Methods in Cell Biology

VOLUME 76

The Zebrafish: Cellular and Developmental Biology

Series Editors

Leslie Wilson

Department of Molecular, Cellular, and Developmental Biology
University of California
Santa Barbara, California

Paul Matsudaira

Whitehead Institute for Biomedical Research
Department of Biology
Division of Biological Engineering
Massachusetts Institute of Technology
Cambridge, Massachusetts

Methods in Cell Biology

VOLUME 76

The Zebrafish: Cellular and Developmental Biology

Edited by

H. William Detrich, III

Department of Biology
Northeastern University
Boston, Massachusetts

Monte Westerfield

Institute of Neuroscience
University of Oregon
Eugene, Oregon

Leonard I. Zon

Division of Hematology/Oncology
Children's Hospital of Boston
Department of Pediatrics and Howard Hughes Medical Institute
Boston, Massachusetts

ELSEVIER
ACADEMIC
PRESS

AMSTERDAM • BOSTON • HEIDELBERG • LONDON
NEW YORK • OXFORD • PARIS • SAN DIEGO
SAN FRANCISCO • SINGAPORE • SYDNEY • TOKYO

Elsevier Academic Press
525 B Street, Suite 1900, San Diego, California 92101-4495, USA
84 Theobald's Road, London WC1X 8RR, UK

This book is printed on acid-free paper. ∞

For all information on all Academic Press publications
visit our Web site at www.books.elsevier.com

ISBN: 0-12-564171-0

PRINTED IN THE UNITED STATES OF AMERICA
04 05 06 07 08 9 8 7 6 5 4 3 2 1

CONTENTS

4. Imaging Blood Vessels in the Zebrafish

Makoto Kamei, Sumio Isogai, and Brant M. Weinstein

5. Zebrafish Apoptosis Assays for Drug Discovery

Chuenlei Parng, Nate Anderson, Christopher Ton, and Patricia McGrath

6. Lipid Metabolism in Zebrafish

Shiu-Ying Ho, Juanita L. Thorpe, Yun Deng, Evelyn Santana, Robert A. DeRose, and Steven A. Farber

PART II   Developmental and Neural Biology

PART III Disease Models

Index   613

Volumes in Series   625

CONTRIBUTORS

Numbers in parenthess indicates the pages on which the authors' contributions begin.

R. Craig Albertson (437), Department of Cytokine Biology, The Forsyth Institute; and Department of Oral and Developmental Biology, Harvard School of Dental Medicine, Boston, Massachusetts 02115

James F. Amatruda (109), Division of Hematology-Oncology, Children's Hospital and Dana-Farber Cancer Institute, Boston, Massachusetts 02115

Nate Anderson (75), Phylonix Pharmaceuticals, Inc., Cambridge, Massachusetts 02139

Andrei Avanesov (333), Department of Ophthalmology, Harvard Medical School. Boston, Massachusetts 02114

Robert Baker (385), Department of Physiology and Neuroscience, New York University School of Medicine, New York, New York 10016

Laure Bally-Cuif (163), Zebrafish Neurogenetics Junior Research Group, Institute of Virology, Technical University-Munich, D-81675 Munich, Germany; and GSF-National Research Center for Environment and Health, Institute of Developmental Genetics, D-85764 Neuherberg, Germany

Serena Banden (261), Department of Biology, Northeastern University, Boston, Massachusetts 02115

James C. Beck (385), Department of Physiology and Neuroscience, New York University School of Medicine, New York, New York 10016

Melissa A. Borla (261), Department of Biology, Northeastern University, Boston, Massachusetts 02115

Robert J. Bryson-Richardson (37), Victor Chang Cardiac Research Institute, Sydney 2010, Australia

Douglas S. Campbell (13), Department of Neurobiology and Anatomy, University of Utah School of Medicine, Salt Lake City, Utah 84132

Stephen P. Chambers (593), Molecular Genetics and Development Group, School of Biological Sciences, University of Auckland, Auckland 1001, New Zealand

Joanne Chan (475), Vascular Biology Program, Children's Hospital of Boston, Boston, Massachusetts 02115

Prisca Chapouton (163), Zebrafish Neurogenetics Junior Research Group, Institute of Virology, Technical University-Munich, D-81675 Munich, Germany; and GSF-National Research Center for Environment and Health, Institute of Developmental Genetics, D-85764 Neuherberg, Germany

Keith C. Cheng (555), Jack Gittlen Cancer Research Institute, Department of Pathology and Department of Biochemistry and Molecular Biology, Pennsylvania State College of Medicine, Hersey, Pennsylvania 17033

Chi-Bin Chien (13), Department of Neurobiology and Anatomy, University of Utah Medical Center, Salt Lake City, Utah 84132

Paul Collodi (151), Department of Animal Sciences, Purdue University, West Lafayette, Indiana 47907

Jennifer Crodian (151), Department of Animal Sciences, Purdue University, West Lafayette, Indiana 47907

Peter D. Currie (37), Victor Chang Cardiac Research Institute, Sydney 2010, Australia

Matthew Cykowski (489), Department of Cellular and Structural Biology, The University of Texas Health Science Center at San Antonio, San Antonio, Texas 78229

Robert A. DeRose (87), Department of Microbiology and Immunology, Kimmel Cancer Center, Thomas Jefferson University, Philadelphia, Pennsylvania 19107

Yun Deng (87), Department of Microbiology and Immunology, Kimmel Cancer Center, Thomas Jefferson University, Philadelphia, Pennsylvania 19107

H. William Detrich III (261), Department of Biology, Northeastern University, Boston, Massachusetts 02115

Andrew Dodd (593), Molecular Genetics and Development Group, School of Biological Sciences, University of Auckland, Auckland 1001, New Zealand

John E. Dowling (315), Department of Molecular and Cellular Biology, Harvard University, Cambridge, Massachusetts 02138

Wolfgang Driever (531), Developmental Biology, University of Freiburg, D-79104 Freiburg, Germany

Iain A. Drummond (501), Department of Medicine, Harvard Medical School and Renal Unit, Massachusetts General Hospital, Charlestown, Massachusetts 02129

James M. Fadool (315), Department of Biological Science, Florida State University, Tallahassee, Florida 32306

Lianchun Fan (151), Department of Animal Sciences, Purdue University, West Lafayette, Indiana 47907

Steven A. Farber★ (87), Department of Microbiology and Immunology, Kimmel Cancer Center, Thomas Jefferson University, Philadelphia, Pennsylvania 19107

Mark C. Fishman (569), Novartis Institutes for Biomedical Research, Cambridge, Massachusetts 02139

Scott E. Fraser (207), Biological Imaging Center, Beckman Institute, California Institute of Technology, Pasadena, California 91125

Ethan Gahtan (261), Department of Biology, Northeastern University, Boston, Massachusetts 02115

Erin E. Gestl (555), Jack Gittlen Cancer Research Institute, Department of Pathology and Department of Biochemistry and Molecular Biology, Pennsylvania State College of Medicine, Hersey, Pennsylvania 17033

Edwin Gilland (385), Department of Physiology and Neuroscience, New York University School of Medicine, New York, New York 10016

Lara Gnügge (531), Developmental Biology, University of Freiburg, D-79104 Freiburg, Germany

★Current address: Department of Embryology, Carnegie Institution of Washington, Baltimore, Maryland 21210

Paul D. Henion (237), Center for Molecular Neurobiology and Department of Neuroscience, Ohio State University, Columbus, Ohio 43210

Shiu-Ying Ho (87), Department of Microbiology and Immunology, Kimmel Cancer Center, Thomas Jefferson University, Philadelphia, Pennsylvania 19107

Lara D. Hutson (13), Department of Biology, Williams College, Williamstown, Massachusetts 01267

Sumio Isogai (51), Department of 1st Anatomy, School of Medicine, Iwate Medical University, Morioka 020-8505, Japan

Pudur Jagadeeswaran (489), Department of Cellular and Structural Biology, The University of Texas Health Science Center at San Antonio, San Antonio, Texas 78229

Yashar Javidan (415), Department of Developmental and Cell Biology, University of California, Irvine, Irvine, California 92697

Makoto Kamei (51), Laboratory of Molecular Genetics, National Institute of Child Health and Human Development, National Institutes of Health, Bethesda, Maryland 20892

John P. Kanki (237), Department of Pediatric Oncology, Dana-Farber Cancer Institute, Boston, Massachusetts 02115

Reinhard W. Köster (207), GSF-National Research Centre for Environment and Health, Institute of Developmental Genetics, 85764 Neuherberg, Germany

Charles A. Lessman★ (285), Department of Microbiology & Molecular Cell Sciences, The University of Memphis, Memphis, Tennessee 38152

A. Thomas Look (237), Department of Pediatric Oncology, Dana-Farber Cancer Institute, Boston, Massachusetts 02115

Donald R. Love (593), Molecular Genetics and Development Group, School of Biological Sciences, University of Auckland, Auckland 1001, New Zealand

Jarema Malicki (333), Department of Ophthalmology, Harvard Medical School, Boston, Massachusetts 02114

Patricia McGrath (75), Phylonix Pharmaceuticals, Inc., Cambridge, Massachusetts 02139

Dirk Meyer (531), Developmental Biology, University of Freiburg, D-79104 Freiburg, Germany

Jessica L. Moore[†] (555), Jake Gittlen Cancer Research Institute, Department of Pathology, Pennsylvania State College of Medicine, Hershey, Pennsylvania 17033

Peter E. Nielsen (593), Center of Biomolecular Recognition, Department of Medical Biochemistry and Genetics, University of Copenhagen, Copenhagen 2200, Denmark

Donald M. O'Malley (261), Department of Biology, Northeastern University, Boston, Massachusetts 02115

Sandra Parker (261), Department of Biology, Northeastern University, Boston, Massachusetts 02115

Chuenlei Parng (75), Phylonix Pharmaceuticals, Inc., Cambridge, Massachusetts 02139

★Current address: Department of Biology, The University of Memphis, Memphis, Tennessee 38152

†Current address: Department of Biology, University of South Florida, Tampa, Florida 33620

Brian D. Perkins[‡] (315), Department of Molecular and Cellular Biology, Harvard University, Cambridge, Massachusetts 02138

Randall T. Peterson (569), Developmental Biology Laboratory, Cardiovascular Research Center, Massachusetts General Hospital, Charlestown, Massachusetts 02129

Kathleen L. Pfaff (109), Division of Hematology-Oncology, Children's Hospital and Dana-Farber Cancer Institute, Boston, Massachusetts 02115

Nagarajan S. Sankrithi (261), Department of Biology, Northeastern University, Boston, Massachusetts 02115

Evelyn Santana (87), Department of Microbiology and Immunology, Kimmel Cancer Center, Thomas Jefferson University, Philadelphia, Pennsylvania 19107

Thomas F. Schilling (415), Department of Developmental and Cell Biology, University of California, Irvine, Irvine, California 92697

Fabrizio C. Serluca (475), Novartis Institutes for Biomedical Research, Cambridge, Massachusetts 02139

Jennifer L. Shepard (109), Division of Hematology-Oncology, Children's Hospital and Dana-Farber Cancer Institute, Boston, Massachusetts 02115

Didier Y. R. Stainier (455), Department of Biochemistry and Biophysics, Programs in Developmental Biology, Genetics, and Human Genetics, University of California, San Francisco, San Francisco, California 94143-0448

Howard M. Stern (109), Division of Hematology-Oncology, Children's Hospital and Dana-Farber Cancer Institute, Boston, Massachusetts 02115; and Department of Pathology, Brigham and Women's Hospital, Boston, Massachusetts 02115

Rodney A. Stewart (237), Department of Pediatric Oncology, Dana-Farber Cancer Institute, Boston, Massachusetts 02115

David W. Tank (385), Departments of Physics and Molecular Biology, Princeton University, Princeton, New Jersey 08544

Bijoy Thattaliyath (489), Department of Cellular and Structural Biology, The University of Texas Health Science Center at San Antonio, San Antonio, Texas 78229

Juanita L. Thorpe (87), Department of Microbiology and Immunology, Kimmel Cancer Center, Thomas Jefferson University, Philadelphia, Pennsylvania 19107

Christopher Ton (75), Phylonix Pharmaceuticals, Inc., Cambridge, Massachusetts 02139

David Traver (127), Section of Cell and Developmental Biology, University of California, San Diego, La Jolla, California 92093

Le A. Trinh (455), Department of Biochemistry and Biophysics, Programs in Developmental Biology, Genetics, and Human Genetics, University of California, San Francisco, San Francisco, California 94143-0448

Brant M. Weinstein (51), Laboratory of Molecular Genetics, National Institute of Child Health and Human Development, National Institutes of Health, Bethesda, Maryland 20892

[‡]Current address: Department of Biology, Texas A & M University, College Station, Texas 77843

Pamela C. Yelick (437), Department of Cytokine Biology, The Forsyth Institute; and Department of Oral and Developmental Biology, Harvard School of Dental Medicine, Boston, Massachusetts 02115

Hao Zhu (3), Division of Hematology/Oncology, Children's Hospital of Boston, Department of Pediatrics and Howard Hughes Medical Institute, Harvard Medical School, Boston, Massachusetts 02115

Leonard I. Zon (3), Division of Hematology/Oncology, Children's Hospital of Boston, Department of Pediatrics and Howard Hughes Medical Institute, Harvard Medical School, Boston, Massachusetts 02115

PREFACE

Research Vessel Nathaniel B. Palmer
Southern Atlantic Ocean, 54° 47' S, 59° 15' W
On the Burdwood Banks
20 May 2004

Monte, Len, and I welcome you to two new volumes of *Methods in Cell Biology* devoted to *The Zebrafish: Cellular and Developmental Biology* and *Genetics, Genomics, and Informatics*. In the five years since publication of the first pair of volumes, *The Zebrafish: Biology* (Vol. 59) and *The Zebrafish: Genetics and Genomics* (Vol. 60), revolutionary advances in techniques have greatly increased the versatility of this system. At the Fifth Conference on *Zebrafish Development and Genetics*, held at the University of Wisconsin in 2003, it was clear that many new and compelling methods were maturing and justified the creation of the present volumes. The zebrafish community responded enthusiastically to our request for contributions, and we thank them for their tremendous efforts.

The new volumes present the post-2000 advances in molecular, cellular, and embryological techniques (Vol. 76) and in genetic, genomic, and bioinformatic methods (Vol. 77) for the zebrafish, *Danio rerio*. The latter volume also contains a section devoted to critical infrastructure issues. Overlap with the prior volumes has been minimized intentionally.

The first volume, *Cellular and Developmental Biology*, is divided into three sections: Cell Biology, Developmental and Neural Biology, and Disease Models. The first section focuses on microscopy and cell culture methodologies. New microscopic modalities and fluorescent reporters are described, the cell cycle and lipid metabolism in embryos are discussed, apoptosis assays are outlined, and the isolation and culture of stem cells are presented. The second section covers development of the nervous system, techniques for analysis of behavior and for screening for behavioral mutants, and methods applicable to the study of major organ systems. The volume concludes with a section on use of the zebrafish as a model for several diseases.

The second volume, *Genetics, Genomics, and Informatics*, contains five sections: Forward and Reverse Genetics, The Zebrafish Genome and Mapping Technologies, Transgenesis, Informatics and Comparative Genomics, and Infrastructure. In the first, forward-genetic (insertional mutagenesis, maternal-effects screening), reverse-genetic (antisense morpholino oligonucleotide and peptide nucleic acid gene knockdown strategies, photoactivation of caged mRNAs), and hybrid (target-selected screening for ENU-induced point mutations) technologies are described. Genetic applications of transposon-mediated transgenesis of zebrafish are presented, and the status of the genetics and genomics of *Medaka*, the honorary zebrafish, is updated. Section 2 covers the zebrafish genome project, the cytogenetics of zebrafish chromosomes, several methods for mapping zebrafish genes and mutations, and the recovery of mutated genes via positional cloning. The third section presents multiple methods for transgenesis in zebrafish and describes the application of nuclear transfer for cloning of zebrafish. Section 4 describes bioinformatic analysis of the zebrafish genome and of microarray data, and emphasizes the importance of comparative analysis of genomes in gene discovery and in the elucidation of gene regulatory elements. The final section provides important, but difficult to find, information on small- and large-scale infrastructure available to the zebrafish biologist.

The attentive reader will have noticed that this Preface was drafted by the first editor, Bill Detrich, while he (I) was at sea leading the sub-Antarctic ICEFISH Cruise (International Collaborative Expedition to collect and study Fish Indigenous to Sub-antarctic Habitats; visit www.icefish.neu.edu). Wearing my second biological hat, I study the adaptational biology of Antarctic fish and use them as a system for comparative discovery of erythropoietic genes. Antarctic fish embryos generally hatch after six months of development, and they reach sexual maturity only after several years. Imagine attempting genetic studies on these organisms! My point is that the zebrafish system and its many advantages greatly inform my research on Antarctic fish, while at the same time I can move genes discovered by study of the naturally evolved, but very unusual, phenotypes of Antarctic fish into the zebrafish for functional analysis. We the editors emphasize that comparative strategies applied to multiple organisms, including the diverse fish taxa, are destined to play an increasing role in our understanding of vertebrate development.

We wish to express our gratitude to the series editors, Leslie Wilson and Paul Matsudaira, and the staff of Elsevier/Academic Press, especially Kristi Savino, for their diligent help, great patience, and strong encouragement as we developed these volumes.

H. William Detrich, III
Monte Westerfield
Leonard I. Zon

These volumes are dedicated to Jose Campos-Ortega and Nigel Holder, departed colleagues whose wisdom and friendship will be missed by the zebrafish community

PART I

Cell Biology

CHAPTER 1

Use of the DsRed Fluorescent Reporter in Zebrafish

Hao Zhu and Leonard I. Zon

Division of Hematology/Oncology
Children's Hospital of Boston
Department of Pediatrics and Howard Hughes Medical Institute
Harvard Medical School
Boston, Massachusetts 02115

Green fluorescent protein (GFP) is firmly established as a fluorescent reporter for the imaging of specific tissues in zebrafish. The employment of other reporters such as DsRed in transgenic zebrafish has made multicolored labeling experiments possible. To date, several DsRed transgenic lines have been generated for lineage labeling, transplantation assays, and commercial applications. Advances in multicolored labeling experiments will depend on the implementation of newly engineered reporters and fusion proteins, as well as on innovative experiments that exploit the power of direct visualization.

I. Introduction

Since the mid-1990s, there has been an explosion in the number of GFP expressing transgenic zebrafish lines reported in the literature. This is primarily true because the "know-how" necessary to make transgenics has become widespread. In addition, transgenic zebrafish can be used in a wide range of experiments. As a testament to this, nearly every organ system is now represented by tissue-specific GFP transgenic lines. The transgenic catalog is comprehensive, and includes cardiac and skeletal muscle, pancreas, erythrocytes and lymphocytes, vessels, alimentary canal, notochord/floor plate, and central nervous system neurons (Goldman *et al.*, 2001; Higashijima *et al.*, 2000; Huang *et al.*, 2001; Ju *et al.*, 1999; Lawson and Weinstein, 2002; Long *et al.*, 1997; Motoike *et al.*, 2000; Udvadia *et al.*, 2001).

Fluorescent proteins have been utilized successfully in mammalian systems, but they are especially useful in the zebrafish. Zebrafish embryos develop *ex vivo* and thus do not require unwieldy culture conditions for microscopic investigation. Furthermore, zebrafish embryos and larvae are optically clear, enabling the observation of even anatomically deep organs throughout the first month of development. These developing tissues are thin enough that they can be penetrated by laser confocal microscopy even in late developmental stages.

Zebrafish expressing fluorescent reporters under the control of constitutive and tissue-specific promoters have been featured in a wide range of experiments. Fluorescent labeling of a specific subset of cells makes it possible to isolate that subpopulation for a variety of purposes. For instance, cells isolated by fluorescence activated cell sorting (FACS) can be transplanted and subsequently followed in living hosts. These cell populations have also been used to generate cDNA libraries and can potentially be used to establish cell lines. In many situations, fluorescent protein expression in living embryos can replace wholemount *in situ* hybridization of fixed specimens. Manipulations of gene levels by ectopic overexpression and morpholino-mediated knockdown experiments can be assayed in fluorescent embryos, a strategy that permits visualization of morphological or gene expression changes in living animals. GFP can also be used as a sentinel marker in genetic or chemical screens. On the most fundamental level, fluorescence allows us to see morphogenetic processes that were previously invisible.

While most of these advances have been made using GFP, there have been attempts to introduce other fluorescent proteins such as DsRed into the zebrafish, as discussed in this chapter. The combinatorial use of these reporters allows the visualization of multiple tissues with multiple colors, making an already powerful cell biological and genetic system more vibrant.

II. DsRed: History and Properties

The introduction of GFP from *Aequorea* jellyfish in 1994 revolutionized cell biology. The applications of GFP have ranged widely, from tracking gene expression and subcellular labeling to monkey transgenesis (Chan *et al.*, 2001) and transgenic artwork (Eduardo Kac's "GFP Bunny"), capturing the imagination of the scientific and public worlds.

In an effort to widen the spectrum of fluorescent proteins, DsRed (originally designated drFP583) was one of several GFP homologs cloned from reef corals (Matz *et al.*, 1999). Though the protein is only 23% homologous to GFP, it has several conserved residues in the vicinity of a virtually identical chromophore (Yarbrough *et al.*, 2001). With easily separable emission wavelengths of 509 and 583 nm, respectively, enhanced green fluorescent protein (EGFP) and DsRed are suited for dual-color labeling with minimal crossover interference. The original DsRed protein was mutagenized to yield a somewhat faster maturing and more soluble variant that is offered commercially as DsRed2 (Clontech Laboratories, Inc., Palo Alto, CA).

GFP is an optimal genetic fusion tag because of its properties of fast maturation, solubility, and existence as a monomer. In contrast, the obligate tetramerization and slow maturation of DsRed posed formidable obstacles to its employment in fusion protein and gene expression experiments (Baird *et al.*, 2000). Also, DsRed maintains high levels of fluorescence for long periods of time after it is produced because of its high extinction coefficient, a property that is both advantageous and yet problematic for short-term gene expression applications. In our experience with DsRed2 in zebrafish embryos, it is detectable approximately 12 hours after the initiation of transcription. As a consequence, we have not been able to use DsRed for the observation or isolation of cells that have activated DsRed transgene expression within a restricted time window. Because DsRed exists predominantly as a stable homotetramer, attempts at using it as a genetic fusion tag have failed. Fortunately, efforts by several groups to develop faster maturing, monomeric DsRed variants have led to the creation of mRFP1, a rapidly maturing (detectable in less than 1 hr) monomeric red fluorescent protein with excitation at 584 nm and emission at 607 nm (Campbell *et al.*, 2002). Although mRPF1 has a lower extinction coefficient, quantum yield, and photostability than DsRed, all of which results in lower signal intensity, it represents a significant improvement over DsRed that can and will be further refined. The use of mRFP1 in zebrafish has not yet been reported.

Even before DsRed was widely used in zebrafish for scientific purposes, DsRed fluorescent zebrafish were marketed commercially as pets. These "GloFish^TM," originally designed to monitor environmental pollution (Knight, 2003), generated considerable controversy because of the potential risks of releasing transgenic animals into the environment. Nevertheless, GloFish^TM catapulted zebrafish into the national spotlight and, for better or worse, stretched the boundaries of utility for fluorescent proteins in fish.

III. DsRed in Transgenic Zebrafish

To date, only a handful of transgenic lines have utilized fluorescent reporters other than GFP. This may be attributed to the drawbacks of DsRed or a relative lack of experience using reporters other than GFP in transgenesis. The earliest report of such use investigated the efficacy of three fluorescent proteins for multi-labeling in zebrafish (Finley *et al.*, 2001). Single-, double-, or triple-labeled transiently expressing embryos were produced by injecting GFP, blue-shifted GFP variant (BFP), or DsRed DNA constructs. Finley *et al.* found that simultaneously expressed fluorescent proteins could be detected independently within the same cells, showing that it would be effective to label zebrafish tissues with these markers. Subsequently, this group generated germline transgenic lines that ubiquitously express GFP, BFP, and DsRed in order to demonstrate that the Sleeping Beauty transposase can enhance transgenesis efficiency (Davidson *et al.*, 2003).

The first lines expressing DsRed or yellow-shifted GFP (YFP) under tissue-specific control were made for ornamental and industrial purposes. Gong *et al.* (2003) used a muscle-specific *mylz2* promoter to drive GFP, YFP, and DsRed expression in zebrafish. These fish were considered suitable for ornamentation because the promoter drove such strong expression that fluorescence was visible without the aid of ultraviolet (UV) light. The authors proposed that the fish might also be used as a source of recombinant proteins because of the high levels of protein production. Experimentally speaking, these and other commercially available lines might also function as markers in transplant or explant assays since donor-derived cells with high levels of fluorescence would be easily identified.

Recently, the *lmo2* promoter was used to create two transgenic lines, *Tg(lmo2:EGFP)* and *Tg(lmo2:DsRed)* (Fig. 1A,B), which both exhibit embryonic blood and endothelial expression (Zhu and Zon, unpublished data). To distinguish hematopoietic from vascular cell populations in living embryos, *Tg(lmo2:DsRed)* was mated to *Tg(fli1:EGFP)*, a line that labels endothelial cells (Lawson and Weinstein, 2002). In these double transgenic embryos, the vasculature was labeled with EGFP, while the vasculature and blood were labeled with DsRed (Fig. 1C). *Tg(lmo2:DsRed)* was also mated to *Tg(gata1:EGFP)*, an erythrocyte-specific line (Long *et al.*, 1997). Since erythroid progenitors and their progeny coexpress *gata1* and *lmo2*, these cells expressed both EGFP and DsRed in this double transgenic embryo (Fig. 1D). Endothelial cells were DsRed+/EGFP−, and could thus be distinguished from fluorescent yellow erythroid cells that expressed both proteins.

From these examples, several points can be made about the combinatorial use of fluorescent proteins. First, an examination of *Tg(lmo2:DsRed)* in the background of GFP expressing vascular and erythroid transgenics demonstrates that lineage-specific cells can be distinguished by fluorescent gene expression even if they are morphologically indistinct. Second, it is theoretically possible to identify or isolate cell populations that are marked by the overlap of gene expression domains similar to a Venn diagram. Third, the interplay between cell types in close physical association can be examined using time-lapse video microscopy. In essence, the techniques that have been used to analyze the interactions between subcellular components within individual cells can be translated onto whole organisms.

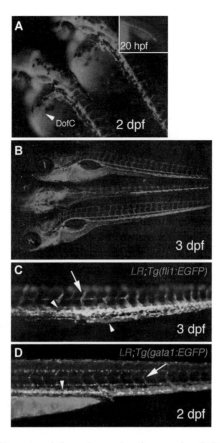

Fig. 1 Visualization of hematopoietic and vascular tissues using DsRed and EGFP transgenic embryos. (A) In *Tg(lmo2:DsRed)* embryos (abbreviated *LR*), DsRed protein is initially detected at 20 hpf (inset); 2 dpf *LR* embryos labeling hematopoietic and endothelial cells in the ducts of Cuvier (DofC). (B) Labeling of the vascular endothelial network of a 3 dpf *LR* embryo. (C) *LR; Tg(fli1:EGFP)* embryos distinctly label hematopoietic (arrowheads) and endothelial cells (arrow) in 3 dpf embryos. (D) In *LR; Tg(gata1:EGFP)* transgenic embryos, green/red erythrocytes (arrowheads) circulate through vessels (arrow) labeled by DsRed in 2 dpf embryos. (Zhu and Zon, unpublished data) (See Color Insert.)

IV. Use of Multiple Fluorescent Reporters in Transplantation Assays

Transplant and chimeric experiments in mice have traditionally taken advantage of Y-chromosomes and the *lacZ* reporter to identify donor cells in recipients. In zebrafish, it is preferable to employ fluorescent markers because they can be used to identify donor-derived cells without sacrificing the recipients. Also consider that GFP or DsRed expressing tissues can be identified by immunohistochemistry if, for example, visualization of deeper tissues is required. The lack of cell-type specific surface antibodies in the zebrafish has made tissue-restricted fluorescent markers important for the isolation of these subpopulations that would be otherwise inaccessible. Another advantage of using fluorescent genetic tags in zebrafish is that breeding multi-labeled animals is easy, making it possible to devise and create sophisticated reagents in a relatively short amount of time.

Traver and colleagues (2003) pioneered the use of multicolored transgenic zebrafish in the setting of hematopoietic cell transplantation. They used whole kidney marrow (WKM), which is the primary site of adult hematopoiesis and bone marrow equivalent in teleost fish, from double-labeled fluorescent fish to follow multilineage, donor-derived hematopoiesis and early homing events (Fig. 2A–D). Two transgenic lines were used to independently label erythrocytes and leukocytes. The first line, *Tg(gata1:DsRed)*, expresses DsRed under the control of the *gata1* promoter and marks the erythroid lineage over an animal's lifetime (Long *et al.*, 1997; Traver *et al.*, 2003). The second, *Tg(β-acting:EGFP)*, expresses GFP in virtually all blood lineages except erythroid, and was used to mark myeloid, precursor, and lymphoid cells. WKM was collected from *Tg(gata1:DsRed); Tg(β-actin:EGFP)* double transgenic adults and injected into the circulation of wild-type embryos, resulting in the appearance of DsRed or GFP positive cells in the hosts. The red fluorescent cells were erythrocytes. Two morphologically distinct types of green fluorescent cells, seen rolling along vessel lumens, were likely lymphocytes and myelomonocytic cells. GFP expressing cells also homed to the sites of the developing thymus and kidney, whereas DsRed cells were only seen in circulation, suggesting that the GFP fraction contains lymphocyte and progenitor cells. A single fluorescent marker, GFP, was able to resolve two distinct cell types by outlining the shape of small round cells, identified as lymphocytes, and amoeboid cells with pseudopodia, identified as myelomonocytic cells.

The double transgenic kidney marrow was also transplanted into *vlad tepes* (*vlt*), a *gata1*$^{-/-}$ mutant lacking erythrocytes (Lyons *et al.*, 2002), and *bloodless* (*bls*), a mutant with an absence of primitive blood cells (Fig. 2B–D) (Liao *et al.*, 2002). While transplantation into *vlt* resulted in robust reconstitution of DsRed erythrocytes, the reconstitution of GFP+ leukocytes occurred at levels comparable to the wild-type recipient setting. In *bls*, the appearance of both DsRed+ erythrocytes and GFP+ cells in the developing thymus and pronephros was fast and robust, suggesting that there was a relative lack of competing host cells during the first few days of *bls* development. In this context, the independent labeling of two cell types with different colors helped to resolve the differences in engraftment kinetics between two mutant recipients. These results demonstrate an elegant way to identify donor cell types in the context of living transplant recipients. This type of experiment provides a bird's eye view of post transplantation hematopoietic cell homing, proliferation, and differentiation, a view that is impossible to obtain in any other system.

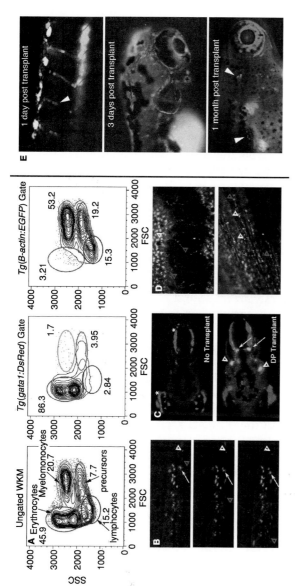

Fig. 2 Use of multiple fluorescent reporters in transplantation assays. Left panel: Transplantation of whole kidney marrow from double transgenic donors allows independent visualization of leukocytes and erythrocytes in recipient embryos. (A) Scatter profile of ungated WKM in a representative *Tg(gata1:DsRed); Tg(β-actin:EGFP)* double transgenic adult (left). DsRed+ cells were contained only within the erythrocyte gate (middle), whereas GFP+ cells were non-erythroid (right). (B–C) Transplantation of 48 hpf recipients showed transient reconstitution of donor-derived erythrocytes and leukocytes. (B) Visualization of the tail vessels in a *gata1⁻/⁻* transplant recipient showed a slow-moving, round leukocyte (arrowhead), a larger leukocyte displaying an end-over-end tumbling migration (arrow), and a rapidly circulating erythrocyte (red arrowhead) at 1 day post-transplantation. Each frame is separated by 300ms (20× magnification, anterior to the left). (C) Dorsal views comparing untransplanted (upper) and transplanted *bloodless* (bls) recipients (lower). *bls* recipients showed rapid and robust engraftment of the pronephros (arrows) and bilateral thymi (arrowheads) by GFP+ leukocytes by day 5 post-transplantation. Asterisks denote autofluorescence of the eyes and swim bladder in the DsRed channel. (D) *bls* recipients display sustained, multilineage hematopoiesis from donor-derived cells. Upper panel shows robust reconstitution of DsRed+ erythrocytes (red arrowheads) and GFP+ leukocytes (white arrowhead) as observed in the dermal capillaries of a *bloodless* recipient at 8 weeks post-transplantation. Lower panel shows similar multilineage reconstitution as observed in the tail capillaries of another *bls* recipient at 8 weeks (20× magnification). From Traver et al., 2003.

Right Panel: Transplantation of primitive wave hematopoietic progenitors from *Tg(lmo2:EGFP)* embryos into *vlad tepes* (vlt) recipients. (E) The EGFP expressing population from 10–12 somite *Tg(lmo2:EGFP); Tg(gata1:DsRed)* transgenic embryos were isolated by FACS and transplanted into 48 hpf *vlt* embryos. One day post transplantation, circulating donor-derived cells could be identified by DsRed and EGFP fluorescence (arrowhead points to GFP+ circulating cell). By 3 days post transplantation, most circulating cells were DsRed+, suggesting that the EGFP+ donor progenitors had differentiated into Gata1+ erythrocytes. One month after transplantation, each of the surviving recipients carried approximately 10–200 DsRed+ circulating cells (arrowheads). (Zhu and Zon, unpublished data.) (See Color Insert.)

The transplantation of fluorescently labeled cells can also be used to assay the cell fate of a specific population. For example, the *Tg(gata1:DsRed)* fish were used to evaluate *lmo2* and EGFP expressing primitive hematopoietic progenitors (Zhu and Zon, unpublished data). In this experiment, EGFP+ donor cells were FACS sorted from 10- to 12-somite staged *Tg(lmo2:EGFP); Tg(gata1:DsRed)* embryos and injected into the circulation of 48 hpf *vlt* embryos that normally die within 14 days (Fig. 2E). One day after transplantation, circulating donor-derived erythrocytes could be identified by DsRed and EGFP fluorescence (Fig. 2E). By three days post transplantation, few EGFP$^+$ circulating cells were present and virtually all observable fluorescence was in the DsRed channel, suggesting that the EGFP$^+$ donor progenitors had differentiated into Gata1/DsRed expressing ery-throcytes. One month after transplantation, surviving fish contained between 10 and 200 DsRed$^+$ circulating cells. This transplantation experiment showed that donor-derived primitive wave hematopoietic cells and their progeny could be detected in the circulation of *vlt* recipients for more than one month. This experi-ment demonstrated how one fluorescent label can be used for donor isolation, while another can be used as a marker of donor cells. It follows that the inclusion of even more reporters within a single transgenic animal will increase the versatility of possible experiments.

V. Fusion Protein Reporters

The number of novel fluorescent proteins is rapidly increasing. Thus, a thor-ough account of these developments is beyond the scope of this chapter. In addition to novel fluorescent proteins, fluorescent fusion proteins that label sub-cellular structures are being applied to fish. The first example was a transgenic line that labels cell nuclei with a histone-GFP fusion protein (Pauls *et al.*, 2001). It is only a matter of time before the tools that had been developed for other animal models are transferred to zebrafish. In *Drosophila*, GFP fused to the C-terminal end of moesin, a protein that localizes to the cortical actin-cytoskeleton, has been used for the analysis of cell shape changes during morphogenesis (Edwards *et al.*, 1997). GFP-*moesin* can potentially be used in zebrafish in order to analyze cell shape changes in processes that have specific relevance to vertebrate organogene-sis. Now that red fluorescent fusion proteins can be made using mRFP1, multiple subcellular components can be labeled with several colors in the context of the whole organism.

Fluorescent fusion proteins have also played a role in zebrafish disease models. Recently, a zebrafish model of T-cell acute lymphoblastic leukemia was made by driving the expression of a mouse c-Myc-GFP fusion transgene under the control of the zebrafish Rag2 promoter (Langenau *et al.*, 2003). The GFP fusion protein was integral in showing that the leukemia arose in the thymus and spread to the gills, retroorbital soft tissue, skeletal muscle and abdominal organs. In this leuke-mia model, it would be advantageous to ectopically express other genes that can modulate cancer progression using fusions with non-GFP fluorescent proteins. This way, cells that are expressing multiple oncogenes can be identified by fluorescence. In general, using fluorescent markers to visualize tumor progression will play a pivotal role in studying cancer in zebrafish.

VI. Conclusion

The creative use of fluorescent reporters in zebrafish is bound only by imagination. Its utility has ranged from the development of pets to the tracking of donor cells in transplants. By introducing a spectrum of fluorescent reporters into zebrafish, we are approaching a living histology. One can envision an embryo that possesses a color for each cell type in any organ of interest. Soon it will be possible to take optical sections through a living embryo that contains GFP labeled cardiomyocytes, DsRed labeled leukocytes, YFP labeled endothelial cells, BFP labeled nephrons, etc. The zebrafish is already equipped for these possibilities: they are optically clear, develop *ex vivo*, and are genetically tractable. Advances in multi-fluorescent labeled experiments will depend on the implementation of newly engineered reporters and experiments that exploit the power of direct visualization.

References

Baird, G. S., Zacharias, D. A., and Tsien, R. Y. (2000). Biochemistry, mutagenesis, and oligomerization of DsRed, a red fluorescent protein from coral. *Proc. Natl. Acad. Sci. USA* **97**, 11984–11989.

Campbell, R. E., Tour, O., Palmer, A. E., Steinbach, P. A., Baird, G. S., Zacharias, D. A., and Tsien, R. Y. (2002). A monomeric red fluorescent protein. *Proc. Natl. Acad. Sci. USA* **99**, 7877–7882.

Chan, A. W., Chong, K. Y., Martinovich, C., Simerly, C., and Schatten, G. (2001). Transgenic monkeys produced by retroviral gene transfer into mature oocytes. *Science* **291**, 309–312.

Davidson, A. E., Balciunas, D., Mohn, D., Shaffer, J., Hermanson, S., Sivasubbu, S., Cliff, M. P., Hackett, P. B., and Ekker, S. C. (2003). Efficient gene delivery and gene expression in zebrafish using the Sleeping Beauty transposon. *Dev. Biol.* **263**, 191–202.

Edwards, K. A., Demsky, M., Montague, R. A., Weymouth, N., and Kiehart, D. P. (1997). GFP-moesin illuminates actin cytoskeleton dynamics in living tissue and demonstrates cell shape changes during morphogenesis in *Drosophila*. *Dev. Biol.* **191**, 103–117.

Finley, K. R., Davidson, A. E., and Ekker, S. C. (2001). Three-color imaging using fluorescent proteins in living zebrafish embryos. *Biotechniques* **31**(72), 66–70.

Goldman, D., Hankin, M., Li, Z., Dai, X., and Ding, J. (2001). Transgenic zebrafish for studying nervous system development and regeneration. *Transgenic Res.* **10**, 21–33.

Gong, Z., Wan, H., Tay, T. L., Wang, H., Chen, M., and Yan, T. (2003). Development of transgenic fish for ornamental and bioreactor by strong expression of fluorescent proteins in the skeletal muscle. *Biochem. Biophys. Res. Commun.* **308**, 58–63.

Higashijima, S., Hotta, Y., and Okamoto, H. (2000). Visualization of cranial motor neurons in live transgenic zebrafish expressing green fluorescent protein under the control of the islet-1 promoter/enhancer. *J. Neurosci.* **20**, 206–218.

Huang, H., Vogel, S. S., Liu, N., Melton, D. A., and Lin, S. (2001). Analysis of pancreatic development in living transgenic zebrafish embryos. *Mol. Cell Endocrinol.* **177**, 117–124.

Ju, B., Xu, Y., He, J., Liao, J., Yan, T., Hew, C. L., Lam, T. J., and Gong, Z. (1999). Faithful expression of green fluorescent protein (GFP) in transgenic zebrafish embryos under control of zebrafish gene promoters. *Dev. Genet.* **25,** 158–167.

Knight, J. (2003). GloFish casts light on murky policing of transgenic animals. *Nature* **426,** 372.

Langenau, D. M., Traver, D., Ferrando, A. A., Kutok, J. L., Aster, J. C., Kanki, J. P., Lin, S., Prochownik, E., Trede, N. S., Zon, L. I., *et al.* (2003). Myc-induced T cell leukemia in transgenic zebrafish. *Science* **299,** 887–890.

Lawson, N. D., and Weinstein, B. M. (2002). *In vivo* imaging of embryonic vascular development using transgenic zebrafish. *Dev. Biol.* **248,** 307–318.

Liao, E. C., Trede, N. S., Ransom, D., Zapata, A., Kieran, M., and Zon, L. I. (2002). Non-cell autonomous requirement for the bloodless gene in primitive hematopoiesis of zebrafish. *Development* **129,** 649–659.

Long, Q., Meng, A., Wang, H., Jessen, J. R., Farrell, M. J., and Lin, S. (1997). GATA-1 expression pattern can be recapitulated in living transgenic zebrafish using GFP reporter gene. *Development* **124,** 4105–4111.

Lyons, S. E., Lawson, N. D., Lei, L., Bennett, P. E., Weinstein, B. M., and Liu, P. P. (2002). A nonsense mutation in zebrafish gata1 causes the bloodless phenotype in vlad tepes. *Proc. Natl. Acad. Sci. USA* **99,** 5454–5459.

Matz, M. V., Fradkov, A. F., Labas, Y. A., Savitsky, A. P., Zaraisky, A. G., Markelov, M. L., and Lukyanov, S. A. (1999). Fluorescent proteins from nonbioluminescent Anthozoa species. *Nat. Biotechnol.* **17,** 969–973.

Motoike, T., Loughna, S., Perens, E., Roman, B. L., Liao, W., Chau, T. C., Richardson, C. D., Kawate, T., Kuno, J., Weinstein, B. M., *et al.* (2000). Universal GFP reporter for the study of vascular development. *Genesis* **28,** 75–81.

Pauls, S., Geldmacher-Voss, B., and Campos-Ortega, J. A. (2001). A zebrafish histone variant H2A.F/Z and a transgenic H2A.F/Z:GFP fusion protein for *in vivo* studies of embryonic development. *Dev. Genes Evol.* **211,** 603–610.

Traver, D., Paw, B. H., Poss, K. D., Penberthy, W. T., Lin, S., and Zon, L. I. (2003). Transplantation and *in vivo* imaging of multilineage engraftment in zebrafish bloodless mutants. *Nat. Immunol.* **4,** 1238–1246.

Udvadia, A. J., Koster, R. W., and Skene, J. H. (2001). GAP-43 promoter elements in transgenic zebrafish reveal a difference in signals for axon growth during CNS development and regeneration. *Development* **128,** 1175–1182.

Yarbrough, D., Wachter, R. M., Kallio, K., Matz, M. V., and Remington, S. J. (2001). Refined crystal structure of DsRed, a red fluorescent protein from coral, at 2.0-A resolution. *Proc. Natl. Acad. Sci. USA* **98,** 462–467.

CHAPTER 2

Analyzing Axon Guidance in the Zebrafish Retinotectal System

Lara D. Hutson, [*] **Douglas S. Campbell,** [†] **and Chi-Bin Chien** [†]

[*]Department of Biology
Williams College
Williamstown, Massachusetts 01267

[†]Department of Neurobiology and Anatomy
University of Utah Medical Center
Salt Lake City, Utah 84132

I. Introduction

The developing visual system has been the subject of intensive study in many model organisms except *Caenorhabditis elegans*, which unfortunately lacks eyes. There are several reasons for its popularity. It is experimentally accessible, its normal anatomy is understood extremely well, and its function is understood equally well. Thus one can observe how the visual system develops, use perturbations to test the mechanisms of its development, and link alterations in development to changes in mature anatomy and, ultimately, to visual behavior. Many laboratories, including ours, have been led by these advantages to study the projections of retinal ganglion cells (RGCs) to the optic tectum. The retinotectal projection in the zebrafish is an especially good system for imaging because of the larva's transparency. In addition, it is a useful system for genetic analysis because of the large number of known mutants that affect its development (reviewed in Culverwell and Karlstrom, 2002; Hutson and Chien, 2002b). Here we describe the strategies that have been used to observe and to perturb retinotectal development in the zebrafish.

Retinal axons originate from the RGCs, which are the primary cell type in the innermost cellular layer of the retina. The axons first navigate radially to the optic nerve head near the center of the retina and exit the eye. This occurs at 32 hours post fertilization (hpf) in zebrafish (Burrill and Easter, 1994). The retinal axons then cross the midline and enter the contralateral optic tract, coursing dorsalward in the diencephalon until they reach the pretectal nuclei and the optic tectum at 48 hpf. The tectum, which is located at the dorsal roof of the midbrain, is the primary visual center in non-mammalian vertebrates and is likely primarily responsible for high spatial-resolution vision. Axons find their topographic targets on the tectum, then arborize and begin to form synapses. The axonal arbors are refined over time via activity-dependent mechanisms (Schmidt, 2004). The first retinal axons reach the tectum before visual function begins, but by about 3 days post fertilization (dpf), larvae begin to show visually evoked behavior, and by 5 dpf, behaviors such as the optokinetic response are robustly displayed (Easter and Nicola, 1996; Neuhauss, 2003).

In the next section, we begin with a brief overview of the genetic control of retinal axon guidance. We then describe methods for labeling and observing retinal axons. We follow with methods for perturbing retinotectal development. Finally, we end with a brief discussion of methods likely to be important in the future. The methods described here for the retinotectal system are also useful, either directly or with some modifications, for studying the development of other axon tracts.

II. Retinotectal Mutants

Among the many zebrafish mutants affecting specific developmental processes that were isolated in large-scale genetic screens (Neumann, 2002), at least 27 are known to affect development of the retinotectal projection (Table I; Haffter *et al.*, 1996; reviewed in Culverwell and Karlstrom, 2002; Hutson and Chien, 2002b). Most of these were discovered in a landmark screen carried out in Friedrich Bonhoeffer's laboratory, which used as an assay the direct labeling of retinal axons with lipophilic dyes (Baier *et al.*, 1996; Karlstrom *et al.*, 1996; Trowe *et al.*, 1996). Screens for visual behavioral defects are a promising avenue for discovering additional mutants (reviewed in Baier, 2000 and Neuhauss, 2003),

Table I
Retinotectal Pathfinding Mutants

Mutant	Region in which pathfinding affected	Gene	Brain defect?	References
acerebellar (ace)	Chiasm, anterior projections, optic tract, topography	*fgf8*	Yes	Picker *et al.*, 1999; Shanmugalingam *et al.*, 2000
astray (ast)	Chiasm, anterior projections, optic tract	*robo2*	No	Karlstrom *et al.*, 1996; Fricke *et al.*, 2001
bashful (bal)	Retinal exit, anterior projections	?	Yes	Karlstrom *et al.*, 1996
belladonna (bel)	Midline crossing	?	Yes	Karlstrom *et al.*, 1996
blowout (blw)	Midline crossing	?	Yes	Karlstrom *et al.*, 1996; Nakano *et al.*, 2004
blumenkohl (blu)	Expanded terminations	?	No	Trowe *et al.*, 1996
boxer (box)	Tract sorting, crossing in posterior commissure	?	No	Karlstrom *et al.*, 1996; Trowe *et al.*, 1996
chameleon (con)	Retinal exit, midline crossing	*dispatched 1*	Yes	Karlstrom *et al.*, 1996; Nakano *et al.*, 2004
cyclops (cyc)	Midline crossing	*nodal related-2 (ndr2)*	Yes	Karlstrom *et al.*, 1996; Rebagliati *et al.*, 1998; Sampath *et al.*, 1998
dackel (dak)	Tract sorting	?	No	Karlstrom *et al.*, 1996; Trowe *et al.*, 1996
detour (dtr)	Midline crossing	*gli1*	Yes	Karlstrom *et al.*, 1996, 2003
esrom (esr)	Midline crossing, termination	?	No	Karlstrom *et al.*, 1996; Trowe *et al.*, 1996
gnarled (gna)	Tectal entry, tectal misrouting	?	No	Trowe *et al.*, 1996; Wagle *et al.* 2004
grumpy (gup)	Anterior projections, midline crossing	*laminin β1*	Yes	Karlstrom *et al.*, 1996; Parsons *et al.*, 2002
iguana (igu)	Midline crossing	?	Yes	Karlstrom *et al.*, 1996
macho (mao)	Expanded terminations	?	No	Gnuegge *et al.*, 2001; Trowe *et al.*, 1996
nevermind (nev)	Tract sorting, D-V topography	?	No	Trowe *et al.*, 1996
no isthmus (noi)	Chiasm, anterior projections, tectal bypass	*pax2a*	Yes	Brand *et al.*, 1996; Macdonald *et al.*, 1997; Trowe *et al.*, 1996
parachute (pac)	Ipsilateral projections, entering chiasm area	*N-cadherin*	Yes	Lele *et al.*, 2002; Masai *et al.*, 2003
pinscher (pic)	Tract sorting, crossing in posterior commissure	?	No	Karlstrom *et al.*, 1996; Trowe *et al.*, 1996
sleepy (sly)	Anterior projections, midline crossing	*laminin γ1*	Yes	Karlstrom *et al.*, 1996; Parsons *et al.*, 2002
smooth muscle omitted (smu)	Midline crossing	*smoothened (smo)*	Yes	Chen *et al.*, 2001; Varga *et al.*, 2001
sonic-you (syu)	Retinal exit, midline crossing	*sonic hedgehog (shh)*	Yes	Brand *et al.*, 1996; Schauerte *et al.*, 1998
space cadet (spc)	Retinal exit, midline crossing	?	No	Karlstrom *et al.*, 1996; Lorent *et al.*, 2001
umleitung (uml)	Midline crossing	?	Yes	Karlstrom *et al.*, 1996
who cares (woe)	Tract sorting, D-V topography	?	No	Trowe *et al.*, 1996
you-too (yot)	Midline crossing	*gli2*	Yes	Karlstrom *et al.*, 1996, 1999

although behavioral defects do not necessarily reflect axon guidance defects, nor vice versa (Neuhauss *et al.*, 1999).

In most of the retinotectal mutants, the axon guidance defect seems to be a secondary consequence of disrupted brain patterning that results in abnormally patterned axon guidance cues (Barresi and Karlstrom, personal communication, 2004; Culverwell and Karlstrom, 2002; Hutson and Chien, 2002b). In others, brain patterning appears to be normal, suggesting that the mutation affects retinal axon guidance more specifically. Of course, one of the most interesting questions is, What are the mutated genes? Eleven retinotectal mutants have been cloned and published (Table I), and most of the remaining have either been cloned or are in the process of being cloned (personal communication from several authors). To date, most of the cloned genes clearly affect brain patterning, either through the Hedgehog pathway [*syu* (*shh*), *smu* (*smo*), *dtr* (*gli1*), *yot* (*gli2*)] or other pathways [*cyc* (*ndr2*), *noi* (*pax2a*), *ace* (*fgf8*)]. The only clear "axon guidance gene" to emerge so far is *astray* (*ast*), which encodes the axon guidance receptor Robo2, a homolog of the *Drosophila* Roundabout receptor first found in *Drosophila* (Fricke *et al.*, 2001; Kidd *et al.*, 1998). RGC axons in *ast* mutants are presumably "blind" to members of the Slit family of axon guidance signals, three of which are expressed in or near the optic pathway (Hutson and Chien, 2002a; Hutson *et al.*, 2003). The lack of Slit-Robo signaling in the retinal axons causes defects in the formation of the optic chiasm and the optic tract. At the current pace, cloning of (nearly) all of the retinotectal mutants will be complete by 2006. This should give us an idea of a large part of the genetic toolkit used to build the visual system. This information will be important for understanding human visual development, since many of these genes have conserved functions in mammals. However, the identification of the genes is only half the battle. We will still need to understand how these genes control retinal axons. It is here that the experimental accessibility of the retinotectal system—the ease of labeling, observing, and perturbing retinal axons—will be especially important.

III. Labeling the Retinotectal System

A. Overview

The discovery of many mutants with specific retinotectal phenotypes (Karlstrom *et al.*, 1996; Trowe *et al.*, 1996) has provided the impetus to develop an array of techniques to visualize retinal axons, using antibody staining, lipophilic dyes, or DNA constructs encoding green fluorescent protein (GFP) or other fluorescent proteins. Each technique is best suited to a particular part of the pathway (inside the retina, near the chiasm, in the optic tract, or on the tectum) and a particular type of sample (few or many axons, live or fixed embryos). A rogues' gallery of labeled embryos is shown in Fig. 1.

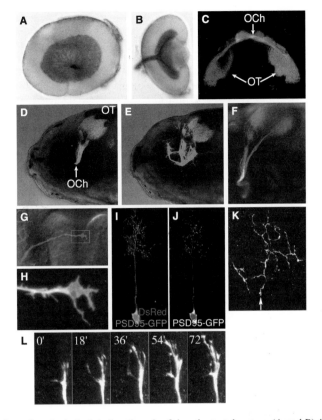

Fig. 1 A variety of methods for labeling the zebrafish retinotectal system. (A and B) 48 hpf zebrafish eye labeled with the zn-8 antibody, which recognizes the cell surface molecule Alcam/Neurolin/DM-GRASP, expressed in all RGCs. Courtesy of Arminda Suli. (A) Lateral view, rostral to the right. (B) Dorsal view, rostral up. (C) Dorsal view of 5 dpf retinotectal projection labeled by intraocular injection (Method 1) with DiO (*green*, left eye) and DiI (*red*, right eye). The eyes have been dissected away. Rostral is up. (D and E) Lateral views of optic tracts labeled with DiO in 5 dpf wild type (D) and *astray* (E) larvae. The eye contralateral to the injected eye was removed in order to facilitate imaging. (F) Lateral view of optic tract and tectum of 6 dpf wild-type larva, after topographic labeling of the eye. DiI (*red*) was injected into dorsonasal retina, and DiO (*green*) was injected into ventrotemporal retina. Courtesy of Jeong-Soo Lee. (G and H) Leading growth cone labeled by intraocular injection of DiI (whole eye fill, Method 1). The *boxed area* in G is shown at higher magnification in H. (G) 20× dry objective. (H) 60× water-immersion objective. (I and J) A single wild-type tectal neuron at 100 hpf, labeled by coinjection of pBSKαtubulin:GAL4 and UAS:PSD95-GFP; UAS:DsRedExpress-1 (Niell *et al.*, 2004; Method 3). The DsRed labels the cytoplasm of the soma and apical dendrite, while PSD95-GFP is preferentially localized to postsynaptic sites. 60×/1.2 water-immersion objective. (I) Merged image. (J) PSD95-GFP image alone. (K) A single retinal axon arbor on the tectum, labeled by injecting plasmid DNA containing the Brn3c promoter driving expression of membrane targeted GFP using GAL4/UAS amplification (Method 3). (L) Time-lapse imaging of a cluster of zebrafish retinal growth cones, labeled with DiI, as they grow through the developing brain over a 72-m period (Method 2). 60× water-immersion objective. (A and B) Obtained on an Olympus BX50WI compound microscope. (C through L) Captured on Olympus FV200 or FV300 confocal microscopes. OCh, optic chiasm; OT, optic tectum. (See Color Insert.)

B. Antibody Labeling

Two antibodies have been widely used to label retinal axons, using standard whole-mount antibody staining techniques. Anti-acetylated tubulin (Sigma St. Louis, MO) recognizes a form of tubulin preferentially found in stable microtubules, and thus labels all axons. This staining has been used to look at the earliest retinal axons crossing the optic chiasm (Karlstrom *et al.*, 1996), but at later stages it becomes hard to distinguish retinal axons amongst the rest of the axon scaffold. Two monoclonal antibodies from the panel generated by Bill Trevarrow (Trevarrow *et al.*, 1990), zn-5 and zn-8 (Zebrafish International Resource Center, Developmental Studies Hybridoma Bank) both recognize Alcam (previously known as either neurolin or DM-GRASP) (Fashena and Westerfield, 1999), Zn-5 and zn-8 are likely duplicate isolates derived from the same hybridoma. Alcam is a cell-surface protein expressed by newly born RGCs and their axons (Kanki *et al.*, 1994; Laessing *et al.*, 1996), which are added in a ring around the periphery of the retina. Since central RGCs have started to turn off Alcam expression by 48 hpf (Laessing *et al.*, 1996), zn-5 or zn-8 staining is especially useful for showing RGC axons navigating within the retina to the optic nerve head (Fig. 1A and B).

C. Lipophilic Dye Labeling

Lipophilic dyes such as DiI, DiO, or DiD are wonderful tools for labeling axons (Honig and Hume, 1989). Structurally, these dyes consist of a fluorophore attached to two long aliphatic tails, which make them lipophilic so that they dissolve easily into cell membranes. Lipophilic dyes applied to live neurons become incorporated into intracellular membrane pools and then are transported both anterogradely and retrogradely by fast axonal transport. In addition, these dyes have the unique ability to label axons in fixed tissue. Since aldehyde fixation crosslinks proteins, but leaves membranes untouched, lipophilic dye applied to a neuron's cell body can dissolve into the plasma membrane, then diffuse through the plane of the membrane, lighting up the entire axon. Because the zebrafish larva is so small, its axons are relatively short, and the required diffusion times for lipophilic dye labeling are only a few hours.

We use mainly DiO and DiI, but sometimes DiD, in several different ways—labeling either the entire retina or specific retinal regions in fixed larvae, or labeling live axons in living embryos. The original retinotectal screen analyzed fixed 5 dpf larvae, using a vibrating-needle injection apparatus (which works much like a tiny tattoo needle), to label the dorsonasal and ventrotemporal quadrants of one eye with DiO and DiI (described in detail by Baier *et al.*, 1996). We have replicated this apparatus and routinely use it for topographic injections (Fig. 1F).

Here we describe a simpler method that can be easily duplicated in any laboratory that uses a pressure injector to label eyes in fixed embryos or larvae (Method 1). Using DiI or DiO dissolved in chloroform, it is quite simple to label the entire retinotectal projection (whole eye fills) in order to analyze axon pathfinding (Fig. 1C–E) and growth cone morphology (Fig. 1G–H). This method can be modified slightly to give reliable topographic labeling of regions of the eye by injecting less dye solution and dissolving the dye in ethanol or dimethylformamide (which gives more localized labeling because the dye precipitates immediately at the point of injection, rather than spreading over the entire RGC layer). Individual ganglion cells can also be retrogradely labeled with DiI, allowing the visualization of their dendritic arbors (Mangrum *et al.*, 2002). We also describe a method for labeling RGCs in live embryos in order to observe axon growth using time-lapse microscopy (Method 2; Hutson and Chien, 2002a).

D. Using GFP or Other Fluorescent Proteins for Labeling

As useful as lipophilic dyes are: (1) In our hands, it is difficult to label single axons using pressure injections (although others have had better success, e.g., Schmidt *et al.*, 2004). Single axon labeling is particularly important for visualizing arbors on the tectum, as neighboring arbors frequently overlap. (2) For large-scale experiments such as screens, lipophilic dye labeling is very time-consuming. These two problems can be overcome by transient expression and stable transgenic expression, respectively, of constructs encoding fluorescent proteins such as GFP. DNA-based labeling methods can also be used to express fusion proteins, for instance to visualize synaptogenesis (Niell *et al.*, 2004). Furthermore, DNA-based methods are not restricted to reporter proteins, but can also be used to express proteins that may perturb axonal development. In Method 3, discussed later in this chapter, we describe how we use these methods to label individual tectal neurons or retinal arbors (Fig. 1I–K).

Transient expression of DNA constructs is comparatively easy in the zebrafish. One can simply inject plasmid DNA (either supercoiled or linearized) for a construct with an enhancer and promoter (usually just abbreviated as "promoter") placed upstream of the desired coding sequence. DNA constructs are expressed mosaically, perhaps because the injected DNA is packaged into "micronuclei," which are inherited at random by a subset of cells in the embryo (Forbes *et al.*, 1983; Westerfield *et al.*, 1992). Several groups have carried out time-lapse imaging of neuronal development using transient GFP expression (Dynes and Ngai, 1998; Köster and Fraser, 2001a) or stable transgenic lines (Bak and Fraser, 2003). For a given experiment, it is necessary to optimize both the promoter and the coding sequence. By using a cell type-specific promoter, it is possible to get expression specifically in the tissue of interest. Promoters from at least four RGC-specific genes have been characterized in zebrafish: *nAChRβ3* (*PAR*) (Tokuoka *et al.*, 2002), *brn3c* (Roeser and Baier, personal communication, 2003), *isl3* (Pittman and Chien, unpublished data, 2003), and *ath5* (Masai, personal communication, 2004). Each of these drives expression starting at a slightly different time and each also drives in a different set of cells outside the retina.

Single axons must express at relatively high levels in order to be visible. Two methods are available for increasing the brightness of GFP expression. First, by changing the coding sequence, GFP localization can be biased to the axon using a membrane-targeting sequence such as the N-terminal palmitoylation sequence from GAP-43 (Moriyoshi *et al.*, 1996). Second, by changing the promoter, either by using a stronger promoter or by using the GAL4/UAS amplification system, in which a tissue-specific promoter drives expression of the strong transcriptional activator GAL4-VP16, which then drives expression from a 14xUAS control element (Köster and Fraser, 2001b). We have successfully used both of these methods to improve labeling of retinal axons. Note however that it is important to control for potential disruptions of cell behavior or viability that may be caused by high levels of either cytoplasmic or membrane-targeted GFP.

Stable transgenic lines can be generated by raising injected founders to sexual maturity, crossing them, and examining their offspring to find those founders with germline integration of the injected DNA construct (Meng *et al.*, 1999). Several stable transgenic lines now exist that express GFP under the control of RGC-specific promoters (Table II; see also review by Udvadia and Linney, 2003). Some of these lines express GFP in the optic pathway or the optic tectum, making it difficult to distinguish retinal axons from surrounding tissue. However, several

Table II
Transgenic Lines that Label Retinal Ganglion Cells

Transgenic line	Retinal expression	Other expression	Reference
ath5:GFP	RGCs, cytoplasmic GFP	Forebrain, tectum	Masai *et al.*, 2003
brn3c:mGFP	Subset of RGCs, membrane-targeted GFP	Inner ear, lateral line neuromasts	Roeser and Baier, personal communication, 2003
isl3:GFP	All RGCs, cytoplasmic GFP	Trigeminal ganglion, Rohon-Beard neurons, a few cells in dorsal midbrain	Pittman and Chien, 2003
nAChRβ3:GFP	RGCs, cytoplasmic GFP	Trigeminal ganglion, Rohon-Beard neurons, some tectal cells	Tokuoka *et al.*, 2002; Yoshida and Mishina, 2003
shh:GFP	RGCs, cytoplasmic GFP	Notochord, floor plate, pharyngeal arch endoderm, ventral forebrain	Neumann and Nuesslein-Volhard, 2000; Roeser and Baier, 2003
[RARE+GATA2 basal promoter]: nGFP	Dorsal retina (weak), ventral retina (strong), nuclear GFP	Caudal hindbrain, spinal cord, notochord	Perz-Edwards *et al.*, 2001
deltaD:GAL4	Drives expression throughout retina from UAS reporter constructs	Not described	Scheer *et al.*, 2001

mGFP, membrane-targeted GFP; nGFP, nuclear-localized GFP; RGC, retinal ganglion cell.

lines allow clear visualization of the retinal projection, making it possible to label large numbers of embryos simply by crossing transgenic carriers. Indeed, we are beginning to use transgenics in new screens for retinotectal mutants.

1. Method 1: Whole Eye Fills in Fixed Embryos

a. Summary

This method uses lipophilic dyes (DiI, DiO, or DiD) to label the entire retinofugal projection in fixed embryos. The dyes are made up at approximately 1% w/v in chloroform (originally suggested by Suresh Jesuthasan). The chloroform evaporates quickly, but the concentration is not critical in chloroform. Since these dyes are very soluble in chloroform, pipette clogging is not usually a problem. When the dye is injected just right into the gap between retina and lens, it spreads out quickly to cover the retinal ganglion cell layer (since chloroform is immiscible with water), not stopping until the dye has dissolved into cell membranes. This labels all of the retinal axons, often so strongly that the optic nerve can even be seen in brightfield.

b. Solutions Needed

4% PFA	4% paraformaldehyde in 0.1 M phosphate buffer, pH 7.4
PBS	phosphate-buffered saline

1.0% DiI, DiO, or DiD in chloroform
  (Molecular Probes, Eugene, OR)
1.5% agarose in 1/3 × PBS*
1.5% low-melt agarose in 1/3 × PBS*
(*At one point it was thought that 1/3 × PBS
  gave better labeling than 1 × PBS, but we
  have not seen a great difference and, at this
  point, its use is historical.)

c. Protocol

1. If the embryos or larvae are pigmented, expose them to bright light from a fiberoptic illuminator for a few minutes before fixation to contract the melanophores so that they do not obscure the tectum. Alternately, raise the embryos in 0.1 mM phenylthiourea (PTU) to prevent melanin formation. PTU also makes it easy to see into the eye being injected and is therefore a good idea when learning the procedure.

2. For fixation, remove as much of the embryo medium as possible and add ice-cold fix (4% PFA), then leave for a few hours at room temperature or overnight at 4 °C. For imaging growth cones and arbors, however, do not use cold fix as it depolymerizes microtubules. Rather, fix for about 30 minutes at room temperature, then store at 4 °C. For lipophilic dye labeling, the

length of fixation is not critical—we often store embryos in fix at 4 °C for months and then have successful injections.

3. For a quick mounting system, make up at least 50 ml of 1.5% agarose in 1/3 × PBS. Do this in an Erlenmeyer flask, so that it will stay hot long enough to work with. Carefully microwave the agarose to melt it. Pipette 1 to 2 ml onto a microscope slide (Fig. 2A).

4. After the agarose cools, use a scalpel or razor blade to slice it into slabs, and separate the slabs slightly (Fig. 2B). Each slab will hold an embryo.

5. Working under a dissecting microscope at 20× magnification, use a pipette or forceps to place the embryos on the agarose, then orient the embryo with forceps or a pin (Fig. 2C). The vertical sides of the cut hold the embryo's tail fin to give a perfect dorsal orientation. Push the agarose slabs together to hold the embryos, then pipette on a bit more agarose to seal the embryos in place. Using 1.5% low-melt agarose for sealing is recommended for easier penetration.

   With a large number of embryos to analyze, an alternate method is to fill the lid of a 60-mm Petri dish with 1.5% agarose in 1/3 × PBS. Once the agarose cools, make several long cuts and arrange embryos head-to-tail in each slab. Each slab can hold up to 25 embryos. Seal in the embryos with 1.5% low-melt agarose as described earlier. Overlay the entire dish with 1/3 × PBS to prevent desiccation.

6. Fill a pulled glass micropipette with 1 to 2 µl of lipophilic dye solution (we use Eppendorf Micro loader tips to back fill). We use a Picospritzer pressure injector for injections, set at 40 psi pressure and 30 ms pulse time

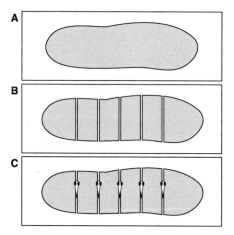

Fig. 2 Embedding of zebrafish embryos or larvae for intraocular injections (whole eye fills), Method 1. (A) Drop of agarose on microscope slide. (B) Agarose with slices made with a scalpel or razor blade. (C) Embryos/larvae inserted into slices. The edges of the cuts hold the tail fin for a perfect dorsal view of the embryo.

(adjust pulse time to change drop size). Working under a dissecting microscope at 40× magnification, eject test droplets into the medium and carefully
break back the tip of the pipette with forceps until you get the desired
drop size.

The eye will be easiest to reach if there is not too much agarose covering
it. Holding the pipette holder either freehand or with a micromanipulator,
pierce the eye with the pipette tip, aiming for the small gap between retina
and lens. Inject a drop of dye. In PTU-treated, *albino*, or *golden* embryos,
you will see the dye instantly fill the RGC layer when the injection is
correct. In pigmented embryos, you may see the eye swell briefly when dye
is injected. If the injection goes awry, dye may go through the eye, back out
of the eye, etc. Stray drops of dye should be teased away with a pin before
they cause background labeling.

7. Allow the dye to diffuse overnight at room temperature or for a few hours at
   28.5 °C in your fish incubator. Then observe on an upright fluorescence
   dissecting scope (no remounting necessary). Overlaying the slides or dishes
   with a little 1/3 × PBS can improve the optics.

8. Use a compound microscope, ideally a confocal microscope, to attain
   higher resolution imaging (Fig. 1C–H). It may be necessary to remount
   your samples. To image the optic chiasm region, use forceps to tease the
   embryos out of the agarose and transfer one at a time to a 24 × 50 mm
   coverslip or coverslip-bottomed Petri dish (Fig. 3). Place a small droplet of
   low-melt agarose over the embryo and hold the embryo in the desired
   orientation until the agarose hardens. To image the optic tract, remove
   the contralateral eye by using an electrolytically sharpened tungsten needle
   to cut away the surrounding skin, sever the optic nerve, and very carefully
   release the eye. We prefer an inverted microscope, but for an upright
   microscope, it helps to sandwich the mounted samples with another coverslip. A 20× objective lens is sufficient to view the entire projection, while a
   40× or 60× objective lens is necessary to image individual growth cones or
   arbors.

2. Method 2: Live Axon Labeling for Time-Lapse

a. Summary

Time-lapse imaging of RGC axons and their arbors has been performed by
many investigators (Gnuegge *et al.*, 2001; Hutson and Chien, 2002a; Kaethner and
Stuermer, 1992; Schmidt *et al.*, 2000). We have obtained better labeling with DiI
than with DiO. However, DiI has been reported to cause greater phototoxicity
than DiO (Kaethner and Stuermer, 1992), so it may be advisable to test both. DiD
also works well, but its long-red emission makes it difficult to screen for labeling
by eye.

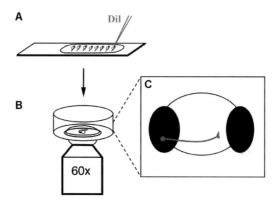

Fig. 3 Time-lapse imaging of retinal axons, Method 2. (A) Injection of DiI into eyes of embryos while immobilized in methylcellulose on a microscope slide. (B) Embryo with DiI-labeled RGCs embedded in agarose in observation chamber (a drilled Petri dish with a coverslip glued on the bottom). (C) Schematic view of ventral diencephalic region of the optic pathway with labeled axon and growth cone. Dorsorostral is up.

b. Solutions and Materials Needed

E2 embryo medium	15.0 mM NaCl, 0.5 mM KCl, 1.0 mM CaCl$_2$, 1.0 mM MgSO$_4$, 0.15 mM KH$_2$PO$_4$, 0.70 mM NaHCO$_3$
E2/GN	10 μg/ml gentamycin in E2
Tricaine stock	0.4% tricaine (MESAB, MS222), 10 mM HEPES, pH 7.4
Methylcellulose	3% methylcellulose, 0.016% tricaine in E2/GN
DiI, DiO, or DiD	0.25% dissolved as well as possible in 95% EtOH
1.5% agarose	1.5% low-melt agarose, 0.016% tricaine in E2/GN

Picospritzer or other microinjection apparatus
Microinjection needles
Microscope slide
Observation chamber (coverslip-bottomed Petri dish, see Fig. 3.)

c. Protocol

1. Label the embryos 3 to 6 h before imaging. This allows time for axons to become well-labeled, without being long enough for dye to enter the blood-stream and give high background fluorescence. The first cohort of axons enters the ventral diencephalon at approximately 32 hpf, enters the contra-lateral optic tract at approximately 38 hpf, and reaches the optic tectum at approximately 48 hpf. Anesthetize the embryos by adding tricaine stock to a final concentration of 0.016%. Carefully transfer the embryos, a few at a time, into a large pool of 3% methylcellulose solution on a microscope slide. Transfer as little embryo medium as possible and remove extra medium using the corner of a tissue.

2. Use forceps or a large-caliber dissecting needle to *carefully* orient the embryos with one eye facing up. The methylcellulose is extremely viscous and can shear the embryos.

3. Microcentrifuge the dye solution briefly to pellet precipitated dye since 0.25% is near saturation for the lipophilic dyes. Fill a micropipette with dye solution, taking care not to transfer any crystals.

   Pressure-inject a very small volume of DiI solution into one eye of each embryo (Fig. 3A). Do not inject between the eye and the lens, as you did for whole eye fills, but instead aim for the body of the neural retina. (The ethanol solution does not penetrate as well as chloroform.) Figure 3 shows the best angle of approach. When injections are complete, place the slide containing the embryos into a Petri dish that contains E2, and transfer the embryos to 28.5 °C for at least 3 h. The E2 will slowly dissolve the methylcellulose, making it easier to liberate the embryos.

4. At the desired stage, release embryos, using a transfer pipette to "blow" away any remaining methylcellulose.

5. Add tricaine to a final concentration of 0.016%. Screen under a fluorescence dissecting microscope for nicely labeled axons.

6. Mount 6 to 8 of your best-labeled embryos. Liquefy the low-melt agarose solution in a glass test tube, then hold it at 40 °C in a heat block. This temperature is close to the maximum that embryos can withstand, but high enough to keep the agarose liquid. Using a fire-polished Pasteur pipette, transfer one embryo at a time into the agarose, then onto a slide (if using an upright microscope) or into the bottom of an observation chamber (if using an inverted microscope; see Fig. 3B). Before the agarose cools, orient the embryo using forceps. If using an upright microscope, you need not coverslip the embryo, and it can be imaged directly using a dipping lens. If using an inverted microscope, partially fill the dish with E2 containing tricaine to keep it wet.

7. Use confocal microscopy because it works well for time-lapse imaging (see Fig. 1L). Phototoxicity is not generally a problem in our hands, although we do try to minimize laser intensity and dwell time. We have been able to take as many as 22 slices per Z-series every 3 minutes, for as long as 8 h (Hutson and Chien, 2002a).

3. Method 3: Single Axon Labeling Using XFP Constructs

a. Summary

This is a method to label small numbers of RGCs by injecting embryos at the 1 through 4 cell stage with plasmids encoding a fluorescent protein (e.g., GFP or DsRed) under the control of an RGC-specific promoter. This gives mosaic expression from the DNA, so that by chance one or a few RGCs will express the construct. The first published report using this method to label single zebrafish RGCs used the nicotinic acetylcholine receptor $\beta3$ subunit promoter (PAR) to drive the expression of soluble enhanced green fluorescent protein (EGFP) to label arbors, and a vesicle associated membrane protein (VAMP)-EGFP fusion construct to label presynaptic sites (Tokuoka *et al.*, 2002). However, in our hands the GFP expression from these constructs is very dim, making it difficult to visualize individual arbors and sites of synapse formation in living embryos (D. S. Campbell and C.-B. Chien, 2004). We find that using a GAL4/UAS amplification cassette (Köster and Fraser, 2001b; Brn3c constructs, gift of Tong Xiao) overcomes this problem and gives sufficiently bright expression in retinal axons (see Fig. 1K). Using a similar set of constructs described by Niell *et al.* (2004) gives very nice labeling of tectal neurons (Fig. 1I–J).

b. Solutions and Materials Needed

DNA constructs	1. pG1 Brn3c:GAL4;UAS:GAP43-GFP (gift of T. Roeser, T. Xiao, M. Smear, and H. Baier). Drives expression of membrane-targeted GFP in RGCs.
	2. pBSKα–tubulin:GAL4 (Köster and Fraser, 2001b). Drives expression in differentiated neurons including tectal cells.
	3. UAS:PSD95-GFP;UAS:DsRedExpress-1 (Niell *et al.*, 2004). Dual construct that labels tectal dendrites with DsRedExpress, and post-synaptic sites with PSD95-GFP, when coinjected with pBSKα–tubulin:GAL4.
Phenol red stock	1% phenol red in deionized water
E2 embryo medium	15.0 mM NaCl, 0.5 mM KCl, 1.0 mM $CaCl_2$, 1.0 mM $MgSO_4$, 0.15 mM KH_2PO_4, 0.70 mM $NaHCO_3$
E2/GN	10 μg/ml gentamycin in E2
E3 medium	5 mM NaCl, 0.17 mM KCl, 0.33 mM $CaCl_2$, 0.33 mM $MgSO_4$
E3/PTU	0.1 mM phenylthiourea (phenylthiocarbamide) in E3
1.5% agarose	1.5% low-melt agarose in E2/GN for injection plates
1% agarose	1% low-melt agarose in E2/GN for mounting embryos

Picospritzer or other microinjection apparatus
Microinjection needles
Coverslip-bottomed dishes

c. Protocol

1. Prepare DNA solution to be injected in deionized water, with a final concentration of 0.1% phenol red to help visualize the injected bolus. For labeling single RGC axons and arbors, the DNA concentration should be approximately 35 ng/μl of pG1 Brn3c:GAL4;UAS:GAP43-GFP. For labeling single tectal cells, the DNA concentration should be approximately

25 ng/μl pBSKα–tubulin:GAL4 and 25ng/μl UAS:PSD95-GFP;UAS:DsRed Express-1. The injected bolus should be approximately 1 nl, or a sphere 120 μm in diameter. This can be measured with an eyepiece micrometer or estimated very roughly as one-sixth the diameter of the yolk cell.

2. Arrange the embryos in grooves in agarose injection trays overlaid with E2/GN and inject DNA into the cell(s) at the 1 to 4 cell stage, using a Picospritzer or similar microinjector.

3. Once injected, transfer the embryos to E3 medium and raise at 28.5 °C.

4. Between 12 and 22 hpf, replace the medium with E3/PTU to prevent melanin formation.

5. At approximately 3 dpf, screen the embryos under a fluorescence dissecting microscope for GFP or DsRed expression. Mount labeled embryos dorsal side down in 1% agarose in a coverslip-bottomed Petri dish for imaging on an inverted confocal microscope. Labeled arbors can be visualized using a 60× water-immersion objective.

6. Amplify the signal (should the GFP appear too dim) by fixing the embryos and staining them with an anti-GFP polyclonal antibody (Molecular Probes, Inc., Eugene, OR) followed by an Alexa 488-labeled ant-rabbit secondary antibody.

7. This method can be extended to examine the effects of perturbing gene function on RGC arbor formation *in vivo*, by coinjection of the pG1 Brn3c:GAL4;UAS:GAP43-GFP construct together with morpholino antisense oligonucleotides.

IV. Perturbing the Retinotectal System

A. Overview

Simple observation of normal embryos can tell us much about development. However, to rigorously test hypotheses about developmental mechanisms, one must perform perturbations. The strategies for perturbing retinotectal development are, broadly, pharmacological manipulation, generation of mosaics by transplantation, loss-of-function perturbation, and gain-of-function perturbation. For pharmacological experiments, drugs have been applied by injection into the eyes (Schmidt *et al.*, 2004; Stuermer *et al.*, 1990), or by bath application (Masai *et al.*, 2003; Schmidt *et al.*, 2000). The penetration of bath-applied agents is presumably aided by the small size of zebrafish larvae.

Transplantation of cells from labeled donors to unlabeled hosts at the blastula or gastrula stage is the classic zebrafish method for creating genetic mosaics to test cell autonomy (Ho and Kane, 1990), and can be used to target the retina quite effectively (Moens and Fritz, 1999). Indeed, using a stable transgenic line expressing GFP under an RGC-specific promoter, it is possible to test cell autonomy of axon guidance mutations (Suli and Chien, 2002). However, such mosaics often have clones of cells in the brain that may make the results difficult to interpret. A method that is more difficult, but perhaps more elegant, is to transplant entire eye primordia (Fig. 4; Method 4). This yields mosaics in which the entire eye comes from the donor, and the rest of the embryo is completely host derived. This method has been used to show that *astray* acts eye-autonomously in retinal axon guidance, while *ace* (*fgf8*) has both eye-autonomous and non-autonomous actions (Fricke *et al.*, 2001; Picker *et al.*, 1999).

Two main methods are used for loss-of-function experiments. The first is obviously, mutant analysis. The second, injection of stable, nontoxic antisense morpholino oligonucleotides (MOs) into one-cell stage embryos, is well established for the short-term disruption of gene function (Nasevicius and Ekker, 2000). MOs directed near a start codon can inhibit protein translation. MOs directed against an exon/intron or intron/exon boundary can inhibit splicing of pre-mRNAs (Draper *et al.*, 2001), and have the advantage that their efficacy can be tested by RT-PCR. A general problem with MOs is that they lose efficacy at later stages, as they become diluted by cell divisions and titrated out by new mRNA synthesis. This is a particular concern for the retinotectal system, which develops relatively late. We have used both translation-blocking and splice-blocking morpholinos to successfully disrupt development of the retinotectal pathway at 48 hpf and in some cases up to 4 dpf (Hutson and Chien, unpublished results, 2002). Another problem is that, for genes with dual roles in early development, as well as in retinotectal development, a potential retinotectal phenotype may be obscured by earlier defects. The discovery that morpholinos appear to be effective when injected into the yolk at later stages suggests a possible method to circumvent this problem (Stenkamp *et al.*, 2000). Blastula- or gastrula-stage cell transplants are another way to avoid early lethality.

It is becoming quite simple to perform gain-of-function experiments (i.e., to incorrectly express genes in particular cells of interest) by using transient or stable transgenic expression driven by cell-specific, tissue-specific, or inducible promoters. The nicotinic acetylcholine receptor-$\beta3$ subunit promoter has been used to drive incorrect expression in RGCs and to analyze the arborization of retinal axons (Tokuoka *et al.*, 2002; Yoshida and Mishina, 2003). The *hsp70* promoter, derived from the heat shock protein 70 gene, allows genes to misexpress either globally using heat shock or locally using focal laser induction (Halloran *et al.*, 2000). Mary Halloran's group (Liu *et al.*, 2004) has used heat shock controlled incorrect expression of Sema3D in order to study its effects on retinal axon guidance. The temporal control provided by the *hsp70* promoter makes it an especially powerful tool, sure to be used widely in the future.

1. Method 4: Eye Transplants

a. *Summary*

This procedure is tricky, but is the most convincing way to determine whether a gene acts eye autonomously or brain autonomously during retinotectal development

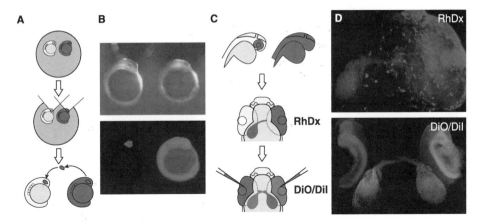

Fig. 4 Eye transplant procedure, Method 4. (A) At 5 somite stage, host embryo (unlabeled) and donor embryo (labeled with rhodamine dextran) are embedded in a small drop of low-melt agarose. Scalpel cuts are used to remove wedges of agarose, giving access to eye primordium, then the host eye is replaced with the donor eye. (B) Brightfield (*top*) and rhodamine fluorescence (*bottom*) micrographs of embedded embryos after removing agarose wedges. (C) Host and donor embryos are raised until 4 to 5 dpf. Rhodamine fluorescence is used to check that the entire eye has been transplanted without any accompanying brain tissue. Lipophilic dye labeling is then used to assay the retinal projections. (D) Confocal images of transplants. *Top*, rhodamine dextran labeling shows the transplanted eye and retinal axons that have grown out into the host brain. Notice a few cells that have migrated out of the transplanted eye. *Bottom*, lipophilic dye labeling shows the host (*red*) and donor (*green*) projections in a wild type-to-wild type transplant. Note that a few donor axons project ipsilaterally (*yellow* labeling in right tectum). (See Color Insert.)

(Fricke *et al.*, 2001). Wild-type to wild-type transplantations can result in certain retinal axon guidance defects (especially ipsilateral projections), so it is important to do transplants in both directions (i.e., wild-type into mutant and mutant into wild-type) and to evaluate the results quantitatively.

b. Solutions and Materials

Fish Ringer's	116 mM NaCl, 3 mM KCl, 4 mM CaCl$_2$, 1 mM MgCl$_2$, 5 mM HEPES, 20 μg/ml gentamycin sulfate. This has been slightly modified from the Ringer's in the Zebrafish Book, (Websterfield, 2003) by addition of magnesium and the raising of calcium to promote healing.
E2 embryo medium	15 mM NaCl, 0.5 mM KCl, 1 mM Ca Cl$_2$, 1 mM Mg SO$_4$, 0.15 mM KH$_2$PO$_4$, 0.05 mM Na$_2$HPO$_4$, 0.70 mM NaHCO$_3$.
E2/PTU	0.1 mM phenylthiourea (phenylthiocarbamide) in E2
Rhodamine dextran	lysine-fixable rhodamine dextran, 3000 MW, 50 mg/ml in water
Low-melt agarose	1.2% low melt agarose in Fish Ringer's

Mineral oil

Picospritzer or other microinjector

18 °C incubator

Agarose injection trays for single-cell injections

Tungsten dissecting needles, electrolytically sharpened

Flame-polished large-bore Pasteur pipette

Scalpel, #11 blade

Nylon crystal-picking loop (0.1–0.2 mm, Hampton Research, Aliso Viejo, CA)

c. Protocol

1. Label donor embryos at the one- to two-cell stage with rhodamine dextran. We use Tübingen-style agarose injecting dishes with long grooves (Gilmour *et al.*, 2002). Hold the pipette holder freehand and, using a Picospritzer, inject the embryos. Use medium forceps to orient the embryos and hold them when withdrawing the needle. It is gentlest to inject dextran into the yolk just below the cell(s). The dextran will be carried upwards by cytoplasmic streaming. Inject just enough to turn the embryo slightly pink. Occasional batches of dextran are said to be much more toxic than usual, presumably because of impurities.

2. Carry out the transplants at 5 to 10 somites (about 12 to 13 hpf), when the eye anlage has just formed a distinct pocket. Before this, the tissue is not firm enough to manipulate; after this, healing is excellent but retinal axons never enter the brain. The reason seems to be that the optic stalk is still quite wide at this stage. At later stages the stalk thins down rapidly, making it impossible to match up the ends of the donor and host stalks. Unless you have fish on a delayed light cycle, the most convenient schedule is to inject dextran the first morning, raise the embryos at 28.5 °C until about 70 to 80% epiboly that afternoon (about 5 P.M.), then shift them to 18 °C to slow their development. They will then reach 5 somites around 9 or 10 A.M. the next morning. At this stage, transfer about 20 donors and about 20 hosts to separate 50-mm glass Petri dishes filled with E2. It is important to match stages well so that the donor and host eyes are the same size. Dechorionate embryos manually with fine forceps. Hereafter, transfer them using a cut-off, flame-polished Pasteur pipette. It is easy to damage the yolks of embryos at this stage, so always use clean tools.

3. Microwave 0.12 g of low-melt agarose in 10 ml of Ringer's in a glass test tube. Hold this at 38 to 40 °C (no hotter) in a heat block. To mount embryos, pick up one donor and one host in the same pipette, drop them briefly in the agarose to wash off the E2, then place them back in the pipette and put them in a drop of agarose on an inverted 35-mm Petri dish lid. Orient them quickly with forceps: lateral view, right eye up, noses in the same direction (see Fig. 4A and B).

4. After orienting several pairs of donors and hosts (typically two or four pairs in each lid), use a clean #11 scalpel to cut wedges in the agarose drops, pointing at the heads of the embryos. Next, flood the dishes with Ringer's. Use a sharpened tungsten needle to flip the wedges of agarose out, starting just underneath the embryo so that no film of agarose is left over the eye (see Fig. 4A and B).

5. Make an entry through the tough embryonic skin. This is a key step. Use a sharpened tungsten needle to nick the skin just dorsal to the eyes of both donor and host. Apply a small drop of mineral oil here. To apply the oil, we use a broken-off micropipette attached to a piece of silastic tubing whose end has been plugged, mounted using a 1-ml tuberculin syringe as a handle. Squeezing the silastic with an index finger gives good control over oil ejection. After the oil droplet wets the skin, it can be removed immediately. After about two minutes, a small bubble of dead, white skin should rise up. This can be flicked away with a needle, leaving a window through which to reach the eye.

6. Use a very sharp tungsten needle with a hooked end to loosen the eye anlage, cutting through the optic stalk and also separating it from overlying skin. Loosen the donor eye first, then the host.

7. Use a fine nylon loop to remove the host eye. (These loops are designed to handle protein crystals for x-ray crystallography. Search for one that is just the right size for the eyes.) It is important to leave the overlying skin to hold the transplanted eye in place. The donor eye should then be moved into place as soon as possible using the loop, and pushed into the pocket formed by the overlying skin.

8. Carefully free the donor and host from the agarose after healing occurs (within 20 to 30 minutes). They can then be raised in E2/PTU to inhibit pigment formation.

9. Label eyes by whole-eye fill at 4 to 5 dpf. It is important to use the rhodamine dextran lineage label to confirm that the entire eye is donor derived, and also to check that no donor cells have been inadvertently transplanted into the brain (see Fig. 4C and D).

V. Future Directions

Two techniques stand out as ones that will undoubtedly be adapted to the study of the zebrafish retinotectal system: *in vivo* electroporation and multiphoton imaging. In *Xenopus* embryos, *in vivo* electroporation is a powerful method for delivering both DNA constructs (Haas *et al.*, 2001) and dextran-coupled indicators (Edwards and Cline, 1999). There seems to be no reason that it should not be just as useful in zebrafish (Teh *et al.*, 2003). Electroporation is especially attractive because it would allow both spatial and temporal control when expressing reporters, gain-of-function constructs, or dominant negative constructs. Indeed, with effective electroporation, it should be possible to use short hairpin constructs driven by Pol III promoters to generate small interfering RNAs (siRNAs) in RGCs, thus specifically inhibiting gene function (see McManus and Sharp, 2002, for review).

Multiphoton excitation has two main advantages (reviewed in Denk and Svoboda, 1997; Helmchen and Denk, 2002): (1) reduced photobleaching and phototoxicity, since multiphoton absorption is restricted to the plane of focus; and (2) improved brightness and resolution in thick or scattering tissues, since infrared light scatters less than visible light and nondescanned detectors can make use of scattered emitted light. Thus it is ideally suited to imaging cell behavior in live thick samples such as the intact zebrafish larva and in particular in the retinotectal system.

Combining these emerging techniques with established methods in the zebrafish retinotectal system promises to make significant contributions to our understanding of the *in vivo* control of axon guidance.

Acknowledgments

We dedicate this chapter to Friedrich Bonhoeffer for his farsighted support of the retinotectal screen. We thank Arminda Suli for Fig. 1A and B, Jeong-Soo Lee for Fig. 1F, and Andrew Pittman and Ichiro Masai for allowing us to mention unpublished transgenic lines. We thank many colleagues for generously providing reagents: Tobias Roeser, Tong Xiao, and Herwig Baier for unpublished Brn3c constructs and transgenic fish; Reinhard Köster for GAL4 and UAS constructs; and Cris Niell and Stephen Smith for PSD95-GFP constructs. Thanks to Nick Marsh-Armstrong and Pam Kainz for suggestions on the eye transplant procedure. This work was supported by NIH F32 EY07017 to LDH, an EMBO Long-Term Fellowship to DSC, and NSF IBN-021385 and NIH R01 EY12873 to CBC.

References

Baier, H. (2000). Zebrafish on the move: Towards a behavior-genetic analysis of vertebrate vision. *Curr. Opin. Neurobiol.* **10,** 451–455.

Baier, H., Klostermann, S., Trowe, T., Karlstrom, R. O., Nusslein-Volhard, C., and Bonhoeffer, F. (1996). Genetic dissection of the retinotectal projection. *Development* **123,** 415–425.

Bak, M., and Fraser, S. E. (2003). Axon fasciculation and differences in midline kinetics between pioneer and follower axons within commissural fascicles. *Development* **130,** 4999–5008.

Brand, M., Heisenberg, C. P., Jiang, Y. J., Beuchle, D., Lun, K., Furutani-Seiki, M., Granato, M., Haffter, P., Hammerschmidt, M., Kane, D. A., *et al.* (1996). Mutations in zebrafish genes affecting the formation of the boundary between midbrain and hindbrain. *Development* **123,** 179–190.

Burrill, J. D., and Easter, S. S., Jr. (1994). Development of the retinofugal projections in the embryonic and larval zebrafish (Brachydanio rerio). *J. Comp. Neurol.* **346,** 583–600.

Chen, W., Burgess, S., and Hopkins, N. (2001). Analysis of the zebrafish smoothened mutant reveals conserved and divergent functions of hedgehog activity. *Development* **128,** 2385–2396.

Culverwell, J., and Karlstrom, R. O. (2002). Making the connection: Retinal axon guidance in the zebrafish. *Semin. Cell Dev. Biol.* **13,** 497–506.

Denk, W., and Svoboda, K. (1997). Photon upmanship: Why multiphoton imaging is more than a gimmick. *Neuron.* **18,** 351–357.

Draper, B. W., Morcos, P. A., and Kimmel, C. B. (2001). Inhibition of zebrafish fgf8 pre-mRNA splicing with morpholino oligos: A quantifiable method for gene knockdown. *Genesis* **30,** 154–156.

Dynes, J. L., and Ngai, J. (1998). Pathfinding of olfactory neuron axons to stereotyped glomerular targets revealed by dynamic imaging in living zebrafish embryos. *Neuron.* **20,** 1081–1091.

Easter, S. S., Jr., and Nicola, G. N. (1996). The development of vision in the zebrafish (Danio rerio). *Dev. Biol.* **180,** 646–663.

Edwards, J. A., and Cline, H. T. (1999). Light-induced calcium influx into retinal axons is regulated by presynaptic nicotinic acetylcholine receptor activity *in vivo. J. Neurophysiol.* **81,** 895–907.

Fashena, D., and Westerfield, M. (1999). Secondary motoneuron axons localize DM-GRASP on their fasciculated segments. *J. Comp. Neurol.* **406,** 415–424.

Forbes, D. J., Kirschner, M. W., and Newport, J. W. (1983). Spontaneous formation of nucleus-like structures around bacteriophage DNA microinjected into Xenopus eggs. *Cell* **34,** 13–23.

Fricke, C., Lee, J. S., Geiger-Rudolph, S., Bonhoeffer, F., and Chien, C.-B. (2001). Astray, a zebrafish roundabout homolog required for retinal axon guidance. *Science* **292,** 507–510.

Gilmour, D. T., Jessen, J. R., and Lin, S. (2002). Manipulating gene expression in the zebrafish. *In* "Zebrafish: A Practical Approach" (C. Nüsslein-Volhard and R. Dahm, eds.), pp. 121–144. Oxford University Press, New York.

Gnuegge, L., Schmid, S., and Neuhauss, S. C. (2001). Analysis of the activity-deprived zebrafish mutant macho reveals an essential requirement of neuronal activity for the development of a fine-grained visuotopic map. *J. Neurosci.* **21,** 3542–3548.

Haas, K., Sin, W. C., Javaherian, A., Li, Z., and Cline, H. T. (2001). Single-cell electroporation for gene transfer *in vivo. Neuron.* **29,** 583–591.

Haffter, P., Granato, M., Brand, M., Mullins, M. C., Hammerschmidt, M., Kane, D. A., Odenthal, J., van Eeden, F. J., Jiang, Y. J., Heisenberg, C. P., *et al.* (1996). The identification of genes with unique and essential functions in the development of the zebrafish, Danio rerio. *Development* **123,** 1–36.

Halloran, M. C., Sato-Maeda, M., Warren, J. T., Su, F., Lele, Z., Krone, P. H., Kuwada, J. Y., and Shoji, W. (2000). Laser-induced gene expression in specific cells of transgenic zebrafish. *Development* **127,** 1953–1960.

Helmchen, F., and Denk, W. (2002). New developments in multiphoton microscopy. *Curr. Opin. Neurobiol.* **12,** 593–601.

Ho, R. K., and Kane, D. A. (1990). Cell-autonomous action of zebrafish spt-1 mutation in specific mesodermal precursors. *Nature* **348,** 728–730.

Honig, M. G., and Hume, R. I. (1989). DiI and diO: Versatile fluorescent dyes for neuronal labeling and pathway tracing. *Trends Neurosci.* **12,** 333–335, 340–341.

Hutson, L. D., and Chien, C.-B. (2002a). Pathfinding and error correction by retinal axons: The role of astray/robo2. *Neuron.* **33,** 205–217.

Hutson, L. D., and Chien, C.-B. (2002b). Wiring the zebrafish: Axon guidance and synaptogenesis. *Curr. Opin. Neurobiol.* **12,** 87–92.

Hutson, L. D., Jurynec, M. J., Yeo, S. Y., Okamoto, H., and Chien, C.-B. (2003). Two divergent slit1 genes in zebrafish. *Dev. Dyn.* **228,** 358–369.

Kaethner, R. J., and Stuermer, C. A. (1992). Dynamics of terminal arbor formation and target approach of retinotectal axons in living zebrafish embryos: A time-lapse study of single axons. *J. Neurosci.* **12,** 3257–3271.

Kanki, J. P., Chang, S., and Kuwada, J. Y. (1994). The molcular cloning and characterization of potential chick DM-GRASP homologs in zebrafish and mouse. *J. Neurobiol* **25,** 831–845.

Karlstrom, R. O., Talbot, W. S., and Schier, A. F. (1999). Comparative synteny cloning of zebrafish you-too: Mutations in the Hedgehog target gli2 affect ventral forebrain patterning. *Genes Dev.* **13,** 388–393.

Karlstrom, R. O., Trowe, T., Klostermann, S., Baier, H., Brand, M., Crawford, A. D., Grunewald, B., Haffter, P., Hoffmann, H., Meyer, S. U., *et al.* (1996). Zebrafish mutations affecting retinotectal axon pathfinding. *Development* **123,** 427–438.

Karlstrom, R. O., Tyurina, O. V., Kawakami, A., Nishioka, N., Talbot, W. S., Sasaki, H., and Schier, A. F. (2003). Genetic analysis of zebrafish gli1 and gli2 reveals divergent requirements for gli genes in vertebrate development. *Development* **130,** 1549–1564.

Kidd, T., Brose, K., Mitchell, K. J., Fetter, R. D., Tessier-Lavigne, M., Goodman, C. S., and Tear, G. (1998). Roundabout controls axon crossing of the CNS midline and defines a novel subfamily of evolutionarily conserved guidance receptors. *Cell* **92,** 205–215.

Köster, R. W., and Fraser, S. E. (2001a). Direct imaging of *in vivo* neuronal migration in the developing cerebellum. *Curr. Biol.* **11,** 1858–1863.

Köster, R. W., and Fraser, S. E. (2001b). Tracing transgene expression in living zebrafish embryos. *Dev. Biol.* **233,** 329–346.

Laessing, U., and Stuermer, C. A. (1996). Spatiotemporal pattern of retinal ganglion cell differetiation revealed by the expression of neurolin in embryonic zebrafish. *J. Neurobiol.* **29,** 65–74.

Lele, Z., Folchert, A., Concha, M., Rauch, G. J., Geisler, R., Rosa, F., Wilson, S. W., Hammerschmidt, M., and Bally-Cuif, L. (2002). Parachute/n-cadherin is required for morphogenesis and maintained integrity of the zebrafish neural tube. *Development* **129,** 3281–3294.

Liu, Y., Berndt, J., Su, F., Tawarayama, H., Shoji, W., Kuwada, J. Y., and Halloran, M. C. (2004). Semaphorin3D guides retinal axons along the dorsoventral axis of the tectum. *J. Neurosci.* **24,** 310–318.

Lorent, K., Liu, K. S., Fetcho, J. R., and Granato, M. (2001). The zebrafish space cadet gene controls axonal pathfinding of neurons that modulate fast turning movements. *Development* **128,** 2131–2142.

Macdonald, R., Scholes, J., Strahle, U., Brennan, C., Holder, N., Brand, M., and Wilson, S. W. (1997). The Pax protein Noi is required for commissural axon pathway formation in the rostral forebrain. *Development* **124,** 2397–2408.

Mangrum, W. I., Dowling, J. E., and Cohen, E. D. (2002). A morphological classification of ganglion cells in the zebrafish retina. *Vis. Neurosci.* **19,** 767–779.

Masai, I., Lele, Z., Yamaguchi, M., Komori, A., Nakata, A., Nishiwaki, Y., Wada, H., Tanaka, H., Nojima, Y., Hammerschmidt, M., *et al.* (2003). N-cadherin mediates retinal lamination, maintenance of forebrain compartments and patterning of retinal neurites. *Development* **130,** 2479–2494.

McManus, M. T., and Sharp, P. A. (2002). Gene silencing in mammals by small interfering RNAs. *Nat. Rev. Genet.* **3,** 737–747.

Meng, A., Jessen, J. R., and Lin, S. (1999). Transgenesis. *Methods Cell Biol.* **60,** 133–148.

Moens, C. B., and Fritz, A. (1999). Techniques in neural development. *Methods Cell Biol.* **59,** 253–272.

Moriyoshi, K., Richards, L. J., Akazawa, C., O'Leary, D. D., and Nakanishi, S. (1996). Labeling neural cells using adenoviral gene transfer of membrane-targeted GFP. *Neuron.* **16,** 255–260.

Nakano, Y., Kim, H. R., Kawakami, A., Roy, S., Schier, A. F., and Ingham, P. W. (2004). Inactivation of dispatched 1 by the chameleon mutation disrupts Hedgehog signaling in the zebrafish embryo. *Dev. Biol.* **269,** 381–92.

Nasevicius, A., and Ekker, S. C. (2000). Effective targeted gene 'knockdown' in zebrafish. *Nat. Genet.* **26,** 216–220.

Neuhauss, S. C. (2003). Behavioral genetic approaches to visual system development and function in zebrafish. *J. Neurobiol.* **54,** 148–160.

Neuhauss, S. C., Biehlmaier, O., Seeliger, M. W., Das, T., Kohler, K., Harris, W. A., and Baier, H. (1999). Genetic disorders of vision revealed by a behavioral screen of 400 essential loci in zebrafish. *J. Neurosci.* **19,** 8603–8615.

Neumann, C. J. (2002). Vertebrate development: A view from the zebrafish. *Semin. Cell Dev. Biol.* **13,** 469.

Neumann, C. J., and Nuesslein-Volhard, C. (2000). Patterning of the zebrafish retina by a wave of sonic hedgehog activity. *Science* **289,** 2137–2139.

Niell, C. M., Meyer, M. P., and Smith, S. J. (2004). *In Vivo* imaging of synapse formation on a growing dendritic arbor. *Nat. Neurosci.* **7,** 254–260.

Parsons, M. J., Pollard, S. M., Saude, L., Feldman, B., Coutinho, P., Hirst, E. M., and Stemple, D. L. (2002). Zebrafish mutants identify an essential role for laminins in notochord formation. *Development* **129,** 3137–3146.

Perz-Edwards, A., Hardison, N. L., and Linney, E. (2001). Retinoic acid-mediated gene expression in transgenic reporter zebrafish. *Dev. Biol.* **229,** 89–101.

Picker, A., Brennan, C., Reifers, F., Clarke, J. D., Holder, N., and Brand, M. (1999). Requirement for the zebrafish mid-hindbrain boundary in midbrain polarisation, mapping and confinement of the retinotectal projection. *Development* **126,** 2967–2978.

Rebagliati, M. R., Toyama, R., Haffter, P., and Dawid, I. B. (1998). Cyclops encodes a nodal-related factor involved in midline signaling. *Proc. Natl. Acad. Sci. USA* **95,** 9932–9937.

Roeser, T., and Baier, H. (2003). Visuomotor behaviors in larval zebrafish after GFP-guided laser ablation of the optic tectum. *J. Neurosci.* **23,** 3726–3734.

Sampath, K., Rubinstein, A. L., Cheng, A. M., Liang, J. O., Fekany, K., Solnica-Krezel, L., Korzh, V., Halpern, M. E., and Wright, C. V. (1998). Induction of the zebrafish ventral brain and floorplate requires cyclops/nodal signalling. *Nature* **395,** 185–189.

Schauerte, H. E., van Eeden, F. J., Fricke, C., Odenthal, J., Strahle, U., and Haffter, P. (1998). Sonic hedgehog is not required for the induction of medial floor plate cells in the zebrafish. *Development* **125,** 2983–2993.

Scheer, N., Groth, A., Hans, S., and Campos-Ortega, J. A. (2001). An instructive function for Notch in promoting gliogenesis in the zebrafish retina. *Development* **128,** 1099–1107.

Schmidt, J. T., Buzzard, M., Borress, R., and Dhillon, S. (2000). MK801 increases retinotectal arbor size in developing zebrafish without affecting kinetics of branch elimination and addition. *J. Neurobiol.* **42,** 303–314.

Schmidt, J. T., Fleming, M. R., and Leu, B. (2004). Presynaptic protein kinase C controls maturation and branch dynamics of developing retinotectal arbors: Possible role in activity-driven sharpening. *J. Neurobiol.* **58,** 328–340.

Schmidt, J. T. (2004). Activity-driven sharpening of the retinotectal projection: The search for retrograde synaptic signaling pathways. *J. Neurobiol.* **59,** 114–133.

Shanmugalingam, S., Houart, C., Picker, A., Reifers, F., Macdonald, R., Barth, A., Griffin, K., Brand, M., and Wilson, S. W. (2000). Ace/Fgf8 is required for forebrain commissure formation and patterning of the telencephalon. *Development* **127,** 2549–2561.

Stenkamp, D. L., Frey, R. A., Prabhudesai, S. N., and Raymond, P. A. (2000). Function for Hedgehog genes in zebrafish retinal development. *Dev. Biol.* **220,** 238–252.

Stuermer, C. A., Rohrer, B., and Munz, H. (1990). Development of the retinotectal projection in zebrafish embryos under TTX-induced neural-impulse blockade. *J. Neurosci.* **10,** 3615–3626.

Teh, C., Chong, S. W., and Korzh, V. (2003). DNA delivery into anterior neural tube of zebrafish embryos by electroporation. *Biotechniques* **35,** 950–954.

Tokuoka, H., Yoshida, T., Matsuda, N., and Mishina, M. (2002). Regulation by glycogen synthase kinase-3beta of the arborization field and maturation of retinotectal projection in zebrafish. *J. Neurosci.* **22,** 10324–10332.

Trevarrow, B., Marks, D. L., and Kimmel, C. B. (1990). Organization of hindbrain segments in the zebrafish embryo. *Neuron.* **4,** 669–679.

Trowe, T., Klostermann, S., Baier, H., Granato, M., Crawford, A. D., Grunewald, B., Hoffmann, H., Karlstrom, R. O., Meyer, S. U., Muller, B., *et al.* (1996). Mutations disrupting the ordering and topographic mapping of axons in the retinotectal projection of the zebrafish, Danio rerio. *Development* **123,** 439–450.

Udvadia, A. J., and Linney, E. (2003). Windows into development: Historic, current, and future perspectives on transgenic zebrafish. *Dev. Biol.* **256,** 1–17.

Varga, Z. M., Amores, A., Lewis, K. E., Yan, Y. L., Postlethwait, J. H., Eisen, J. S., and Westerfield, M. (2001). Zebrafish smoothened functions in ventral neural tube specification and axon tract formation. *Development* **128,** 3497–3509.

Wagle, M., Grunewald, B., Subburaju, S., Barzaghi, C., Le Guyader, S., Chan, J., and Jesuthasan, S. (2004). EphrinB2a in the zebrafish retinotectal system. *J. Neurobiol.* **59,** 57–65.

Westerfield, M., Wegner, J., Jegalian, B. G., DeRobertis, E. M., and Puschel, A. W. (1992). Specific activation of mammalian Hox promoters in mosaic transgenic zebrafish. *Genes Dev.* **6,** 591–598.

Yoshida, T., and Mishina, M. (2003). Neuron-specific gene manipulations to transparent zebrafish embryos. *Methods Cell. Sci.* **25,** 15–23.

CHAPTER 3

Optical Projection Tomography for Spatio-Temporal Analysis in the Zebrafish

Robert J. Bryson-Richardson and Peter D. Currie

Victor Chang Cardiac Research Institute
Sydney 2010, Australia

Information about the spatio-temporal patterning of an embryonic structure, gene, or protein expression pattern can be invaluable in elucidating function. The determination of the three-dimensional (3D) organization of an anatomical structure within a zebrafish embryo has required serial confocal microscopy, followed by reconstruction, or generating a reconstruction from serial sections.

The process of sectioning can damage the morphology of an embryo and cause the loss of 3D organization. Optical sectioning using confocal microscopy removes a lot of these problems in the zebrafish embryo. However, optical section in the zebrafish juvenile and adult stages is prevented by the size of the sample. In addition, for confocal analysis, samples must be fluorescently labeled precluding the analysis of many common labeling techniques such as *in situ* hybridization and *β-Galactosidase* staining. To overcome these problems, we have applied the Optical projection tomography (OPT) technique described by Sharpe *et al.* (2002) and used it to examine phenotype and morphology in juvenile zebrafish.

I. Introduction

Traditionally the analysis of gene expression and anatomy has relied on two-dimensional images, which we try to consider in the 3D context of the organism. While methods do exist for the generation of 3D images, they all have their limitations. These methods include the reconstruction of serial physical sections (Brune *et al.*, 1999; Streicher *et al.*, 2000; Weninger and Mohun, 2002), serial optical sections using confocal microscopy (Cooper *et al.*, 1999), and magnetic resonance imaging (MRI). In the zebrafish, 3D analysis has largely depended on confocal imaging (Bassett *et al.*, 2003; Isogai *et al.*, 2001) or, to a much lesser extent, serial sectioning. Confocal microscopy requires fluorescent labeling of the sample and there are also limits on the size of the sample. Use of two-photon lasers can increase the working depth but it is not feasible to scan entire juvenile fish in this way. Generating 3D information by serial sectioning is a highly time-intensive procedure, and the resulting images do not always join to form a smooth 3D object. While MRI does not require fluorescent labeling, it also has the disadvantage that regular labeling techniques cannot be used to label tissues or expression. Additionally, MRI relies on prohibitively expensive equipment and currently has a limited resolution. A more thorough comparison of OPT to other 3D imaging techniques has been carried out by Sharpe (2003).

Therefore we decided to test the method of OPT described by Sharpe *et al.* (2002) for the characterization of late stage morphology and also for phenotypic analysis of samples beyond the range of confocal analysis. We also tested the ability of OPT to reconstruct early stage embryos (24 h) to determine whether this method is suitable as a single tool for the analysis of zebrafish throughout development.

II. The Principle of Tomography

Tomography is the process by which information about an object is determined from the object's sections or projections. Projection tomography has been used successfully for x-ray scans, positron emission tomography, and electron tomography. In all of these methods, a 3D object is produced from reconstructions using 2D projection images. Examining the projection of an object removes a dimension as demonstrated in Fig. 1, where the projection of a 3D sample is a 2D image. As the light travels through the sample it is reduced in intensity. Where the sample is thickest, very little light passes all the way through. Conversely, where the sample is thinnest, hardly any light is lost. The projection is like a shadow, a 3D object creating a 2D shadow. Except, in this case, the shadow is not uniformly dark but contains information about the thickness and density of the sample.

From a single projection image, we cannot reconstruct the 3D shape of the object. However, if we obtain projection images from a series of angles, it is

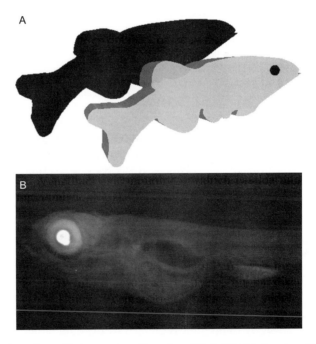

Fig. 1 Projection of an object removes a dimension. (A) A 3D object casts a 2D shadow. In a projection, the shadow is not uniform but contains information about the object. Where the object is thickest, the shadow is darker. Similarly, where the object is thin, there is only a faint shadow. (B) Projection image from a scan of a 2-week-old zebrafish embryo. Using fluorescent imaging, a stronger signal is received from the thicker tissues and those that fluoresce at higher levels. In addition to giving information about the shape of the sample, unlike the shadow, the projection contains information about the internal structures.

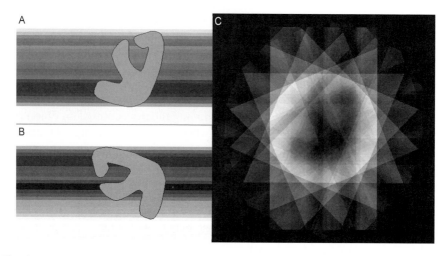

Fig. 2 Back projection. (A) Light is shone through a sample (*blue*) creating a projection (*grey background*). The amount of light passing through the sample is dependent on the thickness of the sample. (B) The sample is rotated 90° counter-clockwise and the process is repeated. (C) To reconstruct the sample, the projection images are rotated to match the angle of the sample and merged together. This image is the result of using nine images of the sample in 22.5° intervals throughout 180°. The contrast of the image has then been adjusted to make the reconstructed object clearer. The shape of the object is clearly visible in the back projection. However, there is a star-shape pattern surrounding the object.

possible to calculate the shape of the object. The object is reconstructed as a series of 2D sections. The sections are subsequently stacked to form a 3D object. The reconstruction process used is a back projection method. In the simplest form of back projection, the projection images are projected back through the object and the combination of the images reconstructs the shape of the original image, as shown in Fig. 2. Such simple reconstruction methods have problems such as the formation of a star-shaped pattern around each object as demonstrated in Fig. 2. To remove these artifacts, a filtered back projection method is used. The Radon transform is a projective transformation that generates a projection from a 2D object. Using the inverse of the Radon transform, we can generate 2D information from the projection. The result of the inverse Radon transformation is still blurred in comparison with the original object and so filtering is carried out to reduce this effect. A ramp filter is applied to the data either before or after back projection. The ramp filter attenuates data according to its frequency, with the greatest effect on low-frequency data and no attenuation of the highest frequency data. The frequency refers to the rate of change of intensities in the projection (i.e., a very bright pixel next to a black pixel has a very high frequency, while a line of pixels with very similar values has a low frequency). This enhances edges in the reconstruction, making the results sharper. In most cases, including OPT, the filter is applied to the projection before reconstruction, as the application of a 1D filter is less computationally intensive than a 2D filter. To reconstruct the object from the series of projections, information must be obtained over at least 180°. As the number of angles used increases, the quality of the reconstruction improves, as the distance between data points is reduced.

III. Problems for OPT in Zebrafish

The technique of OPT was developed using mouse embryos as samples (Sharpe *et al.*, 2002). Obvious differences create problems in applying this technique to zebrafish. Early zebrafish embryos are a fraction of the size of mouse embryos. The higher magnifications used for zebrafish cause a loss of depth of field in the captured images. As a consequence of this, the sample must be orientated along the axis of rotation and carefully adjusted so that as much of the sample as possible stays in focus throughout the rotation of the sample during the scan.

The best results for OPT of whole embryos have been obtained using fluorescent imaging, relying on the autofluorescence of embryonic tissue. In the zebrafish the yolk fluoresces at levels above the embryonic tissue and can reduce the quality of the reconstruction of the embryonic tissue. In antibody-labeled embryos this is not a problem as this fluorescence appears to be quenched during the staining procedure in unstained embryos. However when the morphology is being examined, a fluorescent dye must be used to increase the fluorescence of the embryonic tissue above that of the yolk. The increased fluorescence also results in short exposure times, which in turn reduce background noise in the images.

When examining the early embryos and particularly the juvenile stages, pigmentation blocks the passage of light through the embryo, preventing accurate reconstruction. For this reason, where possible, we use fish with reduced pigmentation such as *golden*. Where this is impossible, we remove pigmentation using the method described in VI. Sample Preparation, B. Removing Pigmentation.

Finally, given the smaller size of zebrafish embryonic structures, we wished to reconstruct images at the highest resolution possible. We therefore captured images approximately four times the size of those routinely used in mouse embryos. While this does result in higher resolution reconstructions, the time taken for reconstruction and subsequent rendering is greatly increased. We have therefore tried to optimize this process where possible.

We describe here the technique we have developed and use for carrying out OPT on zebrafish. This application of OPT to zebrafish samples continues to be developed in our lab, and the OPT technique itself is continually being improved and refined in Edinburgh (J. Sharpe, personal communication, 2004).

IV. Materials

A. Reagents

Methanol series; 100% methanol, 75% methanol, 50% methanol, 25% methanol in H_2O.

2% low melting point agarose in H_2O.

4% paraformaldehyde (PFA) in phosphate-buffered saline (PBS).

Propidium Iodide 0.2 mg/ml.

Bleach solution: 5% Formamide, $0.5 \times SSC$, 10% H_2O_2. This solution is made from a 50% Formamide $5 \times SSC$ stock diluted 1 in 7 and then 3 parts 30% H_2O_2 added. Adding H_2O_2 to the undiluted formamide: SSC solution can be explosive.

Benzyl alcohol: Benzyl Benzoate 2 : 1.

B. Equipment

1-ml syringes with attached 30G 1/2 needles

20-ml glass vials

60- $\times$ 20.3-mm Petri dishes

Charge-coupled device (CCD) camera

Fluorescent stereomicroscope (Leica MZ FLIII, Leica Microsystems, Gladesville, Australia)

IPlab software (Scanalytics, Inc., Fairfax, VA)

Macintosh Computer (OS9)

Mouse Atlas software

OPT reconstruction software

Polyethylene gloves

Reconstruction scripts

MRC OPT Scanner

Visualization Toolkit

Unix or Linux workstation

V. Methods

A. Overview

To carry out OPT, the stained or unstained sample must be fixed and then mounted in agarose before dehydration. The dehydrations and subsequent clearing steps allow the light to pass through the sample. A series of 400 images of the sample are then taken throughout a 360° revolution. These 400 images can then be used to create a 3D model of the sample. This 3D model allows virtual sectioning of the sample in any plane. Alternatively, the sample can be rendered to produce images showing the surface of the sample or to give a 3D representation.

VI. Sample Preparation

A. Fixation

1. Samples should be fixed in 4% PFA at 4 °C at least overnight. In young unstained zebrafish embryos, the autofluorescence of the yolk is much greater than that of embryonic tissue in unstained embryos.
2. To increase embryonic fluorescence above that of the yolk tissue, embryos younger than 72 h should be washed for 1 h at room temperature in propidium iodide at a concentration of 0.2 μg/ml in PBS.
3. The embryos should be washed 5 times in H_2O at room temperature.

B. Removing Pigmentation

For OPT to work, light needs to be able to pass all the way through the embryo. Pigmentation therefore can reduce the quality of, or entirely prevent, reconstruction. For this reason we work, where possible, with strains with reduced or absent pigmentation such as *golden*. This is of course not always possible and, even in these strains, pigmentation still occurs at later stages. Therefore we remove pigmentation using a protocol similar to that described for *Xenopus* (Robinson and Guille, 1999).

1. After fixation, embryos are rapidly dehydrated using 100% methanol for at least 30 m at room temperature.
2. The embryos are then rehydrated through a methanol : water series of 75%, 50%, and 25% methanol and finally H_2O for 10 m at each step.
3. The embryos are then transferred to the bleach solution, illuminated on a light box or microscope stage, and turned every couple of minutes. It should take 10 to 15 m for the pigmentation to disappear, the last pigmentation to go being that of the eye. In older fish the pigmentation is more difficult to remove and may require changing the bleaching solution every 20 m until the pigmentation clears.

C. Mounting the Sample for Scanning

1. Wash the embryos 4 times in H_2O for 5 m to remove any traces of salts that may crystallize during the agarose mounting procedure.
2. Rinse the samples in low melting point (LMP) agarose cooled to 37 °C before mounting. Mount the samples in LMP agarose in a deep 50-mm Petri dish.
3. Orient the sample within the dish using 30G 1/2 needles on a 1-ml syringe. The zebrafish should be vertical within the agarose and such that the center of the sample is approximately 1 cm from the base of the dish. To cool the LMP agarose faster, the mounting can be carried out on a bed of ice to reduce the time until the agarose sets.

4. When the agarose starts to solidify and the sample position is fixed, move the dish to 4 °C for at least 1 h to completely set.

5. Once the agarose is completely set, remove the sample using a plastic borer (made from a cut pipet). Push the borer through the agarose trying to ensure the sample is in the centre of the core to be removed. Cut away the surrounding agarose with a razor blade until the core can be removed.

6. Using a razor blade, trim the agarose to form a cone around the sample.

7. Lower the mounted sample using tweezers into a 20-ml vial containing either 25% methanol 75% H_2O, for samples older than 72 h, or 100% methanol, for samples younger than 72 h and leave at 4 °C overnight. The wash series is used to reduce the amount of shrinking that occurs to the sample during dehydration. If the sample shrinks more as a result of dehydration than the agarose, bubbles are seen around the sample. Samples surrounded by air bubbles cannot be accurately reconstructed. This can be reduced by longer fixation and gradual dehydration. Move samples younger than 72 h to a vial containing benzyl alcohol : benzyl benzoate (BABB) 2 : 1 over night or 5 h or more at 4 °C. Move older samples through a methanol water series of 50%, 75%, 2 × 100% methanol for 5 h or overnight at 4 °C at each step before moving to BABB overnight. Remember to wear polyethylene gloves during all procedures involving benzyl benzoate : benzyl alcohol solutions.

VII. Scanning

1. Using forceps, remove the sample from the glass vial containing BABB and transfer to a glass Petri dish. Plastic dishes can be used but degrade with exposure to BABB. The excess BABB should be removed using tissue and then the sample transferred to the OPT machine. It is best to move the sample in the dish to the machine and then transfer to the OPT scanner to limit the chance of dropping BABB on to the scanner or microscope. If any BABB is spilled, it should be removed immediately to prevent corrosion of the scanner.

2. Once the sample is in the scanner and ready to be scanned, run the OPT scanner software. This guides the user through the process of adjusting the position of the mount, ensuring the specimen remains within the field of view, and is as close as possible to the axis of rotation. Then set the exposure and focus to capture the 400 images throughout the 360° rotation. The OPT scanner software is used in combination with IPlab (Scanalytics, Fairfax, VA). Images can be captured at either the full resolution of the camera (for example 1360 × 1036 pixels) or binning ×2, depending on the resolution required. While the results obtained without binning are of a higher

resolution, there is an increase in the scanning times and a large increase in the time required for all the subsequent reconstruction steps.

3. Save the complete image series as a 12-bit tiff stack and then copied to a CD or DVD for backup and transfer to the Unix workstation used for reconstruction.

4. Where possible, crop the image stack to remove unnecessary blank space around the sample as this reduces the size of the images and significantly reduces the reconstruction times. It is common that the sample will only fill the center of the image, given the shape of fish. When cropping the images, a small space should be left around the sample to prevent edge effects forming too close to the sample.

VIII. Reconstruction

Reconstruction of the data is carried out on a Unix or Linux workstation. The required software is written by the Edinburgh Mouse Atlas project (Baldock *et al.*, 2003) and the Sharpe lab. These programs are run from the command line and require a basic understanding of Unix. The OPT software copies the data to the Unix system and converts the images to the wlz format used by the reconstruction software. Wlz format images can be viewed using the MA3Dview software.

1. The images are analyzed and noise and bad pixels removed using OptPreprocess. The images are then ready to be reconstructed by the OptRecon program.

2. The OptRecon program assumes that the axis of rotation is in the center of the image, which is not always the case. To determine the displacement of the axis of rotation from the center, a series of individual sections throughout the sample are reconstructed using OptRecon to test different displacement values. If an incorrect value is used, the image will become blurred, and objects start to appear as circles as opposed to discrete spots. The optimum value is determined by empirical examination of the reconstructed sections created using a range of displacement values. Once the optimum value has been determined, the entire object is reconstructed using this value.

3. The reconstructed object is a 16-bit grey object. The grey range can be adjusted to reduce background noise and be converted to an 8-bit object using MA3Dview. The object is also cropped to remove excess blank space from around the sample and to reduce the file sizes further still. The resulting object is the end result of the reconstruction process, sections of the object in any plane be can be viewed using the MA3Dview software (Fig. 3). The process of reconstruction is summarized in Fig. 4.

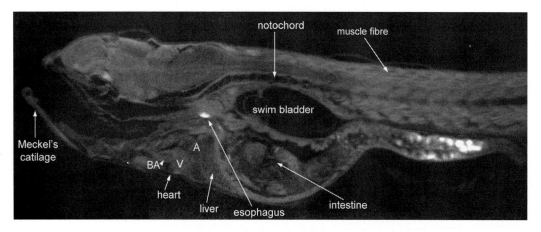

Fig. 3 Parasagittal section through a reconstruction of a 2-week-old juvenile. Many of the organs and tissues are clearly defined, including the heart where the atrium (A), ventricle (V), and bulbus arteriosus (BA) are visible. Further demonstrating the high resolution obtained, individual muscle fibers are discernable within the myotome.

IX. Presentation of Reconstructions

Using MA3Dview software developed by the Edinburgh Mouse Atlas project, the completed reconstruction can be viewed as a series of 2D sections. The data can also now be visualized in 3D by applying surface or volume rendering techniques.

The output from the OPT is a digital 3D object. One of the advantages of this type of data is that it can be visualized in many ways, some of which we present later in this chapter. We can also use rendering techniques to create isosurface models or volume models. Although a thorough description of rendering techniques and software is beyond the scope of this chapter, we present examples of these methods.

A. Surface Rendering

The first step is to convert the 3D object to the .slc format, using the WlzExtFF-Convert program. The .slc format stores the 3D object as a series of contours. Surface rendering produces a solid-appearing object by linking voxels with a gray value above a certain threshold. We use the IsoSurface program that is part of the Visualization Tool Kit (VTK; Schroeder *et al.*, 2003). This creates a surface-rendered object that can be visualized and manipulated using vtkDecimate. For example, if all values in a 3D object of a zebrafish were above 20, creating an

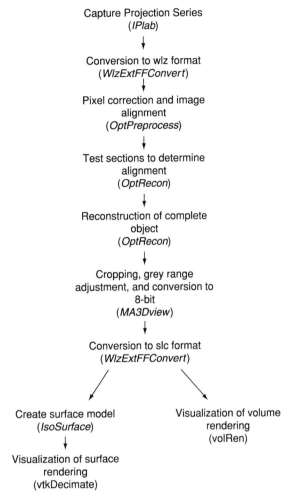

Fig. 4 The process of reconstruction. The major steps in the reconstruction process are shown together with the programs used at those stages (italicized in parenthesis).

isosurface with a threshold of 20 would produce a model of the surface of the whole fish in 3D. If the sample had been labeled with a neuronal-specific antibody, the values in the neurons would be much higher (e.g., 200). The creation of a surface rendering with a cutoff of 200 would therefore produce a model of just the neurons. By reducing the opacity of the model of the whole fish, we can merge the two renderings and have a model showing the 3D pattern within the context of the sample (Fig. 5A).

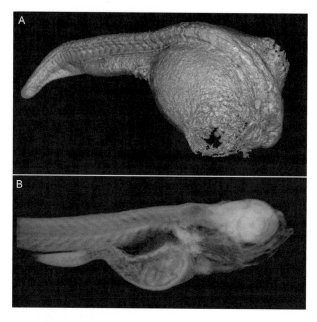

Fig. 5   Volume and surface renderings. (A) Surface rendering of 28-somite stage embryo labeled with an antiacetylated antibody. The 3D reconstruction can be surface rendered to give a better visualization of the 3D object. The rendered object can be rotated to allow any angle to be viewed. Thresholding can be used to identify the antibody signal (*green*) over that of the embryo. Acetylated tubulin is one of the first components of the neuronal microtubule cytoskeleton to form and is a marker of developing post-mitotic neurons (Piperno and Fuller, 1985; Wilson and Easter, Jr., 1991). (B) Volume rendering of a reconstruction of a 2-week-old juvenile zebrafish. Like surface rendering, volume rendering can be used to give a clear 3D representation of the sample. By reducing the opacity of the rendered object, all of the internal features can be visualized. (See Color Insert.)

B. Volume Rendering

The VTK also provides tools for generating volume renderings. Again, using WlzExtFFConvert, the first step is to convert the 3D object to the .slc format. This model can then be viewed using scripts that use functions from VTK, such as volRen. As for surface renderings, thresholds can be applied to remove background noise. The data set is rendered with a reduced opacity so that all of the internal features are still visible compared with surface renderings where a hollow shell representing the shape of embryo is produced (Fig. 5B).

C. Viewing Renderings

While 2D images of these renderings are useful to display the 3D patterns, the best display of a 3D object is achieved when the object can be moved in a movie or by using the software described earlier in this chapter. In addition to direct visualization of these rendered objects, a series of 2D rendered images may be combined to form a movie, which is a convenient way to visualize and distribute the data without specialized software or knowledge.

X. Discussion

We have demonstrated that OPT can be used on zebrafish as small as 26-somite embryos and up to 2-week-old juvenile fish, providing results that are of sufficient quality to present valuable information about a staining pattern or anatomical features. The visualization of morphology of juvenile fish has previously only been possible using serial sectioning, OPT provides a nondestructive method for obtaining the morphology of the sample and additionally provides a complete 3D object that can be further manipulated and rendered.

XI. The Future of OPT

A. Atlases of Morphology and Gene Expression

OPT allows atlases of morphology and gene-expression pattern to be easily generated in a similar way to data that is currently being collected for the mouse. In addition to morphological and phenotypic characterization of zebrafish, these techniques for the generation of high-resolution 3D models at various stages of development will provide valuable staging series and atlases for other species. As only a single embryo would be required at each stage, this would be even more beneficial in the study of organisms where embryonic samples are difficult to obtain.

B. Screening Tool during Larval Stages

The large quantities of information that can be extracted from a single embryo means that OPT is suitable for use in high-throughput screens, characterizing phenotypes or expression patterns at embryonic and larval stages. Recording 3D data from a single embryo prevents the need for repeated sectioning, improving the speed at which screens can be carried out and greatly aiding the analysis of the results. In addition to its use for description, OPT also produces quantitative data in the intensity levels of expression. It may be possible to use this data to compare levels of expression between different tissues in a quantitative manner.

C. Time-Lapse Analysis of Embryonic Development

Given the optical clarity of the early embryo and the ability for embryos to be mounted alive in agarose, it may be possible to image live embryos in 3D throughout development. Although differences in the optical properties of embryonic tissues have currently prevented such an approach, tomographic principles do exist that may allow the imaging of diffracting samples.

XII. OPT Equipment

The OPT technology was invented by J. Sharpe at the MRC Human Genetics Unit in Edinburgh, Scotland, and we performed this work through a collaboration. The MRC is currently redesigning and commercializing the technique (as a fully integrated imaging system), so please contact the Sharpe lab for any requests (James.Sharpe@hgu.mrc.ac.uk).

References

Baldock, R. A., Bard, J. B., Burger, A., Burton, N., Christiansen, J., Feng, G., Hill, B., Houghton, D., Kaufman, M., Rao, J., *et al.* (2003). EMAP and EMAGE: A framework for understanding spatially organized data. *Neuroinformatics* **1,** 309–325.

Bassett, D. I., Bryson-Richardson, R. J., Daggett, D. F., Gautier, P., Keenan, D. G., and Currie, P. D. (2003). Dystrophin is required for the formation of stable muscle attachments in the zebrafish embryo. *Development* **130,** 5851–5860.

Brune, R. M., Bard, J. B., Dubreuil, C., Guest, E., Hill, W., Kaufman, M., Stark, M., Davidson, D., and Baldock, R. A. (1999). A three-dimensional model of the mouse at embryonic day 9. *Dev. Biol.* **216,** 457–468.

Cooper, M. S., D'Amico, L. A., and Henry, C. A. (1999). Confocal microscopic analysis of morphogenetic movements. *In* "Methods in Cell Biology" (H. W. Detrich, M. Westerfield, and L. I. Zon, eds.), pp. 179–204. Academic Press, San Diego.

Isogai, S., Horiguchi, M., and Weinstein, B. M. (2001). The vascular anatomy of the developing zebrafish: An atlas of embryonic and early larval development. *Dev. Biol.* **230,** 278–301.

Piperno, G., and Fuller, M. T. (1985). Monoclonal antibodies specific for an acetylated form of alpha-tubulin recognize the antigen in cilia and flagella from a variety of organisms. *J. Cell Biol.* **101,** 2085–2094.

Robinson, C., and Guille, M. (1999). Immunohistochemistry of Xenopus embryos. *In* "Methods in Molecular Biology" (M. Guille, ed.), pp. 89–97. Humana Press, Totowa, NJ.

Schroeder, W., Martin, K., and Lorenson, B. (2003). "The Visualization Toolkit: An Object Oriented Approach to 3D Graphics." Kitware Inc., New York.

Sharpe, J., Ahlgren, U., Perry, P., Hill, B., Ross, A., Hecksher-Sorensen, J., Baldock, R., and Davidson, D. (2002). Optical projection tomography as a tool for 3D microscopy and gene expression studies. *Science* **296,** 541–545.

Sharpe, J. (2003). Optical projection tomography as a new tool for studying embryo anatomy. *J. Anat.* **202,** 175–181.

Streicher, J., Donat, M. A., Strauss, B., Sporle, R., Schughart, K., and Muller, G. B. (2000). Computer-based three-dimensional visualization of developmental gene expression. *Nat. Genet.* **25,** 147–152.

Weninger, W. J., and Mohun, T. (2002). Phenotyping transgenic embryos: A rapid 3-D screening method based on episcopic fluorescence image capturing. *Nat. Genet.* **30,** 59–65.

Wilson, S. W., and Easter, S. S., Jr. (1991). Stereotyped pathway selection by growth cones of early epiphysial neurons in the embryonic zebrafish. *Development* **112,** 723–746.

CHAPTER 4

Imaging Blood Vessels in the Zebrafish

Makoto Kamei,* Sumio Isogai,† and Brant M. Weinstein*

*Laboratory of Molecular Genetics
National Institute of Child Health and Human Development
National Institutes of Health
Bethesda, Maryland 20892

†Department of 1st Anatomy, School of Medicine
Iwate Medical University
Morioka 020-8505, Japan

I. Introduction

The circulatory system is one of the first organ systems to begin functioning during vertebrate development, and its proper assembly is critical for embryonic survival. Blood vessels innervate all other tissues, supplying them with oxygen, nutrients, hormones, and cellular and humoral immune factors. The heart pumps blood through a complex network of blood vessels composed of an inner single-cell thick endothelial epithelium surrounded by outer supporting pericyte or smooth muscle cells embedded in a fibrillar matrix. The mechanisms of blood vessel growth and morphogenesis are a subject of intensive investigation, and a large number of genes important for blood vessel formation have been identified in recent years. This has been achieved through developmental studies in mice and other animal models. However, our understanding of how these genes work together to orchestrate the proper assembly of the intricate system of blood vessels in the living animal remains limited, in part because of the challenging nature of these studies. The architecture and context of blood vessels are difficult to reproduce *in vitro*, and most developing blood vessels *in vivo* are relatively inaccessible to observation and experimental manipulation. Furthermore, since a properly functioning vasculature is required for embryonic survival and major defects lead to early death and embryonic resorption in amniotes, genetic analysis of blood vessel formation has been largely limited to reverse-genetic approaches.

The zebrafish provides a number of advantages for *in vivo* analysis of vascular development. As noted elsewhere in this book, zebrafish embryos are readily accessible to observation and experimental manipulation. Genetic and experimental tools and methods are available for functional manipulation of the entire organism, vascular tissues, or even single vascular or nonvascular cells. Two features in particular make zebrafish especially useful for studying vascular development. First, developing zebrafish are very small—a 2 dpf embryo is just 2 mm long. Their embryos are so small, in fact, that the cells and tissues of the zebrafish receive enough oxygen by passive diffusion to survive and develop in a reasonably normal fashion for the first 3 to 4 days of development, even in the complete absence of blood circulation. This makes it fairly straightforward to assess the cardiovascular specificity of genetic or experimental defects that affect the circulation. Second, zebrafish embryos and early larvae are virtually transparent. The embryos of zebrafish (and many other teleosts) are telolecithic—yolk is sequestered in a single large cell separate from the embryo proper. The absence of obscuring yolk proteins gives embryos and larvae a high degree of optical clarity. Genetic variants deficient in pigment cells or pigment formation are even more transparent. This remarkable transparency is probably the most valuable feature of the fish for studying blood vessels, facilitating high-resolution imaging *in vivo*.

In this chapter we review some of the methods used to image and assess the pattern and function of the zebrafish vasculature, both in developing animals and in adults. First, we briefly touch on visualizing vascular gene expression (*in situ* hybridization, immunohistochemistry). In the next section we detail methods for imaging blood vessels in fixed developing and adult zebrafish specimens. (Resin and dye injection, alkaline phosphatase staining.) In the final section we describe several methods for imaging blood vessels in living animals (microangiography, time-lapse imaging of transgenic zebrafish with fluorescently tagged blood vessels). Collectively, these methods provide an unprecedented capability to image blood vessels in developing and adult animals.

II. Imaging Vascular Gene Expression

Experimental analysis of blood vessel formation during development requires the use of methods for visualizing the expression of particular genes within blood vessels and their progenitors. Two general methods are available to visualize endogenous gene expression within zebrafish embryos and larvae, *in situ* hybridization and immunohistochemistry. Neither of the methods are specific to the vasculature, and detailed protocols for these methods are available elsewhere (Hauptmann and Gerster, 1994; Westerfield, 2000). *In situ* hybridization is used routinely to assay the spatial and temporal patterns of vascular genes. A variety of different probes are available, some of which are listed in Table I. The *fli1* and *scl* genes are early markers of vascular and hematopoietic lateral mesoderm. The expression of the *fli1* becomes restricted to endothelial cells, a subset of circulating myeloid cells, and cranial neural crest derivatives (Brown *et al.*, 2000; Thompson *et al.*, 1998), while *scl* expression becomes restricted to the hematopoietic lineage at later stages (Gering *et al.*, 1998). The *tie2* gene is a zebrafish ortholog of a mammalian tyrosine kinase receptor for angiopoetin-1 and is expressed in a vascular-specific manner (Lyons *et al.*, 1998). The *flk1* and *flt4* genes (Fouquet *et al.*, 1997; Sumoy *et al.*, 1997; Thompson *et al.*, 1998) are zebrafish orthologs of mammalian endothelial-specific tyrosine kinase receptors for the important vascular signaling molecule *vascular endothelial growth factor (vegf)*. They are initially expressed in hemangiogenic lateral mesoderm then become restricted to angioblasts and endothelium. In the axial vessels of the trunk (dorsal aorta and posterior cardinal vein), *flk1* becomes preferentially expressed in the aorta while *flt4* becomes preferentially expressed in the posterior cardinal vein (similar expression patterns of the corresponding orthologs are observed in mice). Other genes such as *efnb2*, *grl*, and *notch5* are useful as markers of specification of arterial

Table I
Common Vascular Marker Genes Used in Zebrafish Vasculature Research

Marker genes	Expression pattern	Reference
fli1	Pan-endothelial	Brown *et al.*, 2000; Thompson *et al.*, 1998
tie2	Pan-endothelial	Lyons *et al.*, 1998
scl	Hematopoietic	Gering *et al.*, 1998
flk1	Pan-endothelial	Sumoy *et al.*, 1997
flt4	Initially pan-endothelial. Later restricted to vein only	Thompson *et al.*, 1998
efnb2	Artery only	Lawson *et al.*, 2001; Zhong *et al.*, 2000
grl	Artery only	Zhong *et al.*, 2000
notch5	Artery only	Lawson *et al.*, 2001
ephb4	Vein only	Lawson *et al.*, 2001

rather than venous endothelium, although all of these markers also exhibit substantial expression in nonvascular tissues, particularly the nervous system (Lawson *et al.*, 2001; Zhong *et al.*, 2000). Relatively few markers are expressed in venous but not arterial endothelium. The most commonly used is *ephb4*, although *flt4* can also be used to identify venous endothelium as noted earlier.

III. Nonvital Blood Vessel Imaging

A number of methods are available for visualizing the pattern of blood vessels in fixed specimens. Micro-dye injection and micro-resin injection can be used to delineate the patent vasculature (lumenized or open blood vessels connected to the systemic circulation). Both of these methods rely on injection to fill blood vessels with dye or plastic resin that can be visualized in detail after the procedure. Dye injection methods are most useful in embryos and larvae up to a few weeks of age. At later juvenile stages and in adults, tissue opacity and thickness interfere with dye visualization in deeper vessels. Resin injections can thus be more useful. Resin injections are difficult to perform on small specimens (such as embryos) but could be used to visualize vessels at almost any stage of development. While technically challenging, resin injection provides excellent visualization of the adult vasculature, since tissues surrounding the plastic resin are digested away and do not interfere with vessel observation. In addition to these two injection methods for lumenized vessels, staining for the endogenous alkaline phosphatase activity of vascular endothelium can also be used to visualize vessels in fixed specimens. This method is useful for easy, rapid observation of vessel patterns. The method does not require that the vessel be patent, but it cannot be used effectively before approximately 3 dpf because of low signal and high background staining. Even at 3 dpf, the method gives a relatively high background and is not particularly useful for visualizing cranial vessels. This method is also less useful at later stages because of increasing background. We describe the procedures for all of these methods in detail.

A. Micro-Dye and Micro-Resin Injection

Since the nineteenth century, the dye injection method has been the most widely used tool for visualizing the developing circulatory system. Pioneering vascular embryologists such as Florence Sabin carried out their groundbreaking descriptive studies by injecting India ink into blood vessels of vertebrate embryos to reveal their patterns (Evans, 1910; Sabin, 1917). In the 1970s the corrosive resin casting method, previously employed to visualize larger adult blood vessels, was combined with scanning electron microscopy to permit its use for visualizing vessels on a microscopic scale, such as in the developing renal vasculature (Murakami, 1972). Although microangiography and vascular-specific transgenic fish have now become the tools of choice in most cases for visualizing vessels in living zebrafish embryos (discussed later in this chapter), these newer methods have limited usefulness in later stage larvae, juveniles, and adult fish. At these later stages the "classic" dye or resin injection methods still provide the best visualization of the majority of blood vessels. The resin casting method involves injection of a plastic resin that is allowed to harden *in situ*, followed by etching away of tissues to leave behind only the plastic cast. The cast is rotary shadowed and visualized by scanning electron microscopy. The dye injection method described later in this chapter involves injection of Berlin Blue dye followed by fixation and clearing of the embryos or larvae and whole-mount microscopic visualization.

1. Resin Injection Method

a. Materials

- Paraffin bed (Fig. 1)
- Injection apparatus for circulating saline buffer (×2; Fig. 2)
- Injection apparatus for fixative
- Injection apparatus for resin injection (one apparatus per sample to be injected)
- Physiological saline buffer suitable for bony fish
- 2% glutaraldehyde solution in saline buffer (Sigma catalog number G6403, 50% solution in water)
- Methacrylate resin components
- Methyl methacrylate monomer (Aldrich catalog number M55909 or Fluka catalog number 03989)
- Ethyl methacrylate monomer (Aldrich catalog number 234893 or Fluka catalog number 65852)
- 2-hydroxypropy methacrylate monomer (Aldrich catalog number 268542 or Fluka catalog number 17351)

2. Protocol

a. Preparation of the Apparatus

1. The paraffin bed is made in a 9-cm glass Petri dish by pouring molten paraffin wax (Fig. 1). While the wax is solidifying, tilt the dish approximately 15° to create a gentle slope. A depression is made in the middle of the bed for settling a fish.

2. The glass needles are made from stock glass tubing (3-mm outside diameter). The tubes are cut into 10- to 12-cm lengths (Fig. 2). The needles are pulled from the tube by heating the middle of the tube with a Bunsen burner. When the color of the glass tube is changed to red and the tubes feel soft, remove the tube from the heat and pull on both ends. This should produce two injection needles with length of 5 to 6 cm. Let the needles cool down. Then, holding the thick end of the needle by hand and the sharp end of the needle by a pair of forceps, reheat the sharp end

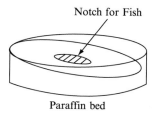

Fig. 1 Paraffin bed used for holding adult zebrafish.

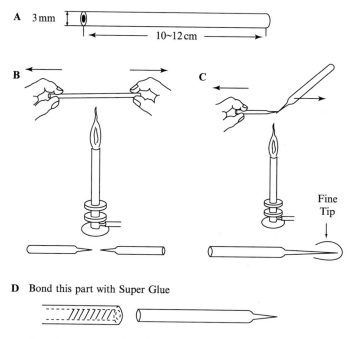

D Bond this part with Super Glue

Fig. 2 Preparation of glass needles for injection of adult zebrafish. Glass stock (A) is pulled on a Bunsen burner (B). The tips are repulled (C). Then the needles are bonded to vinyl tubing with Super Glue (D).

on a Bunsen burner and pull as before. By pulling the needles twice, it is possible to create needles with very fine points. The tip of the needles made in this manner is closed, and the tip needs to be broken open just before use.

3. The apparatus for injecting physiological saline buffer is made by attaching a glass needle (as described in Step 2) to a clear vinyl tubing (3-mm inside diameter, 20 cm in length). When plugging the needle in, a small amount of superglue is applied to reinforce the attachment. For injection of the buffer, a 2- to 10-ml glass syringe is attached, as shown in Fig. 3.

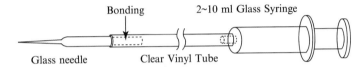

Fig. 3 Apparatus for injecting adult zebrafish with saline and fixative. For injecting resin, the glass syringe is replaced with a disposable plastic syringe.

4. The apparatus for injecting fixative is prepared as described in Step 3.

5. The apparatus for injecting resin is prepared as described in Step 3, except that a 10-ml disposable syringe is attached instead of a glass syringe and heat-resistant silicone tubing is used. Remember, always use caution when working with resin. As resin polymerizes, heat is generated and the viscosity of the resin increases. The heated vinyl tube can detach from the syringe suddenly, causing pressurized, viscous hot resin to splatter. The resin is harmful to skin and mucous membranes, and personal safety measures should always be taken when performing this procedure (e.g., use of goggles, face masks, gloves, and other protective clothing). In addition, make sure that the end of the vinyl tubing is securely attached to the syringe and use heat-resistant silicone tubing for the resin injection apparatus.

6. Preparing resin (methyl methacrylate and ethyl methacrylate monomer). Commercially available methacrylate monomers contain monomethyl ether hydroxy quinone to prevent polymerization, and it is necessary to remove this before use. Prepare 500 ml of 5% NaOH. Pour 100 ml methacrylate monomer and 50 ml 5% NaOH in a separating funnel and shake (Fig. 4A). Wait until the two solutions separate, and then remove the lower 5% NaOH layer (should be brown in color). Repeat until the NaOH solution remains clear, then remove the NaOH by extracting the methacrylate monomer (upper layer) with distilled water. Pour 100 ml of distilled water in the separating funnel containing the methacrylate monomer, shake, and wait until the two layers separate. Remove the lower distilled water layer. Repeat three to four times. Filter methacrylate monomer using double filter paper, and incubate in a 150-ml air-tight container with sodium sulphite overnight (Fig. 4B). Place this container within a desiccator containing silica gel, and store in a refrigerator at 4 °C.

b. Experimental Procedure: Steps 1, 2, and 4 are to be Carried Out Under a Dissecting Microscope

1. Washing the circulatory system with saline buffer. Place anesthetized adult zebrafish on the depressed part of the paraffin bed, ventral side up. Use a pair of watchmaker's forceps and a pair of fine surgical or iridectomy scissors to remove the outer skin and pericardial sac surrounding the heart. Use the forceps to sever the sinus venosus to allow blood to drain. Break the tip of the needle to the size of the ventricle and attach it to a glass syringe containing saline buffer. Stab the

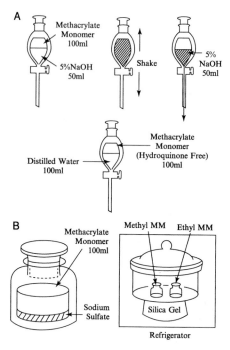

Fig. 4 Preparation of resin for injection. Commercial resin is supplied with monomethyl ether hydroxyquinone to prevent polymerization. This must be extracted before use (A) as described in the text. After extraction, resin is stored refrigerated over sodium sulphate in a desiccator (B).

glass needle into the ventricle in the direction of the head and apply pressure on the syringe to flush the circulatory system with buffer (Fig. 5A). Do not stop until the system is very well flushed out. Flushing should be continued well after the flow from the sinus venosus has become clear saline.

2. Fixation with 2% glutaraldehyde. Break a glass needle as described in Step 2, attach to a syringe containing the 2% glutaraldehyde solution, and start circulating the fixative in the same manner as for the saline buffer (Fig. 5B). Fix well.

3. Mixing the resin. Mix together 3 ml methyl methacrylate monomer, 1.75 ml ethyl methacrylate monomer and 5.25 ml 2-hydroxypropylmethacrylate monomer (to make 10 ml final volume) in a disposable plastic 100-ml cup (Fig. 5C). To this mixture, add 0.15 g benzoyl peroxide (catalyst), 0.15 ml N,N-dimethyl aniline (polymerization agent) and Sudan III (dye), then mix well and sonicate for 2 m.

4. Resin injection. Immediately after sonication, suck the resin mixture up in a 10-ml disposable plastic syringe and attach a glass needle. This time break the needle so that it has a slightly larger bore. Make sure the vinyl tubing is attached to the end of the syringe firmly. Push the needle in the direction of the head of the fish through the same hole used for washing and fixing (Fig. 5D). If possible, push

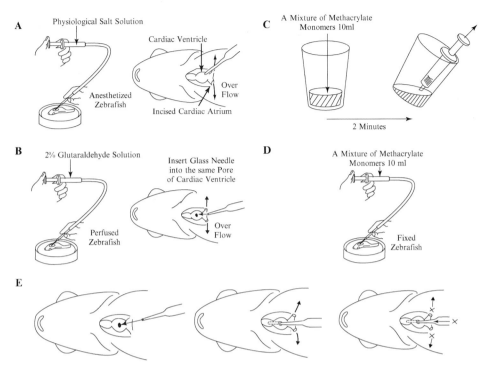

Fig. 5 Resin injection of adult zebrafish. Anesthetized animal is thoroughly flushed with physiological salt solution (A), then with glutaraldehyde solution (B). Mixed resin is taken up into a plastic syringe (C), which is attached to the rest of the resin injection apparatus and the injection is performed (D). The injection should be stopped when the resin hardens and flow ceases (E). See text for details.

the tip of the needle all the way into the arterial cone (back of the ventricle) before beginning the injection. At first, the injection will be easy since the viscosity is low, but after 3 to 5 m, the viscosity will increase and the resin will start to heat up. Keep pushing down on the syringe. After 10 to 12 m, the resin hardens sufficiently for resin flow to stop, at which point the injection should be stopped (Fig. 5E). Wait about 10 m more for resin to fully harden.

5. Digestion of tissue. The injected adult zebrafish is digested with 10 to 20% potassium hydroxide (KOH) for a few days and gently washed with distilled water to remove tissue. The resin cast is dissected out by using watchmaker's forceps and small scissors. Sometimes sonication is used to remove bones and hard-to-remove tissues, but care is required not to destroy the cast itself. For scanning electron microscope (SEM) observation, each local vascular system is divided and trimmed. These procedures must be performed under water, using a dissecting microscope. Each block is frozen in distilled water then freeze dried.

6. SEM. The dried block is mounted on a metal stub and coated with osmium or platinum. Observations are performed with an SEM, using an acceleration voltage of 5 to 10 kV.

3. Dye Injection Method

a. Materials

- Berlin blue dye solution
- Injection apparatus for embryos and early larvae (as described later in this chapter for microangiography), or injection apparatus for juvenile and adult fish (as described earlier in this chapter for resin injection)
- Paraffin bed (for juvenile and adult injections—see resin injection, discussed earlier)

4. Protocol

a. Dye Injection of Embryos and Early Larvae

1. Prepare Berlin blue dye solution by adding 0.5 to 0.75 g Berlin blue powder (Aldrich catalog number 234125, Prussian blue) to 100 ml distilled water and dissolving thoroughly. Filter the solution through double layers of Whatman 3MM filter paper and store in an air-tight bottle.

2. Prepare glass microneedles as for micro angiography (described later).

3. Embed embryos/larvae in agarose as follows (Fig. 6A): Immobilize the embryo in 1X tricaine in embryo media (Westerfield, 2000). Place the embryo on the slide in a drop of tricaine embryo media and remove as much of liquid as possible with a pipette. Place one drop of 1% molten low-melting temperature agarose on the embryo, allowing it to harden and embed the embryo. Attempt to orient the embryo before the agarose hardens so that either its left or right side is facing up.

4. Allow the blood to drain by severing the sinus venosus, using watchmaker's forceps.

5. Remove the agarose covering the caudal half of the trunk or cranial half of the tail with the forceps and add a drop of 1X tricaine.

6. Break the fine tip of a needle to the size of the dorsal aorta or the caudal artery, and, from the needle tip, suck enough Berlin blue solution to cannulate all the vessels thoroughly (Fig. 6B).

7. Target the point just beneath the notochord in order to pierce precisely the dorsal aorta or the caudal artery with the fine tip. To make sure the tip is in the correct position of these vessels, inject the dye for 0.1 sec, using a Picopump. The blood cells move according to the pumping if the tip is in correct position. Inject the dye for 1 to 2 sec and continue the procedure until all the vessels are cannulated thoroughly (Fig. 6C).

8. Fix the dye-injected embryos.

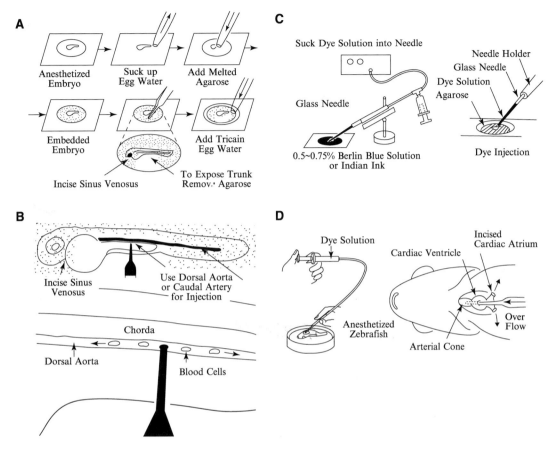

Fig. 6 Mounting and dye injection of developing and adult zebrafish. Embryos and larvae are embedded in agarose on a glass slide for dye injection (A). Injection is performed into the dorsal aorta or caudal artery, with the sinus venosus incised to permit dye to flow through the vasculature (B). The apparatus used for dye injection is shown in (C). For dye injection of adult zebrafish, an apparatus similar to that used for saline and fixative injection before resin injection is employed (D). See text for further details.

b. Dye Injection of Juvenile and Adult Zebrafish: This Method is Similar to the Resin Injection Method Described Earlier

1. Carry out the injection of dye under a dissecting microscope. Place anesthetized adult zebrafish on the depressed part of the paraffin bed ventral side up.

2. Use a pair of watchmaker's forceps and a pair of fine surgical or iridectomy scissors to remove the outer skin and pericardial sac surrounding the heart. Use the forceps to sever the sinus venosus to allow blood to drain during injection (Fig. 6D).

3. Break the tip of the needle to the size of the ventricle and attach it to a glass syringe containing 0.5 to approximately 0.7% Berlin blue solution buffer (apparatus is very similar to that used for saline injection in the resin injection method described earlier). Stab the glass needle into the ventricle in the direction of the head and apply pressure on the syringe to flush the circulatory system with buffer. Continue injection well after the dye begins to flow from sinus venosus. Make sure that the dye is thoroughly injected.

4. Fix the dye-injected sample immediately in either 4% paraformaldehyde or 10% neutral formaldehyde. Store in fixative.

5. For observation, lightly wash the samples and then clear them by passing through 50, 70, 80, 90, and 100% glycerol solution (one solution change per day for a total of five days). Image samples under a dissecting microscope, dissecting away tissues as needed for observation of deeper vessels.

B. Alkaline Phosphatase Staining for 3-dpf Embryos

Zebrafish blood vessels possess endogenous alkaline phosphatase (AP) activity. Endogenous AP activity is not detectable in 24-hpf embryos, but is weakly detectable by 48 hpf and strongly detectable at 72 hpf. Staining vessels by endogenous AP activity is useful for easy and rapid visualization of the vasculature in many specimens, but provides less resolution than many of the other methods. We use protocol modified from Childs *et al.* (2002).

1. Materials

- Fixation buffer: 10 ml 4% paraformaldehyde + 1 ml 10% Triton-X100. Makes 11 ml, scale up or down as needed.
- Rinse buffer: 10 ml 10 × PBS + 5 ml 10% Triton-X100 + 1 ml normal horse serum + 84 ml distilled water. Makes 100 ml, scale up or down as needed.
- Staining buffer: 1 ml 5 M NaCl + 2.5 ml 1 M MgCl2 + 5 ml 1 M Tris pH 9.0–9.5 + 500 μl 10% Tween + 41 ml distilled water. Makes 50 ml, scale up or down as needed.
- Staining solution: 10 ml staining buffer + 45 μl nitro blue tetrazolium (NBT) + 35 μl 5-Bromo-4-chloro-3-indolyl-phosphate (BCIP). Scale up or down as needed.
- NBT: 4-Nitroblue tetrazolium chloride (Boehringer-Mannheim catalog number 1-383-213), 100 mg/ml in 70% dimethylformamide.
- BCIP: X-Phosphate or 5-Bromo-4-chloro-3-indolyl-phosphate (Boehringer-Mannheim catalog number 1-383-221), 50 mg/ml in dimethylformamide.

2. Protocol

    1. Fix at room temperature for 1 h in fixation buffer.

    2. Rinse 1× in rinse buffer.

    3. Wash 5× for 10 m at room temperature in rinse buffer or leave washing in rinse buffer at 4 °C for up to several days. If doing the latter, wash again at room temperature for 10 m before going on to the next step.

    4. Wash 2× for 5 m in staining buffer

    5. Stain in 1 ml of staining solution. Color development takes about 5 to 30 m.

    6. Stop the reaction by washing 3× in rinse buffer without horse serum, then fix in 4% paraformaldehyde for 30 m and store in fixative at 4 °C.

a. Important Notes

1. Avoid putting the embryos in methanol. This destroys endogenous AP activity. If embryos have been placed in methanol, some AP activity can be reconstituted by washing embryos in PBT overnight or even throughout a weekend before starting the staining procedure, although staining will be weaker than in non-methanol-exposed embryos.

2. Complete antibody staining first if combined AP staining and antibody staining [e.g., anti-enhanced green fluorescent protein (EGFP) staining] is desired. After the diaminobenzidine (DAB) staining, do a quick post-fix and go straight into the washes for the AP protocol. AP staining works very well after an antibody stain. Alternatively, stain for exogenous AP phosphatase at a time point when the endogenous vascular form is not active (24 hpf).

IV. Vital Imaging of Blood Vessels

While the methods we have described are useful for visualizing vascular patterns in fixed zebrafish specimens, particularly at later developmental stages and in adults, the zebrafish is perhaps best known for its accessibility to vital imaging methods. A number of vascular imaging methods are available that take advantage of the optical clarity and experimental accessibility of zebrafish embryos and larvae. Confocal microangiography can be used to image blood vessels with active circulation and to detect defects in their patterning or function. Confocal microangiography is performed by injecting fluorescent microspheres into the circulation of living embryos. Then, using a confocal or (preferably) multiphoton microscope, you may collect 3D image "stacks" of the fluorescently labeled vasculature in the living animal. This method can be used from the initiation of circulation at approximately 1-dpf to 10-dpf (or even older) larvae, although injections become progressively more technically challenging to perform after about 2 dpf. Increasing tissue depth makes high-resolution imaging of deep vessels progressively more difficult at later stages. Repeated microangiographic imaging of the same animal and microangiography on animals with impaired circulation may be difficult or impossible to perform.

Recently, transgenic zebrafish with endogenous fluorescently labeled blood vessels have been derived. These animals facilitate high-resolution imaging of the vasculature *in vivo* and made possible long-term time-lapse imaging of the dynamics of blood vessel growth and remodeling. Zebrafish Fli1-EGFP (Lawson and Weinstein, 2002), mTie2-GFP (Motoike *et al.*, 2000), and Flk1-EGFP (Cross *et al.*, 2003) transgenic zebrafish lines have been generated with fluorescently "tagged" endothelial cells and angioblasts. These lines permit imaging of the endothelial cells lining vessels rather than vessel lumens and, as such, they can be used to image vessels that are not carrying circulation, cords of endothelial cells lacking a vascular lumen, or even isolated migrating angioblasts. The Fli-EGFP has already been used to examine mechanisms of cranial (Lawson and Weinstein, 2002) and trunk (Isogai *et al.*, 2003) blood vessel formation. Unlike microangiography, animals can be repeatedly reimaged through an extended period, particularly when multiphoton imaging is employed. Methods for confocal microangiography and time-lapse multiphoton imaging of transgenic animals are described in more detail later in this chapter. Finally, we describe some of the novel insights into *in vivo* vessel formation processes that have already been obtained through use of time-lapse imaging methods.

A. Microangiography

Confocal microangiography is useful for visualizing and assessing the patent circulatory system—blood vessels that are actually carrying blood flow or that at least have open lumens connected to the functioning vasculature. The method facilitates detailed study of the pattern, function, and integrity (leakiness) of vessels. The method is relatively easy to perform, particularly on younger animals and does not require that the animal be of any particular genotype. However, animals with impaired circulation may be difficult or impossible to infuse with fluorescent microspheres.

1. Materials

• Fluoresceinated carboxylated latex beads (0.02 to 0.04 μm), available from Molecular Probes, Eugene, OR. The yellow-green (catalog number F8787), red-orange (catalog number F8794), or dark red (catalog number F8783) beads are suitable for confocal imaging using the laser lines on standard Krypton-Argon laser confocal microscopes. Other colors may be used for when multiphoton imaging is employed. Bead suspension as supplied is diluted 1:1 with 2% BSA (Sigma) in deionized distilled water, sonicated approximately 25 cycles of 1 in each at maximum power on a Branson sonifier equipped with a microprobe, and subjected to centrifugation for 2 m at top speed in an Eppendorf microcentrifuge.

• Omega dot (OD) glass capillaries (1 mm) available from World Precision Instruments (WPI), Sarasota, FL (catalog number TW100-4 without filament or TW100F-4 with an internal filament) for preparing holding and microinjection pipettes.

• Two coarse micromanipulators with magnetic holders and base plates.

• Danieu's solution (30%). To make a 100% Danieu's solution, mix 58 mM NaCl, 0.7 mM KCl, 0.4 mM $MgSO_4$, 0.6 mM $Ca(NO_3)_2$, 5 mM Hepes, pH 7.6 Dilute with reverse osmosis (RO) water to obtain 30% concentration.

• Holding and microinjection pipettes.

• Falcon culture dish (6 cm) available from BD Biosciences, Bedford, MA.

• Micromanipulator and microinjection apparatus.

• Dissecting microscope equipped with epifluorescence optics.

2. Protocol

a. Preparation of the Apparatus

1. The glass microinjection needles are prepared from 1-mm capillaries with internal filaments using a Kopf vertical pipette puller (approximate settings: heat = 12, solenoid = 4.5; see Fig. 7A for desired shape of microneedle). Needles are broken open with a razor blade just behind their tip to give an opening of approximately 5 to 10 μm in width.

2. The holding pipettes are prepared from 1-mm capillaries WITHOUT filaments by carefully partially melting one end of the capillary with a Bunsen burner, such that the opening is narrowed to approximately 0.2 mm (slightly smaller for younger embryos, slightly larger for older larvae). A photographic image of the end of the tip of a holding pipette is shown in Fig. 7B.

3. The apparatus for microinjection is made by attaching a glass microinjection needle (as described in Step 2) to a pipette holder (WPI catalog number MPH6912; adapter for holder and tubing to attach to Pico pump, WPI catalog number 5430). The pipette holder is attached to a controlled air pressure station such as WPI Pneumatic Pico pump (catalog number PV820).

4. The apparatus for holding embryos is made by attaching a glass holding pipette (as described in Step 3) to a pipette holder (WPI catalog number MPH6912). The holding pipettes and their holders are attached via mineral-oil filled tubing (Stoelting Instruments Wood Dale, IL, catalog number 51162) to a manual micro syringe pump (Stoelting Instruments catalog number 51222, with 25 microliter syringe).

5. Holding pipettes and microneedles and their associated holders and other equipment are arranged on either side of a stereo dissecting microscope as diagrammed in Fig. 7C. A photographic image of a typical arrangement is shown in Fig. 7D.

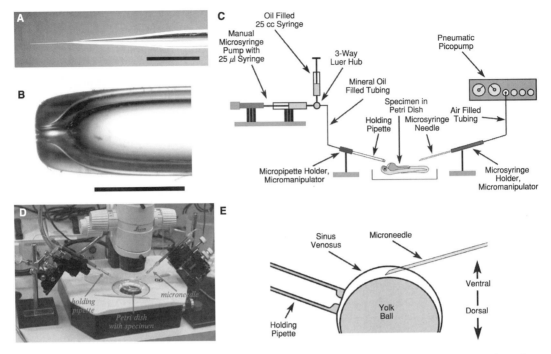

Fig. 7 Microangiography of developing zebrafish embryos and larvae. The desired configurations for injection needles (A) and holding pipettes (B) are shown. A schematic diagram of the apparatus used is shown in (C) and a photographic image of an actual setup is shown in (D). For injection, an embryo is held ventral side up, with suction applied through the holding pipette and injected obliquely through the sinus venosus (E). Older larvae are injected by direct intracardiac injection. See text for details. Scale bars are 3 mm (A) and 1 mm (B).

3. Experimental Procedure

1. Embryos are collected and incubated to the desired stage of development. Use of albino mutant lines or 1-phenyl-2-thiourea (PTU) treatment improves visualization of many vascular beds at later stages. See Westerfield (2000) for PTU treatment protocol.

2. A few mL of fluorescent micro sphere suspension are used to backfill a glass microneedle for injection. The tip should be broken off to the desired diameter just before use.

3. Embryos are dechorionated and anesthetized with tricaine in embryo media.

4. Embryos and larvae (1 to 3 days old) are held ventral side up for injection using a holding pipette applied to the side of the yolk ball (Fig. 7E), with suction applied via a micro syringe driver. Care should be taken to prevent the holding pipette from rupturing the yolk ball. Larvae (4 to 7 days old) are held ventral side up for injection by embedding in 0.5% low melting temperature agarose.

5. For 1- to 3-day-old embryos, a broken glass microneedle is inserted obliquely into the sinus venosus (as diagrammed in Fig. 7E). For 4- to 7-day-old larvae, a broken glass microneedle is inserted through the pericardium directly into the ventricle.

6. After microneedle insertion, many (20+) small boluses of bead suspension are delivered during the course of up to a minute. Smaller numbers of overly large boluses can cause temporary or permanent cardiac arrest. The epifluorescence attachment on the dissecting microscope can be used to monitor the success of the injection.

7. Embryos or larvae are allowed to recover from injection briefly (approximately 1 m) in tricaine-free embryo media, then rapidly mounted in 5% methyl cellulose (Sigma) or low-melt agarose (both made up in embryo media with tricaine). For short-term imaging (generally one stack of images) methylcellulose is applied to the bottom of a thick depression well slide. The rest of the well is carefully filled with 30% Danieu's solution containing 1× tricaine, trying not to disturb the methylcellulose layer below. The injected zebrafish embryo is placed in the well, moved on top of the methylcellulose, and then gently pushed into the methylcellulose in the desired orientation to fully immobilize. Methylcellulose is only useful for short-term mounting because the embryo gradually sinks in the methylcellulose (which also loses viscosity by absorbing additional water over time). For longer term or repeated imaging, animals can be mounted in agarose, using methods such as that described later.

8. Injected, mounted animals are imaged on a confocal or multiphoton microscope using the appropriate laser lines/wavelengths. Although the fluorescent beads are initially distributed uniformly throughout the vasculature of the embryo, within minutes they began to be phagocytosed by and concentrate in selected cells lining the vessels (cf. "tail reticular cells" in Westerfield, 2000). Because of this, specimens must be imaged as rapidly as possible, generally within 15 m after injection. Generally between 20 and 50 frame-averaged (5 frames) optical sections are collected with a spacing of 2 to 5 μm between sections, depending on the magnification (smaller spacing at higher magnifications). Reconstructions (3D) can be generated using a variety of commercial packages, discussed later.

B. Imaging Blood Vessels in Transgenic Zebrafish

Confocal microangiography is a valuable tool for imaging developing blood vessels, but it has limitations. The method is well suited for delineating the luminal spaces of functional blood vessels, but those that lack circulation, vessels that have not yet formed open lumens, and isolated endothelial progenitor cells are essentially invisible. Much of the "action" of early blood vessel formation occurs before the initiation of circulation through the relevant vessels. The first major axial vessels of the zebrafish trunk—the dorsal aorta and cardinal vein—coalesce as defined cords of cells at the trunk midline with distinct molecular arterial-venous identities many hours before they actually begin to carry circulation. Vessels that develop later generally form by sprouting and by the migration of strings of endothelial cells or even individual cells that are likewise undetectable by angiography until well after their initial growth has been completed. Furthermore, because of the leakage of low molecular weight dyes or the pinocytic clearance of microspheres, injected animals can only be imaged for a short time (up to 0.5 h) after injection and repeated imaging requires reinjection of microspheres with different excitation and emission spectra. Thus, for most practical purposes, dynamic imaging of blood vessel growth using this method is not possible. What is needed is a specific and durable fluorescent "tag" for endothelial cells and their angioblast progenitors.

Autofluorescent proteins such as green fluorescent protein (GFP) have been used to mark a variety of tissues in transgenic zebrafish embryos and larvae. Methods for generating germline transgenic zebrafish are now widely used, and their application and resulting lines have been thoroughly reviewed elsewhere (Lin, 2000; Udvadia and Linney, 2003). Tissue-specific expression of fluorescent (or other) proteins in germline transgenic animals is achieved through the use of tissue-specific promoters, and a number of different promoters have been used to drive fluorescent protein expression in zebrafish vascular endothelium. Murine Tie2 promoter (a tyrosine kinase receptor expressed in endothelium activated by angiopoietin ligands) constructs drive GFP expression in endothelial cells in mice and zebrafish, and stable germline transgenic lines have been prepared in both species (Motoike *et al.*, 2000). The usefulness of mTie2-GFP has been limited in fish by the fact that germline transgenic zebrafish show substantial non-vascular expression of GFP in the hindbrain and more posterior neural tube. The overall level of expression from the Murine promoter is also relatively low in fish compared with mice. Germline transgenic zebrafish have also been generated expressing EGFP in the vasculature under the control of the zebrafish Fli1 promoter (Lawson and Weinstein, 2002). Fli1 is a transcription factor expressed in the presumptive hemangioblast lineage and later restricted to vascular endothelium, cranial neural crest derivative, and a small subset of myeloid derivatives (Brown *et al.*, 2000). These lines express abundant EGFP in the vasculature, faithfully recapitulating the expression pattern of the endogenous *fli1* gene and permitting resolution of very fine cellular features of vascular endothelial cells *in vivo*. The Fli1-EGFP transgenic lines have become the most widely used resource for transgenic visualization of blood vessels and have already been used in a variety of different published studies examining the developing trunk and cranial vessels (Isogai *et al.*, 2003; Lawson and Weinstein, 2002; Roman *et al.*, 2002) and regenerating vessels in the adult fin (Huang *et al.*, 2003). Most recently, transgenic zebrafish with fluorescently labeled blood vessels have also been generated by using the promoter for the *flk1* receptor tyrosine kinase to drive EGFP expression in endothelium (Cross *et al.*, 2003).

Here, we review methods for exploiting what is perhaps the most important feature of these transgenics: They permit repeated and continuous imaging of the fluorescently labeled blood vessels. This has made it possible, for the first time, to image the dynamics of blood vessel growth and development of vascular networks in living animals. We describe methods for mounting embryos and larvae for long-term observation and for time-lapse multiphoton microscopy of blood vessels within these animals. The mounting of animals for time-lapse imaging is much more difficult and, in some ways, more critical to the success of the experiment than the actual imaging, which is relatively straightforward to set up on most imaging systems.

1. Long–Term Mounting for Time-lapse Imaging

For time-lapse imaging of blood vessels in transgenics through the course of hours or even days, the animals must be carefully mounted in a way that maintains the region of the animal being imaged in a relatively fixed position, yet keeps the animal alive and developing normally throughout the course of the experiment. This task is complicated further by the fact that developing zebrafish are continuously growing and undergoing morphogenetic movements, which must be accommodated in whatever scheme is used to hold them in place. We describe next a relatively simple mounting method that is adaptable to imaging different areas of embryos or larvae while holding them in place through the course of hours. For time-lapse experiments that run up to a day or more, imaging chambers with buffer circulation are employed. However, we will not describe these chambers and the procedures for mounting animals in them in this chapter.

a. Materials
- Imaging vessels (described later).
- Low-melting temperature agarose (2%) made up in 30% Danieu's solution containing $1\times$ PTU (if non-albino animals are used) and $1.25 \times$ tricaine (if nonparalyzed animals are used).
- Danieu's solution (30%) with or without $1 \times$ PTU and $1.25 \times$ tricaine (see earlier).
- Fine forceps (Dumont #55).

2. Method

a. Preparation of Imaging Chambers
1. Imaging vessels are prepared from 6-cm polystyrene culture dishes (Falcon 3002) and 14-ml polypropylene tubes (Falcon 2059). Model cement is also required for assembly.

2. The polypropylene tube is sliced into 5-mm segments (rings) using a heated razor blade. One ring is glued to the bottom plate of the culture dish, using model cement. Care should be taken to glue the slice of the polypropylene tube to the center of the dish and to avoid smearing the glue inside the polypropylene ring (to avoid obscuring the optical clarity). See Fig. 8A.

3. The glue should be allowed to dry overnight before use.

4. Just before use, the polypropylene ring in the imaging chamber should be slightly overfilled with the low-melting temperature agarose to make a slightly convex dome (Fig. 8A).

b. Mounting Animals in Imaging Chambers

These procedures are done on a dissecting microscope.

1. Dechorionate and select embryos for mounting. Only a single embryo is generally mounted per imaging chamber.

2. Fill the imaging vessel with the 30% Danieu's solution to just below the rim of the Petri dish. If pigment-free, albino mutant embryos are used, the PTU can be left out. If paralyzed mutant embryos are used, the tricaine can be left out of the Danieu's solution.

3. Pick an embryo for mounting and place it in the middle of the agarose bed. (It may begin to roll off because of the convex surface.)

4. Using the fine forceps blades, make a shallow, narrow trench in the center of the agarose dome. This trench should be slightly wider than the dimensions of the embryo in its desired orientation for imaging, with the animal below the surface of the agarose but not too deep (Fig. 8B). It is critical that the trench be carved out carefully to make a space that holds the animal relatively motionless at rest. For imaging most portions of later stage embryos and larvae, the animal should lie on its side in the trench, for lateral view. A larger cavity should be carved out posterior to the tail to accommodate additional increases in trunk or tail length. Additional space should also be left around the head to accommodate shifting and growth, particularly in younger animals.

5. Two additional large cavities should also be carved out perpendicular to the trench on either side (Fig. 8B). These cavities act as anchor points for the agarose layer that is overlaid on top of the embryo.

6. Place the embryo in the trench and slowly overlay with molten agarose. It should be warm enough to freely flow in the buffer, but not too hot to kill the embryo. We typically use glass Pasteur pipette for this since it offers more precise control. Start from one well next to the trench. Apply the agarose at steady rate and, once it fills the well, move over to the other well by moving the pipette over the embryo. Once over the other well, fill up this well (Fig. 8C). This should create an agarose bridge over the embryo and should hold down the embryo.

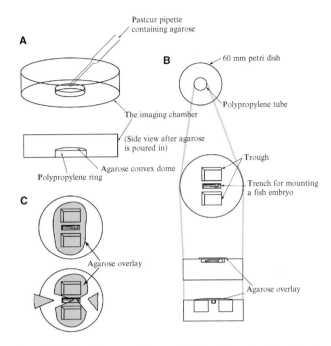

Fig. 8 Mounting zebrafish embryos and larvae for time-lapse imaging. Imaging chambers are prepared from a ring of polypropylene tube glued to a 60-mm Petri dish (A). The ring is filled with low-melt agarose, and cavities carved out to accommodate the animal and to act as anchor points for top agarose (B). After covering the animal with top agarose much of the agarose over the animal is carved away in wedges (C). See text for further details.

7. Cut excess agarose away by using the blades of a pair of fine forceps. For imaging the trunk, we slice away a triangle of agarose over the trunk and tail and over the rostral region, leaving a "bridge" of agarose over the yolk sac, posterior head, and anterior-most trunk sufficient to hold the embryo firmly in place (Fig. 8C). This ensures the optical clarity of the trunk vessels. These cuts are necessary since the embryo is growing in anterior–posterior axis and straightening. Without removing these wedges of agarose, continued growth and straightening of the embryo/larva could not be accommodated.

3. Multiphoton Time–Lapse Imaging

Once animals are properly mounted, they can be imaged by relatively straight-forward time-lapse imaging methods. It is strongly recommended that a multiphoton microscopy system be used for this rather than a standard confocal microscope. The advantages of multiphoton microscopy and its use in developmental studies have been reviewed elsewhere (Denk and Svoboda, 1997;

Weinstein, 2002). Multiphoton imaging reduces photo damage during the course of long imaging experiments, improves resolution of fluorescent structures deep in tissues, and improves the "three dimensionality" of resulting image reconstructions. Most imaging systems designed for or adaptable to multiphoton imaging (Bio-Rad, Leica, Zeiss, etc.) have software interfaces that allow for simple implementation of "4D imaging" (x, y, z, time) experiments. Next we provide a cursory experimental description with some of the important parameters and experimental considerations.

1. Very carefully transfer the imaging chamber with the mounted animal to the stage of a multiphoton microscope, taking care not to dislodge the animal. If the animal is very easily dislodged in transfer, then it was likely not well enough mounted and should be more securely held when remounted.

2. After locating the field to be imaged, the time-lapse parameters should be set. The following guidelines offer a few of the important parameters for multiphoton transgenic blood vessel imaging:

Maximal imaging depth	Approximately 250 μm for best image quality
Objectives	10 to 100×, must pass long wavelength light
Spacing between planes	1 to 5 μm, depending on magnification
Number of planes imaged	10 to 60, depending on region and magnification
Interval between time points	1 to 15 m (5 m is most typical)
Length of time lapse	Up to 24 h, longer if a chamber is used with circulation of warmed buffer. The chamber described earlier is mainly used for shorter time-lapse experiments, and some developmental delay may be noted in longer runs
Laser power setting	Minimum necessary to obtain good images. If possible, increase power with greater depth and use sensitive detectors to permit further decreases in required power
Frames averaged	5 frames averaged/plane

3. Once an imaging run has been initiated, the images being collected should be checked frequently for shifting of the field being imaged. Often some shifting occurs because of growth or morphogenetic movements of the developing animal. Some initial shifting is also sometimes seen at the beginning of a time-lapse run as the animal "settles in." The field being imaged will sometimes need to be adjusted a number of times during the course of an experiment to maintain the vessels being imaged within the field. The stage can be adjusted to reset the X and Y positions. The Z positions (bottom and top of the image stack) may also need to be reset, usually by stopping and restarting the time-lapse program. If excessive shifting occurs because the embryo has been improperly mounted, a new animal should be remounted.

V. Conclusion

Recent evidence suggests that genetic factors are critical in the formation of major vessels formed during early development. Understanding the mechanisms behind the emergence of these early vascular networks will require the use of genetically and experimentally accessible model vertebrates. The accessibility and optical clarity of the zebrafish embryo and larva make it particularly useful for studies of vascular development. Studies of developing vessels are likely to have far-reaching implications for human health, since understanding mechanisms underlying the growth and morphogenesis of blood vessels has become critical for a number of important emerging clinical applications. Proangiogenic and antiangiogenic therapies show great promise for treating ischemia and cancer, respectively, and a great deal of effort is currently going into uncovering and characterizing factors that can be used to promote or inhibit vessel growth *in vivo*. Since many of the molecules that play key roles in developing vessels carry out analogous functions during postnatal angiogenesis, it seems likely that the zebrafish will yield many important clinically applicable insights in the future.

References

Brown, L. A., Rodaway, A. R., Schilling, T. F., Jowett, T., Ingham, P. W., Patient, R. K., and Sharrocks, A. D. (2000). Insights into early vasculogenesis revealed by expression of the ETS-domain transcription factor Fli-1 in wild-type and mutant zebrafish embryos. *Mech. Dev.* **90**, 237–252.

Childs, S., Chen, J. N., Garrity, D. M., and Fishman, M. C. (2002). Patterning of angiogenesis in the zebrafish embryo. *Development* **129**, 973–982.

Cross, L. M., Cook, M. A., Lin, S., Chen, J. N., and Rubinstein, A. L. (2003). Rapid analysis of angiogenesis drugs in a live fluorescent zebrafish assay. *Arterioscler. Thromb. Vasc. Biol.* **23**, 911–912.

Denk, W., and Svoboda, K. (1997). Photon upmanship: Why multiphoton imaging is more than a gimmick. *Neuron* **18**, 351–357.

Evans, H. (1910). "Manual of Human Embryology." J. B. Lippincott & Company, Philadelphia.

Fouquet, B., Weinstein, B. M., Serluca, F. C., and Fishman, M. C. (1997). Vessel patterning in the embryo of the zebrafish: Guidance by notochord. *Dev. Biol.* **183**, 37–48.

Gering, M., Rodaway, A. R., Gottgens, B., Patient, R. K., and Green, A. R. (1998). The SCL gene specifies haemangioblast development from early mesoderm. *Embo. J.* **17**, 4029–4045.

Hauptmann, G., and Gerster, T. (1994). Two-color whole-mount in situ hybridization to vertebrate and Drosophila embryos. *Trends Genet.* **10**, 266.

Huang, C. C., Lawson, N. D., Weinstein, B. M., and Johnson, S. L. (2003). Reg6 is required for branching morphogenesis during blood vessel regeneration in zebrafish caudal fins. *Dev. Biol.* **264**, 263–274.

Isogai, S., Lawson, N. D., Torrealday, S., Horiguchi, M., and Weinstein, B. M. (2003). Angiogenic network formation in the developing vertebrate trunk. *Development* **130**, 5281–5290.

Lawson, N. D., Scheer, N., Pham, V. N., Kim, C. H., Chitnis, A. B., Campos-Ortega, J. A., and Weinstein, B. M. (2001). Notch signaling is required for arterial-venous differentiation during embryonic vascular development. *Development* **128**, 3675–3683.

Lawson, N. D., and Weinstein, B. M. (2002). *In vivo* imaging of embryonic vascular development using transgenic zebrafish. *Dev. Biol.* **248**, 307–318.

Lin, S. (2000). Transgenic zebrafish. *Methods Mol. Biol.* **136**, 375–383.

Lyons, M. S., Bell, B., Stainier, D., and Peters, K. G. (1998). Isolation of the zebrafish homologues for the tie-1 and tie-2 endothelium-specific receptor tyrosine kinases. *Dev. Dyn.* **212,** 133–140.

Motoike, T., Loughna, S., Perens, E., Roman, B. L., Liao, W., Chau, T. C., Richardson, C. D., Kawate, T., Kuno, J., Weinstein, B. M., *et al.* (2000). Universal GFP reporter for the study of vascular development. *Genesis* **28,** 75–81.

Murakami, T. (1972). Vascular arrangement of the rat renal glomerulus: A scanning electron microscope study of corrosion casts. *Arch. Histol. Jap.* **34,** 87–107.

Roman, B. L., Pham, V. N., Lawson, N. D., Kulik, M., Childs, S., Lekven, A. C., Garrity, D. M., Moon, R. T., Fishman, M. C., Lechleider, R. J., *et al.* (2002). Disruption of acvrl1 increases endothelial cell number in zebrafish cranial vessels. *Development* **129,** 3009–3019.

Sabin, F. R. (1917). Origin and development of the primitive vessels of the chick and of the pig. *Carnegie Contrib. Embryol.* **6,** 61–124.

Sumoy, L., Keasey, J. B., Dittman, T. D., and Kimelman, D. (1997). A role for notochord in axial vascular development revealed by analysis of phenotype and the expression of VEGR-2 in zebrafish flh and ntl mutant embryos. *Mech. Dev.* **63,** 15–27.

Thompson, M. A., Ransom, D. G., Pratt, S. J., MacLennan, H., Kieran, M. W., Detrich, H. W., 3rd, Vail, B., Huber, T. L., Paw, B., Brownlie, A. J., *et al.* (1998). The cloche and spadetail genes differentially affect hematopoiesis and vasculogenesis. *Dev. Biol.* **197,** 248–269.

Udvadia, A. J., and Linney, E. (2003). Windows into development: Historic, current, and future perspectives on transgenic zebrafish. *Dev. Biol.* **256,** 1–17.

Weinstein, B. (2002). Vascular cell biology in vivo: A new piscine paradigm? *Trends Cell Biol.* **12,** 439–445.

Westerfield, M. (2000). "The zebrafish book. A guide for the laboratory use of zebrafish (Danio rerio)." University of Oregon Press, Eugene.

Zhong, T. P., Rosenberg, M., Mohideen, M. A., Weinstein, B., and Fishman, M. C. (2000). Gridlock, an HLH gene required for assembly of the aorta in zebrafish. *Science* **287,** 1820–1824.

CHAPTER 5

Zebrafish Apoptosis Assays for Drug Discovery

Chuenlei Parng, Nate Anderson, Christopher Ton, and Patricia McGrath

Phylonix Pharmaceuticals, Inc.,
Cambridge, Massachusetts 02139

I. Introduction

Development of agents that modulate apoptosis is a major focus of biopharmaceutical research. Morphologically, apoptosis is characterized by a series of structural changes in dying cells: blebbing of the plasma membrane, condensation of the cytoplasm and nucleus, and cellular fragmentation into membrane apoptotic bodies (Ellis *et al.*, 1991; Hale *et al.*, 1996; Kerr *et al.*, 1972; William and Smith, 1993). Physiologically, cellular apoptotic pathways function to ensure proper cell numbers and cell fates during development and to prevent the propagation of virus-infected or DNA-damaged cells (Majno and Joris, 1995; Song and Steller, 1999). Pathologically, apoptosis, which is programmed cell death, plays

METHODS IN CELL BIOLOGY, VOL. 76
0091-679X/04 $35.00

an important role in regulating tissue homeostasis. Imbalances between cell death and survival may result in premature death, uncontrolled proliferation, tumor formation, or disease states that include cancer, neurodegeneration, autoimmunity, and heart and renal diseases (Cecconi and Gruss, 2001; Friedlander, 2003). The ability to manipulate apoptosis could become a major strategy for therapeutic intervention in numerous diseases.

The transparency of the zebrafish embryo permits visual observation of *ex utero* development. The organs, including the gut, liver, kidney, and vasculature, are in place by 48 hours. By 72 hours, the hatched embryo has nearly completed morphogenesis. A high degree of evolutionary conservation in the molecular mechanisms of development and cellular physiology exists between zebrafish and mammals (Chen and Fishman, 1996; Granato and Nusslein-Volhard, 1996), underscoring the potential of the system for use as a model for studying human disease. Zebrafish embryos, which are permeable to small molecules (Ho *et al.*, 2003; Milan *et al.*, 2003; Zhang *et al.*, 2003), provide easy access for drug administration and vital dye staining. Small molecules, including peptides, dyes, and drugs, can be dissolved directly in fish water and freely diffuse into zebrafish in the absence or presence of carrier [e.g., dimethyl sulfoxide (DMSO)] (Seng and McGrath, 2002). In addition, our preliminary drug toxicity results demonstrate that the zebrafish embryo exhibits a dose/response comparable to mouse and human and is a good model for toxicity assessment (Zhang *et al.*, 2003).

Here, we describe the zebrafish apoptosis gene, caspase substrate recognition, and *in situ* apoptosis assays, including monitoring caspase activity, and visual and quantitative analysis to assess drug effects on apoptosis for research and therapeutic discovery.

II. Zebrafish Apoptosis Genes

Compared with molecular studies in mice, worms, and flies, analyses using zebrafish as a model system are still in their infancy. However, the zebrafish expressed sequence tags (EST) project has elucidated that zebrafish exhibit most metazoan apoptosis genes, including: 9 Bcl-2 family members (Mcl-1a, Mcl-1b, BLP1, Bcl-xL, Bax, Bad, Nip3, Nip3L), 8 caspases [Caspy (Caspase-1), Caspy2 (Caspase-2), -3, -4, -6, -8, -9, -13], 2 Ced-4-like molecules (Apaf-1 and Nod1), 4 IAP (inhibitor of apoptosis) (IAP1, XIAP, survivin1, survivin 2), 4 death receptors (NGFR1, TNFR, DR6, ZH-DR), the death ligand, TRAIL, 10 apoptosis-related kinases (Ask1a, 1b, Akt1, Akt2, DAP kinase, KRAK1, 2, RICK, RIP3, ZIPK), transcriptional factors (PML1, p53, c-Myc, mdm2, etc.), and many other death-related molecules (Inohara and Nunez, 2000; Lamkanfi *et al.*, 2002). Functional studies by ectopic expression or gene knockdown (induced by antisense morpholinos) have shown that some apoptotic genes, including p53, mdm2, bax, and caspases exhibit conserved proapoptotic or antiapoptotic functions (Hsieh *et al.*, 2003; Langheinrich *et al.*, 2002; Masumoto *et al.*, 2003).

III. Caspase Substrate Recognition in Zebrafish

Caspase inhibitors are useful tools for studying the relationship between caspases and other factors involved in apoptosis. Caspase inhibitors bind to the active site of caspases and form either a reversible or irreversible linkage. Caspases are cysteine-dependent proteases and cleave at the carboxyl side of aspirate residues (Howard *et al.*, 1991; Sleath *et al.*, 1990). Activation of caspases requires cleavage at Asp-X sequences in the proenzymes to yield the large and small subunits that are present in the active $\alpha2\beta$ tetramer (Donepudi and Grutter, 2002; Nicholson, 1999; Takahashi and Earnshaw, 1996). Generally, the structure of a caspase inhibitor includes a peptide recognition sequence attached to a functional group such as an aldehyde (CHO), methylcoumaryl-7-amide (MCA), dimethybenzoyl-oxy-methane (DMB), chloromethylketone (CMK), fluoromethylketone (FMK), or fluoroacyloxy-methylketone (FAOM). Caspase inhibitors that have a CHO group are reversible and those which have an MCA, DMB, CMK, FMK, and FAOM group are irreversible and cell-permeable. Modified N-terminal acetyl (Ac) or carbobenzoxy (Z) groups and a methyl ester amino acid (OME) are designed to enhance membrane permeability. Zebrafish apoptotic proteins exhibit pharmacological responses similar to such responses in mammals. Caspase peptide inhibitors, functional for vertebrate homologues of caspases 1, 4, and 5 (Ac-YVAD-CHO and Ac-YVAD-CMK) (Y = Tyr; V = Val; A = Ala; D = Asp), or caspases 2, 3, and 7 (Ac-DEVD-CHO) (E = Glu) have been shown to inhibit zebrafish DNA fragmentation *in vitro*, thereby modeling apoptosis and preventing nocodazole-induced apoptosis in early development (Chan *et al.*, 1998; Ikegami *et al.*, 1997). Caspase-1, 4, 5 inhibitor Ac-YVAD-CHO prevents camptothecin (a topoisomerase I inhibitor)-induced apoptosis (Ikegami *et al.*, 1999), but caspase-2, 3, 7 inhibitor Ac-DEVD-CHO does not. In addition, recombinant zebrafish caspase-3, which hydrolyzed Ac-DEVD-4-MCA, was inhibited by z-DEVD-FMK and showed low enzymatic activity against other caspase substrates (caspase-1, 6, 8, and 9) and exhibited similar substrate specificity to the mammalian caspase-3 subfamily (Yabu *et al.*, 2001). These data suggest that zebrafish caspases share identical substrate recognition sites with their mammalian homologues and likely target the proteins with the same catalytic ability *in vivo*.

IV. Detection of Apoptosis *In Situ*

Apoptosis can be visualized in live zebrafish embryos using acridine orange, fluorogenic caspase substrate, and inhibitor staining. These reagents can be incubated with dechorionated zebrafish directly or injected into the yolk (Furutani-Seiki *et al.*, 1996). Other cyanine dyes such as YO-PRO-1 and SYTOX, which have been shown to be permeable to apoptotic cells (Daly *et al.*, 1997; Idziorek *et al.*, 1995), did not work well for us in live zebrafish.

A. Acridine Orange Staining

Acridine orange, which is a metachromatic intercalator sensitive to DNA conformation, has been widely used to detect apoptosis in wild-type, mutant, and drug-treated zebrafish (Brand *et al.*, 1996; Furutani-Seiki *et al.*, 1996; Parng *et al.*, 2002). Similar to other animals, in zebrafish embryos, programmed cell death maintains the homeostasis of organ development, removes surplus cells to form proper morphology, and ablates neurons that make inappropriate connections in zebrafish embryos. Consistent patterns of apoptosis can be clearly observed in: dorsal neural tube, hatching glands, retina, lens, cornea, inner ear, olfactory organs, notochord, somites, muscle, tail bud, fins, Rohon-Beard neurons, and lateral-line neuromasts (Cole and Ross, 2001; Furutani-Seiki *et al.*, 1996; Parng *et al.*, 2002; Svoboda *et al.*, 2001; Yamashita, 2003). These organs and tissues are rapidly developing at early stages of embryogenesis and exhibit complex mechanisms for cell differentiation, proliferation, and apoptosis. Inappropriate cell death creates abnormal patterns of apoptosis. In response to proapoptotic agents, cell death increases in zebrafish embryos. In contrast, antiapoptotic drugs decrease cell death. As shown in Fig. 1, neomycin-induced apoptosis in the liver and retinoic acid-induced apoptosis in the brain were observed by acridine orange staining.

B. Quantitative Acridine Orange Assay

The overall approach for this quantitative acridine orange assay is depicted in Fig. 2. Since acridine orange bound to chromatin can be easily extracted from zebrafish embryos after lysis or permeabilization with 100% ethanol (Parng *et al.*, 2002), as a general index for apoptosis, we evaluated the use of a fluorescent microplate reader to quantitate acridine orange. At Ex.490 nm/Em.530, the level of fluorescence increased in a linear relationship to acridine orange concentration (0.01 μg/ml to to 1 μg/ml), indicating that acridine orange can be quantitated in solution (Fig. 3A). In addition, to ascertain that the amount of fluorescence increases proportional to the number of acridine-orange stained cells, we performed dose response experiments. As shown in Fig. 3B, the increase in fluorescence was proportional to the number of embryos per well, indicating that changes in apoptosis in zebrafish can be quantitated. As shown in Fig. 4, apoptosis increased in a dose-dependent manner after neomycin and staurosporine treatment. This vital dye assay has signification potential for high throughput quantitative drug assessment using a conventional liquid handling workstation and a microplate format.

C. Fluorogenic Caspase Substrate

Fluorogenic caspase substrates can be used to determine the sites of apoptosis. After cleavage, the shift in emission of AMC (7-amino-4-methoxy coumarin) (Ex. 380; Em 460) or AFC (7-amino-4-trifluoromethyl coumarin)

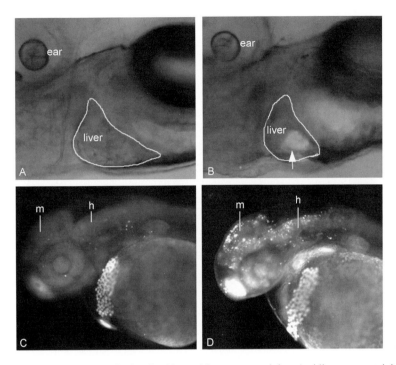

Fig. 1 Drug-induced apoptosis visualized by acridine orange staining. Acridine orange staining was performed, as described previously (Furutani-Seiki *et al.*, 1996). Untreated zebrafish (A), 4-day-old zebrafish were treated with 1 μg/ml neomycin for 24 h (B). Liver apoptosis can be observed after neomycin treatment. Embryos, (20 hpf) were treated with vehicle alone (C) or with 1 μM retinoic acid (D). Aberrant apoptosis was observed in the midbrain (m) and hindbrain (h) regions. (See Color Insert.)

(Ex. 400; Em 505)-conjugated peptides can be visualized by fluorescence microscopy. As shown in Fig. 5, developmental apoptosis in the retina, otolith, and pronephros was detected by fluorescent caspase-3 substrate (Ac-DEVD-AMC). Fluorometric and colormetric caspase substrates can be used to detect caspase activity using whole zebrafish lysate. Zebrafish are homogenized in a lysis buffer (0.5% Triton X-100, 0.2% SDS, 1 mM EDTA, 1× protease inhibitor in phosphate buffer saline (PBS, pH 7.2) with a homogenizer. Lysate is centrifuged at 12,000*g* for 30 m and supernatant is used for caspase activity assay. Additional protein concentration measurement is required to normalize the enzyme activity.

D. Fluorochrome Inhibitor of Caspases

Fluorochrome inhibitor of caspases (FLICA) has been widely used for *in vivo* staining (Bender *et al.*, 2000; Smolewski *et al.*, 2001). Fluorescein isothiocyanate (FITC), carboxyfluoresceir (FAM), or sulforhodamine (SR) group has

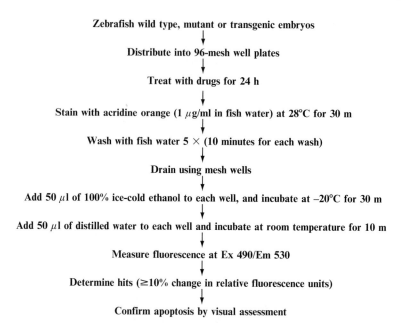

Zebrafish wild type, mutant or transgenic embryos

Distribute into 96-mesh well plates

Treat with drugs for 24 h

Stain with acridine orange (1 μg/ml in fish water) at 28°C for 30 m

Wash with fish water 5 × (10 minutes for each wash)

Drain using mesh wells

Add 50 μl of 100% ice-cold ethanol to each well, and incubate at –20°C for 30 m

Add 50 μl of distilled water to each well and incubate at room temperature for 10 m

Measure fluorescence at Ex 490/Em 530

Determine hits (≥10% change in relative fluorescence units)

Confirm apoptosis by visual assessment

Fig. 2 Flow diagram of quantitative acridine orange assay. In this format, zebrafish embryos can be sorted and deposited into 96-well plate by a large particle sorter, COPAS (complex object parametric analyzer and sorter, Union Biometrica, Somerville, MA) (Schneider, 2003). A robotic workstation can be used to dispense drug into wells, and mesh well plate (MultiScreen-Mesh well plates, Millipore, Billerica, MA) can be used to assist washing and draining.

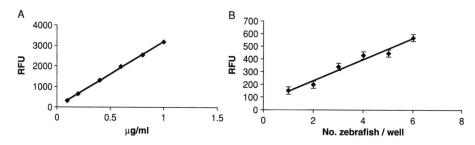

Fig. 3 Quantitative acridine orange assay. The soluble acridine orange solution exhibited a linear increase in fluorescence units at Ex. 490/Em. 530 (A). In addition, a linear relationship between the number of zebrafish per microwell and relative fluorescence units (RFU) was observed. Varying numbers of zebrafish (from 1 to 6) were deposited into individual wells of a 96-well plate and processed. RFU was measured. Using 5-day-old embryos, a linear response was observed. Each data point represents the average of 5 experiments.

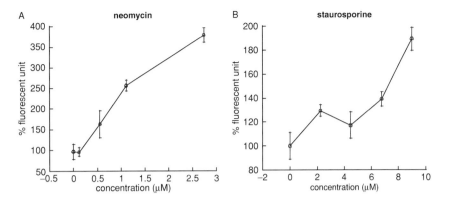

Fig. 4 Dose-dependent apoptosis can be quantitated using a microplate reader. Four-day-old embryos were treated with neomycin (A) and staurosporine (B) for 24 h and fluorescence was measured. Neomycin- and staurosporine-induced apoptosis occurred in a dose-dependent manner.

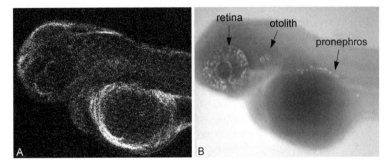

Fig. 5 Fluorescent caspase substrate staining. Three-day-old zebrafish larvae were incubated with vehicle alone (A) or fluorometric caspase-3 substrate (Ac-DEVD-AMC, Calbiochem, A Brand of EMD, Biosciences, Inc., San Diego, CA) (10 μg/ml) (B) for 2 h at 28 °C, and visualized by fluorescent microscopy. Developmental apoptosis was observed in the retina, otolith, and pronephros. To visualize the outline of zebrafish, the exposure time in (A) was increased from 20 msec (B) to 100 msec.

been substituted for the carbobenzoxy (Z) N-terminal blocking group in the peptide inhibitor to create fluorescent markers. FITC and FAM produce green fluorescence, and SR produces red fluorescence. The fluorochrome label allows for a single-reagent addition to assess caspase activity *in situ* by fluorescence microscopy. Additionally, a flow cytometer or a fluorescence microplate reader can be used to quantitate FLICA signals (Grabarek *et al.*, 2002). These fluorescently labeled caspase inhibitors irreversibly bind to caspases and fluoresce. Caspase inhibitors can be administered by incubating embryos in medium or by microinjection. Generally, for incubation, a minimum amount of 10 μM of caspase inhibitor is required to induce inhibitory effects. To ensure maximum inhibition, caspase inhibitors must be replenished at 24-h intervals. For microinjection, approximately 1 mM of inhibitor is injected into the yolk region.

E. TUNEL Staining

To locate apoptotic cells *in situ*, the TUNEL (terminal deoxynucleotidyl transferase-mediated dUTP-biotin nick end-labeling) method is widely used on tissue sections or in whole embryos (Dong *et al.*, 2001; Gavrieli *et al.*, 1992; Knight *et al.*, 2003; Link *et al.*, 2001). TUNEL staining detects cells that exhibit DNA fragmentation using fluorescein deoxyuridine triphosphate (dUTP) or digoxigenin dUTP. Dechorionated embryos are fixed in cold paraformaldehyde in PBS for 6 h and dehydrated in a graded ethanol series (50, 70, 85, 95, 100%), followed by 20 m in acetone at $-20\,^\circ$C. The embryos are further permeabilized by incubating in a solution containing 0.5% Triton X-100 and 0.1% sodium citrate in PBS for 15 m and in a 20 μg/ml Proteinase K for 5 to 10 m. The embryos are refixed with 4% paraformaldehyde before incubation with Terminal deoxynucleotidyl Transferase (TdT) solution according to manufacture's instruction (*In Situ* Cell Death Detection Kit from Roche, Basel, Switzerland). Sectioning may be performed to get more detailed staining. After TUNEL staining, embryos are dehydrated through a graded ethanol series, then embedded and sectioned (Prince *et al.*, 1998).

For paraffin embedded sections, section specimens (approximately 10 μM) are deparaffinized by immersing slides twice in fresh xylene for 5 m at room temperature. The slides are then washed with 100% ethanol for 5 m and rehydrated through graded ethanol washes (100, 95, 85, 70, and 50%) for 3 m each at room temperature, incubated in 0.85% NaCl for 5 m and rinsed with PBS. As described earlier, the sections are then permeabilized with Triton X-100 and proteinase K and refixed with 4% paraformaldehyde for TdT staining.

V. Perspectives

The apoptosis pathway has been an important target for drug invention. Zebrafish exhibit inherent attributes including transparency, conserved apoptotic machinery, easy manipulation, and permeability to vital dye and small molecules, which make it possible to perform throughput drug screening, using a quantitative microplate reader. Fluorochrome caspase substrates or inhibitors can be used to visualize apoptosis in zebrafish, permitting mapping of caspase activation and dissecting apoptotic functions *in vivo*. Mutant or transgenic zebrafish exhibiting defects in specific apoptosis pathway can further assist target screening for proapoptic or antiapoptotic agents.

Acknowledgment

This publication was made possible by support from NIH grants:1R43CA83256-01; 2R44CA83256-02A1; 1R43CA86181-01.

References

Bender, E., Smolewski, P., Amstad, P., and Darzynkiewicz, Z. (2000). Activation of caspases measured *in situ* by binding of fluorochrome-labeled inhibitors of caspases (FLICA): Correlation with DNA fragmentation. *Exp. Cell Res.* **259**, 308–313.

Brand, M., Heisenberg, C. P., Warga, R. M., Pelegri, F., Karlstrom, R. O., Beuchle, D., Picker, A., Jiang, Y. J., Furutani-Seiki, M., van Eeden, F. J., *et al.* (1996). Mutations affecting development of the midline and general body shape during zebrafish embryogenesis. *Development* **123**, 129–142.

Chan, D. W., and Yager, T. D. (1998). Preparation and imaging of nuclear spreads from cells of the zebrafish embryos: Evidence for large degradation intermediates in apoptosis. *Chromosoma* **107**, 39–60.

Chen, J. N., and Fishman, M. C. (1996). Zebrafish tinman homolog demarcates the heart field and initiates myocardial differentiation. *Development* **123**, 293–302.

Cecconi, F., and Gruss, P. (2001). Apaf1 in developmental apoptosis and cancer: How many ways to die? *Cell Mol. Life Sci.* **58**, 1688–1697.

Cole, L. K., and Ross, L. S. (2001). Apoptosis in the developing zebrafish embryo. *Dev. Biol.* **240**, 123–142.

Daly, J. M., Jannot, C. B., Beerli, R. R., Graus-Porta, D., Maurer, F. G., and Hynes, N. E. (1997). Neu differentiation factor induces ErbB2 down-regulation and apoptosis of ErbB2 overexpressing breast tumor cells. *Cancer Res.* **57**, 3804–3811.

Donepudi, M., and Grutter, M. G. (2002). Structure and zymogen activation of caspases. *Biophy. Chem.* **101**, 145–153.

Dong, H., Teraoka, H., Kondo, S., and Hiraga, T. (2001). 2, 3, 7, 8-tetrachlorodibenzo-p-dioxin induces apoptosis in the dorsal midbrain of zebrafish embryos by activation of arylhydrocarbon receptor. *Neurosci. Lett.* **303**, 169–172.

Ellis, A. E. (1977). Ontogeny of the immune response in *Salmo salar*. Histogenesis of the lymphoid organs and appearance of membrane immunoglobulin and mixed leucocyte reactivity. *In* "Developmental Immunobiology" (J. B. Solomon, ed.), pp. 225–231. Elsevier/North Holland Biomedical Press, Amsterdam.

Friedlander, R. M. (2003). Apoptosis and caspases in neurodegenerative diseases. *New Engl. J. Med.* **348**, 1365–1375.

Furutani-Seiki, M., Jiang, Y.-J., Brand, M., Heisenberg, C.-P., Houart, C., Beuchle, D., van Eeden, F. J. M., Granato, M., Haffter, P., Hammerschmidt, M., *et al.* (1996). Neural degeneration mutants in the zebrafish, Danio rerio. *Development* **123**, 229–239.

Gavrieli, Y., Sherman, Y., and Ben-Sasson, S. A. (1992). Identification of programmed cell death *in situ* via specific labeling of nuclear DNA fragmentation. *J. Cell Biol.* **119**, 493–501.

Grabarek, J., Amstad, P., and Darzynkiewicz, Z. (2002). Use of fluorescently labeled caspase inhibitors as affinity labeled to detect activated caspases. *Hum. Cell* **15**, 1–12.

Granato, M., and Nusslein-Volhard, C. (1996). Fishing for genes controlling development. *Curr. Opin. Gen. Dev.* **6**, 461–468.

Hale, A. J., Smith, C. A., Sutherland, L. C., Stoneman, V. E., Longthorne, V. L., Culhane, A. C., and Williams, G. T. (1996). Apoptosis: Molecular regulation of cell death. *Eur. J. Biochem.* **236**, 1–26.

Hsieh, Y. C., Chang, M. S., Chen, J. Y., Yen, J. J., Lu, I. C., Chou, C. M., and Huang, C. J. (2003). Cloning of zebrafish BAD, a BH-3 only proapoptotic protein, whose expression leads to apoptosis in Cos-1 cells and zebrafish embryos. *Biochem. Biophys. Res. Commun.* **304**, 667–675.

Ho, S. Y., Pack, M., and Farber, S. A. (2003). Analysis of small molecule metabolism in zebrafish. *Methods Enzymol.* **364**, 408–426.

Howard, A. D., Kostura, M. J., Thornberry, N., Ding, G. J., Limjuco, G., Weidner, J., Salley, J. P., Hogquist, K. A., Chaplin, D. D., and Mumford, R. A. (1991). IL-1-converting enzyme requires aspartic acid residues for processing of the IL-1 beta precursor at two distinct sites and does not cleave 31-kDa IL-1 alpha. *J. Immunol.* **147**, 2964–2969.

Idziorek, T., Estaquier, J., De Bels, F., and Ameisen, J. C. (1995). YOPRO-1 permits cytofluorimetric analysis of programmed cell death without interfering with cell viability. *J. Immunolog. Methods* **185**, 249–258.

Ikegami, R., Zhang, J., Rivera-Bennetts, A. K., and Yager, T. D. (1997). Activation of the metaphase checkpoint and an apoptosis programme in the early zebrafish embryo, by treatment with the spindle-destabilising agent nocodazole. *Zygote* **5**, 329–350.

Ikegami, R., Hunter, P., and Yager, T. D. (1999). Developmental activation of the capability to undergo checkpoint-induced apoptosis in the early zebrafish embryo. *Dev. Biol.* **209**, 409–433.

Inohara, N., and Nunez, G. (2000). Genes with homology to mammalian apoptosis regulators identified in zebrafish. *Cell Death Differ.* **7**, 509–510.

Kerr, J. F., Wyllie, A. H., and Currie, A. R. (1972). Apoptosis: Abasic biological phenomenon with wide-ranging implication in tissue kinetics. *Br. J. Cancer* **26**, 239–257.

Knight, R. D., Nair, S., Nelson, S. S., Afshar, A., Javidan, Y., Geisler, R., Rauch, G. J., and Schilling, T. F. (2003). Lockjaw encodes a zebrafish tfap2a required for early neural crest development. *Development* **130**, 5755–5768.

Lamkanfi, M., Declercq, W., Kalai, M., Saelens, X., and Vandenabeele, P. (2002). Alice in caspase land. A phylogenetic analysis of caspases from worm to man. *Cell Death Differ.* **9**, 358–361.

Langheinrich, U., Hennen, E., Scott, G., and Vacun, G. (2002). Zebrafish as a model organism for the identification and characterization of drugs and genes affecting p53 signalling. *Curr. Biol.* **12**, 2023–2028.

Link, B. A., Kainz, P. M., Ryou, T., and Dowling, J. E. (2001). The perplexed and confused mutations affect distinct stages during the transition from proliferating to post-mitotic cells within the zebrafish retina. *Dev. Biol.* **15**, 436–453.

Majno, G., and Joris, I. (1995). Apoptosis, oncosis, and necrosis. An overview of cell death. *Am. J. Pathol.* **146**, 3–15.

Masumoto, J., Zhou, W., Chen, F. F., Su, F., Kuwada, J. Y., Hidaka, E., Katsuyama, T., Sagara, J., Taniguchi, S., Ngo-Hazelett, P., *et al.* (2003). Caspy, a zebrafish caspase, activated by ASC oligomerization is required for pharyngal arch development. *J. Biol. Chem.* **2003**, 4268–4276.

Milan, D. J., Peterson, T. A., Ruskin, J. N., Peterson, R. T., and MacRae, C. A. (2003). Drugs that induce repolarization abnormalities cause bradycardia in zebrafish. *Circulation* **107**, 1355–1358.

Nicholson, D. W. (1999). Caspase structure, proteolytic substrates, and function during apoptotic cell death. *Cell Death Differ.* **6**, 1028–1042.

Parng, C., Seng, W. L., Semino C., and McGrath, P. (2002). Zebrafish: A preclinical model for drug screening. *Assay and Drug Devel. Technol.* **1**, 41–48.

Prince, V. E., Price, A. L., and Ho, R. K. (1998). Hox gene expression reveals regionalization along the anteroposterior axis of the zebrafish notochord. *Dev. Genes Evol.* **208**, 517–522.

Schneider, I. (2003). Novel high-throughput screening technologies. *Genetic Engineering News* **23**, 10–13.

Seng, W. L., and McGrath, P. (2002). A high throughput *in vivo* zebrafish angiogenesis assay. *Drug Plus International* **1**, 22–24.

Sleath, P. R., Hendrickson, R. C., Kronheim, S. R., March, C. J., and Black, R. A. (1990). Substrate specificity of the protease that processes human interleukin-1 beta. *J. Biol. Chem.* **265**, 14526–14528.

Smolewski, P., Bedner, E. L. D., Hsieh, T. C., Wu, J. M., Phelps, D. J., and Darzynkiewicz, Z. (2001). Detection of caspases activation by fluorochrome-labeled inhibitors: Multiparameter analysis by laser scanning cytometry. *Cytometry* **44**, 73–82.

Song, Z., and Steller, H. (1999). Death by design: Mechanism and control of apoptosis. *Trends Cell Biol.* **9**, M49–M52.

Svoboda, K. R., Linares, A. E., and Ribera, A. B. (2001). Activity regulates programmed cell death of zebrafish Rohon-Beard neurons. *Development* **128**, 3511–3520.

Takahashi, A., and Earnshaw, W. C. (1996). ICE-related proteases in apoptosis. *Curr. Opin. Gen. Dev.* **6**, 50–55.

William, G. T., and Smith, C. A. (1993). Molecular regulation of apoptosis: Genetic controls on cell death. *Cell* **74,** 777–779.

Yabu, T., Kishi, S., Okazaki, T., and Yamashita, M. (2001). Characterization of zebrafish caspase-3 and induction of apoptosis through ceramide generation in fish fathead minnow tailbud cells and zebrafish embryos. *Biochem. J.* **360,** 39–47.

Yamashita, M. (2003). Apoptosis in zebrafish development. Comparative Biochemistry and Physiology Part B. *Biochem. Mol. Biol. Int.* **136,** 731–742.

Zhang, C., Fremgen, T., and Willett, C. (2003). Zebrafish: An animal model for toxilogical studies. *In* "Current Protocols in Toxicology," John W. Wiley and Sons, Inc., NY. Unit 1.7, Supplement 17.

CHAPTER 6

Lipid Metabolism in Zebrafish

Shiu-Ying Ho, Juanita L. Thorpe, Yun Deng, Evelyn Santana, Robert A. DeRose, and Steven A. Farber

Department of Microbiology and Immunology
Kimmel Cancer Center
Thomas Jefferson University
Philadelphia, Pennsylvania 19107

I. Introduction

Lipids play multiple and essential roles in all living cells. A variety of human diseases that affect many individuals worldwide are associated with abnormal lipid metabolism, such as atherosclerosis and diabetes mellitus (Joffe *et al.*, 2001; McNeely *et al.*, 2001). Although intensive studies have demonstrated how lipids act as signaling molecules, components of cell membranes, and as a rich energy source, there remain many unanswered questions at both the genetic and molecular level.

The zebrafish, *Danio rerio*, as a model vertebrate system was initially developed primarily by George Streisinger and colleagues (1981) in the 1980s. It became a popular system not only for studies of vertebrate patterning and development, but recently as a model for human disease as illustrated by the positional cloning of a number of zebrafish mutations whose phenotypes mimic many aspects of human disease (Donovan *et al.*, 2000; Dooley and Zon, 2000; Fisher *et al.*, 2003; Wang *et al.*, 1998). Zebrafish contain from 30,000 to 80,000 genes, more than 80% of

METHODS IN CELL BIOLOGY, VOL. 76

which are similar to those in mouse and human (Lander *et al.*, 2001). The conservation between the two species makes zebrafish a powerful genetic system to study vertebrate physiology.

Lipid metabolism has been examined intensively in teleostean fish. Both endogenous lipids transport and lipolysis in fish is similar to that observed in mammals, with slightly different absorptional and depositional processes (Sheridan, 1988). We have shown that zebrafish process dietary lipids similar to humans and that pharmaceuticals used to treat hypercholesterolemia in humans are effective in this vertebrate model (Farber *et al.*, 2001). In addition, zebrafish and mammals have similar pathways for prostanoid synthesis as evidenced by the conserved cyclooxygenases required for prostaglandins synthesis (Grosser *et al.*, 2002). The formation of lipid signaling molecules, such as the prostaglandins and thromboxanes, can be inhibited by commonly used nonsteroidal antiinflammatory drugs (Grosser *et al.*, 2002).

The zebrafish is an ideal vertebrate model not only because its genes are similar to human genes, but because of the ease with which both morphological and functional processes can be monitored. To date, several large-scale mutagenesis screening projects were initiated or completed. In this chapter, we focus on how we applied optical biosensors to screen and study lipid metabolism in live zebrafish larvae.

II. Genetic Screen—Forward, Reverse, and Targeted–Mutagenesis

The ability to perform forward genetic studies in zebrafish by mutagenizing the entire genome and screening for particular phenotypes has made this vertebrate model popular (Driever *et al.*, 1996; Haffter *et al.*, 1996). The mutagenesis methods utilized in the zebrafish community include the generation of large deletions and translocations by gamma ray radiation (Fisher *et al.*, 1997), point mutations by soaking founder fish in a variety of chemicals (most frequently ethylnitrosourea [ENU]) (Driever *et al.*, 1996; Haffter *et al.*, 1996), and retrovirus or transposon insertions (Chen *et al.*, 2002; Ivics and Izsvak, 2004). We have initiated an ENU-mutagenized screening effort and searched for mutations that perturbed lipid processing. This is accomplished by soaking larval fish in several fluorescent lipid reporters that are swallowed and ultimately allow the visualization of lipid metabolism *in vivo* (Farber *et al.*, 2001). One fluorescent phosphoethanolamine analogue, PED6,   [N-((6-(2,4-dinitro-phenyl)amino)hexanoyl)-1-palmitoyl-2-BODIPY-FL-pentanoyl-*sn*-glycerol-3-phosphoethanolamine], (Fig. 1) results in the release of a fluorescent BODIPY labeled acyl chain upon phospholipase A_2 (PLA_2) cleavage (Farber *et al.*, 1999). When 5-dpf zebrafish larvae are immersed in media containing PED6, an intense bright green fluorescence is observed in the intestine, gall bladder, and liver (Fig. 1E). NBD-cholesterol (22-[N-(7-nitrobenz-2-oxa-1,3-diazol-4-yl) amino]-23,24-bisnor-5-cholen-3-ol) (Fig. 2) was also used to visualize

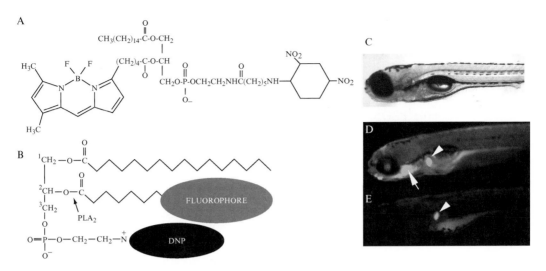

Fig. 1 PED6 as a biosensor to visualize lipid metabolism in live zebrafish larva. (A) The structure of PED6. The emission of the BODIPY-labeled acyl chain at the *sn*-2 position is quenched by the dinitrophenol group at the *sn*-3 position when this molecule is intact. (B) Upon PLA$_2$ cleavage at the *sn*-2 position, the BODIPY-labeled acyl chain is liberated and can emit a green fluorescence (515 nm). (C) Bright field view of a 5-dpf embryo. (D) Embryo soaked in unquenched phospholipids (2 ug/ml, D3803). Arrowhead marks gall bladder and the arrow marks the pharynx. (E) Embryo (5 dpf) soaked in PED6 (3 ug/ml, 6 h). Arrowhead indicates the gall bladder. (See Color Insert.)

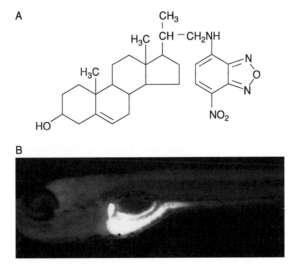

Fig. 2 NBD-Cholesterol labeled live zebrafish larva. (A) Structure of NBD-Cholesterol. (B) Larval zebrafish (5 dpf) was immersed into embryo media containing NBD-Cholesterol (3 μg/ml, solubilized with fish bile) for 2 h. (See Color Insert.)

cholesterol absorption in live larvae. Both fluorescent reporters provide a rapid readout for digestive organ morphology, as well as lipid processing, and have successfully enabled the identification of several mutant lines.

One mutation identified with these lipid reporters, *fat-free*, was a recessive lethal mutation, whose digestive system appears morphologically normal despite exhibiting significantly diminished fluorescence in the intestine and the gall bladder after PED6 or NBD-cholesterol ingestion (Fig. 3). Phenotypic analysis indicated that fish harboring this mutation have normal swallowing function and PLA_2 activity (Farber *et al.*, 2001). When we immersed *fat-free* mutant larvae in the media containing radioactive oleic acid (250 μCi/mmole, 20 h), followed by whole embryo lipid extraction and thin layer chromatography (TLC) analysis, we found that *fat-free* larvae have significantly reduced radioactivity incorporated into phosphatidylcholine (PC) fraction ($p < 0.05$) (Fig. 4). However, when *fat-free* mutant larvae were incubated with BODIPY FL-C5, a short chain fatty acid analogue, they had nearly normal digestive organ fluorescence. Because short chain fatty acids are less hydrophobic and do not depend on emulsifiers (such as bile) for absorption, we therefore proposed that the *fat-free* mutation might

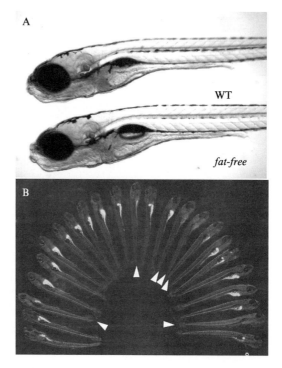

Fig. 3 PED6 labeled larvae. (A) *Fat-free* is morphologically normal as wild type (WT). (B) *Fat-free* has diminished fluorescence in intestinal lumen and gall bladder. Arrowheads mark mutant embryos. (See Color Insert.)

attenuate biliary synthesis or secretion (Ho *et al.*, 2003). Indeed, histology of *fat-free* larvae revealed that these mutants have impaired hepatic canuliculi. The positional cloning of the *fat-free* gene indicated this gene encodes a protein with no known function. Further sequence analysis revealed *fat-free* to be well conserved from invertebrates to mammals. Domain analyses and transcriptional profiling of *fat-free* mutants indicated that this gene is likely to modulate vesicular trafficking.

This example illustrates that the forward genetic screens can assist in identifying the functions of vertebrate genes by identifying novel genes that regulate lipid metabolism. However, the positional cloning of an ENU-induced mutation generally takes months to years. On the other hand, a reverse genetic screen can be used to more rapidly study presumptive lipid modulators. The development of targeted antisense morpholino (MO) or gripNA based knockdown technologies (Nasevicius and Ekker, 2000; Urtishak *et al.*, 2003) has enabled genome-wide, sequence based, reverse genetic screens in this model vertebrate. These technologies are now widely used in the zebrafish community, and numerous studies have successfully pheno copied mutants by targeting particular genes with these reagents (Dutton *et al.*, 2001; Karlen and Rebagliati, 2001; Urtishak *et al.*, 2003).

Targeted antisense reagents are injected into the yolk at the 1- to 4-cell stage to assess the role of a particular gene on early larval lipid metabolism. This injection is followed by a second yolk injection of a BODIPY-labeled fatty acid (at 24 hpf). Embryos are allowed to develop for some period, and the lipids extracted to assay fluorescent lipid metabolite levels by TLC. This methodology allows one to screen for abnormal lipid metabolites at stages before the mouth opens and swallowing begins. For example, the lipid profile for MO (TC82778L32) injected embryos is clearly abnormal (Fig. 5). The injected dosage of MO correlates well with the altered lipomic profile. The more MO injected, the less fluorescence incorporated into the phospholipids (PL) and diacylglycerol lipid classes. These results indicate that this MO-targeted gene is possibly involved in PL rather than triacylglycerol metabolism. We have so far screened approximately 90 genes using this approach and have identified 12 with altered lipid profiles. Further studies are ongoing to illuminate the mechanism by which particular genes regulate the utilization of the labeled fatty acid. The drawback of these knockdown technologies is that they only generate transient gene-silencing that persist for only 3 to 5 days.

Recently, efficient target-selected mutagenesis in zebrafish has been reported (Wienholds *et al.*, 2003). This method provides a new approach to permanently knockout specific genes through the selection of randomly ENU-mutagenized fish. This method could enable rapid target-selected mutagenesis generating mutations in known lipid mediators. These lines could be extremely useful to study the physiological functions of specific genes. Nevertheless, the efforts of trying to establish zebrafish embryonic stem cells in order to generate transgenic zebrafish cannot be undervalued (Shafizadeh *et al.*, 2002), and may indeed become the main methodology to generate transgenic zebrafish in the near future.

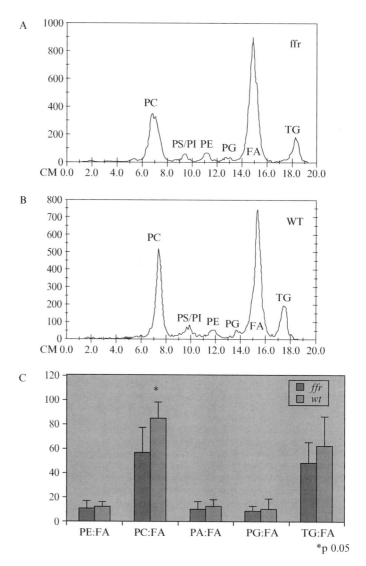

Fig. 4 Lipomics of *fat-free* and wild type (Ho *et al.*, 2003). Both *fat-free* and wild-type larvae (4 dpf) were incubated with radioactive oleic acid for 20 h, followed by lipid extraction and thin layer chromatography (TLC). The solvent chloroform:ethanol:triethylamine:water (30:34:30:8) was used to develop the TLC plate. The radioactivities were then scanned. The major metabolites derived from oleic acid (FA) are phosphatidylcholine (PC), phosphatidylserine (PS), phosphatidylinositol (PI), phosphatidylethanolamine (PE), phosphatidyl glycerol (PG), and triacylglycerol (TG). (A) Representative lipomics of *fat-free* larva. (B) Representative lipomics of wild-type larva. (C) Comparison of lipomics between *fat-free* and wild-type larva. *Fat-free* has significantly decreased radioactives incorporated into PC fraction (n = 9, mean ± SD).

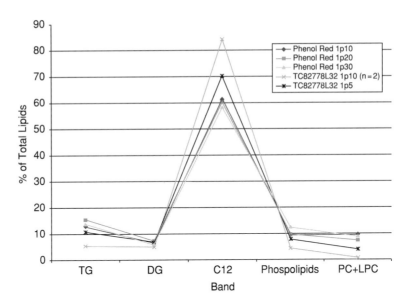

Fig. 5 Lipid profiles after MO injection. The embryos (1- to 4-cell stage) were injected with phenol red (control) or MO TC82778L32 at different doses, then BODIPY-fatty acids (C12) were injected at 1 dpf. The embryos were collected at 3 dpf, lipids were extracted (chloroform:methanol:water—2:1:1) and subject to thin layer chromatography (TLC). TLC plates were analyzed using a laser scanner and the fluorescences incorporated into different lipid classes were determined. Relative fluorescence intensity values were normalized by total fluorescence in all lipid classes. TG: triacylglycerol; DG: diacylglycerol; C12: BODIPY-fatty acids; PC: phosphatidylcholine; LPC: LysoPC; PL: phospholipids except PC and lysoPC.

III. Pharmacological Studies Using Zebrafish

Lipid metabolism in zebrafish larvae can be successfully modulated by a number of pharmacological means. We have previously shown that atorvastatin (Lipitor) can block the accumulation of fluorescent lipids in the gall bladder of larval zebrafish (Farber *et al.*, 2001). Lipitor is a widely prescribed medication for the treatment of hypercholesterolemia, and is one of the most economically successful drugs ever developed. Lipitor belongs to the statin family of 3-hydroxyl-3-methylglutaryl-CoA reductase (HMGCoAR) inhibitors (Fig. 6). It has been reported that in *Drosophila*, reduced HMGCoAR activity results in primordial germ cell (PGCs) migration defects (Van Doren *et al.*, 1998). We therefore examined both zebrafish overall morphological development and PGC migration following statin-mediated HMGCoAR inhibition. PGCs migration in zebrafish embryos can be visualized in late somite-stage embryos by injecting *in vitro* transcribed capped *gfp-nanos* mRNA at the one-cell stage. Both the *gfp-nanos*

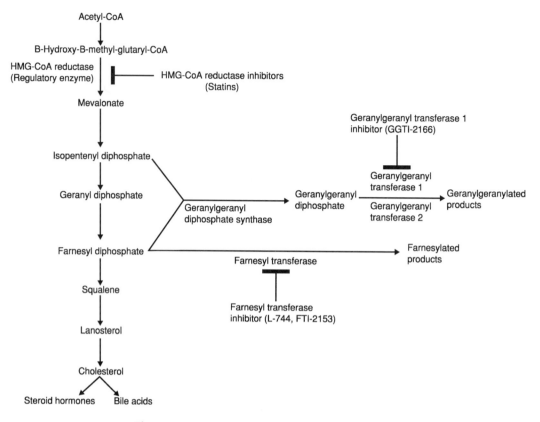

Fig. 6   Cholesterol and isoprenoid biosynthesis pathway (Thorpe *et al.*, 2004).

message and protein are stabilized only in PGCs such that they maintain their fluorescence throughout early development, facilitating detailed study of PGC migratory behavior and ultimately gonad formation (Doitsidou *et al.*, 2002). When zebrafish embryos are soaked in mevinolin (Lovastatin) and simvastatin (Zocor), embryos exhibited developmental arrest, blunted axis elongation, misshapen somites, and head and axial necrosis. Further, PGC migration was profoundly perturbed with a similar dose-response profile (Fig. 7). Embryos treated with the more hydrophobic statin, Lipitor, exhibited PGCs migration defects and only mild morphologic abnormalities (Fig. 8).

To determine which downstream products of HMGCoAR mediate these defects, we pioneered a "block in rescue" approach by injecting putative biochemical pathway intermediates downstream of the pharmacologically inhibited enzyme. We initially established the validity of this approach by demonstrating that an injection of mevalonate, the product of HMGCoAR, completely rescued all the

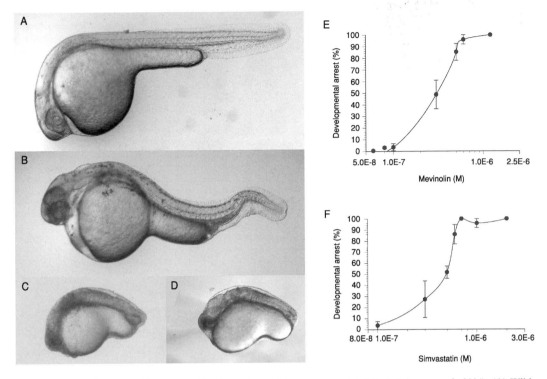

Fig. 7 Embryos treated with statins exhibit developmental defects (Thorpe *et al.*, 2004). (A) Wild-type embryos (24 hpf). (B) Embryos soaked in a low dose of mevinolin (0.06 μM) have a slight tail kink. (C) Exposure to higher doses of mevinolin (1.2 μM) results in blunted axis elongation, necrosis, and developmental arrest. (D) Simvastatin-treated embryos (2.0 μM) exhibit similar defects as in (C). (E) Dose-response of mevinolin on developmental arrest (mean $\pm$ SEM, n = 3). (F) Dose-response of simvastatin on developmental arrest (mean $\pm$ SEM, n = 3).

phenotypes associated with statin treatment (Fig. 9). Similar experiments using intermediates downstream of mevalonate (see Fig. 6) indicated that the prenylation pathway was mediating the effect of statins on PGCs. We tested this hypothesis by treating *gfp-nanos* mRNA injected embryos with selective geranylgeranylation or farnesylation inhibitors. High doses of the selective farnesyl transferase inhibitors L-744 or FTI-2153 (Crespo *et al.*, 2001; Sun *et al.*, 1999) had no effect on PGC migration. Embryos treated with geranylgeranyl transferase I (GGTI) inhibitor (GGTI-2166) exhibited a strong PGC migration phenotype with only mild morphological defects manifested as a slight disruption of the notochord (Fig. 10). These data suggest that protein prenylation, specifically by GGTI, is required for correct PGC migration (Thorpe *et al.*, 2004).

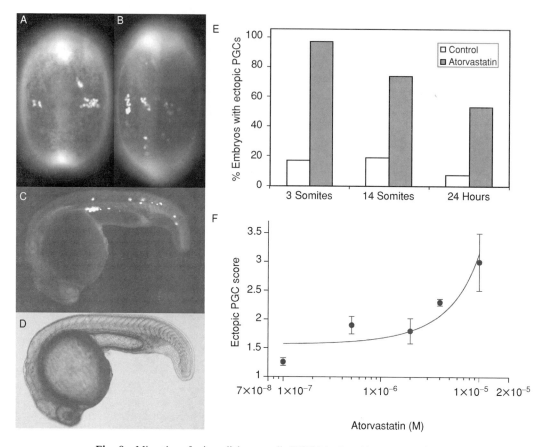

Fig. 8 Migration of primordial germ cells (PGCs) is altered by atorvastatin treatment (Thorpe *et al.*, 2004). (A) A single fluorescent image from a time lapse (1-somite stage, first capture) of the movement of PGCs in a wild-type embryo injected at the 1 cell stage with *gfp-nanos* mRNA (1–2 nl, 60 ng/ul). (B) Embryos (1-somite stage) soaked in atorvastatin (10 mM) and injected with *gfp-nanos* mRNA exhibit ectopic PGCs that fail to cluster. (C) Later stage embryos (22-somite stage) exhibit many ectopic germ cells after atorvastatin (10 mM) treatment. (D) Brightfield image of an embryo (24 hpf) injected with *gfp-nanos* mRNA and treated with atorvastatin (10 mM) that exhibits only a mild tail defect. (E) Analysis of the effect of atorvastatin treatment on PGC migration at different developmental stages. The number of embryos at a given stage with more than 2 ectopic PGCs were determined and expressed as a percent of the total number of embryos examined (n = 64 – 84 for each stage). (F) The effect of atorvastatin on numbers of ectopic PGCs is dose dependent. Embryos were soaked in atorvastatin (10 μM). At 24 hpf, embryos were each scored on the degree of ectopic PGCs (level 1 embryo had a wild-type single gonadal cluster and a level 4 embryo had no discernible cluster). Data represent the mean $\pm$ SEM from 3 to 4 experiments/dose.

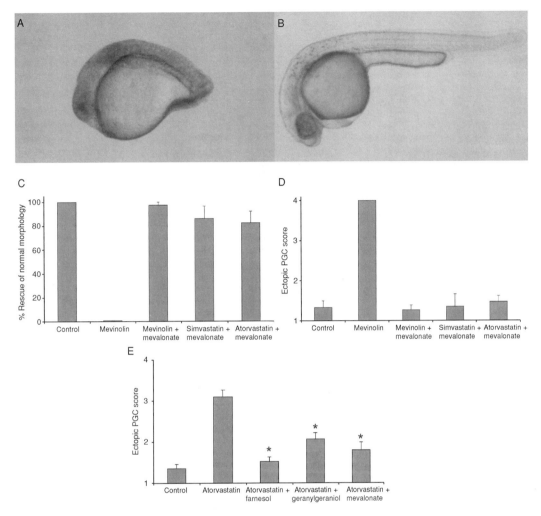

Fig. 9 Injection of isoprenoid intermediates abrogates the effects of statins (Thorpe *et al.*, 2004). (A) Uninjected embryos show severe developmental defects after 24 h of mevinolin (1.2 μM) treatment. (B) Embryos injected at early cell stages (1 to 16 cell stage) with mevalonate (1 to 2 nl, 0.5 M) and then soaked in mevinolin (1.2 μM, 24 h) exhibit normal morphology. (C) Mevalonate injection rescues the somatic defects observed in embryos treated with statins. Embryos at early cell stages were injected with mevalonate as in (B) and soaked overnight in statin drugs: mevinolin (1.2 μM), simvastatin (2.0 μM), and atorvastatin (10 μM). At 24 hpf, developmental defects were scored. Data represent the MEAN $\pm$ SEM from 3 to 4 experiments. (D) The appearance of ectopic PGCs after statin treatment is prevented by mevalonate injections. Embryos at early cell stages were injected with *gfp-nanos* mRNA and mevalonate as in (B) and soaked overnight in statin drugs [mevinolin (1.2 μM), simvastatin (2.0 μM), and atorvastatin (10 μM)]. At 24 hpf, the PGC score was determined. Data represent the Mean $\pm$ SEM from 3 to 4 experiments. (E) Embryo's injected at the 1- to 16-cell stage with farnesol (1 to 2 nl, 1 M) , geranyl geraniol (0.5 to 1.0 nl, 1 M) or mevalonate (1 to 2 nl, 0.5 M) then soaked overnight with atorvastatin (10 μM) show normal PGC migration. Statistics were performed using ANOVA with a post hoc test that utilizes a Bonferroni correction. Data represents MEAN $\pm$ SEM, *p < 0.01 difference from atorvastatin alone.

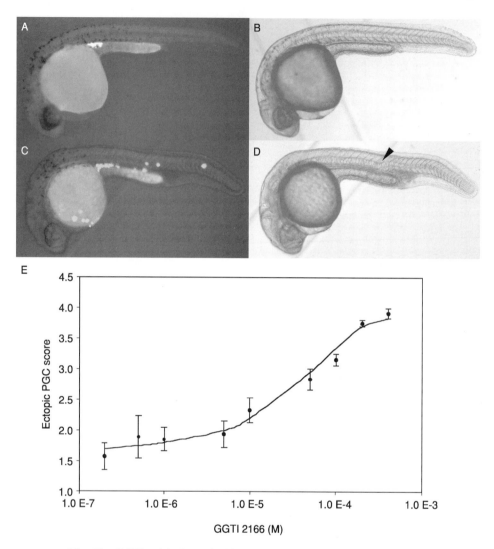

Fig. 10 GGT1 activity is required for PGC migration (Thorpe *et al.*, 2004). (A) A fluorescent image of a wild-type embryo (24 hpf) injected with *gfp-nanos* mRNA at the 1-cell stage reveals a cluster of PGCs. (B) A brightfield image of the embryo in (A). (C) A fluorescent image of an embryo (24 hpf) that was injected with *gfp-nanos* mRNA at the 1-cell stage and immediately soaked in the geranylgeranyl transferase inhibitor (GGTI-2166; 40 μM), resulting in ectopic migration of many PGCs. (D) A brightfield image of the embryo in (C) showing a mild notochord defect (arrowhead). (E) Dose response of GGTI-2166 on PGC migration defect. Data represent the MEAN $\pm$ SEM from 3 to 6 experiments for each dose.

IV. Mechanism of Action Studies Using Zebrafish: The Annexin2-Caveolin1 Complex

In addition to PGC migration, we have focused extensive effort to identify the components of a putative intestinal sterol transport. Caveolin 1 (CAV1) is a 22 kDa protein that forms the cytoplasmic coat associated with caveolae (Kurzchalia *et al.*, 1992; Rothberg *et al.*, 1992; Smart *et al.*, 1999). However, CAV1 is not restricted to caveolae and the protein has been shown to be involved in the intracellular trafficking of sterol (Uittenbogaard *et al.*, 1998; Uittenbogaard and Smart, 2000; Uittenbogaard *et al.*, 2002). CAV1 can form at least two distinct chaperone complexes. One complex consists of CAV1, heat shock protein 56, cyclophilin A, and cyclophilin 40 and this complex traffics newly synthesized cholesterol from the endoplasmic reticulum to caveolae (Uittenbogaard *et al.*, 1998). The other complex consists of CAV1, Annexin 2 (ANX2), cyclophilin A, and cyclophilin 40 and this complex traffics exogenous cholesterol from caveolae to the endoplasmic reticulum (Uittenbogard *et al.*, 2002). In mammalian cell culture systems, these two CAV1 chaperone complexes appear to play roles in regulating both total cellular and caveolar cholesterol levels.

The annexins are a family of calcium-dependent phospholipid-binding proteins that are widespread in eukaryotes, but, despite extensive study, the true *in vivo* functions of the annexins have remained unclear. While many studies performed *in vitro* and in cultured cells implicate ANXs in a variety of cellular processes, the physiological significance of these results is often far from evident. Previous studies have reported the involvement of ANX2 in membrane fusion (a feature exploited by cytomegalovirus) (Raynor *et al.*, 1999) and in actin-dependent "rocketing" transport of macro pinocytic vesicles (Merrifield, *et al.*, 2001), as well as intracellular cholesterol traffic (Uittenbogaard *et al.*, 2002). We have previously cloned 11 zebrafish *anx* genes and analyzed their expression (Farber *et al.*, 2003).

Using both targeted MO injections and immunoprecipitation experiments coupled with mass spectrometry analysis, we determined that ANX2b complexes with CAV1 in the intestine of the zebrafish (Smart *et al.*, 2004). This complex is heat stable and unaffected by sodium dodecyl sulphate (SDS) or reducing conditions. *In situ* hybridization showed that both *anx2b* and *cav1* mRNA are expressed in the larval zebrafish intestinal epithelium (Fig. 11). Knockdown of either *anx2b* or *cav1* by MO injection prevents the formation of the protein heterocomplex (Fig. 12). Further, *anx2b* MO injection prevents processing of a fluorescent cholesterol reporter and results in a reduction in sterol mass (Fig. 13). During the second half of 2003 and the first half of 2004, the FDA approved Zetia (ezetimibe) alone and in conjunction with statins to reduce serum cholesterol levels. A significant body of data is related to the mechanism of action of ezetimibe, which has been demonstrated to block the uptake of cholesterol in the small intestines of a variety of mammals, including rats, hamsters, and monkeys (Smart *et al.*, 2004). Binding of ezetimibe has been localized to the brush border of the small intestine, its likely

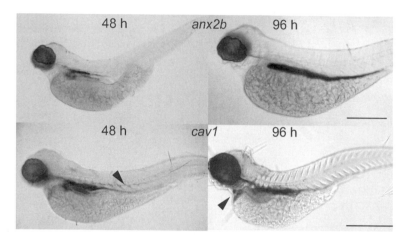

Fig. 11 CAV1 and ANX2b can form a stable heterocomplex (Smart *et al.*, 2004). Expression of *cav1* and *anx2b* in zebrafish larvae. Embryos were fixed in 4% paraformaldehyde and probed with digoxigenin-labeled antisense RNA probe. Top, lateral views of embryos probed for *anx2b* at 48 hpf (left) and 96 hpf (right). Note strong expression in the intestinal epithelium. Scale bar: 500 μm. Bottom, lateral views of embryos probed for *cav1* at 48 hpf (left) and 96 hpf (right). Expression is concentrated in the intestinal epithelium, but *cav1* can also be seen in the somite boundaries at 48 hpf (left, arrowhead) and in the heart ventricle (96 hpf). Scale bar: 500 μm.

site of action. Pharmacological treatment with ezetimibe disrupts the ANX2b-CAV1 heterocomplex, suggesting this heterocomplex is a target of the drug ezetimibe (Fig. 14). The observation that CAV1 and ANX2b form a 55 kD heterocomplex stabilized, most likely, by a covalent interaction is clearly an exciting observation unprecedented in either the CAV or ANX literature.

The uptake and efflux of sterol is tightly regulated at the cellular and organismal level by multiple mechanisms (Connelly *et al.*, 2003; Repa *et al.*, 2002; Rogler *et al.*, 1991; Wang *et al.*, 2003). While it is well established that the intestine is the primary site of dietary cholesterol absorption, the mechanisms involved in the transport of cholesterol across the intestinal epithelium are poorly understood. A better understanding of these mechanisms of sterol transport may have direct translational impact on developing treatment strategies for addressing diet-induced obesity, diabetes, and cardiovascular disease. These data suggest that the ANX2-CAV1 heterocomplex are components of an intestinal sterol transport complex (Smart *et al.*, 2004). Further, we have provided the first clear evidence that this complex is a target of ezetimibe potentially opening the way for new drug development efforts.

Studies have shown that the zebrafish system can be used to screen angiogenic drugs (Chan *et al.*, 2002) and is suggested to be a good model for anti-cancer drug screening (Amatruda *et al.*, 2002). We have demonstrated that Lipitor can block

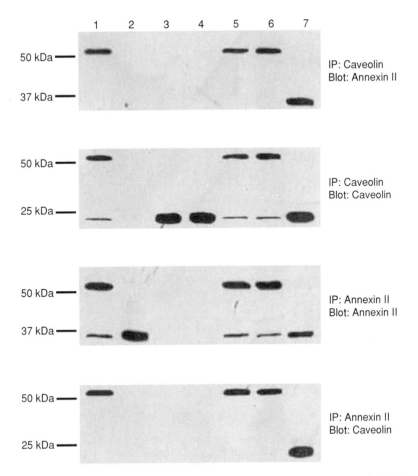

Fig. 12 Disruption and reformation of the CAV1-ANX2b heterocomplex (Smart *et al.*, 2004). Effect of *cav1* and *anx* morpholinos on the formation of the CAV1-ANX2b heterocomplex. Embryos (1- to 8-cell stage) were injected with the following morpholinos: (1) uninjected, (2) *cav1*, (3) *anx2b* synthesis 1, (4) *anx2b* synthesis 2, (5) *anx2b* mismatched, (6) *anx2a*. 3T3 cell lysate (20 µg) was loaded directly onto the gel as a positive control for ANX2 and CAV1 (Lane 7). The embryos were then allowed to develop for 48 h. Larvae were processed to generate lysates (approximately 20 embryos/sample) and 50 µg of protein were used for immunoprecipitation with CAV1 IgG or ANX2 IgG as indicated. The precipitates were resolved by SDS-PAGE and immuno blotted with ANX2 IgG or caveolin IgG as indicated. The data are representative of 3 to 4 independent experiments.

fluorescent lipid absorption, and these processes can be seen in live zebrafish larvae (Farber *et al.*, 2001). This method provides a rapid readout to screen antilipid drugs *in vivo*. We can perform large-scale drug screening simply by arraying zebrafish larvae into multi well plates that contain different chemical compounds and fluorescent lipid reporters, and identify potential compounds that block the lipid absorption.

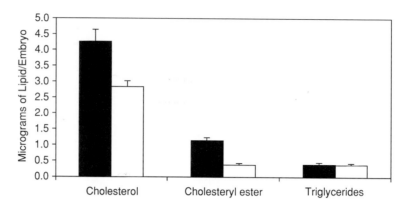

Fig. 13 Effect of reducing ANX2b protein in zebrafish larvae (Smart *et al.*, 2004). Newly fertilized embryos (1- to 8-cell stage) were injected with *anx2b* MO and allowed to develop for 72 h. Embryos were then collected and extracted and the total amount of cholesterol, cholesteryl ester, and triglycerides was determined for injected (white bars) and control uninjected (black bars) embryos. Each bar represents the mean of 6 measurements, 20 embryos per measurement. Differences between injected and control embryos for both cholesterol and cholesteryl ester are statistically significant, $p < 0.05$.

V. Gene-Specific Studies: The Acyl-CoA Synthetase Enzymes

Acyl-CoA Synthetase (ACS) family members are responsible for the addition of coenzyme A (CoA) to fatty acids (Chang *et al.*, 1997; Sillero and Sillero, 2000). This essential step in fatty acid metabolism is necessary in processes such as β-oxidation, desaturation and elongation of fatty acid, triacylglycerol synthesis, and membrane PL synthesis (Faergeman and Knudsen, 1997; Fulgencio *et al.*, 1996; Prentki and Corkey, 1996). ACS activity is also involved in protein acylation, gene regulation, signal transduction, vesicle trafficking, and eicosanoid production (Corkey *et al.*, 2000; Faergeman and Knudsen, 1997; Sakuma *et al.*, 2001; Stremmel *et al.*, 2001). Moreover, recent data implicate ACS in the segregation of fatty acids into acyl-CoA pools that channel specific fatty acids to either fat storage or fat utilization (Muoio *et al.*, 2000). ACS family members are classified according to the carbon chain length of the fatty acid substrate, although considerable overlap in substrate preference has been reported for a number of family members. To date, four families have been identified, including short (C2-C4), medium (C6-C12), long (C14-C20), and very long ($\geq$C22) chain fatty acyl-CoA synthetases (Coleman *et al.*, 2000). In addition, ACS families are further classified into subfamilies and different isoforms have been documented (Cao *et al.*, 2002; Steinberg *et al.*, 2000; Suzuki *et al.*, 1995).

We have identified and cloned seven ACS genes by using the zebrafish expressed sequence tags (EST) database (http://zfish.wustl.edu/). All clones contain the consensus adenosine monophosphate (AMP)-binding motif characteristic of ACS genes. One clone shares high homology with bubblegum, a fatal disorder

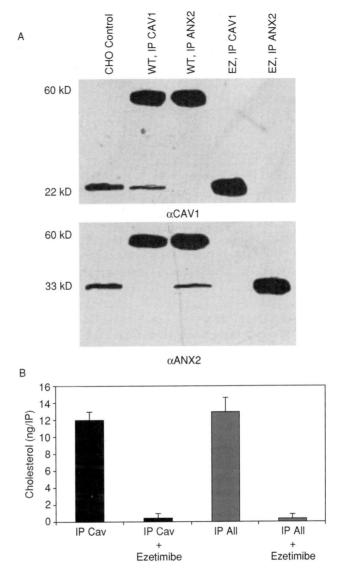

Fig. 14 Effect of ezetimibe on CAV1/ANX2 complex (Smart *et al.*, 2004). (A) Effects of ezetimibe on complex formation in zebrafish embryos. Top, immunoblot with CAV IgG. Bottom, immunoblot with ANX2 IgG. Lanes are identical in both blots. Embryos (48 hpf) were lysed, immunoprecipitated, subjected to SDS-PAGE (20 embryos per lane), and immuno blotted. Lane 1, CHO cells control. Lane 2, untreated embryos, IP with CAV IgG. Lane 3, untreated embryos, IP with ANX2 IgG. Lane 4, embryos soaked in 100 μM ezetimibe from 24 hpf to 48 hpf, IP with CAV IgG. Lane 5, embryos soaked in ezetimibe as in lane 4, IP with ANX2 IgG. (B) Enterocytes from C57BL/6 mice fasted for 4 h were isolated and treated with 10 μM ezetimibe for 2 h at 37 °C. The cytosol was immunoprecipitated and the amount of cholesterol determined by gas chromatography.

characterized by neurodegeneration attributed to the demyelination of white matter and accumulation of very long-chain fatty acids (Kemp *et al.*, 2001; Percy and Rutledge, 2001; Petroni, 2002). Further lipid and transcriptional analyses of these ACS MO knockdowns are ongoing and should reveal the diverse functions of these ACS genes.

VI. Summary

Forward genetics is an unbiased methodology to discover new genes or functions of genes. At the present, the zebrafish is one of the few vertebrate systems where large-scale forward genetic studies are practical. Fluorescent lipid labeling of zebrafish larvae derived from families created from ENU-mutagenized fish enabled us to perform a large scale *in vivo* screen to identify mutants with perturbed lipid processing. With the aid of the zebrafish genome project, positional cloning of mutated genes with abnormal lipid metabolism can be accelerated. MO- and gripNA-based transient gene silencing is feasible in zebrafish embryos and provides a reverse genetic screening strategy to search for important lipid regulators. The advantages of using zebrafish as a vertebrate model to study lipid metabolism include its rapid external development and its optical clarity that enables the monitoring of biological processes. Large scale, high-throughput drug screening *in vivo*, especially for drugs that inhibit lipid absorption, can be easily achieved in this model. These zebrafish-based assays are important tools to understand aspects of lipid biology with significant clinical implications.

References

Amatruda, J. F., Shepard, J. L., Stern, H. M., and Zon, L. I. (2002). Zebrafish as a cancer model system. *Cancer Cell.* **1,** 229–231.

Cao, Y., Murphy, K. J., McIntyre, T. M., Zimmerman, G. A., and Prescott, S. M. (2002). Expression of fatty acid-CoA ligase 4 during development and in brain. *FEBS Lett.* **467,** 263–267.

Chan, J., Bayliss, P. E., Wood, J. M., and Roberts, T. M. (2002). Dissection of angiogenic signaling in zebrafish using a chemical genetic approach. *Cancer Cell.* **1,** 257–267.

Chang, K. H., Xiang, H., and Dunaway-Mariano, D. (1997). Acyl-adenylate motif of the acyl-adenylate/thioester-forming enzyme superfamily: A site-directed mutagenesis study with the Pseudomonas sp. strain CBS3 4-chlorobenzoate: Coenzyme Coenzyme A ligase. *Biochemistry* **36,** 15650–15659.

Chen, W., Burgess, S., Golling, G., Amsterdam, A., and Hopkins, N. (2002). High-throughput selection of retrovirus producer cell lines leads to markedly improved efficiency of germ line-transmissible insertions in zebra fish. *J. Virol.* **76,** 2192–2198.

Coleman, R. A., Lewin, T. M., and Muoio, D. M. (2000). Physiological and nutritional regulation of enzymes of triacylglycerol synthesis. *Annu. Rev. Nutr.* **20,** 77–103.

Connelly, M. A., De La Llera-Moya, M., Peng, Y., Drazul-Schrader, D., Rothblat, G. H., and Williams, D. L. (2003). Separation of lipid transport functions by mutations in the extracellular domain of scavenger receptor class B, type I. *J. Biol. Chem.* **278,** 25773–25782.

Corkey, B. E., Deeney, J. T., Yaney, G. C., Tornheim, K., and Prentki, M. (2000). The role of long-chain fatty acyl-CoA esters in beta-cell signal transduction. *J. Nutr.* **130,** 299S–304S.

Crespo, N. C., Ohkanda, J., Yen, T. J., Hamilton, A. D., and Sebti, S. M. (2001). The farnesyltransferase inhibitor, FTI-2153, blocks bipolar spindle formation and chromosome alignment and causes prometaphase accumulation during mitosis of human lung cancer cells. *J. Biol. Chem.* **276,** 16161–16167.

Doitsidou, M., Reichman-Fried, M., Stebler, J., Koprunner, M., Dorries, J., Meyer, D., Esguerra, C. V., Leung, T., and Raz, E. (2002). Guidance of primordial germ cell migration by the chemokine SDF-1. *Cell* **111,** 647–659.

Donovan, A., Brownlie, A., Zhou, Y., Shepard, J., Pratt, S. J., Moynihan, J., Paw, B. H., Drejer, A., Barut, B., Zapata, A., Law, T. C., Brugnara, C., Lux, S. E., Pinkus, G. S., Pinkus, J. L., Kingsley, P. D., Palis, J., Fleming, M. D., Andrews, N. C., and Zon, L. I. (2000). Positional cloning of zebrafish ferroportin1 identifies a conserved vertebrate iron exporter. *Nature* **403,** 776–781.

Dooley, K., and Zon, L. I. (2000). Zebrafish: A model system for the study of human disease. *Curr. Opin. Genet. Dev.* **10,** 252–256.

Driever, W., Solnica-Krezel, L., Schier, A. F., Neuhauss, S. C., Malicki, J., Stemple, D. L., Stainier, D. Y., Zwartkruis, F., Abdelilah, S., Rangini, Z., Belak, J., and Boggs, C. (1996). A genetic screen for mutations affecting embryogenesis in zebrafish. *Development* **123,** 37–46.

Dutton, K., Dutton, J. R., Pauliny, A., and Kelsh, R. N. (2001). A morpholino phenocopy of the colourless mutant. *Genesis* **30,** 188–189.

Faergeman, N. J., and Knudsen, J. (1997). Role of long-chain fatty acyl-CoA esters in the regulation of metabolism and in cell signaling. *Biochem. J.* **323,** 1–12.

Farber, S. A., Olson, E. S., Clark, J. D., and Halpern, M. E. (1999). Characterization of Ca2+dependent phospholipase A2 activity during zebrafish embryogenesis. *J. Biol. Chem.* **274,** 19338–19346.

Farber, S. A., Pack, M., Ho, S. Y., Johnson, I. D., Wagner, D. S., Dosch, R., Mullins, M. C., Hendrickson, H. S., Hendrickson, E. K., and Halpern, M. E. (2001). Genetic analysis of digestive physiology using fluorescent phospholipid reporters. *Science* **292,** 1385–1388.

Farber, S. A., De Rose, R. A., Olson, E. S., and Halpern, M. E. (2003). The zebrafish annexin gene family. *Genome. Res.* **13,** 1082–1096.

Fisher, S., Jagadeeswaran, and Halpern, M. E. (2003). Radiographic analysis of zebrafish skeletal defects. *Dev. Biol.* **264,** 64–76.

Fisher, S., Amacher, S. L., and Halpern, M. E. (1997). Loss of cerebrum function ventralizes the zebrafish embryo. *Development* **124,** 1301–1311.

Fulgencio, J. P., Kohl, C., Girard, J., and Pegorier, J. P. (1996). Troglitazone inhibits fatty acid oxidation and esterification, and gluconeogenesis in isolated hepatocytes from starved rats. *Diabetes* **45,** 1556–1562.

Grosser, T., Yusuff, S., Cheskis, E., Pack, M. A., and FitzGerald, G. A. (2002). Developmental expression of functional cyclooxygenases in zebrafish. *Proc. Natl. Acad. Sci. USA* **99,** 8418–8423.

Haffter, P., Granato, M., Brand, M., Mullins, M. C., Hammerschmidt, M., Kane, D. A., Odenthal, J., van Eeden, F. J., Jiang, Y. J., Heisenberg, C. P., Kelsh, R. N., Furutani-Seiki, M., Vogelsang, E., Beuchle, D., Schach, U., Fabian, C., and Nusslein-Volhard, C. (1996). The identification of genes with unique and essential functions in the development of the zebrafish, Danio rerio. *Development* **123,** 1–36.

Ho, S. Y., Pack, M., and Farber, S. A. (2003). Analysis of small molecule metabolism in zebrafish. *Methods Enzymol.* **364,** 408–426.

Ivics, Z., and Izsvak, Z. (2004). Transposable elements for transgenesis and insertional mutagenesis in vertebrates: A contemporary review of experimental strategies. *Methods Mol. Biol.* **260,** 255–276.

Joffe, B. I., Panz, V. R., and Raal, F. J. (2001). From lipodystrophy syndromes to diabetes mellitus. *Lancet* **357,** 1379–1381.

Karlen, S., and Rebagliati, M. (2001). A morpholino phenocopy of the cyclops mutation. *Genesis* **30,** 126–128.

Kemp, S., Pujol, A., Waterham, H. R., van Geel, B. M., Boehm, C. D., Raymond, G. V., Cutting, G. R., Wanders, R. J., and Moser, H. W. (2001). ABCD1 mutations and the X-linked adrenoleukodystrophy mutation database: Role in diagnosis and clinical correlations. *Hum. Mutat.* **18**, 499–515.

Kurzchalia, T. V., Dupree, P., Parton, R. G., Kellner, R., Virta, H., Lehnert, M., and Simons, K. (1992). VIP21, a 21-kD membrane protein is an integral component of trans-Golgi-network-derived transport vesicles. *J. Cell Biol.* **118**, 1003–1014.

Lander, E. S., Linton, L. M., Birren, B., Nusbaum, C., Zody, M. C., Baldwin, J., Devon, K., Lander, E. S., Linton, L. M., Birren, B., Nusbaum, C., Zody, M. C., Baldwin, J., Devon, K., Dewar, K., Doyle, M., FitzHugh, W., Funke, R., Gage, D., Harris, K., Heaford, A., Howland, J., Kann, L., Lehoczky, J., LeVine, R., McEwan, P., McKernan, K., Meldrim, J., Mesirov, J. P., Miranda, C., Morris, W., Naylor, J., Raymond, C., Rosetti, M., Santos, R., Sheridan, A., Sougnez, C., Stange-Thomann, N., Stojanovic, N., Subramanian, A., Wyman, D., Rogers, J., Sulston, J., Ainscough, R., Beck, S., Bentley, D., Burton, J., Clee, C., Carter, N., Coulson, A., Deadman, R., Deloukas, P., Dunham, A., Dunham, I., Durbin, R., French, L., Grafham, D., Gregory, S., Hubbard, T., Humphray, S., Hunt, A., Jones, M., Lloyd, C., McMurray, A., Matthews, L., Mercer, S., Milne, S., Mullikin, J. C., Mungall, A., Plumb, R., Ross, M., Shownkeen, R., Sims, S., Waterston, R. H., Wilson, R. K., Hillier, L. W., McPherson, J. D., Marra, M. A., Mardis, E. R., Fulton, L. A., Chinwalla, A. T., Pepin, K. H., Gish, W. R., Chissoe, S. L., Wendl, M. C., Delehaunty, K. D., Miner, T. L., Delehaunty, A., Kramer, J. B., Cook, L. L., Fulton, R. S., Johnson, D. L., Minx, P. J., Clifton, S. W., Hawkins, T., Branscomb, E., Predki, P., Richardson, P., Wenning, S., Slezak, T., Doggett, N., Cheng, J. F., Olsen, A., Lucas, S., Elkin, C., Uberbacher, E., Frazier, M., Gibbs, R. A., Muzny, D. M., Scherer, S. E., Bouck, J. B., Sodergren, E. J., Worley, K. C., Rives, C. M., Gorrell, J. H., Metzker, M. L., Naylor, S. L., Kucherlapati, R. S., Nelson, D. L., Weinstock, G. M., Sakaki, Y., Fujiyama, A., Hattori, M., Yada, T., Toyoda, A., Itoh, T., Kawagoe, C., Watanabe, H., Totoki, Y., Taylor, T., Weissenbach, J., Heilig, R., Saurin, W., Artiguenave, F., Brottier, P., Bruls, T., Pelletier, E., Robert, C., Wincker, P., Smith, D. R., Doucette-Stamm, L., Rubenfield, M., Weinstock, K., Lee, H. M., Dubois, J., Rosenthal, A., Platzer, M., Nyakatura, G., Taudien, S., Rump, A., Yang, H., Yu, J., Wang, J., Huang, G., Gu, J., Hood, L., Rowen, L., Madan, A., Qin, S., Davis, R. W., Federspiel, N. A., Abola, A. P., Proctor, M. J., Myers, R. M., Schmutz, J., Dickson, M., Grimwood, J., Cox, D. R., Olson, M. V., Kaul, R., Raymond, C., Shimizu, N., Kawasaki, K., Minoshima, S., Evans, G. A., Athanasiou, M., Schultz, R., Roe, B. A., Chen, F., Pan, H., Ramser, J., Lehrach, H., Reinhardt, R., McCombie, W. R., de, la, Bastide, M., Dedhia, N., Blocker, H., Hornischer, K., Nordsiek, G., Agarwala, R., Aravind, L., Bailey, J. A., Bateman, A., Batzoglou, S., Birney, E., Bork, P., Brown, D. G., Burge, C. B., Cerutti, L., Chen, H. C., Church, D., Clamp, M., Copley, R. R., Doerks, T., Eddy, S. R., Eichler, E. E., Furey, T. S., Galagan, J., Gilbert, J. G., Harmon, C., Hayashizaki, Y., Haussler, D., Hermjakob, H., Hokamp, K., Jang, W., Johnson, L. S., Jones, T. A., Kasif, S., Kaspryzk, A., Kennedy, S., Kent, W. J., Kitts, P., Koonin, E. V., Korf, I., Kulp, D., Lancet, D., Lowe, T. M., McLysaght, A., Mikkelsen, T., Moran, J. V., Mulder, N., Pollara, V. J., Ponting, C. P., Schuler, G., Schultz, J., Slater, G., Smit, A. F., Stupka, E., Szustakowski, J., Thierry-Mieg, D., Thierry-Mieg, J., Wagner, L., Wallis, J., Wheeler, R., Williams, A., Wolf, Y. I., Wolfe, K. H., Yang, S. P., Yeh, R. F., Collins, F., Guyer, M. S., Peterson, J., Felsenfeld, A., Wetterstrand, K. A., Patrinos, A., Morgan, M. J., Szustakowki, J., de, Jong, P., Catanese, J. J., Osoegawa, K., Shizuya, H., Choi, S., and Chen, Y. J. (2001). Initial sequencing and analysis of the human genome. *Nature* **409**, 860–921.

McNeely, M. J., Edwards, K. L., Marcovina, S. M., Brunzell, J. D., Motulsky, A. G., Austin, M. A. (2001). Lipoprotein and apolipoprotein abnormalities in familial combined hyperlipidemia: A 20-year prospective study. *Atherosclerosis* **159**, 471–481.

Merrifield, C. J., Rescher, U., Almers, W., Proust, J., Gerke, V., Sechi, A. S., and Moss, S. E. (2001). Annexin 2 has an essential role in actin-based macro pinocytic rocketing. *Curr. Biol.* **11**, 1136–1141.

Muoio, D. M., Lewin, T. M., Wiedmer, P., and Coleman, R. A. (2000). Acyl-CoAs are functionally channeled in liver: Potential role of acyl-CoA synthetase. *Am. J. Physiol. Endocrinol. Metab.* **279**, E1366–E1373.

Nasevicius, A., and Ekker, S. C. (2000). Effective targeted gene 'knockdown' in zebrafish. *Nat. Genet.* **26,** 216–220.

Percy, A. K., and Rutledge, S. L. (2001). Adrenoleukodystrophy and related disorders. *Ment. Retard. Dev. Disabil. Res. Rev.* **7,** 179–189.

Petroni, A. (2002). Androgens and fatty acid metabolism in X-linked Adrenoleukodystrophy. *Prostaglandins Leukot. Essent. Fatty Acids* **67,** 137–139.

Prentki, M., and Corkey, B. E. (1996). Are the beta-cell signaling molecules malonyl-CoA and cystolicsystolic long-chain acyl-CoA implicated in multiple tissue defects of obesity and NIDDM? *Diabetes* **45,** 273–283.

Raynor, C. M., Wright, J. F., Waisman, D. M., and Pryzdial, E. L. (1999). Annexin II enhances cytomegalovirus binding and fusion to phospholipid membranes. *Biochemistry* **38,** 5089–5095.

Repa, J. J., Dietschy, J. M., and Turley, S. D. (2002). Inhibition of cholesterol absorption by SCH 58053 in the mouse is not mediated via changes in the expression of mRNA for ABCA1, ABCG5, or ABCG8 in the enterocyte. *J. Lipid Res.* **43,** 1864–1874.

Rogler, G., Herold, G., and Stange, E. F. (1991). HDL3-retroendocytosis in cultured small intestinal crypt cells: A novel mechanism of cholesterol efflux. *Biochim. Biophys. Acta* **1095,** 30–38.

Rothberg, K. G., Heuser, J. E., Donzell, W. C., Ying, Y. S., Glenney, J. R., and Anderson, R. G. (1992). Caveolin, a protein component of caveolae membrane coats. *Cell* **68,** 673–682.

Sakuma, S., Fujimoto, Y., Katoh, Y., Kitao, A., and Fujita, T. (2001). The effects of fatty acyl CoA esters on the formation of prostaglandin and arachidonoyl-CoA formed from arachidonic acid in rabbit kidney medulla microsomes. *Prostaglandins Leukot. Essent. Fatty Acids* **64,** 61–65.

Shafizadeh, E., Huang, H., and Lin, S. (2002). Transgenic zebrafish expressing green fluorescent protein. *Methods Mol. Biol.* **183,** 225–233.

Sheridan, M. A. (1988). Lipid dynamics in fish: Aspects of absorption, transportation, deposition and mobilization. *Comp. Biochem. Physiol. B.* **90,** 679–690.

Sillero, A., and Sillero, M. A. (2000). Synthesis of dinucleotide polyphosphates catalyzed by firefly luciferase and several ligases. *Pharmacol. Ther.* **87,** 91–102.

Smart, E. J., Graf, G. A., McNiven, M. A., Sessa, W. C., Engelman, J. A., Scherer, P. E., Okamoto, T., and Lisanti, M. P. (1999). Caveolins, liquid-ordered domains, and signal transduction. *Mol. Cell. Biol.* **19,** 7289–7304.

Smart, E. J., De Rose, R. A., and Farber, S. A. (2004). Annexin 2-caveolin 1 complex is a target of ezetimibe and regulates intestinal cholesterol transport. *Proc. Natl. Acad. Sci. USA* **101,** 3450–3455.

Steinberg, S. J., Morgenthaler, J., Heinzer, A. K., Smith, K. D., and Watkins, P. A. (2000). Very long-chain acyl-CoA synthetases. Human "bubblegum" represents a new family of proteins capable of activating very long-chain fatty acids. *J. Biol. Chem.* **275,** 35162–35169.

Streisinger, G., Walker, C., Dower, N., Knauber, D., and Singer, F. (1981). Production of clones of homozygous diploid zebra fish (Brachydanio rerio). *Nature* **291,** 293–296.

Stremmel, W., Pohl, L., Ring, A., and Herrmann, T. (2001). A new concept of cellular uptake and intracellular trafficking of long-chain fatty acids. *Lipids* **36,** 981–989.

Sun, J., Blaskovich, M. A., Knowles, D., Qian, Y., Ohkanda, J., Bailey, R. D., Hamilton, A. D., and Sebti, S. M. (1999). Antitumor efficacy of a novel class of non-thiol-containing peptidomimetic inhibitors of farnesyltransferase and geranylgeranyl transferase I: Combination therapy with the cytotoxic agents cisplatin, Taxol, and gemcitabine. *Cancer Res.* **59,** 4919–4926.

Suzuki, H., Watanabe, M., Fujino, T., and Yamamoto, T. (1995). Multiple promoters in rat acyl-CoA synthetase gene mediate differential expression of multiple transcripts with 5′-end heterogeneity. *J. Biol. Chem.* **270,** 9676–9682.

Thorpe, J. L., Doitsidou, M., Ho, S. Y., Raz, E., Farber, S. A. (2004). Germ cell migration in zebrafish is dependent on HMGCoA reductase activity and prenylation. *Dev. Cell.* **6,** 295–302.

Uittenbogaard, A., and Smart, E. J. (2000). Palmitoylation of caveolin-1 is required for cholesterol binding, chaperone complex formation, and rapid transport of cholesterol to caveolae. *J. Biol. Chem.* **275,** 25595–25599.

Uittenbogaard, A., Everson, W. V., Matveev, S. V., and Smart, E. J. (2002). Cholesteryl ester is transported from caveolae to internal membranes as part of a caveolin-annexin II lipid-protein complex. *J. Biol. Chem.* **277**, 4925–4931.

Uittenbogaard, A., Ying, Y., and Smart, E. J. (1998). Characterization of a cytosolic heat-shock protein-caveolin chaperone complex. Involvement in cholesterol trafficking. *J. Biol. Chem.* **273**, 16525–16532.

Urtishak, K. A., Choob, M., Tian, X., Sternheim, N., Talbot, W. S., Wickstrom, E., and Farber, S. A. (2003). Targeted gene knockdown in zebrafish using negatively charged peptide nucleic acid mimics. *Dev. Dyn.* **228**, 405–413.

Van Doren, M., Broihier, H. T., Moore, L. A., and Lehmann, R. (1998). HMG-CoA reductase guides migrating primordial germ cells. *Nature* **396**, 466–469.

Wang, H., Long, Q., Marty, S. D., Sassa, S., and Lin, S. (1998). A zebrafish model for hepatoerythropoietic porphyria. *Nat. Genet.* **20**, 239–243.

Wang, L., Connelly, M. A., Ostermeyer, A. G., Chen, H. H., Williams, D. L., Brown, D. A. (2003). Caveolin-1 does not affect SR-BI-mediated cholesterol efflux or selective uptake of cholesteryl ester in two cell lines. *J. Lipid Res.* **44**, 807–815.

Wienholds, E., van Eeden, F., Kosters, M., Mudde, J., Plasterk, R. H., and Cuppen, E. (2003). Efficient target-selected mutagenesis in zebrafish. *Genome Res.* **13**, 2700–2707.

CHAPTER 7

Analysis of the Cell Cycle in Zebrafish Embryos

Jennifer L. Shepard,★ Howard M. Stern,★,† Kathleen L. Pfaff,★ and James F. Amatruda★

★Division of Hematology-Oncology
Children's Hospital and Dana-Farber Cancer Institute
Boston, Massachusetts 02115

†Department of Pathology
Brigham and Women's Hospital
Boston, Massachusetts 02115

I. Introduction: The Cell Cycle in Zebrafish

Genetic dissection of the molecular pathways that regulate cell division in eukaryotes has revealed that many of the genes involved in cell cycle progression are evolutionarily conserved. Because cell division is intimately tied to cancer, cell cycle analysis in non-mammalian organisms has proven relevant to human cancer. Forward genetic screens in yeast and *Drosophila* have been invaluable for gene discovery and have made important contributions to understanding pathways regulating cell proliferation. Importantly, it has been found that the human orthologues of some genes identified in these organisms are expressed incorrectly in human tumors (Hariharan and Haber, 2003). The zebrafish model offers the opportunity to combine gene discovery with carcinogenesis analysis in one organism. Zebrafish develop tumors spontaneously, after exposure to carcinogens and by transgenic overexpression of oncogenes. They develop a variety of tumor types including carcinomas, sarcomas, germ cell tumors, and hematologic malignancies. Many zebrafish neoplasms histologically resemble those in humans (Langenau *et al.*, 2003; Spitsbergen and Kent, 2003; Spitsbergen *et al.*, 2000a,b). The zebrafish system has great promise as a model organism in which to genetically dissect the vertebrate cell cycle and potentially glean a better understanding of uncontrolled cell division and tumor development *in vivo*.

The advantages of the zebrafish system that make it a powerful organism to study vertebrate development include external fertilization of oocytes, transparent embryos, and rapid embryonic development. These features also provide the opportunity to study early cell divisions and tissue-specific cellular proliferation. Studies of the developing zebrafish embryo have revealed similarities to the early cell divisions of other vertebrates, such as *Xenopus*. In the zebrafish, the first seven cell divisions are synchronous and cycle rapidly between DNA replication (S phase) and mitosis (M phase) without the intervening gap phases, G1 and G2 (Kimmel *et al.*, 1995). The mid blastula transition (MBT) ensues during the tenth cell division, which is approximately 3 hpf. MBT is accompanied by loss of division synchrony, increased cell cycle duration, activation of zygotic transcription, and an onset of cellular motility (Kane and Kimmel, 1993). Embryonic cells first exhibit a G1 gap phase between the M and S phases during MBT. Onset of the G1 phase is dependent on the activation of zygotic transcription, which raises the possibility that some of the early zygotic transcripts may be cell cycle regulators that are directly responsible for slowing the cell cycle (Zamir *et al.*, 1997). Greater understanding of cell cycle regulation in zebrafish embryos was obtained by studying their responses to various cell cycle inhibitors, including aphidicolin, hydroxyurea, etoposide, camptothecin, and nocodazole (Ikegami *et al.*, 1997a,b, 1999). Exposure to these agents after MBT induces cell cycle arrest, sometimes accompanied by initiation of an apoptotic program. However, before MBT, the embryonic cells continue to divide, often with deleterious effects, after exposure to cell cycle inhibitors. These studies indicate that zebrafish embryos do possess cell cycle checkpoints, but they are not functional until after MBT. These findings are supported by a mutation in the maternal-effect gene *futile cycle*, which prevents mitotic spindle assembly in the first cell division (Dekens *et al.*, 2003). Despite this defect, the early cleavage divisions proceed and produce anucleate cells, indicating that cell cycle checkpoints are not present during the earliest cell divisions.

Later developmental stages of zebrafish embryogenesis provide the opportunity to study the cell cycle in distinct tissue types. Studies of cell cycle regulation in older embryos (10 to 36 hpf) have focused on the developing eyes and central nervous system (CNS). Lineage analysis of CNS progenitor cells revealed a correlation between morphogenesis and cell cycle number, implying that the nervous system development may be at least partially regulated by the cell cycle (Kimmel *et al.*, 1994). Whereas most developing vertebrate embryos exhibit a constant lengthening of the cell cycle duration throughout development, meticulous analysis of cell number in the developing zebrafish retina revealed a surprising mechanism of modulated cell cycle control. Li and colleagues (2000) reported that the retinal cell cycle duration temporarily slows between 16 and 24 hpf, followed by an abrupt change to more rapid cell divisions.

A large scale N-ethyl-N-nitrosourea (ENU) mutagenesis screen conducted in Tübigen uncovered a class of mutants that are hypothesized to bear mutations in cell cycle regulatory genes (Kane *et al.*, 1996). These early arrest mutants, such as *speed bump, ogre, zombie, specter, and poltergeist*, have phenotypes that include mitotic bridges, mitotic arrest, absence of cytokinesis, and abnormal nuclei accompanied by cell lysis. As these characteristics typically result from cell division defects, it is likely that the responsible mutations are in genes required for cell cycle progression.

In this chapter, we provide protocols for assays that characterize the various phases of cell division in zebrafish embryos. We discuss the design of screens for mutations affecting embryonic cell proliferation and a method to detect novel compounds affecting the zebrafish cell cycle. Assays discussed in this chapter include: DNA content analysis by flow cytometry, whole-mount embryonic antibody staining, mitotic spindle analysis, 5-bromo-2-deoxyuridine (BrdU) incorporation, cell death analysis, and *in situ* hybridization with cell cycle regulatory genes. Each assay targets different phases of the cell cycle and, in total, create a detailed picture of zebrafish embryo cell proliferation. Although our studies have focused on embryonic assays for cell cycle characterization, it is likely that these protocols can be modified to study adult tissues. These protocols can be applied to a variety of experiments, such as characterization of the cell cycle phenotypes of mutants or the analysis of RNA overexpression and morpholino knockdown of cell cycle regulatory genes. Furthermore, the genetic tractability of the zebrafish system (Patton and Zon, 2001) makes it an excellent organism in which to pursue forward genetic screens for mutations or chemical screens for novel compounds that alter cell division using one or more of these cell cycle assays.

II. Zebrafish Embryo Cell Cycle Protocols

A. DNA Content Analysis

A profile of the cell cycle in disaggregated zebrafish embryos or adult tissue can be obtained through DNA content analysis. In this technique, cells are stained with a dye that fluoresces on DNA binding, such as Hoechst 33342 or propidium iodide. The intensity of fluorescence is proportional to the amount of DNA in each cell (Krishan, 1975). Analysis by fluorescence activated cell sorting (FACS) generates a histogram showing the proportion of cells that have a complement of DNA (G1 phase) that is not replicated, those that have a fully replicated complement of DNA (G2 or M phase) and those that have an intermediate amount of DNA (S phase).

Protocol:

All steps are performed on ice except for the dechorionation (Step 1) and RNAse incubation (Step 9).

1. Dechorionate embryos and wash with **E3**. (Items in bold are reagents and supplies found in Section VI.) Analysis of single embryos is possible, though in practice we typically pool approximately 20 embryos/tube.
2. Disaggregate embryos (using small **pellet pestle**) in 500 μl of DMEM (or other tissue culture medium) + 10% fetal calf serum in a matching homogenizing tube.
3. Bring volume to 1 ml with DMEM/serum and remove aggregates by passing cell suspension sequentially through **105 μm mesh and 40 μm mesh**.
4. Count a sample using a hemocytometer.
5. Place volume containing at least 2×10^6 cells in a 15-ml conical tube, and bring volume to 5 ml with 1X **PBS**.
6. Spin at 1200 rpm for 10 m at 4 °C.
7. Carefully aspirate off liquid and gently resuspend cell pellet in 2 ml **Propidium Iodide solution**.
8. Add 2 μg of DNAse-free RNAse (Roche, Basel, Switzerland).
9. Incubate in the dark at room temperature for 30 m.
10. Place samples on ice and analyze on FACS machine.

B. Whole Mount Immunohistochemistry with Mitotic Marker Phospho Histone H3

Histone H3 phosphorylation is considered to be a crucial event for the onset of mitosis and an antibody against this protein has been widely used in *Drosophila* and mammalian cell lines as a mitotic marker (Hendzel *et al.*, 1997). Two members of the Aurora/AIK kinase family, Aurora A and Aurora B, phosphorylate histone H3 at the serine 10 residue (Chadee *et al.*, 1999; Crosio *et al.*, 2002). Increased serine 10 phosphorylation of histone H3 has been seen in transformed fibroblasts (Chadee *et al.*, 1999), suggesting that this antibody could make an excellent marker for cell proliferation in the zebrafish, as well as detecting cell cycle mutations that may result in transformed phenotypes. In zebrafish, the phosphohistone H3 antibody (pH3) stains mitotic cells throughout the embryo (Fig. 1A). In developing organs like the nervous system, pH3 staining increases as zebrafish embryos undergo proliferation during distinct developmental stages.

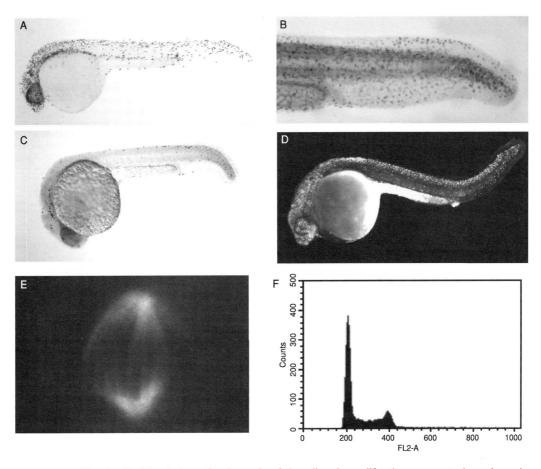

Fig. 1 Useful techniques for the study of the cell cycle, proliferation, or apoptosis as shown in zebrafish embryos. (A) Antibody staining against phosphorylated histone H3 in wild-type 24-hpf embryos. (B) BrdU incorporation to mark cells in S phase in the tail of a wild-type 28-hpf embryo. (C) and (D) Apoptotic cells can be visualized by TUNEL (in C, wild-type 24-hpf embryos) or acridine orange (in D, 24-hpf *crash&burn* mutant embryo [Shepard *et al.*, submitted]). (E) Anti-alpha tubulin can be used to examine mitotic spindle formation. (F) DNA content analysis shows the population of embryonic cells present in all phases of the cell cycle.

Protocol:

1. Fix embryos overnight at 4 °C in 4% paraformaldehyde (**PFA**).
2. Permeabilize embryos for 7 m in −20 °C acetone.
3. Wash embryos in H_2O followed by 2 × 5 m washes in **PBST**.
4. Incubate for 30 m at room temperature in **block**.

5. Incubate overnight at 4 °C in rabbit anti-phospho histone H3 at a concentration of 1.33 μg/ml in **block**. Two different sources of antibody have been used: Santa Cruz Biotechnology (Santa Cruz, CA) and an anti phospho peptide polyclonal antibody to the sequence (ARKS[PO$_4$]TGGKAPRKQLC) made and affinity purified by Genemed Synthesis (South San Francisco, CA).

6. Wash 4 × 15 m in **PBST**.

7. Incubate 2 h at room temperature in horseradish peroxidase-conjugated secondary goat anti-rabbit IgG (Jackson ImmunoResearch Laboratories, Inc., West Grove, PA) at a concentration of 3 μg/ml in block.

8. Wash 4 × 15 m in **PBST**.

9. Develop in the dark for 5 m at room temperature in diaminobenzidine (DAB) solution (0.67 mg/ml in 15 ml of PBST) and 12 μl 30% H$_2$O$_2$.

10. Wash in **PBST** and store embryos at 4 °C in **PFA**.

C. Mitotic Spindle/Centrosome Detection

Study of the mitotic spindle and centrosomes is an important step in understanding mutants with cell cycle defects, particularly those whose phenotypes appear to be related to problems in mitosis. Genomic instability is one of the main alterations seen in human cancers and such unequal segregation of chromosomes can be caused by problems in mitotic spindle formation or centrosome number (Kramer *et al.*, 2002). In this protocol, anti-α-tubulin labels the mitotic spindle, anti-γ-tubulin detects the centrosome, and 4′,6-diamidino-2-phenylindole (DAPI) labels the DNA.

Protocol:

1. Fix embryos in **PFA** for 4 h at room temperature.

2. Dehydrate in methanol at −20 °C for at least 30 m.

3. Rehydrate embryos in graded methanol:**PBST** series (3:1, 1:1, 1:3) for 5 m each.

4. Wash 1 × 5 m in **PBST**.

5. Place in −20 °C acetone for 7 m.

6. Wash 3 × 5 m in **PBST**.

7. Incubate 1 h at room temperature in **block**.

8. Incubate in monoclonal mouse α-tubulin antibody (Sigma, St. Louis, MO) at a concentration of 1:500 and in polyclonal rabbit γ-tubulin antibody (Sigma, St. Louis, MO) at a concentration of 1:1000 (both diluted in block) at 4 °C overnight.

9. Wash 4 × 15 m in **PBST**.

10. Incubate in rhodamine-conjugated goat anti-mouse secondary (Molecular Probes, Eugene, OR) at 1:600 dilution and fluorescein-conjugated goat

anti-rabbit secondary (Jackson ImmunoResearch Laboratories, Inc.) at 1:600 dilution for 2 h room temperature.

11. Wash 2 × 15 m in **PBST**.
12. Include a 1:500 dilution of 100 μM **DAPI** during the third wash to stain DNA.
13. Wash 2 × 15 m in **PBST**.

Mount whole embryos (if younger than 18 hpf) or just tails of embryos (if older than 18 hpf) on glass slides with VectaShield mounting media (Vector Laboratories, Burlingame, CA).

D. BrdU Incorporation

BrdU (5-bromo-2-deoxyuridine) is a nucleoside analog that is specifically incorporated into DNA during S-phase (Meyn *et al.*, 1973) and can subsequently be detected with an anti-BrdU specific antibody. This technique has been used to label replicating cells in zebrafish embryos (Larison and Bremiller, 1990) and adults (Rowlerson *et al.*, 1997). The following protocol is designed to label a fraction of proliferating cells in zebrafish embryos, to allow comparison of the replication fraction of different embryos (Fig. 1B). If the embryos are chased for varying amounts of time after the BrdU pulse, then fixed and stained for both BrdU and pH3 (as described in Section B), the transit of cells from S phase into G2/M can be assessed. This is useful in analyzing mutants with mitotic phenotypes.

Protocol:

1. Dechorionate and chill 15 m on ice in **E3**.
2. Prepare cold 10 mM BrdU/15% Dimethylsulfoxide (DMSO) in **E3** and chill on ice. Place embryos in BrdU solution and incubate 20 m on ice to allow uptake of BrdU.
3. Change into warm **E3** and incubate exactly 5 m, 28.5 °C. Note: Longer incubation times will result in more cells being labeled.
4. Fix 2 h at room temperature in **PFA**. Longer fixation may decrease the staining.
5. Transfer to methanol at −20 °C overnight. All subsequent steps are performed at room temperature unless otherwise noted.
6. Rehydrate in graded methanol:**PBST** series (3:1, 1:1, 1:3) for 5 m each.
7. Wash 2× in **PBST,** 5 m.
8. Digest embryos in 10 ug/ml Proteinase K, 10 m.
9. Wash **PBST**. Refix in PFA for not more than 20 m.
10. Wash quickly 3× in H$_2$O, then 2× in 2N HCl.
11. Incubate 1 h in 2N HCl. This step denatures the labeled DNA to expose the BrdU epitope.

12. Rinse several times in **PBST**. It is important to bring the pH back up to approximately 7 before adding blocking solution.

13. Block for 30 m in **BrdU blocking solution**.

14. Incubate in monoclonal **anti-BrdU antibody** at a dilution of 1:100 in BrdU block for 2 h at room temperature or overnight at 4 °C. (If carrying out simultaneous BrdU/pH3 staining, add the primary anti-pH3 antibody as described in Section B, except that BrdU block is used.)

15. Wash 5 × 10 m in **PBST**.

16. Incubate 2 h at room temperature with horseradish peroxidase or fluorophore-conjugated anti-mouse secondary antibody. (For simultaneous BrdU/pH3 stain, add a fluorescent anti-rabbit antibody as well.)

17. Wash 5 × 10 m in **PBST**. Develop with diaminobenzidine if using HRP-conjugated secondary antibody.

E. Apoptosis Detection by TUNEL Staining

Apoptosis is a form of programmed cell death that eliminates damaged or unneeded cells. It is controlled by multiple signaling pathways that mediate responses to growth, survival, or death signals. Cell cycle checkpoint controls are linked to apoptotic cascades and these connections can be compromised in diseases, including cancer. The defining characteristics of apoptosis are membrane blebbing, cell shrinkage, nuclear condensation, segmentation, and division into apoptotic bodies that are phagocytosed (Wyllie, 1987). The DNA strand breaks that occur during apoptosis can be detected by enzymatically labeling the free ends with modified nucleotides, which can then be detected with antibodies (Gavrieli *et al.*, 1992).

Protocol:

1. Embryos are fixed overnight at 4 °C in **PFA**.

2. Wash in **PBS** and transfer to methanol for 30 m at −20 °C.

3. Rehydrate embryos in a graded methanol:**PBST** series (3:1, 1:1, 1:3) for 5 m each.

4. Wash 1 × 5 m in **PBST**.

5. Digest embryos in proteinase K (10 μg/ml) at room temperature (1 m for embryos younger than 16 hpf, 2 m for embryos older than 16 hpf).

6. Wash twice in **PBST**.

7. Postfix in **PFA** for 20 m room temperature.

8. Wash 5 × 5 m in **PBST**.

9. Postfix for 10 m at −20 °C with pre chilled Ethanol:Acetic Acid 2:1.

10. Wash 3 × 5 m in **PBST** at room temperature.

11. Incubate for 1 h at room temperature in 75 ul equilibration buffer (TdT-ApopTag Peroxidase *In Situ* Apoptosis Detection Kit from Serologicals, Norcross, GA Corporation).

12. Add a small volume of working-strength Terminal deoxynucleotidyl Transferase (TdT)—reaction buffer and TdT at a ratio of 2:1 plus 0.3% Triton—from Serologics Corporation.

13. Incubate overnight at 37 °C.

14. Stop reaction by washing in working-strength stop/wash buffer (1 ml concentrated buffer from Serologics Kit with 34 ml water) for 3 to 4 h at 37 °C.

15. Wash 3 × 5 m in **PBST**.

16. Block with 2 mg/ml BSA, 5% sheep serum in **PBST** for 1 h at room temperature.

17. Incubate in anti-digoxigenin peroxidase antibody included in kit (full strength).

18. Wash 4 × 30 m **PBST** at room temperature.

19. Develop in the dark for 5 m at room temperature in DAB solution (0.67 mg/ml in 15 ml of **PBST**) and 12 μl 30% H_2O_2.

20. Wash in **PBST** and store embryos at 4 °C in PFA.

F. Apoptosis Detection by Acridine Orange

Another method of apoptotic cell detection that can be performed on living embryos is acridine orange staining. The basis of this method is that the ATP-dependent lysosomal proton pump is preserved in apoptotic cells but not in necrotic cells. Therefore, apoptotic cells will take up the acridine orange dye; whereas living or necrotic cells will not (Darzynkiewicz *et al.*, 1992). This method is useful for identifying mutants based on an apoptotic phenotype in order to further characterize them in living assays.

Protocol:

1. Live dechorionated embryos are incubated in a 2 μg/ml solution of acridine orange (Sigma, St. Louis, MO) in 1 × **PBS** for 30 m at room temperature.

2. Embryos are washed 5 × quickly in **E3** 5 × and then visualized on a stereo dissecting microscope equipped for fluoresceinisothiocynate (FITC) epifluorescence.

G. *In Situ* Hybridization

RNA expression analysis by *in situ* hybridization of antisense probes in whole-mount zebrafish embryos is a commonly used technique to localize expression of developmental regulatory genes. While the technique is not exceptionally quantitative, it can reveal stark differences in gene expression. More quantitative analysis of gene expression, such as Northern blotting, reverse transcriptase-polymerase chain reaction (RT-PCR), or real-time polymerase chain reaction (PCR) do not permit the examination of alterations in tissue-specific expression or an expression pattern.

Cell division is a highly controlled process that involves regulation at both the transcriptional and posttranslational stages. Cyclins are a class of proteins that play critical roles in guiding cells through the G1, S, G2, and M phases of the cell cycle by regulating the activity of the cyclin-dependent kinases. The name cyclin alludes to the fact that their expression levels oscillate between peaks and nadirs that are coordinated with particular phases of the cell cycle (reviewed in Murray, 2004). The tightly regulated expression of these important cell cycle genes incorporates transcriptional, translational, and posttranslational controls. Many genes involved in cell cycle regulation are specifically expressed during the cell cycle phase in which they act.

Protein expression and activity are the most direct ways to understand a particular gene's role in a process such as the cell cycle. Unfortunately, most mammalian antibodies do not cross-react with the zebrafish orthologs, prohibiting analysis by either whole-mount immunohistochemistry or immunoblotting. While many groups are working to obtain functional zebrafish antibodies to expand zebrafish experiments into biochemical understanding, most current zebrafish studies focus on gene expression.

Zebrafish orthologs of cell cycle regulatory genes such as PCNA and cyclins have been found to possess similar expression patterns throughout the proliferative tissues of developing zebrafish embryos (www.zfin.org). A subset of these genes has altered expression levels in response to cell cycle perturbations, such as exposure to the microtubule poison nocodazole (K. Pfaff and L. Zon, unpublished, 2004). These cell cycle regulatory genes also exhibit expression changes in zebrafish mutants that have abnormal cell proliferation (Shepard *et al.*, 2004, submitted; and unpublished data, 2004). *In situ* hybridization for cell cycle regulatory genes can be performed using previously published *in situ* hybridization protocols (Jowett, 1999; Thisse *et al.*, 1993, 1994).

III. Screening for Mutations Affecting the Zebrafish Cell Cycle

Genetic systems such as flies and yeast have provided vital information about the identity of the genes involved in proliferation and cell cycle control. Many of the genes uncovered in these model organisms have been linked to cancer through work in vertebrate systems. Zebrafish, with its strengths as both a genetic system and its anatomical similarity to humans, combines the strengths of the other models and can build on the knowledge they have already generated. A forward genetic screen in zebrafish has uncovered mutations that affect embryonic cellular proliferation, and these mutants have been used to establish a link between the embryonic phenotype and adult cancer susceptibility (Shepard *et al.*, submitted).

Many different screens can be envisioned using the cell cycle assays described as markers earlier in this chapter to detect defects in pathways controlling many aspects of cell proliferation. As an example of one such screen, our laboratory has performed a haploid screen for mutations in cell proliferation using the antibody against phosphorylated histone H3 (pH3) as a marker for mitotic cells (Shepard *et al.*, 2004, submitted and K. Pfaff, unpublished, 2004). ENU males were generated using published techniques (Solnica-Krezel *et al.*, 1994) and haploid embryos were created from the F1 females (Corley-Smith *et al.*, 1999). These haploid clutches were fixed at 36 hpf and stained with the pH3 antibody. The 36 hpf time point was chosen because it is well past the mid-blastula transition and maternal transcripts should be largely absent by this time. Therefore, any embryos with mutations in necessary housekeeping genes should be lethal by 36 hpf. Additionally, by 36 hpf almost all of the organ systems have begun to form. Therefore, any cell type or organ-specific cell cycle changes caused by the mutations should be detectable. Choosing a time point that is appropriate for the chosen method of detection is a very important step. It is necessary to consider both the developmental stage of wild-type embryos and at which stage mutations in the targeted pathway are likely to become apparent. After screening haploid clutches from more than 1000 F1 females for changes in the number of mitotic cells as seen by pH3 staining, more than 20 diploid mutants were identified that had varying defects in mitotic cell number. Our finding of these mutants demonstrates the ability to use zebrafish screens to identify mutants with interesting cell cycle phenotypes.

IV. Screening for Chemical Suppressors of Zebrafish Cell Cycle Mutants

Another way to probe the cell cycle is via chemical agents. Chemical screens could identify novel compounds that are useful tools for studying the cell cycle. Furthermore, mutations in cell cycle genes are commonly found in human cancer. Given the need to improve on current cancer therapy, one approach is to identify small molecule suppressors that bypass the consequences of specific cell cycle gene mutations. Akin to the use of genetic modifier screens to identify secondary mutations that enhance or suppress a primary defect (St Johnston, 2002), chemical suppressor screens would directly identify small molecules that rescue a genetic phenotype. If the phenotype is disease related, such compounds might represent lead therapeutic agents.

Zebrafish have recently been utilized in chemical screens to identify compounds that perturb specific aspects of development (Khersonsky *et al.*, 2003; Peterson *et al.*, 2000). The zebrafish system offers several advantages for chemical screens, providing information on tissue specificity and toxicity, and accounting for compound activation via drug metabolism. Furthermore, cells are not transformed and are in their normal physiological milieu of cell-cell and cell-extracellular matrix interactions. For these reasons, we developed a zebrafish chemical suppressor screening technology using a recessive lethal cell cycle mutant that has a fourfold increase in the number of mitotic cells as detected by pH3-staining (Stern *et al.*, 2004, submitted). This technology could easily be applied to other cell cycle mutants and could be modified to use cell cycle assays other than pH3 staining. In addition, such chemical suppressor screens could be applied to any zebrafish model of human disease (Dooley and Zon, 2000).

The following protocol can be repeated weekly, giving a throughput of more than 1000 compounds per week for a recessive lethal mutation. In the case of homozygous viable mutants, the throughput could be improved by using fewer embryos (3 to 5) per well in 96-well plates.

Protocol:

1. Generate large numbers of embryos for a chemical screen. The embryos must be at approximately the same developmental stage. Set up 100 heterozygote pair wise matings with fish separated by a divider. The next morning, remove the divider, allow the fish to mate, and collect the embryos.

2. Dilute chemicals into screening medium. The screen is conducted in 48-well plates with a volume of 300 μl per well. Individual chemicals could be added to each well, but to improve throughput, we devised a matrix pooling strategy: The chemical library (courtesy of the Institute of Chemistry and Cell Biology, Harvard Medical School) was arrayed in 384 well plates with the last 4 columns empty, thus containing 320 compounds per plate. Given this plate geometry, 8 by 10 matrix pools were created. A hit detected in both a horizontal and a vertical pool identified the individual compound (Fig. 2A).

   a. Transfer 80 μl of **screening medium** to each well of four 384-well plates using a Tecan liquid handling robot (Tecan, Durham, NC).

   b. Pin transfer 1 μl of each compound (arrayed at 5 mg/ml in DMSO) into each well of screening medium by performing 10 transfers with a 100 nl 384-pin array for each of the four 384-well plates (total of 320 $\times$ 4 = 1280 compounds).

   c. Pipette the diluted chemicals from the 384-well plates to 48-well plates to carry out pooling, using a Tecan liquid handling robot. For vertical pools, 30 μl was transferred from each of 8 wells plus an additional 60 μl of screening medium to bring the total volume to 300 μl. For horizontal pools, 30 μl was transferred from each of 10 wells.

3. Aliquot embryos to the 48-well plates at 50% epiboly.

   a. Before aliquoting embryos to wells, examine them under a dissecting microscope and discard all dead, delayed, or deformed embryos.

A

1	2	3	4	5	6	7	8	9	10	A
11	12	13	14	15	16	17	18	19	20	B
21	22	23	24	25	26	27	28	29	30	C
31	32	33	34	35	36	37	38	39	40	D
41	42	43	44	45	46	47	48	49	50	E
51	52	53	54	55	56	57	58	59	60	F
61	62	63	64	65	66	67	68	69	70	G
71	72	73	74	75	76	77	78	79	80	H
I	J	K	L	M	N	O	P	Q	R	

B

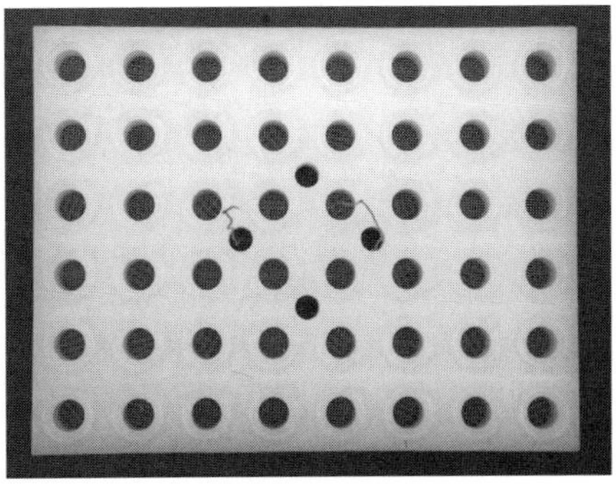

Fig. 2   Chemical screen methods. (A) Example of matrix pooling. In an 8-by-10 matrix pool, 80 compounds (numbered 1 to 80) are screened in 18 pools (A to R). Each compound is thus tested in 2 distinct pools. Individual active compounds are identified by deconvoluting the matrix. For example, if a hit is identified in pools B and P, the active compound would be number 18. (B) A 48-well staining grid.

   b. Pool embryos in a single 100-mm tissue culture dish or a 50-ml conical tube.

   c. Decant the embryo medium and remove as much liquid from the embryo suspension as possible with a transfer pipette. Pressing the transfer pipette tip to the bottom of the tube or dish allows most liquid to be removed without aspirating the embryos.

   d. Add approximately 20 embryos to each well by scooping them with a small chemical weighing spatula. With 20 embryos per well and a Mendelian recessive inheritance, there is a 0.3% chance of a well having no mutants. Since a hit requires detection in both a horizontal and a vertical pool, each with 20 embryos, the false-positive rate for identification of complete suppressors is 0.001%.

4. Place 48-well plates into an incubator at 28.5 °C.

5. One to 2 h later, clean out any dead embryos from each well using a long glass Pasteur pipette bent at a 90° angle.

6. Incubate at 28.5 °C overnight.

7. Dechorionate embryos by adding 150 μl of a 5 mg/ml pronase solution to each well. After 10 m, gently shake plates until embryos come out of the chorions.

8. Using a transfer pipette fitted with a 10 μl tip, remove as much of the pronase/chemical mixture as possible from each well.

9. Rinse the embryos once in fresh embryo medium and remove, as in Step 8.

10. Add 500 μl of **PFA** to each well.

11. Parafilm the edges of the plates to prevent evaporation and fix at 4 °C at least overnight but not longer than a week.

12. Using a transfer pipette, move embryos to 48-well staining grids made of acetone-resistant plastic with a wire mesh bottom (Fig. 2B).

13. Perform pH3 staining protocol by placing staining grids into 11- by 8.5-cm reservoirs containing 20 to 30 ml of the appropriate solution. To change solutions, the grid can be lifted out of one reservoir and placed into another reservoir with the next solution. For overnight antibody incubations, the reservoir should be sealed with parafilm to prevent evaporation.

14. After staining is complete, move embryos with a transfer pipette back into 48-well plates that have been precoated with 100 μl of 1% agarose in 1 × **PBS**. The agarose forms a meniscus that keeps embryos in the center of the well where they are easier to score.

15. Score for absence of mutants or for partial suppression without effect on wild types. In addition to suppressors and enhancers, one can identify compounds that affect both wild types and mutants, thus having a more general effect.

16. Deconvolute the matrix pool to identify individual chemicals.

V. Conclusions

Given the power of zebrafish in forward vertebrate genetics and organism-based small molecule screens, the system will nicely complement traditional model organisms for studying the cell division cycle. Many of the assays that are commonly used to probe the cell cycle in systems such as yeast, *Drosophila*, and mammalian cells can be used in the zebrafish. The protocols outlined in this chapter can be utilized to characterize known mutants for alterations in cell proliferation or, alternatively, can be used to screen for more cell cycle mutants. Given that zebrafish embryos are amenable to gene knockdown via antisense morpholino-modified oligonucleotides and overexpression by mRNA injection, these protocols can also be used to study cell cycle genes in the zebrafish without generating a mutant.

One of the remaining challenges in the field is a deficit of antibodies against zebrafish cell cycle regulatory proteins. More effort will be necessary to identify existing antibodies that cross-react with zebrafish proteins or to develop such antibodies from zebrafish peptides so that genetic and small molecule discoveries can be more easily characterized via biochemistry. Regardless, one of the real advantages of the zebrafish system in cell cycle biology is the ability to link genetic discoveries directly to cancer. Zebrafish develop cancer that is histologically similar to human neoplasms, and large numbers of fish can be put through carcinogenesis assays (Spitsbergen *et al.*, 2000a,b) to determine whether cell cycle mutations confer altered cancer susceptibility. Similarly, small molecule suppressors of cell cycle mutations could be tested in zebrafish for chemotherapeutic or chemopreventative activity. Ongoing development of tools to study the cell cycle in zebrafish combined with the methods described here will forward our understanding of the vertebrate cell division cycle and cancer.

VI. Reagents and Supplies

Anti-BrdU	Roche, Basel Switzerland (catalog number 1170 376).
Block	2% blocking reagent (Roche catalog number 1096 176), 10% fetal calf serum, 1% dimethylsulfoxide in PBST.
BrdU Block	0.2% blocking reagent (Roche catalog number 1096 176), 10% fetal calf serum, 1% dimethylsulfoxide in PBST. The lower concentration of blocking reagent improves detection.
DAPI	4′,6-Diamidino-2-phenylindole.
E3	5 mM NaCl, 0.17 mM KCl, 0.33 mM $CaCl_2$, 0.33 mM $MgSO_4$.
Mesh	Small Parts, Inc., Miami Lakes, FL 105-μm mesh (catalog number U-CMN-105D). 40-μm mesh (catalog number U-CMN-40D).
PBS	Phosphate-buffered saline, pH 7.5.
PBST	1X PBS with 0.1% (v/v) Tween-20.
Pellet pestle & tubes	Fisher Scientific International, Hampton, NH (catalog number K749520-0090).
PFA	4% paraformaldehyde buffered with $1 \times$ PBS.
Propidium Iodide	0.1% Sodium Citrate, 0.05 mg/ml propidium iodide, 0.0002% Triton X-100 (added fresh)
Screening medium	**E3** supplemented with 1% DMSO, 20 μM metronidazole, 50 units/mL penicillin, 50 μg/ml streptomycin, and 1 mM Tris pH 7.4.

Acknowledgments

We thank Len Zon and members of the Zon laboratory for useful discussions, Oninye Onyekwere for advice on FACS protocols, and Cassandra Belair and Caroline Burns for critical reading of the manuscript. Supported by NIH grants 1R01 DK55381 and 1 R01 HD044930 (Leonard I. Zon), 5 K08 HL04082 (J.F.A), 5 K08 DK061849 (H.M.S.) and the Albert J. Ryan Fellowship (J.L.S and K.L.P.).

References

Chadee, D. N., Hendzel, M. J., Tylipski, C. P., Allis, C. D., Bazett-Jones, D. P., Wright, J. A., and Davie, J. R. (1999). Increased Ser-10 phosphorylation of histone H3 in mitogen-stimulated and oncogene-transformed mouse fibroblasts. *J. Biol. Chem.* **274**, 24914–24920.

Corley-Smith, G. E., Brandhorst, B. P., Walker, C., and Postlethwait, J. H. (1999). Production of haploid and diploid androgenetic zebrafish (including methodology for delayed *in vitro* fertilization). *Methods Cell Biol.* **59**, 45–60.

Crosio, C., Fimia, G. M., Loury, R., Kimura, M., Okano, Y., Zhou, H., Sen, S., Allis, C. D., and Sassone-Corsi, P. (2002). Mitotic phosphorylation of histone H3: Spatio-temporal regulation by mammalian Aurora kinases. *Mol. Cell. Biol.* **22**, 874–885.

Darzynkiewicz, Z., Bruno, S., Del Bino, G., Gorczyca, W., Hotz, M. A., Lassota, P., and Traganos, F. (1992). Features of apoptotic cells measured by flow cytometry. *Cytometry* **13**, 795–808.

Dekens, M. P., Pelegri, F. J., Maischein, H. M., and Nusslein-Volhard, C. (2003). The maternal-effect gene futile cycle is essential for pronuclear congression and mitotic spindle assembly in the zebrafish zygote. *Development* **130**, 3907–3916.

Dooley, K., and Zon, L. I. (2000). Zebrafish: A model system for the study of human disease. *Curr. Opin. Genet. Dev.* **10**, 252–256.

Gavrieli, Y., YSherman, Y., and Ben-Sasson, S. A. (1992). Identification of programmed cell death *in situ* via specific labeling of nuclear DNA fragmentation. *J. Cell Biol.* **119**, 493–501.

Hariharan, I. K., and Haber, D. A. (2003). Yeast, flies, worms, and fish in the study of human disease. *N. Engl. J. Med.* **348**, 2457–2463.

Hendzel, M. J., Wei, Y., Mancini, M. A., Van Hooser, A., Ranalli, T., Brinkley, B. R., Bazett-Jones, D. P., and Allis, C. D. (1997). Mitosis-specific phosphorylation of histone H3 initiates primarily within peri centromeric heterochromatin during G2 and spreads in an ordered fashion coincident with mitotic chromosome condensation. *Chromosoma* **106**, 348–360.

Ikegami, R., Hunter, P., and Yager, T. D. (1999). Developmental activation of the capability to undergo checkpoint-induced apoptosis in the early zebrafish embryo. *Dev. Biol.* **209**, 409–433.

Ikegami, R., Rivera-Bennetts, A. K., Brooker, D. L., and Yager, T. D. (1997a). Effect of inhibitors of DNA replication on early zebrafish embryos: Evidence for coordinate activation of multiple intrinsic cell-cycle checkpoints at the mid-blastula transition. *Zygote* **5**, 153–175.

Ikegami, R., Zhang, J., Rivera-Bennetts, A. K., and Yager, T. D. (1997b). Activation of the metaphase checkpoint and an apoptosis programme in the early zebrafish embryo, by treatment with the spindle-destabilising agent nocodazole. *Zygote* **5**, 329–350.

Jowett, T. (1999). Analysis of protein and gene expression. *Methods Cell Biol.* **59**, 63–85.

Kane, D. A., and Kimmel, C. B. (1993). The zebrafish midblastula transition. *Development* **119**, 447–456.

Kane, D. A., Maischein, H. M., Brand, M., van Eeden, F. J., Furutani-Seiki, M., Granato, M., Haffter, P., Hammerschmidt, M., Heisenberg, C. P., Jiang, Y. J., *et al.* (1996). The zebrafish early arrest mutants. *Development* **123**, 57–66.

Khersonsky, S. M., Jung, D. W., Kang, T. W., Walsh, D. P., Moon, H. S., Jo, H., Jacobson, E. M., Shetty, V., Neubert, T. A., and Chang, Y. T. (2003). Facilitated forward chemical genetics using a tagged triazine library and zebrafish embryo screening. *J. Am. Chem. Soc.* **125**, 11804–11805.

Kimmel, C. B., Ballard, W. W., Kimmel, S. R., Ullmann, B., and Schilling, T. F. (1995). Stages of embryonic development of the zebrafish. *Dev. Dyn.* **203**, 253–310.

Kimmel, C. B., Warga, R. M., and Kane, D. A. (1994). Cell cycles and clonal strings during formation of the zebrafish central nervous system. *Development* **120**, 265–276.

Kramer, A., Neben, K., and Ho, A. D. (2002). Centrosome replication, genomic instability and cancer. *Leukemia* **16**, 767–775.

Krishan, A. (1975). Rapid flow cytofluorimetric analysis of mammalian cell cycle by propidium iodide staining. *J. Cell Biol.* **66**, 188–193.

Langenau, D. M., Traver, D., Ferrando, A. A., Kutok, J. L., Aster, J. C., Kanki, J. P., Lin, S., Prochownik, E., Trede, N. S., Zon, L. I., *et al.* (2003). Myc-induced T cell leukemia in transgenic zebrafish. *Science* **299**, 887–890.

Larison, K. D., and Bremiller, R. (1990). Early onset of phenotype and cell patterning in the embryonic zebrafish retina. *Development* **109**, 567–576.

Li, Z., Hu, M., Ochocinska, M. J., Joseph, N. M., and Easter, S. S., Jr. (2000). Modulation of cell proliferation in the embryonic retina of zebrafish (Danio rerio). *Dev. Dyn.* **219**, 391–401.

Meyn, R. E., Hewitt, R. R., and Humphrey, R. M. (1973). Evaluation of S phase synchronization by analysis of DNA replication in 5-bromodeoxyuridine. *Exp. Cell Res.* **82**, 137–142.

Murray, A. W. (2004). Recycling the cell cycle: Cyclins revisited. *Cell* **116**, 221–234.

Patton, E. E., and Zon, L. I. (2001). The art and design of genetic screens: Zebrafish. *Nat. Rev. Genet.* **2**, 956–966.

Peterson, R. T., Link, B. A., Dowling, J. E., and Schreiber, S. L. (2000). Small molecule developmental screens reveal the logic and timing of vertebrate development. *Proc. Natl. Acad. Sci. USA* **97,** 12965–12969.

Rowlerson, A., Radaelli, G., Mascarello, F., and Veggetti, A. (1997). Regeneration of skeletal muscle in two teleost fish: Sparus aurata and Brachydanio rerio. *Cell Tissue Res.* **289,** 311–322.

Solnica-Krezel, L., Schier, A. F., and Driever, W. (1994). Efficient recovery of ENU-induced mutations from the zebrafish germline. *Genetics* **136,** 1401–1420.

Spitsbergen, J. M., and Kent, M. L. (2003). The state of the art of the zebrafish model for toxicology and toxicologic pathology research—advantages and current limitations. *Toxicol. Pathol.* **31**(Suppl.), 62–87.

Spitsbergen, J. M., Tsai, H. W., Reddy, A., Miller, T., Arbogast, D., Hendricks, J. D., and Bailey, G. S. (2000a). Neoplasia in zebrafish (Danio rerio) treated with 7,12-dimethylbenz[a]anthracene by two exposure routes at different developmental stages. *Toxicol. Pathol.* **28,** 705–715.

Spitsbergen, J. M., Tsai, H. W., Reddy, A., Miller, T., Arbogast, D., Hendricks, J. D., and Bailey, G. S. (2000b). Neoplasia in zebrafish (Danio rerio) treated with N-methyl-N′ nitro-N-nitroso guanidine by three exposure routes at different developmental stages. *Toxicol. Pathol.* **28,** 716–725.

St Johnston, D. (2002). The art and design of genetic screens: *Drosophila melanogaster. Nat. Rev. Genet.* **3,** 176–188.

Thisse, C., Thisse, B., Halpern, M. E., and Postlethwait, J. H. (1994). Goosecoid expression in neurectoderm and mesentoderm is disrupted in zebrafish cyclops gastrulas. *Dev. Biol.* **164,** 420–429.

Thisse, C., Thisse, B., Schilling, T. F., and Postlethwait, J. H. (1993). Structure of the zebrafish snail1 gene and its expression in wild-type, spade tail and no tail mutant embryos. *Development* **119,** 1203–1215.

Wyllie, A. H. (1987). Apoptosis: Cell death in tissue regulation. *J. Pathol.* **153,** 313–316.

Zamir, E., Kam, Z., and Yarden, A. (1997). Transcription-dependent induction of G1 phase during the zebra fish midblastula transition. *Mol. Cell. Biol.* **17,** 529–536.

CHAPTER 8

Cellular Dissection of Zebrafish Hematopoiesis

David Traver

Section of Cell and Developmental Biology
University of California, San Diego
La Jolla, California 92093

I. Introduction

From the mid-1990s to early in the twenty-first century, the development of forward genetic approaches in the zebrafish system has provided unprecedented power in understanding the molecular basis of vertebrate blood cell development. Establishment of cellular and hematological approaches to better understand the biology of resulting blood mutants, however, has lagged behind these efforts. In this chapter, recent advances in zebrafish hematology are reviewed, with an emphasis on prospective isolation strategies for both embryonic and adult hematopoietic stem cells and the development of assays to rigorously test their function.

II. Zebrafish Hematopoiesis

A. Primitive Hematopoiesis

Fate mapping experiments in the early zebrafish embryo have demonstrated that blood cells derive from ventral mesoderm (Ho, 1992; Ho and Kane, 1990; Kozlowski *et al.*, 1997; Lee, 1994; Stainier *et al.*, 1993). The isolation of zebrafish homologues of genes, known to be important in the early specification of murine blood cells, such as *scl* and *lmo2*, have been shown to be expressed in bilateral stripes as mesodermal derivatives become patterned (reviewed in Zon, 1995). At early somite stages, expression of *scl* and *lmo2* appears to mark precursors of both the hematopoietic and vascular lineages. Blood and endothelial cell development has been extensively characterized in the mouse, where hematopoiesis and vasculogenesis are first observed in blood islands of the extraembryonic yolk sac (YS) at day 7.5 post-coitum (E7.5). The blood islands consist of nucleated erythroid cells surrounded by endothelial cells. Early blood and endothelial precursors share a similar pattern of gene expression, with both lineages expressing *flk1*, *cd34*, *tie2*, *gata-2*, *lmo2* and *scl* (reviewed in Keller *et al.*, 1999). Mouse embryonic stem (ES) cells cultured under specific conditions express *flk1* and produce clonal populations of blood and endothelial cells (Keller *et al.*, 1999; Nishikawa, 2001). Targeted disruption of the murine *flk1* gene results in catastrophic defects in both YS hematopoiesis and vasculogenesis (Shalaby *et al.*, 1995), suggesting that signaling through *flk1* is important for hemangioblast development. Although not formally proven from clonal studies, these data support the development of YS blood and blood vessels from a common hemangioblastic progenitor. As I discuss later in this chapter, the development of transgenic zebrafish lines expressing fluorescent markers under the control of promoters such as *flk1*, *lmo2*, and *tie2* may now permit testing of hemangioblast potential by prospective isolation strategies.

While the existence of clonogenic hemangioblasts has yet to be demonstrated *in vivo*, it is generally believed that bipotent hemangioblasts are present only as transient intermediates in early development that give rise to blood-specific hematopoietic stem cells (HSCs). Hematopoiesis in many organisms can be divided into two phases, a primitive embryonic phase followed by a definitive phase where all adult cell types are produced. In the mouse, distinct tissues support each major wave of blood cell production, with the embryonic YS producing the initial wave of primitive erythrocytes followed by definitive, multilineage hematopoiesis in the fetal liver (FL). In the zebrafish, two stripes of mesodermal cells express *scl*, *lmo2*, and *gata1* at early somite stages (Fig. 1A) and later converge toward the midline and coalesce into a structure termed the *intermediate cell mass* (ICM) (Al-Adhami and Kunz, 1977; Detrich *et al.*, 1995). Based on the expression of embryonic *globin* genes and that the vast majority of cells within this region are erythroid, the ICM appears to be the equivalent of the mammalian YS. By approximately 20 hpf, erythroid precursors within the ICM have become mature erythrocytes, and, by 24 hpf, they are found within circulation.

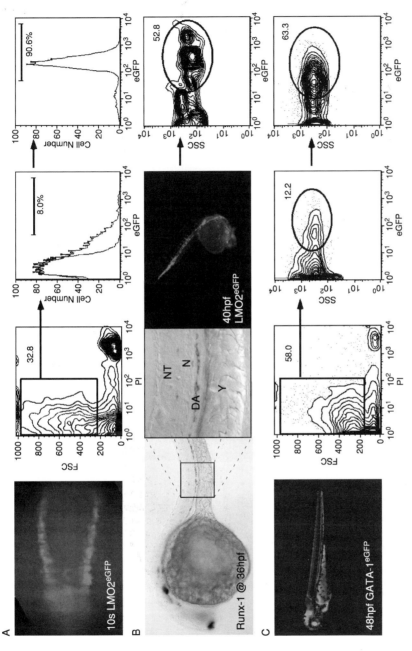

Fig. 1 Zebrafish embryonic hematopoiesis. (A) The LMO2 promoter drives expression of GFP in the primitive blood stripes of 10s transgenic embryos. Staining of single cell suspensions from 10s embryos with propidium iodide (PI) allows live cells to be gated from dead cells and debris (second panel from left) for further analysis. Analysis of GFP expression in live cells (third panel from left) shows a GFP+ population (right histogram) compared with wild-type embryos (left histogram). GFP^{high} cells (8% of starting live population) were sorted twice for purity and reanalyzed as shown in the right panel. Two rounds of sorting yields more than 90% purity of LMO2$^{eGFP\ high}$ cells. (B) Initiation of definitive hematopoiesis may be marked by expression of Runx-1 and LMO2 in the AGM. Inset shows expression of *runx1* in the ventral wall of the dorsal aorta (left panels; courtesy of C. E. Burns). GFP+ cells can be isolated from 40-hpf LMO2^{eGFP} embryos for functional testing via transplantation. (Photo courtesy of H. Zhu.) Right panel shows profile after one round of sorting. (C) Primitive erythroid cells are marked by a GATA-1^{eGFP} transgene. Approximately 12% of live cells express GFP in a 48-hpf transgenic embryo (third panel from left). Purity of GATA-1+ cells is enriched approximately five-fold after one round of sorting (right panel).

B. Initiation of Definitive Hematopoiesis

In mammals, definitive, multilineage hematopoiesis is first observed within the embryo proper. Beginning around E10.5 in the mouse, hematopoietic activity can be found within a region containing the aorta, gonad, and mesonephros (AGM) (Medvinsky et al., 1993). In situ hybridization studies in mice, chick, and zebrafish have revealed that c-myb and runx-1 (aml-1) are expressed in the ventral wall of the dorsal aorta (reviewed in Galloway and Zon, 2003). Murine transplantation studies have shown that the AGM contains HSCs capable of repopulating multilineage hematopoiesis in conditioned adult recipients (Medvinsky and Dzierzak, 1996). Experiments in the mouse comparing transplant reconstitution potentials of pre circulation YS and AGM explant cultures showed that the AGM could contribute to definitive hematopoiesis in lethally irradiated adult mice, whereas YS transplants could not (Cumano and Godin, 2001). When YS cells are transplanted into conditioned newborn animals, however, they too can generate long-term, multilineage hematopoiesis (Yoder et al., 1997a,b). This suggests that the YS contains HSCs with definitive potential, but that multipotency is normally latent when in the YS environment. After interaction with stromal layers derived from the AGM (Matsuoka et al., 2001; Xu et al., 1998), with the fetal liver (Palis and Yoder, 2001), or with environmental aspects present in newborn but not adult animals (Yoder, 1997a), YS HSCs become competent to generate definitive blood cell types. These findings suggest that shifting hematopoietic sites may provide new environments to differentially support migrating HSC populations (reviewed in Traver and Zon, 2002).

While these experiments suggest that the first HSCs with definitive potential arise in the mouse YS, they remain controversial because of the difficulty of isolating YS cells before the onset of circulation, and because of findings in other model systems. Transplantation studies performed in frogs (Ciau-Uitz et al., 2000; Turpen et al., 1997), in birds (Dieterlen-Lievre, 1975), and in fruit flies (Holz et al., 2003) all suggest that the embryonic and adult hematopoietic programs derive from distinct HSC subsets that likely arise independently in different anatomical locations.

The murine YS and AGM regions only transiently harbor HSCs and, by E12, both are devoid of hematopoietic activity. Around E11, the FL is populated by circulating HSCs (Houssaint, 1981; Johnson and Moore, 1975) and becomes the predominant site of blood production during midgestation, producing the first full complement of definitive, adult-type effector cells. Shortly afterward, hematopoiesis is evident in the fetal spleen and ultimately resides in bone marrow (Keller et al., 1999). Shifting hematopoietic sites are also seen in zebrafish where blood precursors are found in a site resembling the mammalian AGM after primitive hematopoiesis in the ICM. Zebrafish blood cells appear in intimate contact with the dorsal aorta and can be distinguished as blood precursors by the expression of definitive hematopoietic genes such as c-myb and runx-1 (Burns et al., 2002; Thompson et al., 1998; Fig. 1B). The zebrafish AGM appears to similarly generate blood cells for only a short period, from approximately 30 to 40 hpf. After this time, blood production shifts to the kidney, which serves as the definitive hematopoietic organ for the remainder of adult life.

C. Adult Hematopoiesis

Previous genetic screens in zebrafish were extremely successful in identifying mutants in primitive erythropoiesis. The screening criteria used in these screens scored visual defects in circulating blood cells at early time points in embryogenesis. Mutants defective in definitive hematopoiesis but displaying normal primitive blood cell development were therefore likely missed. Current screens aimed at identifying mutants with defects in the generation of definitive HSCs in the AGM should reveal new genetic pathways required for multilineage hematopoiesis. An understanding of the biology of mutants isolated through the use of these approaches, however, first requires the characterization of normal, definitive hematopoiesis and the development of assays to more precisely study the biology of zebrafish blood cells. To this end, we have established several tools to characterize the definitive blood-forming system of adult zebrafish.

Blood production in adult zebrafish, like other teleosts, occurs in the kidney, which supports both renal functions and multilineage hematopoiesis (Zapata, 1979). Similar to mammals, T lymphocytes develop in the thymus (Trede and Zon, 1998; Willett *et al.*, 1999; Fig. 2A), which exists in two bilateral sites in zebrafish (Hansen and Zapata, 1998; Willett *et al.*, 1997). The teleostean kidney is a sheath of tissue that runs along the spine (Fig. 2B and E). The anterior portion, or *head kidney*, shows a higher ratio of blood cells to renal tubules than does the posterior portion (Zapata, 1979), termed the *trunk kidney* (Fig. 2B and C). All mature blood cell types are found in the kidney and morphologically resemble their mammalian counterparts (Figs. 2G and 3), with the exceptions that erythrocytes remain nucleated and thrombocytes perform the clotting functions of platelets (Jagadeeswaran *et al.*, 1999). Histologically, the zebrafish spleen (Fig. 2D) has a simpler structure than its mammalian counterpart in that germinal centers are not observed (Zapata and Amemiya, 2000). The absence of immature precursors in the spleen suggests that the kidney is the predominant hematopoietic site in adult zebrafish. The cellular composition of *whole kidney marrow* (WKM), spleen, and blood are shown in Fig. 1F to H. Morphological examples of all kidney cell types are shown in Fig. 3.

Analysis of WKM by flow cytometry (FACS) showed that several distinct populations could be resolved by light scatter characteristics (Fig. 4A). Forward scatter (FSC) is directly proportional to cell size, and side scatter (SSC) is proportional to cellular granularity (Shapiro, 2002). Using combined scatter profiles, the major blood lineages can be isolated to purity from WKM after two rounds of cell sorting (Traver *et al.*, 2003b). Mature erythroid cells were found exclusively within two FSC^{low} fractions (Populations R1 and R2, as shown in Fig. 4A and D), lymphoid cells within a $FSC^{int} SSC^{low}$ subset (Population R3, as shown in Fig. 4A

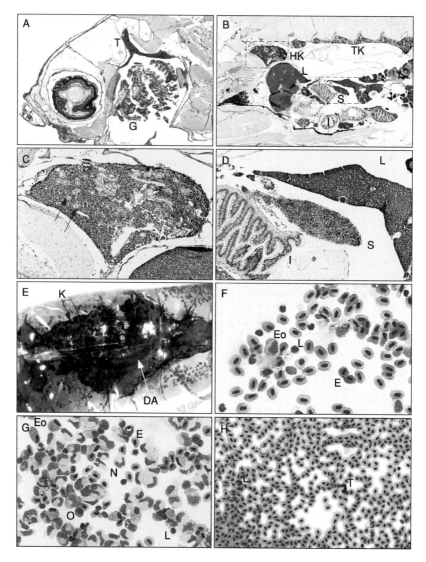

Fig. 2 Histological analyses of adult hematopoietic sites. (A) Sagittal section showing location of the thymus (T), which is dorsal to the gills (G). (B) Midline sagittal section showing location of the kidney, which is divided into the head kidney (HK) and trunk kidney (TK), and spleen (S). The head kidney shows a higher ratio of blood cells to renal tubules (black arrows), as shown in a close up view of the HK in (C). Close up view of the spleen, which is positioned between the liver (L) and the intestine (I). (E) Light microscopic view of the kidney (K), over which passes the dorsal aorta (DA, white arrow). (F) Cytospin preparation of splenic cells, showing erythrocytes (E), lymphocytes (L), and an eo/basophil (Eo). (G) Cytospin preparation of kidney cells showing cell types as already noted plus neutrophils (N) and erythroid precursors (O, orthochromic erythroblast). (H) Peripheral blood smear showing occasional lymphocytes and thrombocyte (T) clusters amongst mature erythrocytes. (A to D) Hematoxylin and Eosin stains, (F to H) May-Grünwald/Giemsa stains. (See Color Insert.)

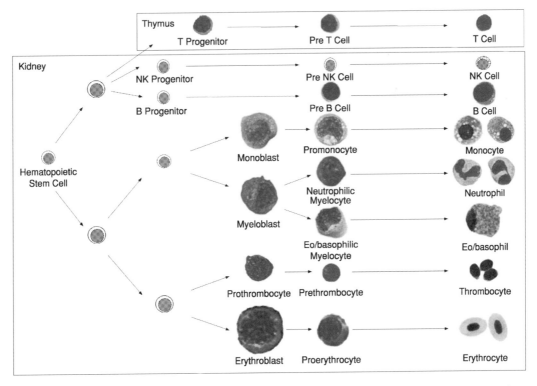

Fig. 3 Proposed model of definitive hematopoiesis in zebrafish. Shown are actual cells types from adult kidney marrow. All cells were photographed with a 100× oil objective from cytospin preparations. Proposed lineage relationships are based on those demonstrated using clonogenic murine progenitor cells. (See Color Insert.)

and E), immature precursors within a FSC^{high} SSC^{int} subset (Population R4, as shown in Fig. 4A and F), and myelomonocytic cells within only a FSC^{high}, SSC^{high} population (Population R5, as shown in Fig. 4A and G). Interestingly, two distinct populations of mature erythroid cells exist (Fig. 4A, R1, R2 gates). Attempts at sorting either of these subsets reproducibly resulted in approximately equal recovery of both (Fig. 4D). This likely resulted due to the elliptical nature of zebrafish red blood cells, since sorting of all other populations yielded cells that fell within the original sorting gates upon reanalysis. Examination of splenic (Fig. 4B) and peripheral blood (Fig. 4C) suspensions showed each to have distinct profiles from WKM, each being predominantly erythroid. Each scatter population sorted from spleen and blood showed each to contain only erythrocytes, lymphocytes, or myelomonocytes in a manner identical to those in the kidney. Immature precursors were not observed in either tissue. Percentages of cells within each

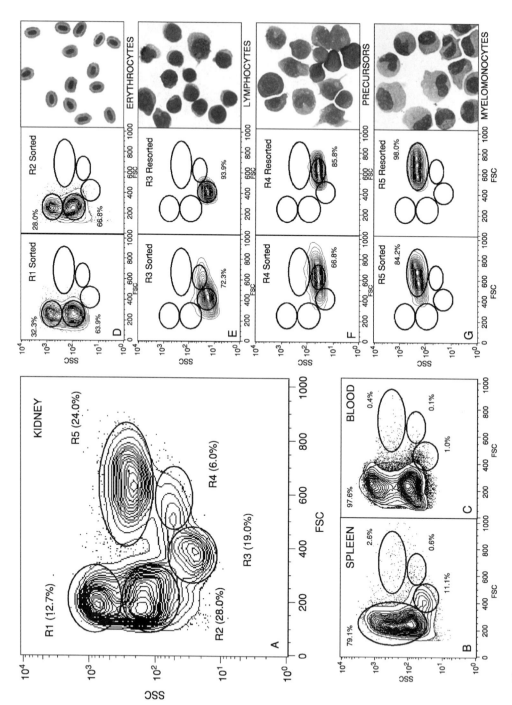

Fig. 4 Each major blood lineage can be isolated by size and granularity using flow cytometry. (A) Scatter profile for WKM. Mature erythrocytes are found within R1 and R2 gates, lymphocytes within the R3 gate, immature precursors within the R4 gate, and myeloid cells within the R5 gate. Mean percentages of each population within WKM are shown. Scatter profiling can also be used for splenocytes (B) and peripheral blood (C).

scatter population closely matched those obtained by morphological cell counts, demonstrating that this flow cytometric assay is accurate in measuring the relative percentages of each of the major blood lineages.

Many transgenic zebrafish lines have been created using proximal promoter elements from genes that demonstrate lineage-affiliated expression patterns in the mouse. These include GATA-1^{eGFP} (Long *et al.*, 1997), GATA-2^{eGFP} (Jessen *et al.*, 1998; Traver et al., 2003b), RAG-2^{eGFP} (Langenau *et al.*, 2003), lck^{eGFP} (Langenau *et al.*, 2004), PU.1^{eGFP} (Hsu *et al.*, 2004; Ward *et al.*, 2003), and CD41^{eGFP} (Traver *et al.*, 2003b) stable transgenic lines. In the adult kidney, we have demonstrated that each of these animals expresses green fluorescent protein (GFP) in the expected kidney scatter fractions (Traver *et al.*, 2003b). For example, all mature erythrocytes express GFP in GATA-1^{eGFP} transgenic animals, as do erythroid progenitors within the precursor population. Expression of GATA-2 is seen only within eosinophils that are contained within the myeloid population, RAG-2 and lck only within cells in the lymphoid fraction, and PU.1 in both myeloid cells and rare lymphoid cells. The development of CD41^{eGFP} transgenic animals has demonstrated that rare thrombocytic cells are found within the kidney, with thrombocyte progenitors appearing in the precursor scatter fraction and mature thrombocytes in the lymphoid fraction. Without fluorescent reporter genes, rare populations such as thrombocytes cannot be resolved by light scatter characteristics alone. By combining the simple technique of scatter separation with fluorescent transgenesis, specific hematopoietic cell subpopulations can now be isolated to a high degree of purity for further analyses.

FACS profiling can also serve as a diagnostic tool in the examination of zebrafish blood mutants. The majority of blood mutants identified to date are those displaying defects in embryonic erythrocyte production (reviewed in Traver *et al.*, 2003a). Most of these mutants are recessive and many are embryonic lethal when homozygous. Most have not been examined for subtle defects as heterozygotes. Several heterozygous mutants, such as *retsina, riesling*, and *merlot* showed haploinsufficiency as evidenced by aberrant kidney erythropoiesis (Traver *et al.*, 2003b). All mutants displayed anemia with concomitant increases in erythroid precursors. These findings suggest that many of the gene functions required to make embryonic erythrocytes are similarly required in their adult counterparts at full gene dosage for normal function.

Purification of each WKM fraction by FACS (D to G). (D) Sorting of populations R1 or R2 yields both upon reanalysis. This appears to be due to the elliptical shape of erythrocytes (right panel). (E) Isolation of lymphoid cells. (F) Isolation of precursor fraction. (G) Isolation of myeloid fraction. FACS profiles after one round of sorting are shown in left panels, after two rounds in middle panels, and morphology of double sorted cells shown in right panels (E to G).

III. Hematopoietic Cell Transplantation

In mammals, transplantation has been used extensively to functionally test putative hematopoietic stem and progenitor cell populations, precursor/progeny relationships, and cell autonomy of mutant gene function. To address similar issues in zebrafish, we have developed several different varieties of hematopoietic cell transplantation (HCT; Fig. 5).

A. Embryonic Donor Cells

While scatter profiling has proven very useful in analyzing and isolating specific blood lineages from the adult kidney, it cannot be used to enrich blood cells from the developing embryo. To study the biology of the earliest blood-forming cells in the embryo, we have made use of transgenic zebrafish expressing fluorescent proteins. As already discussed, hematopoietic precursors appear to be specified from mesodermal derivatives that express *lmo2, flk1*, and *gata2*. The proximal promoter elements from each of these genes are sufficient to recapitulate their endogenous expression patterns. Using germline transgenic animals expressing

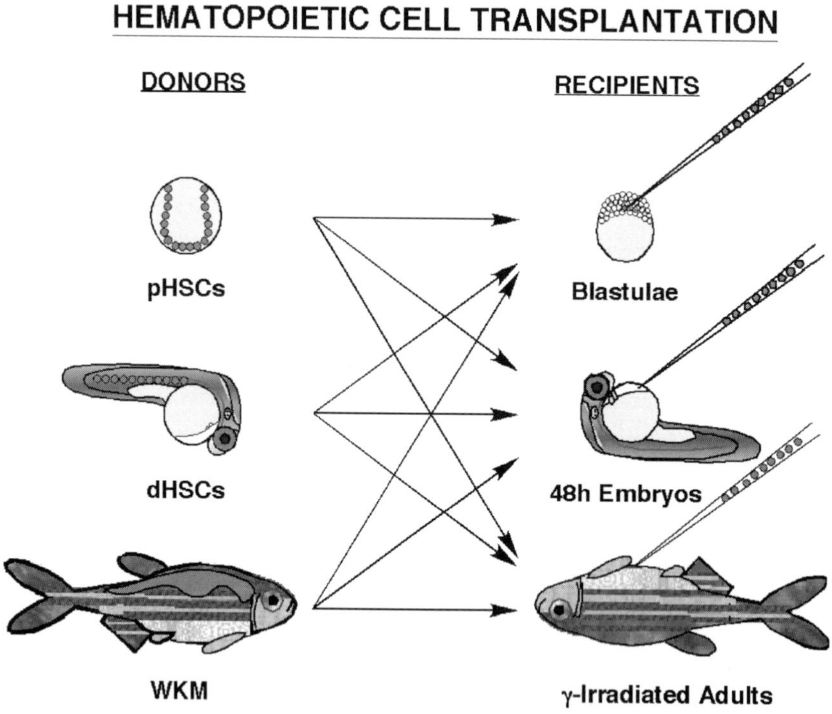

HEMATOPOIETIC CELL TRANSPLANTATION

Fig. 5 Methods of hematopoietic cell transplantation in the zebrafish. See text for details.

GFP under the control of each of these promoters, blood cell precursors can be isolated by flow cytometry from embryonic and larval animals for transplantation into wild-type recipients. For example, GFP$^+$ cells in LMO-2^{eGFP} embryos can be visualized by FACS by 8 to 10 somites (see Fig. 1A). These cells can be sorted to purity and tested for functional potential in a variety of transplantation assays (Fig. 5).

We have used two types of heterochronic transplantation strategies to address two fundamental questions in developmental hematopoiesis. The first is whether cells that express LMO-2 at 8 to 12 somites have hemangioblastic potential (i.e., can generate both blood and vascular cells). We reasoned that purified cells should be placed into a relatively naive environment to provide the most permissive conditions to read out their full fate potentials. Therefore, we attempted transplantation into 1000-cell stage blastulae recipients. Transplanted cells appear to survive this procedure well and GFP$^+$ cells could be found over several days in developing embryos and larvae. By isolating GFP$^+$ cells from LMO-2^{eGFP} animals also carrying a GATA-1^{dsRED} transgene, both donor-derived endothelial and erythroid cells can be independently visualized in green and red, respectively. Using this approach, we have shown that LMO-2$^+$ cells from 8 to 12ss embryos can generate robust regions of donor endothelium and intermediate levels of circulating erythrocytes (Traver, Burns, Zhu, and Zon, unpublished results). We are currently generating additional transgenic lines that express dsRED under ubiquitous promoters to test the full fate potentials of LMO-2$^+$ cells upon transplantation. Additionally, while these studies demonstrate that LMO-2$^+$ cells can generate at least blood and endothelial cells at the population level, single-cell fate-mapping studies need to be performed to assess whether clonogenic hemangioblasts can be identified *in vivo*.

The second question addressed through transplantation is whether the earliest identifiable primitive blood precursors can generate the definitive hematopoietic cells that arise later in embryogenesis. It has been previously reported that the embryonic lethal *vlad tepes* mutant dies from erythropoietic failure because of a defect in the *gata-1* gene (Lyons *et al.*, 2002). This lethality can be rescued by transplantation of WKM from wild-type adults into mutant recipients at 48 hpf (Traver *et al.*, 2003b). We therefore tested whether cells isolated from 8 to 12ss LMO-2^{eGFP} embryos could give rise to definitive cell types and rescue embryonic lethality in *vlad tepes* recipients. After transplantation of GFP$^+$ cells at 48 hpf, approximately half of the cells in circulation were GFP$^+$ and the other half was dsRED$^+$ 1 day posttransplantation. Three days later, analyses of the same animals showed that the vast majority of cells in circulation were dsRED$^+$, apparently because of the differentiation of LMO-2$^+$ precursors to the erythroid fates. Compared with control animals that had not been transplanted (which all died by 12 dpf), some mutant recipients survived for 1 to 2 months after transplantation. We observed no proliferation of donor cells at any time point after transplantation, however, and survivors analyzed more than 1 month posttransplantation showed no remaining cells in circulation (Traver, Burns, Zhu, and Zon, unpublished results). Therefore, these data indicate that mutant survivors were only transiently rescued by short-lived, donor-derived erythrocytes. Thus, within the context of this transplantation setting, it does not appear that primitive hematopoietic precursors can seed definitive hematopoietic organs to give rise to enduring repopulation of the host blood forming system.

1. Protocol for Isolating Hematopoietic Cells from Embryos

This simple physical dissociation procedure is effective in producing single cell suspensions from early embryos (8 to 12ss), as well as from embryos as old as 48 hpf.

1. Stage and collect embryos. We estimate that approximately 200 cells can be isolated per 10 to 12ss LMO-2^{eGFP} embryo. It is recommended that as many embryos as possible be collected since subsequent transplantation efficiency depends largely on cell concentration. At least 500 to 1000 embryos are recommended.

2. Transfer embryos to 1.5 ml Eppendorf centrifuge tubes. Add embryos until they sediment to the 0.5-ml mark. Remove embryo medium since it is not optimal for cellular viability.

3. Wash 2× with 0.9× Dulbecco's PBS (GIBCO, Carlsbad, CA; 500 ml 1× Dulbecco's PBS + 55 mls ddH2O).

4. Remove 0.9× PBS and add 750 ul ice-cold staining media (SM; 0.9× Dulbecco's PBS + 5% FCS). Keep cells on ice from this point onward.

5. Homogenize with blue plastic pestle and pipette a few times with a p1000 tip.

6. Strain resulting cellular slurry through a 40-μm nylon cell strainer (Falcon 2340) atop a 50-ml conical tube. Rinse with additional SM to flush cells through the filter.

7. Gently mash remaining debris atop strainer with a plunger removed from a 28-g insulin syringe.

8. Rinse with more SM until conical is filled to 25 ml mark (helps remove yolk).

9. Centrifuge for 5 m at 200g at 4 °C. Remove supernatant until 1 to 2 ml remain.

10. Add 2 to 3 ml SM, resuspend by pipetting.

11. Strain again through 40-μm nylon mesh into a 5-ml Falcon 2054 tube. It is important to filter the cell suspension at least twice before running the sample by FACS. Embryonic cells are sticky and will clog the nozzle if clumps are not properly removed beforehand.

12. Centrifuge again for 5 m at 200g at 4 °C. Repeat Steps 10 to 12 if necessary.

13. Remove supernatant, resuspend with 1 to 2 ml SM depending on number of embryos used.

14. Propidium Iodide (PI) may be added at this point to $1 \mu g/ml$ to exclude dead cells and debris on the flow cytometer. When using, however, bring samples having PI only and GFP only to set compensations properly. Otherwise, the signal from PI may bleed into the GFP channel, resulting in false positives.

Embryonic cells are now ready for analysis or sorting by flow cytometry. As shown in Fig. 1A, it is often difficult to visualize GFP$^+$ cells when the expression is low or target population is rare. One should therefore always prepare age-matched GFP negative embryos in parallel with transgenic embryos. It is then apparent where the sorting gates should be drawn to sort bona fide GFP$^+$ cells. If highly purified cells are desired, one must perform two successive rounds of sorting. In general, sorting GFP$^+$ cells once yields populations of approximately 50 to 70% purity (Fig. 1B, right panel; Fig. 1C, right panel). Two rounds of cell sorting generally yields greater than 90% purity, as observed with 10ss LMO2^{eGFP} cells (Fig. 1A, right panel). Cells should be kept ice-cold during the sorting procedure.

2. Transplanting Purified Cells into Embryonic Recipients

After sorting, centrifuge cells for 5 m at 200g at 4°C. Carefully remove all supernatant. Resuspend cell pellet in 5 to 10 μl of ice-cold SM that contains 3U Heparin and 1U DnaseI to prevent coagulation and lessen aggregation. Preventing the cells from aggregating or adhering to the glass capillary needle used for transplantation is critical. Mix the cells by gently pipetting with a 10-μl pipette tip. Keep on ice. For transplantation, we use the same needle-pulling parameters used to make needles for nucleic acid injections, the only difference being the use of filament-free capillaries to maintain cell viability. We also use the standard air-powered injection stations used for nucleic acid injections.

3. Transplanting Cells into Blastula Recipients

1. Stage embryonic recipients to reach the 500- to 1000-cell stage at the time of transplantation.

2. Prepare plates for transplantation by pouring a thin layer of 2% agarose made in E3 embryo medium into a 6-cm Petri dish. Drop transplantation mold (similar to the embryo injection mold described in chapter 5 of the Zebrafish book (Westerfield, University of Oregon Press, 1993), but having individual depressions rather than troughs) atop molten agarose and let solidify.

3. Dechorionate blastulae in 1 to 2% agarose-coated Petri dishes by light pronase treatment or manually with watchmaker's forceps.

4. Transfer individual blastulae into individual wells of a transplantation plate that has been immersed in $1 \times$ HBSS (GIBCO). Position the animal pole upward.

5. Using glass, filament-free, fine-pulled capillary needles (1.0-mm OD), back load 3 to 6 μl of cell suspension after breaking the needle on a bevel to an opening of approximately 20 μm. Load into needle holder and force cells to injection end by positive pressure using a pressurized air injection station.

6. Gently insert needle into the center of the embryo and expel cells using either very gentle pressure bursts or slight positive pressure. Transplanting cells near the marginal zone of the blastula leads to higher blood cell yields since embryonic fate maps show blood cells to derive from this region in later gastrula stage embryos.

7. Carefully transfer embryos to agarose-coated Petri dishes, using glass transfer pipettes.

8. Place into E3 embryo medium and incubate at 28.5 °C. Many embryos will not survive the transplantation procedure, so clean periodically to prevent microbial outgrowth.

9. Monitor by fluorescence microscopy for donor cell types.

4. Transplanting Cells into 48 hpf Embryos

1. All procedures are performed as already described, except that dechorionated 48-hpf embryos are staged and used as transplant recipients.

2. Fill transplantation plate with 1× HBSS containing 1× Penicillin/Streptomycin and 1X buffered tricaine, pH 7.0. Do not use E3 as it is suboptimal for cellular viability. Anesthetize recipients in tricaine, then array individual embryos into individual wells of transplantation plate. Position the head at the bottom of well, yolk side up.

3. Load cells as described earlier. Insert injection needle into the sinus venosus/duct of Cuvier and gently expel cells by positive pressure or gentle pressure bursts. Take care not to rupture the yolk sac membrane. A very limited volume can be injected into each recipient. It is thus important to use very concentrated cell suspensions to reconstitute the host blood-forming system. If using WKM as donor cells, concentrations of 5×10^5 cells/μl can be achieved if care is taken to filter and anticoagulate the sample.

4. Allow animals to recover at 28.5 °C in E3. Keep clean and visualize daily by microscopy for the presence of donor-derived cells.

B. Adult Donor Cells

Whereas the initiation of definitive hematopoiesis appears to occur within the embryonic AGM, multilineage hematopoiesis is not fully apparent until the kidney becomes the site of blood cell production. The kidney appears to be the only site of adult hematopoiesis, and we have previously demonstrated that it contains HSCs capable of long-term repopulation of embryonic (Traver et al., 2003b) and adult (Langenau et al., 2004) recipients. For HSC enrichment strategies, both high-dose transplants and limiting dilution assays are required to gauge the purity of input cell populations. In embryonic recipients, we estimate that the maximum number of cells that can be transplanted is approximately 5×10^3, and precise quantitation of transplanted cell numbers is difficult. To circumvent both issues, we have developed HCT into adult recipients.

For transplantation into adult recipients, myeloablation is necessary for successful engraftment of donor cells. We have found γ-irradiation to be the most consistent way to deplete zebrafish hematopoietic cells. The minimum lethal dose (MLD) of 40 Gy specifically ablates cells of the blood-forming system and can be rescued by transplantation of one kidney equivalent (10^6 WKM cells). Of transplanted recipients, 30-day survival is approximately 75% (Traver et al., 2004). We have not yet performed long-term survival studies at this dose. An irradiation dose of 20 Gy is sublethal, and the vast majority of animals survive this treatment despite having nearly total depletion of all leukocyte subsets 1 week after irradiation (Traver et al., 2004). We have shown that this dose is necessary and sufficient for transfer of a lethal T-cell leukemia (Traver et al., 2004), and for long-term (more than 6 month) engraftment of thymus repopulating cells (Langenau et al., 2004). We do not yet know the average relative chimerism of donor to host cells when transplantation is performed after 20 Gy. That this dose is sufficient for robust engraftment, for long-term repopulation, and yields extremely high survival suggests that 20 Gy may be the optimal dose for myeloablative conditioning before transplantation.

1. Protocols for Isolating Hematopoietic Cells from Adult Zebrafish

Anesthetize adult animals in 0.02% tricaine in fish water.

For blood collection, dry animal briefly on tissue, then place on a flat surface with head to the left, dorsal side up. Coat a 10-μl pipette tip with heparin (3 u/μl) then insert tip just behind the pectoral fin and puncture the skin. Direct the tip into the heart cavity, puncture the heart and aspirate up to 10 μl blood by gentle suction. Immediately perform blood smears or place into 0.9 $\times$ PBS containing 5% FCS and 1 u/μl heparin. Mix immediately to prevent clotting. Blood from several animals may be pooled in this manner for later use by flow cytometry. Red cells may be removed using red blood cell hypotonic lysis solution (Sigma, St. Louis, MO; 8.3 g/l ammonium chloride in .01M Tris-HCl, pH7.5) on ice for 5 m. Add 10 volumes of ice-cold SM, then centrifuge at $200g$ for 5 m at $4\,^\circ$C. Resuspended blood leukocytes can then be analyzed by flow cytometry or cytocentrifuge preparations.

For collection of other hematopoietic tissues, place fish on ice for several minutes after tricaine. Make a ventral, midline incision using fine scissors under a dissection microscope.

For spleen collection, locate spleen just dorsal to the major intestinal loops and tease out with watchmaker's forceps. Place into ice-cold SM. Dissect any non-splenic tissue away and place on a 40-μm nylon cell strainer (Falcon 2340) atop a

50-ml conical tube. Gently mash the spleen, using a plunger removed from a 28-g insulin syringe and rinse with SM to flush cells through the filter. Up to 10 spleens can be processed through each filter. Centrifuge at 200g for 5 m at 4 °C. Filter again through 40-μm nylon mesh if using for FACS.

For kidney collection, remove all internal organs, using forceps and a dissection microscope. Take care in dissection as ruptured intestines or gonads will contaminate the kidney preparation. Using watchmaker's forceps, tease the entire kidney away from the body wall starting at the head kidney and working towards the rear. Place into ice-cold SM. Aspirate vigorously with a 1-ml pipette to separate hematopoietic cells from renal cells. Filter through 40-μm nylon mesh, wash, centrifuge, and repeat. Perform last filtration step into a Falcon 2054 tube if using for FACS. It is important to filter the cell suspension at least twice before running the sample. PI may be added at this point to 1 μg/ml to exclude dead cells and debris on the flow cytometer. When using, however, bring samples having PI only and GFP only (if using) to set compensations properly. Otherwise, the signal from PI may bleed into the GFP channel resulting in false positives.

2. Transplanting Whole Kidney Marrow

After filtering and washing WKM suspension three times, centrifuge cells for 5 m at 200g at 4 °C. Carefully remove all supernatant. Resuspend the cell pellet in 5 to 10 μl of ice-cold SM that contains 3U Heparin and 1U DnaseI to prevent coagulation and lessen aggregation. Preventing the cells from aggregating or adhering to the glass capillary needle used for transplantation is critical. Mix the cells by gently pipetting with a 10-μl pipette tip. Keep on ice. For blastulae and embryo transplantation, perform following previous protocols. Between 5×10^2 and 5×10^3 cells can be injected into each 48-hpf embryo if the final cell concentration is approximately $5 \times 10^5/\mu$l.

3. Transplanting Cells into Irradiated Adult Recipients

For irradiation of adult zebrafish, we have used a ^{137}Cesium source irradiator typically used for the irradiation of cultured cells (Gammacell 1000). We lightly anesthetize five animals at a time then irradiate in sealed Petri dishes filled with fish water (without tricaine). We performed careful calibration of the irradiator using calibration microchips to obtain the dose rate at the height within the irradiation chamber nearest to the ^{137}Cesium point source. We found the dose rate to be uniform amongst calibration chips placed within euthanized animals under water, under water alone, or in air alone, verifying that the tissue dosage via total body irradiation (TBI) was accurate.

Transplantation into circulation is most efficiently performed by injecting cells directly into the heart. We perform intracardiac transplantation using pulled filament-free capillary needles as already described, but we break needles at a larger bore size of approximately 50 μm. The needle assembly can be handheld and used with a standard gas-powered microinjection station. We have also had limited success transplanting cells intraperitoneally using a 10-μl Hamilton syringe. Engraftment efficiency for WKM is only marginal using this method, but transplantation of T-cell leukemia or solid tumor suspensions is highly efficient after irradiation at 20 Gy (Traver *et al.*, 2004).

4. Irradiation

1. Briefly anesthetize adult zebrafish in 0.02% tricaine in fish water.
2. Place 5 at a time into 60- by 15-mm Petri dishes (Falcon) containing fish water. Wrap the dish with parafilm and irradiate for the length of time necessary to achieve the desired dose.
3. Return irradiated animals to clean tanks that contain fish water. We have successfully transplanted irradiated animals from 12 to 72 h after irradiation. Using a 20-Gy dose, the nadir of host hematopoietic cell numbers occurs at approximately 72 h postirradiation.

5. Transplantation

1. Prepare cells to be transplanted as already described, taking care to remove particulates or contaminants by multiple filtration and washes. When using WKM as donor cells, we typically make final cell suspensions at 2×10^5 cells/μl. Keep cells on ice.

2. Anesthetize an irradiated animal in 0.02% tricaine in fish water.

3. Transfer ventral side up into a well cut into a sponge wetted with fish water. Under a dissection microscope, use fine forceps to remove the scales that cover the pericardial region.

4. Fill an injection needle with approximately 20 μl of cell suspension. Force the cells to the end of the needle with positive pressure and adjust pressure balance to be neutral. Hold the needle assembly in your right hand, then place gentle pressure on the abdomen of the recipient, using your left index finger. This will position the heart adjacent to the skin and allow visualization of the heartbeat. Insert the needle through the skin and into the heart. If the needle is positioned within the heart and the pressure balance is neutral, blood from the heart will enter the needle and the meniscus will rise and fall with the heartbeat. Inject approximately 5 to 10 μl of the cell suspension by gentle pressure bursts.

5. Return recipient to fresh fish water. Repeat for each additional recipient. Do not feed until the next day to lessen chance of infection.

IV. Enrichment of Hematopoietic Stem Cells

The development of many different transplantation techniques now permits the testing of cell autonomy of mutant gene function, oncogenic transformation, and stem cell enrichment strategies in the zebrafish. For HSC enrichment strategies, fractionation techniques can be used to divide WKM into distinct subsets for functional testing via transplantation. The most successful means of HSC enrichment in the mouse has resulted from the sub fractionation of whole bone marrow cells with monoclonal antibodies (mAbs) and flow cytometry (Spangrude et al., 1988). We have attempted to generate mAbs against zebrafish leukocytes by repeated mouse immunizations using both live WKM and purified membrane fractions followed by standard fusion techniques. Many resulting hybridoma supernatants showed affinity to zebrafish WKM cells in FACS analyses (Fig. 6A). All antibodies showed one of two patterns, however. The first showed binding to all WKM cells at similar levels. The second pattern showed binding to all kidney leukocyte subsets but not to kidney erythrocytes, similar to the pattern shown in the left panel of Fig. 6A. We found no mAbs that specifically bound only to myeloid cells, lymphoid cells, etc. when analyzing positive cells by their scatter profiles. We reasoned that these nonspecific binding affinities may be attributed to different oligosaccharide groups present on zebrafish blood cells. If the glycosylation of zebrafish membrane proteins is different from the mouse, then the murine immune system would likely mount an immune response against these epitopes. To test this idea, we removed both O- and N-linked sugars from WKM using a deglycosylation kit (Prozyme, San Leandro, CA), then incubated these cells with previously positive mAbs. All mAbs tested in this way showed a time-dependent decrease in binding, with nearly all binding disappearing after 2 h of deglycosylation (Fig. 6A). It thus appears that standard immunization approaches using zebrafish WKM cells elicit a strong immune response against oligosaccharide epitopes. This response is likely extremely robust, since we did not recover any mAbs that reacted with specific blood cell lineages. Similar approaches by other investigators using blood cells from frogs or other teleost species have yielded similar results (du Pasquier, Flajnik, personal communication). In an attempt to circumvent the glycoprotein issue, we are now preparing to perform a new series of immunizations using kidney cell membrane preparations that have been deglycosylated.

Previous studies have shown that specific lectins can be used to enrich hematopoietic stem and progenitor cell subsets in the mouse (Huang and Auerbach, 1993; Lu et al., 1996; Visser et al., 1984). In preliminary studies, we have shown that FITC-labeled lectins such as peanut agglutinin (PNA) and potato lectin (PTL) differentially bind to zebrafish kidney subsets. As shown in Fig. 6B, PNA binds to a subset of cells both within the lymphoid and precursor kidney scatter fractions. Staining with PTL also shows that a minor fraction of lymphoid cells binds PTL;

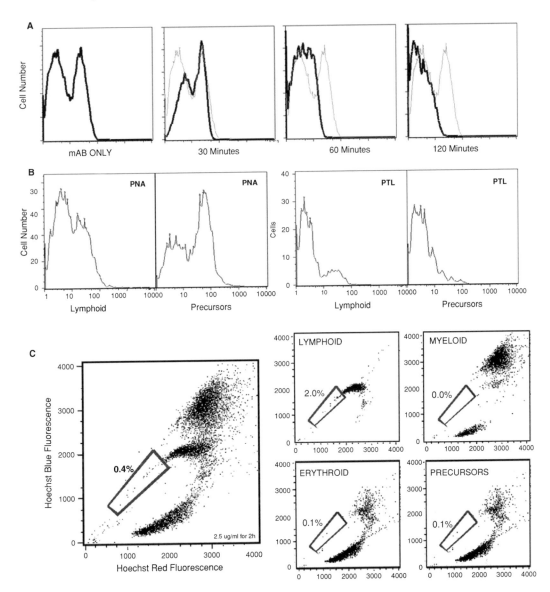

Fig. 6 Potential methods of stem cell enrichment. (A) Mouse monoclonal antibodies generated against zebrafish WKM cells react against oligosaccharide epitopes. Deglycosylation enzymes result in time-dependent loss of antibody binding (bold histograms) compared with no enzyme control (left panel and gray histograms). (B) Differential binding of lectins to WKM scatter fractions. Peanut agglutinin splits both the lymphoid and precursor fraction into positive and negative populations (left panels). Potato lectin shows a minor positive fraction only within the lymphoid fraction (right panels). (C) Hoechst 33342 dye reveals a side population (SP) within WKM. 0.4% of WKM cells appear within the verapamil-sensitive SP gate (left panel). Only the lymphoid fraction, where kidney HSCs reside, contains appreciable numbers of SP cells (right panels).

whereas the precursor (and other) scatter fractions are largely negative (Fig. 6B). We are currently testing both positive and negative fractions in transplantation assays to determine whether these differential binding affinities can be used to enrich HSCs.

We have previously demonstrated that long-term HSCs reside in the adult kidney (Traver *et al.*, 2003b). We therefore isolated each of the kidney scatter fractions from GATA-1^{eGFP} transgenic animals and transplanted cells from each into 48-hpf recipients to determine which subset contains HSC activity. The only population that could generate GFP$^+$ cells for more than 3 weeks in wild-type recipients was the lymphoid fraction. This finding is in accordance with mouse and human studies that have shown purified HSCs to have the size and morphological characteristics of inactive lymphocytes (reviewed in Morrison *et al.*, 1995).

Another method that has been extremely useful in isolating stem cells from whole bone marrow is differential dye efflux. Dyes such as rhodamine 123 (Mulder and Visser, 1987; Visser and de Vries, 1988) or Hoechst 33342 (Goodell *et al.*, 1996) allow the visualization and purification of a "side population" (SP) that is highly enriched for HSCs. This technique appears to take advantage of the relatively high activity of multidrug resistance transporter proteins in HSCs that actively pump each dye out of the cell in a verapamil-sensitive manner (Goodell *et al.*, 1996). Other cell types lack this activity and become positively stained, allowing isolation of the negative SP fraction by FACS. Our preliminary studies of SP cells in the zebrafish kidney have demonstrated a typical SP profile when stained with 2.5 ug/ml of Hoechst 33342 for 2 h at 28 °C (Fig. 6C). This population disappears when verapamil is added to the incubation. Interestingly, the vast majority of SP cells appear within the lymphoid scatter fraction (Fig. 6C). Further examination of whether this population is enriched for HSC activity in transplantation assays is warranted.

Finally, there are many other ways that hematopoietic stem and progenitor cells can be enriched from WKM including sublethal irradiation, cytoreductive drug treatment, and use of transgenic lines expressing fluorescent reporter genes. We have shown that, after 20-Gy doses of γ−irradiation, nearly all hematopoietic lineages are depleted within 1 week (Traver *et al.*, 2004). Examination of kidney cytocentrifuge preparations at this time shows that the vast majority of cells are immature precursors. That this dose does not lead to death of the animals demonstrates that HSCs are spared and are likely highly enriched 5 to 8 days after exposure. We have also shown that cytoreductive drugs such as Cytoxan and 5-Fluorouracil have similar effects on kidney cell depletion, although the effects were more variable than those achieved with sublethal irradiation (Winzeler, Traver, and Zon unpublished). Since HSCs are contained within the kidney lymphoid fraction, they can be further enriched by HSC-specific or lymphocyte-specific transgenic markers. Possible examples of transgenic promoters are *lmo-2, gata-2,* or *c-myb* to positively mark HSCs and *rag-2, lck*, or B cell receptor genes to exclude lymphocytes from this subset.

V. Conclusions

From the mid-1990s through early in the twenty-first century, the zebrafish has rapidly become a powerful model system in which to elucidate the molecular mechanisms of vertebrate blood development through forward genetic screens. In this review, we have described the cellular characterization of the zebrafish blood-forming system and provided detailed protocols for the isolation and transplantation of hematopoietic cells. Through the development of lineal sub fractionation techniques and transplantation technology, a hematological framework now exists for the continued study of the genetics of hematopoiesis. By adapting these experimental approaches that have proven to be powerful in the mouse, the zebrafish should be uniquely positioned to address fundamental questions regarding the biology of hematopoietic stem and progenitor cells.

References

Al-Adhami, M. A., and Kunz, Y. W. (1977). Ontogenesis of haematopoietic sites in *Brachydanio rerio*. *Develop. Growth. Differ.* **19**, 171–179.

Burns, C. E., DeBlasio, T., Zhou, Y., Zhang, J., Zon, L., and Nimer, S. D. (2002). Isolation and characterization of runxa and runxb, zebrafish members of the runt family of transcriptional regulators. *Exp. Hematol.* **30**, 1381–1389.

Ciau-Uitz, A., Walmsley, M., and Patient, R. (2000). Distinct origins of adult and embryonic blood in Xenopus. *Cell* **102**, 787–796.

Cumano, A., and Godin, I. (2001). Pluripotent hematopoietic stem cell development during embryogenesis. *Curr. Opin. Immunol.* **13**, 166–171.

Detrich, H. W., III, Kieran, M. W., Chan, F. Y., Barone, L. M., Yee, K., Rundstadler, J. A., Pratt, S., Ransom, D., and Zon, L. I. (1995). Intraembryonic hematopoietic cell migration during vertebrate development. *Proc. Natl. Acad. Sci. USA* **92**, 10713–10717.

Dieterlen-Lievre, F. (1975). On the origin of haemopoietic stem cells in the avian embryo: An experimental approach. *J. Embryol. Exp. Morphol.* **33**, 607–619.

Galloway, J. L., and Zon, L. I. (2003). Ontogeny of hematopoiesis: Examining the emergence of hematopoietic cells in the vertebrate embryo. *Curr. Top. Dev. Biol.* **53**, 139–158.

Goodell, M. A., Brose, K., Paradis, G., Conner, A. S., and Mulligan, R. C. (1996). Isolation and functional properties of murine hematopoietic stem cells that are replicating in vivo. *J. Exp. Med.* **183**, 1797–1806.

Hansen, J. D., and Zapata, A. G. (1998). Lymphocyte development in fish and amphibians. *Immunol. Rev.* **166**, 199–220.

Ho, R. K. (1992). Cell movements and cell fate during zebrafish gastrulation. *Dev. Suppl.,* 65–73.

Ho, R. K., and Kane, D. A. (1990). Cell-autonomous action of zebrafish spt-1 mutation in specific mesodermal precursors. *Nature* **348**, 728–730.

Holz, A., Bossinger, B., Strasser, T., Janning, W., and Klapper, R. (2003). The two origins of hemocytes in Drosophila. *Development* **130**, 4955–4962.

Houssaint, E. (1981). Differentiation of the mouse hepatic primordium. II. Extrinsic origin of the haemopoietic cell line. *Cell Differ.* **10**, 243–252.

Hsu, K., Traver, D., Kutok, J. L., Hagen, A., Liu, T. X., Paw, B. H., Rhodes, J., Berman, J., Zon, L. I., Kanki, J. P., *et al.* (2004). The pu.1 promoter drives myeloid gene expression in zebrafish. *Blood* **104**, 1291–1297.

Huang, H., and Auerbach, R. (1993). Identification and characterization of hematopoietic stem cells from the yolk sac of the early mouse embryo. *Proc. Natl. Acad. Sci. USA* **90**, 10110–10114.

Jagadeeswaran, P., Sheehan, J. P., Craig, F. E., and Troyer, D. (1999). Identification and characterization of zebrafish thrombocytes. *Br. J. Haematol.* **107,** 731–738.

Jessen, J. R., Meng, A., McFarlane, R. J., Paw, B. H., Zon, L. I., Smith, G. R., and Lin, S. (1998). Modification of bacterial artificial chromosomes through chi-stimulated homologous recombination and its application in zebrafish transgenesis. *Proc. Natl. Acad. Sci. USA* **95,** 5121–5126.

Johnson, G. R., and Moore, M. A. (1975). Role of stem cell migration in initiation of mouse foetal liver haemopoiesis. *Nature* **258,** 726–728.

Keller, G., Lacaud, G., and Robertson, S. (1999). Development of the hematopoietic system in the mouse. *Exp. Hematol.* **27,** 777–787.

Kozlowski, D. J., Murakami, T., Ho, R. K., and Weinberg, E. S. (1997). Regional cell movement and tissue patterning in the zebrafish embryo revealed by fate mapping with caged fluorescein. *Biochem. Cell Biol.* **75,** 551–562.

Langenau, D. M., Ferrando, A. A., Traver, D., Kutok, J. L., Hezel, J. P., Kanki, J. P., Zon, L. I., Look, A. T., and Trede, N. S. (2004). *In vivo* tracking of T cell development, ablation, and engraftment in transgenic zebrafish. *Proc. Natl. Acad. Sci. USA* **101**(19), 7369–7374.

Langenau, D. M., Traver, D., Ferrando, A. A., Kutok, J., Aster, J. C., Kanki, J. P., Lin, H. S., Prochownik, E., Trede, N. S., Zon, L. I., *et al.* (2003). Myc-induced T-Cell Leukemia in Transgenic Zebrafish. *Science* **299,** 887–890.

Lee, R. K., Stainier, D. Y., Weinstein, B. M., and Fishman, M. C. (1994). Cardiovascular development in the zebrafish. II. Endocardial progenitors are sequestered within the heart field. *Development* **120,** 3361–3366.

Long, Q., Meng, A., Wang, H., Jessen, J. R., Farrell, M. J., and Lin, S. (1997). GATA-1 expression pattern can be recapitulated in living transgenic zebrafish using GFP reporter gene. *Development* **124,** 4105–4111.

Lu, L. S., Wang, S. J., and Auerbach, R. (1996). *In vitro* and *in vivo* differentiation into B cells, T cells, and myeloid cells of primitive yolk sac hematopoietic precursor cells expanded >100-fold by coculture with a clonal yolk sac endothelial cell line. *Proc. Natl. Acad. Sci. USA* **93,** 14782–14787.

Lyons, S. E., Lawson, N. D., Lei, L., Bennett, P. E., Weinstein, B. M., and Liu, P. P. (2002). A nonsense mutation in zebrafish gata1 causes the bloodless phenotype in vlad tepes. *Proc. Natl. Acad. Sci. USA* **99,** 5454–5459.

Matsuoka, S., Tsuji, K., Hisakawa, H., Xu, M., Ebihara, Y., Ishii, T., Sugiyama, D., Manabe, A., Tanaka, R., Ikeda, Y., *et al.* (2001). Generation of definitive hematopoietic stem cells from murine early yolk sac and paraaortic splanchnopleures by aorta-gonad-mesonephros region-derived stromal cells. *Blood* **98,** 6–12.

Medvinsky, A., and Dzierzak, E. (1996). Definitive hematopoiesis is autonomously initiated by the AGM region. *Cell* **86,** 897–906.

Medvinsky, A. L., Samoylina, N. L., Muller, A. M., and Dzierzak, E. A. (1993). An early pre-liver intraembryonic source of CFU-S in the developing mouse. *Nature* **364,** 64–67.

Morrison, S. J., Uchida, N., and Weissman, I. L. (1995). The biology of hematopoietic stem cells. *Annu. Rev. Cell Dev. Biol.* **11,** 35–71.

Mulder, A. H., and Visser, J. W. M. (1987). Separation and functional analysis of bone marrow cells separated by Rhodamine-123 fluorescence. *Exp. Hematol.* **15,** 99–104.

Nishikawa, S. I. (2001). A complex linkage in the developmental pathway of endothelial and hematopoietic cells. *Curr. Opin. Cell. Biol.* **13,** 673–678.

Palis, J., and Yoder, M. C. (2001). Yolk-sac hematopoiesis: The first blood cells of mouse and man. *Exp. Hematol.* **29,** 927–936.

Shalaby, F., Rossant, J., Yamaguchi, T. P., Gertsenstein, M., Wu, X. F., Breitman, M. L., and Schuh, A. C. (1995). Failure of blood-island formation and vasculogenesis in Flk-1-deficient mice. *Nature* **376,** 62–66.

Shapiro, H. M. (2002). "Practical Flow Cytometry." Wiley-Liss, New York.

Spangrude, G. J., Heimfeld, S., and Weissman, I. L. (1988). Purification and characterization of mouse hematopoietic stem cells. *Science* **241,** 58–62.

Stainier, D. Y., Lee, R. K., and Fishman, M. C. (1993). Cardiovascular development in the zebrafish. I. Myocardial fate map and heart tube formation. *Development* **119,** 31–40.

Thompson, M. A., Ransom, D. G., Pratt, S. J., MacLennan, H., Kieran, M. W., Detrich, H. W., III, Vail, B., Huber, T. L., Paw, B., Brownlie, A. J., *et al.* (1998). The cloche and spade tail genes differentially affect hematopoiesis and vasculogenesis. *Dev. Biol.* **197,** 248–269.

Traver, D., Herbomel, P., Patton, E. E., Murphy, R. D., Yoder, J. A., Litman, G. W., Catic, A., Amemiya, C. T., Zon, L. I., and Trede, N. S. (2003a). "The zebrafish as a model organism to study development of the immune system." Academic Press, Philadelphia.

Traver, D., Paw, B. H., Poss, K. D., Penberthy, W. T., Lin, S., and Zon, L. I. (2003b). Transplantation and *in vivo* imaging of multilineage engraftment in zebrafish bloodless mutants. *Nat. Immunol.* **4,** 1238–1246.

Traver, D., Winzeler, E. A., Stern, H. M., Mayhall, E. A., Langenau, D. M., Kutok, J. L., Look, A. T., and Zon, L. I. (2004). Effects of lethal irradiation and rescue by hematopoietic cell transplantation. *Blood* **104,** 1296–1305.

Traver, D., and Zon, L. I. (2002). Walking the walk: Migration and other common themes in blood and vascular development. *Cell* **108,** 731–734.

Trede, N. S., and Zon, L. I. (1998). Development of T-cells during fish embryogenesis. *Dev. Comp. Immunol.* **22,** 253–263.

Turpen, J. B., Kelley, C. M., Mead, P. E., and Zon, L. I. (1997). Bipotential primitive-definitive hematopoietic progenitors in the vertebrate embryo. *Immunity* **7,** 325–334.

Visser, J. W., Bauman, J. G., Mulder, A. H., Eliason, J. F., and de Leeuw, A. M. (1984). Isolation of murine pluripotent hemopoietic stem cells. *J. Exp. Med.* **159,** 1576–1590.

Visser, J. W., and de Vries, P. (1988). Isolation of spleen-colony forming cells (CFU-s) using wheat germ agglutinin and rhodamine 123 labeling. *Blood Cells* **14,** 369–384.

Ward, A. C., McPhee, D. O., Condron, M. M., Varma, S., Cody, S. H., Onnebo, S. M., Paw, B. H., Zon, L. I., and Lieschke, G. J. (2003). The zebrafish spi1 promoter drives myeloid-specific expression in stable transgenic fish. *Blood* **102,** 3238–3240.

Willett, C. E., Cortes, A., Zuasti, A., and Zapata, A. G. (1999). Early hematopoiesis and developing lymphoid organs in the zebrafish. *Dev. Dyn.* **214,** 323–336.

Willett, C. E., Zapata, A. G., Hopkins, N., and Steiner, L. A. (1997). Expression of zebrafish rag genes during early development identifies the thymus. *Dev. Biol.* **182,** 331–341.

Xu, M. J., Tsuji, K., Ueda, T., Mukouyama, Y. S., Hara, T., Yang, F. C., Ebihara, Y., Matsuoka, S., Manabe, A., Kikuchi, A., *et al.* (1998). Stimulation of mouse and human primitive hematopoiesis by murine embryonic aorta-gonad-mesonephros-derived stromal cell lines. *Blood* **92,** 2032–2040.

Yoder, M., Hiatt, K., and Mukherjee, P. (1997a). *In vivo* repopulating hematopoietic stem cells are present in the murine yolk sac at day 9.0 postcoitus. *Proc. Nat. Acad. Sci. USA* **94,** 6776.

Yoder, M. C., Hiatt, K., Dutt, P., Mukherjee, P., Bodine, D. M., and Orlic, D. (1997b). Characterization of definitive lymphohematopoietic stem cells in the day 9 murine yolk sac. *Immunity* **7,** 335–344.

Zapata, A. (1979). Ultrastructural study of the teleost fish kidney. *Dev. Comp. Immunol.* **3,** 55–65.

Zapata, A., and Amemiya, C. T. (2000). Phylogeny of lower vertebrates and their immunological structures. *Curr. Top. Microbiol. Immunol.* **248,** 67–107.

Zon, L. I. (1995). Developmental biology of hematopoiesis. *Blood* **86,** 2876–2891.

CHAPTER 9

Culture of Embryonic Stem Cell Lines from Zebrafish

Lianchun Fan, Jennifer Crodian, and Paul Collodi

Department of Animal Sciences
Purdue University
West Lafayette, Indiana 47907

I. Introduction

Despite its many advantages for studies of embryo development and human disease, one deficiency of the zebrafish model has been the lack of methods for targeted mutagenesis using embryonic stem (ES) cells. In mice, ES cell-mediated gene targeting has provided a powerful approach to the study of gene function (Lui *et al.*, 1993; Zhang *et al.*, 1995). A similar strategy applied to zebrafish would complement other genetic methods currently available such as large-scale random mutagenesis (Currie, 1996; Golling *et al.*, 2002; Van Eeden *et al.*, 1999), antisense-based gene knockdown, (Nasevicius and Ekker, 2000) and target selected muta-genesis (Wienholds *et al.*, 2002) approaches to increase the utility of this model system. To address this problem, our laboratory has been working to establish zebrafish ES cell lines that are suitable for use in a gene targeting approach (Fan *et al.*, 2004; Ghosh and Collodi, 1994; Ma *et al.*, 2001). Successful ES cell-mediated gene targeting requires the use of pluripotent ES cell lines that possess the capacity to contribute to the germ-cell lineage of a host embryo (Evans and Kaufman, 1981; Gossler *et al.*, 1986). The germline competent ES cells are genetically altered in culture by targeted incorporation of foreign DNA by homologous recombination followed by *in vitro* selection of cell colonies that have undergone the targeting event (Fig. 1). The selected colonies are expanded in culture and introduced into host embryos to generate germline chimeras carrying the targeted mutation (Capecchi, 1989; Doetschman *et al.*, 1987). Once the chimeras are sexually mature, they are used to establish the knockout line. This chapter describes methods for the derivation of germline competent zebrafish ES cell cultures along with a protocol for the efficient introduction of plasmid DNA into the cells by electroporation and *in vitro* selection of homologous recombinants. Chapter 6 of volume 77 describes methods for the introduction of ES cells into host embryos to generate germline chimeras.

II. Methods

A. General Characteristics of Zebrafish ES Cell Cultures

Although pluripotent, germline competent ES cell cultures have been derived from both blastula- and gastrula-stage zebrafish embryos (Fan *et al.*, 2004), the blastula, which is composed of non-differentiated cells, is the optimal stage for use in initiating the cultures. To maintain pluripotency and germline competency, the ES cell cultures are initiated and maintained on a feeder layer of growth-arrested rainbow trout spleen cells derived from the established RTS34st cell line (Ganassin and Bols, 1999; Ma *et al.*, 2001). In the presence of the feeder layer, the ES cell cultures remain germline competent for at least 6 passages (6 weeks) in culture, a sufficient period for electroporation and selection of homologous

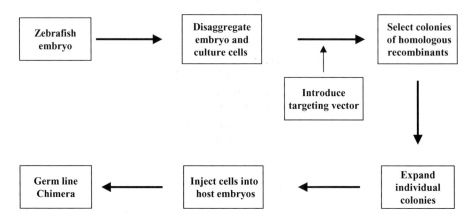

Fig. 1   Strategy for ES cell-mediated gene targeting in zebrafish.

recombinants (Fan *et al.*, 2004). When introduced into a recipient embryo, the cultured ES cells contribute to multiple tissues including the germ cell lineage of the host (Fan and Collodi, 2002). Zebrafish germline chimeras have been generated from ES cells maintained for multiple passages in culture (Fan *et al.*, 2004).

When a primary cell culture is initiated from zebrafish blastulas and maintained on an RTS34st feeder layer, within 24 h after plating, the embryo cells form dense homogeneous aggregates present throughout the culture (Fig. 2A). As the zebrafish embryo cells proliferate, the aggregates become larger while continuing to exhibit a homogeneous appearance without showing morphological indication of differentiation. The primary culture must be passaged at day 4 to 6 to prevent the cell aggregates from becoming too large and differentiating. During the first passage, the aggregates are partially dissociated and added to a fresh feeder layer of growth-arrested RTS34st cells. The aggregates become easier to dissociate with each passage and eventually grow to form a monolayer by the fourth passage (Fig. 2B).

The same strategy can be used to derive pluripotent ES cell cultures initiated from embryos at the germ-ring stage of development (approximately 6 hpf) (Ma *et al.*, 2001). Since cell differentiation has begun to occur at this stage of development, the majority of the cell aggregates in the primary culture will be composed of differentiated cell types including pigmented melanocytes, neural cells, and fibroblasts. Even though cell differentiation is pervasive in the gastrula-derived primary culture, pluripotent ES cells can be obtained by manually selecting the small number of cell aggregates that possess an ES-like morphology characterized by a compact and homogeneous appearance. The ES-like cell aggregates are dissociated and passaged to initiate the long-term culture.

Plasmid DNA is efficiently introduced into the ES cells by electroporation and drug selection is used to isolate colonies of stable transformants. Gene targeting in the ES cells is accomplished by introducing a vector containing sequences that are

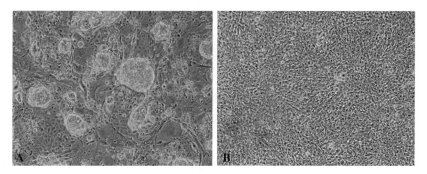

Fig. 2 Zebrafish embryo cell culture initiated from blastula-stage embryos on a feeder layer of RTS34st cells. (A) The primary culture is composed of ES-like cell aggregates growing on the monolayer of feeder cells. (B) By passage 4, the embryo cells grow to form a confluent monolayer in close association with the feeder cells.

homologous to the gene that is being targeted for disruption (Fig. 3A). Zebrafish ES cells that have incorporated the plasmid DNA in a targeted fashion by homologous recombination are selected using a strategy that is based on the positive/negative selection method commonly employed with mouse ES cells (Capecchi, 1989). To use this selection strategy, a targeting vector is constructed that contains *neo* flanked on each side by 3-kb or longer arms that are homologous to the gene being targeted (Fig. 3A). The vector also contains the red fluorescence protein (RFP) gene with its own promoter located outside of the homologous region. After electroporation, the ES cells that have incorporated the vector are selected in G418 and drug-resistant colonies are examined by fluorescence microscopy. The colonies that have incorporated the plasmid by random insertion will possess the entire vector sequence including RFP; whereas the colonies of homologous recombinants will lack RFP (Fig. 3B). Approximately 5 weeks after electroporation, individual RFP-negative colonies are manually picked from the plate using a drawn-out Pasteur pipette or a micropipetor and transferred to individual wells of a 24-well plate (one colony/well) and eventually expanded into 25 cm^2 flasks. Disruption of the targeted gene is confirmed by polymerase chain reaction

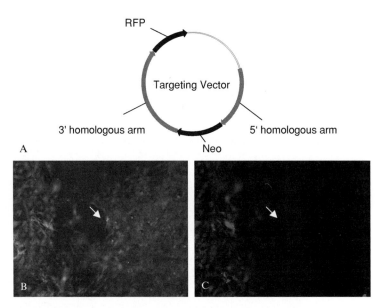

Fig. 3   (A) The general design of a targeting vector that contains *neo* flanked by sequences that are homologous to the gene being targeted along with *RFP* located outside of the homologous region. The targeting vector was introduced into an ES cell line that constitutively expresses the enhanced green fluorescent protein (EGFP). (B) After G418 selection, two GFP$^+$ colonies are shown by fluorescence microscopy. (C) The same two colonies shown in (B) are visualized using a rhodamine filter to detect RFP expression. The colony on the right (*arrow*) is RFP$^-$, indicating that it is a homologous recombinant.

(PCR) and Southern blot analysis. We have found that the frequency of homologous recombination in the ES cells is approximately 1 out of 400 G418-resistant colonies. Using this selection strategy, approximately 50% of the manually selected RFP-negative colonies were confirmed to be homologous recombinants.

B. Derivation of ES Cell Cultures from Blastula-Stage Embryos

1. Collect the embryos in a fine mesh net and rinse several times with water to remove debris. Transfer the embryos into a Petri dish containing egg water and incubate at 28 °C until they reach the blastula-stage of development (approximately 4 hpf). Divide the embryos into groups of approximately 50 individuals, and transfer each group into a 2.5-ml Eppendorf Microfuge tube that had the conical-shaped bottom cut off and replaced with fine mesh netting attached with a rubber band. Submerge the bottom of the tube containing the embryos sitting on the net into 70% ethanol for 10 sec and then immediately submerge the embryos into a beaker of sterile egg water. Transfer the embryos from the tube into a 60-mm Petri dish containing egg water and remove any dead individuals. Remove the egg water and rinse the embryos three times with LDF medium (Collodi et al., 1992). Remove the LDF and add approximately 2 ml of bleach solution to the dish. Incubate the embryos for 2 m in the bleach solution and then remove the bleach and immediately rinse with LDF. Repeat the bleach treatment and rinse 2 additional times. It is important to avoid exposing the embryos to the bleach solution for periods longer than 2 m without rinsing. After the final bleach treatment, rinse the embryos 3 additional times with LDF medium.

2. Remove the chorions by incubating each group of embryos in 3 ml of pronase solution for approximately 15 m or until the chorions begin to break apart. Gently swirl the suspended embryos in the dish to release them from the digested chorion. Use a pipetter to remove the pronase solution containing the floating chorions and gently rinse the dechorionated embryos with LDF medium. Incubate each group of embryos in 3 ml of trypsin/EDTA for 1 to 2 m and gently pipet to dissociate the cells. Collect the cells by centrifugation ($500 \times g$, 5 m) and resuspend the cell pellet obtained from each group of approximately 50 embryos in 1.8 ml LDF medium. Transfer the cell suspension (1.8 ml) to a single well of a 6-well tissue culture plate (Falcon) containing a confluent monolayer of growth-arrested RTS34st cells. Let the plate sit undisturbed for 30 m to allow the embryo cells to attach to the RTS34st monolayer. After the cells have attached, add the following factors to each well: 150 μl of fetal bovine serum (FBS), 15 μl of zebrafish embryo extract, 30 μl of trout plasma, 30 μl of insulin stock solution, 15 μl of epidermal growth factor (EGF) stock solution, 15 μl of basic fibroblast growth factor (bFGF) stock solution, and 945 μl of RTS34st conditioned medium. If the cells are not attached after 30 m, the plate can be incubated for 1 to 2 h before adding the factors. Also, EGF can be added immediately after plating the cells to enhance cell attachment. Incubate the culture for 5 days (22 °C). During this time, the cell aggregates should increase in size while maintaining a

homogeneous, non-differentiated appearance. Although zebrafish cell cultures are normally propagated at 26 °C (Collodi *et al.*, 1992), the embryo cell cultures are maintained at 22 °C to accommodate the feeder layer of trout spleen cells.

3. To passage the primary culture, harvest the cells from each well by adding 2 ml of trypsin/ethylene diamine tetra acetic acid (EDTA) solution per well and incubating 30 sec before transferring the cell suspension to a 15-ml polypropylene centrifuge tube (Owens Corning, Toledo, OH). Pipet the cell suspension up and down several times to partially dissociate the cell aggregates and add 0.2 ml of FBS to stop the action of the trypsin. The cell aggregates cannot be completely dissociated during the first passage. Collect the cells by centrifugation ($500 \times g$, 5 m) and resuspend the pellet in 3.6 ml of LDF medium. Add 1.8 ml of the cell suspension to each of 2 wells of a 6-well plate (Falcon) containing a confluent monolayer of growth-arrested RTS34st cells and add the factors listed in Step 2 to each well. Incubate the 6-well plate for 5 days (22 °C) and harvest the cells as described in Step 2. Combine the cells harvested from 2 wells, collect the cells by centrifugation and resuspend the cell pellet in 3.6 ml of LDF medium. The suspension will still contain a large number of cell aggregates.

4. Add the cell suspension to a 25-cm^2 tissue culture flask (Falcon) containing a confluent monolayer of growth-arrested RTS34st cells. Let the flask sit undisturbed for 1 to 3 h to allow the cells to attach to the feeder layer.

5. Add the following factors to the flask: 300 μl of FBS, 30 μl of zebrafish embryo extract, 60 μl of trout serum, 60 μl of bovine insulin stock solution, 30 μl of EGF stock solution, 30 μl of bFGF stock solution, and 1.890 μl of RTS34st conditioned medium.

6. Incubate the flask for 7 days (22 °C) and then harvest the cells in trypsin/ EDTA as already described and seed them into 2 flasks that each contains a confluent monolayer of growth-arrested RTS34st cells. With each passage, the cell aggregates become easier to dissociate and fewer aggregates are present in the culture. Continue to passage the culture approximately every 7 days.

7. Cryopreserve the cultures beginning at passage 4, when the cells begin to grow as a monolayer. A portion of the culture is frozen at each passage. Harvest the cultures in trypsin/EDTA as described in Step 3 and resuspend the cell pellet obtained from a 25-cm^2 flask in 1.2 ml freezing medium. Transfer the cell suspension to a cryo vial (Nalgene Labware, Rochester, NY), place the vial in Styrofoam insulation and incubate the vial at 4 °C for 10 m, followed by -80 °C for at least 1 h, and then submerge and store the vial in liquid nitrogen.

C. Derivation of ES Cell Cultures from Zebrafish Gastrula-Stage Embryos

1. Initiate primary cell cultures from embryos at the germ-ring stage of development (approximately 6 hpf) (Westerfield, 1995), using the methods already described. After the primary culture has been growing for approximately 5 days, use a drawn-out Pasteur pipet or a micropipetor (Rainin, Oakland, CA) to remove

aggregates of densely packed cells that appear homogeneous without morphological indications of differentiation. Combine 30 to 50 of the isolated cell aggregates in a sterile 2-ml centrifuge tube containing LDF medium.

2. Collect the cell aggregates by centrifugation (500 × g, 5 m), resuspend the pellet in 1.0 ml of trypsin/EDTA solution, and incubate 2 m while occasionally pipetting the cell suspension through a 5-ml pipette to partially dissociate the aggregates. Add 0.1 ml of FBS to stop the action of the trypsin and collect the cells by centrifugation (500 × g, 5 m).

3. Resuspend the cell pellet in 1.8 mL of LDF medium and add to a single well of a 6-well plate containing a monolayer of growth-arrested RTS34st feeder cells.

4. Let the plate sit undisturbed for 5 h to allow the embryo cells to attach and add the factors listed Section B, Step 2. The culture consists of small cell aggregates and some single embryo cells attached to the RTS34st cells.

5. Incubate the plate (22 °C) for 7 days. As the cells proliferate, the aggregates should become larger without exhibiting morphological indications of differentiation. The culture is passaged every 7 days as described in Section B. The cell aggregates will become easier to dissociate and eventually grow as a monolayer after approximately 4 passages.

D. Electroporation of Plasmid DNA into the ES Cell Cultures

1. Once the ES cell culture begins to grow as a monolayer (passage 4) (see Fig. 2B), the cells can be efficiently transformed with plasmid DNA by electroporation, and colonies of stable transformants selected. To prepare the ES cells for electroporation, harvest the cells by trypsinization, wash 2 times with PBS and suspend 6×10^6 cells in 0.75 ml of PBS in a 0.4-cm electroporation cuvette.

2. Add 50 μg of sterile, linearized, plasmid DNA dissolved in 50 μl TE buffer. In addition to the gene of interest, the plasmid should contain a selectable marker gene such as *neo* under the control of a constitutively expressed promoter.

3. Electroporate the cells (950 μF, 300 V) and measure cell mortality by trypan blue staining 0.5 h after electroporation. Cell mortality should be approximately 50%. Plate the cells into 2 100-mm diameter culture dishes containing a confluent layer of growth-arrested RTS34st (*neo*) cells and add the medium and supplements described in Section B. The next day, add 5 μl/ml of the G418 stock solution and change the medium every 2 days, adding fresh G418. Colonies will begin to appear 2 to 3 weeks after G418 selection is initiated.

4. Gene targeting by homologous recombination can be accomplished in the cells using a targeting vector containing *neo* flanked by 5′ and 3′ arms that are homologous to the targeted gene along with *RFP* located outside of the homologous region (see Fig. 3A). The targeting vector is electroporated into the cells and the cells are selected in G418 as described in Step 3. After G418 selection, the

colonies are examined by fluorescence microscopy and the homologous re-combinants are identified by the absence of RFP expression (see Fig. 3B). The RFP-negative colonies are manually removed from the plate using a Pipetman micropipetor (Rainin) approximately 5 weeks after the start of G418 selection. The individual selected colonies are transferred to single wells of a 24-well plate containing growth arrested RTS34st (*neo*) feeder cells. The individual colonies are cultured for 2 to 3 weeks before passaging into single wells of a 12-well plate. During passage, a portion of the cells from each colony are harvested for PCR analysis to confirm homologous recombination.

III. Materials

A. Reagents

1. Cell culture media: Leibowitz's L-15 (catalog number 41300-039), Ham's F12 (catalog number 21700-075), and Dulbecco's modified Eagle's media (catalog number 12100-046) are available from GIBCO-BRL, Grand Island, NY. One liter of each medium is prepared separately by dissolving the powder in ddH$_2$O and adding HEPES buffer (final concentration 15 mM, pH 7.2), penicillin G (120 μg/ml), ampicillin (25 μg/ml), and streptomycin sulfate (200 μg/ml). LDF medium is prepared by combining Leibowitz's L-15, Dulbecco's modified Eagles, and Ham's F12 media (50:35:15) and supplementing with sodium bicarbonate (0.180 g/L) and sodium selenite (10^{-8} M). The medium is filter sterilized before use.

2. Phosphate buffered saline (PBS) (catalog number. 21600-010) is available from GIBCO.

3. TE buffer: 10 mM Tris-HCl, 1 mM EDTA, pH 8.0.

4. Fetal bovine serum (FBS) (catalog number BT-9501-500) is available from Harlan Laboratories, Indianapolis, IN.

5. Calf serum (catalog number 26170-043) is available from GIBCO.

6. Trout plasma (SeaGrow) is available from East Coast Biologics, Inc., North Berwick, ME. The plasma is sterile filtered and heat treated (56 °C, 25 m) and centrifuged (10,000 × g, 10 m) before use.

7. Trypsin/EDTA solution (2mg/ml trypsin, 1 mM EDTA) is prepared in PBS. The solution is filter sterilized before use. Trypsin (catalog number T-7409) and EDTA (catalog number E-6511) are available from Sigma, St. Louis, MO.

8. Human epidermal growth factor (EGF) (catalog number 13247-051) is available from Invitrogen, Carlsbad, CA. Stock EGF solution is prepared at 10 μg/ml in ddH$_2$O.

9. Human basic fibroblast growth factor (bFGF) (catalog number 13256-029) is available from Invitrogen. Stock bFGF solution is prepared at 10 μg/ml in 10 mM Tris-HCl, pH 7.6.

10. Bovine insulin (catalog number I-5500) is available from Sigma. Stock insulin is prepared at 1 mg/ml in 20 mM HCl.

11. Bleach (Chlorox) solution is prepared fresh at 0.5% in ddH$_2$O from a newly opened bottle.

12. Zebrafish embryo extract is prepared by homogenizing approximately 500 embryos in 0.5 ml of LDF medium and centrifuging (20,000 × g, 10 m) to remove the debris. The supernatant is collected, filter sterilized, and the protein measured. The extract is diluted to 10 mg protein/ml and stored frozen ($-20\,°C$) in 0.2-ml aliquots.

13. Geneticin (G418 sulfate, catalog number 11811-031) is available from GIBCO-BRL. G418 stock solution is prepared at 100 mg/ml in ddH$_2$O and filter sterilized before use.

14. Pronase (catalog number P6911) is available from Sigma and is prepared at 0.5 mg/ml in Hanks solution.

15. Egg water: 60 μg/ml aquarium salt.

16. Freezing medium: 80% FD medium (1:1 mixture of Ham's F12 and DMEM), 10% FBS, 10% dimethyl sulfoxide (DMSO).

B. Feeder Cell Lines

1. Growth-Arrested Feeder Cells

RTS34st cells (Ganassin and Bols, 1999) are cultured (18 °C) in Leibowitz's L-15 medium (Sigma, catalog number L5520) supplemented with 30% calf serum. To prepare growth-arrested cells, a confluent culture of RTS34st cells contained in a flask (25 cm^2) or dish (100-mm diameter) is irradiated (3000 RADS) and then harvested by trypsinization and frozen in liquid nitrogen within 24 h after irradiation. To recover the frozen growth-arrested cells, the vial is thawed briefly in a water bath (37 °C) and the cells are collected by centrifugation and resuspending in L-15 medium. The cells from 1 frozen vial are distributed into 2 25-cm^2 flasks or 4 wells of a 6-well plate. After the cells have attached to the culture surface, the medium is supplemented with calf serum (30%). After 24 h, the growth-arrested cells should be spread on the culture surface and used immediately as feeder layers. Before using the growth-arrested cells as a feeder layer, the L-15 medium is removed and the cells are rinsed 1 time.

2. Drug-Resistant Feeder Cells

A feeder cell line that is resistant to G418 was prepared by transfecting RTS34st with the pBKRSV plasmid, which contains the aminoglycoside phosphotransferase gene (*neo*) under the control of Roussarcoma virus (RSV) promoter. Colonies of cells that stably express *neo* were selected in G418 (500 μg/ml). The *neo*-resistant cell line, RTS34st (*neo*), is cultured in L15 medium supplemented with 30% calf serum plus 200 μg/ml G418. Growth-arrested RTS34st (*neo*) cells are prepared using the same methods described for RTS34st.

3. RTS34st Cell-Conditioned Medium

Conditioned medium is prepared by adding fresh L-15 plus 30% FBS to a confluent culture of RTS34st cells and incubating for 3 days (18 °C). The medium is removed, filter sterilized, and stored frozen (−20 °C).

Acknowledgments

This work was supported by grants from the U.S. Dept. of Agriculture NRI 01-3242, Illinois-Indiana SeaGrant R/A-03-01 and the National Institutes of Health R01-GM069384-01.

References

Capecchi, M. (1989). Altering the genome by homologous recombination. *Science* **244,** 1288–1292.

Collodi, P., Kamei, Y., Ernst, T., Miranda, C., Buhler, D., and Barnes, D. (1992). Culture of cells from zebrafish embryo and adult tissues. *Cell. Biol. Toxicol.* **8,** 43–61.

Currie, P. D. (1996). Zebrafish genetics: Mutant cornucopia. *Curr. Biol.* **6,** 1548–1552.

Doetschman, T., Gregg, R. G., Maeda, N., Hooper, M. L., Melton, D. W., Thompson, S., and Smithies, O. (1987). Targeted correction of a mutant HPRT gene in mouse embryonic stem cells. *Nature* **330,** 576–578.

Evans, M. J., and Kaufman, M. H. (1981). Establishment in culture of pluripotential cells from mouse embryos. *Nature* **292,** 154–156.

Fan, L., Alestrom, A., Alestrom, P., and Collodi, P. (2004). Development of zebrafish cell cultures with competency for contributing to the germ line. *Crit. Rev. Eukaryot. Gene. Expr.* **14,** 43–51.

Fan, L., and Collodi, P. (2002). Progress towards cell-mediated gene transfer in zebrafish. *Brie. Functional Genom. Proteom.* **1,** 131–138.

Ganassin, R., and Bols, N. C. (1999). A stromal cell line from rainbow trout spleen, RTS34st, that supports the growth of rainbow trout macrophages and produces conditioned medium with mitogenic effects on leukocytes. *In Vitro Cell Dev. Biol. Animal* **35,** 80–86.

Ghosh, C., and Collodi, P. (1994). Culture of cells from zebrafish blastula-stage embryos. *Cytotechnology* **14,** 21–26.

Golling, G., Amsterdam, A., Sun, Z., Antonelli, M., Maldonado, E., Chen, W., Burgess, S., Haldi, M., Artzt, K., Farrington, S., *et al.* (2002). Insertional mutagenesis in zebrafish rapidly identifies genes essential for early vertebrate development. *Nat. Genet.* **31,** 135–140.

Gossler, A., Doetschman, T., Korn, R., Serfling, E., and Kemler, R. (1986). Transgenesis by means of blastocyst-derived embryonic stem cell lines. *Proc. Natl. Acad. Sci. USA* **83,** 9065–9069.

Lui, J., Baker, J., Perkins, A. S., Robertson, E. J., and Efstratiadis, A. (1993). Mice carrying null mutations of the genes encoding insulin-like growth factor 1 (Igf-1) and type 1 IGF receptor (Igf1r). *Cell* **75,** 59–72.

Ma, C., Fan, L., Ganassin, R., Bols, N., and Collodi, P. (2001). Production of zebrafish germ-line chimeras from embryo cell cultures. *Proc. Natl. Acad. Sci. USA* **98,** 2461–2466.

Nasevicius, A., and Ekker, S. C. (2000). Effective targeted gene "knockdown" in zebrafish. *Nat. Genet.* **26,** 216–220.

Van Eeden, F. J. M., Granato, M., Odenthal, J., and Haffter, P. (1999). Developmental mutant screens in the zebrafish. *In* "Methods in Cell Biology" (H. W. Detrich, M. Westerfield, and L. I. Zon, eds.), Vol. 60, pp. 21–41. Academic Press, San Diego.

Westerfield, M. (1995). "The Zebrafish Book, ed. 3." University of Oregon Press, Eugene, OR.

Wienholds, E., Schulte-Merker, S., Walderich, B., and Plasterk, R. (2002). Target-selected inactivation of the zebrafish rag1 gene. *Science* **297,** 99–101.

Zhang, W., Behringer, R. R., and Olson, E. N. (1995). Inactivation of the myogenic bHLH gene MRF4 results in up-regulation of myogenin and rib anomalies. *Genes Dev.* **9,** 1388–1399.

PART II

Developmental and Neural Biology

CHAPTER 10

Neurogenesis

Prisca Chapouton and Laure Bally-Cuif

Zebrafish Neurogenetics Junior Research Group
Institute of Virology
Technical University-Munich
D-81675 Munich, Germany

GSF–National Research Center for Environment and Health
Institute of Developmental Genetics
D-85764 Neuherberg, Germany

METHODS IN CELL BIOLOGY, VOL. 76
Copyright 2004, Elsevier Inc. All rights reserved.
0091-679X/04 $35.00

I. Introduction

We will adopt here the broad definition of neurogenesis, shared by most authors, as the multistep process that begins with neural induction and leads to the differentiation of functional neurons (Appel and Chitnis, 2002). During this process, cells evolve from a precursor state (also referred to as *precursor cell, progenitor cell*, or *neuroblast*) to a committed and freshly postmitotic state (also referred to as *early differentiating neuron* or *early postmitotic neuron*) to the differentiated state. We restrict our review to the building of the zebrafish central nervous system, but include current knowledge on the formation of glia (sensu stricto *gliogenesis*), which can influence neuronal selection, birth, and differentiation, as well as, at least in rodents, take an active part in late neurogenesis (Alvarez-Buylla and Garcia-Verdugo, 2002; Gotz *et al.*, 2002; Taupin and Gage, 2002). For a discussion of the zebrafish peripheral sympathetic nervous system, see chapter 12.

Neurogenesis in zebrafish, as in other lower vertebrates, occurs in two successive (primary and secondary) waves (Kimmel, 1993). Primary neurogenesis begins during late gastrulation and continues during embryogenesis to produce early-born, big neurons with long axons such as the brain epiphyseal and post-optic clusters, Mauthner cells, Rohon-Beard (RB) sensory neurons, and the three types of primary spinal motoneurons (CaP, MiP, RoP) (Kimmel and Westerfield, 1990). Axonogenesis starts between 14 and 24 hpf. Primary neurons build the first functional embryonic and early larval neuronal scaffold. Secondary neurogenesis occurs particularly from post-embryonic stages (2 dpf) onward at all rostrocaudal levels, to take over the primary system through a refined and increasingly complex network (Mueller and Wullimann, 2003). Because the zebrafish grows throughout life, zebrafish neurogenesis is believed to extend into adulthood. As we will discuss, the mechanisms and factors controlling primary and secondary neurogenesis are not intrinsically different.

II. The Primary Neuronal Scaffold

The initiation of neuronal differentiation in the zebrafish embryo has traditionally been revealed through the use acetylcholinesterase (AChE) activity (Hanneman and Westerfield, 1989; Ross *et al.*, 1992; Wilson *et al.*, 1990), antibodies against acetylated tubulin (Chitnis and Kuwada, 1990), or HNK1 (Metcalfe *et al.*, 1990). Axon tracts have also been visualized after DiI applications, facilitated by the large size of zebrafish primary neurons (Myers *et al.*, 1986; Wilson and Easter, 1991a,b). These techniques, later complemented by *in situ* hybridization approaches with neurogenesis markers and by the production of large sets of antibodies (Trewarrow *et al.*, 1990), highlighted defined differentiation centers and stereotyped axonal pathways within the early neural tube (Kimmel, 1993) (Fig. 1).

The embryonic brain neuronal network (Ross *et al.*, 1992; Wilson and Easter, 1991a,b) is composed of several nuclei (dorsorostral, ventrorostral, epiphyseal, ventrocaudal, and nucleus of the posterior commissure) that extend longitudinal or transverse (commissural) axon tracts that connect the clusters with each other, with the spinal cord, or with the peripheral system.

Reticulospinal neurons are long projection interneurons whose cell bodies are located in a rhombomere-specific pattern, within the hindbrain or within the basal mesencephalon (nMLF). The axons of reticulospinal neurons extend into the spinal cord (Mendelson, 1986a,b). These neurons can be labeled by backfilling with fluorescent dyes from the spinal cord at trunk levels. An extensively studied reticulospinal neuron is the Mauthner cell, located in rhombomere 4, which links sensory information originating from trigeminal ganglia to contralateral motoneurons, mediating the escape response (Kimmel *et al.*, 1981, 1982; O'Malley *et al.*, 1996).

Spinal cord neurons (Bernhardt *et al.*, 1990; Eisen *et al.*, 1990; Kuwada *et al.*, 1990b; Westerfield *et al.*, 1986) compose motoneurons, intermediate neurons, and sensory neurons, arranged in a ventral to dorsal sequence. Three types of primary motoneurons (CaP, MiP, RoP), regularly arranged opposite each somite, differ in their position and projection pattern to the somite-derived fast muscles (Eisen *et al.*, 1990; Westerfield *et al.*, 1986). Eight types of interneurons have been identified (Bernhardt *et al.*, 1990; Hale *et al.*, 2001) that, at least in part, participate in distinct motor behaviors (Ritter *et al.*, 2001). Among those, at least CoPA and VeLD have both primary and secondary components. RB mechanosensory neurons are the only type of primary sensory neurons. Their central axons join the lateral longitudinal fascicle and extend into the hindbrain and spinal cord. The peripheral axons of RBs exit the spinal cord and innervate the skin. Of all six primary spinal neurons, RBs are the only transient cell type. They gradually die by apoptosis around 3 dpf, coinciding with the emergence of the dorsal root ganglia (Reyes *et al.*, 2004; Svoboda *et al.*, 2001; Williams *et al.*, 2000).

Together, these results highlight a number of interesting features. First, at all locations, the basal (ventral) plate generally matures earlier than the alar (dorsal) plate. Second, the location and connection pattern of all zebrafish early clusters/ neuronal groups are very reminiscent of the early neuronal organization observed in other vertebrates, such as the mouse or chick (Easter *et al.*, 1994). This suggests that establishment of the early neuronal differentiation profile in vertebrates is controlled in time and space in a strict and evolutionarily conserved manner. Finally, anteriorly, the position of these clusters is generally related to neuromeric organization of the brain. Thus patterning cues that regulate the development of neuromeres might also play a role in the spatial regulation of neurogenesis onset.

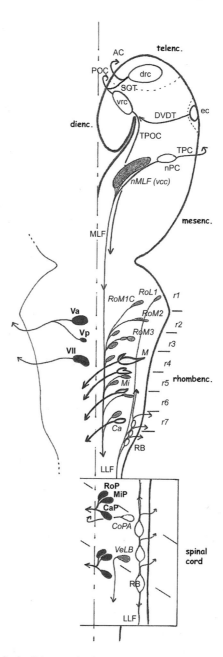

Fig. 1 Earliest clusters of zebrafish central primary neurons, schematized on an "open" preparation of the neural tube at 24 to 48 hpf (anterior up, spinal cord—in box—represented at a higher magnification for clarity). Brain clusters in the tel-, di- and mesencephalon (drc, ec, nMLF, nPC, vrc) build a scaffold of orthogonally oriented tracts (AC, DVDT, POC, SOT, TPC). More posterior

III. Early Development of the Zebrafish Neural Plate

A. Morphogenesis

Direct lineage tracings using fluorescent markers and time-lapse videomicroscopy have localized presumptive neural plate cells to the central 4 blastomeres of the 16-celled embryos (Helde *et al.*, 1994; Wilson *et al.*, 1995) and the dorsal aspect of the blastoderm at the shield stage (Fig. 2A) (Woo and Fraser, 1995). Starting at gastrulation, neural plate cells are then rearranged along the anteroposterior and mediolateral axes (Fig. 2B) by convergence-extension movements (Solnica-Krezel and Cooper, 2002) or active migration (Varga *et al.*, 1999). Zebrafish neurulation has been studied in detailed histological and lineage analyses (Kimmel and Warga, 1986; Papan and Campos-Ortega, 1994; Schmitz *et al.*, 1993). Zebrafish neuralation is initiated at early somatogenesis, when characteristic thickenings form along the lateral and medial aspects of the neural plate (Fig. 2C). The lateral thickenings progressively converge toward the midline while the neural plate folds inward, giving rise around the 6- to 10-somite stage to a solid neural keel (Fig. 2D), which later detaches from the adjacent epidermis and forms a neural rod (Fig. 2E). The neural lumen starts becoming visible around the 17-somite stage (Fig. 2F) and has been hypothesized to result from secondary cavitation (Kimmel, 1993; Papan and Campos-Ortega, 1994; Schmitz *et al.*, 1993). An important specific feature of zebrafish neurulation is the fact that dividing cells frequently contribute progeny to both sides of the neural tube (Papan and Campos-Ortega, 1997). This bilateral distribution is permitted by the long-lasting keel/rod structure and the perpendicular orientation of mitoses at that stage (Fig. 2D and E) (Geldmacher-Voss *et al.*,

---

neuronal groups can be subdivided into motoneurons (black cell bodies, bold labeling), intermediate neurons (italics), and sensory neurons (Roman). Primary motoneurons of the rhombencephalon first differentiate in rhombomeres 2 to 4 and produce cranial nerves V and VII that innervate the branchial arches. Primary motoneurons of the spinal cord are of three types (RoP, MiP, CaP) that differ in their location relative to somitic boundaries (diagonal lines) and axonal arborization. Primary interneurons compose the reticulospinal system of the mes-(nMLF) and rhombencephalon (Ca, M, Mi, Ro) (for a full nomenclature see Mendelson, 1986a,b), and CoPA and VeLP spinal neurons. They project ipsilaterally (dark gray cell bodies) or contralaterally (light gray cell bodies). Finally, the primary sensory system is composed of the RB cells, which axons form the LLF and also innervate the skin. Abbreviations—Clusters/neurons: Ca; caudal reticulospinal neurons; CaP: caudal primary motoneurons; CoPA: commissural primary ascending interneuron; drc: dorso-rostral cluster; ec: epiphyseal cluster; M: Mauthner cell; Mi: middle reticulospinal neurons; MiP: middle primary motoneuron; nMLF (or vcc): nucleus of the MLF; nPC: nucleus of the PC; RB: RB neurons; RoM: rostral medial reticulospinal neurons; RoP: rostral primary motoneurons; RoL: rostral lateral reticulospinal neurons; VeLD: ventral longitudinal descending interneurons; vcc (or nMLF): ventro-caudal cluster; vrc; ventro-rostral cluster; tracts—AC: anterior commissure; DVDT: dorsoventral diencephalic tract; LLF: lateral longitudinal fascicle; MLF: medial longitudinal fascicle; POC: post-optic commissure; TPC: tract of the posterior commissure; TPOC: tract of the post-optic commissure; V: fifth cranial nerve; VII: seventh cranial nerve; territories—dienc.: diencephalon; mesenc.: mesencephalon (midbrain): telencephalon.

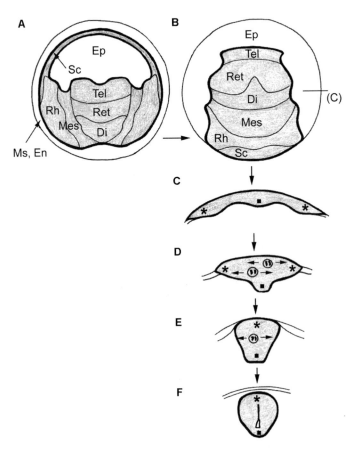

Fig. 2 Fate map and morphogenesis of the zebrafish embryonic neural plate. A and B, Fate map of the neural plate (gray shading) at the shield (A) and tail-bud (B) stages (dorsal views), schematized from (Woo and Fraser, 1995). The different neural territories are organized into coherent (and partially overlapping) domains that align along AP during gastrulation. C through F, Morphogenesis of the neural tube (gray shading) at tail-bud (C), 5 somites (D), 14 somites (E), and 20 somites (cross sections at the level indicated in B, schematized from Schmitz *et al.*, 1993). Lateral neural plate bulges converge toward the midline, which itself folds inward, leading to the formation of a compact neural keel/rod that later cavitates. However, as in other vertebrates, prospective dorsal neural tube cells originate from lateral neural plate domains (stars) while ventral cells originate at the midline (square). At the keel/rod stage, dividing cells often contribute progeny to both sides of the future neural tube (arrows in D and E). Abbreviations—Di: diencephalon; En: endoderm; Ep: epidermis; Mes: mesencephalon; Ms: mesoderm; Ret: retina; Rh: rhombencephalon; Sc: spinal cord; Tel: telencephalon.

2003). Sister cells distributing bilaterally are not related in fate and mostly give rise to different neuronal or glial types (Papan and Campos-Ortega, 1997). Outside this specific feature, zebrafish neurulation is comparable to that of other vertebrates, in particular with the distribution of medial versus lateral cells of the neural plate to the ventral versus midline, respectively, of the mature neural tube (Fig. 2C through F).

B. Neural Induction

Markers of neural induction, or *preneural* genes, best established in *Xenopus* and chicken, include the Sox family members (*Sox2* and *Sox3*) and *ERNI*. Their zebrafish homologs have not been studied to date, although expression of several zebrafish *sox* genes was reported to delimit the early neural plate at gastrulation (de Martino *et al.*, 2000; Rimini *et al.*, 1999; Vriz *et al.*, 1996). Traditionally, early markers of neural anteroposterior (AP) and dorsoventral (DV) patterning are used to assess neural induction in zebrafish. These include *otx2*, expressed within the anterior neural plate (presumptive forebrain and midbrain), and *hoxb1b* (prev. *hoxb1, hoxa-1*) (rhombomere 4 and beyond), from early gastrulation (Koshida *et al.*, 1998; Li *et al.*, 1994; Prince *et al.*, 1998). Induction of the anterior neural plate border (placodal fields) is revealed by *dlx3b* (prev. *dlx3*)/*dlx5a* (prev. *dlx4*) and *foxi1* at early somatogenesis stages (Akimenko *et al.*, 1994; Lee *et al.*, 2003; Nissen *et al.*, 2003; Quint *et al.*, 2000; Solomon *et al.*, 2003). Finally, the posterior neural plate border (future neural crests and RB neurons) (Cornell and Eisen, 2002) expresses *foxd3* (previously *fkd6*), *sox9b*, *sox10* (*cll*) or *snail1b* (prev. *snail2*) at late gastrulation (Dutton *et al.*, 2001; Li *et al.*, 2002; Odenthal and Nusslein-Volhard, 1998; Thisse *et al.*, 1995). As in other vertebrates, the initial specification of the zebrafish neural plate is anterior in character, later progressively posteriorized by signals originating from the nonaxial mesoderm (Erter *et al.*, 2001; Lekven *et al.*, 2001; Woo and Fraser, 1997, 1998).

Until recently, largely based on studies in amphibians, neural induction was viewed as a default pathway, where ectodermal cells would become neural unless driven toward an epithelial fate by bone morphogenetic protein (Bmp) signaling during gastrulation (Hemmati-Brivanlou and Melton, 1997). In this model, neural development on the dorsal embryonic side was permitted by the diffusion of Bmp inhibitors (such as Noggin and Chordin) from the Spemann organizer (or its equivalents the mouse node, chicken Hensen's node, and zebrafish shield). Along this line, the size of the neural plate is impaired in zebrafish mutants affected in Bmp2/4 signaling (Hammerschmidt and Mullins, 2002), or in the absence of organizer tissue [e.g., in *bozozok* (*dharma*)] (Fekany *et al.*, 1999; Koos and Ho, 1999; Leung *et al.*, 2003; Yamanaka *et al.*, 1998), *squint;cyclops* (*ndr1;ndr2*), or maternal zygotic *one-eyed pinhead* [(*oep*) mutants] (Gritsman *et al.*, 1999). However, even in these cases, a neural plate forms, suggesting that blocking Bmp is neither required nor sufficient for neural induction.

Findings in chicken and mice strongly reinforce this view and further point to two other major signaling pathways as the prime neural inducers, acting before gastrulation occurs: Fgf and Wnt (Bainter *et al.*, 2001; Stern, 2002; Wilson and Edlund, 2001). For instance, chemical interference with Fgf signaling abolishes neural induction in chicken embryos (Streit *et al.*, 2000; Wilson *et al.*, 2000). In support of these hypotheses, selective targets of Fgf in the neural induction process, such as Churchill, have recently been identified in chicken (Sheng *et al.*, 2003). A role for Fgf in initiating neural induction has also been demonstrated in ascidians (Bertrand *et al.*, 2003). To date however, in zebrafish, early Fgf and Wnt signaling have been mostly implicated in neural patterning rather than in neural induction (Furthauer *et al.*, 1997; Kudoh *et al.*, 2002; Reifers *et al.*, 1998; Shiomi *et al.*, 2003; Werter *et al.*, 2001). Expressed at the dorsal embryonic margin in the late blastula, *fgf3* participates in induction of the posterior neural plate; however, this function requires Bmp antagonists and thus is not independent of organizer activity (Koshida *et al.*, 2002). The zebrafish maternal Wnt pathway is involved in organizer induction (Hammerschmidt and Mullins, 2002) and regionalization of the yolk syncytial layer (Ho *et al.*, 1999), but a direct role in neural induction has not been shown. Searching for homologs of Fgf or Wnt targets that mediate neural competence in other chordates (Bertrand *et al.*, 2003; Sheng *et al.*, 2003) might help implicating early Fgf and Wnt signaling in zebrafish neural induction. Bmp signaling is important, however, in the definition of the neural plate border and dorsal neural cell types at all anteroposterior (AP) levels of the zebrafish neural plate (Barth *et al.*, 1999; Houart *et al.*, 2002; Wilson *et al.*, 2002).

C. Delimitation of Proneural Fields by Prepattern Factors

Our understanding of neurogenesis in vertebrates, including zebrafish, stems largely from key original findings in *Drosophila* (Campos-Ortega, 1993). The definition of neurogenesis-competent domains (*proneural clusters* or *proneural fields*) within the fly is achieved by the combinatorial expression of neurogenesis activators (*proneural* factors) and neurogenesis inhibitors. The former compose Achaete-Scute proteins, the latter include Hairy-like factors. All belong to the basic helix-loop-helix (bHLH) class of transcription factors and are expressed in direct response to the early embryonic patterning machinery, to establish a neurogenesis "prepattern" within the neuroectoderm.

As we have already discussed, neurogenesis in the early vertebrate neural plate occurs at stereotyped loci and avoids others, suggesting the existence of a prepattern of proneural/incompetent fields similar to that observed in *Drosophila*. Immediately downstream of neural induction, several transcription factors that promote the neural fate have been recently identified. The most studied of these belong to the Sox, Gli, POU, and Iroquois families. They are expressed across broad domains of the neural plate (Bainter *et al.*, 2001; Bally-Cuif and Hammerschmidt, 2003). In zebrafish, *iro1, iro7*, and *pou5f1* (prev. *pou2*) are expressed across the presumptive midbrain and hindbrain areas (Hauptmann and Gerster, 1995; Itoh *et al.*, 2002; Lecaudey *et al.*, 2001). Iro1 and 7 and Pou5f1 are required for *neurog1* (prev. *ngn1*) expression in their respective expression domains. At least Iro1 and 7 are sufficient, when expressed incorrectly, to induce ectopic *neurog1* expression within non-neural ectoderm (Belting *et al.*, 2001; Itoh *et al.*, 2002). More anteriorly, expression of *flh* (*znot*) defines the epiphyseal proneural field and permits expression of the proneural factors Neurog1 and Ash1a, driving neurogenesis in this area (Cau and Wilson, 2003).

Recent evidence in zebrafish and *Xenopus* highlights that proneural fields are also defined as the domains that do not express active neurogenesis inhibitors. These domains include the anterior neural plate (prospective telencephalon, diencephalon, and eyes), the longitudinal spinal cord stripes that separate the columns of sensory, motoneurons and interneurons, and the midbrain-hindbrain boundary. They are generally characterized by the expression of transcription factors such as Zic1-3, Iro3, Anf and BF1, as well as members of Hairy/E(Spl) family (so-called Hes, Hey, Her, or Hairy in different species), which have been best studied in *Xenopus* (Andreazzoli *et al.*, 2003; Bellefroid *et al.*, 1998; Brewster *et al.*, 1998; Hardcastle and Papalopulu, 2000). In zebrafish, *zic1* (prev. *opl*) and *zic3* expression highlight anterior and posterior neural plate domains, respectively (Grinblat and Sive, 2001; Grinblat *et al.*, 1998) *hesx1* (prev. *anf*) and *foxg1* (prev. *bf1*) are labeling the anterior neural plate (Houart *et al.*, 2002). *her5* is labeling the midbrain-hindbrain boundary (Müller *et al.*, 1996). Gain- and loss-of-function experiments recently established Her5 as a crucial inhibitor of neurogenesis at the midbrain-hindbrain boundary, acting upstream of the definition of proneural fields (Geling *et al.*, 2003, 2004).

In all examples studied, expression of neurogenesis prepattern factors is directly controlled by the early embryonic patterning machinery. This expression of neurogenesis prepattern factors thus links early patterning information to the definition of the first neurogenesis sites in a manner analogous to the definition of proneural fields in *Drosophila*. However, the molecular pathways activated or inhibited by these factors in vertebrates remain poorly understood. These factors might directly inhibit expression of proneural genes, control the cell cycle machinery, or both.

As a result of prepatterning, early neurogenesis occurs at discrete sites, which can be revealed at the 3-somite stage by the characteristic expression profile of the proneural gene *neurog1* (see the section on primary neurogenesis, discussed later) (Fig. 3A). Anteriorly, prominent *neurog1* expression highlights precursors of the olfactory neurons, the ventrocaudal cluster (vcc), and trigeminal ganglia. Posteriorly, longitudinal columns of presumptive motoneurons, interneurons, and sensory neurons are visible. These proneural clusters are organized around zones where inhibitory prepatterning takes place, such as the anterior neural plate, the midbrain-hindbrain boundary, and domains that separate the posterior longitudinal stripes in the presumptive hindbrain and spinal cord (Fig. 3A).

D. Neural Tube Organizers as Proliferation Signals?

The expression of early neurogenesis prepatterning genes and the positioning of neurogenic zones at later stages are mainly influenced by local secreted factors produced within or adjacent to the neural plate. Members of the Wnt, Fgf, Bmp,

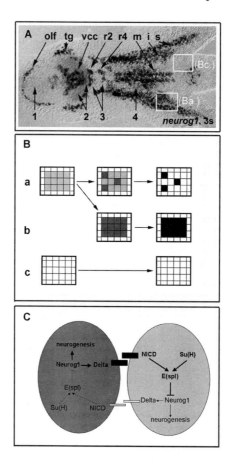

Fig. 3 Early proneural clusters and the principles of lateral inhibition. A, Expression of *neurog1* revealed at the 3-somite stage by whole-mount *in situ* hybridization, dorsal view of a flat-mounted embryo, anterior left. *neurog1* expression highlights a discrete distribution of early proneural clusters (olf: olfactory neuron precursors, tg: trigeminal ganglion, vcc: ventro-caudal cluster, r2: motoneuron precursors in rhombomere 2, r4: motoneuron precursors in rhombomere 4, m: spinal motoneuron precursors, i: spinal interneuron precursors, s: spinal sensory neuron precursors). These clusters are separated by prepatterned zones of neuronal inhibition (1: ANP, 2: MHB, 3: longitudinal stripes in r2 and r4, 4: longitudinal stripes in the spinal cord). B and C, Lateral inhibition selects a few competent precursors for further neurogenesis. Ba, Precursors selection within a proneural cluster (boxed in A.) occurs in three steps: (i) weak and ubiquitous expression of an early proneural factor (e.g., Neurog1; light grey; left panel). (ii) Reinforcement of stochastic differences in the expression of *neurog1* between adjacent cells by Notch-Delta interactions. C, Leading to the juxtaposition of strongly and weakly *neurog1*-positive cells (dark and light grey, respectively; middle panel). (iii) Further commitment of strongly *neurog1*-expressing cells (black) and extinction of *neurog1* expression in their neighbors (white; right panel). Bb, In the absence of lateral inhibition, all early precursors commit to differentiation. Bc, In prepatterned domains of lateral inhibitions (as boxed in A), early expression of proneural genes is blocked and lateral inhibition does not take place.

and Hh families are involved in this process. In addition to patterning control, these factors have been directly implicated in the control of cell proliferation/ differentiation in many systems. Thus, in addition to positioning the expression of prepatterning genes, these factors may also play a direct role on neurogenesis by influencing cellular status.

Along the anterior-posterior (AP) axis, several local signaling centers have been identified that influence neural plate patterning and correlate with specific neurogenesis status. In zebrafish, at gastrulation stages, a signal originating from the nonaxial blastoderm margin and possibly encoded by *wnt8* posteriorizes the neural plate. This may partly account for the anterior-to-posterior gradient of CNS maturation. At late gastrulation, the first cell row of the neural plate acts as a signaling source, the anterior neural border, to permit telencephalic maintenance (Houart *et al.*, 1998). Anterior neural border activity is encoded by Sfrp (prev. Tlc), a secreted Wnt inhibitor of the Frizzled family, proposed to antagonize Wnt8b signaling from the posterior diencephalon (Houart *et al.*, 2002). A role of the anterior neural border on neurogenesis has not been studied, but its position is suggestive: It lies at the junction between prospective placodal fields (which undergo early neurogenesis) and the anterior neural plate (where neurogenesis is actively prevented by prepatterning inhibitors). At later stages, forebrain neurogenesis is influenced by Fgf3 and Fgf8 (Walshe and Mason, 2003). The midbrain-hindbrain boundary is the source of secreted factors of the Wnt (Wnt1, Wnt10b, Wnt8, Wnt3) and Fgf (Fgf8, Fgf15, Fgf17, Fgf18) families (Kelly and Moon, 1995; Kelly *et al.*, 1995; Lekven *et al.*, 2003; Reifers *et al.*, 1998, 2000), which (at least in part) encode activity of the isthmic organizer (Martinez, 2001; Rhinn and Brand, 2001; Wurst and Bally-Cuif, 2001) and are required for growth and patterning of the mid-hindbrain and anterior hindbrain. It positions expression of the prepatterning neurogenesis inhibitor Her5 (Geling *et al.*, 2003). A possible direct role of midbrain-hindbrain boundary secreted factors in neurogenesis control has not been studied; however, both Wnt1 and Fgf8 have been shown to influence cell proliferation within the midbrain-hindbrain area in the mouse _(Panhuysen *et al.*, 2004; Xu *et al.*, 2000). Finally, at somitogenesis stages, expression of Fgf3 and Fgf8 in rhombomere 4 (r4) permits development of r5-6 (Maves *et al.*, 2002). Perhaps meaningfully, r5-6 also exhibit delayed maturation compared with r4.

Signals influencing neurogenesis along the DV axis include Bmp, Wnt, and Hh. Bmp's are expressed by the dorsal ectoderm and dorsal neural tube and can distinctly affect neurogenesis in different contexts. In the mouse embryo, Bmp signaling sequentially promotes proliferation, then cell cycle exit of dorsal neural tube cells via alternative Bmp receptor usage (Panchision *et al.*, 2001). Expression of *Wnt1* and *Wnt3a* at the dorsal midline has also been proposed to establish a dorso-ventral gradient promoting proliferation (and blocking differentiation) across the neural tube (Megason and McMahon, 2002). Interestingly, sensitivity to this gradient might be influenced by ventrally derived Shh (Ishibashi and McMahon, 2002). Hh signaling itself was associated with several events of late neural proliferation (Ruiz i Altaba *et al.*, 2002), as well as with the regulation of expression of Zic and Gli factors to define longitudinal stripes of neurogenesis in the *Xenopus* spinal cord (Brewster *et al.*, 1998). The role of Bmp, Wnt, and Hh signals in controlling the neural proliferation/differentiation process in zebrafish has not been directly studied. Hh signaling controls the mediolateral pattern of *neurog1* expression (Blader *et al.*, 1997), but a direct regulation of the *neurog1* enhancer (as opposed to general neural plate repatterning) has not been reported (Blader *et al.*, 2003). However, it is likely that these pathways are involved in neurogenesis control in a manner similar to other vertebrates.

IV. Lateral Inhibition and the Neurogenesis Cascade

A. Primary Neurogenesis

Competent neurogenic domains (or proneural fields) express low levels of proneural factors, which need to be reinforced to drive neurogenesis, and only some precursors will further commit to neurogenesis or gliogenesis. In *Drosophila*, the selection of these precursors relies on the process of lateral inhibition, where cells expressing high levels of the Notch ligand Delta will commit to differentiation and at the same time, via Delta-Notch binding, prohibit their neighboring cells from doing so (Simpson, 1997). After binding of Delta, the intracellular domain of Notch (NICD) translocates to the nucleus and activates transcription of downstream effectors, among which bHLH transcriptional repressors of the Enhancer-of-Split [E(Spl)] family. E(Spl) proteins prevent expression or activity of proneural factors. Similarly, cells expressing high levels of Delta maintain expression of proneural factors and Delta transcription. Thus, initial differences in Delta expression are amplified, leading to the reinforcement of neurogenic predispositions in a selection of precursors (Bray and Furriols, 2001; Mumm and Kopan, 2000) (see Fig. 3C).

Current evidence suggests that similar molecular and cellular mechanisms are at play in vertebrates (Appel and Chitnis, 2002; Lewis, 1998). Studies in zebrafish have revealed expression of *notch* (*notch1a, notch1b, notch3* -prev. *notch5*-, and *notch2*- prev. *notch6*-) (Westin and Lardelli, 1997), delta (*dlA, B, D*) (Haddon, 1998), and *E(spl)* (*her4*) (Takke *et al.*, 1999) genes, as well as expression of proneural genes (*neurog1, neurog3* -prev. *ngn2, ngn3*-, *ash1a, ash1b, atoh1* -prev. *ath1*-, neurod4 -prev. *ath3*-, and *coe2*) (Allende and Weinberg, 1994; Bally-Cuif *et al.*, 1998; Blader *et al.*, 1997; Koster and Fraser, 2001; Liao *et al.*, 1999; Park *et al.*, 2003; Wang *et al.*, 2001) in proneural domains of the neural plate. The relative expression of these factors obeys the mechanics of lateral inhibition. For instance, NICD induces *her4* and prevents *neurog1* expression (Takke *et al.*, 1999), while an increase in Delta function maintains high levels of *neurog1* throughout early proneural fields (Appel and Eisen, 1998; Appel *et al.*, 2001; Dornseifer *et al.*, 1997; Haddon, 1998). In addition, a number of zebrafish mutants with altered *neurog1* expression proved invaluable to our understanding of vertebrate neurogenesis and

Notch/Delta function in this process. Deficiencies in Neurog1 function were produced by insertional mutagenesis into the *neurog1* locus (*neurod3^{hi1059}*) (Golling *et al.*, 2002) and injection of *neurog1* morpholino into zebrafish eggs (Cornell and Eisen, 2002; Park *et al.*, 2003). Resulting embryos display a severe reduction of cranial ganglia and of the number of nucleus of the medial longitudinal fasciculus (nMLF) and spinal sensory neurons, while spinal motoneurons and interneurons and epiphysial neurons are less affected (Cau and Wilson, 2003; Cornell and Eisen, 2002; Geling *et al.*, 2004; Golling *et al.*, 2002; Park *et al.*, 2003). These results point to Neurog1 as a crucial proneural factor of zebrafish primary neurogenesis and highlight differential spatial requirements for Neurog1 function in this process. Mutations *after-eight-aei-* (*delD*), *deadly-seven-des-*(*notch1a*), *delA^{dx2}* (*dlA*), and *mind-bomb* (*mib*) all directly affect Notch/Delta signaling (Appel *et al.*, 1999; Holley *et al.*, 2002; Jiang *et al.*, 1996; van Eeden *et al.*, 1996). Mutants *aei*, *des*, and *delA* display a moderate excess of primary neurons, as predicted from a lateral inhibition defect (Gray *et al.*, 2001; Jiang *et al.*, 1996). Redundancy between Notch and Delta proteins within the zebrafish neural plate might explain the relatively mild effects of these mutations. In contrast, *mib* mutants suffer from a severe neurogenic phenotype. The mutant *mib* encodes a new ubiquitin ligase that interacts with the intracellular domain of Delta to promote Notch signaling and neurogenesis inhibition in neighboring cells (Itoh *et al.*, 2003). This function is strongly reminiscent of *Drosophila* Neuralized (Deblandre *et al.*, 2001; Lai *et al.*, 2001; Pavlopoulos *et al.*, 2001), and further analysis of zebrafish *mib* mutants should provide crucial insight into the mechanisms controlling the effectiveness of Notch/Delta signaling in vertebrates. Finally, impairment of Notch/Delta signaling or the mutation *narrow-minded* (Artinger *et al.*, 1999) all affect concomitantly the development of trunk neural crest and RB cells, suggesting that both cell types are regulated by the same genetic pathways. Further, RB and neural crest cells likely originate from a common pool of precursors, since Notch/Delta signaling influences development of crest versus RB identities. Deficiencies in Notch signaling expand the RB population at the expense of neural crests (Cornell and Eisen, 2000; Jiang *et al.*, 1996). Recently, a zebrafish non-HLH-containing factor expressed in primary neuron progenitors, Onecut1 (prev. Onecut), was also isolated (Hong *et al.*, 2002). Its function in the neuronal specification process is unknown.

Following lateral inhibition, the progenitors displaying increased levels of Delta and of proneural expression further commit to differentiation. In zebrafish, as in *Drosophila* and other vertebrates, this is accompanied by cell cycle exit and the transcription of a new set of proneural bHLH genes, such as *neurod2* (prev. *ndr2*) (Korzh *et al.*, 1998; Liao *et al.*, 1999). In addition, a late Delta factor, DlB (Haddon *et al.*, 1998), and RNA-binding proteins of the Hu family (Mueller and Wullimann, 2002a; Park *et al.*, 2000a), are expressed in committed progenitors. The transition from proliferating to postmitotic cells involves a number of cell cycle regulators, which perhaps have been best studied in *Xenopus* (Ohnuma *et al.*, 2001, 2002). Several zebrafish mutants however have been isolated where this transition is impaired. To date, all of those affect retinal differentiation, perhaps because cell cycle exit in this system is technically easier to score. In *young* mutants, affecting the *smarca4* gene, cell cycle withdrawal across the retina is initiated normally but is slower than in wild type, and retinal cells do not fully differentiate (Gregg *et al.*, 2003; Link *et al.*, 2000). The mutations *perplexed* and *confused*, respectively, cause generalized or restricted apoptosis of retinal precursors during the transition phase from proliferating to postmitotic cells (Link *et al.*, 2001). Molecular identification of these mutations promises to provide great insight into the mechanisms that link cell cycle arrest and neuronal differentiation in vertebrates.

B. Molecular Control of Secondary Neurogenesis

The mechanisms underlying secondary neurogenesis remain (to date) less understood. Early findings suggest that at least some of these might differ from primary neurogenesis control, possibly at late stages of neuronal maturation (Grunwald *et al.*, 1988). Molecular data, however, strongly support a view where, at least for early steps of the neurogenesis cascade, the same early processes are used reiteratively throughout embryonic and larval development to generate both primary and secondary neurons. Indeed, the early neurogenesis program (*notch1a, ash1a, ash1b, neurog1, neurod2,* and *dlA*) is expressed in the postembryonic zebrafish brain, in a similar relative distribution at embryonic stages, and with a comparable correlation to proliferation zones (Mueller and Wullimann, 2002a,b, 2003; Wullimann and Knipp, 2000; Wullimann and Mueller, 2002). For instance, *notch1a* expression generally coincides with proliferating cells, *neurod2* with immediately postmitotic neurons, and *neurog1* transiently covers both populations. Note, however, that the expression of these factors at postembryonic stages is not always widespread, suggesting the involvement of alternative cascades in some neural tube areas. It is likely that these different genetic networks, nevertheless, rely on bHLH factors, yet to be identified.

V. Establishment of Neuronal Identity

Important parameters in the choice of neuronal identity are combinations of positional cues and differentiation timing. Positional cues are extrinsic (e.g., distance to a signaling factor) or intrinsic (e.g., expression of a given set of transcriptional regulators by the precursor cell). Positional cues are significant in relation to a specific differentiation status, generally the moment of cell cycle exit (birth date) or of the last few rounds of cell division that precede cell cycle exit. The identification of mutants and functional tests in zebrafish made important contribution to this field in vertebrates by leading to the identification of a number of pathways or factors attributing specific CNS neuronal identities.

At gastrulation, as already discussed, Bmp levels determine the extent of non-neural versus neural ectoderm. However, extensive studies of zebrafish Bmp signaling mutants revealed an additional role of Bmp's in the specification of intermediate and dorsal neural cell types. Indeed, embryos in which Bmp signaling is strongly reduced, such as *swirl* (*bmp2b*) or strong alleles of *somitabun* (*madh5*, prev. *smad5*), develop an excess of interneurons at the expense of neural crests/ RBs. In contrast, weak *somitabun* alleles or *snailhouse* (*bmp7*) mutants, in which Bmp signaling is less reduced, display an excess of dorsal neuronal types (Barth *et al.*, 1999; Nguyen *et al.*, 2000; Schmid *et al.*, 2000). Similarly, development of the telencephalic fate occurs between thresholds of Bmp activity (Houart *et al.*, 2002). Thus, graded Bmp levels might be required *in vivo* for the specification of neuronal identities along the neural plate border. In a series of single-cell RNA injection experiments, canonical Wnt signaling from the dorsal midline (possibly encoded by Wnt1 or Wnt3a) was also directly implicated in neural versus pigment cell fate choice of neural crests (Dorsky *et al.*, 1998). Hh signaling from the ventral midline has been implicated in the specification of cranial motoneurons. Specifically, branchio-motoneurons, but not spinal motoneurons, are eliminated in a cell-autonomous manner when failing to receive or process Sonic and Tiggy Winkle Hedgehog information (Bingham *et al.*, 2001; Chandrasekhar *et al.*, 1998, 1999). Hh and Nodal midline signaling, as well as signaling from the isthmic organizer, also control development of zebrafish catecholaminergic or serotonergic neurons located in the forebrain or rhombomere 1. However, these effects are likely associated with general patterning control (Holzschuh *et al.*, 2003b; Teraoka *et al.*, 2004).

Zebrafish studies also lead to the identification of a number of transcription factors crucially involved in the specification of distinct neuronal subtypes. As in other vertebrates, LIM family members are expressed in a combinatorial manner in different spinal cord neuronal populations. For example, all primary motoneurons express *lhx3* (prev. *lim3*), while RoP and MiP express *isl1*, and CaP and VaP express *isl2* (Appel *et al.*, 1995; Inoue *et al.*, 1994; Korzh *et al.*, 1993; Tokumoto *et al.*, 1995). In addition, *isl2* is expressed in RB cells. Perturbation of Isl2 function leads to a partial transformation of CaP and VaP neurons into ventrolateral descending (VeLD) interneurons and impairs peripheral axon outgrowth from RB neurons (Segawa *et al.*, 2001). Recent results suggest that, in chicken, the attribution of motoneuron subtype identity by the LIM-homeodomain proteins Isl1 and Lhx3 involves synergistic activity of these factors and the bHLH proteins Ngn2 and NeuroM (Lee and Pfaff, 2003).

The bHLH transcription factor Olig2 is expressed in zebrafish spinal cord precursors of motoneurons and oligodendrocytes and is required for the specification of these two neuronal types (Park and Appel, 2003; Park *et al.*, 2002). Recent results in mammalian cells suggest that this function might involve the physical association of Olig2 with the homeodomain transcription factor Nkx2.2 (Sun *et al.*, 2003). Together with results on trunk motoneuron specification (as discussed earlier), these findings support the more general possibility that a functional linkage of bHLH factors and homeodomain proteins might be used to synchronize neurogenesis and neuronal subtype specification.

Development of zebrafish motoneurons and reticulospinal neurons of the hind-brain relies on a combinatorial code of Hox proteins and partners, in a manner that controls both neuronal identity and hindbrain patterning (Deflorian *et al.*, 2004; McClintock *et al.*, 2002; Popperl *et al.*, 2000; Waskiewicz *et al.*, 2002).

Finally, the specification of more anterior neuronal subtypes has been studied, in particular, through a genetic screen for mutations affecting catecholaminergic development (Guo *et al.*, 1999b). Several of these mutations have now been cloned. Development of hypothalamic dopaminergic and serotonergic neurons is controlled (non-autonomously) by the zinc finger protein Fezl (mutation *too few*), expressed in the forebrain (Levkowitz *et al.*, 2003), as well as by the dose of the transcription elongation factor Supth5 (prev. Spt5, mutation *foggy*) (Guo *et al.*, 2000; Keegan *et al.*, 2002). A role for the paired-box transcription factor Pax6a (prev. Pax6.1) in the development of diencephalic dopaminergic neurons is suggested by its selective expression in this neuronal type (Wullimann and Rink, 2001). Noradrenergic neurons of the locus coeruleus require, as in other vertebrates, the homeodomain protein Phox2a (mutation *soulless*) (Guo *et al.*, 1999a), as well as the transcriptional activator Tfap2a (mutation *lockjaw*) (Holzschuh *et al.*, 2003a; Knight *et al.*, 2003). How the expression of these different neuronal identity factors is regulated and encodes neuronal subtype remains to be investigated.

Great progress has been made on this aspect in the zebrafish retinal system, where the generation of distinct neuronal types in a strict temporal sequence from common precursors permits to directly address the combined roles of differentiation timing and positional information in cell type specification (Cepko *et al.*, 1996; Easter and Malicki, 2002; Hu and Easter, 1999; Malicki, 1999; Neumann, 2001). Several factors have been isolated that are necessary for the development of one retinal cell subtype. An instructive role for Notch signaling has been proposed for the specification of Müller glia, the latest born retinal cell type, probably by orienting ganglion cell precursors toward a glial fate (Scheer *et al.*, 2001). The bHLH factor Atoh7 (prev. Ath5, mutation *lakritz*) controls specification of the earliest born cells, ganglion cells, by permitting the first wave of retinal neurogenesis (Kay *et al.*, 2001). In the absence of Atoh7, progenitors fail to appropriately exit the cell cycle and participate to later neurogenesis waves that produce later cell types (Kay *et al.*, 2001). The molecular connection between the encoding of cellular diversity by Notch or Ath5 and cell cycle control remains a most interesting avenue to explore (Ohnuma *et al.*, 2001, 2002). In addition to temporal control, retinal differentiation is strictly controlled in space as it progresses across the retinal field from the ventronasal retina (Neumann, 2001). This differentiation wave would be initiated by Nodal signaling from anterior midline tissues (Masai *et al.*, 2000), and propagated by Hh signaling from the Nodal-induced optic stalk and first differentiated retinal neurons (Neumann and Nuesslein-Volhard, 2000; Stenkamp and Frey, 2003). A number of additional zebrafish retinal mutants deficient in selective subtypes are still to be cloned, holding promise for in-depth understanding of the molecular bases for retinal cell subtype encoding (Neuhauss, 2003; Neuhauss *et al.*, 1999).

VI. Aspects of Neuronal Differentiation

Concomitant to or after the withdrawal from mitosis and the specification to particular cell types by multipotent progenitors, key steps in the generation of functional neurons include their migration to the appropriate position, their molecular and morphological differentiation (with the establishment of characteristic projections), and their survival. These aspects encompass a vast body of recent zebrafish literature, which we briefly survey in the next sections.

A. Neuronal Migration

As in other vertebrates, neuronal migration responds to a number of general, as well as neuronal, subtype-specific cues. For instance, the positioning of most neurons is affected in the absence of neural adhesion molecules such as Cdh2 (prev. Ncad) (Erdmann et al., 2003; Lele et al., 2002; Malicki et al., 2003; Masai et al., 2003). The *oko meduzi* mutation causes comparable non-cell-autonomous retinal lamination defects (Malicki and Driever, 1999). Selective migration defects of the facial (nVII) neurons are observed in the absence of Pbx4 (mutation *lazarus*) function (Cooper et al., 2003) and nVII, together with glossopharyngeal (nIX) neurons, fail to migrate to the posterior hindbrain when the transmembrane protein Vangl2 (prev. Strabismus, mutation *trilobite*) or its genetic partner Prickle1 are absent (Bingham et al., 2002; Carreira-Barbosa et al., 2003; Jessen et al., 2002). The latter proteins are also involved in controlling cell convergence and extension at gastrulation, demonstrating their requirement in at least two spatio-temporally distinct types of directional movements during zebrafish development. Finally, primary and secondary trunk motoneurons exhibit migration defects in the absence of a functional Gli2 (mutation *you-too*) Hh target, a phenotype associated with, and possibly resulting from, aberrant axonal pathfinding (Zeller et al., 2002).

B. Axonal Pathfinding

Axons can be easily visualized after fluorescent labeling on live zebrafish specimen. Thus the zebrafish model is particularly amenable to study the establishment of axonal projections. A large number of mutations have been recovered and cloned that deepened our understanding of this process. These mostly concern retinotectal and motor nerves pathfinding (Baier et al., 1996; Beattie and Eisen, 1997; Beattie et al., 2000; Eisen and Pike, 1991; Granato et al., 1996; Karlstrom et al., 1996; Trowe et al., 1996) (reviewed in Beattie et al., 2002; Culverwell and Karlstrom, 2002; Hjorth and Key, 2002; Hutson and Chien, 2002a). The cues controlling axonal growth of other neuronal groups have been addressed in a candidate approach and remain more hypothetical (Bernhardt et al., 1992a,b; Hjorth and Key, 2001; Kanki and Kuwada, 2000; Kuwada et al., 1990a,b; Marx et al., 2001; Roth et al., 1999; Xiao et al., 2003).

Retinal ganglion cells in lower vertebrates project in a topographically ordered fashion to the contralateral optic tectum, and zebrafish mutations have been recovered that affect all choice points along this path (i.e., exit of the optic nerve from the eye, crossing of the ventral midline at the level of the optic stalk, projection toward posterior, and branching in the correct tectal area). These mutations can be related to patterning defects (e.g., midline defects in Hh pathway mutants) (Culverwell and Karlstrom, 2002). However, in several instances, these mutations are likely to affect attractive or repulsive receptor/ligand guidance systems more directly implicated in pathfinding, such as Robo/Slit (Robo2, mutation *astray*) (Fricke *et al.*, 2001; Hutson and Chien, 2002b), ephrin/Eph (in the mutants *no-isthmus* and *acerebellar*) (Picker *et al.*, 1999), Semaphorins (in *ace/fgf8*) (Picker *et al.*, 1999; see also Liu *et al.*, 2004), or Tenascin (Becker *et al.*, 2003a). A number of mutations affecting retinotectal projections remain to be cloned, holding promise for a deeper understanding of this process. Among them is the interesting gene *space cadet*, which also controls axonal pathfinding in a small population of hindbrain commissural neurons, the spiral fiber neurons, which synapse on the Mauthner axon and modulate fast turning movements (Lorent *et al.*, 2001).

Primary motoneurons all exit the spinal cord in a common pathway that reaches the muscle pioneer cells, after which they diverge to distinct targets. Secondary motoneurons later follow the same paths. A large body of evidence suggests that pathfinding of both motoneuron classes requires multiple myotome-derived cues, which can affect either motor axon extension or target choice (Zeller *et al.*, 2002). Transplantation experiments have identified a crucial cell population controlling this aspect, the muscle pioneers, although the factors involved remain hypothetical (Eisen and Melancon, 2001; Lewis and Eisen, 2004; Melancon *et al.*, 1997). Among identified molecules are chondroitin sulfates, integrins, and semaphorins (Becker *et al.*, 2003b; Bernhardt and Schachner, 2000; Halloran *et al.*, 2000; Roos *et al.*, 1999). In addition, several zebrafish mutations that affect motor axon growth of trajectory once these reach the periphery have been isolated. Both the *unplugged* and *diwanka* genes are necessary for groups of adaxial mesodermal cells to guide all three types of motor axons at the level of muscle pioneer cells (Zeller and Granato, 1999; Zhang and Granato, 2000; Zhang *et al.*, 2001). The path of CaP primary motoneurons is selectively affected in *stumpy*, both in a cell-autonomous and non-cell-autonomous manner (Beattie *et al.*, 2000). Molecular identification of these mutations will undoubtedly provide crucial information on the processes of axonal guidance as they contribute to the wiring of the zebrafish nervous system. Finally, the ubiquitously expressed Survival motor neuron (Smn) protein appears cell-autonomously necessary for motor axon growth and pathfinding. Its potential targets and mode of action remain unknown (McWhorter *et al.*, 2003).

C. Neuronal Survival

Programmed cell death is a general feature of normal animal development that, within the nervous system, is often associated with neurons failing to establish the right connections. Thus, the neurons do not receive proper trophic support or electrical activity. These processes can be enhanced in pathological conditions where given neuronal groups or environmental factors are missing.

Zebrafish RB neurons have provided a tractable model to address some processes underlying programmed neural cell death. RB neurons build up the sensory system of the late embryo and early larva, after which time they progressively degenerate and their function is taken over by dorsal root ganglion (DRG) neurons (Reyes *et al.*, 2004). However, RB death has not been functionally correlated with DRG formation. It matches with reduced Ntrk3a (prev. TrkC1)/ Nt3 signaling (Williams *et al.*, 2000) and sustained Na^+ current activity. The latter conclusion is particularly supported by analysis of zebrafish *macho* mutants, where RB cells lack voltage-gated Na^+ current. The RB cells thus cannot fire action potentials in response to tactile stimulation and they survive longer than normal (Svoboda *et al.*, 2001). In addition, RB neurons degenerate prematurely in embryos lacking acetylcholinesterase activity and where the level of transsynaptic acetylcholine is increased (*ache* mutants) (Behra *et al.*, 2002). Together, these results suggest that RB death is promoted by a combination of mechanisms involving trophic and activity parameters.

A number of zebrafish mutations have also been recovered that selectively display cell death in the CNS. These most often affect the optic tectum and retina, but can be as widespread as including the entire brain and spinal cord (Abdelilah *et al.*, 1996; Furutani-Seiki *et al.*, 1996; Golling *et al.*, 2002; Matsuda and Mishina, 2004; Rodriguez and Driever, 1997). Selective retinal degeneration processes have been characterized in more detail in some of these (Bahadori *et al.*, 2003; Doerre and Malicki, 2001; Goldsmith *et al.*, 2003; Kainz *et al.*, 2003). (See chapters 16 through 18 in this volume for extensive reviews.) Selective retinal degeneration processes point to a crucial role of cell-cell interactions in the survival of most retinal cell types (Doerre and Malicki, 2001; Goldsmith *et al.*, 2003). To date, one mutation causing apoptosis of most photoreceptors, *not really finished*, has been molecularly identified (Becker *er al.*, 1998b). It encodes a transcription factor showing high similarity to human nuclear factor erythroid-related factor 1 (NRF1) and is widely expressed within the developing zebrafish retina (Becker *et al.*, 1998b). Its downstream regulatory targets, as well as its cell-autonomous or non-cell-autonomous requirements, remain to be determined. In-depth analysis of zebrafish retinal mutations are likely to provide important information relevant to human retinal pathologies (Goldsmith and Harris, 2003; Malicki *et al.*, 2002; Neuhauss, 2003).

VII. Useful Tools for the Study of Zebrafish Neurogenesis

Tables I to IV provide information on the study of zebrafish neurogenesis. Table I presents information on zebrafish markers and tools you can use to assess proliferation status and neuronal differentiation stages. Table II discusses zebrafish markers and tools you can use to assess neuronal identity. Table III presents zebrafish glial markers. Table IV provides information about zebrafish mutants of lateral inhibition or neuronal identity.

Table I

Zebrafish Markers and Tools to Assess Proliferation Status and Neuronal Differentiation Stages

Name	Ab/RNA	Type of cells labeled	Reference	Transgenic line
Proliferation status				
Phosphohistone 3	Ab	Cells in late G2 and M-phase	Wei and Allis, 1998	
PCNA	Ab	All proliferating cells (the antigen remains present for about 24 h)	Wullimann and Knipp, 2000	
Histone H1	Ab	Cells in G1 phase	Huang and Sato, 1998; Tarnowka *et al.*, 1978	
BrdU	mAb	Cells in S-Phase after BrdU treatment		
Histone H2A		Visualization of mitotic figures		*H2A.F/Z:gfp* (Pauls *et al.*, 2001)
Precursors cells				
notch1a	RNA	Proliferative cells during primary and secondary neurogenesis	Mueller and Wullimann, 2002a	*UAS:notch1a^{ac}* (Scheer *et al.*, 2001)
dlA	RNA	Neuronal precursors	Haddon, 1998; Mueller and Wullimann, 2002a	
dlD	RNA	Neuronal precursors	Dornseifer *et al.*, 1997; Haddon, 1998	*deltaD:gfp* (Hans and Campos-Ortega, 2002)
Precursors and differentiating postmitotic neurons				
dlB	RNA	Subpopulation of *deltaA* and *deltaD* expressing cells: singled-out primary neurons	Haddon, 1998	
onecut1	RNA	Neuronal precursors, early differentiating neurons, except in the telencephalon	Hong *et al.*, 2002	
coe2		Neuronal precursors, early postmitotic neurons	Bally-Cuif *et al.*, 1998	
neurog1 (ngn1)	RNA	Neuronal precursors, early postmitotic neurons	Blader *et al.*, 1997; Mueller and Wullimann, 2002a	Several *neurog1:gfp* lines (Blader *et al.*, 2003)
neurod4 (zath3)	RNA	Neuronal precursors, early postmitotic neurons	Park *et al.*, 2003; Wang *et al.*, 2003	
ash1a	RNA	Subpopulation of neuronal precursors	Allende and Weinberg, 1994	
ash1b	RNA	Subpopulation of neuronal precursors	Allende and Weinberg, 1994	
pou3 (tai-ji)	RNA	Precursor cells	Huang and Sato, 1998	
pou5f1 (pou2)	RNA	Subpopulation of precursors	Hauptmann and Gerster, 1996	
sox19	RNA	Subpopulation of precursors	Vriz *et al.*, 1996	
sox31	RNA	Subpopulation of precursors	Girard *et al.*, 2001	
sox11a	RNA	Subpopulation of precursors	de Martino *et al.*, 2000	

(*continues*)

Table I *continued*

Name	Ab/RNA	Type of cells labeled	Reference	Transgenic line
sox11b	RNA	Subpopulation of precursors	de Martino *et al.*, 2000	
sox21	RNA	Subpopulation of precursors	Rimini *et al.*, 1999	
neurod2	RNA	Differentiating neurons	Korzh *et al.*, 1998; Mueller and Wullimann, 2002b	
dcc	RNA	First neuronal clusters	Hjorth *et al.*, 2001	
β-thymosin	RNA	Early differentiating neurons	Roth *et al.*, 1999	
Polysialic acid (PSA)	Ab	Differentiating neurons, expression on cell bodies	Marx *et al.*, 2001	
cntn2 (tag1)	RNA	Outgrowing and migrating neurons	Warren *et al.*, 1999	
L2/HNK1	mAb Zn12	Outgrowing neurons	Metcalfe *et al.*, 1990	
elavl3 (HuC)	RNA	Early differentiating and mature neurons. Start at 1 somite	Kim *et al.*, 1996; Mueller and Wullimann, 2002a; Park *et al.*, 2000a	*huC:gfp* (Park *et al.*, 2000b)
elavl4 (HuD)	RNA	Subset of postmitotic neurons. Start at 10 somites	Park *et al.*, 2000a	
tuba1 (α1-Tubulin)	RNA	Early differentiating and regenerating neurons	Hieber *et al.*, 1998	*α1-Tubulin:egfp* (Goldman *et al.*, 2001)
Elavl3 + 4 (HuC+D)	mAb 16A11	Early postmitotic and mature neurons	Mueller and Wullimann, 2002a	
β1-Tubulin	RNA	Early differentiating, start at 24 hpf	Oehlmann *et al.*, 2004	
gap43	RNA	Postmitotic neurons in the phase of axonal growth, and regenerating neurons start at 17 hpf	Reinhard *et al.*, 1994	(rat) *gap43:gfp* (Udvadia *et al.*, 2001)
Mature neurons				
α2-Tubulin	RNA	Mature neurons	Hieber *et al.*, 1998	
Acetylated Tubulin	Ab	Membrane staining of all differentiated neurons		
Neurofilament	Ab RMO-44	Mature neurons, reticulospinal neurons (bodies + axons)	Gray *et al.*, 2001; Lee *et al.*, 1987	
Gefiltin (intermediate filament)			Leake *et al.*, 1999	
plasticin (intermediate filament)	Ab RNA	Subset of neurons extending axons	Canger *et al.*, 1998	
nadl1.1 (L1.1)	RNA	Reticulospinal neurons during axonogenesis	Becker *et al.*, 1998a	
E587 (L1 related)	mAb E17	Axons of primary tracts and commissures. start at 17 hpf	Weiland *et al.*, 1997	
	CON1	Subset of axons	Bernhardt *et al.*, 1990	
	zn-1	All neurons	Trewarrow *et al.*, 1990	

Table II
Zebrafish Markers and Tools to Assess Neuronal Identity

Identity	Marker	Ab/RNA	Reference	Transgenic line
GABAergic neurons				
		Antibodies	Extensive list in Marc and Cameron, 2001; Yazulla and Studholme, 2001	
	gad2 (gad65)	RNA	Martin et al., 1998	
	gad1 (gad67)	RNA	Martin et al., 1998	
Catecholaminergic neurons (dopamine, noradrenalin, adrenalin)				
Locus Coeruleus	phox2a	RNA	Guo et al., 1999a	
	TH (tyrosyne hydroxylase)	RNA Ab (Chemicon)	Guo et al., 1999b	
Dopaminergic	Uch-L1	RNA	Son et al., 2003	
	slc6a3 (dat)	RNA	Holzschuh et al., 2001	
Noradrenergic and adrenergic neurons	dbh (dopamine β hydroxylase)	RNA	Guo et al., 1999b	
Serotonergic neurons				
	5HT (serotonin)	Ab (Chemicon)	Bellipanni et al., 2002	
	tph (tphD1)	RNA	Bellipanni et al., 2002	
	tph2 (tphD2)	RNA	Teraoka et al., 2004	
	tphR	RNA		
Glutamatergic neurons				
		Antibodies	Extensive list in Marc and Cameron, 2001; Yazulla and Studholme, 2001	
Cholinergic neurons				
		Antibodies	Extensive list in Marc and Cameron, 2001; Yazulla and Studholme, 2001	
Glycinergic neurons				
	glra1, glra4a, glrb (glyR subunits)	RNA	Imboden et al., 2001	
Motoneurons				
Primary motoneurons, early processes	znp-1	mAb	Melancon et al., 1997; Trewarrow et al., 1990	

Cell/neuron type	Marker	Method	Reference	Transgenic line
Secondary motoneurons during axonal growth	Alcam (DM-GRASP/ neurolin)	Ab zn-5	Fashena and Westerfield, 1999; Ott et al., 2001	
Primary (RoP, MiP, VaP, CaP) and secondary motoneurons	lhx3 (lim3)	RNA, pAb	Appel et al., 1995; Glasgow et al., 1997	
MiP + RoP, secondary motoneurons and cranial motoneurons	islet1	RNA Ab(DSHB, 40.2D6 and 39.5D5)	Inoue et al., 1994; Korzh et al., 1993; Segawa et al., 2001	islet1:gfp (Higashijima et al., 2000) (only for cranial motoneurons)
CaP and VaP	islet2	RNA	Appel et al., 1995	
Primary motoneurons	pnx	RNA	Bae et al., 2003	
Primary motoneurons	olig2	RNA	Park et al., 2002	olig2.egfp (Shin et al., 2003)
Cranial motoneurons	tbx20	RNA	Ahn et al., 2000	
Occulomotor and trochlear motoneurons	phox2a	RNA	Guo et al., 1999a	
Spinal motoneurons (transiently) + vagal motor nucleus(nX)	sst1 (ppss1) (somatostatin)	RNA	Devos et al., 2002	
	cxcr4b	RNA	Chong et al., 2001	
	appa (app)	RNA	Musa et al., 2001	
Reticulospinal interneurons				
	tlx3a (tlxA)	RNA	Andermann and Weinberg, 2001	
Mauthner neuron	3A10	Ab	Hatta, 1992	
Mauthner neuron	RMO-44	mAb	Gray et al., 2001; Lee et al., 1987	
Mauthner neuron	glra1, glra4a, glrb (glyR subunits)	RNA	Imboden et al., 2001	
Spinal cord commissural interneurons				
	pax2a (pax2.1)	RNA, Ab	Mikkola et al., 1992	pax2.1:gfp (Picker et al., 2002)
Interneurons of the spinal cord				
Small number of interneurons	pnx	RNA	Bae et al., 2003	
	dbx1a (hlx1)	RNA	Fjose et al., 1994	
	cxcr4a	RNA	Chong et al., 2001	
	islet1	RNA, Ab	Korzh et al., 1993	

(continues)

Table II *continued*

Identity	Marker	Ab/RNA	Reference	Transgenic line
Rohon-Beard primary sensory neurons				
	pnx	RNA	Bae et al., 2003	
	cxcr4b	RNA	Chong et al., 2001	
	islet1	RNA, Ab	Korzh et al., 1993	
	islet2	RNA	Segawa et al., 2001	
	tlx3a (tlxA)	RNA	Andermann and Weinberg, 2001; Langenau et al., 2002	
	tlx3b	RNA	Langenau et al., 2002	
	cbfb	RNA	Blake et al., 2000	
	runx3	RNA	Kalev-Zylinska et al., 2003	
Subpopulation of RB	ntrk3a (trkC)	RNA	Williams et al., 2000	
Cerebellar granule cells	atoh1 (ath1)	RNA	Koster and Fraser, 2001	
Purkinje cells	zebrin2	mAb	Jaszai et al., 2003	
	M1	mAb	Miyamura and Nakayasu, 2001	
Retinal cells	See extensive reviews	Antibodies	Marc and Cameron, 2001; Yazulla and Studholme, 2001	
Amacrine cells	pax6a (pax6.1)	RNA	Kay et al., 2001	pax6:gfp (Kay et al., 2001)
Ganglion cells (RGC)	atoh7 (ath5)	RNA	Kay et al., 2001; Masai et al., 2000;	
RGC and INL	islet 1	RNA, Ab	Korzh et al., 1993; Masai et al., 2000	
	islet3		Kikuchi, 1997	
RGC and INL	lhx3 (lim3)	RNA, Ab	Glasgow et al., 1997; Masai et al., 2000	
RGC	cxcr4b	RNA	Chong et al., 2001	
RGC	dacha	RNA	Hammond et al., 2002	
RGC	zn-5	mAb	Kawahara et al., 2002	
Epiphyseal neurons	Serotonin-N-acetyltransferase-2 (aanat2)	RNA	Gothilf et al., 2002	zfAANAT-2:gfp (Gothilf et al., 2002)

	islet1	RNA, Ab	Korzh et al., 1993
	lhx3 (lim3)	RNA, pAb	Glasgow et al., 1997
	tph (tphD1)	RNA	Bellipanni et al., 2002
	cxcrb	RNA	Chong et al., 2001
Pituitary gland neurons			
	tbx20	RNA	Ahn et al., 2000
	lhx3 (lim3)	RNA, pAb	Glasgow et al., 1997
Hypothalamic clusters			
	Histamine	Ab	Kaslin and Panula, 2001
	neurog3 (ngn2, ngn3), expression start at 24 hpf	RNA	Wang et al., 2003
	cxcrb	RNA	Chong et al., 2001
	sst1 (ppss1), start at 5 dpf	RNA	Devos et al., 2002
	tph (tphD1)	RNA	Bellipanni et al., 2002
	dacha	RNA	Hammond et al., 2002

Table III
Zebrafish Glial Markers

Identity	Marker	Ab/RNA	Reference
Radial glial cells			
	zrf-1	mAb	Trewarrow *et al.*, 1990
	zrf-3	mAb	Trewarrow *et al.*, 1990
	gfap GFAP	RNA, Ab	Marcus and Easter, 1995; Nielsen and Jorgensen, 2003
	FABP (fatty acid binding protein)/ BLBP	RNA, Ab	Denovan-Wright *et al.*, 2000; Hartfuss *et al.*, 2001; Liu *et al.*, 2003; Our own observations
	RC2	mAb	Hartfuss *et al.*, 2001; Our own observations
	zrf-1 = zns-2	mAb	Trewarrow *et al.*, 1990; Wullimann and Rink, 2002
	sox11a	RNA	de Martino *et al.*, 2000
	sox11b	RNA	de Martino *et al.*, 2000
	C-4 subset of gfap-expressing cells, tested in adult	mAb	Tomizawa *et al.*, 2000b
Mueller glial cells	HNK1	mAb zn12	Peterson *et al.*, 2001
Mueller glial cells	Glutamine synthetase	mAb	Kay *et al.*, 2001; Peterson *et al.*, 2001
Mueller glial cells	Carbonic anhydrase		Peterson *et al.*, 2001
Oligodendrocytes			
	olig2	RNA	Park *et al.*, 2002
	mpz (myelin protein zero)	RNA	Brosamle and Halpern, 2002
	plp1a (DM20)	RNA	Brosamle and Halpern, 2002
	myelin basic protein (MBP)	RNA, Ab	Brosamle and Halpern, 2002; Tomizawa *et al.*, 2000a
	34 kDa protein band	mAb	Tomizawa *et al.*, 2000a
	A-20	mAb	Arata and Nakayasu, 2003
Astrocytes			
Reticular astrocytes (restricted to the optic nerve)	cytokeratin	Ab	Macdonald *et al.*, 1997; Maggs and Scholes, 1990
Reticular astrocytes (restricted to the optic nerve)	*pax2a (pax2.1)*	RNA Ab	Macdonald *et al.*, 1997; Mikkola *et al.*, 1992
Subpopulation of astrocytes (tested only in adults)	A-22	mAb	Kawai *et al.*, 2001

Table IV
Zebrafish Mutants of Lateral Inhibition or Neuronal Identity

Name	Abbreviation	Gene mutated	Phenotype	Reference
Neurogenesis mutants				
narrowminded	nrd		Complete loss of Rohon-Beard neurons + reduction of early neural crest cells	Artinger et al., 1999
mind bomb (or white tail)	mib (wit)	Ubiquitin ligase (required for notch signaling)	Neurons produced in excessive number within the neural plate. Increase in primary neurons and reduction of secondary motor neurons and eye and hindbrain radial glial cells	Haddon, 1998; Itoh et al., 2003; Jiang et al., 1996; Schier et al., 1996
after eight	aei	dlD	Increase in primary sensory neurons	Holley et al., 2000
deadly seven	des	notch1a	Increase in reticulospinal neurons and slight increase in motoneurons	Gray et al., 2001; Liu et al., 2003
heart and soul	has	prkci (aPKCλ)	Defects in cell polarity leading to failure of brain ventricles to inflate, defects in retina (+ other defects in endodermal organs)	Horne-Badovinac et al., 2001; Schier et al., 1996
dlA^{dx2}	dx2	dlA (hypermorph)	Excess of RB neurons	Appel et al., 1999
Neuronal specification mutants				
motionless	mot		Less dopamiergic hypothalamic neurons, lacking brain ventricles, cell death in telencephalon and lens by 50 hpf.	Guo et al., 1999b
foggy	fog	supt5h (spt5)	Lack of dopaminergic neurons in hypothalamus, telencephalon and retina, and lack of noradrenergic neurons in the locus Coeruleus (+ cardiovascular defects)	Guo et al., 1999b; Guo et al., 2000
too few	tof	fezl	Reduction of hypothalamic dopaminergic neurons	Guo et al., 1999b
soulless	sll	phox2a	Loss of locus coeruleus and arch associated catecholaminergic neurons	Guo et al., 1999a
lakritz	lak	atoh7 (ath5)	Loss of retinal ganglion cells	Kay et al., 2001
young	yng	smarca4	Blocked final differentiation of retinal cells	Gregg et al., 2003; Link et al., 2000
perplexed	plx		Cell death of retinal cells before exiting the cell cycle	Link et al., 2001
confused	cfs		Cell death in a subset of retinal postmitotic cells	Link et al., 2001

VIII. Conclusion

Our understanding of zebrafish early neurogenesis has made enormous progress with the unraveling of several patterning mechanisms, a large number of proneural factors, and some of their interactions. Interestingly, however, in most cases, expression of different proneural factors are complementary or only partially overlapping rather than identical. This suggests that proneural factors might exert shared, as well as specific, functions, the latter being achieved either by each factor alone or as a combinatorial "code." Such functional subdivision may explain the maintenance of related proneural factors during zebrafish evolution. It will now be most important to analyze these factors in more detail and understand their exact contribution to generic, as well as cell- or stage-specific, aspects of zebrafish neurogenesis.

Our understanding of the zebrafish embryonic and larval brain also remains relatively fragmentary. Comparably fewer molecular and biochemical tools are available for zebrafish than those which are available for other vertebrate models to characterize the different brain neuronal types. Their networks of axonal connections remain largely unknown. It is our hope that future selective green fluorescent protein (GFP) transgenic lines will permit us to extend detailed projection tracing to the unwiring of most brain circuits, as well as to isolate and molecularly characterize most neuronal types.

Acknowledgments

We are grateful to Jovica Ninkovic for Fig. 3A, and to all lab members for regular discussions. Many thanks to Drs. J. A. Campos-Ortega, M. Hammerschmidt, and U. Straehle for their input on some aspects of this review. Work in Laure Bally-Cuif's laboratory, Laure Bally-Cuif, and Prisca Chapouton are funded by the VolkswagenStiftung, with additional funds from the DFG (BA2024/1-2) and the European Commissions as part of the *ZF-MODELS* Integrated Project in the 6th Framework Programme (Contract No. LSHG-CT-2003-503496).

References

Abdelilah, S., Mountcastle-Shah, E., Harvey, M., Solnica-Krezel, L., Schier, A. F., Stemple, D. L., Malicki, J., Neuhauss, S. C., Zwartkruis, F., Stainier, D. Y., *et al.* (1996). Mutations affecting neural survival in the zebrafish Danio rerio. *Development* **123,** 217–227.

Ahn, D., Ruvinsky, I., Oates, A., Silver, L., and Ho, R. (2000). tbx20 a new vertebrate T-box gene expressed in the cranial motor neurons and developing cardiovascular structures in zebrafish. *Mech. Dev.* **95,** 253–258.

Akimenko, M. A., Ekker, M., Wegner, J., Lin, W., and Westerfield, M. (1994). Combinatorial expression of three zebrafish genes related to distal-less: Part of a homeobox gene code for the head. *J. Neurosci.* **14,** 3475–3486.

Allende, M. L., and Weinberg, E. S. (1994). The expression pattern of two zebrafish achaete-scute homolog (ash) genes is altered in the embryonic brain of the cyclops mutant. *Dev. Biol.* **166,** 509–530.

Alvarez-Buylla, A., and Garcia-Verdugo, M. (2002). Neurogenesis in adult subventricular zone. *J. Neurosci.* **22,** 629–634.

Andermann, P., and Weinberg, E. (2001). Expression of zTlxA, a Hox11-like gene, in early differentiating embryonic neurons and cranial sensory ganglia of the zebrafish embryo. *Dev. Dynam.* **222**, 595–610.

Andreazzoli, M., Gestri, G., Cremisi, F., Casarosa, S., Dawid, I. B., and Barsacchi, G. (2003). Xrx1 controls proliferation and neurogenesis in Xenopus anterior neural plate. *Development* **130**, 5143–5154.

Appel, B., and Chitnis, A. (2002). Neurogenesis and specification of neuronal identity. *Results Probl. Cell Differ.* **40**, 237–251.

Appel, B., and Eisen, J. S. (1998). Regulation of neuronal specification in the zebrafish spinal cord by Delta function. *Development* **125**, 371–380.

Appel, B., Fritz, A., Westerfield, M., Grunwald, D. J., Eisen, J. S., and Riley, B. B. (1999). Delta-mediated specification of midline cell fates in zebrafish embryos. *Curr. Biol.* **9**, 247–256.

Appel, B., Givan, L. A., and Eisen, J. S. (2001). Delta-Notch signaling and lateral inhibition in zebrafish spinal cord development. *BMC Dev. Biol.* **1**, 13.

Appel, B., Korzh, V., Glasgow, E., Thor, S., Edlund, T., Dawid, I. B., and Eisen, J. S. (1995). Motoneuron fate specification revealed by patterned LIM homeobox gene expression in embryonic zebrafish. *Development* **121**, 4117–4125.

Arata, N., and Nakayasu, H. (2003). A periaxonal net in the zebrafish central nervous system. *Brain. Res. Dev. Brain Res.* **961**, 179–189.

Artinger, K., Chitnis, A., Mercola, M., and Driever, W. (1999). Zebrafish *narrowminded* suggests a genetic link between formation of neural crest and primary sensory neurons. *Development* **126**, 3969–3979.

Bae, Y. K., Shimizu, T., Yabe, T., Kim, C. H., Hirata, T., Nojima, H., Muraoka, O., Hirano, T., and Hibi, M. (2003). A homeobox gene, pnx, is involved in the formation of posterior neurons in zebrafish. *Development* **130**, 1853–1865.

Bahadori, R., Huber, M., Rinner, O., Seeliger, M. W., Geiger-Rudolph, S., Geisler, R., and Neuhauss, S. C. (2003). Retinal function and morphology in two zebrafish models of oculo-renal syndromes. *Eur. J. Neurosci.* **18**, 1377–1386.

Baier, H., Klostermann, S., Trowe, T., Karlstrom, R. O., Nusslein-Volhard, C., and Bonhoeffer, F. (1996). Genetic dissection of the retinotectal projection. *Development* **123**, 415–425.

Bainter, J., Boos, A., and Kroll, K. L. (2001). Neural Induction takes a transcriptional twist. *Dev. Dyn.* **222**, 315–327.

Bally-Cuif, L., Dubois, L., and Vincent, A. (1998). Molecular cloning of Zcoe2, the zebrafish homolog of Xenopus Xcoe2 and mouse EBF-2, and its expression during primary neurogenesis. *Mech. Dev.* **77**, 85–90.

Bally-Cuif, L., and Hammerschmidt, M. (2003). Induction and patterning of neuronal development, and its connection to cell cycle control. *Curr. Opin. Neurobiol.* **13**, 16–25.

Barth, K., Kishimoto, Y., Rohr, K., Seydler, C., Schulte-Merker, S., and Wilson, S. (1999). Bmp activity establishes a gradient of positional information throughout the entire neural plate. *Development* **126**, 4977–4987.

Beattie, C. E., and Eisen, J. S. (1997). Notochord alters the permissiveness of myotome for pathfinding by an identified motoneuron in embryonic zebrafish. *Development* **124**, 713–720.

Beattie, C. E., Granato, M., and Kuwada, J. Y. (2002). Cellular, genetic and molecular mechanisms of axonal guidance in the zebrafish. *Results Probl. Cell Differ.* **40**, 252–269.

Beattie, C. E., Melancon, E., and Eisen, J. S. (2000). Mutations in the stumpy gene reveal intermediate targets for zebrafish motor axons. *Development* **127**, 2653–2662.

Becker, C., Schweitzer, J., Feldner, J., Becker, T., and Schachner, M. (2003a). Tenascin-R as a repellant guidance molecule for developing optic axons in zebrafish. *J. Neurosci.* **23**, 6232–6237.

Becker, T., Bernhardt, R. R., Reinhard, E., Wullimann, M. F., Tongiorgi, E., and Schachner, M. (1998a). Readiness of zebrafish brain neurons to regenerate a spinal axon correlates with differential expression of specific cell recognition molecules. *J. Neurosci.* **18**, 5789–5803.

Becker, T. S., Burgess, S. M., Amsterdam, A. H., Allende, M. L., and Hopkins, N. (1998b). Not really finished is crucial for development of the zebrafish outer retina and encodes a transcription factor highly homologous to human Nuclear Respiratory Factor-1 and avian Initiation Binding Repressor. *Development* **125**, 4369–4378.

Becker, T., McLane, M., and Becker, C. (2003b). Integrin antagonists affect growth and pathfinding of ventral motor nerves in the trunk of embryonic zebrafish. *Mol. Cell Neurosci.* **23**, 54–68.

Behra, M., Cousin, X., Bertrand, C., Vonesch, J. L., Biellmann, D., Chatonnet, A., and Strahle, U. (2002). Acetylcholinesterase is required for neuronal and muscular development in the zebrafish embryo. *Nat. Neurosci.* **5**, 111–118.

Bellefroid, E. J., Kobbe, A., Gruss, P., Pieler, T., Gurdon, J. B., and Papalopulu, N. (1998). Xiro3 encodes a Xenopus homolog of the Drosophila Iroquois genes and functions in neural specification. *EMBO J.* **17**, 191–203.

Bellipanni, G., Rink, E., and Bally-Cuif, L. (2002). Cloning of two tryptophane hydroxylase genes expressed in the diencephalon of the developing zebrafish brain. *Mech. Dev.* **119S**, S215–S220.

Belting, H. G., Hauptmann, G., Meyer, D., Abdelilah-Seyfried, S., Chitnis, A., Eschbach, C., Soll, I., Thisse, C., Thisse, B., Artinger, K. B., *et al.* (2001). Spiel ohne grenzen/pou2 is required during establishment of the zebrafish midbrain-hindbrain boundary organizer. *Development* **128**, 4165–4176.

Bernhardt, R., and Schachner, M. (2000). Chondroitin sulfates affect the formation of the segmental motor nerves in zebrafish embryos. *Dev. Biol.* **221**, 206–219.

Bernhardt, R. R., Chitnis, A. B., Lindamer, L., and Kuwada, J. Y. (1990). Identification of spinal neurons in the embryonic and larval zebrafish. *J. Comp. Neurol.* **302**, 603–616.

Bernhardt, R. R., Nguyen, N., and Kuwada, J. Y. (1992a). Growth cone guidance by floor plate cells in the spinal cord of zebrafish embryos. *Neuron* **8**, 869–882.

Bernhardt, R. R., Patel, C. K., Wilson, S. W., and Kuwada, J. Y. (1992b). Axonal trajectories and distribution of GABAergic spinal neurons in wild type and mutant zebrafish lacking floor plate cells. *J. Comp. Neurol.* **326**, 263–272.

Bertrand, V., Hudson, C., Caillol, D., Popovici, C., and Lemaire, P. (2003). Neural tissue in ascidian embryos is induced by FGF9/16/20, acting via a combination of maternal GATA and Ets transcription factors. *Cell* **115**, 615–627.

Bingham, S., Higashijima, S., Okamoto, H., and Chandrasekhar, A. (2002). The zebrafish trilobite gene is essential for tangential migration of branchiomotor neurons. *Dev. Biol.* **242**, 149–160.

Bingham, S., Nasevicius, A., Ekker, S. C., and Chandrasekhar, A. (2001). Sonic hedgehog and tiggy-winkle hedgehog cooperatively induce zebrafish branchiomotor neurons. *Genesis* **30**, 170–174.

Blader, P., Fischer, N., Gradwohl, G., Guillemont, F., and Strahle, U. (1997). The activity of neurogenin1 is controlled by local cues in the zebrafish embryo. *Development* **124**, 4557–4569.

Blader, P., Plessy, C., and Strahle, U. (2003). Multiple regulatory elements with spatially and temporally distinct activities control neurogenin1 expression in primary neurons of the zebrafish embryo. *Mech. Dev.* **120**, 211–218.

Blake, T., Adya, N., Kim, C. H., Oates, A. C., Zon, L., Chitnis, A., Weinstein, B. M., and Liu, P. P. (2000). Zebrafish homolog of the leukemia gene CBFB: Its expression during embryogenesis and its relationship to scl and gata-1 in hematopoiesis. *Blood* **96**, 4178–4184.

Bray, S., and Furriols, M. (2001). Notch pathway: Making sense of Suppressor of Hairless. *Curr. Biol.* **11**, R217–R221.

Brewster, R., Lee, J., and Ruiz i Altaba, A. (1998). Gli/Zic factors pattern the neural plate by defining domains of cell differentiation. *Nature* **393**, 579–583.

Brosamle, C., and Halpern, M. (2002). Characterization of myelination in the developing zebrafish. *Glia* **39**, 47–57.

Campos-Ortega, J. A. (1993). Mechanisms of early neurogenesis in *Drosophila melanogaster*. *J. Neurobiol.* **10**, 1305–1327.

Canger, A., Passini, M., Asch, W., Leake, D., Zafonte, B., Glasgow, E., and Schechter, N. (1998). Restricted expression of the neuronal intermediate filament protein plasticin during zebrafish development. *J. Comp. Neurol.* **399**, 561–572.

Carreira-Barbosa, F., Concha, M., Takeuchi, M., Ueno, N., Wilson, S., and Tada, M. (2003). Prickle 1 regulates cell movements during gastrulation and neuronal migration in zebrafish. *Development* **130**, 4037–4046.

Cau, E., and Wilson, S. W. (2003). Ash1a and Neurogenin1 function downstream of Floating head to regulate epiphysial neurogenesis. *Development* **130**, 2455–2466.

Cepko, C. L., Austin, C. P., Yang, X., Alexiades, M., and Ezzeddine, D. (1996). Cell fate determination in the vertebrate retina. *Proc. Natl. Acad. Sci. USA* **93**, 589–595.

Chandrasekhar, A., Schauerte, H. E., Haffter, P., and Kuwada, J. Y. (1999). The zebrafish detour gene is essential for cranial but not spinal motor neuron induction. *Development* **126**, 2727–2737.

Chandrasekhar, A., Warren, J. T., Jr., Takahashi, K., Schauerte, H. E., van Eeden, F. J., Haffter, P., and Kuwada, J. Y. (1998). Role of sonic hedgehog in branchiomotor neuron induction in zebrafish. *Mech. Dev.* **76**, 101–115.

Chitnis, A., and Kuwada, J. Y. (1990). Axonogenesis in the brain of zebrafish embryos. *J. Neurosci.* **10**, 1892–1905.

Chong, S., Emelyanov, A., Gong, Z., and Korzh, V. (2001). Expression pattern of two zebrafish genes, cxcr4a and cxcr4b. *Mech. Dev.* **109**, 347–354.

Cooper, K., Leisenring, W., and Moens, C. (2003). Autonomous and nonautonomous functions for Hox/Pbx in branchiomotor neuron development. *Dev. Biol.* **253**, 200–213.

Cornell, R. A., and Eisen, J. S. (2000). Delta signaling mediates segregation of neural crest and spinal sensory neurons from zebrafish lateral neural plate. *Development* **127**, 2873–2882.

Cornell, R. A., and Eisen, J. S. (2002). Delta/Notch signaling promotes formation of zebrafish neural crest by repressing Neurogenin1 function. *Development* **129**, 2639–2648.

Culverwell, J., and Karlstrom, R. O. (2002). Making the connection: Retinal axon guidance in the zebrafish. *Semin. Cell Dev. Biol.* **13**, 497–506.

de Martino, S., Yan, Y. L., Jowett, T., Postlethwait, J. H., Varga, Z. M., Ashworth, A., and Austin, C. A. (2000). Expression of sox11 gene duplicates in zebrafish suggests the reciprocal loss of ancestral gene expression patterns in development. *Dev. Dyn.* **217**, 279–292.

Deblandre, G. A., Lai, E. C., and Kintner, C. (2001). *Xenopus* neuralized is a ubiquitin ligase that interacts with XDelta1 and regulates Notch signaling. *Dev. Cell* **1**, 795–806.

Deflorian, G., Tiso, N., Ferretti, E., Meyer, D., Blasi, F., Bortolussi, M., and Argenton, F. (2004). Prep1.1 has essential genetic functions in hindbrain development and cranial neural crest cell differentiation. *Development* **131**, 613–627.

Denovan-Wright, E., Pierce, M., and Wright, J. (2000). Nucleotide sequence of cDNA clones coding for a brain-type fatty acid binding protein and its tissue-specific expression in adult zebrafish (Danio rerio). *BBA-Gene Structure and Expression* **1492**, 221–226.

Devos, N., Deflorian, G., Biemar, F., Bortolussi, M., Martial, J., Peers, B., and Argenton, F. (2002). Differential expression of two somatostatin genes during zebrafish embryonic development. *Mech. Dev.* **115**, 133–137.

Doerre, G., and Malicki, J. (2001). A mutation of early photoreceptor development, mikre oko, reveals cell–cell interactions involved in the survival and differentiation of zebrafish photoreceptors. *J. Neurosci.* **21**, 6745–6757.

Dornseifer, P., Takke, C., and Campos-Ortega, J. A. (1997). Overexpression of a zebrafish homologue of the Drosophila neurogenic gene Delta perturbs differentiation of primary neurons and somite development. *Mech. Dev.* **63**, 159–171.

Dorsky, R. I., Moon, R. T., and Raible, D. W. (1998). Control of neural crest cell fate by the Wnt signalling pathway. *Nature* **396**, 370–373.

Dutton, K. A., Pauliny, A., Lopes, S. S., Elworthy, S., Carney, T. J., Rauch, J., Geisler, R., Haffter, P., and Kelsh, R. N. (2001). Zebrafish colourless encodes sox10 and specifies non-ectomesenchymal neural crest fates. *Development* **128**, 4113–4125.

Easter, S. S., Jr., Burrill, J., Marcus, R. C., Ross, L., Taylor, J. S. H., and Wilson, S. W. (1994). Initial tract formation in the vertebrate brain. *Prog. Brain Res.* **102**, 79–93.

Easter, S. S., Jr., and Malicki, J. J. (2002). The zebrafish eye: Developmental and genetic analysis. *Results Probl. Cell Differ.* **40**, 346–370.

Eisen, J. S., and Melancon, E. (2001). Interactions with identified muscle cells break motoneuron equivalence in embryonic zebrafish. *Nat. Neurosci.* **4**, 1065–1070.

Eisen, J. S., and Pike, S. H. (1991). The spt-1 mutation alters segmental arrangement and axonal development of identified neurons in the spinal cord of the embryonic zebrafish. *Neuron* **6**, 767–776.

Eisen, J. S., Pike, S. H., and Romancier, B. (1990). An identified motoneuron with variable fates in embryonic zebrafish. *J. Neurosci.* **10**, 34–43.

Erdmann, B., Kirsch, F. P., Rathjen, F. G., and More, M. I. (2003). N-cadherin is essential for retinal lamination in the zebrafish. *Dev. Dyn.* **226**, 570–577.

Erter, C. E., Wilm, T. P., Basler, N., Wright, C. V., and Solnica-Krezel, L. (2001). Wnt8 is required in lateral mesendodermal precursors for neural posteriorization *in vivo*. *Development* **128**, 3571–3583.

Fashena, D., and Westerfield, M. (1999). Secondary motoneuron axons localize DM-GRASP on their fasciculated segments. *J. Comp. Neurol.* **406**, 415–424.

Fekany, K., Yamanaka, Y., Leung, T., Sirotkin, H. I., Topczewski, J., Gates, M. A., Hibi, M., Renucci, A., Stemple, D., Radbill, A., *et al.* (1999). The zebrafish bozozok locus encodes Dharma, a homeodomain protein essential for induction of gastrula organizer and dorsoanterior embryonic structures. *Development* **126**, 1427–1438.

Fjose, A., Izpisua-Belmonte, J. C., Fromental-Ramain, C., and Duboule, D. (1994). Expression of the zebrafish gene hlx-1 in the prechordal plate and during CNS development. *Development* **120**, 71–81.

Fricke, C., Lee, J. S., Geiger-Rudolph, S., Bonhoeffer, F., and Chien, C. B. (2001). Astray, a zebrafish roundabout homolog required for retinal axon guidance. *Science* **292**, 507–510.

Furthauer, M., Thisse, C., and Thisse, B. (1997). A role for FGF-8 in the dorsoventral patterning of the zebrafish gastrula. *Development* **124**, 4253–4264.

Furutani-Seiki, M., Jiang, Y. J., Brand, M., Heisenberg, C. P., Houart, C., Beuchle, D., van Eeden, F. J., Granato, M., Haffter, P., Hammerschmidt, M., *et al.* (1996). Neural degeneration mutants in the zebrafish, Danio rerio. *Development* **123**, 229–239.

Geldmacher-Voss, B., Reugels, A. M., Pauls, S., and Campos-Ortega, J. A. (2003). A 90-degree rotation of the mitotic spindle changes the orientation of mitoses of zebrafish neuroepithelial cells. *Development* **130**, 3767–3780.

Geling, A., Itoh, M., Tallafuss, A., Chapouton, P., Tannhäuser, B., Kuwada, J. Y., Chitnis, A. B., and Bally-Cuif, L. (2003). bHLH transcription factor Her5 links patterning to regional inhibition of neurogenesis at the midbrain-hindbrain boundary. *Development* **130**, 1591–1604.

Geling, A., Plessy, C., Rastegar, S., Straehle, U., and Bally-Cuif, L. (2004). Her5 acts as a prepattern factor that blocks neurogenin1 and coe2 expression upstream of Notch to inhibit neurogenesis at the midbrain-hindbrain boundary. *Development* **131**, 1591–1604.

Girard, F., Cremazy, F., Berta, P., and Renucci, A. (2001). Expression pattern of the Sox31 gene during Zebrafish embryonic development. *Mech. Dev.* **100**, 71–73.

Glasgow, E., Karavanov, A., and Dawid, I. (1997). Neuronal and neuroendocrine expression of LIM3, a LIM class homeobox gene, is altered in mutant zebrafish with axial signaling defects. *Dev. Biol.* **192**, 405–419.

Goldman, D., Hankin, M., Li, Z., Dai, X., and Ding, J. (2001). Transgenic zebrafish for studying nervous system development and regeneration. *Trans. Res.* **10**, 21–33.

Goldsmith, P., Baier, H., and Harris, W. A. (2003). Two zebrafish mutants, ebony and ivory, uncover benefits of neighborhood on photoreceptor survival. *J. Neurobiol.* **57**, 235–245.

Goldsmith, P., and Harris, W. A. (2003). The zebrafish as a tool for understanding the biology of visual disorders. *Semin. Cell Dev. Biol.* **14**, 11–18.

Golling, G., Amsterdam, A., Sun, Z., Antonelli, M., Maldonado, E., Chen, W., Burgess, S., Haldi, M., Artzt, K., Farrington, S., *et al.* (2002). Insertional mutagenesis in zebrafish rapidly identifies genes essential for early vertebrate development. *Nat. Genet.* **31**, 135–140.

Gothilf, Y., Toyama, R., Coon, S., Du, S., Dawid, I., and Klein, D. (2002). Pineal-specific expression of green fluorescent protein under the control of the serotonin-N-acetyltransferase gene regulatory regions in transgenic zebrafish. *Dev. Dynam.* **225**, 241–249.

Gotz, M., Hartfuss, E., and Malatesta, P. (2002). Radial glial cells as neuronal precursors: A new perspective on the correlation of morphology and lineage restriction in the developing cerebral cortex of mice. *Brain Res. Bull.* **57**, 777–788.

Granato, M., van Eeden, F. J., Schach, U., Trowe, T., Brand, M., Furutani-Seiki, M., Haffter, P., Hammerschmidt, M., Heisenberg, C. P., Jiang, Y. J., *et al.* (1996). Genes controlling and mediating locomotion behavior of the zebrafish embryo and larva. *Development* **123**, 399–413.

Gray, M., Moens, C. B., Amacher, S. L., Eisen, J. S., and Beattie, C. E. (2001). Zebrafish deadly seven functions in neurogenesis. *Dev. Biol.* **237**, 306–323.

Gregg, R. G., Willer, G. B., Fadool, J. M., Dowling, J. E., and Link, B. A. (2003). Positional cloning of the young mutation identifies an essential role for the Brahma chromatin remodeling complex in mediating retinal cell differentiation. *Proc. Natl. Acad. Sci. USA* **100**, 6535–6540.

Grinblat, Y., Gamse, J., Patel, M., and Sive, H. (1998). Determination of the zebrafish forebrain: Induction and patterning. *Development* **125**, 4403–4416.

Grinblat, Y., and Sive, H. (2001). Zic gene expression marks anteroposterior pattern in the presumptive neurectoderm of the zebrafish gastrula. *Dev. Dyn.* **222**, 688–693.

Gritsman, K., Zhang, J., Cheng, S., Heckscher, E., Talbot, W. S., and Schier, A. F. (1999). The EGF-CFC protein one-eyed pinhead is essential for nodal signaling. *Cell* **97**, 121–132.

Grunwald, D. J., Kimmel, C. B., Westerfield, M., Walker, C., and Streisinger, G. (1988). A neural degeneration mutation that spares primary neurons in the zebrafish. *Dev. Biol.* **126**, 115–128.

Guo, S., Brush, J., Teraoka, H., Goddard, A., Wilson, S. W., Mullins, M. C., and Rosenthal, A. (1999a). Development of noradrenergic neurons in the zebrafish hindbrain requires BMP, FGF8, and the homeodomain protein soulless/Phox2a. *Neuron* **24**, 555–566.

Guo, S., Wilson, S. W., Cooke, S., Chitnis, A. B., Driever, W., and Rosenthal, A. (1999b). Mutations in the zebrafish unmask shared regulatory pathways controlling the development of catecholaminergic neurons. *Dev. Biol.* **208**, 473–487.

Guo, S., Yamaguchi, Y., Schilbach, S., Wada, T., Lee, J., Goddard, A., French, D., Handa, H., and Rosenthal, A. (2000). A regulator of transcriptional elongation controls vertebrate neuronal development. *Nature* **408**, 366–369.

Haddon, C., Smithers, L., Schneider-Maunoury, S., Coche, T., and Henrique, D. and Lewis, J. (1998). Multiple delta genes and lateral inhibition in zebrafish primary neurogenesis. *Development* **125**, 359–370.

Hale, M., Ritter, D., and Fetcho, J. (2001). A confocal study of spinal interneurons in living larval zebrafish. *J. Comp. Neurol.* **437**, 1–16.

Halloran, M. C., Sato-Maeda, M., Warren, J. T., Su, F., Lele, Z., Krone, P. H., Kuwada, J. Y., and Shoji, W. (2000). Laser-induced gene expression in specific cells of transgenic zebrafish. *Development* **127**, 1953–1960.

Hammerschmidt, M., and Mullins, M. C. (2002). Dorsoventral patterning in the zebrafish: Bone morphogenetic proteins and beyond. *Results Probl. Cell Differ.* **40**, 72–95.

Hammond, K. L., Hill, R. E., Whitfield, T. T., and Currie, P. D. (2002). Isolation of three zebrafish dachshund homologues and their expression in sensory organs, the central nervous system and pectoral fin buds. *Mech. Dev.* **112**, 183–189.

Hanneman, E., and Westerfield, M. (1989). Early expression of acetylcholinesterase activity in functionally distinct neurons of the zebrafish. *J. Comp. Neurol.* **284**, 350–361.

Hans, S., and Campos-Ortega, J. A. (2002). On the organisation of the regulatory region of the zebrafish deltaD gene. *Development* **129**, 4773–4784.

Hardcastle, Z., and Papalopulu, N. (2000). Distinct effects of XBF-1 in regulating the cell cycle inhibitor p27(XIC1) and imparting a neural fate. *Development* **127**, 1303–1314.

Hartfuss, E., Galli, R., Heins, N., and Gotz, M. (2001). Characterization of CNS precursor subtypes and radial glia. *Dev. Biol.* **229**, 15–30.

Hatta, K. (1992). Role of the floor plate in axonal patterning in the zebrafish CNS. *Neuron* **9**, 629–642.

Hauptmann, G., and Gerster, T. (1995). Pou-2—a zebrafish gene active during cleavage stages and in the early hindbrain. *Mech. Dev.* **51**, 127–138.

Hauptmann, G., and Gerster, T. (1996). Complex expression of the zp-50 pou gene in the embryonic zebrafish brain is altered by overexpression of sonic hedgehog. *Development* **122**, 1769–1780.

Helde, K., Wilson, E., Cretekos, C., and Grunwald, D. (1994). Contribution of early cells to the fate map of the zebrafish gastrula. *Science* **265**, 517–520.

Hemmati-Brivanlou, A., and Melton, D. (1997). Vertebrate embryonic cells will become nerve cells unless told otherwise. *Cell* **88**, 13–17.

Hieber, V., Dai, X., Foreman, M., and Goldman, D. (1998). Induction of alpha-1-tubulin gene expression during development and regeneration of the fish central nervous system. *J. Neurobiol.* **37**, 429–440.

Higashijima, S., Hotta, Y., and Okamoto, H. (2000). Visualization of cranial motor neurons in live transgenic zebrafish expressing green fluorescent protein under the control of the islet-1 promoter/enhancer. *J. Neurosci.* **20**, 206–218.

Hjorth, J., and Key, B. (2001). Are pioneer axons guided by regulatory gene expression domains in the zebrafish forebrain? High-resolution analysis of the patterning of the zebrafish brain during axon tract formation *Dev. Biol.* **229**, 271–286.

Hjorth, J., and Key, B. (2002). Development of axon pathways in the zebrafish central nervous system. *Int. J. Dev. Biol.* **46**, 609–619.

Hjorth, J. T., Gad, J., Cooper, H., and Key, B. (2001). A zebrafish homologue of deleted in colorectal cancer (zdcc) is expressed in the first neuronal clusters of the developing brain. *Mech. Dev.* **109**, 105–119.

Ho, C., Houart, C., Wilson, S., and Stainier, D. Y. (1999). A role for the extraembryonic yolk syncytial layer in patterning the zebrafish embryo suggested by properties of the hex gene. *Curr. Biol.* **9**, 1131–1134.

Holley, S. A., Geisler, R., and Nusslein-Volhard, C. (2000). Control of her1 expression during zebrafish somitogenesis by a delta-dependent oscillator and an independent wave-front activity. *Genes Dev.* **14**, 1678–1690.

Holley, S. A., Julich, D., Rauch, G. J., Geisler, R., and Nusslein-Volhard, C. (2002). her1 and the notch pathway function within the oscillator mechanism that regulates zebrafish somitogenesis. *Development* **129**, 1175–1183.

Holzschuh, J., Barrallo-Gimeno, A., Ettl, A. K., Durr, K., Knapik, E. W., and Driever, W. (2003a). Noradrenergic neurons in the zebrafish hindbrain are induced by retinoic acid and require tfap2a for expression of the neurotransmitter phenotype. *Development* **130**, 5741–5754.

Holzschuh, J., Hauptmann, G., and Driever, W. (2003b). Genetic analysis of the roles of Hh, FGF8, and nodal signaling during catecholaminergic system development in the zebrafish brain. *J. Neurosci.* **23**, 5507–5519.

Holzschuh, J., Ryu, S., Aberger, F., and Driever, W. (2001). Dopamine transporter expression distinguishes dopaminergic neurons from other catecholaminergic neurons in the developing zebrafish embryo. *Mech. Dev.* **101**, 237–243.

Hong, S., Kim, C., Yoo, K., Kim, H., Kudoh, T., Dawid, I., and Huh, T. (2002). Isolation and expression of a novel neuron-specific onecut homeobox gene in zebrafish. *Mech. Dev.* **112**, 199–202.

Horne-Badovinac, S., Lin, D., Waldron, S., Schwarz, M., Mbamalu, G., Pawson, T., Jan, Y., Stainier, D., and Abdelilah-Seyfried, S. (2001). Positional cloning of heart and soul reveals multiple roles for PKC lambda in zebrafish organogenesis. *Curr. Biol.* **11**, 1492–1502.

Houart, C., Caneparo, L., Heisenberg, C., Barth, K., Take-Uchi, M., and Wilson, S. (2002). Establishment of the telencephalon during gastrulation by local antagonism of Wnt signaling. *Neuron* **35**, 255–265.

Houart, C., Westerfield, M., and Wilson, S. W. (1998). A small population of anterior cells patterns the forebrain during zebrafish gastrulation. *Nature* **391**, 788–792.

Hu, M., and Easter, S. S. (1999). Retinal neurogenesis: The formation of the initial central patch of postmitotic cells. *Dev. Biol.* **207**, 309–321.

Huang, S., and Sato, S. (1998). Progenitor cells in the adult zebrafish nervous system express a Brn-1-related Pou gene, Tai-ji. *Mech. Dev.* **71**, 23–35.

Hutson, L., and Chien, C.-B. (2002a). Wiring the zebrafish: Axon guidance and synaptogenesis. *Curr. Opin. Neurobiol.* **12**, 87–92.

Hutson, L. D., and Chien, C.-B. (2002b). Pathfinding and error correction by retinal axons: The role of astray/robo2. *Neuron* **33**, 205–217.

Imboden, M., Devignot, V., Korn, H., and Goblet, C. (2001). Regional distribution of glycine receptor messenger RNA in the central nervous system of zebrafish. *Neuroscience* **103**, 811–830.

Inoue, A., Takahashi, M., Hatta, K., Hotta, Y., and Okamoto, H. (1994). Developmental regulation of islet-1 mRNA expression during neuronal differentiation in embryonic zebrafish. *Dev. Dyn.* **199**, 1–11.

Ishibashi, M., and McMahon, A. P. (2002). A sonic hedgehog-dependent signaling relay regulates growth of diencephalic and mesencephalic primordia in the early mouse embryo. *Development* **129**, 4807–4819.

Itoh, M., Kim, C. H., Palardy, G., Oda, T., Jiang, Y. J., Maust, D., Yeo, S. Y., Lorick, K., Wright, G. J., Ariza-McNaughton, L., *et al.* (2003). Mind bomb is a ubiquitin ligase that is essential for efficient activation of Notch signaling by Delta. *Dev. Cell* **4**, 67–82.

Itoh, M., Kudoh, T., Dedekian, M., Kim, C. H., and Chitnis, A. B. (2002). A role for iro1 and iro7 in the establishment of an anteroposterior compartment of the ectoderm adjacent to the midbrain-hindbrain boundary. *Development* **129**, 2317–2327.

Jaszai, J., Reifers, F., Picker, A., Langenberg, T., and Brand, M. (2003). Isthmus-to-midbrain transformation in the absence of midbrain-hindbrain organizer activity. *Development* **130**, 6611–6623.

Jessen, J., Topczewski, J., Bingham, S., Sepich, D., Marlow, F., Chandrasekhar, A., and Solnica-Krezel, L. (2002). Zebrafish trilobite identifies new roles for Strabismus in gastrulation and neuronal movements. *Nat. Cell Biol.* **4**, 610–615.

Jiang, Y. J., Brand, M., Heisenberg, C. P., Beuchle, D., Furutani-Seiki, M., Kelsh, R. N., Warga, R. M., Granato, M., Haffter, P., Hammerschmidt, M., *et al.* (1996). Mutations affecting neurogenesis and brain morphology in the zebrafish, Danio rerio. *Development* **123**, 205–216.

Kainz, P., Adolph, A., Wong, K., and Dowling, J. (2003). Lazy eyes zebrafish mutation affects Muller glial cells, compromising photoreceptor function and causing partial blindness. *J. Comp. Neurol.* **463**, 265–280.

Kalev-Zylinska, M. L., Horsfield, J. A., Flores, M. V., Postlethwait, J. H., Chau, J. Y., Cattin, P. M., Vitas, M. R., Crosier, P. S., and Crosier, K. E. (2003). Runx3 is required for hematopoietic development in zebrafish. *Dev. Dyn.* **228**, 323–336.

Kanki, J., and Kuwada, J. Y. (2000). Growth cones utilize both widespread and local directional cues in the zebrafish brain. *Dev. Biol.* **219**, 364–372.

Karlstrom, R. O., Trowe, T., Klostermann, S., Baier, H., Brand, M., Crawford, A. D., Grunewald, B., Haffter, P., Hoffmann, H., Meyer, S. U., *et al.* (1996). Zebrafish mutations affecting retinotectal axon pathfinding. *Development* **123**, 427–438.

Kaslin, J., and Panula, P. (2001). Comparative anatomy of the histaminergic and other aminergic systems in zebrafish (Danio rerio). *J. Comp. Neurol.* **440**, 342–377.

Kawahara, A., Chien, C. B., and Dawid, I. B. (2002). The homeobox gene mbx is involved in eye and tectum development. *Dev. Biol.* **248**, 107–117.

Kawai, H., Arata, N., and Nakayasu, H. (2001). Three-dimensional distribution of astrocytes in zebrafish spinal cord. *Glia* **36**, 406–413.

Kay, J., Finger-Baier, K., Roeser, T., Staub, W., and Baier, H. (2001). Retinal ganglion cell genesis requires lakritz, a zebrafish atonal homolog. *Neuron* **30**, 725–736.

Keegan, B. R., Feldman, J. L., Lee, D. H., Koos, D. S., Ho, R. K., Stainier, D. Y., and Yelon, D. (2002). The elongation factors Pandora/Spt6 and Foggy/Spt5 promote transcription in the zebrafish embryo. *Development* **129**, 1623–1632.

Kelly, G. M., Greenstein, P., Erezyilmaz, D. F., and Moon, R. T. (1995). Zebrafish wnt8 and wnt8b share a common activity but are involved in distinct developmental pathways. *Development* **121**, 1787–1799.

Kelly, G. M., and Moon, R. T. (1995). Involvement of wnt1 and pax2 in the formation of the midbrain-hindbrain boundary in the zebrafish gastrula. *Dev. Genet.* **17**, 129–140.

Kikuchi, Y., Segawa, H., Tokumoto, M., Tsubokawa, T., Hotta, Y., Uyemura, K., and Okamoto, H. (1997). Ocular and cerebellar defects in zebrafish induced by overexpression of the LIM domains of the islet-3 LIM/homeodomain protein. *Neuron* **18**, 369–382.

Kim, C. H., Ueshima, E., Muraoka, O., Tanaka, H., Yeo, S. Y., Huh, T. L., and Miki, N. (1996). Zebrafish elav/HuC homologue as a very early neuronal marker. *Neurosci. Lett.* **216**, 109–112.

Kimmel, C., and Warga, R. (1986). Tissue specific cell lineages originate in the gastrula of the zebrafish. *Science* **231**, 356–368.

Kimmel, C., and Westerfield, M. (1990). Primray neurons of the zebrafish. *In* "Signal and sense: Local and global order in perceptual maps" (G. M. Edelman, W. E. Gall, and M. W. Cowan, eds.), pp. 561–588.

Kimmel, C. B. (1993). Patterning the brain of the zebrafish embryo. *Annu. Rev. Neurosci.* **16**, 707–732.

Kimmel, C. B., Powell, S. L., and Metcalfe, W. K. (1982). Brain neurons which project to the spinal cord in young larvae of the zebrafish. *J. Comp. Neurol.* **205**, 112–127.

Kimmel, C. B., Sessions, S. K., and Kimmel, R. J. (1981). Morphogenesis and synaptogenesis of the zebrafish Mauthner neuron. *J. Comp. Neurol.* **198**, 101–120.

Knight, R. D., Nair, S., Nelson, S. S., Afshar, A., Javidan, Y., Geisler, R., Rauch, G. J., and Schilling, T. F. (2003). Lockjaw encodes a zebrafish tfap2a required for early neural crest development. *Development* **130**, 5755–5768.

Koos, D. S., and Ho, R. K. (1999). The nieuwkoid/dharma homeobox gene is essential for bmp2b repression in the zebrafish pregastrula. *Dev. Biol.* **215**, 190–207.

Korzh, V., Edlund, T., and Thor, S. (1993). Zebrafish primary neurons initiate expression of the LIM homeodomain protein Isl-1 at the end of gastrulation. *Development* **118**, 417–425.

Korzh, V., Sleptsova, I., Liao, J., He, J., and Gong, Z. (1998). Expression of zebrafish bHLH genes ngn1 and nrd defines distinct stages of neural differentiation. *Dev. Dynam.* **213**, 92–104.

Koshida, S., Shinya, M., Mizuno, T., Kuroiwa, A., and Takeda, H. (1998). Initial anteroposterior pattern of the zebrafish central nervous system is determined by differential competence of the epiblast. *Development* **125**, 1957–1966.

Koshida, S., Shinya, M., Nikaido, M., Ueno, N., Schulte-Merker, S., Kuroiwa, A., and Takeda, H. (2002). Inhibition of BMP activity by the FGF signal promotes posterior neural development in zebrafish. *Dev. Biol.* **244**, 9–20.

Koster, R. W., and Fraser, S. E. (2001). Direct imaging of *in vivo* neuronal migration in the developing cerebellum. *Curr. Biol.* **11**, 1858–1863.

Kudoh, T., Wilson, S. W., and Dawid, I. B. (2002). Distinct roles for Fgf, Wnt and retinoic acid in posteriorizing the neural ectoderm. *Development* **129**, 4335–4346.

Kuwada, J. Y., Bernhardt, R. R., and Chitnis, A. B. (1990a). Pathfinding by identified growth cones in the spinal cord of zebrafish embryos. *J. Neurosci.* **10**, 1299–1308.

Kuwada, J. Y., Bernhardt, R. R., and Nguyen, N. (1990b). Development of spinal neurons and tracts in the zebrafish embryo. *J. Comp. Neurol.* **302**, 617–628.

Lai, E. C., Deblandre, G. A., Kintner, C., and Rubin, G. M. (2001). Drosophila neuralized is a ubiquitin ligase that promotes the internalization and degradation of delta. *Dev. Cell* **1**, 783–794.

Langenau, D., Palomero, T., Kanki, J., Ferrando, A., Zhou, Y., Zon, L., and Look, A. (2002). Molecular cloning and developmental expression of Tlx (Hox11) genes in zebrafish (Danio rerio). *Mech. Dev.* **117**, 243–248.

Leake, D., Asch, W., Canger, A., and Schechter, N. (1999). Gefiltin in zebrafish embryos: Sequential gene expression of two neurofilament proteins in retinal ganglion cells. *Differentiation* **65**, 181–189.

Lecaudey, V., Thisse, C., Thisse, B., and Schneider-Maunoury, S. (2001). Sequence and expression pattern of ziro7, a novel, divergent zebrafish iroquois homeobox gene. *Mech. Dev.* **109**, 383–388.

Lee, S., and Pfaff, S. (2003). Synchronization of neurogenesis and motor neuron specification by direct coupling of bHLH and homeodomain transcription factors. *Neuron* **38**, 731–745.

Lee, S. A., Shen, E. L., Fiser, A., Sali, A., and Guo, S. (2003). The zebrafish forkhead transcription factor Foxi1 specifies epibranchial placode-derived sensory neurons. *Development* **130**, 2669–2679.

Lee, V. M., Carden, M. J., Schlaepfer, W. W., and Trojanowski, J. Q. (1987). Monoclonal antibodies distinguish several differently phosphorylated states of the two largest rat neurofilament subunits (NF-H and NF-M) and demonstrate their existence in the normal nervous system of adult rats. *J. Neurosci.* **7**, 3474–3488.

Lekven, A. C., Buckles, G. R., Kostakis, N., and Moon, R. T. (2003). Wnt1 and wnt10b function redundantly at the zebrafish midbrain-hindbrain boundary. *Dev. Biol.* **254**, 172–187.

Lekven, A. C., Thorpe, C. J., Waxman, J. S., and Moon, R. T. (2001). Zebrafish wnt8 encodes two wnt8 proteins on a bicistronic transcript and is required for mesoderm and neurectoderm patterning. *Dev. Cell* **1**, 103–114.

Lele, Z., Folchert, A., Concha, M., Rauch, G. J., Geisler, R., Rosa, F., Wilson, S. W., Hammerschmidt, M., and Bally-Cuif, L. (2002). Parachute/n-cadherin is required for morphogenesis and maintained integrity of the zebrafish neural tube. *Development* **129**, 3281–3294.

Leung, T., Bischof, J., Soll, I., Niessing, D., Zhang, D., Ma, J., Jackle, H., and Driever, W. (2003). Bozozok directly represses bmp2b transcription and mediates the earliest dorsoventral asymmetry of bmp2b expression in zebrafish. *Development* **130**, 3639–3649.

Levkowitz, G., Zeller, J., Sirotkin, H. I., French, D., Schilbach, S., Hashimoto, H., Hibi, M., Talbot, W. S., and Rosenthal, A. (2003). Zinc finger protein controls the development of monoaminergic neurons. *Nat. Neurosci.* **6**, 28–33.

Lewis, J. (1998). Notch signling and the control of cell fate choices in vertebrates. *Semin. Cell Dev. Biol.* **9**, 583–589.

Lewis, K. E., and Eisen, J. S. (2004). Paraxial mesoderm specifies zebrafish primary motoneuron subtype identity. *Development* **131**, 891–902.

Li, M., Zhao, C., Wang, Y., Zhao, Z., and Meng, A. (2002). Zebrafish sox9b is an early neural crest marker. *Dev. Genes Evol.* **212**, 203–206.

Li, Y., Allende, M. L., Finkelstein, R., and Weinberg, E. S. (1994). Expression of two zebrafish orthodenticle-related genes in the embryonic brain. *Mech. Dev.* **48**, 229–244.

Liao, J., He, J., Yan, T., Korzh, V., and Gong, Z. (1999). A class of neuroD-related basic helix-loop-helix transcription factors expressed in developing central nervous system in zebrafish. *DNA Cell Biol.* **18**, 333–344.

Link, B., Fadool, J., Malicki, J., and Dowling, J. (2000). The zebrafish young mutation acts non-cell-autonomously to uncouple differentiation from specification for all retinal cells. *Development* **127**, 2177–2188.

Link, B., Kainz, P., Ryou, T., and Dowling, J. (2001). The perplexed and confused mutations affect distinct stages during the transition from proliferating to post-mitotic cells within the zebrafish retina. *Dev. Biol.* **236**, 436–453.

Liu, K. S., Gray, M., Otto, S. J., Fetcho, J. R., and Beattie, C. E. (2003). Mutations in deadly seven/notch1a reveal developmental plasticity in the escape response circuit. *J. Neurosci.* **23**, 8159–8166.

Liu, Y., Berndt, J., Su, F., Tawarayama, H., Shoji, W., Kuwada, J., and Halloran, M. (2004). Semaphorin3D guides retinal axons along the dorsoventral axis of the tectum. *J. Neurosci.* **24**, 310–318.

Lorent, K., Liu, K. S., Fetcho, J. R., and Granato, M. (2001). The zebrafish space cadet gene controls axonal pathfinding of neurons that modulate fast turning movements. *Development* **128**, 2131–2142.

Macdonald, R., Scholes, J., Strahle, U., Brennan, C., Holder, N., Brand, M., and Wilson, S. W. (1997). The Pax protein Noi is required for commissural axon pathway formation in the rostral forebrain. *Development* **124**, 2397–2408.

Maggs, A., and Scholes, J. (1990). Reticular astrocytes in the fish optic nerve: Macroglia with epithelial characteristics form an axially repeated lacework pattern, to which nodes of Ranvier are apposed. *J. Neurosci.* **10**, 1600–1614.

Malicki, J. (1999). Development of the retina. *Meth. Cell Biol.* **59**, 273–299.

Malicki, J., and Driever, W. (1999). Oko meduzi mutations affect neuronal patterning in the zebrafish retina and reveal cell–cell interactions of the retinal neuroepithelial sheet. *Development* **126**, 1235–1246.

Malicki, J., Jo, H., and Pujic, Z. (2003). Zebrafish N-cadherin, encoded by the glass onion locus, plays an essential role in retinal patterning. *Dev. Biol.* **259**, 95–108.

Malicki, J., Jo, H., Wei, X., Hsiung, M., and Pujic, Z. (2002). Analysis of gene function in the zebrafish retina. *Methods* **28**, 427–438.

Marc, R., and Cameron, D. (2001). A molecular phenotype atlas of the zebrafish retina. *J. Neurocytol.* **30**, 593–654.

Marcus, R., and Easter, S. J. (1995). Expression of glial fibrillary acidic protein and its relation to tract formation in embryonic zebrafish (danio rerio). *J. Comp. Neurol.* **359**, 365–381.

Martin, S., Heinrich, G., and Sandell, J. H. (1998). Sequence and expression of glutamic acid decarboxylase isoforms in the developing zebrafish. *J. Comp. Neurol.* **396**, 253–266.

Martinez, S. (2001). The isthmic organizer and brain regionalization. *Int. J. Dev. Biol.* **45**, 367–371.

Marx, M., Rutishauser, U., and Bastmeyer, M. (2001). Dual function of polysialic acid during zebrafish central nervous system development. *Development* **128**, 4949–4958.

Masai, I., Lele, Z., Yamaguchi, M., Komori, A., Nakata, A., Nishiwaki, Y., Wada, H., Tanaka, H., Nojima, Y., Hammerschmidt, M., *et al.* (2003). N-cadherin mediates retinal lamination, maintenance of forebrain compartments and patterning of retinal neurites. *Development* **130**, 2479–2494.

Masai, I., Stemple, D., Okamoto, H., and Wilson, S. (2000). Midline signals regulate retinal neurogenesis in zebrafish. *Neuron* **27**, 251–263.

Matsuda, N., and Mishina, M. (2004). Identification of chaperonin CCT (gamma) subunit as a determinant of retinotectal development by whole-genome subtraction cloning from zebrafish tectal neuron mutant. *Development* **131**, 1913–1925.

Maves, L., Jackman, W., and Kimmel, C. B. (2002). FGF3 and FGF8 mediate a rhombomere 4 signaling activity in the zebrafish hindbrain. *Development* **129**, 3825–3837.

McClintock, J. M., Kheirbek, M. A., and Prince, V. E. (2002). Knockdown of duplicated zebrafish hoxb1 genes reveals distinct roles in hindbrain patterning and a novel mechanism of duplicate gene retention. *Development* **129**, 2339–2354.

McWhorter, M. L., Monani, U. R., Burghes, A. H., and Beattie, C. E. (2003). Knockdown of the survival motor neuron (Smn) protein in zebrafish causes defects in motor axon outgrowth and pathfinding. *J. Cell Biol.* **162**, 919–931.

Megason, S. G., and McMahon, A. P. (2002). A mitogen gradient of dorsal midline Wnts organizes growth in the CNS. *Development* **129**, 2087–2098.

Melancon, E., Liu, D. W., Westerfield, M., and Eisen, J. S. (1997). Pathfinding by identified zebrafish motoneurons in the absence of muscle pioneers. *J. Neurosci.* **17**, 7796–7804.

Mendelson, B. (1986a). Development of reticulospinal neurons of the zebrafish. I. Time of origin. *J. Comp. Neurol.* **251**, 160–171.

Mendelson, B. (1986b). Development of reticulospinal neurons of the zebrafish. II. Early axonal outgrowth and cell body position. *J. Comp. Neurol.* **251**, 172–284.

Metcalfe, W., Myers, P., Trevarrow, B., Bass, M., and Kimmel, C. (1990). Primary neurons that express the L2/HNK-1 carbohydrate during early development in the zebrafish. *Development* **110**, 491–504.

Mikkola, I., Fjose, A., Kuwada, J. Y., Wilson, S., Guddal, P. H., and Krauss, S. (1992). The paired domain-containing nuclear factor pax[b] is expressed in specific commissural interneurons in zebrafish embryos. *J. Neurobiol.* **23,** 933–946.

Miyamura, Y., and Nakayasu, H. (2001). Zonal distribution of Purkinje cells in the zebrafish cerebellum: Analysis by means of a specific monoclonal antibody. *Cell. Tissue Res.* **305,** 299–305.

Mueller, T., and Wullimann, M. F. (2002a). BrdU-, neuroD (nrd)- and Hu-studies reveal unusual non-ventricular neurogenesis in the postembryonic zebrafish forebrain. *Mech. Dev.* **117,** 123–135.

Mueller, T., and Wullimann, M. F. (2002b). Expression domains of neuroD (nrd) in the early postembryonic zebrafish brain. *Brain Res. Bull.* **57,** 377–379.

Mueller, T., and Wullimann, M. F. (2003). Anatomy of neurogenesis in the early zebrafish brain. *Brain Res. Dev. Brain Res.* **140,** 137–155.

Müller, M. V., Weizsäcker, E., and Campos-Ortega, J. A. (1996). Transcription of a zebrafish gene of the hairy-Enhancer of split family delineates the midbrain anlage in the neural plate. *Dev. Genes Evol.* **206,** 153–160.

Mumm, J., and Kopan, R. (2000). Notch signaling: From the outside in. *Dev. Biol.* **228,** 151–165.

Musa, A., Lehrach, H., and Russo, V. (2001). Distinct expression patterns of two zebrafish homologues of the human APP gene during embryonic development. *Dev. Genes Evol.* **211,** 563–567.

Myers, P. Z., Eisen, J. S., and Westerfield, M. (1986). Development and axonal outgrowth of identified motoneurons in the zebrafish. *J. Neurosci.* **6,** 2278–2289.

Neuhauss, S., Biehlmaier, O., Seeliger, M., Das, T., Kohler, K., Harris, W., and Baier, H. (1999). Genetic disorders of vision revealed by a behavioral screen of 400 essential loci in zebrafish. *J. Neurosci.* **19,** 8603–8615.

Neuhauss, S. C. (2003). Behavioral genetic approaches to visual system development and function in zebrafish. *J. Neurobiol.* **54,** 148–160.

Neumann, C. (2001). Pattern formation in the zebrafish retina. *Cell Dev. Biol.* **12,** 485–490.

Neumann, C., and Nuesslein-Volhard, C. (2000). Patterning of the zebrafish retina by a wave of sonic hedgehog activity. *Science* **289,** 2137–2139.

Nguyen, V., Trout, J., Connors, S., Andermann, P., Weinberg, E. S., and Mullins, M. (2000). Dorsal and intermediate neuronal cell types of the spinal cord are established by a BMP signaling pathway. *Development* **127,** 1209–1220.

Nielsen, A., and Jorgensen, A. (2003). Structural and functional characterization of the zebrafish gene for glial fibrillary acidic protein, GFAP. *Gene* **3310,** 123–132.

Nissen, R. M., Yan, J., Amsterdam, A., Hopkins, N., and Burgess, S. M. (2003). Zebrafish foxi one modulates cellular responses to Fgf signaling required for the integrity of ear and jaw patterning. *Development* **130,** 2543–2554.

O'Malley, D. M., Kao, Y. H., and Fetcho, J. R. (1996). Imaging the functional organization of zebrafish hindbrain segments during escape behaviors. *Neuron* **17,** 1145–1155.

Odenthal, J., and Nusslein-Volhard, C. (1998). Fork head domain genes in zebrafish. *Dev. Genes Evol.* **208,** 245–258.

Oehlmann, V., Berger, S., Sterner, C., and Korsching, S. (2004). Zebrafish beta tubulin expression is limited to the nervous system throughout development, and in the adult brain is restricted to a subset of proliferative regions. *Gene Expr. Patterns* **4,** 191–198.

Ohnuma, S., Hopper, S., Wang, K. C., Philpott, A., and Harris, W. A. (2002). Co-ordinating retinal histogenesis: Early cell cycle exit enhances early cell fate determination in the Xenopus retina. *Development* **129,** 2435–2446.

Ohnuma, S., Philpott, A., and Harris, W. A. (2001). Cell cycle and cell fate in the nervous system. *Curr. Opin. Neurobiol.* **11,** 66–73.

Ott, H., Diekmann, H., Stuermer, C., and Bastmeyer, M. (2001). Function of neurolin (DM-GRASP/SC-1) in guidance of motor axons during zebrafish development. *Dev. Biol.* **235,** 86–97.

Panchision, D. M., Pickel, J. M., Studer, L., Lee, S. H., Turner, P. A., Hazel, T. G., and McKay, R. D. (2001). Sequential actions of BMP receptors control neural precursor cell production and fate. *Genes Dev.* **15,** 2094–2110.

Panhuysen, M., Vogt-Weisenhorn, D., Blanquet, V., Brodski, C., Heinzmann, U., Beister, W., and Wurst, W. (2004). Effects of Wnt signaling on proliferation in the developing mid-hindbrain region. *Mol. Cell. Neurosci.* **26,** 101–111.

Papan, C., and Campos-Ortega, J. A. (1994). On the formation of the neural keel and neural tube in the zebrafish Danio (Brachydanio) rerio. *Roux's Arch Dev. Biol.* **203,** 178–186.

Papan, C., and Campos-Ortega, J. A. (1997). A clonal analysis of spinal cord development in the zebrafish. *Dev. Genes Evol.* **207,** 71–81.

Park, H., and Appel, B. (2003). Delta-Notch signaling regulates oligodendrocyte specification. *Development* **130,** 3747–3755.

Park, H., Hong, S., Kim, H., Kim, S., Yoon, E., Kim, C., Miki, N., and Huh, T. (2000a). Structural comparison of zebrafish Elav/Hu and their differential expressions during neurogenesis. *Neurosci. Lett.* **279,** 81–84.

Park, H. C., Kim, C. H., Bae, Y. K., Yeo, S. Y., Kim, S. H., Hong, S. K., Shin, J., Yoo, K. W., Hibi, M., Hirano, T., *et al.* (2000b). Analysis of upstream elements in the HuC promoter leads to the establishment of transgenic zebrafish with fluorescent neurons. *Dev. Biol.* **227,** 279–293.

Park, H. C., Mehta, A., Richardson, J. S., and Appel, B. (2002). olig2 is required for zebrafish primary motor neuron and oligodendrocyte development. *Dev. Biol.* **248,** 356–368.

Park, S. H., Yeo, S. Y., Yoo, K. W., Hong, S. K., Lee, S., Rhee, M., Chitnis, A. B., and Kim, C. H. (2003). Zath3, a neural basic helix-loop-helix gene, regulates early neurogenesis in the zebrafish. *Biochem. Biophys. Res. Commun.* **308,** 184–190.

Pauls, S., Geldmacher-Voss, B., and Campos-Ortega, J. A. (2001). A zebrafish histone variant H2A.F/Z and a transgenic H2A.F/Z:GFP fusion protein for *in vivo* studies of embryonic development. *Dev. Genes Evol.* **211,** 603–610.

Pavlopoulos, E., Pitsouli, C., Klueg, K. M., Muskavitch, M. A., Moschonas, N. K., and Delidakis, C. (2001). Neuralized encodes a peripheral membrane protein involved in delta signaling and endocytosis. *Dev. Cell* **1,** 807–816.

Peterson, R., Fadool, J., McClintock, J., and Linser, P. (2001). Muller cell differentiation in the zebrafish neural retina: Evidence of distinct early and late stages in cell maturation. *J. Comp. Neurol.* **429,** 530–540.

Picker, A., Brennan, C., Reifers, F., Clarke, J. D., Holder, N., and Brand, M. (1999). Requirement for the zebrafish mid-hindbrain boundary in midbrain polarisation, mapping and confinement of the retinotectal projection. *Development* **126,** 2967–2978.

Picker, A., Scholpp, S., Bohli, H., Takeda, H., and Brand, M. (2002). A novel positive transcriptional feedback loop in midbrain–hindbrain boundary development is revealed through analysis of the zebrafish pax2.1 promoter in transgenic lines. *Development* **129,** 3227–3239.

Popperl, H., Rikhof, H., Chang, H., Haffter, P., Kimmel, C. B., and Moens, C. B. (2000). Lazarus is a novel pbx gene that globally mediates hox gene function in zebrafish. *Mol. Cell* **6,** 255–267.

Prince, V. E., Moens, C. B., Kimmel, C. B., and Ho, R. K. (1998). Zebrafish hox genes: Expression in the hindbrain region of wild-type and mutants of the segmentation gene, valentino. *Development* **125,** 393–406.

Quint, E., Zerucha, T., and Ekker, M. (2000). Differential expression of orthologous Dlx genes in zebrafish and mice: Implications for the evolution of the Dlx homeobox gene family. *J. Exp. Zool.* **288,** 235–241.

Reifers, F., Adams, J., Mason, I. J., Schulte-Merker, S., and Brand, M. (2000). Overlapping and distinct functions provided by fgf17, a new zebrafish member of the Fgf8/17/18 subgroup of Fgfs. *Mech. Dev.* **99,** 39–49.

Reifers, F., Bohli, H., Walsh, E. C., Crossley, P. H., Stainier, D. Y., and Brand, M. (1998). Fgf8 is mutated in zebrafish acerebellar (ace) mutants and is required for maintenance of midbrain-hindbrain boundary development and somitogenesis. *Development* **125,** 2381–2395.

Reinhard, E., Nedivi, E., Wegner, J., Skene, J. H., and Westerfield, M. (1994). Neural selective activation and temporal regulation of a mammalian GAP-43 promoter in zebrafish. *Development* **120**, 1767–1775.

Reyes, R., Haendel, M., Grant, D., Melancon, E., and Eisen, J. S. (2004). Slow degeneration of zebrafish Rohon-Beard neurons during programmed cell death. *Dev. Dyn.* **229**, 30–41.

Rhinn, M., and Brand, M. (2001). The midbrain–hindbrain boundary organizer. *Curr. Opin. Neurobiol.* **11**, 34–42.

Rimini, R., Beltrame, M., Argenton, F., Szymczak, D., Cotelli, F., and Bianchi, M. E. (1999). Expression patterns of zebrafish sox11A, sox11B and sox21. *Mech. Dev.* **89**, 167–171.

Ritter, D., Bhatt, D., and Fetcho, J. (2001). *In vivo* imaging of zebrafish reveals diffeences in the spinal networks for escape and swimming movements. *J. Neurosci.* **21**, 8956–8965.

Rodriguez, M., and Driever, W. (1997). Mutations resulting in transient and localized degeneration in the developing zebrafish brain. *Biochem. Cell Biol.* **75**, 579–600.

Roos, M., Schachner, M., and Bernhardt, R. R. (1999). Zebrafish semaphorin Z1b inhibits growing motor axons *in vivo*. *Mech. Dev.* **87**, 103–117.

Ross, L., Parrett, T., and Easter, S. J. (1992). Axonogenesis and morphogenesis in the embryonic zebrafish brain. *J. Neurosci.* **12**, 467–482.

Roth, L., Bormann, P., Bonnet, A., and Reinhard, E. (1999). Beta-thymosin is required for axonal tract formation in developing zebrafish brain. *Development* **126**, 1365–1374.

Ruiz i Altaba, A., Palma, V., and Dahmane, N. (2002a). Hedgehog-Gli signalling and the growth of the brain. *Nat. Rev. Neurosci.* **3**, 24–33.

Ruiz i Altaba, A., Sanchez, P., and Dahmane, N. (2002b). Gli and hedgehog in cancer: Tumours, embryos and stem cells. *Nat. Rev. Cancer* **2**, 361–372.

Scheer, N., Groth, A., Hans, S., and Campos-Ortega, J. A. (2001). An instructive function for Notch in promoting gliogenesis in the zebrafish retina. *Development* **128**, 1099–1107.

Schier, A. F., Neuhauss, S. C., Harvey, M., Malicki, J., Solnica-Krezel, L., Stainier, D. Y., Zwartkruis, F., Abdelilah, S., Stemple, D. L., Rangini, Z., *et al.* (1996). Mutations affecting the development of the embryonic zebrafish brain. *Development* **123**, 165–178.

Schmid, B., Furthauer, M., Connors, S. A., Trout, J., Thisse, B., Thisse, C., and Mullins, M. C. (2000). Equivalent genetic roles for bmp7/snailhouse and bmp2b/swirl in dorsoventral pattern formation. *Development* **127**, 957–967.

Schmitz, B., Papan, C., and Campos-Ortega, J. A. (1993). Neurulation in the anterior trunk region of the zebrafish Brachydanio rerio. *Roux's Arch. Dev. Biol.* **202**, 250–259.

Segawa, H., Miyashita, T., Hirate, Y., Higashijima, S., Chino, N., Uyemura, K., Kikuchi, Y., and Okamoto, H. (2001). Functional repression of Islet-2 by disruption of complex with Ldb impairs peripheral axonal outgrowth in embryonic zebrafish. *Neuron* **30**, 423–436.

Sheng, G., dos Reis, M., and Stern, C. D. (2003). Churchill, a zinc finger transcriptional activator, regulates the transition between gastrulation and neurulation. *Cell* **115**, 603–613.

Shin, J., Park, H. C., Topczewska, J. M., Mawdsley, D. J., and Appel, B. (2003). Neural cell fate analysis using olig2 BAC transgenics. *Meth. Cell Sci.* **25**, 7–14.

Shiomi, K., Uchida, H., Keino-Masu, K., and Masu, M. (2003). Ccd1, a novel protein with a DIX domain, is a positive regulator in the Wnt signaling during zebrafish neural patterning. *Curr. Biol.* **13**, 73–77.

Simpson, P. (1997). Notch signaling in development: On equivalence groups and asymmetric developmental potential. *Curr. Opin. Genet. Dev.* **7**, 537–542.

Solnica-Krezel, L., and Cooper, M. S. (2002). Cellular and genetic mechanisms of convergence and extension. *Results Probl. Cell Differ.* **40**, 136–165.

Solomon, K. S., Kudoh, T., Dawid, I. B., and Fritz, A. (2003). Zebrafish foxi1 mediates otic placode formation and jaw development. *Development* **130**, 929–940.

Son, O., Kim, H., Ji, M., Yoo, K., Rhee, M., and Kim, C. (2003). Cloning and expression analysis of a Parkinson's disease gene, uch-L1, and its promoter in zebrafish. *BBRC* **312**, 601–607.

Stenkamp, D., and Frey, R. (2003). Extraretinal and retinal hedgehog signaling sequentially regulate retinal differentiation in zebrafish. *Dev. Biol.* **258**, 349–363.

Stern, C. (2002). Induction and initial patterning of the nervous system—The chick embryo enters the scene.. *Curr. Opin. Genet. Dev.* **12**, 447–451.

Streit, A., Berliner, A., Papnayotou, C., Sirulnik, A., and Stern, C. (2000). Initiation of neural induction by FGF signaling before gastrulation. *Nature* **406**, 74–78.

Sun, T., Dong, H., Wu, L., Kane, M., Rowitch, D., and Stiles, C. (2003). Cross-repressive interaction of the Olig2 and Nkx2.2 transcription factors in developing neural tube associated with formation of a specific physical complex. *J. Neurosci.* **23**, 9547–9556.

Svoboda, K. R., Linares, A. E., and Ribera, A. B. (2001). Activity regulates programmed cell death of zebrafish Rohon-Beard neurons. *Development* **128**, 3511–3520.

Takke, C., Dornseifer, P., v Weizsacker, E., and Campos-Ortega, J. A. (1999). her4, a zebrafish homologue of the Drosophila neurogenic gene E(spl), is a target of NOTCH signalling. *Development* **126**, 1811–1821.

Tarnowka, M. A., Baglioni, C., and Basilico, C. (1978). Synthesis of H1 histones by BHK cells in G1. *Cell* **15**, 163–171.

Taupin, P., and Gage, F. (2002). Adult neurogenesis and neural stem cells of the central nervous system in mammals. *J. Neurosci. Res.* **69**, 745–749.

Teraoka, H., Russell, C., Regan, J., Chandrasekhar, A., Concha, M., Yokoyama, R., Higashi, K., Take-Uchi, M., Dong, W., Hiraga, T., *et al.* (2004). Hedgehog and Fgf signaling pathways regulate the development of tphR-expressing serotonergic raphe neurons in zebrafish embryos. *J. Neurobiol.* **60**, 275–288.

Thisse, C., Thisse, B., and Postlethwait, J. H. (1995). Expression of snail2, a second member of the zebrafish snail family, in cephalic mesendoderm and presumptive neural crest of wild-type and spadetail mutant embryos. *Dev. Biol.* **172**, 86–99.

Tokumoto, M., Gong, Z., Tsubokawa, T., Hew, C. L., Uyemura, K., Hotta, Y., and Okamoto, H. (1995). Molecular heterogeneity among primary motoneurons and within myotomes revealed by the differential mRNA expression of novel islet-1 homologs in embryonic zebrafish. *Dev. Biol.* **171**, 578–589.

Tomizawa, K., Inoue, Y., Doi, S., and Nakayasu, H. (2000a). Monoclonal antibody stains oligodendrocytes and Schwann cells in zebrafish (Danio rerio). *Anat. Embryol.* **201**, 399–406.

Tomizawa, K., Inoue, Y., and Nakayasu, H. (2000b). A monoclonal antibody stains radial glia in the adult zebrafish (Danio rerio) CNS. *J. Neurocytol.* **29**, 119–128.

Trewarrow, B., Marks, D., and Kimmel, C. B. (1990). Organization of hindbrain segments in the zebrafish embryos. *Neuron* **4**, 669–679.

Trowe, T., Klostermann, S., Baier, H., Granato, M., Crawford, A. D., Grunewald, B., Hoffmann, H., Karlstrom, R. O., Meyer, S. U., Muller, B., *et al.* (1996). Mutations disrupting the ordering and topographic mapping of axons in the retinotectal projection of the zebrafish, Danio rerio. *Development* **123**, 439–450.

Udvadia, A. J., Koster, R. W., and Skene, J. H. (2001). GAP-43 promoter elements in transgenic zebrafish reveal a difference in signals for axon growth during CNS development and regeneration. *Development* **128**, 1175–1182.

van Eeden, F. J., Granato, M., Schach, U., Brand, M., Furutani-Seiki, M., Haffter, P., Hammerschmidt, M., Heisenberg, C. P., Jiang, Y. J., Kane, D. A., *et al.* (1996). Mutations affecting somite formation and patterning in the zebrafish, Danio rerio. *Development* **123**, 153–164.

Varga, Z. M., Wegner, J., and Westerfield, M. (1999). Anterior movement of ventral diencephalic precursors separates the primordial eye field in the neural plate and requires cyclops. *Development* **126**, 5533–5546.

Vriz, S., Joly, C., Boulekbache, H., and Condamine, H. (1996). Zygotic expression of the zebrafish Sox-19, an HMG box-containing gene, suggests an involvement in central nervous system development. *Brain Res. Mol. Brain Res.* **40**, 221–228.

Walshe, J., and Mason, I. (2003). Unique and combinatorial functions of Fgf3 and Fgf8 during zebrafish forebrain development. *Development* **130,** 4337–4349.

Wang, X., Chu, L. T., He, J., Emelyanov, A., Korzh, V., and Gong, Z. (2001). A novel zebrafish bHLH gene, neurogenin3, is expressed in the hypothalamus. *Gene* **275,** 47–55.

Wang, X., Emelyanov, A., Korzh, V., and Gong, Z. (2003). Zebrafish atonal homologue zath3 is expressed during neurogenesis in embryonic development. *Dev. Dyn.* **227,** 587–592.

Warren, J. T., Jr., Chandrasekhar, A., Kanki, J. P., Rangarajan, R., Furley, A. J., and Kuwada, J. Y. (1999). Molecular cloning and developmental expression of a zebrafish axonal glycoprotein similar to TAG-1. *Mech. Dev.* **80,** 197–201.

Waskiewicz, A. J., Rikhof, H. A., and Moens, C. B. (2002). Eliminating zebrafish pbx proteins reveals a hindbrain ground state. *Dev. Cell* **3,** 723–733.

Wei, Y., and Allis, C. (1998). Pictures in cell biology. *Trends Cell Biol.* **8,** 266.

Weiland, U., Ott, H., Bastmeyer, M., Schaden, H., Giordano, S., and CAO, S. (1997). Expression of an LI-related cell adhesion molecule on developing CNS fiber tracts in zebrafish and its functional contribution to axon fasciculation. *Mol. Cell Neurosci.* **9,** 77–89.

Werter, C., Wilm, T., Basler, N., Wright, C., and Solnica-Krezel, L. (2001). Wnt8 is required in lateral mesendodermal precursors for neural posteriorization *in vivo*. *Development* **128,** 3571–3583.

Westerfield, M., McMurray, J. V., and Eisen, J. S. (1986). Identified motoneurons and their innervation of axial muscles in the zebrafish. *J. Neurosci.* **6,** 2267–2277.

Westin, J., and Lardelli, M. (1997). Three novel Notch genes in zebrafish: Implications for vertebrate Notch gene evolution and function. *Dev. Genes Evol.* **207,** 51–63.

Williams, J. A., Barrios, A., Gatchalian, C., Rubin, L., Wilson, S. W., and Holder, N. (2000). Programmed cell death in zebrafish rohon beard neurons is influenced by TrkC1/NT-3 signaling. *Dev. Biol.* **226,** 220–230.

Wilson, E., Cretekos, C., and Helde, K. (1995). Cell mixing during early epiboly in the zebrafish embryo. *Dev. Genet.* **17,** 6–15.

Wilson, S., and Edlund, T. (2001). Neural induction: Toward a unifying mechanism. *Nat. Neurosci.* **4,** 1161–1168.

Wilson, S., Graziano, E., Harland, R. M., Jessell, T., and Edlund, T. (2000). An early requirement for FGF signaling in the acquisition of neural cell fate in the chick embryo. *Curr. Biol.* **10,** 421–429.

Wilson, S., Ross, L., Parrett, T., and Easter, S. (1990). The development of a simple scaffold of axon tracts in the brain of the embryonic zebrafish Brachydanio rerio. *Development* **108,** 121–145.

Wilson, S. W., Brand, M., and Eisen, J. S. (2002). Patterning the zebrafish central nervous system. *Results Probl. Cell Differ.* **40,** 181–215.

Wilson, S. W., and Easter, S. S., Jr. (1991a). A pioneering growth cone in the embryonic zebrafish brain. *Proc. Natl. Acad. Sci. USA* **88,** 2293–2296.

Wilson, S. W., and Easter, S. S., Jr. (1991b). Stereotyped pathway selection by growth cones of early epiphysial neurons in the embryonic zebrafish. *Development* **112,** 723–746.

Woo, K., and Fraser, S. E. (1995). Order and coherence in the fate map of the zebrafish nervous system. *Development* **121,** 2595–2609.

Woo, K., and Fraser, S. E. (1997). Specification of the zebrafish nervous system by nonaxial signals. *Science* **277,** 254–257.

Woo, K., and Fraser, S. E. (1998). Specification of the hindbrain fate in the zebrafish. *Dev. Biol.* **197,** 283–296.

Wullimann, M., and Knipp, S. (2000). Proliferation pattern changes in the zebrafish brain from embryonic through early postembryonic stages. *Anat. Embryol.* **202,** 385–400.

Wullimann, M., and Rink, E. (2002). The teleostean forebrain: A comparative and developmental view based on early proliferation, Pax6 activity and catecholaminergic organization. *Brain Res. Bull.* **57,** 363–370.

Wullimann, M. F., and Mueller, T. (2002). Expression of Zash-1a in the postembryonic zebrafish brain allows comparison to mouse Mash1 domains. *Brain Res. Gene Expr. Patterns* **1,** 187–192.

Wullimann, M. F., and Rink, E. (2001). Detailed immunohistology of Pax6 protein and tyrosine hydroxylase in the early zebrafish brain suggests role of Pax6 gene in development of dopaminergic diencephalic neurons. *Brain Res. Dev. Brain Res.* **131,** 173–191.

Wurst, W., and Bally-Cuif, L. (2001). Neural plate patterning: Upstream and downstream of the isthmic organizer. *Nat. Rev. Neurosci.* **2,** 99–108.

Xiao, T., Shoji, W., Zhou, W., Su, F., and Kuwada, J. Y. (2003). Transmembrane sema4E guides branchiomotor axons to their targets in zebrafish. *J. Neurosci.* **23,** 4190–4198.

Xu, J., Liu, Z., and Ornitz, D. M. (2000). Temporal and spatial gradients of Fgf8 and Fgf17 regulate proliferation and differentiation of midline cerebellar structures. *Development* **127,** 1833–1843.

Yamanaka, Y., Mizuno, T., Sasai, Y., Kishi, M., Takeda, H., Kim, C. H., Hibi, M., and Hirano, T. (1998). A novel homeobox gene, dharma, can induce the organizer in a non-cell autonomous manner. *Genes Dev.* **12,** 2345–2353.

Yazulla, S., and Studholme, K. (2001). Neurochemical anatomy of the zebrafish retina as determined by immunocytochemistry. *J. Neurocytol.* **30,** 551–592.

Zeller, J., and Granato, M. (1999). The zebrafish diwanka gene controls an early step of motor growth cone migration. *Development* **126,** 3461–3472.

Zeller, J., Schneider, V., Malayaman, S., Higashijima, S., Okamoto, H., Gui, J., Lin, S., and Granato, M. (2002). Migration of zebrafish spinal motor nerves into the periphery requires multiple myotome-derived cues. *Dev. Biol.* **252,** 241–256.

Zhang, J., and Granato, M. (2000). The zebrafish unplugged gene controls motor axon pathway selection. *Development* **127,** 2099–2111.

Zhang, J., Malayaman, S., Davis, C., and Granato, M. (2001). A dual role for the zebrafish unplugged gene in motor axon pathfinding and pharyngeal development. *Dev. Biol.* **240,** 560–573.

CHAPTER 11

Time-Lapse Microscopy of Brain Development

Reinhard W. Köster⋆ and Scott E. Fraser[†]

⋆GSF–National Research Center
for Environment and Health
Institute of Developmental Genetics
85764 Neuherberg, Germany

[†]Biological Imaging Center
Beckman Institute
California Institute of Technology
Pasadena, California 91125

I. Introduction—Why and When to Use Intravital Imaging

During embryogenesis, the vertebrate brain undergoes dramatic morphological changes. On the cellular level, neural cells have to undergo fate decisions, need to control proliferation and phenotypic differentiation, must sense and interpret positional signals, establish polarity along their cell body, regulate motility and adherence, communicate with homotypic and heterotypic cell types, and form axonal connections throughout the brain. Therefore, the behavior of neural cells is highly dynamic and requires dynamic analytical methods to be fully understood. Moreover, a functional brain is unlikely to result from the independent execution of these developmental processes. Instead, they must be orchestrated carefully in a proper spatial and temporal order. All these complex cellular interactions can hardly be recapitulated in isolated environments of cellular cultures. Noninvasive observation by means of intravital imaging of neural cells and their behavior within their natural environment, the embryonic brain, can thus offer a wealth of information about the interactions and decisions that lead to the formation of a vertebrate brain.

Zebrafish, as a see-through vertebrate model organism for embryogenesis with its external development and rapid brain formation, represents an ideal model organism to tackle these demanding experimental questions of continuous *in vivo* imaging. Furthermore, the accessibility of zebrafish for genetic methods, transient gene knock-down (Nasevicius and Ekker, 2000), and ectopic gene expression, as well as the possibility to establish mutant (Driever *et al.*, 1996; Haffter *et al.*, 1996) and stable transgenic lines (Udvadia and Linney, 2003), allow patterning and signal transduction events to be linked spatially and temporally with their morphological consequences on the tissue, and the cellular and molecular level *in vivo*.

Initially, dyes were used to mark cells and follow their developmental time course. The advent of the green fluorescent protein (GFP) (Chalfie *et al.*, 1994) and its derived and related variants has propelled intravital imaging into new dimensions. These intravital, genetically encoded dyes are regulated and can thus be controlled by cellular transcription and translation. They can be fused to endogenous proteins as fluorescent tags and even be made functional so that they can report intracellular molecular events.

This chapter introduces some of the techniques that can be used for intravital, noninvasive, time-lapse confocal microscopy of zebrafish embryonic brain development. No single experimental approach covers all the questions of brain development. But the different approaches have in common that they all follow the basic line of labeling the embryo and embedding it, followed by data recording and data analysis. Thus, the following sections keep this order, introducing and weighing different known techniques for each experimental step.

II. Techniques for Vital Staining of the Nervous System

Labeling of the specimen is the most important and often the technically most challenging part of intravital imaging. In most experiments efficient labeling determines the project's success or failure. That is why some careful thinking should be dedicated to the labeling approach, the question that is asked, and the specificity of the labeling that is desired before the experiment is started. Different answers can be obtained from different types of labeling.

A. Overview of Techniques

1. Ubiquitous Labeling

Labeling all or almost all cells throughout the embryo is helpful to obtain information about cellular morphologies (e.g., cell polarization) to address morphological changes of tissues, to distinguish between static and motile areas within a tissue of interest, or simply to create a histological counterstain. The easiest labeling method is soaking a dye into zebrafish embryos by adding it to the rearing medium. Most of these vital dyes however are non fixable for further immunohistochemical applications. In addition, dyes can be pressure-injected into zebrafish embryos at the single-cell stage to get distributed to all daughter cells during the subsequent cell divisions. Alternatively, messenger ribonucleic acid (mRNA) encoding for variants of GFP can be injected as it gets evenly distributed among dividing cells. Injecting dyes has the advantage that successful labeling can be scored and recorded right away. Injecting GFP-mRNA causes a delay in labeling because of the required time for translation and protein maturation within the cells; however, the label is amplified manifold through rounds of translation and virtually any subcellular structure can be targeted with proper GFP-fusion proteins.

2. Labeling Cell Clusters Randomly

In contrast to ubiquitous labeling, a mosaic labeling is favorable if individual cellular behavior is to be studied. The very good contrast between the labeled cell or small cell clusters and the unlabeled cellular environment allows even membrane protrusions to be visualized (Köster and Fraser, 2001a). Such labeling can be achieved by pressure injection of dyes or mRNA encoding GFP into individual blastomeres during early cleavage stages. Only descendants of the injected cell will be labeled; as distinct cellular fates are not yet established during these early stages of embryogenesis, the labeled cells will be distributed randomly.

Injected dyes get diluted through cell divisions and mRNA is degraded through time, thus these labeling approaches are mostly limited to the first 30 hours of embryonic development. In contrast, injected GFP-DNA expression vectors are stable through weeks. But even when injected at the single-cell stage, they get distributed in a random mosaic manner and thus result in a mosaic labeling usually marking fewer cells than obtained by dye or mRNA injection. It has been reported, though, that a higher degree of labeling can be achieved by either enhancing the integration frequency of the delivered DNA during early blastula stages by using the meganuclease I-SceI (Thermes et al., 2002), by flanking the enhancer-transgene unit with insulating sequences (Hsiao et al., 2001), or by internal amplification through strong transcriptional activators (Köster and Fraser, 2001b).

3. Targeted Labeling of Specific Cell Clusters

If the lineage of a distinct cell population or the behavior of a distinct cell type of interest is to be studied, a targeted labeling procedure has to be chosen. Using dyes or GFP-mRNA, one can inject embryos at the one-cell stage and later transplant labeled cells from injected donor embryos into gastrula stage embryos (Chen and Schier, 2001) into regions of known fate. (If you would like to see a fate map of the nervous system, turn to Woo and Fraser, 1995.) Alternatively, photo-activatable dyes such as caged fluorescein-dextran (Concha et al., 2003) or photo-convertible GFP (Ando et al., 2002; Patterson and Lippincott-Schwartz, 2002) or photoactivatable caged GFP-mRNA (Ando et al., 2001) can be injected at the single-cell stage and subsequently converted into active fluorophores by focused ultraviolet (UV) light onto the cells of interest.

Plasmid-DNA can be injected at the one-cell stage when the expression construct contains a suitable enhancer driving expression reliably inside the neural cells of interest. Expression, however, remains mosaic within this cell population. Alternatively, using electroporation plasmid-DNA, as well as mRNA, can be electroporated right into the tissue or cells of interest (Concha et al., 2003; Teh et al., 2003).

4. Targeted Single-Cell Labeling

Fate mapping requires that single, individual cells be labeled. This can be achieved via single-cell iontophoresis during which a dye is directly applied to a cell by an electrical field (Woo and Fraser, 1995). This requires that the cells are accessible for capillary injection and limits this technique mostly to surface cells. Single-cell labeling deeper inside tissues of interest is manageable by uncaging photoactivatable dyes with excitation light focused down to sizes smaller than one-cell diameter. Similarly, it has been shown that expression vectors driven by heat-shock inducible promoters can be activated in individual cells of interest by local warming, using focused laser beams (Halloran *et al.*, 2000).

5. Targeted Labeling of Entire Cell Populations

When the mosaic labeling achieved with most transient transgenic methods is to overcome cell type specific enhancers can be used to establish stable transgenic GFP-expressing lines. Usually, established imaging conditions, such as excitation intensity, can be transferred from one labeled specimen to the next, as the labeling intensity remains relatively unchanged. Although most laborious, this nonrandom, targeted, cell-type-specific, genetic labeling offers a broad range of applications such as mutant analysis (Neumann and Nüsslein-Volhard, 2000), screening for mutants, and visualization of regeneration processes (Udvadia *et al.*, 2001). A detailed overview of strategies used to generate stable transgenic zebrafish lines containing a list of currently established transgenic strains has been reported (Udvadia and Linney, 2003).

B. Details of Techniques

1. Soaking Embryos in Dyes for Ubiquitous Labeling

The easiest way of vital staining is by soaking the zebrafish embryo in fluorescent dyes. The most versatile dye for cytoplasmic labeling is the quite photostable boron dipyrromethane (BODIPY). Its lipophilic derivative BODIPY-Ceramide is more specific in labeling only the cytoplasmic membrane, the Golgi apparatus and the interstitial fluid (Cooper *et al.*, 1999; Dynes and Ngai, 1998). As these BODIPY dyes leave the cellular nucleus unlabeled, individual cells can easily be identified. BODIPY comes in different colors (Table I) and can thus be used in combination with other labeling methods such as GFP, DsRed, or monomeric red fluorescent protein (mRFP) expressing cells (Cooper *et al.*, 1999; Dynes and Ngai, 1998; Köster and Fraser, 2001b). Depending on the attempted depth of imaging, soaking for optimal labeling can vary between 30 min and 24 h. BODIPY and its derivatives are non fixable dyes. Therefore, labeling is lost during histological staining procedures such as fluorescent immunostaining. As these dyes work almost as specific on fixed tissue as on living samples BODIPY dyes can be reapplied to immunostained samples to recover the labeling.

a. Protocol

1. Prepare the stock solution by dissolving BODIPY in dimethyl sulfoxide (DMSO) at a concentration of 2.5 μg/μl (store at $-20\,°C$).

2. Place the dechorionated embryo in a small container (e.g., Nunc, Portsmouth, NH, produces 4-well dishes, each of which holds 1 ml).

Table I

Characteristics of Dyes for Vital Stainings in Developing Zebrafish Embryos

Dye (Exc.$_{max}$/Em.$_{max}$)	Delivery	Range	Labeled structure
BODIPY (505/515) BODIPY (548/578)	Soaking	Ubiquitous	Cytoplasm, interstitial fluid

BODIPY Ceramide (505/511)
BODIPY CeramideTR (589/617)
	Soaking	Ubiquitous	Cytoplasmic membrane,
			   Golgi apparatus,
			   interstitial fluid

Hoechst-33342 (350/461)	Soaking	Ubiquitous	Nucleus
YO-PRO (491/509)	Soaking	Apoptotic cells	Nucleus
TO-PRO (515/531)	Soaking	Apoptotic cells	Nucleus

FM 4–64 (515/640)	Soaking	Endocytotic cells	Membrane,
			   synaptic vesicles,
			   endosome

Fluorescein (494/521)	Pressure
	   injection
		Stays within
		   injected cells
			Entire cell

Rhodamine (555/580)	Pressure
	   injection
		Stays within
		   injected cells
			Entire cell

DiO (484/510)
DiI (549/565)
DiD (644/665)
	Pressure
	injection into
	cytoplasm/
	extracellular
	region or
	body cavities
		Stays within
		   injected cell/
		   is carried along
		   by motile cells
			Cytoplasm and
			   cellular membrane/
			   cellular membrane

3. Replace the embryo medium with a 1:1000 dilution of BODIPY (2.5 μg/ml) in 30% Danieau solution and incubate the embryo at 28 °C for the desired time.

4. Rinse the excessive dye off the embryo's skin and image the specimen.

Vital nuclear staining can be achieved using Hoechst 33342 (1:1000 dilution of 10 μg/ml stock in 30% Danieau solution). Detection of the fluorescence of this dye, like many other DNA-binding dyes, requires excitation with UV light (350 nm). The recent incorporation of UV-diode lasers in confocal microscopy systems, as well as multi-photon excitation microscopes, although still expensive equipment, will broaden the applicability of this vital nuclear stain. Other nuclear dyes such as YO-PRO and TO-PRO (see Table I) and their related DNA-binding dyes (1:300 dilution of 1 μg/ml DMSO stock in 30% Danieau solution) share the advantage that they absorb and emit in the visible spectrum. In contrast to Hoechst, these dyes cannot penetrate the cytoplasmic membrane of living cells and stain dying or dead cells only. They can therefore be used as fluorescent apoptotic markers.

To counterstain enhanced green fluorescent protein (EGFP)-labeled cells FM 4-64 (1 mM stock in ethanol, 10 μM in 30% Danieau solution) can be used. This dye enters cells via endocytosis and thus labels the cytoplasmic membrane and endosomes of endocytotically active cells.

2. Pressure Injection of Dyes for Labeling Groups of Cells

To achieve a more cell-specific label, fluorescent dyes can be injected into groups of cells or individual cells. This is commonly used for cell tracing in mapping approaches, to address cell autonomy questions, or to analyze projection patterns of neurons. Two different types of dyes are usually used for this approach, hydrophilic fluorescent dye-coupled dextrans (fluorescein- or rhodamine-coupled) or lipophilic carbocyanines (DiO, DiI, or DiD). Aqueous solutions of fluorescein- or rhodamine-coupled dextrans are poured into glass capillaries and injected through mild air pressure pulses into the cytoplasm of the cells of interest. Here, the dye diffuses quickly and labels the entire cytoplasm and nucleus, resulting in brightly marked cells, but subcellular structures cannot be visualized. The size of the dextran determines whether the dye is allowed to cross gap junctions (molecular weight less than 1 kD) or remains within the labeled cell (usually dextrans around 10.000 and 15.000 kD are used). Furthermore, fluorescein can also be acquired as a photo-caged compound for light inducible labeling approaches. Remember that injection and subsequent incubation and handling of embryos has to occur in the dark to prevent premature fluorophore uncaging (Concha et al., 2003). It is important to note, however, that fluorescein and rhodamine are bleached easily, thus extensive excitation before image recording should be avoided. Because of their dextran residue, these dyes can be fixed in place after live imaging procedures and the analyzed embryo can be subjected to immunohistochemical double labeling techniques.

a. Protocol

1. Dissolve fluorescein/rhodamine dextran at 100 mg/ml.
2. Apply the solution to a Micro-Centricon ultrafiltration tube (Millipore, Billerica, MA), threshold of membrane: 3 kDa, and spin at maximum speed until most of the liquid has drained.
3. Wash several times with distilled water to remove synthetic byproducts.
4. Dissolve the remaining supernatant in water at 100 mg/ml, aliquot into 100 μl volumes, and store at $-20\,^{\circ}$C until needed for injection.

As DiO, DiI, or DiD are lipophilic dyes, stock solutions are usually prepared in 70 or 100% ethanol at a concentration of 0.5% w/v. Depending on the method of administration, different types of labeling can be achieved. Similar to the dextran-coupled dyes, these carbocyanines can be pressure-injected into cells of interest through the use of glass capillaries, resulting in individually labeled cells. Because of immediate precipitation in the aqueous environment, these dyes stay within the injected cell and its descendants; however, the insolubility of DiI-derivatives in water makes vital labeling of zebrafish embryonic cells challenging. Cells labeled by injection can be monitored with regard to their behavior immediately after being labeled. In contrast, focal injections into the extracellular space of tissue regions containing motile cells (e.g., migrating neuronal precursors in the hindbrain) result in a bright depository of DiI or its derivatives. Cell labeling and subsequent observation relies on the fact that precipitating dye crystals attach to the cell membranes. When the motile cells leave the injection site, they carry the dye crystals with them, resulting in bright cellular membrane labeling in an unlabeled environment. Alternatively, large amounts of DiI and its derivatives can be injected as isotonic sucrose solutions into body cavities such as the lumen of the neural tube. Again, cells migrating away from the ventricular surface carry with them the dye crystals as a bright fluorescent label. While the ventricle is labeled too brightly to be imaged, the cells leaving the ventricular region enter a non-fluorescent environment that provides a good contrast. Once attached to a cell, the DiI crystals are not passed onto other cells during the course of cellular migration, ensuring that the same individual cells can be observed during their developmental time course. The introduction of Celltracker (Molecular Probes, Inc., Eugene, Oregon), CM-DiI, which binds the dye through a thiol-reactive moiety to cellular compounds, retains the label within the cell throughout fixation and permeabilization and thus allows immunostainings to be performed on top of DiI-labeling.

b. Protocol

1. Dissolve DiD, DiI, and DiO in 100% ethanol at a concentration of 0.5% w/v.
2. Use self-pulled glass capillaries to inject 1:10 dilution (in 100% ethanol) into cells of interest, **or**
   Focally inject 1:10 dilution (in 100% ethanol) in tissue region of interest, **or**
   Inject 1:10 dilution (in isotonic sucrose) into lumen of neural tube or other body cavities.
3. Record individually labeled cells or motile cells leaving the brightly labeled injection site carrying with them DiI-crystals attached to their membranes.

3. Iontophoretic Labeling of Single Cells

Labeling of individual cells can be achieved by single-cell injection, using iontophoresis at later embryonic stages. Instead of pressure injection, the charged dye (dye-coupled dextrans or DiI-derivatives) is transferred to the pierced cell through directed migration in an electrical field. Intracellular injection, using electrophysiology equipment, permits single cells to be labeled with certainty (cf. Woo and Fraser, 1995) by recording the change in potential that accompanies the penetration of the cell membrane. After injection, the presence of a single labeled cell can be confirmed by brief inspection of the fluorescence signal. Care must be taken to minimize exposure of the newly-injected cells to the exciting wavelengths for the dye, as the bleaching that would otherwise result will create toxic byproducts. The needed components for intracellular iontophoretic labeling and the key features of successful injections are presented elsewhere (Fraser, 1996).

If single cell labeling is not absolutely required, a much simpler and inexpensive approach can be employed. Use visual guidance and a simple battery power supply to label cell groups as small as single cells (often doublets or small clusters of cells) with lipophilic dyes, such as DiI. The carbocyanine dye, diluted in an organic solvent, is ejected from the very sharp pipette tip by current, allowing much greater control than possible with pressure injections. A simple experimental setup and procedure has been described in detail (Fraser, 1996).

4. Quantum Dots

Recently, quantum dots have captured much attention as new biological labeling agents. In principle, they behave like other dyes with extreme photostability. They consist of cadmium selenide, with only the crystal size determining the excitation and emission characteristics. Therefore, with the same chemistry, a wide range of colors can be generated (Jaiswal *et al.*, 2003). Furthermore, their small size allows high fluorophore concentrations to be achieved. When coated by micelles (Dubertret *et al.*, 2002), zinc sulfide (Mattoussi *et al.*, 2001) or silica (Bruchez *et al.*, 1998; Chan and Nie, 1998) these crystals are nontoxic to the organism (zebrafish: Köster *et al.*, in press) and can be further functionalized through added adapter molecules (Lidke *et al.*, 2004). Therefore, quantum dots may soon become very useful for intravital imaging purposes.

5. Genetic Labeling

In dye labeling techniques that use dyes which are not genetically encoded, the dye dilutes with cell growth and mitosis and is lost by photo-bleaching. In contrast, genetically encoded dyes such as GFP, DsRed, or mRFP can be produced continuously by the cell through ongoing transcription and translation, thereby replacing photo-damaged fluorophores. A battery of different GFP variants has been produced and a list, certainly incomplete, of useful GFP variants for intravital zebrafish imaging is provided in Table II.

When used in stable transgenic lines, GFP can label distinct cell types. Some useful strains for imaging brain development are listed in Table III.

These transgenic strains are powerful tools to analyze the dynamics and progression of phenotypes when crossed into mutant strains. But, for many genes, mutant strains have not been identified. In these cases interference strategies with the gene product of interest have to be established. For the first 24 to 30 h of embryogenesis, RNA-based strategies are commonly used with the injection of antisense-morpholinos for loss-of-function (Nasevicius and Ekker, 2000) and

Table II

Different Colors, Subcellular Localizations and Functional Properties of Fluorescent Proteins that are Useful for Intravital Imaging

	GFP Variant	Reference
Different Colors	$Exc._{max}/Em._{max}$	
	EBFP (383 nm/447 nm)	(Finley et al., 2001)
	ECFP (439 nm/476 nm)	(Heim and Tsien, 1996)
	EGFP (484 nm/510 nm)	(Cormack et al., 1996)
	EYFP (512 nm/529 nm)	(Heim and Tsien, 1996)
	DsRed (558 nm/583 nm)	(Matz et al., 1999)
	mRFP (584 nm/607 nm)	(Campbell et al., 2002)
Targeted Fusions		
	NLS-GFP (nuclear)	(Linney et al., 1999)
	H2B-GFP (nuclear)	(Kanda et al., 1998)
	Lyn-GFP (membrane)	(Teruel et al., 1999)
	Unc-GFP (neurites)	(Dynes and Ngai, 1998)
	Gap43-GFP (neurites)	(Moriyoshi et al., 1996)
	Mito-GFP (mitochondria)	(Rizzuto et al., 1995)
	Actin-GFP (actin skeleton)	(Westphal et al., 1997)
	Tubulin-GFP (microtubules)	(Clontech, 1998)
Functionalized Variants		
	Venus (fast folding YFP)	(Nagai et al., 2002)
	Timer (time-dependent green-to-red conversion)	(Verkhusha et al., 2001)
	Kaede (UV-induced green-to-red conversion)	(Ando et al., 2002)
	PA-GFP (photoactivatable)	(Patterson and Lippincott-Schwartz, 2002)
	Flash-pericam (Ca^{2+}-indicator)	(Nagai et al., 2001)
	Inverse pericam (Ca^{2+}-indicator)	(Nagai et al., 2001)
	Ratiometric pericam (Ca^{2+}-indicator)	(Nagai et al., 2001)
	Protein tyrosine kinase activity reporters	(Ting et al., 2001)

capped mRNA for gain-of-function studies. When antisense morpholinos are injected into stable transgenic GFP lines or dye-soaked embryos, intravital imaging approaches can be used as dynamic analytical method. With progressing embryogenesis, however, injected RNA becomes degraded, making the injection of plasmid-DNA, which is stable throughout several weeks, the interference method of choice (Hammerschmidt et al., 1999). As injected plasmid-DNA leads to a mosaic distribution of vector-containing and expressing cells, the challenge is to label the transgene-expressing cells. This can be achieved by creating

Table III

Stable Transgenic Zebrafish Lines with GFP Expression inside the Developing Nervous System that Are Particularly Useful for *in vivo* Imaging Approaches

Transgenic Line	Labeled Tissue	Reference
HuC:gfp	Postmitotic neuronal precursors	(Park *et al.*, 2000)
Gap43:gfp	Neuronal precursors	(Udvadia *et al.*, 2001)
α−tubulin:gfp	Neuronal precursors	(Hieber *et al.*, 1998)
Netrin:gfp	Floor plate, hypo chord	(Rastegar *et al.*, 2002)
Shh:gfp	Retinal ganglion cells	(Neumann and Nüsslein-Volhard, 2000)
Her5:gfp	Neuroectodermal cells at mid-hindbrain boundary	(Tallafuss and Bally-Cuif, 2003)
Pax2.1:gfp	Dorsoposterior telencephalon, anteroposterior diencephalon, posterior tectum, tegmentum, anterior cerebellum, rhombomere 3 and 5, otic vesicle and spinal cord interneurons	(Picker *et al.*, 2002)
Islet1:gfp	Cranial motor neurons	(Higashijima *et al.*, 2000)
Gata2:gfp	Neuronal precursors of vrc in the diencephalon	(Bak and Fraser, 2003; Meng *et al.*, 1997)
Omp:gfp	Neuronal precursors in the olfactory placode	(Yoshida *et al.*, 2002)
Flh:egpf	Neuronal precursors of epithalamus in the telencephalon	(Concha *et al.*, 2003)
Ngn1:gfp, ngn1:rfp	Neuronal precursors throughout neural plate (except spinal cord interneurons)	(Blader *et al.*, 2003)

GFP-fusion proteins, but their functional integrity has to be ensured. Alternatively, simultaneous coexpression of at least two transgenes—with one of them being a marker (GFP)—allows the transgene-expressing cells to be identified and imaged. Two different strategies can be employed. One is the use of internal ribosomal entry sites (IRES), where an expression vector gives rise to a single bicistronic mRNA that gets translated into two different proteins, the protein of interest and GFP. The efficiency at which the IRES-dependent transgene is being translated varies strongly with the IRES that is being used. However, in general the IRES-dependent cistron is expressed at lower levels than the cap-dependent cistron (Fahrenkrug *et al.,* 1999; Köster *et al.,* 1996). A second type of vectors relies on the indirect expression of two cistrons by an enhancer/promoter through the diffusible transcriptional activator Gal4Vp16. Once transcribed and translated,

Gal4Vp16 binds back to the expression vector to several Gal4-specific binding sites, thereby activating the expression of the Gal4-dependent transgene of interest and Gal4-dependent GFP (Köster and Fraser, 2001b). Because of their long-lasting strong fluorescence, these vectors are of use when long-term time-lapse imaging or observation of small size subcellular structures are attempted (Niell et al., 2004).

III. Preparation of the Zebrafish Specimen

A. Imaging Chambers

For imaging, the labeled embryo has to be mounted in an imaging chamber that provides the embryo with the necessary aqueous medium and oxygen but also establishes optimal conditions for the light path. Usually zebrafish embryos are raised in plastic dishes, but most objectives are corrected for glass coverslips. In addition, imaging through plastic strongly scatters the light, diminishing the picture quality. A simple but very efficient imaging chamber can be produced by drilling a hole into the bottom of a cell culture dish (Fig. 1A). A coverslip is subsequently "glued" to the dish bottom through the use of silicon grease to seal the dish again, providing the desired optics of glass for non-immersive imaging (Fig. 1B and C). When an inverted confocal microscope is used, simply covering the imaging dish with the plastic dish cover efficiently restricts evaporation (Fig. 1E). In case an upright microscope is used, plastic rings obtained from culture dishes can be used in a manner similar to an imaging chamber. After sealing one side of the plastic ring with a coverslip, the embryo is embedded on this coverslip, the dish is filled with 30% Danieau solution, and sealed on the other side with a second coverslip. Finally, the entire dish can be flipped over and mounted on the scope.

Immersion objectives can be dipped right into the embryonic rearing medium just above the tissue that is to be imaged. As these objectives are corrected for aqueous solutions, their use does not require specific imaging dishes. However, in case long-term time-lapse recording experiments are being performed, excessive evaporation from this open dish has to be prevented. Sealing the space between the objective and the imaging chamber can be achieved by plastic wrapping of the space between the rim of the dish and the objective. Using a condom is one easy approach to reduce evaporation (Potter, 2000).

Time-lapse imaging, especially at high magnifications, requires that the embryo is lying still, fixed in a desired position as any embryonic movement would move the imaged cells out of focus and field of vision. Therefore, embryos are mounted in a matrix that allows them to further develop but prevents or restricts movements to a minimum. Several mounting techniques, varying mostly in the matrix media, have been found useful.

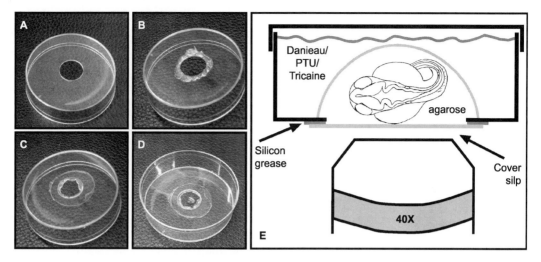

Fig. 1 Agarose embedding for zebrafish time-lapse recordings. To ensure a light path that accounts for the coverslip correction of most objectives and ensures optimal working distances, embryos are embedded in special imaging chambers. (A) These chambers are custom-made by drilling a hole into the bottom of a round 5-cm Petri dish. (B) Silicon grease is applied to the rim of this hole from the bottom of the dish. (C) This allows a glass coverslip to be glued underneath the dish, sealing the hole and providing glass optics for imaging. (D) Subsequently, a zebrafish embryo is transferred—within a drop of ultra-low gelling agarose—to the coverslip and mounted in the proper position before the agarose cools and solidifies. (E) Finally, the embedded embryo is overlaid with 30% Danieau/PTU/ Tricaine, the dish is covered to avoid evaporation, and the entire imaging chamber is mounted on the stage of an inverted confocal laser scanning microscope.

B. Stabilizing the Embryo

1. Methylcellulose Embedding

Methylcellulose in Danieau solution provides a very viscous matrix into which the dechorionated embryo can be transferred and oriented using manipulation needles or forceps (Westerfield, 1995). As the matrix remains pliable, repositioning of the embryo in the desired orientation can be performed. The matrix confers no constraints onto the embryo. It is thus a favorable embedding method for young, fragile embryos during gastrulation or neurulation stages with extensive axis growth. As the matrix remains pliable, movements of the embryo cannot be fully prevented. Therefore, methylcellulose embedding is usually chosen for shorter time-lapse sessions lasting a couple of hours.

a. Protocol
1. Prepare a 3% methylcellulose solution in 30% Danieau solution by stirring for at least 24 h.
2. Pipette a small droplet of methylcellulose solution into the imaging chamber.

3. Transfer the labeled and dechorionated embryo into a methylcellulose droplet and orient it using micromanipulation needles, forceps, or a hair loop.

4. Overlay the droplet and fill the imaging chamber with 30% Danieau solution.

2. Agarose Embedding

High-magnification time-lapse imaging and long-term imaging approaches lasting through several days require that the embryo is restricted from moving. Ultralow gelling agarose provides a matrix that solidifies and thus keeps the embryo still in any desired position (Fig. 1E). It is porous enough to provide the embryo with oxygen, also allowing diffusion of added chemicals such as phenylthiourea (PTU) to prevent pigmentation of the imaged embryo. As the agarose gels at 17 °C, mounting can be performed at temperatures that do not harm the embryo. To some extent, axial growth is still allowed by the agarose. If extensive axial growth of the embryo is expected, the agarose matrix can be removed along the trunk and tail after it has solidified, holding down the head only.

a. Protocol

1. Dissolve 1.2% of ultra-low gelling agarose in 30% Danieau solution by boiling it and subsequently cooling the solution to 28 °C. (A stock of melted agarose/30% Danieau solution can be kept inside the embryo incubator.)

2. Pipette the labeled and dechorionated embryo into the agarose solution.

3. Transfer a drop of agarose containing the embryo onto the glass coverslip of the imaging chamber.

4. Place the imaging chamber onto a cooled surface (e.g., a cell culture dish on ice water) and orient the embryo with micromanipulation needles into the desired position until the agarose solidifies.

5. Remove the imaging chamber from the cooled surface and overlay the embryo with 30% Danieau solution containing 0.010 to 0.005% Tricaine (MS22, 3-aminobenzonic acid ethylester) for sedation. PTU at a 0.75-mM concentration should be added to the Danieau solution if pigmentation interferes with imaging.

3. Plasma Clot Embedding

The embedding techniques already described do not require further manipulation of the labeled embryo. Proper orientation of the specimen before imaging provides convenient access to the dorsal and lateral sides of the brain. The ventral side, instead, is difficult to access and requires immense imaging depth. Recently, an embedding method has been described that allows the embryo to be freed from the yolk and interfering tissue to be dissected away from the ventral side of the brain. Mechanical forces exerted by healing processes that would extensively reshape the dissected brain are inhibited by the paralyzing agent adenosine $5'(\beta,\alpha$-imido) triphosphate (AMP-PNP). Subsequently the dissected brain is held in position by a plasma clot (Langenberg *et al.*, 2003). Although not strictly live imaging of an intact embryo, the entire brain or head is being cultured, providing the cells with their complex natural environment and giving access to regions of the brain that cannot be obtained by other embedding methods.

a. Protocol

1. Drop ca. 20 μl of reconstituted bovine plasma onto a coverslip and let dry.
2. Wet the plasma-covered area with thrombin (100 U/ml).
3. Cover the plasma clot with L15 amphibian culture medium containing penicillin/streptomycin/antimycotics (at a final concentration of 100 U/ml penicillin).
4. Inject 8 nl of 40 mM AMP-PNP into the yolk cell of the labeled embryo.
5. Remove the yolk with a microneedle and dissect the head or brain.
6. Place the tissue on a coverslip and mount the embryo in the desired position. Further stabilization by addition of reconstituted plasma is optional. For a detailed description of this technique, please refer to Langenberg *et al.*, 2003.

IV. The Microscopic System

A. Heating Chamber

To guarantee proper temporal development of the embryo during the image-recording period, the environmental temperature of the microscopic stage needs to be adjusted to 29 or 30 °C. Microscope companies often provide incubation chambers for their microscopic setup.

Constructing your own incubation chamber can help you avoid these additional costs. Cardboard covered with insulating foil can be used to design custom-fit chambers around the microscopic stage, still leaving enough space to move the stage for focusing purposes (Fig. 2A and B). Strips of velcro are helpful to hold the individual pieces of the chamber together and to mount the chamber on the confocal microscope. For image quality, the heating chamber should not cover the photomultipliers as their signal-to-noise ratio correlates inversely with the temperature.

B. Heater

Depending on the available space, the heater can either be positioned outside (Fig. 2A) or included into the heating chamber (Fig. 2B). When mounted outside, the heated air can be fed into the chamber by flexible plastic or aluminum tubes that can be purchased at almost any hardware store. The heated air should not

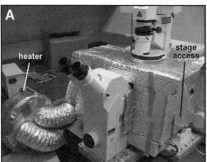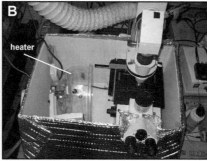

Fig. 2 Heating chambers during time-lapse recording sessions provide the desired environmental temperature for proper temporal development of the imaged specimen. Efficient heating chambers can be built from cardboard covered with insulating foil (A and B). Individual chamber parts can be held together with tape or Velcro. By sealing the microscopic stage (A), they serve both to stabilize the interior temperature and prevent environmental light from passing through the objective to the detectors. Thus, they provide the imaged object with its ideal developmental temperature, minimize evaporation and condensation, and improve the signal-to-noise ratio, respectively. Depending on the available space, the heater can feed the incubation chamber from outside by flexible tubes (A) or the heater can be included into the chamber (B) with the restriction that it should not blow the heated air directly across the microscopic stage. Note that the detectors should not be included in the chamber, as their signal-to-noise performance decreases with increasing temperatures.

be blown directly across the microscopic stage but rather heat the environmental space surrounding the stage. This prevents temperature differences from building up within the imaging chamber and avoids movements of the medium inside the dish. Therefore, heaters with a weak fan are preferable. Also, heaters with low heating power but sensitive temperature sensors should be chosen. Although requiring some preparatory time for heating the interior of the chamber to the desired temperature, overshooting and extensive fluctuation around the aimed temperature is avoided. This is important as temperature changes during the image-recording period lead to shrinkage or expansion of the mounting medium holding the specimen and therefore result in a change of the focal plane. We found that small incubators used for breeding of chicken eggs can fulfill these requirements within affordable costs. To keep the temperature inside the heating chamber stable, the incubator can be controlled by a thermostat with its sensor placed inside the heating chamber next to the microscope stage.

V. Data Recording

A. Preparations

Before starting to record actual images, the specimen should be mounted and the imaging chamber placed on the microscope for equilibration purposes for at least an hour up to several hours. Therefore time-lapse experiments require proper planning to ensure that the actual developmental process that is to be observed does not commence until stable image recording is achieved. Warming of the microscopic stage, the objectives, and the optical parts inside the confocal microscope extensively influence the position of the optical plane.

Three-dimensional imaging through time is performed by recording a stack of images for a field of interest along its z-axis, with repetition of this procedure after a certain time interval, for the duration of the entire session. Every experimental time-lapse approach requires its specific imaging conditions that have to be established by test experiments. Imaging depth, magnification, laser power, size of pinhole, distance of individual sections along the z-axis, recording period, and recording interval have to be balanced. These careful adjustments are necessary to avoid heating of the observed tissue and cells by laser light (causing photo-toxicity effects), to minimize bleaching of the label, and to choose the proper mounting setup (e.g., methylcellulose versus agarose embedding). The following suggestions and rules of thumb might help you find the proper conditions faster.

B. Mounting

Long time-lapse recording periods, longer than 24 h, require that the embryo be mounted in a stable setting such as in agarose. To prevent the agarose from floating off the coverslip during long imaging periods, a plastic mesh can be "glued" onto the coverslip with silicon grease before the application of the embryo within the liquid agarose. Incorporation of the mesh into the solidifying agarose will hold down the mounted specimen for several days.

C. Choice of Specimen

A brightly labeled cell requires less laser power to return a bright signal than a moderately labeled one. In the long run, however, the brightly labeled cell tends to be more sensitive to heating and photo-toxicity. Therefore, the brightest embryo is not necessarily the best for time-lapse imaging.

D. Saturation

To ensure that the best resolution of the recorded images is achieved, their histogram should be checked regularly. For best results, the number of gray levels used should be maximized, avoiding underexposure or oversaturation of the sample as much as possible by adjusting the gain and black level of the detector. Usually these conditions do not vary extensively during a time-lapse recording session but they may need to be adjusted regularly, as bleaching of the label (or probably developmental increase in GFP-expression levels), growth, and movement of the tissue (change in imaging depth) influence these parameters.

E. Pinhole

The size of the pinhole is tightly correlated to the distance at which individual confocal z-sections need to be recorded. The best axial resolution will be achieved if the pinhole is closed down to one Airy unit, which itself is dependent on the objective that is used and its numerical aperture. At this setting (one Airy unit), photons from outside the focal plane are prevented from reaching the detector while all the photons from within the focal plane are allowed to pass. Closing the pinhole below one Airy unit does not yield much further resolution along the z-axis but severely decreases the number of photons from the focal plane and thus the brightness and contrast of the image. In most cases, however, closing the pinhole down to one Airy unit will require too many z-sections to cover the entire image depth without running into photo-toxicity problems. Therefore, a compromise between laser power, number of z-sections for a given imaging depth, and gain of the detector have to be found. Keeping the cells alive is better than recording the perfect image. A proper 3D reconstruction can still be achieved if an object (e.g., cell, nucleus, growth cone) appears in at least two consecutive z-sections.

F. Time Interval

Choosing the proper time interval, after which image recording along the z-axis is repeated to generate time-lapse movies, strongly depends on the developmental or cellular process that is being analyzed and on the experimental question that is being asked. For example, fate-mapping approaches require the continuous identification of the same individual cell through time. To unambiguously trace individual cells from images of one time point to the next, the cells should not have moved by more than half of their cellular diameter. Therefore, the maximal time interval that can be chosen between two recording time points depends on the speed with which the observed cell moves. Table IV gives some approximate time intervals that can be chosen as starting conditions for different cellular processes when time-lapse imaging brain development.

Table IV

Dependence of the Approximate Recording Interval on the Observed Cellular or Subcellular Process when Recording Images for Generating Time-Lapse Movies

Process	Time interval
Brain morphogenesis	15–20 min
Neuronal migration	10–15 min
Axonal pathfinding	1–5 min
Nuclear translocation in migrating cells	1–5 min
Calcium currents	20 sec
Mitochondrial translocations	3–5 sec
Actin dynamics	1–5 sec

G. Localizing Cells

Ideally, a transmitted light picture of the imaged tissue is taken at every recording time point to locate the imaged cells within the context of the surrounding tissue. At least at the beginning, the end, and several times during the image-recording period such a transmitted light picture should be taken for orientation purposes. Alternatively, a fluorescent counterstain with different excitation/emission properties than the actual label can be used (e.g., GFP-labeling and BODIPY548/578-counterstain; Dynes and Ngai, 1998).

H. Refocusing

Long time-lapse recording sessions require occasional refocusing. Sometimes the imaged specimen slowly moves out of the field of vision because of continued growth and ongoing morphogenetic processes. This requires the moving of the embryo back into the field of vision and the capturing of a new transmitted light image/counterstain image before continuing the time-lapse recording. Image realignment during data analysis procedures can correct for the sudden "jump" of the imaged cells that have been caused by moving the microscopic stage.

VI. Data Analysis

A single time-lapse recording session usually yields a large amount of data in the range of several hundred megabytes or even gigabytes. To get a first impression of the quality of the data, the developmental processes that the data cover and the cellular behavior that the data reveal, an image analysis and rendering software are needed. They should not be too time-consuming to obtain first movie animations.

A. NIH *Image*

In our hands, National Institutes of Health (NIH) *Image* (rsb.info.nih.gov/nih-image/) has been found to be very suitable for this purpose. Developed by Wayne Rasband (Rasband and Bright, 1995) at NIH, it is a public domain program and thus free of costs, which is significant since commercial image analysis products can cost tens of thousands of U.S. dollars. *Image* was originally developed for Macintosh. A Windows version, Scion Image, is now available (Scion Corporation, Frederick, MD; www.scioncorp.com). Meanwhile, a more versatile version of NIH *Image*, Object-Image (University of Virginia, Charlottesville; simon.bio.uva.nl/object-image.html) and a Java version, ImageJ (National Institutes of Health; rsb.info.nih.gov/ij/) that runs on any Java-supporting platform, have been released. Besides allowing basic image rendering, measuring, and analysis operations, the strength of NIH *Image* and its updated versions is that user-developed macros can be applied. This enables almost any image rendering and analysis step to be automated, speeding up the initial analysis conveniently. Furthermore, a NIH *Image* user group has been set up that allows a huge community of NIH *Image* users to be reached and addressed with specific image recording and analysis questions. To sign up, use "subscribe nih-image" as the subject and send to (LISTSERV@LIST.NIH.GOV). Besides writing to the community for help, all the messages distributed in the past have been archived, which can be searched for possible solutions. Therefore, this community represents a very valuable tool to help you find any kind of macro you need to automate the initial data analysis. Another option for getting help with image recording difficulties using laser-scanning microscopy is provided by a user group called Confocal Microscopy List, which also contains a searchable archive (listserv.buffalo.edu/archives/confocal.html).

B. Projections

When working with data from 3D time-lapse recordings, a first quick analysis can be obtained by projecting the individual image stacks of each recorded time point into a single plain (Fig. 3A). This can be done either by a mean value projection or a brightest point projection. A mean value projection will calculate an average brightness from all pixels with the same xy coordinates within a given z-stack and project it to the single image formed from the z-stack. A brightest point projection will instead use the brightest level found along the z–axis for given xy coordinates (Fig. 3A). Both types of projections can be useful depending on the dynamic process that is to be observed and should be tried. This flattening of the 3D data into two dimensions results in a single image for each recorded time point (Fig. 3C), and these images can be immediately animated into a movie. If the embryo had to be moved during the image recording session as already described, a sudden "jump" of labeled structures will appear in the movie. This movement can be corrected by using image rendering programs such as Adobe Photoshop (Adobe Systems Incorporated, San Jose, CA) to apply a movement in the opposite direction to the images that were recorded after the microscopic stage had been moved. This, however, requires that either the labeled structures only move minutely during two subsequent time points (which you must keep in mind when choosing the recording interval) or that static points are available in the pictures. For example, autofluorescing nonmotile skin cells could be used as orientation for realigning pictures. Using such programs as NIH *Image* and Photoshop, for the described movie-rendering process has the advantage that all rendering steps can be automated, giving a relatively quick access to a first movie.

C. LSM Software

Recently, the laser scanning microscopy (LSM) manufacturers have started to provide their own image-rendering software packages that are quite powerful (LaserPix from Bio-Rad Laboratories, Hercules, CA; LCS—Leica Confocal

Software from Leica Microsystems, Inc., Exton, PA; GUI—Graphical User Interface from Nikon, Melville, NY; FluoView software from Olympus, Melville, NY; and LSM510—Laser Scanning Microscope Software from Zeiss, Thornwood, NY). An advantage of using these software tools is that the compatibility of the recorded data with the accepted format of the rendering and analysis software is not an issue. Furthermore, many useful data such as the applied imaging conditions are being stored together with the recorded images, facilitating measurements and scaling. In addition, these software packages offer some advanced data analysis features that go beyond the capabilities of NIH *Image* and can only be found otherwise in software packages that are specialized in image rendering, such as Amira (Indeed—Visual Concepts GmbH, Berlin, Germany), Metamorph (Universal Imaging Corporation, Downingtown, PA), Imaris (Bitplane AG, St. Paul, MN), Slidebook (Intelligent Imaging Innovations, Santa Monica, CA), Volocity (Improvision, Lexington, MA) or VoxelView (Vital Images, Inc., Plymouth, MN). For more information about these specialized software products, a comparison of their different properties, advantages, and disadvantages can be found in Megason and Fraser, 2003. For some of these software packages, demonstration versions can be downloaded and tested free of charge:

Amira: http://www.amiravis.com/download.html

Imaris: http://www.bitplane.com/download/archive.shtml

Volocity: http://www.improvision.com/downloads/default.lass

D. Three-Dimensional Renderings Through Time

To maintain the 3D nature of the recorded image stacks, the obtained data can be visualized when using these advanced software products by creating stereooptic pairs of projections from each recorded z-stack of pictures for every individual time point. An alternative is represented by red-green rendering of the projections, creating a 3D impression to the viewer when wearing red-green glasses. As helpful as these movies are, they have the disadvantage of being difficult to publish and of being inaccessible to people with red-green blindness, which affects about 10% of all humans. A further option that allows presenting the 3D data in two dimensions can be obtained by color coding the different z-values (Fig. 3B and D). Therefore, structures such as cells, axons, and mitochondria that move along the z-axis will change their color through time. This allows one, for example, to address whether cells cross each other's pathways during migration, migrate toward each other, or disperse. To obtain information about absolute changes of imaged structures along the z-axis, like cells moving deeper into the brain or toward its surface, one has to make sure that the embedded embryo itself is not moving along the z-axis. Fix points such as nonmotile skin cells have to stay in the same z-level (i.e., maintain their color) within an analyzed stack through time.

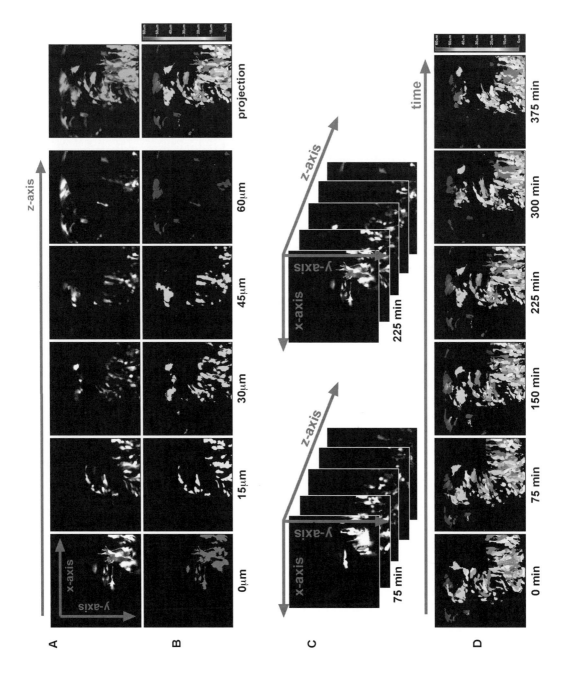

E. Cell Tracking

A challenge for most software is the automated tracking of cells or other labeled structures in 3D time-lapse recordings within a living embryo. For tracking in two dimensions through time using projections of 3D-image stacks, several software packages exist with the dynamic image analysis system (DIAS) software (http://www.geocities.com/solltech/dias/) being capable of analyzing movies in Quick-Time format. From such tracking data, a wealth of information can be obtained, such as the directionality of movements or changes in migratory speeds (Glickman *et al.*, 2003; Kulesa and Fraser, 1998, 2000). However, most software products appear to fail to reliably track individual cells through time automatically under these *in vivo* conditions. In addition, the tracking has to be done manually, image by image, which can be quite laborious. Improved labeling techniques and multi-color labeling of different subcellular structures might help to solve this problem by providing tracking software with more individual information for every single cell.

VII. Pitfalls to Avoid

Time-lapse imaging, although not difficult per se, requires some practice to identify general pitfalls and the specific pitfalls for each approach, depending on the labeling procedure and quality, the age of the embryo, the tissue, and the process that is to be observed. Therefore, an exhaustive list of potential pitfalls cannot be provided and will not be attempted here. But some of the following general rules of thumb and suggestions might be helpful.

A. Technical Pitfalls

1. DNA Purification

Successful mosaic labeling by the injection of plasmid DNA is very dependent on the purity of the DNA and the embryonic stage at which the DNA is being injected. In our hands, further cleaning of maxi prep DNA with the Geneclean Turbo Kit from Qbiogene, Carlsbad, CA, before injection worked best. In addition, the plasmid DNA should be injected as early as possible at the one-cell stage just when the blastodisk of the fertilized egg becomes visible.

Fig. 3 Analysis of 3D time-lapse recordings. (A) A fast and efficient approach to analyze z-stacks of images that have been recorded through time is to project the individual images of every z-stack/time point into a single plane by either mean value or brightest point projection (shown here). (B) If the 3D nature of the data is to be retained, the recorded signal can be coded with pseudo colors, reflecting the individual z-values within each z-stack. (C) Once every z-stack of each recorded time point is projected into a single plain, these projections can be animated into a movie to allow individual cells to be followed. (D) In the case of color-coded z-levels, movements of cells along the z-axis can be observed by cells changing colors while migrating. (See Color Insert.)

2. Fixation of the Mounted Embryo

Time-lapse recordings that last through several days can be demanding on the individual embedding techniques. Agarose mounted on a coverslip can come loose during these long recordings, with the embedded embryo floating out of focus. This can be prevented by incorporating a small-size plastic mesh into the agarose by gluing the mesh with silicon grease onto the coverslip of the imaging chamber before applying the embryo within the liquid agarose drop. The mesh will hold down the agarose and prevent it from floating away.

3. Colocalization of Fluorophores

For colocalization of two different fluorophores, the excitation and detection settings have to separate both emissions properly. Ideally, controls within the same scanning plane should be attempted to ensure that the overlapping signal is obtained from the presence of both fluorescent probes instead of one fluorophore being detected by both detection channels. Also, remember that the pinhole has to be adjusted differently for both dyes to record data from an optical section of identical thickness because the thickness of an optical slice for a given pinhole depends on the wavelength of the recorded light.

4. Storage during Data Recording

Data obtained from time-lapse recordings are often stored in the temporary memory of the computer. Once the temporary memory is filled, the confocal microscope will stop recording and tends to crash. Therefore, if long time-lapse recording sessions are attempted, one should make sure that the recorded images are being saved directly to the hard drive. This option is usually available in the settings menu of the confocal microscope software. Also, ensure that the hard drive contains enough empty space for saving the expected amount of data.

5. Image Resolution

This point leads immediately to the resolution at which images are being recorded. Modern confocal microscopes allow data to be recorded not only at a 512- × 512-pixel resolution but also at 1024 × 1024 and 2048 × 2048 pixels. This means that an individual gray-scale image increases in size from 256 KB to 1 MB to 4 MB. A 24-h 3D time-lapse recording at 2048- × 2048-pixel resolution will lead to large data files for which enough space on the hard drive has to be available. In addition, later handling of these files might be difficult and imaging software is often unable to perform 4D time-lapse rendering processes with these large amounts of data. Therefore, it is important to test the resolution that is required to sufficiently answer the experimental question with short time-lapse recordings to make sure that one does not run into later analysis problems with the obtained data.

B. Analytical Pitfalls

1. Annotation of Recorded Data

Three-dimensional time-lapse data that have been projected into a single plane need to be reanalyzed in three dimensions if answers about cell to cell contacts are to be obtained. Interacting cells need to be localized within the same plane instead of moving above and below each other. Often it is helpful to generate rocking or rotating projections of z-stacked pictures to look at the 3D volume from different angles.

2. Reference Points

Depending on the angle at which a specimen is being optically sectioned, different results can be obtained. Two different cell types can appear within the same plane while in the next specimen they do not share the same z-value. Similarly, distances between structures can vary from specimen to specimen as the position at which the embryo is embedded; therefore, the angle of sectioning changes every time. To ensure that meaningful data are obtained, independent reference points with which to correlate the data are needed.

3. Identifying Individual Cells

In tracking experiments, it is required that individual cells be followed with certainty from one time point to the next. As a rule of thumb, cells should not move further than half of their cell diameter during the time interval of two subsequent recordings to allow an unambiguous correlation of cells.

VIII. Summary

Zebrafish embryos represent an ideal vertebrate model organism for noninvasive intravital imaging because of their optical clarity, external embryogenesis, and fast development. Many different labeling techniques have been adopted from other model organisms or newly developed to address a wealth of different developmental questions directly inside the living organism. The parallel advancements in the field of optical imaging let us now observe dynamic processes at the cellular and subcellular resolution. Combined with the repertoire of available surgical and genetic manipulations, zebrafish embryos provide the powerful and almost unique possibility to observe the interplay of molecular signals with cellular, morphological, and behavioral changes directly within a living and developing vertebrate organism. A bright future for zebrafish is yet to come, let there be light.

Acknowledgments

We thank Veronika Zapilko for critically reading the manuscript and commenting on it. We are grateful to Dr. Laure Bally-Cuif for the schematic drawing in Figure 1. R. W. K. was supported by the NIH and the BMBF Biofuture-Award (0311889). S. E. F. was supported by the Biological Imaging Center of the Beckman Institute, Caltech and the NIH.

References

Ando, H., Futura, T., Stein, R. Y., and Okamoto, H. (2001). Photo-mediated gene activation using caged RNA/DNA in zebrafish embryos. *Nature Genetics* **28,** 317–325.

Ando, R., Hama, H., Yamamoto-Hino, M., Mizuno, H., and Miyawaki, A. (2002). An optical marker based on the UV-induced green-to-red photoconversion of a fluorescent protein. *Proc. Natl. Acad. Sci. USA* **99,** 12651–12656.

Bak, M., and Fraser, S. E. (2003). Axon fasciculation and differences in midline kinetics between pioneer and follower axons within commissural fascicles. *Development* **130,** 4999–5008.

Blader, P., Plessy, C., and Strahle, U. (2003). Multiple regulatory elements with spatially and temporally distinct activities control neurogenin1 expression in primary neurons of the zebrafish embryo. *Mechanisms Development* **120,** 211–218.

Bruchez, M. J., Moronne, M., Gin, P., Weiss, S., and Alivisatos, A. P. (1998). Semiconductor nanocrystals as fluorescent biological labels. *Science* **281,** 2013–2016.

Campbell, R. E., Tour, O., Palmer, A. E., Steinbach, P. A., Baird, G. S., Zacharias, D. A., and Tsien, R. Y. (2002). A monomeric red fluorescent protein. *Proc. Natl. Acad. Sci. USA* **99,** 7877–7882.

Chalfie, M., Tu, Y., Euskirchen, G., Ward, W. W., and Prasher, D. C. (1994). Green fluorescent protein as a marker for gene expression. *Science* **263,** 802–805.

Chan, W. C., and Nie, S. (1998). Quantum dot bioconjugates for ultrasensitive nonisotopic detection. *Science* **281,** 2016–2018.

Chen, Y., and Schier, A. F. (2001). The zebrafish nodal signal squint functions as a morphogen. *Nature* **411,** 607–610.

Clontech Laboratories. (1998). Living Colors subcellular localization vectors. CLONTECHniques, Palo Alto, CA, **XIII,** 8-9

Concha, M. L., Russel, C., Regan, J. C., Tawk, M., Sidi, S., Gilmour, D. T., Kapsimali, M., Sumoy, L., Goldstone, K., Amaya, E., *et al.* (2003). Local tissue interactions across the dorsal midline of the forebrain establish CNS laterality. *Neuron* **39,** 423–438.

Cooper, M. S., D'Amico, L. A., and Henry, C. A. (1999). Confocal microscopy analysis of morphogenetic movements. *Meth. Cell Biol.* **59,** 179–204.

Cormack, B. P., Valdivia, R. H., and Falkow, S. (1996). FACS-optimized mutants of the green fluorescent protein (GFP). *Gene* **173,** 33–38.

Driever, W., SolnicaKrezel, L., Schier, A. F., Neuhauss, S. C. F., Malicki, J., Stemple, D. L., Stainier, D. Y. R., Zwartkruis, F., Abdelilah, S., Rangini, Z., *et al.* (1996). A genetic screen for mutations affecting embryogenesis in zebrafish. *Development* **123,** 37–46.

Dubertret, B., Skourides, P., Norris, D. J., Noireaux, V., Brivanlou, A. H., and Libchaber, A. (2002). *In vivo* imaging of quantum dots encapsulated in phospholipid micelles. *Science* **298,** 1759–1762.

Dynes, J. L., and Ngai, J. (1998). Pathfinding of olfactory neuron axons to stereotyped glomerular targets revealed by dynamic imaging in living zebrafish embryos. *Neuron* **20,** 1081–1091.

Fahrenkrug, S. C., Clark, K. J., Dahlquist, M. O., and Hackett, P. B. (1999). Dicistronic gene expression in developing zebrafish. *Marine Biotechnol.* **1,** 552–561.

Finley, K. R., Davidson, A. E., and Ekker, S. C. (2001). Three-color imaging using fluorescent proteins in living zebrafish embryos. *Biotechniques* **31,** 66–73.

Fraser, S. E. (1996). Iontophoretic Dye Labeling of Embryonic Cells. *Meth. Cell Biol.* **51,** 147–160.

Glickman, N. S., Kimmel, C. B., Jones, M. A., and Adams, R. J. (2003). Shaping the zebrafish notochord. *Development* **130,** 873–887.

Haffter, P., Granato, M., Brand, M., Mullins, M. C., Hammerschmidt, M., Kane, D. A., Odenthal, J., vanEeden, F. J. M., Jiang, Y. J., Heisenberg, C. P., *et al.* (1996). The identification of genes with unique and essential functions in the development of the zebrafish, Danio rerio. *Development* **123**, 1–36.

Halloran, M. C., Sato-Maeda, M., Warren, J. T., Su, F. Y., Lele, Z., Krone, P. H., Kuwada, J. Y., and Shoji, W. (2000). Laser-induced gene expression in specific cells of transgenic zebrafish. *Development* **127**, 1953–1960.

Hammerschmidt, M., Blader, P., and Strahle, U. (1999). Strategies to perturb zebrafish development. *Meth. Cell Biol.* **59**, 87–115.

Heim, R., and Tsien, R. Y. (1996). Engineering green fluorescent protein for improved brightness, longer wavelengths and fluorescence resonance energy transfer. *Current Biol.* **6**, 178–182.

Hieber, V., Dai, X. H., Foreman, M., and Goldman, D. (1998). Induction of alpha 1-tubulin gene expression during development and regeneration of the fish central nervous system. *J. Neurobiol.* **37**, 429–440.

Higashijima, S., Hotta, Y., and Okamoto, H. (2000). Visualization of cranial motor neurons in live transgenic zebrafish expressing green fluorescent protein under the control of the islet-1 promoter/enhancer. *J. Neurosci.* **20**, 206–218.

Hsiao, C.-D., Hsieh, F.-J., and Tsai, H.-J. (2001). Enhanced expression and stable transmission of transgenes flanked by inverted terminal repeats from adeno-associated virus in zebrafish. *Developmental Dynamics* **220**, 323–336.

Jaiswal, J. K., Mattoussi, H., Mauro, J. M., and Simon, S. M. (2003). Long-term multiple color imaging of live cells using quantum dot bioconjugate. *Nature Biotechnol.* **21**, 47–51.

Kanda, T., Sullivan, K. F., and Wahl, G. M. (1998). Histone-GFP fusion enables sensitive analysis of chromosome dynamics in living mammalian cells. *Current Biol.* **8**, 377–385.

Köster, R., Götz, R., Altschmied, J., Sendtner, R., and Schartl, M. (1996). Comparison of monocistronic and bicistronic constructs for neurotrophin transgene and reporter gene expression in fish cells. *Molec. Marine Biol. Biotechnol.* **5**, 1–8.

Köster, R. W., and Fraser, S. E. (2001a). Direct imaging of in vivo neuronal migration in the developing cerebellum. *Current Biol.* **11**, 1858–1863.

Köster, R. W., and Fraser, S. E. (2001b). Tracing transgene expression in living zebrafish embryos. *Developmental Biol.* **233**, 329–346.

Kulesa, P. M., and Fraser, S. E. (1998). Neural crest cell dynamics revealed by time-lapse video microscopy of whole embryo chick explant cultures. *Developmental Biology* **204**, 327–344.

Kulesa, P. M., and Fraser, S. E. (2000). In ovo-time-lapse analysis of chick hindbrain neural crest cell migration shows cell interactions during migration to the branchial arches. *Development* **127**, 1161–1172.

Langenberg, T., Brand, M., and Cooper, M. S. (2003). Imaging brain development and organogenesis in zebrafish using immobilized embryonic explants. *Developmental Dynamics* **228**, 464–474.

Lidke, D. S., Nagy, P., Heintzmann, R., Arndt-Jovin, D. J., Post, J. N., Grecco, H. E., Jares-Erijman, J., and Jovin, T. M. (2004). Quantum dot ligands provide new insights into erbB/HER receptro-mediated signal transduction. *Nature Biotechnol.* **22**, 198–203.

Linney, E., Hardison, N. L., Lonze, B. E., Lyons, S., and DiNapoli, L. (1999). Transgene expression in zebrafish: A comparison of retroviral-vector and DNA-injection approaches. *Developmental Biol.* **213**, 207–216.

Mattoussi, H., Mauro, J. M., Goldman, E. R., Green, T. M., Anderson, G. P., Sundar, V. C., and Bawendi, M. G. (2001). Bioconjugation of highly luminescent colloidal CdSe-ZnS quantum dots with an engineered two-domain recombinant protein. *Physica Status Solidi* **224**, 277–283.

Matz, M. V., Fradkov, A. F., Labas, Y. A., Savitsky, A. P., Zaraisky, A. G., Markelov, M. L., and Lukyanov, S. A. (1999). Fluorescent proteins from nonbioluminescent Anthozoa species. *Nature Biotechnol.* **17**, 969–973.

Megason, S. G., and Fraser, S. E. (2003). Digitizing life at the level of the cell: High-performance laser-scanning microscopy and image analysis for in toto imaging of development. *Mechanisms of Development* **120**, 1407–1420.

Meng, A., Tang, H., Ong, B. A., Farrell, M. J., and Lin, S. (1997). Promoter analysis in living zebrafish embryos identifies a cis-acting motif required for neuronal expression of GATA-2. *Proc. Natl. Acad. Sci. USA* **94**, 6267–6272.

Moriyoshi, K., Richards, L. J., Akazawa, C., O'Leary, D. D. M., and Nakanishi, S. (1996). Labeling neural cells using adenoviral gene transfer of membrane-targeted GFP. *Neuron* **16**, 255–260.

Nagai, T., Iabata, K., Park, E. S., Kubota, M., Mikoshiba, K., and Miyawaki, A. (2002). A variant of yellow fluorescent protein with fast and efficient maturation for cell-biological applications. *Nature Biotechnol.* **20**, 87–90.

Nagai, T., Sawano, A., Park, E. S., and Miyawaki, A. (2001). Circularly permuted green fluorescent proteins engineered to sense Ca2+. *Proc. Natl. Acad. Sci. USA* **98**, 3197–3202.

Nasevicius, A., and Ekker, S. C. (2000). Effective targeted gene 'knockdown' in zebrafish. *Nature Genetics* **26**, 216–220.

Neumann, C. J., and Nüsslein-Volhard, C. (2000). Patterning of the zebrafish retina by a wave of sonic hedgehog activity. *Science* **289**, 2137–2139.

Niell, C. M., Meyer, M. P., and Smith, S. J. (2004). *In vivo* imaging of synapse formation on a growing dendritic arbor. *Nature Neurosci.* **7**, 254–260.

Park, H. C., Kim, C. H., Bae, Y. K., Yeo, S. Y., Kim, S. H., Hong, K. S., Shin, J., Yoo, K. W., Hibi, M., Hirano, T., *et al.* (2000). Analysis of upstream elements in the HuC promoter leads to the establishment of transgenic zebrafish with fluorescent neurons. *Developmental Biol.* **227**, 279–293.

Patterson, G. H., and Lippincott-Schwartz, J. (2002). A photoactivatable GFP for selective photolabeling of proteins and cells. *Science* **297**, 1873–1877.

Picker, A., Scholpp, S., Böhli, H., Takeda, H., and Brand, M. (2002). A novel positive transcriptional feedback loop in midbrain-hindbrain boundary development is revealed through analysis of the zebrafish pax2.1 promoter in transgenic lines. *Development* **129**, 3227–3239.

Potter, S. M. (2000). Two-Photon Microscopy for 4D Imaging of Living Neurons. *In* "Imaging Neurons: A Laboratory Manual." (L. A. K. Yuste, ed.), pp. 20.1–20.16. Cold Spring Harbor Laboratory Press, Cold Spring Harbor, NY.

Rasband, W. S., and Bright, D. S. (1995). NIH Image: A public domain image processing program for the Macintosh. *Microbeam Analysis Soc. J.* **4**, 137–149.

Rastegar, S., Albert, S., Le Roux, I., Fischer, N., Blader, P., Müller, F., and Strähle, U. (2002). A floor plate enhancer of the zebrafish netrin1 gene requires Cyclops (Nodal) signaling and the winged helix transcription factor FoxA2. *Developmental Biol.* **252**, 1–14.

Rizzuto, R., Brini, M., Pizzo, P., Murgia, M., and Pozzan, T. (1995). Chimeric green fluorescent protein as a tool for visualizing subcellular organelles in living cells. *Current Biol.* **5**, 635–642.

Tallafuss, A., and Bally-Cuif, L. (2003). Tracing of her5 progeny in zebrafish transgenics reveals the dynamics of midbrain-hindbrain neurogenesis and maintenance. *Development* **130**, 4307–4323.

Teh, C., Chong, S. W., and Korzh, V. (2003). DNA delivery into anterior neural tube of zebrafish embryos by electroporation. *Biotechniques* **35**, 950–954.

Teruel, M. N., Blanpied, T. A., Shen, K., Augustine, G. J., and Meyer, T. (1999). A versatile microporation technique for the transfection of cultured CNS neurons. *J. Neurosci. Methods* **93**, 37–48.

Thermes, V., Grabher, C., Ristoratore, F., Bourrat, F., Choulika, A., Wittbrodt, J., and Joly, J.-S. (2002). *I-SceI* meganuclease mediates highly efficient transgenesis in fish. *Mechanisms of Development* **118**, 91–98.

Ting, A. Y., Kain, K. H., Klemke, R. L., and Tsien, R. Y. (2001). Genetically encoded fluorescent reporters of protein tyrosine kinase activities in living cells. *Proc. Natl. Acad. Sci. USA* **98**, 15003–15008.

Udvadia, A. J., Koster, R. W., and Skene, J. H. P. (2001). GAP-43 promoter elements in transgenic zebrafish reveal a difference in signals for axon growth during CNS development and regeneration. *Development* **128,** 1175–1182.

Udvadia, A. J., and Linney, E. (2003). Windows into development: Historic, current, and future perspectives on transgenic zebrafish. *Developmental Biol.* **256,** 1–17.

Verkhusha, V. V., Otsuna, H., Awasaki, T., Oda, H., Tsukita, S., and Ito, K. (2001). An enhanced mutant of red fluorescent protein DsRed for double labeling and developmental timer of neural fiber bundle formation. *J. Biological Chemistry* **276,** 29621–29624.

Westerfield, M. (1995). "The Zebrafish Book." University of Oregon Press, Eugene, OR.

Westphal, M., Jungbluth, A., Heidecker, M., Muhlbauer, B., Heizer, C., Schwartz, J. M., Marriott, G., and Gerisch, G. (1997). Microfilament dynamics during cell movement and chemotaxis monitored using a GFP-actin fusion protein. *Current Biol.* **7,** 176–183.

Woo, K., and Fraser, S. E. (1995). Order and coherence in the fate map of zebrafish nervous system. *Development* **121,** 2595–2609.

Yoshida, T., Ito, A., Matsuda, N., and Mishina, M. (2002). Regulation by protein kinase A switching of axonal pathfinding of zebrafish olfactory sensory neurons through the olfactory placoe-olfactory bulb boundary. *J. Neurosci.* **22,** 4964–4972.

Further Reading

Inoué, S., and Spring, K. R. (1997). "Video Microscopy, ed. 2." Plenum Press, New York.

James, B., and Pawley, J. B. (1995). "Handbook of Biological Confocal Microscopy, ed. 2." Plenum Press, New York.

Mason, W. T. (1999). "Fluorescent and Luminescent Probes for Biological Activity, ed. 2." Academic Press, London.

Neher, R., and Neher, E. (2003). Optimizing imaging parameters for the separation of multiple labels in a fluorescence image. *J. Microscopy* **213,** 46–62.

CHAPTER 12

Development of the Peripheral Sympathetic Nervous System in Zebrafish

Rodney A. Stewart,* A. Thomas Look,* John P. Kanki,* and Paul D. Henion[†]

*Department of Pediatric Oncology
Dana-Farber Cancer Institute
Boston, Massachusetts 02115

[†]Center for Molecular Neurobiology and Department of Neuroscience
Ohio State University
Columbus, Ohio 43210

I. Introduction

The combined experimental attributes of the zebrafish model system accommodate cellular, molecular, and genetic approaches, making it particularly well suited for studying mechanisms underlying normal developmental processes, as well as disease states, such as cancer. The ability to analyze developing tissues in the optically clear embryo, combined with the unbiased nature of forward genetic screens, allows one to identify previously elusive genes and to analyze their *in vivo* function. In this chapter, we describe the advantages of using the zebrafish for identifying genes and their functional regulation of the developing peripheral sympathetic nervous system (PSNS). We provide a brief overview of the genetic pathways regulating vertebrate PSNS development, the rationale for developing a zebrafish model, and our current understanding of zebrafish PSNS development. We also include examples that illustrate the potential of mutant analysis in zebrafish PSNS research. Finally, we explore the potential of the zebrafish system for discovering genes that are disrupted in neuroblastoma, a highly malignant cancer of the PSNS.

II. The Peripheral Autonomic Nervous System

A. Overview

The internal organs, smooth muscles, skin, and exocrine glands of the vertebrate body are innervated by the peripheral autonomic nervous system (ANS), which comprises the PSNS and the parasympathetic (PAS) and enteric nervous systems (ENS). These three components are structurally and functionally distinguished by differences in the characteristic locations of their cell bodies, the targets they innervate, the neurotransmitters they utilize, and the molecular pathways controlling their development (Brading, 1999). In the sympathetic nervous system, the cell bodies of the preganglionic neurons are generally found in the thoracic and lumbar areas of the spinal cord. Their axons exit ventrally and innervate peripheral ganglia lying near the spinal cord where they synapse with post-ganglionic neurons, which in turn, innervate target organs. Sympathetic neurons are predominantly adrenergic, producing the neurotransmitter noradrenalin along with one or more neuropeptides. In contrast, the cell bodies of the central component of the PAS lie in the brain and sacral regions of the spinal cord, and synapse with peripheral cell bodies located within the immediate vicinity or in the target organs. PAS neurons are cholinergic, producing acetylcholine along with other neuromodulators (Thexton, 2001). These structural and functional differences allow the sympathetic and parasympathetic systems to function largely in opposition to each other to maintain homeostasis, such as vascular tone, and to generate the fight-or-flight response. Although the ANS normally consists of central preganglionic and peripheral ganglionic neurons to regulate the function of a target organ, an exception can be found in the sympathetic nervous system where chromaffin cells in the adrenal medulla are directly innervated by central preganglionic neurons, but do not innervate a target organ. Chromaffin cells are neural crest-derived endocrine cells closely related lineally and neurochemically to postganglionic neurons of the PSNS. However, rather than functioning in neural transmission, chromaffin cells exhibit an endocrine function secreting hormones, such as adrenalin (Gabella, 2001).

The function of the enteric nervous system is relatively independent of the CNS and other components of the ANS. In most vertebrates, the enteric neurons form 2 layers of ganglionic plexuses located along the entire length of the gastrointestinal tract, consisting of a microscopic meshwork of ganglia connected to each other by short nerve trunks. The inner myenteric plexus, situated between the longitudinal and circular muscle layers, is mainly responsible for muscle contraction. The submucosal plexus controls motility, secretion, and microcirculation processes in the gut. The functions of the enteric nervous system are complex, and 17 different neuronal types have been identified that produce a variety of different neurotransmitters including, nitric oxide, adenosine triphosphate (ATP), and 5-hyroxytryptamine. Enteric ganglia function locally to integrate sensory and reflex activities for the coordination of the sequential relaxation and contractions occurring during peristalsis (Hansen, 2003).

B. Molecular Pathways Underlying PSNS Development

The early development of the PSNS can be divided into four overlapping stages, based on both morphologic and molecular criteria:

1. Formation and fate specification of neural crest cells that will develop into sympathoadrenal (SA) progenitors.

2. Bilateral migration of SA cells and their coalescence in regions adjacent to the dorsal aorta.

3. Neuronal and noradrenergic differentiation of SA progenitors.

4. Maintenance of PSNS neurons in fully developed ganglia and the establishment of their efferent synaptic connections.

Considerable progress has been made elucidating the cellular and molecular mechanisms underlying PSNS development, which represents one of the best described genetic pathways establishing vertebrate neuronal and neurotransmitter identity (Anderson, 1993; Francis and Landis, 1999; Goridis and Rohrer, 2002; Fig. 1). Briefly, neural crest progenitors form at the border between the neural and non-neural ectoderm through a process regulated by bone morphogenetic proteins (BMPs) and Wnt signaling (Knecht and Bronner-Fraser, 2002). These neural crest progenitor cells express genes such as *slug, snail, tfap2α,* and *foxd3* that appear to play roles in their induction or early development (reviewed in Knecht and Bronner-Fraser, 2002). Subsequent to the morphogenetic movements that result in neural tube closure, neural crest progenitors are localized to the most dorsal aspect of the neural tube (Aybar *et al.*, 2003; Bronner-Fraser, 2002; Kos *et al.*, 2001). The premigratory neural crest progenitors then undergo an epithelial-mesenchymal transition and begin to migrate away from the neural tube. Neural crest cells first migrate out ventromedially and later, others follow a dorsolateral pathway (Goridis and Rohrer, 2002; LeDouarin and Kalcheim, 1999; Fig. 1A). During migration, precursors of the SA lineage are exposed to signaling

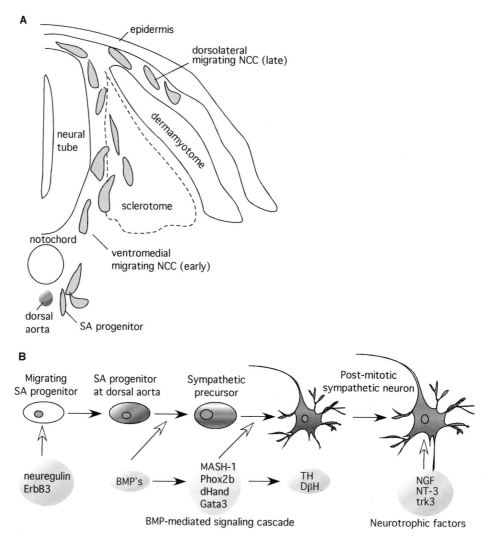

Fig. 1 Neural crest-derived SA progenitors migrate along a ventromedial pathway to bilateral regions adjacent to the dorsal aorta. (A) Schematic diagram of a transverse section through the trunk of a vertebrate embryo (embryonic day 10.5 in the mouse, 2.5 in the chick, and approximately 28 hpf in the zebrafish embryo). In avian and rodent embryos, presumptive SA progenitor cells derived from the neural crest migrate ventromedially within the sclerotome region of the somites and ultimately cease migration in the region of the dorsal aorta. In zebrafish, neural crest-derived SA precursors migrate ventromedially between the neural tube and somites to the dorsal aorta region. (B) Molecular pathways governing sympathetic neuron development. During migration, signaling via the neuregulin-1 growth factor is required for the development of at least some SA progenitors. Once the SA progenitor cells arrive at the dorsal aorta, BMP signaling activates the transcriptional regulators MASH-1 and Phox2b that ultimately lead to the expression of the transcription factors Phox2a, GATA-3 and dHand. Together, these factors are responsible for differentiation of SA progenitors into noradrenergic neurons. Fully differentiated neurons express biosynthetic enzymes responsible for the synthesis of noradrenalin, such as tyrosine hydroxylase and dopamine-β-hydroxylase. Survival of the differentiated sympathetic neurons is governed by a number of neurotrophic factors, such as NGF and NT-3.

factors from the neural tube and notochord, such as sonic hedgehog (Shh) and Neuregulin-1, an epidermal growth factor (EGF)-like growth factor (Crone and Lee, 2002; Krauss et al., 1993; Patten and Placzek, 2000; Williams et al., 2000b). Neuregulin-1 expression is associated with the origin, migration, and target site of SA progenitors. Mice lacking components of the neuregulin-1 pathway, such as the ErbB3 receptor, exhibit severe hypoplasia of the primary sympathetic ganglion chain (Britsch et al., 1998). Further analysis of these mice indicates the requirement for neuregulin signaling for the normal migration of some SA progenitors, rather than their subsequent differentiation into sympathetic neurons (Britsch et al., 1998; Crone and Lee, 2002; Murphy et al., 2002). As the SA precursors aggregate in the vicinity of the dorsal aorta (around embryonic day 10 in the mouse and E2.5 in the chick), a molecular signaling cascade is initiated in response to BMPs that are secreted by dorsal aorta cells (Fig. 1B) (Reissmann et al., 1996; Schneider et al., 1999; Shah et al., 1996; Varley et al., 1995). BMP signaling induces the expression of the proneural gene *MASH-1*, an *achaete-scute* homolog, and the homeodomain transcription factor *phox2b* by sympathetic neuroblasts (Ernsberger et al., 1995; Groves et al., 1995; Guillemot et al., 1993; Hirsch et al., 1998). Several other critical transcription factors are then activated, including the homeobox proteins Phox2a, the basic Helix-Loop-Helix (bHLH) transcription factor dHand, and the zinc-finger protein GATA-3 (Hirsch et al., 1998; Howard et al., 2000; Lim et al., 2000; Lo et al., 1998; Pattyn et al., 1997, 1999; Schneider et al., 1999). Together, these regulatory factors drive SA differentiation further by activating the expression of pan-neural genes and genes encoding enzymes for the synthesis of catecholaminergic neurotransmitters, such as dopamine-beta-hydroxylase (DβH) and tyrosine hydroxylase (TH) (Ernsberger et al., 2000; Kim et al., 2001; Seo et al., 2002). The specification of the neuronal and noradrenergic phenotype of sympathetic neuron precursors by these transcription factors remains incompletely understood, involving a complex interaction of their regulatory pathways, rather than a strictly linear developmental progression (see Goridis and Rohrer, 2002).

A later stage of PSNS development consists of modeling of the sympathetic ganglia through the regulation of cell proliferation and maintenance. The neurotrophic factors, nerve growth factor (NGF) and NT-3, have been shown to control sympathetic neuron survival and the maintenance of their synaptic connections (Birren et al., 1993; DiCicco-Bloom et al., 1993; Francis and Landis, 1999). In the embryo, NGF is secreted from sympathetic target tissues when sympathetic neurons arrive (Chun and Patterson, 1977; Heumann et al., 1984; Korsching and Thoenen, 1983; Shelton and Reichardt, 1984). Analysis of NGF and its high affinity receptor, tyrosine kinase A (TrkA), in mouse mutants confirmed their requirement for the *in vivo* survival of sympathetic neurons (Fagan et al., 1996; Smeyne et al., 1994). In their absence, sympathetic neuron development proceeds normally, but is then followed by neuronal cell death. A similar phenotype is observed in NT-3 mouse mutants, although unlike mutants with NGF-signaling loss, neuronal death occurs at later embryonic stages (Ernfors et al., 1994; Farinas *et al.*, 1994; Francis and Landis, 1999; Wyatt *et al.*, 1997). Therefore, in chick and rodents the action of neurotrophic factors, such as NT-3 and NGF, are largely responsible for establishing and maintaining mature ganglion neuronal cell numbers during embryonic or early postnatal development.

Although many of the inductive signaling pathways affecting different stages of neural crest development have been identified, the regulatory mechanisms controlling these pathways remains poorly understood. While substantial evidence links BMP signaling with the induction of SA progenitor cell development, the genetic control of SA cell responsiveness to BMP signals and their specification remain unclear. While sympathetic precursors are competent to express MASH1 and *phox2b* in response to BMP signaling near the dorsal aorta, they do not respond to BMPs present in the overlying ectoderm during earlier pre migratory stages. Furthermore, the molecular mechanisms regulating the interactions of transcription factors and downstream pathways specifying neuronal and noradrenergic differentiation, are incompletely understood. Other signaling pathways, such as cAMP (Lo *et al.*, 1999), may also contribute to this process. Finally, proliferation of sympathoblasts and differentiated sympathetic neurons occurs throughout embryogenesis (Birren *et al.*, 1993; Marusich *et al.*, 1994; Rohrer and Thoenen, 1987; Rothman *et al.*, 1978). However, little is known about the genetic pathways that actively promote or inhibit SA progenitor cell proliferation. An inability to control cell proliferation in sympathetic ganglia is of particular medical interest, since it can lead to neuroblastoma, the most common human cancer in infants younger than 1 year of age. While the study of PSNS development in tetrapods will continue to contribute to our knowledge of PSNS development, a goal of this chapter is to demonstrate the power of the zebrafish, *Danio rerio*, exploiting its forward genetic potential and advantages as an embryologic system, for making significant contributions to PSNS research. We also propose that the study of zebrafish PSNS development will contribute to our understanding of both normal and abnormal PSNS development and may ultimately provide a genetic model for human neuroblastoma.

III. The Zebrafish as a Model System for Studying PSNS Development

A. Overview

One of the most powerful attributes of the zebrafish system is its capacity for large-scale genetic screens. The unbiased nature of phenotype-based genetic screens enables new genes to be identified without prior knowledge of their function or expression in the tissue of interest. This approach is particularly attractive for study of the PSNS, as many signaling components involved in determining sympathetic fate are either unknown or incompletely understood. Also, most of our current understanding of PSNS development has relied on functional assays on isolated sympathetic cells in culture or analyses of genes that have been expressed incorrectly (Francis and Landis, 1999; Goridis and Rohrer, 2002). While these studies can determine whether certain genes are sufficient to direct sympathetic development, they do not address whether those genes are normally required for PSNS development. Murine gene knockout models have been used in loss-of-function studies to confirm the *in vivo* requirement for particular genes in PSNS development (Guillemot *et al.*, 1993; Lim *et al.*, 2000; Morin *et al.*, 1997; Pattyn *et al.*, 1999). For example, although both Phox2a and Phox2b can induce a sympathetic phenotype when expressed incorrectly in the chick, only the selective knockout of *Phox2b* appears to be necessary for sympathetic development *in vivo*, as the PSNS in *Phox2a*$^{-/-}$ mutant mice appears relatively normal (Morin *et al.*, 1997; Pattyn *et al.*, 1999).

Together, these studies provide valuable insights into the regulatory pathways directing sympathetic neuron development and emphasize the advantages of using mutants to dissect genetic pathways *in vivo*. The capacity for experimental mutagenesis and manipulation of gene expression in the developing embryo is a major strength of the zebrafish system. Forward genetic zebrafish screens can be especially valuable for identifying genes affecting complex signaling pathways that rely on interactions between the developing PSNS and surrounding tissues, which may be impossible to address using *in vitro* assays. Critical roles for extrinsic factors are particularly evident in PSNS development and SA progenitors migrate past a number of tissues expressing different signaling molecules, such as the neural tube, notochord, and floor-plate. In addition, zebrafish mutant embryos can often survive for a longer period of time during embryogenesis than knockout mice lacking orthologous genes. This may be attributed to the external development of the zebrafish embryo and can allow the analysis of the PSNS to extend through later stages of sympathetic neuron differentiation and maintenance. Finally, the zebrafish system offers the most amenable vertebrate model for performing large-scale mutagenesis screens to identify novel genes affecting all aspects of PSNS development, as the time, space, and expense associated with mutagenesis techniques in mice can be prohibitive.

Establishing the zebrafish as a useful vertebrate model for identifying new genes important for PSNS development will require: (1) an analysis of zebrafish sympathetic neuron development and its comparison with other vertebrates, (2) an analysis of the genetic programs regulating zebrafish PSNS development and their conservation in other vertebrates, and (3) the generation of efficient mutagenesis protocols and screening assays for the isolation of PSNS mutants. Each of these areas is addressed in the next sections.

B. Development of the PSNS in Zebrafish

1. Neural Crest Development and Migration

The different stages of PSNS development in the zebrafish have been recently analyzed (An *et al.*, 2002; Raible and Eisen, 1994). The findings show that the morphogenesis and differentiation of sympathetic neurons in zebrafish is qualitatively very similar to other vertebrates. The migration and cell fate specification of trunk sympathetic precursors in zebrafish was analyzed by labeling single neural crest cells with vital dyes and following their subsequent development (Raible and Eisen, 1994). In the trunk, neural crest migration begins around 16 hpf, at the level of somite 7, and sympathetic neurons are only derived from neural crest cells undergoing early migration along the ventromedial pathway. Hence, the ventromedial migration of SA precursor cells is conserved in zebrafish. These studies also demonstrated the existence of both multipotent and fate-restricted neural crest precursors that generate a limited number of neural crest derivatives, such as sympathetic neurons, before or during the initial stages of neural crest migration (Raible and Eisen, 1994). Although little is known about the molecular mechanisms underlying such fate decisions, the ability to analyze the fate restriction of sympathetic neurons in zebrafish, using different neural crest mutants, affords a powerful method to dissect the genetic pathways underlying this process.

2. Gene Expression in Migrating SA Progenitors

Many of the genes capable of inducing the development SA progenitors in birds and rodents have been identified in zebrafish (Table I). Furthermore, the expression of some of these genes in dorsal aorta cells and in neural crest-derived cells in its vicinity, where noradrenergic neurons form, is consistent with their role in fish PSNS development. A number of *BMP* homologues have been identified in zebrafish and some have been shown to be expressed by the dorsal aorta (Dick *et al.*, 2000; Nguyen *et al.*, 1998). Several zebrafish mutants exhibit midline defects affecting structures that may be responsible for BMP signaling. In *flh* mutant embryos, the notochord and dorsal aorta fail to develop (Fouquet *et al.*, 1997; Talbot *et al.*, 1995) and *bmp4* expression is absent. The loss of the local source of BMP signaling corresponds with a failure of sympathetic neurons (Hu$^+$/TH$^+$) to form (Henion, unpublished data, 2002). Interestingly, neural crest-derived cells (crestin$^+$) continue to populate this region where sympathetic neurons would normally develop. These observations suggest that, like other vertebrates, dorsal aorta-derived BMPs are required for SA development in zebrafish (Henion, unpublished data). In contrast, floor-plate cells, which are lacking in the *cyclops* mutant, do not appear to be required for dorsal aorta development, BMP expression, or SA development. All these functions appear to be normal in these mutant embryos (Henion, unpublished data). Interestingly, SA development appears normal in *no tail* mutants (Fig. 2A and B) even though dorsal aorta development is impaired (Fouquet *et al.*, 1997) and BMP expression is reduced. It is possible that weak BMP persists in *ntl* because of the continued presence of notochord precursor cells that fail to differentiate properly in this mutant (Melby *et al.*, 1997). However, whether the notochord is directly responsible for BMP expression and dorsal aorta development is unclear.

Most of the described transcription factors known to direct the development of the sympathetic precursors in other species are present in the zebrafish and exhibit

Table I
Expression of Conserved PSNS Genes in Zebrafish

Genes	Expression during PSNS Development	Reference
ErbB3	Not available	Lo *et al.*, 2003
BMP4	Dorsal aorta	Fig. 2A; Dick *et al.*, 2000; Nguyen *et al.*, 1998
Crestin	Migrating SA, nascent SCG	Luo *et al.*, 2001
Zash1a	SCG	Allende and Weinberg, 1994; Stewart, unpublished data
Phox2a	SCG	Guo *et al.*, 1999a; Holzschuh *et al.*, 2003
Phox2b	Migrating SA, SCG	Guo, Stewart, unpublished data
HuC/D	SCG, trunk sympathetic ganglia	An *et al.*, 2002
GATA3	ND	Neave *et al.*, 1995
dHand	SCG	Fig. 2C and D; Yelon *et al.*, 2000
TH	SCG, trunk sympathetic ganglia	An *et al.*, 2002; Guo *et al.*, 1999b; Holzschuh *et al.*, 2001
DβH	SCG, trunk sympathetic ganglia	An *et al.*, 2002; Holzschuh *et al.*, 2003
PNMT	SCG	Stewart, unpublished data
Trk receptors	ND	Martin *et al.*, 1995; Williams *et al.*, 2000a

appropriate gene expression patterns. The zebrafish *Zash1a* gene, a homolog of *MASH-1*, is transiently expressed in cells near the dorsal aorta by 48 hpf (Allende and Weinberg, 1994; Stewart, unpublished data). Preliminary gene knockdown experiments using antisense morpholinos to specifically target the *Zash1a* gene, resulted in the loss of *TH*-expressing noradrenergic neurons in the developing PSNS (Yang and Stewart, unpublished data). The *Phox2a, Phox2b, GATA-3,* and *dHand* genes have also been cloned in zebrafish. The *dHand* and both *Phox2* genes are also expressed in the first PSNS neurons that develop by 2 dpf (Fig. 2C and D; Guo *et al.*, 1999a; Holzschuh *et al.*, 2003; Neave *et al.*, 1995; Yelon *et al.*, 2000). Therefore, although the expression of *GATA-3* in the developing PSNS has not yet been characterized, many of these early markers of SA precursors appear to be conserved in zebrafish. The future analysis of *Zash1a, Phox2b, Phox2a, GATA-3,* and *dHand* expression, together with the examination of *TH* and *DβH* expression in SA precursors, will provide insight into the functions of these genes with respect to sympathetic neuron differentiation (discussed later). Importantly, analysis of compound mutants utilizing existing zebrafish mutants such as *soulless* (*phox2a*) (Guo *et al.*, 1999a) and *hands down* (*dHand*) (Miller *et al.*, 2000), together with other mutants described in this chapter, will contribute to our understanding of the functional roles of these genes in sympathetic neuron development. The combination of transient gene knockdown techniques using morpholinos and the identification of new zebrafish mutants affecting other regulators of sympathetic

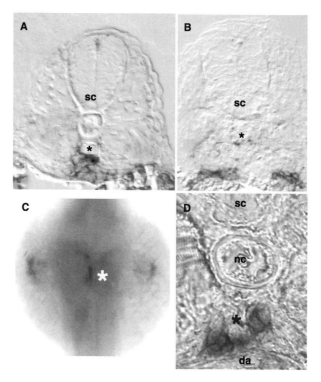

Fig. 2 Expression of *bmp* by the dorsal aorta and dHand by sympathetic neurons. (A) Expression of *bmp4* in transverse sections at the mid-trunk level in wild type (*top*) and *ntl* mutant embryos at 48 hpf. In *bmp4*, expression is evident in cells of the dorsal aorta (*asterisk*) of wild-type embryos. (B) In *ntl* mutants, the dorsal aorta does not form completely (*asterisk*), although expression of *bmp4* is present albeit to a reduced extent. (C) Expression of zebrafish dHand at 58 hpf in cervical sympathetic neurons (asterisk). (D) Coexpression of dHand and TH in sympathetic neurons (*asterisk*) in a transverse section in the anterior trunk region of a 3 dpf embryo. sc = spinal cord; nc = notochord; da = dorsal aorta. Top is dorsal in (A, B, and D) and anterior in (C).

neuron development, should also contribute to novel insights relevant to PSNS development across vertebrate species.

3. Neuronal Differentiation and Coalescence into Sympathetic Ganglia

The timing of overt neuronal differentiation of sympathetic precursors and their transition to fully differentiated noradrenalin (NA)-producing neurons has been described in detail in zebrafish (An *et al.*, 2002). The pan-neuronal antibody 16A11 recognizes members of the Hu family of ribonucleic acid (RNA)-binding proteins (Marusich *et al.*, 1994) and labels sympathetic precursors located ventrolateral to the notochord and adjacent to the dorsal aorta (Fig. 3; An *et al.*, 2002). Sympathetic neurons were found to differentiate at different times in the zebrafish embryo and two populations of sympathetic ganglion neurons were defined. The most rostral population develops at 2 dpf and comprises the superior cervical ganglion (SCG) complex that consists of two separate ganglia arranged in an hourglass shape. Several days later, more caudal trunk sympathetic neurons develop as irregular, bilateral rows of single neurons adjacent to the dorsal aorta, presumably analogous to the primary sympathetic chain in other vertebrates. These neurons differentiate in an anterior to posterior temporal progression, extending caudally as far as the level of the anus, and eventually form regular arrays of segmentally distributed sympathetic ganglia (An *et al.*, 2002).

The reason for the delay in the differentiation of the caudal sympathetic neurons is not known, since the formation of the dorsal aorta (Fouquet *et al.*, 1997) and its expression of *bmps* (see Fig. 2A; Martinez-Barbera *et al.*, 1997), occur well before the differentiation of sympathetic neurons is observed. Importantly, ventrally migrating neural crest-derived cells populate the region adjacent to the dorsal aorta between 24 and 36 hpf. Therefore, the delay in caudal PSNS development may be attributed to a delay in neural crest-derived cells becoming competent to respond to BMP signaling. Since the expression of some zebrafish BMPs have not been examined in the dorsal aorta, it remains possible that different types of BMPs may be selectively expressed by dorsal aorta cells or that SA progenitors exhibit differential responsiveness to different BMPs.

4. Differentiation of Noradrenergic Neurons

One of the key events in PSNS differentiation is the acquisition of the NA-neurotransmitter phenotype, indicated by the expression of noradrenalin and the genes, such as TH and DβH, which are required for its synthesis through the enzymatic conversion of the amino acid L-tyrosine. In zebrafish, *TH* expression has been used as the principal marker for the presence and formation of fully differentiated sympathetic neurons, although it is also expressed by other cate-cholaminergic neurons in the central nervous system (CNS) (Fig. 3A to D; An *et al.*, 2002; Guo *et al.*, 1999b; Holzschuh *et al.*, 2001). Both the TH protein and mRNA are readily detectable in the SCG complex beginning at 48 hpf. Consistent with the expression of Hu proteins, most sympathetic neurons located posterior to the SCG complex do not begin to express *TH* mRNA until approximately 5 days of development, in a few of the more rostral trunk segments (Fig. 3D; An *et al.*, 2002). By 10 dpf, all of the sympathetic ganglia contain neurons expressing *TH*, although some neurons within the nascent sympathetic ganglia do not express TH protein. However, by 28 dpf, all of the neurons uniformly express TH, suggesting the complete maturation of sympathetic ganglia by this time. The expression of DβH protein and mRNA have also been used as markers of PSNS differentiation (Fig. 4A; An *et al.*, 2002; Holzschuh *et al.*, 2003). Because of its requirement for the conversion of dopamine to noradrenaline, *DβH* is expressed in sympathetic neurons of the PSNS, and a subset of catecholaminergic neurons in the CNS (An *et al.*, 2002; Holzschuh *et al.*, 2003). The expression of DβH is generally observed

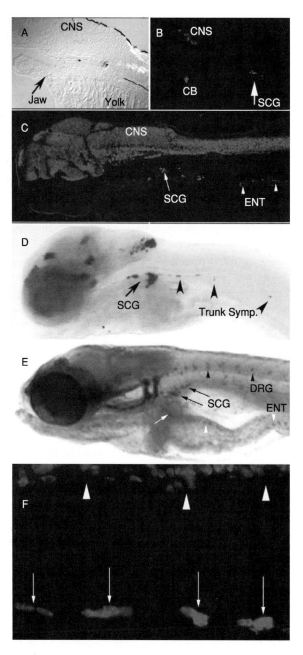

Fig. 3 Development of the peripheral sympathetic nervous system in zebrafish embryos. (A, B, and C) Parasagittal section of 3.5-dpf embryo. (A) High magnification DIC and (B) fluorescence of the same field, showing TH-IR (*red*) in the SCG (*arrow*), carotid body (CB), and a group of anterior cells in the midbrain (CNS). (C) Low magnification view of 3.5-dpf embryo labeled with anti-Hu to reveal

slightly later than TH in differentiating sympathetic neurons (An *et al.*, 2002). However, DβH is expressed along with TH, as early as 2 dpf in the SCG complex (Holzschuh *et al.*, 2003).

Another element of PSNS function is the regulated release of adrenalin and noradrenalin by chromaffin cells of the adrenal gland, which form in or around the developing kidney. Chromaffin cells represent a specialized component of the PSNS and express in the catecholaminergic pathway an additional enzyme, called phenylethanolamine N-methyltransferase (PNMT), which converts noradrenaline into adrenaline (Kalcheim *et al.*, 2002; Schober *et al.*, 2000). In mammals, chromaffin cells are localized to the adrenal medulla, which is located within the cortex of the adrenal glands overlying the kidneys; whereas in zebrafish, both adrenocortical and chromaffin cells are interspersed within the anterior region of the kidney, referred to as the interrenal gland (Hsu *et al.*, 2003; Liu *et al.*, 2003). As in other species, both noradrenergic and adrenergic chromaffin cells have been described in zebrafish. Noradrenergic cells have heterogeneous vesicles containing asymmetrically localized electron-dense granules, while adrenergic cells form smaller vesicles with homogenous electron-lucent granules (Hsu *et al.*, 2003). Initial observations indicated that non-neuronal (16A11$^-$), TH, and DβH-positive cells are present in the SCG complex at 2 dpf, which continue to migrate ventrally to the anterior pronephros (also referred to as the head kidney; Fig. 4; An *et al.*, 2002). Preliminary mRNA *in situ* hybridization assays using *PNMT* expression as a marker for chromaffin cells, support these observations and suggest that these non-neuronal SCG cells represent chromaffin cells (Stewart, unpublished data). Further analysis of chromaffin cell development, using *PNMT* as a marker, should provide a more complete understanding of how SA progenitors form chromaffin cells in the zebrafish.

5. Modeling of Sympathetic Ganglia

In rodents and birds, neurotrophic factors, such as NGF and NT-3, control sympathetic neuron cell numbers through the regulation of their survival and continued maintenance of their synaptic connections (Francis and Landis, 1999; Schober and Unsicker, 2001). In zebrafish, the ability of these factors to control the survival of sympathetic neurons remains unknown. However, NT-3 has been

---

all neurons. A subset of cervical sympathetic neurons are indicated by the arrow and enteric neurons (ENT) by arrowheads. (D) Lateral view of whole-mount *TH* RNA *in situ* preparation at 5 dpf. *TH* RNA is strongly expressed in the SCG (*arrow*) at this stage and is beginning to be expressed in the trunk sympathetic chain (*arrowheads*). A description of *TH* RNA expression in the head is described in Guo *et al.*, 1999. (E) Whole-mount antibody preparation of a 7-dpf larvae labeled with anti-Hu to reveal neurons. Black arrows indicate SCG, black arrowheads indicate dorsal root ganglion (DRG) sensory neurons, and white arrow and white arrowheads indicate enteric neurons (ENT). (F) Parasagittal section in the mid-trunk region of a 17-dpf embryo labeled with anti-Hu. Ventral spinal cord neurons are evident at the top (*arrowheads*). Four segmental sympathetic ganglia (*arrows*) are located ventral to the notochord adjacent to the dorsal aorta. (See Color Insert.)

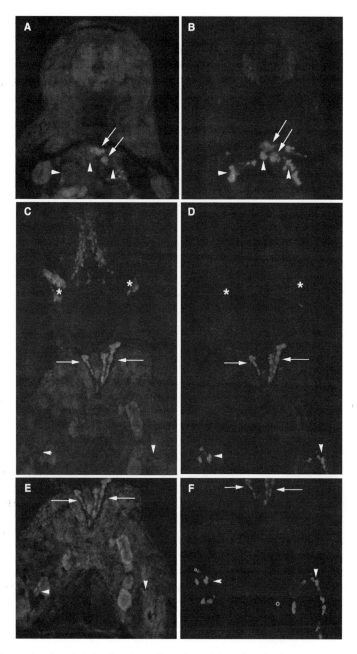

Fig. 4 Sympathoadrenal derivatives in embryonic and juvenile zebrafish. (A and B) Transverse section of a 3.5-dpf embryo double-labeled with anti-Hu (green) and DβH (red). Arrows indicate Hu$^+$/DβH$^+$ sympathetic neurons of the cervical ganglion. Arrowheads indicate Hu$^-$/DβH$^+$ presumptive chromaffin cells. (C and D) Transverse section through the mid-trunk region at 28 dpf, double-labeled

shown to act as a neurotrophic factor regulating cell death in Rohon-Beard sensory neurons (Williams *et al.*, 2000a). In teleost sympathetic ganglia, the proliferation of cells occurs during early development and may possibly continue throughout adult life (An *et al.*, 2002; Weis, 1968). Analysis of 5-bromo-2-deoxyuridine (BrdU) incorporation and the expression of phospho-histone H3 indicate that cells proliferate within the developing sympathetic ganglia (An *et al.*, 2002). Interestingly, some of these cells also expressed the pan-neuronal marker, 16A11, suggesting that preexisting neuronal cells proliferated within the ganglia, a process that has also been observed in both chick and mouse PSNS (Birren *et al.*, 1993; Cohen, 1974; DiCicco-Bloom *et al.*, 1993; Marusich *et al.*, 1994; Rohrer and Thoenen, 1987; Rothman *et al.*, 1978). However, in chick and rodents, sympathetic neurons become postmitotic during embryonic development, while they may remain competent to divide throughout life in the zebrafish (Weis, 1968).

C. Mutations Affecting PSNS Development

1. Introduction

PSNS development, from the induction of neural crest through the overt differentiation of sympathetic ganglia, can be readily observed within the first 5 days of zebrafish development in the SCG (see Fig. 3D; An *et al.*, 2002). During this time, dynamic changes in both the numbers and distribution of sympathetic cells within the SCG can be easily visualized by *TH* mRNA whole-mount *in situ* hybridization. At 2 dpf, bilateral rows containing approximately 5 TH-positive cells are ventrally located near the dorsal aorta. By 5 dpf, the number of *TH*-positive cells have increased five-fold and coalesced into a V-shaped ganglia, including some appearing to migrate ventrally toward the kidney, which may represent putative adrenal chromaffin cells (An *et al.*, 2002). Therefore, the evaluation of the SCG at 5 dpf represents an excellent assay for early PSNS development that can be used in genetic screens to detect mutations affecting any stage of PSNS development. Assaying the SCG would also be able to confirm mutations found to affect very early neural crest development. Such mutagenesis screens are currently being performed and some examples of the different mutant classes that have been isolated thus far are discussed in the next sections.

2. Mutations Affecting Early PSNS Development

Mutations affecting early PSNS development can fail to form either neural crest precursors or SA progenitor cells. They may also disrupt the specification of these cells within the premigratory crest. One example is *sympathetic mutant 1* (*sym1*),

with anti-Hu (green) and anti-TH (red). (E and F) Higher magnification of C and D, including a slightly more ventral region. Arrows indicate sympathetic neurons, arrowheads indicate chromaffin cells, and asterisks denote dorsal root ganglia. (See Color Insert.)

which was discovered in a diploid gynogenetic screen designed to identify mutations disrupting *TH* expression in the SCG complex at 5 dpf (Fig. 5A and C). The *sym1* mutation causes a severe reduction or absence of *TH*- and *PNMT*-expressing cells in the SCG complex of the PSNS, but *TH* expression is not affected in other regions of the CNS. Analysis of the *sym1* phenotype indicates that early neural crest development is disrupted, and the cells exhibit a reduction in the expression of early neural crest markers, such as *crestin* (*cnt*) (Luo *et al.*, 2001). Another example of a mutant within this class is *colourless* (*cls*), which was isolated in screens for neural crest mutants affecting pigment cell development (Kelsh and Eisen, 2000). The zebrafish *cls* mutation disrupts the *sox10* gene, which is required for the development of most non-ectomesenchymal neural crest lineages, including the PSNS (Dutton *et al.*, 2001; Kelsh and Eisen, 2000). The *sox10* gene is expressed in early premigratory neural crest cells and then undergoes rapid downregulation. Analysis of the *cls* phenotype revealed that premigratory neural crest cells fail to migrate and subsequently undergo apoptosis (Dutton *et al.*, 2001). Interestingly, the *sym1* and *cls* mutations differentially affect complementary sets of neural crest

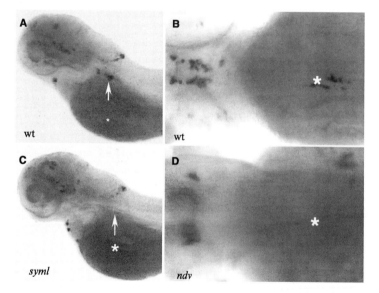

Fig. 5 Isolation of PSNS mutants in zebrafish. Whole-mount *TH in situ* preparation showing expression of *TH* mRNA in (A and B) wild type, (C) *sym1*,, and (D) *nosedive* mutant embryos. (A and C) Lateral view of *TH* expression at 5 dpf. (C) The *sym1* mutant phenotype was identified in an *in situ* screen at 5 dpf for mutations that specifically lack *TH* expression in the SCG (C, *asterisk*), but leave others areas of TH expression in the CNS and carotid body unaffected. (B and D) Dorsal view of *TH* expression in the SCG region at 3 dpf in wild-type embryos (B, *asterisk*). Expression of *TH* is absent in the SCG in *nosedive* mutants (D, *asterisk*). Analysis of Hu immunoreactivity revealed that the lack of *TH* expression in the cervical region in *nosedive* mutant embryos is attributed to the absence of sympathetic neurons (data not shown).

derivatives. For example, unlike *cls*, *sym1* mutants have severe defects in craniofacial cartilage development, while unlike *sym1*, the *cls* mutant lacks pigment cells (Kelsh and Eisen, 2000). The genetic mechanisms specifying the fate restriction of subsets of premigratory neural crest are unknown, although evidence suggests that they exist in the zebrafish (Raible and Eisen, 1994). It is possible that the genes disrupted by the *sym1* and *cls* mutations may contribute to the generation of fate-restricted subsets of neural crest precursors, such as SA progenitors.

In a screen for mutations affecting neural crest derivatives (Henion *et al.*, 1996), three mutants were identified based on the absence of neural crest-derived dorsal root ganglion (DSG) sensory neurons, based on the expression of 16A11 immunoreactivity. Subsequently, it has been found that these mutants also completely lack sympathetic and enteric neurons based on the expression of TH mRNA and 16A11, respectively (Fig. 5B and D; Henion, unpublished data, 2002). In contrast, the development of trunk neural crest-derived chromatophores and glia appears normal in all three mutants. These observations suggest that the genes disrupted by these mutations do not affect all neural crest derivatives, but selectively function during the development of crest-derived neurons. It is tempting to speculate that the phenotypes of these mutants further indicate an early lineage segregation of neurogenic precursors or suggest the function of these genes in the development of multiple neural crest-derived neuronal subtypes.

3. Mutations Affecting Later Stages of PSNS Development

A number of genes are expressed in SA precursors once they have migrated to the region adjacent to the dorsal aorta, indicating further differentiation toward mature sympathetic neurons (see Fig. 1). Therefore, mutations that disrupt later stages of neuronal differentiation may result in normal neural crest migration and expression of early neural crest markers, but exhibit a failure of sympathetic neuron differentiation. The *lockjaw* or *mount blanc* mutant (*tfap2a*) is an example of this mutant class. It shows a lack of *TH*- and *DβH*-positive neurons in the region of the SCG complex (Holzschuh *et al.*, 2003; Knight *et al.*, 2003). The *tfap2a* gene is normally expressed both in the pre migratory neural crest and in the region of the developing SCG. Analysis of neurogenesis, using the 16A11 antibody, revealed normal migration and neural differentiation of SA progenitors at the SCG in *tfap2a* mutants (Holzschuh *et al.*, 2003; Knight *et al.*, 2003). Furthermore, the expression of *Phox2a* in cells in the region of the SCG in these mutants indicates that the signaling cascade required to induce the initial stages of noradrenergic differentiation at the dorsal aorta is intact in these mutants (Holzschuh *et al.*, 2003). The failure of *tfap2a* mutants to express the noradrenergic differentiation markers *TH* and *DβH*, is likely attributed to the requirement for Tfap2a to activate these genes, as Tfap2a has conserved DNA binding regions in both *TH* and *DβH* promoters (Holzschuh *et al.*, 2003; Seo *et al.*, 2002). Retinoic acid (RA) signaling pathways may possibly function upstream of *tfap2a* in the differentiation of noradrenergic neurons because incubation of wild-type zebrafish embryos in

RA induces ectopic *TH*-positive cells in the region of the SCG. This effect is blocked in *tfap2a* mutants (Holzschuh *et al.*, 2003). In addition, mutations in *neckless/rald2*, which disrupt the biosynthesis of RA from vitamin A, have fewer *TH*-expressing cells in the SCG (Holzschuh *et al.*, 2003). These studies demonstrate mutants affecting late stages of sympathetic neuron differentiation and further highlight the advantages of zebrafish mutant analysis for identifying novel genes and signaling pathways affecting PSNS development. Studies in other organisms have not previously evaluated the *in vivo* role of either *tfap2a* or RA signaling in PSNS development.

4. Mutations Affecting Sympathetic Ganglia Modeling

Mutations of this class affect the maintenance or survival of the PSNS. These mutants display abnormal *TH* expression in the SCG because of a disruption of genes affecting the number, morphology, or survival of *TH*-positive cells. Many dynamic processes occur during modeling of the SCG. In addition, mutations affecting the coalescence of differentiated sympathetic neurons into the discrete ganglia, or cell proliferation within the ganglia, can be identified in the zebrafish embryo by 4 to 5 dpf. Another PSNS mutant isolated from a diploid gynogenetic screen, is the *sympathetic2* (*sym2*) mutant, which displays abnormal *TH* expression in the SCG (Stewart, unpublished data). The earlier stages of PSNS development and differentiation proceed normally in *sym2* mutants. These include neural crest induction, migration, aggregation, and noradrenergic differentiation. Although the numbers of sympathetic neurons appear normal in this mutant, changes in the size and shape of the SCG cells become evident at 3 dpf, which worsen until the embryos die at 6 dpf. Preliminary data suggest that *sym2* plays a role in the coalescence of sympathetic neurons into discrete ganglia once they reach the region of the dorsal aorta. Furthermore, the analysis of pigment cells in *sym2* shows that the cells are capable of migration and differentiate normally, but fail to aggregate into discrete lateral or ventral stripes in the developing embryo. Therefore, *sym2* may possibly function in the formation of ganglia from differentiated sympathetic neurons, a critical process that has yet to be addressed at the molecular level.

IV. Zebrafish as a Novel Model for Studying Neuroblastoma

Neuroblastoma (NB) is an embryonic tumor of the PSNS and the most common extracranial solid tumor of children, often arising in the adrenal medulla (40% of cases; Maris and Matthay, 1999). NB affects 650 children in the United States each year and is the leading cause of cancer deaths in children 1 to 4 years of age (Goodman, 1999). Clinically, NB manifests diverse behavior and is one of the few cancers that can spontaneously regress. Often in infants younger than 1 year of age, NB tumors may regress or differentiate without receiving treatment.

However, older children with advanced disease account for 70% of all NB patients, and their poor long-term survival rate has not risen above 20 to 30% (Maris and Matthay, 1999). Despite intensive research efforts during the mid-1970s, 1980s, 1990s, and early in the millennium, many of the genetic pathways disrupted in this cancer remain unknown, posing a major obstacle to understanding its molecular pathology and to the development of effective therapies for this devastating disease.

Analysis of NB by cytogenetic criteria has identified a number of chromosomal regions that are consistently deleted in NB cells (Maris and Matthay, 1999; Westermann and Schwab, 2002). These regions include allelic losses of 1p36.1, 2q, 3p, 4p, 5q, 9p, 11q23, 14q23-qter, 16p12-13, and 18q–, which have been identified in 15 to 44% of primary neuroblastomas (Brodeur, 2003; Brodeur *et al.*, 1997; Maris and Matthay, 1999). Although the critical target genes within these chromosomal regions remain to be identified, the findings suggest that multiple tumor suppressor genes may function during different stages of NB pathogenesis. Such genes may include those that normally suppress cell proliferation, induce developmentally regulated apoptosis, or promote cell differentiation, therefore preventing the uncontrolled proliferation and survival of sympathetic neuroblasts in NB. Of the tumor suppressor genes found in all types of human tumors, we hypothesize that those contributing to NB are among the most likely to also play critical roles in the normal embryologic development of the target tissue, in this case the developing PSNS. Population-based, neonatal screening of infants with elevated levels of catecholamines demonstrated NB with allelic losses in eight chromosomal regions, indicating that loss of heterozygosity of at least one allele of key tumor suppressor genes, occurred within sympathetic neuroblasts during embryologic development (Takita *et al.*, 1995). For these reasons, we postulate that the identification of genes affecting normal embryonic PSNS development in zebrafish, through large-scale genetic screens, should be particularly relevant to NB and provide important insights into the molecular pathways that are dysfunctional in this disease. Such novel approaches are clearly needed to foster the development of therapies to successfully target this deadly cancer.

V. Conclusion and Future Directions

Impressive advances have been made in understanding the genetic mechanisms that regulate PSNS development through studies in birds and rodents. Zebrafish studies that further define the anatomic and morphologic aspects of the developing PSNS should accelerate the pace of discovery in this field. Exploiting the forward genetics of the zebrafish system can contribute significantly to the identification of new genes and pathways that regulate PSNS development. These mutants also provide the means to genetically dissect PSNS developmental processes *in vivo*. Finally, we postulate that knowledge of the genes responsible for normal PSNS development in the zebrafish will help identify the molecular pathways that are affected in neuroblastoma. Ultimately, second generation suppressor screens based on established PSNS mutants exhibiting proliferative abnormalities, typical of NB, can be used to identify potential genes and pathways that may be relevant to the development of effective therapies for this disease.

Acknowledgments

We would like to thank Hermann Rohrer for comments on the manuscript and Su Guo and Hong Wei Yang for unpublished data. R.A.S. was supported by the Hope Street Kids foundation. This work was supported in part by NIH grant CA 104605 to A.T.L., and NIH grant NS38115 and NSF grant IBN0315765 to P.D.H.

References

Allende, M. L., and Weinberg, E. S. (1994). The expression pattern of two zebrafish achaete-scute homolog (ash) genes is altered in the embryonic brain of the cyclops mutant. *Dev. Biol.* **166,** 509–530.

An, M., Luo, R., and Henion, P. D. (2002). Differentiation and maturation of zebrafish dorsal root and sympathetic ganglion neurons. *J. Comp. Neurol.* **446,** 267–275.

Anderson, D. J. (1993). Cell fate determination in the peripheral nervous system: The sympathoadrenal progenitor. *J. Neurobiol.* **24,** 185–198.

Aybar, M. J., Nieto, M. A., and Mayor, R. (2003). Snail precedes slug in the genetic cascade required for the specification and migration of the Xenopus neural crest. *Development* **130,** 483–494.

Birren, S. J., Lo, L., and Anderson, D. J. (1993). Sympathetic neuroblasts undergo a developmental switch in trophic dependence. *Development* **119,** 597–610.

Brading, A. (1999). "The Autonomic Nervous System and its Effectors." Blackwell Science, Malden, MA.

Britsch, S., Li, L., Kirchhoff, S., Theuring, F., Brinkmann, V., Birchmeier, C., and Riethmacher, D. (1998). The ErbB2 and ErbB3 receptors and their ligand, neuregulin-1, are essential for development of the sympathetic nervous system. *Genes Dev.* **12,** 1825–1836.

Brodeur, G. M. (2003). Neuroblastoma: Biological insights into a clinical enigma. *Nat. Rev. Cancer* **3,** 203–216.

Brodeur, G. M., Maris, J. M., Yamashiro, D. J., Hogarty, M. D., and White, P. S. (1997). Biology and genetics of human neuroblastomas. *J. Pediatr. Hematol. Oncol.* **19,** 93–101.

Bronner-Fraser, M. (2002). Molecular analysis of neural crest formation. *J. Physiol. Paris* **96,** 3–8.

Chun, L. L., and Patterson, P. H. (1977). Role of nerve growth factor in the development of rat sympathetic neurons *in vitro*. I. Survival, growth, and differentiation of catecholamine production. *J. Cell Biol.* **75,** 694–704.

Cohen, A. M. (1974). DNA synthesis and cell division in differentiating avian adrenergic neuroblasts. *In* "Wenner-Gren Center International Symposium Series: Dynamics of Degeneration and Growth in Neurons," (K. Fuxe, L. Olson, and Y. Zotterman, eds.), pp. 359–370. Pergamon, Oxford.

Crone, S. A., and Lee, K. F. (2002). Gene targeting reveals multiple essential functions of the neuregulin signaling system during development of the neuroendocrine and nervous systems. *Ann. N. Y. Acad. Sci.* **971,** 547–553.

DiCicco-Bloom, E., Friedman, W. J., and Black, I. B. (1993). NT-3 stimulates sympathetic neuroblast proliferation by promoting precursor survival. *Neuron* **11,** 1101–1111.

Dick, A., Hild, M., Bauer, H., Imai, Y., Maifeld, H., Schier, A. F., Talbot, W. S., Bouwmeester, T., and Hammerschmidt, M. (2000). Essential role of Bmp7 (snailhouse) and its prodomain in dorsoventral patterning of the zebrafish embryo. *Development* **127,** 343–354.

Dutton, K. A., Pauliny, A., Lopes, S. S., Elworthy, S., Carney, T. J., Rauch, J., Geisler, R., Haffter, P., and Kelsh, R. N. (2001). Zebrafish colourless encodes sox10 and specifies non-ectomesenchymal neural crest fates. *Development* **128,** 4113–4125.

Ernfors, P., Lee, K. F., Kucera, J., and Jaenisch, R. (1994). Lack of neurotrophin-3 leads to deficiencies in the peripheral nervous system and loss of limb proprioceptive afferents. *Cell* **77**, 503–512.

Ernsberger, U., Patzke, H., Tissier-Seta, J. P., Reh, T., Goridis, C., and Rohrer, H. (1995). The expression of tyrosine hydroxylase and the transcription factors cPhox-2 and Cash-1: Evidence for distinct inductive steps in the differentiation of chick sympathetic precursor cells. *Mech. Dev.* **52**, 125–136.

Ernsberger, U., Reissmann, E., Mason, I., and Rohrer, H. (2000). The expression of dopamine beta-hydroxylase, tyrosine hydroxylase, and Phox2 transcription factors in sympathetic neurons: Evidence for common regulation during noradrenergic induction and diverging regulation later in development. *Mech. Dev.* **92**, 169–177.

Fagan, A. M., Zhang, H., Landis, S., Smeyne, R. J., Silos-Santiago, I., and Barbacid, M. (1996). TrkA, but not TrkC, receptors are essential for survival of sympathetic neurons *in vivo. J. Neurosci.* **16**, 6208–6218.

Farinas, I., Jones, K. R., Backus, C., Wang, X. Y., and Reichardt, L. F. (1994). Severe sensory and sympathetic deficits in mice lacking neurotrophin-3. *Nature* **369**, 658–661.

Fouquet, B., Weinstein, B. M., Serluca, F. C., and Fishman, M. C. (1997). Vessel patterning in the embryo of the zebrafish: Guidance by notochord. *Dev. Biol.* **183**, 37–48.

Francis, N. J., and Landis, S. C. (1999). Cellular and molecular determinants of sympathetic neuron development. *Annu. Rev. Neurosci.* **22**, 541–566.

Gabella, G. (2001). "Autonomic Nervous System, vol. 2004." Nature Publishing Group, New York.

Goodman, N. W. (1999). An open letter to the Director General of the Cancer Research Campaign. *J. R. Coll. Physicians. Lond.* **33**, 93.

Goridis, C., and Rohrer, H. (2002). Specification of catecholaminergic and serotonergic neurons. *Nat. Rev. Neurosci.* **3**, 531–541.

Groves, A. K., George, K. M., Tissier-Seta, J. P., Engel, J. D., Brunet, J. F., and Anderson, D. J. (1995). Differential regulation of transcription factor gene expression and phenotypic markers in developing sympathetic neurons. *Development* **121**, 887–901.

Guillemot, F., Lo, L. C., Johnson, J. E., Auerbach, A., Anderson, D. J., and Joyner, A. L. (1993). Mammalian achaete-scute homolog 1 is required for the early development of olfactory and autonomic neurons. *Cell* **75**, 463–476.

Guo, S., Brush, J., Teraoka, H., Goddard, A., Wilson, S. W., Mullins, M. C., and Rosenthal, A. (1999a). Development of noradrenergic neurons in the zebrafish hindbrain requires BMP, FGF8, and the homeodomain protein soulless/Phox2a. *Neuron* **24**, 555–566.

Guo, S., Wilson, S. W., Cooke, S., Chitnis, A. B., Driever, W., and Rosenthal, A. (1999b). Mutations in the zebrafish unmask shared regulatory pathways controlling the development of catecholaminergic neurons. *Dev. Biol.* **208**, 473–487.

Hansen, M. B. (2003). The enteric nervous system I: Organisation and classification. *Pharmacol. Toxicol.* **92**, 105–113.

Henion, P. D., Raible, D. W., Beattie, C. E., Stoesser, K. L., Weston, J. A., and Eisen, J. S. (1996). Screen for mutations affecting development of Zebrafish neural crest. *Dev. Genet.* **18**, 11–17.

Heumann, R., Korsching, S., Scott, J., and Thoenen, H. (1984). Relationship between levels of nerve growth factor (NGF) and its messenger RNA in sympathetic ganglia and peripheral target tissues. *EMBO J.* **3**, 3183–3189.

Hirsch, M. R., Tiveron, M. C., Guillemot, F., Brunet, J. F., and Goridis, C. (1998). Control of noradrenergic differentiation and Phox2a expression by MASH1 in the central and peripheral nervous system. *Development* **125**, 599–608.

Holzschuh, J., Barrallo-Gimeno, A., Ettl, A. K., Durr, K., Knapik, E. W., and Driever, W. (2003). Noradrenergic neurons in the zebrafish hindbrain are induced by retinoic acid and require tfap2a for expression of the neurotransmitter phenotype. *Development* **130**, 5741–5754.

Holzschuh, J., Ryu, S., Aberger, F., and Driever, W. (2001). Dopamine transporter expression distinguishes dopaminergic neurons from other catecholaminergic neurons in the developing zebrafish embryo. *Mech. Dev.* **101**, 237–243.

Howard, M. J., Stanke, M., Schneider, C., Wu, X., and Rohrer, H. (2000). The transcription factor dHAND is a downstream effector of BMPs in sympathetic neuron specification. *Development* **127**, 4073–4081.

Hsu, H. J., Lin, G., and Chung, B. C. (2003). Parallel early development of zebrafish interrenal glands and pronephros: Differential control by wt1 and ff1b. *Development* **130**, 2107–2116.

Kalcheim, C., Langley, K., and Unsicker, K. (2002). From the neural crest to chromaffin cells: Introduction to a session on chromaffin cell development. *Ann. N. Y. Acad. Sci.* **971**, 544–546.

Kelsh, R. N., and Eisen, J. S. (2000). The zebrafish colourless gene regulates development of non-ectomesenchymal neural crest derivatives. *Development* **127**, 515–525.

Kim, H. S., Hong, S. J., LeDoux, M. S., and Kim, K. S. (2001). Regulation of the tyrosine hydroxylase and dopamine beta-hydroxylase genes by the transcription factor AP-2. *J. Neurochem.* **76**, 280–294.

Knecht, A. K., and Bronner-Fraser, M. (2002). Induction of the neural crest: A multigene process. *Nat. Rev. Genet.* **3**, 453–461.

Knight, R. D., Nair, S., Nelson, S. S., Afshar, A., Javidan, Y., Geisler, R., Rauch, G. J., and Schilling, T. F. (2003). Lockjaw encodes a zebrafish tfap2a required for early neural crest development. *Development* **130**, 5755–5768.

Korsching, S., and Thoenen, H. (1983). Nerve growth factor in sympathetic ganglia and corresponding target organs of the rat: Correlation with density of sympathetic innervation. *Proc. Natl. Acad. Sci. USA* **80**, 3513–3516.

Kos, R., Reedy, M. V., Johnson, R. L., and Erickson, C. A. (2001). The winged-helix transcription factor FoxD3 is important for establishing the neural crest lineage and repressing melanogenesis in avian embryos. *Development* **128**, 1467–1479.

Krauss, S., Concordet, J. P., and Ingham, P. W. (1993). A functionally conserved homolog of the Drosophila segment polarity gene hh is expressed in tissues with polarizing activity in zebrafish embryos. *Cell* **75**, 1431–1444.

LeDouarin, N., and Kalcheim, C. (1999). "The Neural Crest." Cambridge University Press, New York.

Lim, K. C., Lakshmanan, G., Crawford, S. E., Gu, Y., Grosveld, F., and Engel, J. D. (2000). Gata3 loss leads to embryonic lethality due to noradrenaline deficiency of the sympathetic nervous system. *Nat. Genet.* **25**, 209–212.

Liu, Y. W., Gao, W., Teh, H. L., Tan, J. H., and Chan, W. K. (2003). Prox1 is a novel coregulator of Ff1b and is involved in the embryonic development of the zebra fish interrenal primordium. *Mol. Cell. Biol.* **23**, 7243–7255.

Lo, J., Lee, S., Xu, M., Liu, F., Ruan, H., Eun, A., He, Y., Ma, W., Wang, W., Wen, Z., *et al.* (2003). 15000 unique zebrafish EST clusters and their future use in microarray for profiling gene expression patterns during embryogenesis. *Genome Res.* **13**, 455–466.

Lo, L., Morin, X., Brunet, J. F., and Anderson, D. J. (1999). Specification of neurotransmitter identity by Phox2 proteins in neural crest stem cells. *Neuron* **22**, 693–705.

Lo, L., Tiveron, M. C., and Anderson, D. J. (1998). MASH1 activates expression of the paired homeodomain transcription factor Phox2a, and couples pan-neuronal and subtype-specific components of autonomic neuronal identity. *Development* **125**, 609–620.

Luo, R., An, M., Arduini, B. L., and Henion, P. D. (2001). Specific pan-neural crest expression of zebrafish Crestin throughout embryonic development. *Dev. Dyn.* **220**, 169–174.

Maris, J. M., and Matthay, K. K. (1999). Molecular biology of neuroblastoma. *J. Clin. Oncol.* **17**, 2264–2279.

Martin, S. C., Marazzi, G., Sandell, J. H., and Heinrich, G. (1995). Five Trk receptors in the zebrafish. *Dev. Biol.* **169**, 745–758.

Martinez-Barbera, J. P., Toresson, H., Da Rocha, S., and Krauss, S. (1997). Cloning and expression of three members of the zebrafish Bmp family: Bmp2a, Bmp2b and Bmp4. *Gene* **198**, 53–59.

Marusich, M. F., Furneaux, H. M., Henion, P. D., and Weston, J. A. (1994). Hu neuronal proteins are expressed in proliferating neurogenic cells. *J. Neurobiol.* **25**, 143–155.

Melby, A. E., Kimelman, D., and Kimmel, C. B. (1997). Spatial regulation of floating head expression in the developing notochord. *Dev. Dyn.* **209**, 156–165.

Miller, C. T., Schilling, T. F., Lee, K., Parker, J., and Kimmel, C. B. (2000). Sucker encodes a zebrafish Endothelin-1 required for ventral pharyngeal arch development. *Development* **127**, 3815–3828.

Morin, X., Cremer, H., Hirsch, M. R., Kapur, R. P., Goridis, C., and Brunet, J. F. (1997). Defects in sensory and autonomic ganglia and absence of locus coeruleus in mice deficient for the homeobox gene Phox2a. *Neuron* **18**, 411–423.

Murphy, S., Krainock, R., and Tham, M. (2002). Neuregulin signaling via erbB receptor assemblies in the nervous system. *Mol. Neurobiol.* **25**, 67–77.

Neave, B., Rodaway, A., Wilson, S. W., Patient, R., and Holder, N. (1995). Expression of zebrafish GATA 3 (gta3) during gastrulation and neurulation suggests a role in the specification of cell fate. *Mech. Dev.* **51**, 169–182.

Nguyen, V. H., Schmid, B., Trout, J., Connors, S. A., Ekker, M., and Mullins, M. C. (1998). Ventral and lateral regions of the zebrafish gastrula, including the neural crest progenitors, are established by a bmp2b/swirl pathway of genes. *Dev. Biol.* **199**, 93–110.

Patten, I., and Placzek, M. (2000). The role of Sonic hedgehog in neural tube patterning. *Cell Mol. Life Sci.* **57**, 1695–1708.

Pattyn, A., Morin, X., Cremer, H., Goridis, C., and Brunet, J. F. (1997). Expression and interactions of the two closely related homeobox genes Phox2a and Phox2b during neurogenesis. *Development* **124**, 4065–4075.

Pattyn, A., Morin, X., Cremer, H., Goridis, C., and Brunet, J. F. (1999). The homeobox gene Phox2b is essential for the development of autonomic neural crest derivatives. *Nature* **399**, 366–370.

Raible, D. W., and Eisen, J. S. (1994). Restriction of neural crest cell fate in the trunk of the embryonic zebrafish. *Development* **120**, 495–503.

Reissmann, E., Ernsberger, U., Francis-West, P. H., Rueger, D., Brickell, P. M., and Rohrer, H. (1996). Involvement of bone morphogenetic protein-4 and bone morphogenetic protein-7 in the differentiation of the adrenergic phenotype in developing sympathetic neurons. *Development* **122**, 2079–2088.

Rohrer, H., and Thoenen, H. (1987). Relationship between differentiation and terminal mitosis: Chick sensory and ciliary neurons differentiate after terminal mitosis of precursor cells, whereas sympathetic neurons continue to divide after differentiation. *J. Neurosci.* **7**, 3739–3748.

Rothman, T. P., Gershon, M. D., and Holtzer, H. (1978). The relationship of cell division to the acquisition of adrenergic characteristics by developing sympathetic ganglion cell precursors. *Dev. Biol.* **65**, 322–341.

Schneider, C., Wicht, H., Enderich, J., Wegner, M., and Rohrer, H. (1999). Bone morphogenetic proteins are required *in vivo* for the generation of sympathetic neurons. *Neuron* **24**, 861–870.

Schober, A., Krieglstein, K., and Unsicker, K. (2000). Molecular cues for the development of adrenal chromaffin cells and their preganglionic innervation. *Eur. J. Clin. Invest.* **30**, 87–90.

Schober, A., and Unsicker, K. (2001). Growth and neurotrophic factors regulating development and maintenance of sympathetic preganglionic neurons. *Int. Rev. Cytol.* **205**, 37–76.

Seo, H., Hong, S. J., Guo, S., Kim, H. S., Kim, C. H., Hwang, D. Y., Isacson, O., Rosenthal, A., and Kim, K. S. (2002). A direct role of the homeodomain proteins Phox2a/2b in noradrenaline neurotransmitter identity determination. *J. Neurochem.* **80**, 905–916.

Shah, N. M., Groves, A. K., and Anderson, D. J. (1996). Alternative neural crest cell fates are instructively promoted by TGFbeta superfamily members. *Cell* **85**, 331–343.

Shelton, D. L., and Reichardt, L. F. (1984). Expression of the beta-nerve growth factor gene correlates with the density of sympathetic innervation in effector organs. *Proc. Natl. Acad. Sci. USA* **81**, 7951–7955.

Smeyne, R. J., Klein, R., Schnapp, A., Long, L. K., Bryant, S., Lewin, A., Lira, S. A., and Barbacid, M. (1994). Severe sensory and sympathetic neuropathies in mice carrying a disrupted Trk/NGF receptor gene. *Nature* **368**, 246–249.

Takita, J., Hayashi, Y., Kohno, T., Shiseki, M., Yamaguchi, N., Hanada, R., Yamamoto, K., and Yokota, J. (1995). Allelotype of neuroblastoma. *Oncogene* **11**, 1829–1834.

Talbot, W. S., Trevarrow, B., Halpern, M. E., Melby, A. E., Farr, G., Postlethwait, J. H., Jowett, T., Kimmel, C. B., and Kimelman, D. (1995). A homeobox gene essential for zebrafish notochord development. *Nature* **378,** 150–157.

Thexton, A. (2001). "Vertebrate Peripheral Nervous System, vol. 2001." Nature Publishing Group, New York.

Varley, J. E., Wehby, R. G., Rueger, D. C., and Maxwell, G. D. (1995). Number of adrenergic and islet-1 immunoreactive cells is increased in avian trunk neural crest cultures in the presence of human recombinant osteogenic protein-1. *Dev. Dyn.* **203,** 434–447.

Weis, J. S. (1968). Analysis of the development of nervous system of the zebrafish, Brachydanio rerio. I. The normal morphology and development of the spinal cord and ganglia of the zebrafish. *J. Embryol. Exp. Morphol.* **19,** 109–119.

Westermann, F., and Schwab, M. (2002). Genetic parameters of neuroblastomas. *Cancer Lett.* **184,** 127–147.

Williams, J. A., Barrios, A., Gatchalian, C., Rubin, L., Wilson, S. W., and Holder, N. (2000a). Programmed cell death in zebrafish rohon beard neurons is influenced by TrkC1/NT-3 signaling. *Dev. Biol.* **226,** 220–230.

Williams, Z., Tse, V., Hou, L., Xu, L., and Silverberg, G. D. (2000b). Sonic hedgehog promotes proliferation and tyrosine hydroxylase induction of postnatal sympathetic cells *in vitro. Neuroreport* **11,** 3315–3319.

Wyatt, S., Pinon, L. G., Ernfors, P., and Davies, A. M. (1997). Sympathetic neuron survival and TrkA expression in NT3-deficient mouse embryos. *EMBO J.* **16,** 3115–3123.

Yelon, D., Ticho, B., Halpern, M. E., Ruvinsky, I., Ho, R. K., Silver, L. M., and Stainier, D. Y. (2000). The bHLH transcription factor hand2 plays parallel roles in zebrafish heart and pectoral fin development. *Development* **127,** 2573–2582.

CHAPTER 13

Optical Physiology and Locomotor Behaviors of Wild-Type and *Nacre* Zebrafish

Donald M. O'Malley, Nagarajan S. Sankrithi, Melissa A. Borla, Sandra Parker, Serena Banden, Ethan Gahtan, and H. William Detrich III

Department of Biology
Northeastern University
Boston, Massachusetts 02115

A growing variety of optical techniques are increasing the power and precision with which the larval zebrafish central nervous system (CNS) can be recorded and manipulated. This review focuses on recent advances in the investigation of neural circuitry in brainstem and spinal cord. Calcium imaging approaches are emphasized because they allow for the noninvasive acquisition of neurophysiological data and thereby complement electrophysiological approaches. The use of both genetically encoded and membrane-permeant calcium indicators offers new options for a more extensive mapping of neural activity than has been achieved to date. Issues of phototoxicity, calcium buffering, and imaging mode are considered in this context. More "global" mapping approaches, such as the c-fos-GFP transgenic fish, are also discussed. In addition, the optical accessibility and relative simplicity of zebrafish larvae permit lesioning of defined neural elements with a precision and efficacy that is currently impossible in higher vertebrates. In conjunction with high-speed behavioral recordings, it should therefore be possible to directly establish the neural basis of the larva's locomotive behaviors. Transparency at the larval stage is of critical importance for these optical approaches and so we also discuss several reduced-pigmentation approaches, including use of the *nacre* mutant line.

I. Introduction

The larval zebrafish is gaining momentum as a model organism for the study of neural circuitry. This heightened interest is attributed in part to an increasingly powerful set of optical techniques that exploit the transparency of larval animals. These techniques allow recording of neural activity, precise optical dissection of neural systems, and fine analyses of the impact of perturbations on behavior (reviewed in Fetcho and Liu, 1998; Fetcho *et al.*, 1998; Gahtan and Baier, 2004; and O'Malley *et al.*, 2003a). These approaches are leading toward an increasingly precise mapping between neural activity, neural systems architecture, and behavior (see, for example, Fuss and Korsching, 2001; Gahtan *et al.*, 2002; Hale *et al.*, 2004; Higashijima *et al.*, 2003; Li *et al.*, 2003; Ritter *et al.*, 2001; Takahashi *et al.*, 2002). Indeed, during the coming years, it is possible that larval zebrafish will become a premier system in which the neural architectures underlying vertebrate behaviors are established. To see why, we should first consider what is involved in establishing such relationships.

Most studies of motor control examine a system with regard to one specific behavior. While this provides the critical focus needed to identify circuit elements, it does not provide sufficient information to understand the functional organization of the neural architectures underlying the animal's behaviors. For example, if many cells known to be active during behavior "A" are subsequently found to be active in a very different behavior "B," that would necessitate a quite different interpretation than had the cells been active exclusively during behavior "A." For this reason, one must characterize neural activity in relation to some significant portion of the animal's behavioral (e.g., locomotive) repertoire. That is, one must determine whether a neuron is involved in multiple behaviors, a single behavior, or just a single component of a behavior. Such information is needed to make conclusive findings regarding both the functional roles of individual neurons and the overall organization of the neural architectures generating the behaviors (Morton and Chiel, 1994). This is an extremely difficult challenge for mammalian motor system studies to overcome, since it is difficult to know, from experiment to experiment, that one is studying the exact same type of neuron. It is also difficult to demonstrate that one has systematically examined all (or at least most) of the neurons that might participate in the set of behaviors of interest. In the case of the larval zebrafish, because of its transparency, and because many neurons can be individually identified, such a comprehensive approach can be contemplated. This chapter describes this approach and, while it focuses on the locomotor control system of the larval zebrafish, the techniques described herein are applicable (in principle) to any neural system in the larva.

II. Optical Physiology—Basics

The field of systems neuroscience suffers from paralyzing challenges posed by the tremendous complexity and neuronal diversity inherent in the mammalian brain (MacNeil and Masland, 1998; Stevens, 1998). Optical techniques, in conjunction with the transparency of the larval zebrafish, offer the potential to systematically survey and characterize neurons and their circuit properties, thus arriving at more comprehensive descriptions of the neural architectures underlying the behaviors of interest (see, for example, Gahtan and O'Malley, 2003; Gahtan et al., 2002; Hale et al., 2001; Higashijima et al., 2001; Ritter et al., 2001). The examples that follow regard neurons in the larval brainstem and spinal cord. These neurons have been the subject of numerous anatomical, physiological, and developmental studies (see, for example, Bernhardt et al., 1990; Eaton et al., 2001; Eisen, 1999; Kimmel et al., 1982, 1985; Lewis and Eisen, 2003; Liu and Fetcho, 1999; Liu et al., 2003; Metcalfe et al., 1986; Zottoli and Faber, 2000) and can often be individually identified, which allows the establishment of precise linkages between very different kinds of neuronal information obtained under a variety of conditions. Such linkages are deemed critical to understanding the complex neural architectures that drive behavior. We conclude our discussion of optical physiology with an example of a yet more comprehensive approach, where neural activation patterns can be acquired from the entire CNS while preserving single-cell resolution.

A. Synergism between Electrical and Calcium Recordings

The most direct means of examining a neuron's activity is to record electrical events such as synaptic and action potentials. Recording of electrical activity in the CNS of intact larval zebrafish is difficult, although by immobilizing and partially dissecting the animals, electrical data can be obtained (Ali et al., 2000; Buss et al., 2003; Gamkrelidze et al., 2000; Liu and Westerfield, 1988; Rigo et al., 2003). An alternative means is to study homologous circuitry in the goldfish, which has somewhat larger neurons and a substantial electrophysiological record (see, for example, Aksay et al., 2001, 2003; Faber et al., 1989; Fetcho, 1991).

Goldfish are cyprinid hearing specialists that are closely related to the zebrafish. The two species exhibit striking similarities in the brainstem circuits underlying locomotion and sensory information processing (Bass and Baker, 1997; Lee and Eaton, 1991; Lee *et al.*, 1993; Zottoli *et al.*, 1999). In the zebrafish, aside from the Mauthner cell, there are few published electrophysiological recordings from conclusively identified spinal or brainstem neurons during the larval stages discussed here. A recent sharp-electrode study in adult goldfish, however, characterized the auditory responses of the Mauthner cell's segmental homologues (Nakayama and Oda, 2004). This study supported the results of optical recordings from larval zebrafish (O'Malley *et al.*, 1996), and went considerably further, revealing novel details of the cells' firing patterns. Therefore, a useful synergism exists between these two closely related cyprinids and specifically between the techniques that can be employed in each. These synergies should advance our understanding of the neural systems in both species.

While electrical data are extremely valuable, some types of information cannot be practically or efficiently obtained by such methods and for which imaging methods are better suited. In particular, calcium imaging has been an important complement to electrophysiology because measurable calcium signals generally accompany neural electrical activity, and, more specifically, the firing of action potentials (Fig. 1). Calcium imaging has revealed neural activity patterns in many identified zebrafish neurons for which there are no published electrical recordings (see, for example, Fetcho and O'Malley, 1995; Fuss and Korsching, 2001; Gahtan *et al.*, 2002; Hale *et al.*, 2004; O'Malley *et al.*, 1996; Ritter *et al.*, 2001). In these studies, confocal calcium imaging is generally able to detect single action potentials in neuronal somata inside awake, restrained larvae. Sub-threshold activity does not appear to produce measurable somatic calcium responses. Calcium indicators have excellent spatial resolution and are also suitable for detecting the onset of electrical activity (Fetcho and O'Malley, 1995; Gahtan *et al.*, 2002). They can be used to estimate duration and relative intensity of bouts of electrical activity (Fetcho *et al.*, 1998), but cannot resolve temporal events at frequencies much above 10 Hz (O'Malley *et al.*, 2003a). These approaches to imaging neural activity have been in place since the mid-1990s (O'Donovan *et al.*, 1993), but an exciting new set of approaches is expanding the scope and power of optical recording methods.

B. Simultaneous Recording of Neural Activity and Behavior

The noninvasive nature of optical recording allows neurons to be recorded throughout the larval CNS in awake, intact animals. One advance has been the simultaneous recording of neural activity and high-speed behavioral imaging, which was used to distinguish identified spinal neurons involved in escape versus swimming behaviors (Ritter *et al.*, 2001). Our lab has also begun using a similar approach to evaluate the involvement of brainstem neurons in locomotive

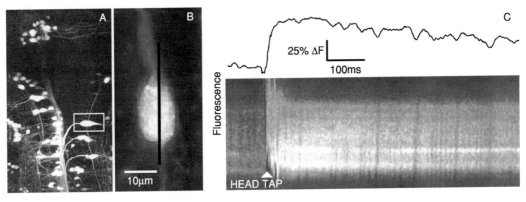

Fig. 1 Line scan of a Mauthner cell. (A) Mauthner cell (M) in rhombomere 4 is scanned in an awake, restrained larva using a BioRad MRC600 confocal microscope. (B) The image of the cell is rotated 90° and magnified. The laser beam is scanned repeatedly across the cell at a fixed location indicated by the vertical bar. (C) Each successive scan line (acquired at 2 msec intervals) is plotted from left to right. Fluorescence intensity is represented by a color scale with orange to yellow indicating the highest fluorescence and purple to blue indicating the lowest. Relative fluorescence intensity is plotted above the line scan. Line scans reveal the latency of a cell's response to a given stimulus (Gahtan *et al.*, 2002). While the increase in calcium is abrupt, the recovery to resting calcium levels takes several seconds because calcium extrusion mechanisms are slower than initial influx. The calcium indicator itself also slows the response dynamics. Because the Mauthner cell is believed to fire only a single action potential per stimulus (Faber *et al.*, 1989), the size of the calcium response is unexpectedly large. This has been a frequent observation with Mauthner cells and may reflect a calcium potentiation mechanism (O'Malley and Fetcho, 1996). Differences in resting fluorescence values are attributed to regional variation in the intracellular behavior of the indicator, rather than persistent resting calcium gradient (O'Malley, 1994; O'Malley *et al.*, 1999). Relative changes in fluorescence intensity are indicative of the magnitude of the calcium response and, to some extent, the firing rate of action potentials (O'Malley *et al.*, 2003a). (See Color Insert.)

behaviors. We include here two examples of how this approach can strengthen the linkages between neural activity and behavior.

Figure 2 shows a neuron recorded during a visually evoked swimming (VES) behavior that had not been previously described for larval zebrafish (Sankrithi, 2003). Larvae are partially immobilized in agar (by embedding just their head) and placed on the stage of the confocal microscope. Exposing these larvae to a step increase in illumination (in a relatively dark room) results in a moderately sustained bout of locomotive activity, in comparison with the relatively brief swim bouts observed spontaneously and during prey capture (Borla *et al.*, 2002). Aside from duration, the behavior appears kinematically similar to the slow swim bouts described by Budick and O'Malley, 2000a. Simultaneous recording of tail movements and of calcium responses in the four largest neurons in the nucleus of the medial longitudinal fasciculus (nMLF) revealed that three of the four were consistently activated during the VES behavior (O'Malley and Sankrithi, 2002). While it is not certain that the visually evoked tail movements seen in restrained animals

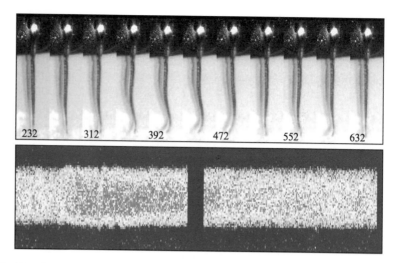

Fig. 2 Visually evoked swimming (VES). While neuronal activity is being recorded in the brainstem, using an inverted confocal microscope, the behavior of partially restrained larvae can be simultaneously recorded. The larva's head (not visible here) is embedded in agar (*dark region at top of image*). The larva's trunk/tail is free to move only in the lighter region at the bottom. The behavioral sequence was recorded using a Redlake Imaging MotionScope high-speed camera mounted on a dissecting scope positioned above the stage of the inverted microscope. The laser beam, which is scanning neurons within the brainstem, is visible as a white dot at the top of the image. The relative time values within the swim bout are indicated (in milliseconds) for alternating frames in the behavioral sequence. Below the larvae are two consecutive "slow" line scans of an nMLF neuron in the midbrain of the same larvae. The line scans were acquired at 6 msec/scan line. The gap in the scan is attributed to the time-lapse between the end of the first scan and the onset of the second scan (the two line scans fill the frame buffer). By the end of the second scan, about 10 sec after the initial stimulus, calcium levels have returned close to their resting value. The gentle VES-like tail movements in restrained larvae are similar to such movements exhibited in free-swimming larvae and are dramatically different from the vigorous tail movements associated with escape and struggling behaviors as shown in Fig. 3. (See Color Insert.)

are identical to the VES behaviors observed in free-swimming animals, the comparable frequencies and amplitudes of the bending patterns suggest that they are the same behavior. These swim bouts are distinct from other elements of the larva's locomotive repertoire, which includes escapes, routine turns, burst swims, capture swims, and J-turns.

A gentle tap to the head of a larva evokes a very different movement, the well-known escape behavior (Kimmel *et al.*, 1974). Free-swimming larvae that are induced to escape show a consistent sequence of movements, specifically a C-bend, counter-turn, and burst swim (Budick and O'Malley, 2000a). While the magnitude of each component can vary, providing adaptive variability in escape direction (Foreman and Eaton, 1993), the pattern of the sequence is quite reproducible. Examination of this behavior in partially restrained animals revealed an unexpected complexity. Tail movements recorded from partially restrained larvae frequently showed "partial-escape" behaviors (i.e., an isolated C-bend like movement or a double-bend analogous to a *C-bend + counter-turn* [Fig. 3A and B]). "Full-escape" behaviors, consisting of two large bends and subsequent burst-swim-like movements, were also frequently observed (Fig. 3C). Taps infrequently elicited a struggling behavior (Fig. 3D) that was readily distinguished from the escape-like behaviors. Because the activity of hindbrain neurons is recorded during these tail movements, this should help establish the role of specific neurons in the different stages of the escape behavior. Earlier optical recordings had indicated that most descending neurons were activated during escape behaviors (Gahtan *et al.*, 2002), but it is not known, for example, whether some neurons are specifically activated during the initial C-bend with perhaps others being activated by the final burst-swimming phase. Sankrithi and O'Malley (2000) showed that all four large nMLF neurons are often active bilaterally during escape behaviors and these same neurons were consistently activated during trials that included swimming. In addition, laser ablation of these neurons had little effect on the escape behavior (Budick and O'Malley, 2000b). Together with the VES results, this suggests that these nMLF neurons may play a greater role in swimming as opposed to turning behaviors. More generally, the ability to simultaneously, and systematically, image neural activity and behavior should facilitate efforts to establish the neural architectures generating vertebrate behaviors.

III. Optical Physiology—Alternative Approaches

A number of alternative approaches to labeling zebrafish neurons with fluorescent calcium indicators are becoming available. Our initial approach consisted of gross injections of fluorescent dextrans into nerve tracts. We recently showed that this labeling was a consequence of the injection pipette severing axons, after which the 10,000 molecular weight (MW) dextrans would enter cut axons and then be trapped when the axons resealed (Gahtan and O'Malley, 2001). Such labeling persists for weeks if not longer (O'Malley, unpublished observations). The retrogradely labeled neurons are often sparsely distributed, providing high contrast and sufficient morphological details to unambiguously identify individual neurons (Hale *et al.*, 2001; O'Malley *et al.*, 2003a). This approach also provides information about axon trajectory since the labeled neurons, wherever they may be, were labeled by the projection of their axon (or possibly dendrite) into the injection site. Some newer techniques offer features that are complementary to this approach, giving interested users a number of choices. For example, injection of membrane-permeant calcium indicator into the zebrafish spinal cord, in conjunction with 2-photon imaging, provides highly stable calcium signals permitting continuous recording for appreciable periods of time (Brustein *et al.*, 2003). This approach also labels large numbers of neurons relatively quickly and does not require that axonal projections be specifically targeted. Presumably, this labeling approach would be well suited for both confocal and 2-photon studies. One potential difficulty, however, is that the morphological details visualized with this approach

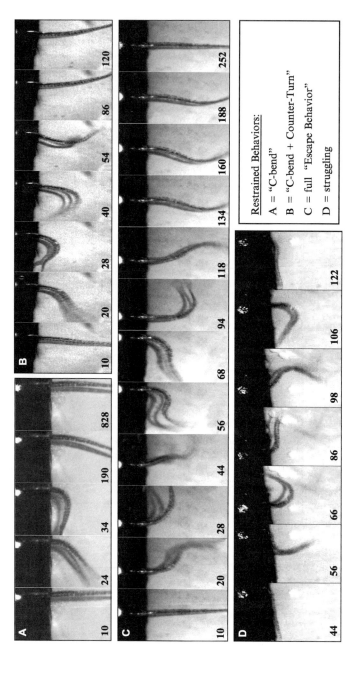

Fig. 3 Components of the escape behavior. Partially restrained larvae, unlike free-swimming larvae, exhibit head-tap responses that appear to consist of portions of the overall escape behavior. (A) A tap elicits an isolated C-bend-like movement: a fast, strong flexion to one side followed by relaxation. (B) A "C-bend + Counter-Turn"-like movement, where the tail flexes left, then right, and then relaxes. (C) The movements are suggestive of a complete escape behavior with two large turns and subsequent undulations indicative of a burst-swim bout. (D) Struggling behavior. The trunk is bent and vigorously whipped from side to side. In the first and last frame, the trunk is pressed up against the agar so firmly that it is barely visible. Selected frames are shown for each behavioral sequence and relative times in each trial are indicated in milliseconds. As in Fig. 2, neuronal activity is being simultaneously recorded and we are in the process of remapping the brainstem network to these locomotive elements of the escape behavior. Larvae used in these studies are usually between 5 and 10 days old, during which time their behavioral capabilities seem fairly constant. They are typically about 4 mm in total length.

do not seem sufficient to identify particular individual neurons. This could be problematic in certain instances, so the choice of which calcium imaging approach to use will depend on the particulars of the experimental question at hand.

A. Molecular Imaging Techniques

A more intriguing approach to visualizing neural activity involves use of the green fluorescent protein (GFP) to construct novel *molecular imaging* tools. By combining GFP with calcium binding proteins, it is possible to construct genetically encoded calcium indicators called "cameleons" (Miyawaki *et al.*, 1999). The cameleon construct contains a calmodulin molecule and M13 domain sandwiched between two different color GFP molecules. When it binds with calcium, calmodulin binds to M13, altering the overall conformation of the construct such that the GFP molecules move closer together and thereby increase the fluorescence energy transfer between them. This altered energy transfer can be quantified using ratiometric imaging. Higashijima *et al.* (2003) have generated transgenic zebrafish in which the cameleon indictor is driven by the pan-neuronal promoter HuC. In this confocal calcium imaging study, distinct calcium responses were recorded through extended periods, comparable to those reported in the 2-photon imaging study we have already cited (Brustein *et al.*, 2003). While there are some theoretical reasons for expecting 2-photon imaging to produce lower phototoxicity than visible-wavelength confocal imaging, it has not been demonstrated that 2-photon imaging actually increases the durability of calcium recordings in larval zebrafish. In contrast, confocal calcium imaging has been used to record Mauthner cell responses, in time-lapse mode, through a 2-day period (Fetcho *et al.*, 1998).

Another issue relating to *in vivo* calcium imaging is the possible side effects of the calcium indicators. For example, millimolar levels of calcium buffer appear to perturb some developmental processes (Ashworth *et al.*, 2001), which raises the question of how longer term labeling of neurons with calcium indicators (i.e., calcium buffers) might influence neural development, neural networks, or larval behaviors. However, the presence of significant amounts of pan-neuronally expressed cameleon protein, beginning from about 3 days post-fertilization, did not impair the establishment of a stable transgenic line (Higashijima *et al.*, 2003). In our experiments, we found that the presence of calcium indicators in many neurons, for days or longer, had no obvious impact on either the observed neuronal activity patterns or the larval behaviors.

The more significant aspect of the molecular imaging approach is that it opens an essentially new frontier for tailoring optical probes to address specific neural questions. While the pan-neuronal HuC-cameleon line may offer similar utility to the membrane-permeant approach (and also avoids the injection step and any injection-related damage), Higashijima *et al.* (2003) emphasize that this approach can be most useful by targeting the reporter to specific classes of neurons, as they did using the *Islet-1* promoter to target the cameleon protein to Rohon-Beard cells. Their group has isolated genes specific to other classes of zebrafish neurons, including glutamatergic, glycinergic, and GABAergic neurons (Higashijima *et al.*, 2001). The future targeting of cameleon proteins to such specific classes of neurons should accelerate characterization of neural circuits underlying different larval behaviors. Alternative genetically encoded calcium indicators are available including, gCaMP, which is also constructed from GFP; however, unlike the cameleon protein, gCaMP does not require ratiometric imaging. It may produce more robust fluorescence responses (Wang *et al.*, 2003). Another possibility would be to combine these various calcium imaging techniques, by, for example, injecting membrane-permeant or retrograde-tracing based indicators into a zebrafish line expressing cameleon or gCaMP in a specific neuronal population. By using two different color fluorophores, one could obtain specific information about the genetically targeted population and simultaneously obtain broader population data from neighboring neurons. Such experiments would ideally be performed on a 2-photon system (if available), because the broad 2-photon absorption profiles of many fluorophores should facilitate the separation and recording of the emissions from multiple calcium indicators.

B. Global Mapping

Calcium dynamics and electrical events are just two activities associated with neuronal stimulation. A third approach to identifying those neural elements involved with particular neural operations or behaviors is called *global mapping* (O'Malley *et al.*, 2003b). This approach is designed to generate a comprehensive spatial map of the neural activation patterns that might be produced by sensory stimulation, repetitive elicitation of a behavior, or any other protocol or condition that affects the pattern of neural activity. It is based on a class of genes called immediate early genes (IEGs) that are rapidly induced in response to sustained periods of neural activity. By simply attaching a GFP to an IEG promoter, it becomes possible to use such "natural" activity reporters to visualize neuronal activation patterns. We have constructed a reporter in which the zebrafish c-fos promoter drives an enhanced GFP (EGFP) gene. C-fos was chosen because it is expressed in virtually every part of the CNS, but in a regionally specific manner appropriate for the stimulation or behavioral task (Bosch *et al.*, 2001; Herdegen and Leah, 1998). In transient expression assays, we have found widespread expression of fos-GFP in response to a combined tactile/vestibular stimulation protocol (Fig. 4). Global mapping thus has the potential to provide a kind of "snapshot" of the global pattern of neural activity that was induced in response to the foregoing stimulation protocol. Because these activity patterns are readout from living animals, two distinct controls are available. A record of the resting GFP pattern can be acquired before the stimulation protocol and in the exact same animal in which fos-GFP will be induced. The second control consists of animals that are not subjected to the stimulation protocol, but are otherwise handled identically.

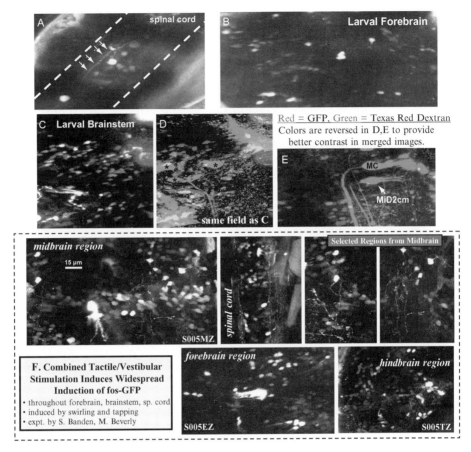

Fig. 4 Fos-GFP mapping of neural activity. 50 head-taps elicited repetitive escape behaviors and induced fos-GFP expression in the forebrain, brainstem, and spinal cord of some larvae containing fos-GFP DNA. (A) In spinal cord (*between dashed lines*), fos-GFP was induced in several large neurons and is present in neuronal processes as well (*arrows*). Fos-GFP is also induced in forebrain (B) and in brainstem (C). (D) Some larvae were double-labeled by injecting Texas-red dextran into the spinal cord, which labels reticulospinal neurons in brainstem (*green)* and fos-GFP neurons (*red*; same fos-GFP field as shown in C). (E) Another larva shows fewer brainstem neurons with fos-GFP induction (*red*). Two large neurons (Mauthner cell, MiD2cm) were retrogradely labeled but were negative for fos-GFP. Induction is expected to be mosaic in these injected animals, because the fos-GFP DNA is not expected to be present in all neurons. The Mauthner cell normally shows c-fos protein expression in response to this stimulus (Bosch *et al.*, 2001). (F) More widespread induction resulted from combined vestibular/tactile stimulation. Numerous cell bodies, fine processes, and varicosities are labeled. Images acquired pre-stimulation never showed basal expression of fos-GFP. In these larvae, fos-GFP DNA (about 25 ng) was injected into one- or two-cell stage zebrafish embryos and the larvae were imaged at 5 to 7 dpf. Batches of injected larvae were group stimulated by repetitive tapping (and for some animals by periodic swirling) of the 24-well trays in which they were maintained. Typically, 5 to 10% exhibited varying degrees of fos-GFP induction. (See Color Insert.)

In our efforts to develop this new technology, we have thus far only examined transient expression, which would typically be mosaic in nature, but establishment of transgenic lines such as these would be of potentially great value to the zebrafish community. Our lab and collaborators are interested in a range of larval and juvenile behaviors including prey capture, aggressive and circadian behaviors, and responses to pharmacological agents that alter these behaviors. Fos-GFP fish have the potential to map the global activation pattern in all such experiments by simply collecting a small number (about 7) of high-resolution image stacks that can span essentially the entire CNS, while providing single-cell resolution. This "memory" of the neural activation pattern would only be read out after allowing the cells in which the c-fos promoter was activated time to synthesize the fully mature (i.e., fluorescent) GFP. This memory trace will be quite persistent, which is necessary for our purposes because the fluorescence needs to persist long enough for acquisition of all confocal (or 2-photon) image stacks. Acquisition of such systematic and complete maps of neuronal activity across the larval CNS (with its approximately 147,000 neurons, by our very rough estimate) is difficult or impossible with any other histological or physiological approach. Also, once the neurons are fluorescently labeled, they could be subjected to a variety of other experiments, including cell filling to establish cell morphology and identity (Gahtan and O'Malley, 2003), laser ablation (Roeser and Baier, 2003), or electrical recording. But simple generation of a spatial map of all neurons active during a behavior or in response to a particular stimulus, would provide an essential context that might dramatically alter the interpretation of more spatially restricted physiological studies.

Perhaps the most exciting application of global mapping concerns the numerous mutant lines of zebrafish in which either the CNS or larval behaviors are affected. For example, an embryo or larva could be chemically or electrically stimulated at a developmental time of interest, and the global activation pattern recorded. This could then be repeated on a mutant/transgenic line made by crossing homozygous fos-GFP fishes to the mutant line. With this hybrid line, the activation pattern in the mutant background could be acquired and then compared with the wild-type pattern. Such data have the potential to sharply accelerate investigations into the locus of action of the many neural and behavioral mutations that to this date remain enigmatic.

IV. Options for Improving the Transparency of Larval Zebrafish

Zebrafish have made a splash in the field of developmental genetics because of the transparency of their embryos and larvae. During this developmental period there are, however, increasing amounts of pigmentation from multiple cell types and this significantly degrades the efficacy of optical techniques, especially in the larval stages. This is problematic not only for the imaging and ablation techniques discussed here, but also for other imaging and visualization methodologies discussed throughout this volume. Being able to use reduced pigmentation animals would thus be of considerable importance, but any time one deviates from the use of wild-type animals, this raises concerns about whether the biological phenomenon or mechanism at hand is in some way altered in the reduced pigmentation animal.

A variety of options are available for increasing the transparency of zebrafish. A widespread approach is to treat embryos with the compound 1-phenyl-2-thio-urea (PTU) to inhibit melanin synthesis. This is very effective in reducing pigmentation throughout the animal, especially in the retina, where the pigmentation may be unaffected by other approaches. By raising embryos in 0.003% PTU, cells in the retina can be very clearly visualized throughout the embryonic stage and into the larval stage (Perkins et al., 2002; Westerfield, 1995). However, this dose of PTU can affect thyroid development and higher doses can cause greater abnormalities (Elsalini and Rohr, 2003). Another common approach is to use the less pigmented *golden* Danio variant, which comes in both short-fin and long-fin varieties. We have used long-fin gold (LFG) danios extensively in our study of locomotor systems to facilitate the imaging of neural circuitry. Since the late 1990s, we have not noticed any differences in either reticulospinal/spinal circuitry or in the behavior of LFG versus wild-type larval fishes. However, we have not made explicit comparisons. It should be noted that adult LFG zebrafish differ from wild type in terms of swimming kinematics, which is an expected result of their exaggerated fin morphology (Plaut, 2000). Differences in aggressive interactions have also been observed (H. Schneider, personal communication).

Other possibilities to enhance transparency include a variety of albino and reduced-pigment lines (Kelsh et al., 1996, 2000). We have begun studying the *nacre* line, which has exceptionally low pigmentation because of the mutation of a microphtalmia-related protein. This results in the absence of melanophores (Lister et al., 1999). In confocal imaging experiments we have observed neuroanatomical details of the brainstem motor systems in *nacre* fish that appear similar to that of the LFG and wild-type strains. Figure 5 shows examples of different neural structures as they appear in the *nacre* fish. The only obvious difference between our wild-type (or LFG) data and the images acquired from *nacre* animals, is that the *nacre* images are considerably brighter. For example, after labeling reticulospinal neurons in *nacre* larvae with Alexa 488 dextran (Molecular Probes, Eugene, OR), the fluorescence intensity is so bright that it presented an unusual problem of saturating the confocal photomultiplier (PMT) tube even with both the highest neutral density filter in place (99% attenuation) and with the confocal aperture reduced to its minimum pinhole size (for the MRC600). This is problematic because fluorophore and PMT saturation both degrade image quality and gain reduction only corrects the PMT problem. However, the BioRad MRC600 can be operated in a low-power mode that is normally used for "standby" purposes. This offers a lower lasing intensity level without the inconvenience of replacing a standard neutral density filter with one of higher attenuation. The variety of structures highlighted in Fig. 5 can be compared with neuroanatomical images

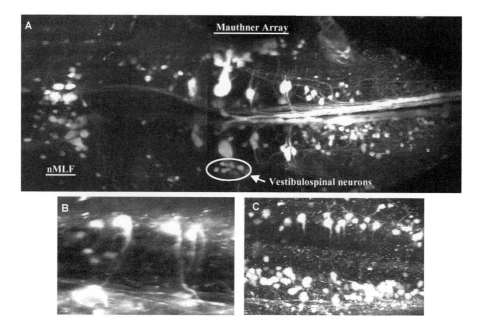

Fig. 5 Neuroanatomy of *nacre* zebrafish. Brainstem neurons were retrogradely labeled by injection of Alexa 488 dextran (10,000 MW) into the spinal cord. Examination of numerous fish revealed normal-looking labeling patterns including, for example, the montage in (A), which shows nMLF labeling, as well as an array of reticulospinal neurons. (B) A higher magnification view of cells with crossing axons. (C) A bilateral population of laterally placed neurons in caudal brainstem that normally give rise to the lateral longitudinal fasciculus. Because we are unable to label all descending neurons in single larvae, it is difficult to make quantitative comparisons of the neuronal populations in *nacre* versus wild-type larvae. We have not yet attempted to compare *nacre* neuroanatomy with that of adult wild-type zebrafish (Wullimann *et al.*, 1996).

shown in the original histological material (Kimmel *et al.*, 1982 and 1985) and with confocal images of these same cells observed in living wild-type and LFG zebrafish (Fetcho and O'Malley, 1997; Fetcho *et al.*, 1998; O'Malley *et al.*, 1996 and 2003a).

Equally important for our interests in establishing the neural basis of locomotive behaviors, is confirming that the *nacre* zebrafish (or any low-pigment animal being used), has behavioral patterns that are essentially identical to the wild-type animals. Figure 6 illustrates a set of locomotive behaviors that provides a representative sampling of the larva's known locomotive repertoire. The slow swim and routine turn shown are similar to those observed in wild-type larvae (Budick and O'Malley, 2000a), while the prey capture behavior shows complex control elements that are similar to those described in Borla and O'Malley, 2002 and Borla *et al.*, 2002. In the next section (on perturbation techniques), we discuss experimental perturbations of the descending control systems and the resulting behavioral deficits that were obtained using the *nacre* line.

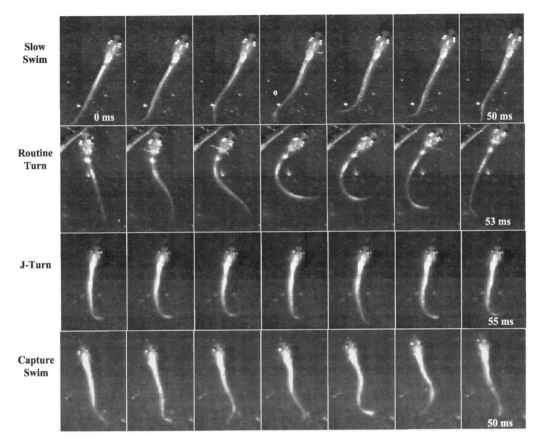

Fig. 6 Locomotive repertoire of *nacre* zebrafish. 7-day old *nacre* larvae were observed (using a Reticon high-speed camera, 1000 frames/sec max frame rate), performing a largely complete repertoire of larval locomotive behaviors. Each behavior appears similar to those previously described for wild-type fish, including the slow swim and routine turn (Budick and O'Malley, 2000a), the J-turn (Borla and O'Malley, 2002) and the capture swim (Borla *et al.*, 2002). The time at the end of each row indicates elapsed time (in milliseconds) from start of each image series. Selected frames are shown from each trial to illustrate the prominent bending features associated with each behavior. The slow swim, J-turn, and capture swim were all recorded from the same larva. Images were collected at between 600 to 800 frames per second. Collection of data at lower frame rates allows less intense illumination to be used.

The agreement between these *nacre* studies and previous experiments on wild-type and LFG animals is useful, especially since new optical methods and applications are continually appearing in the zebrafish literature. This agreement is, however, only anecdotal at this point, as we have not attempted rigorous analytical comparisons. Such comparisons would entail considerable time and expense. It is not expected that we would find much difference between *nacre* and wild-type animals in our studies, since the predominant effect of *nacre* seems to be on the generation of melanophores, with iridophores and xanthophores being produced in substantial numbers (Lister *et al.*, 1999), and specifically on melanophores of neural crest origin. Effects on other crest-derived cell types have not, however, been ruled out. Even if one carried out numerous comparison experiments, it would still be difficult to rule out possibly significant deviations from wild-type characteristics in terms of such traits as neural activity patterns, behavioral repertoire, or gene expression profiles. As optical methods become increasingly important in examining myriad aspects of zebrafish biology (see, for example, Ashworth and Bolsover, 2002; Bak and Fraser, 2003; Carvan *et al.*, 2001; Chen *et al.*, 2002; Das *et al.*, 2003; Downes *et al.*, 2002; Gahtan and O'Malley, 2001, 2003; Goldman *et al.*, 2001; Kato *et al.*, 2004; Langenberg *et al.*, 2003; Li *et al.*, 2003; Niell *et al.*, 2004; Ritter *et al.*, 2001; van der Sar *et al.*, 2003; Webb and Miller, 2003), it becomes all the more important to rigorously evaluate lower pigmentation animals—either lines or treatment protocols. It might be of general use to the zebrafish community if there was a systematic effort to collect and organize information about such reduced pigmentation methods, for example, via the Zebrafish Information Network (ZFIN; Sprague *et al.*, 2003).

V. Behavioral Perturbation Studies

An especially daunting problem in the study of vertebrate neural systems is the establishment of causal roles for neurons—either in the generation of behaviors or in terms of their contributions to neural computations. Traditional lesioning approaches such as tract cutting, neurotoxic cell killing, and electrolytic lesions generally provide "regional" level information since they interfere with large numbers of axons or cells that are likely of diverse anatomy, projection, and function. Establishing causal roles for specific neural circuits in mammals constitutes a further challenge because of the vast number of neurons, the countless interconnections between systems (including reciprocal connections), and the currently unclear extent of neuronal phenotypic diversity. It is therefore imperative to examine simpler vertebrate systems, such as the larval zebrafish CNS, if one is to establish causal relationships between neural circuits and behavior. In this section we discuss several behavioral perturbation approaches that are intended to further the establishment of such linkages.

A. Laser Ablation

Laser ablation of individually identified neurons is perhaps the most straightforward perturbation approach (Fetcho and Liu, 1998; Liu and Fetcho, 1999). In this approach, the confocal imaging laser, when set at full power, can be used to precisely dissect specific elements out of the neural system. Larvae are first embedded in agar and the laser beam is focused on the center of the cell of interest (which has been labeled retrogradely or perhaps by GFP). After killing a specific set of cells, high-speed behavioral recording is used to evaluate the larval locomotive behaviors to discover any behavioral deficits that may have been produced. In theory, this approach can be used to create any desired neural "mutation," in that we could conceivably remove any or all cellular elements of interest. In practice, there are three major limitations:

1. The challenge of labeling all of the neurons that one wishes to laser ablate.
2. The time required for laser ablation.
3. The requirement to confirm that lased cells have in fact been disabled or killed.

For our motor system experiments, a general approach is to inject several dozen or more larvae with calcium-indicating dextrans and subsequently (the next day usually) screen these animals for the desired labeling patterns (O'Malley and Fetcho, 2000). Screening such numbers of larvae can be done fairly quickly and usually results in at least a few larvae with the labeling pattern desired for the experiment at hand. A much bigger obstacle is the 10-minute period that is typically required to laser ablate each individual neuron using the MRC600 confocal imaging laser (a 15-mW Krypton-Argon laser). This is not problematic if only a few cells are being targeted (Liu and Fetcho, 1999), but if many dozens of cells are to be targeted, the required time quickly becomes prohibitive.

The problem of confirming cell kills is also crucial. Liu and Fetcho (1999) used a double labeling approach, where delayed loss of the non-lased (i.e., non-bleached) dye's fluorescence was taken as evidence that the lasing had inflicted lethal damage on the cell after which the cell lost its membrane integrity and died. In single-wavelength experiments, we often observe after laser-ablation attempts, at time scales ranging from minutes to days after lasing, that a lased and "fully bleached" cell will recover to a lower but clear level of fluorescence, as if fluorescent dextrans from remote reaches of the cell had gradually diffused back into the soma where the lasing had taken place. Subsequent re-lasing of the cell will often result in a permanent loss of fluorescence, which we regard as indicating cell death. This is based on our attempts to relabel such cells by injecting dextrans into spinal cord at a site where its axon should be intact. In such attempts at relabeling cells that had met the "permanently photobleached" criterion, we have never observed relabeling of cells presumed dead. Other approaches to confirming cell kills, such as the labeling of dying or dead cells with acridine orange, have not proved reliable, perhaps because doing this in living animals requires that one "catch" the lased cells in the act of dying (i.e., before macrophages or other digestive processes have removed the remnants of the cell).

Liu and Fetcho (1999) showed that killing just three cells, the Mauthner cell and its two homologues, dramatically increased the latency of the larval escape behavior. However, the delayed behavior was of high angular velocity and was kinematically well formed, indicating that the remaining hindbrain neurons were capable of producing escape-like behaviors. Our early attempts to eliminate or degrade this delayed behavior by laser ablating much larger numbers of neurons were unsuccessful (Budick and O'Malley, 2000b; Gahtan and O'Malley, 2000), which turned out to be consistent with our subsequent finding that a large fraction of the entire population of descending neurons is active during the escape behavior (Gahtan *et al.*, 2002). We quickly ran into the problem of cell numbers: 100 cells × 10 min/cell = intolerable amount of lasing time for fish and researcher. Use of higher powered lasers suffers from the problem that such lasers often damage blood vessels in the larval CNS, rendering such experiments unable to be interpreted. (These and other issues are discussed further in O'Malley *et al.*, 2003a.) Two additional strategies offer the potential for extending lesioning applications to take advantage of the accessibility and identifiability of zebrafish neurons.

B. Photosensitizing Agents and Labeled Lesion

In the first approach, a photosensitizing agent (tin chlorin e6) is conjugated to 10,000 MW dextran and this chlorin-dextran is co-injected with fluorescent dextrans to load the agent into retrogradely labeled cells (Sankrithi *et al.*, 2002). This approach has shown promise in preliminary photo-killing experiments, in that (1) the required lasing time is reduced from 10 minutes to 2 minutes, and (2) the lasing is accomplished using a red laser line (either 635 or 647 nm), which inflicts lethal damage without photobleaching the fluorescent label, thus allowing us to use the "delayed-loss-of-fluorescence" criterion to confirm cell kills. With such an approach, it therefore appears feasible to kill precise but large subsets of the 300 known neurons that project from the brain into the spinal cord of the larval zebrafish. This should not only allow us to disable substantial elements of the descending control system, but, because there is a complete template of all known descending neurons (O'Malley *et al.*, 2003a), one can "check-off" the laser-ablated cells in the template and then analyze the behavioral data in terms of the specific neurons that have been "spared." This neuronal sparing experiment will reveal which spared neurons underlie the observed spared behaviors.

A second approach to creating larger scale perturbations of populations of identified neurons is to create "labeled lesions," where neurons are simultaneously axotomized and labeled with a fluorescent dextran (Gahtan and O'Malley, 2001). This occurs normally in our calcium imaging experiments, but can generally be ignored because the dextran is injected into the far caudal spinal cord and produces no significant behavioral deficit on its own. If, however, the injection is made into the far-rostral spinal cord, near the juncture between spinal cord and brainstem, then all of the labeled neurons are effectively disconnected from their spinal targets, hence the term *labeled lesion*. It is not possible to determine exactly which sets of axons will be struck on any given injection, but as a practical matter we simply make many injections and then screen the larvae for labeled-lesion patterns of interest. Unilateral labeling of large numbers of descending neurons resulted in a noticeable deficit in which escapes on the lesioned side showed an unusually rostral bend, causing the larvae to assume an S-shape that is not observed in control fishes or on the control side of the injected fishes (Fig. 7A). More recently we have begun performing these rostral axotomies on *nacre* fish,

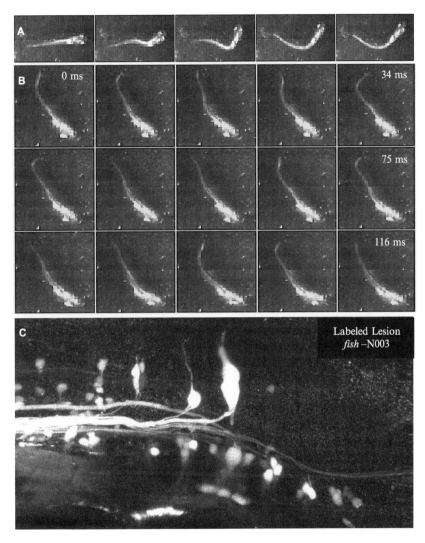

Fig. 7 Behavioral deficits produced by rostral axotomy (in *nacre* fish). (A) S-bends are observed after rostral unilateral axotomy of descending tracts. Selective bending of the most rostral portion of the fish is thought to be attributed to residual function of axotomized neurons that may be able to communicate with the most rostral musculature (Gahtan and O'Malley, 2001). (B) Fish with substantial rostral axotomy exhibit defective locomotion during both tracking and capture swim portions of the prey capture behavior. This larva is making poorly formed J-turn movements during its efforts to track a paramecium. In other recordings from such larvae, they are unable to hold a fixed orientation during the final capture swim and repeatedly fail to capture paramecia (in contrast, control larvae succeed a majority of the time). Such control of orientation and stiffness during locomotion was considered to be more demanding of coordinated descending controls (Behrend, 1984) and so it was predicted that prey capture would be more vulnerable to disruption, as was observed. (C) Confocal image of the *labeled lesion* present in the larvae shown in (B). Larvae are analyzed behaviorally within 24 to 48 hours after making the labeled lesion. At later time points, axonal regeneration may result in functional recovery (DePoister *et al.*, 2001).

where the lesion can be reconstructed more precisely. While it has been difficult to disrupt the larva's swimming behaviors, we expected that axotomy of a modest fraction of the descending projection would have a more pronounced effect on prey-capture locomotion because such behaviors are more complex and presumably place more demands on the descending control system. This was indeed observed (Fig. 7B).

VI. Conclusions

The most important element in these experimental endeavors is the transparency of the larval zebrafish. While optical techniques are individually becoming increasingly powerful, combining techniques may yield important synergies. For example, GFP labeling can be used to guide laser ablation of specific neural structures (Roeser and Baier, 2003), and, if one were to use *global mapping* results (using e.g., fos-GFP) to guide GFP-based photoablation, this might yield a supremely powerful technique for optically dissecting neural architectures that could be characterized as Guided Optical Dissection (or *G.O.D.*). While this chapter has focused on locomotor control systems, the approaches described herein should be applicable to all neural systems for as late in development as the larval CNS remains sufficiently transparent. If optical approaches continue their rapid advance, it is conceivable that, in coming years, zebrafish neural-system studies will have as great an impact on neuroscience as zebrafish mutational analysis has had on the study of vertebrate developmental genetics.

Acknowledgments

This work was supported by NIH grants NS37789 (DMO) and NS11127 (EG) and by NSF grant OPP-0089451 (HWD). We are grateful for use of the *nacre* mutant line, which was generated by J. A. Lister, D. W. Raible, and co-workers and provided by J. Kanki. Excellent technical assistance was provided by Ms. Jeanette Campos and Mr. Matthew Beverly.

References

Aksay, E., Gamkrelidze, G., Seung, H. S., Baker, R., and Tank, D. W. (2001). *In vivo* intracellular recording and perturbation of persistent activity in a neural integrator. *Nature Neurosci.* **4,** 184–193.

Aksay, E., Baker, R., Seung, H. S., and Tank, D. W. (2003). Correlated discharge among cell pairs within the oculomotor horizontal velocity-to-position integrator. *J. Neurosci.* **23,** 10852–10858.

Ali, D. W., Buss, R. R., and Drapeau, P. (2000). Properties of miniature glutamatergic EPSCs in neurons of the locomotor regions of the developing zebrafish. *J. Neurophysiol.* **83,** 181–191.

Ashworth, R., and Bolsover, S. R. (2002). Spontaneous activity-independent intracellular calcium signals in the developing spinal cord of the zebrafish embryo. *Brain Res. Dev. Brain Res.* **139,** 131–137.

Ashworth, R., Zimprich, F., and Bolsover, S. R. (2001). Buffering intracellular calcium disrupts motoneuron development in intact zebrafish embryos. *Brain Res. Dev. Brain Res.* **129,** 169–179.

Bak, M., and Fraser, S. E. (2003). Axon fasciculation and differences in midline kinetics between pioneer and follower axons within commissural fascicles. *Development* **130,** 4999–5008.

Bass, A. H., and Baker, R. (1997). Phenotypic specification of hindbrain rhombomeres and the origins of rhythmic circuits in vertebrates. *Brain Behav. Evol.* **50**, 3–16.

Behrend, K. (1984). Cerebellar influence on the time structure of movement in the electric fish. *Eigenmannia. Neurosci.* **13**, 171–178.

Bernhardt, R. R., Chitnis, A. B., Lindamer, L., and Kuwada, J. Y. (1990). Identification of spinal neurons in the embryonic and larval zebrafish. *J. Comp. Neurol.* **302**, 603–616.

Borla, M. A., and O'Malley, D. M. (2002). High-speed imaging of tracking swims used in the larval zebrafish prey capture behavior. *Soc. Neurosci. Abs.* **28**(18), 361.

Borla, M. A., Palecek, B., Budick, S. A., and O'Malley, D. M. (2002). Prey capture by larval zebrafish: Evidence for fine axial motor control. *Brain Behavior and Evolution* **60**, 207–229.

Bosch, T. J., Maslam, S., and Roberts, B. L. (2001). Fos-like immunohistochemical identification of neurons active during the startle response of the rainbow trout. *J. Comp. Neurol.* **439**, 306–314.

Brustein, E., Marandi, N., Kovalchuk, Y., Drapeau, P., and Konnerth, A. (2003). *In vivo* monitoring of neuronal network activity in zebrafish by two-photon calcium imaging. *Pflugers Arch.* **446**, 766–773.

Budick, S. A., and O'Malley, D. M. (2000a). Locomotive repertoire of the larval zebrafish: Swimming, turning and prey capture. *J. Exp. Biol.* **203**, 2565–2579.

Budick, S. A., and O'Malley, D. M. (2000b). Minimal behavioral deficits are observed after laser-ablation of the nMLF in larval zebrafish. *Soc. Neurosci. Abs.* **26**, 158.

Buss, R. R., Bourque, C. W., and Drapeau, P. (2003). Membrane properties related to the firing behavior of zebrafish motoneurons. *J. Neurophysiol.* **89**, 657–664.

Carvan, M. J., Sonntag, D. M., Cmar, C. B., Cook, R. S., Curran, M. A., and Mille, G. L. (2001). Oxidative stress in zebrafish cells: Potential utility of transgenic zebrafish as a deployable sentinel for site hazard ranking. *Sci. Total Environ.* **274**, 183–196.

Chen, M. C., Zhou, Y., and Detrich, H. W., III. (2002). Zebrafish mitotic kinesin-like protein 1 (Mklp1) functions in embryonic cytokinesis. *Physiol. Genomics* **8**, 51–66.

Das, T., Payer, B., Cayouette, M., and Harris, W. A. (2003). *In vivo* time-lapse imaging of cell divisions during neurogenesis in the developing zebrafish retina. *Neuron* **37**, 597–609.

DePoister, B., Bhatt, D. H., Palma, S. A., and Fetcho, J. R. (2001). Monitoring the regeneration and functional recovery of injured spinal cord in larval zebrafish. *Soc. Neurosci. Abs.* **27**, 961.

Downes, G. B., Waterbury, J. A., and Granato, M. (2002). Rapid *in vivo* labeling of identified zebrafish neurons. *Genesis* **34**, 196–202.

Eaton, R. C., Lee, R. K. K., and Foreman, M. B. (2001). The Mauthner cell and other identified neurons of the brainstem escape network of fish. *Prog. Neurobiol.* **63**, 467–485.

Eisen, J. S. (1999). Patterning motoneurons in the vertebrate nervous system. *Trends Neurosci.* **22**, 321–326.

Elsalini, O. A., and Rohr, K. B. (2003). Phenylthiourea disrupts thyroid function in developing zebrafish. *Dev. Genes Evol.* **212**, 593–598.

Faber, D. S., Fetcho, J. R., and Korn, H. (1989). Neuronal networks underlying the escape response in goldfish: General implications for motor control. *Ann. N.Y. Acad. Sci.* **563**, 11–33.

Fetcho, J. R. (1991). Spinal network of the Mauthner cell. *Brain Behav. Evol.* **37**, 298–316.

Fetcho, J. R., and Liu, K. S. (1998). Zebrafish as a model system for studying neuronal circuits and behavior. *Annals N.Y. Acad. Sci.* **860**, 333–345.

Fetcho, J. R., and O'Malley, D. M. (1995). Visualization of active neural circuitry in the spinal cord of intact zebrafish. *J. Neurophysiology* **73**, 399–406.

Fetcho, J. R., and O'Malley, D. M. (1997). Imaging neuronal networks in behaving animals. *Curr. Opin. Neurobiol.* **7**, 832–838.

Fetcho, J. R., Cox, K., and O'Malley, D. M. (1998). Monitoring activity in neuronal populations with single-cell resolution in a behaving vertebrate. *Histochem. J.* **30**, 153–167.

Foreman, M. B., and Eaton, R. C. (1993). The direction change concept for reticulospinal control of goldfish escape. *J. Neurosci.* **13**, 4101–4113.

Fuss, S. H., and Korsching, S. I. (2001). Odorant feature detection: Activity mapping of structure response relationships in the zebrafish olfactory bulb. *J. Neurosci.* **21,** 8396–8407.

Gahtan, E., and Baier, H. (2004). Of lasers, mutants, and see-through brains: Functional neuroanatomy in zebrafish. *J. Neurobiol.* **59,** 147–161.

Gahtan, E., and O'Malley, D. M. (2000). Analysis of spontaneous activity in control and reticulospinal (RS) ablated zebrafish larvae. *Soc. Neurosci. Abs.* **26,** 158.

Gahtan, E., and O'Malley, D. M. (2001). Rapid lesioning of large numbers of identified vertebrate neurons: Applications in zebrafish. *J. Neurosci. Meth.* **108,** 97–110.

Gahtan, E., and O'Malley, D. M. (2003). Visually-guided injection of identified reticulospinal neurons in zebrafish: A survey of spinal arborization patterns. *J. Comp. Neurol.* **459,** 186–200.

Gahtan, E., Sankrithi, N., Campos, J. B., and O'Malley, D. M. (2002). Evidence for a widespread brainstem escape network in larval zebrafish. *J. Neurophys.* **87,** 608–614.

Gamkrelidze, G., Baker, R., Aksay, E., Seung, H. S., and Tank, D. W. (2000). Persistent activity of hindbrain neurons during eye fixations in larval zebrafish *Danio rerio. Soc. Neurosci. Abs.* **26,** 196.

Goldman, D., Hankin, M., Li, Z., Dai, X., and Ding, J. (2001). Transgenic zebrafish for studying nervous system development and regeneration. *Transgenic Res.* **10,** 21–33.

Hale, M. E., Ritter, D. A., and Fetcho, J. R. (2001). A confocal study of spinal interneurons in living larval zebrafish. *J. Comp. Neurol.* **437,** 1–16.

Hale, M. E., Kheirbek, M. A., Schriefer, J. E., and Prince, V. E. (2004). Hox gene misexpression and cell-specific lesions reveal functionality of homeotically transformed neurons. *J. Neurosci.* **24,** 3070–3076.

Herdegen, T., and Leah, J. D. (1998). Inducible and constitutive transcription factors in the mammalian nervous system: Control of gene expression by Jun, Fos and Krox, and CREB/ATF proteins. *Brain Res. Reviews* **28,** 370–490.

Higashijima, S., Bhatt, D. H., Mandel, G, and Fetcho, J. R. (2001). Neurotransmitter properties of spinal interneurons in embryonic/larval zebrafish. *Soc. Neurosci. Abs.* **27**(4), 830.

Higashijima, S., Masino, M. A., Mandel, G., and Fetcho, J. R. (2003). Imaging neuronal activity during zebrafish behavior with a genetically encoded calcium indicator. *J. Neurophysiol.* **90,** 3986–3997.

Kato, S., Nakagawa, T., Ohkawa, M., Muramoto, K., Oyama, O., Watanabe, A., Nakashima, H., Nemoto, T., and Sugitani, K. (2004). A computer image processing system for quantification of zebrafish behavior. *J. Neurosci. Meth.* **134,** 1–7.

Kelsh, R. N., Brand, M., Jiang, Y. J., Heisenberg, C. P., Lin, S., Haffter, P., Odenthal, J., Mullins, M. C., van Eeden, F. J., Furutani-Seiki, M., *et al.* (1996). Zebrafish pigmentation mutations and the processes of neural crest development. *Development* **123,** 369–389.

Kelsh, R.N., Schmid, B., and Eisen, J. S. (2000). Genetic analysis of melanophore development in zebrafish embryos. *Dev. Biol.* **225,** 277–293.

Kimmel, C. B., Patterson, J., and Kimmel, R. O. (1974). The development and behavioral characteristics of the startle response in the zebrafish. *Devel. Psychobiol.* **7,** 47–60.

Kimmel, C. B., Powell, S. L., and Metcalfe, W. K. (1982). Brain neurons which project to the spinal cord in young larvae of the zebrafish. *J. Comp. Neurol.* **205,** 112–127.

Kimmel, C. B., Metcalfe, W. K., and Schabtach, E. (1985). T-reticular interneurons: A class of serially repeating cells in the zebrafish hindbrain. *J. Comp. Neurol.* **233,** 365–376.

Langenberg, T., Brand, M., and Cooper, M. S. (2003). Imaging brain development and organogenesis in zebrafish using immobilized embryonic explants. *Dev. Dyn.* **228,** 464–474.

Lee, R. K. K., and Eaton, R. C. (1991). Identifiable reticulospinal neurons of the adult zebrafish, *Brachydanio rerio. J. Comp. Neurol.* **304,** 34–52.

Lee, R. K. K., Eaton, R. C., and Zottoli, S. J. (1993). Segmental arrangement of reticulospinal neurons in the goldfish hindbrain. *J. Comp. Neurol.* **329,** 539–556.

Lewis, K. E., and Eisen, J. S. (2003). From cells to circuits: Development of the zebrafish spinal cord. *Prog. Neurobiol.* **69,** 419–449.

Li, W., Ono, F., and Brehm, P. (2003). Optical measurements of presynaptic release in mutant zebrafish lacking postsynaptic receptors. *J. Neurosci.* **23**, 10467–10474.

Lister, J. A., Robertson, C. P., Lepage, T., Johnson, S. L., and Raible, D. W. (1999). *Nacre* encodes a zebrafish microphthalmia related protein that regulates neural-crest-derived pigment cell fate. *Development* **126**, 3757–3767.

Liu, K. S., Gray, M., Otto, S. J., Fetcho, J. R., and Beattie, C. E. (2003). Mutations in deadly seven/ notch1a reveal developmental plasticity in the escape response circuit. *J. Neurosci.* **23**, 8159–8166.

Liu, K., and Fetcho, J. R. (1999). Laser ablation reveals functional relationships of segmental hindbrain neurons in zebrafish. *Neuron* **23**, 325–335.

Liu, D. W., and Westerfield, M. (1988). Function of identified motoneurones and co-ordination of primary and secondary motor systems during zebrafish swimming. *J. Physiol. (Lond.)* **403**, 73–89.

MacNeil, M. A., and Masland, R. H. (1998). Extreme diversity among amacrine cells: Implications for function. *Neuron* **20**, 971–982.

Metcalfe, W.K., Mendelson, B., and Kimmel, C. B. (1986). Segmental homologies among reticulospinal neurons in the hindbrain of the zebrafish larva. *J. Comp. Neurol.* **251**, 147–159.

Miyawaki, A., Griesbeck, O., Heim, R., and Tsien, R. Y. (1999). Dynamic and quantitative Ca21 measurements using improved cameleons. *Proc. Natl. Acad. Sci.* **96**, 2135–2140.

Morton, D. W., and Chiel, H. J. (1994). Neural architectures for adaptive behavior. *Trends Neurosci.* **17**, 413–420.

Nakayama, H., and Oda, Y. (2004). Common sensory inputs and differential excitability of segmentally homologous reticulospinal neurons in the hindbrain. *J. Neurosci.* **24**, 3199–3209.

Niell, C. M., Meyer, M. P., and Smith, S. J. (2004). *In vivo* imaging of synapse formation on a growing dendritic arbor. *Nature Neurosci.* **7**, 254–260.

O'Donovan, M. J., Ho, S., Sholomenko, G., and Yee, W. (1993). Real-time imaging of neurons retrogradely and anterogradely labeled with calcium sensitive dyes. *J. Neurosci. Meth.* **46**, 91–106.

O'Malley, D. M. (1994). Calcium permeability of the neuronal nuclear envelope: Evaluation using confocal volumes and intracellular perfusion. *J. Neurosci.* **14**, 5741–5758.

O'Malley, D. M., and Fetcho, J. R. (1996). Calcium signals in the zebrafish Mauthner cell: Large size and potentiation with repetitive stimulation. *Soc. Neurosci. Abs.* **22**, 795.

O'Malley, D. M., and Fetcho, J. R. (2000). The zebrafish hindbrain: A transparent system for imaging motor circuitry. *In* "Imaging Living Cells" (R. Yuste, F. Lanni, and A. Konnerth, eds.), pp. 14.1–14.12. Cold Spring Harbor Laboratory Press, Cold Spring Harbor, NY.

O'Malley, D. M., and Sankrithi, N. S. (2002). The nMLF neurons of the larval zebrafish are multimodal and multifunctional. *Soc. Neurosci. Abs.* **28**(17), 361.

O'Malley, D. M., Kao, Y.-H., and Fetcho, J. R. (1996). Imaging the functional organization of zebrafish hindbrain segments. *Neuron* **17**, 1145–1155.

O'Malley, D. M., Burbach, B. J., and Adams, P. R. (1999). Fluorescent calcium indicators: Subcellular behavior and use in confocal imaging. *In* "Confocal Microscopy: Methods and Protocols" (S. Paddock, ed.), pp. 261–303. Humana Press, Totowa, NJ.

O'Malley, D. M., Zhou, Q., and Gahtan, E. (2003a). Probing neural circuits in the zebrafish: A suite of optical techniques. *Methods* **30**, 49–63.

O'Malley, D. M., Beverly, M. H., Parker, S. K., Sicoli, J., Zhdanova, I. V., and Detrich, H. W. (2003b). Design of molecular tools for the global mapping of neural activity. *Soc. Neurosci. Abs.* **29**(4), 759.

Perkins, B. D., Kainz, P. M., O'Malley, D. M., and Dowling, J. E. (2002). Transgenic expression of a GFP-rhodopsin C-terminal fusion protein in zebrafish rod photoreceptors. *Visual Neurosci.* **19**, 257–264.

Plaut, I . (2000). Effects of fin size on swimming performance, swimming behaviour and routine activity of zebrafish *Danio rerio*. *J. Exp. Biol.* **203**, 813–820.

Rigo, J. M., Badiu, C. I., and Legendre, P. (2003). Heterogeneity of postsynaptic receptor occupancy fluctuations among glycinergic inhibitory synapses in the zebrafish hindbrain. *J. Physiol.* **553**, 819–832.

Ritter, D. A., Bhatt, D. H., and Fetcho, J. R. (2001). *In vivo* imaging of zebrafish reveals differences in the spinal networks for escape and swimming movements. *J. Neurosci.* **21,** 8956–8965.

Roeser, T., and Baier, H. (2003). Visuomotor behaviors in larval zebrafish after GFP-guided laser ablation of the optic tectum. *J. Neurosci.* **23,** 3726–3734.

Sankrithi, N. (2003). The nMLF neurons of larval zebrafish are multimodal and multifunctional. Ph.D. thesis. Northeastern University, Boston, MA.

Sankrithi, N. S., and O'Malley, D. M. (2000). Optical recordings of nMLF neurons in larval zebrafish demonstrate asymmetrical responses to sensory stimuli. *Soc. Neurosci. Abs.* **26,** 158.

Sankrithi, N. S., Purohit, A., Jones, G. B., and O'Malley, D. M. (2002). Enhanced photoablation of reticulospinal neurons with a dextran conjugated photosensitizer. *Soc. Neurosci. Abs.* **28**(11), 208.

Sprague, J., Clements, D., Conlin, T., Edwards, P., Frazer, K., Schaper, K., Segerdell, E., Song, P., Sprunger, B., and Westerfield, M. (2003). The Zebrafish Information Network (ZFIN): The zebrafish model organism database. *Nucleic Acids Res.* **31,** 241–243.

Stevens, C. F. (1998). Neuronal diversity: Too many cell types for comfort? *Curr. Biol.* **8,** R708–R710.

Takahashi, M., Narushima, M., and Oda, Y. (2002). *In vivo* imaging of functional inhibitory networks on the Mauthner cell of larval zebrafish. *J. Neurosci.* **22,** 3929–3938.

van der Sar, A. M., Musters, R. J., van Eeden, F. J., Appelmelk, B. J., Vandenbroucke-Grauls, C. M., and Bitter, W. (2003). Zebrafish embryos as a model host for the real time analysis of Salmonella typhimurium infections. *Cell Microbiol.* **5,** 601–611.

Wang, J. W., Wong, A. M., Flores, J., Vosshall, L. B., and Axel, R. (2003). Two-photon calcium imaging reveals an odor-evoked mp of activity in the fly brain. *Cell* **112,** 271–282.

Webb, S. E., and Miller, A. L. (2003). Imaging intercellular calcium waves during late epiboly in intact zebrafish embryos. *Zygote* **11,** 175–182.

Westerfield, M. (1995). "The Zebrafish Book: A Guide for the Laboratory Use of Zebrafish (*Brachydanio rerio*), 3rd ed." University of Oregon Press, Eugene, OR.

Wullimann, M. F., Rupp, B., and Reichert, H. (1996). "Neuroanatomy of the Zebrafish Brain: A Topological Atlas." Birkhauser-Verlag, Basel, Switzerland.

Zottoli, S. J., and Faber, D. S. (2000). The Mauthner cell: What has it taught us? *The Neuroscientist* **6,** 26–38.

Zottoli, S. J., Newman, B. C., Rieff, H. I., and Winters, D. C. (1999). Decrease in occurrence of fast startle responses after selective Mauthner cell ablation in goldfish (*Carassius auratus*). *J. Comp. Physiol. [A]* **184,** 207–218.

CHAPTER 14

Computer-Aided Screening for Zebrafish Embryonic Motility Mutants

Charles A. Lessman

Department of Microbiology & Molecular Cell Sciences
The University of Memphis
Memphis, Tennessee 38152

METHODS IN CELL BIOLOGY, VOL. 76
Copyright 2004, Elsevier Inc. All rights reserved.
0091-679X/04 $35.00

I. Introduction

The zebrafish, *Danio rerio*, has become an important vertebrate model organism, especially because of the ease with which large-scale mutagenesis screens may be carried out. However, finding the mutant phenotypes in such a screen is a daunting task since multitudes of offspring must be scrutinized throughout their 48-h developmental period and beyond. This problem becomes even more acute when the phenotype involves a non-morphological trait such as behavior. Zebrafish embryos develop movement behaviors soon after somite formation. These behaviors include trunk movements and somersaulting within the chorion confines. The movements are intermittent and transient, thus simple investigator observation, which would reveal morphological phenotypes readily, is much less likely to detect mutant motility phenotypes. Therefore tools that aid the screener are needed to better detect and characterize motility mutants. A method called Computer-Aided-Screening (CAS) has been developed in our laboratory as a tool for detecting motility mutants in embryonic zebrafish (Lessman, 2002). CAS utilizes transparency scanners, coupled with computers and macro programming, to automate image capture (on the scanner bed) of embryo arrays throughout their development. The CAS technique incorporates a longitudinal (temporal) design to analyze spatial arrays of embryos simultaneously from multiple matings.

A second technique called Computer-Aided Larval Motility Screening (CALMS) has been derived from CAS and extends the screening procedure into the post-hatch larval period. CALMS is also based on the use of transparency scanners controlled by macro programming to automate collection of scans. Larvae are arrayed in standard 96-well plates.

Both CAS and CALMS use inexpensive, yet reliable, flatbed transparency scanners to image the zebrafish arrays at preset intervals. It should be noted that scanners use a linear charged-coupled-device (CCD), which is scanned beneath the bed and can produce images resulting from effective mega pixel resolution. For example, a 1200 dots-per-inch (dpi) linear CCD, found in many average scanners, has a maximum optical resolution of 1200 dpi or pixels per 2.54 cm. Scanning a standard 96-well plate on such a scanner will result in an image derived from an effective CCD of 21.4 mega pixels. High-end scanners have CCDs with dots per inch at 2400, 3600, or higher optical resolution. Therefore, the resulting image resolution will depend on the optical resolution of the scanner used. Nevertheless, for the detection of movement (or lack of it) high resolution is not really necessary. In fact high resolution demands a cost in terms of larger file sizes and prolonged scan times.

CAS uses plastic screening affixed to optically clear culture dishes to produce high-density embryo arrays that physically separate individual embryos into discrete, yet fluid-connected, wells. A novel feature of this technology is the maintenance of physiological conditions on the scanner bed by perfusion of solutions that control temperature, pH, ionic strength, and parasites (Lessman,

2002). CAS allows screening of numerous spawns at one time and hundreds of embryos may be compared simultaneously. The images are assessed for movement in two different ways: (1) Region-of-interest (ROI) over each embryo is analyzed for mean signal intensity through time. (2) An image stack is animated in ImageJ (National Institutes of Health, Bethesda, MD) and for direct observation of movement. Statistical analysis may then be used to look for significant differences.

CAS-CALMS include a number of benefits:

1. High-throughput screening,
2. Automatic archiving of image files for each mating,
3. Simple operation of screening system,
4. Scanning intervals may be changed to suit experiment,
5. Data in a format suitable for statistical analysis, and
6. Low cost.

The traditional method of screening involves investigator observation with a dissecting microscope for identification of mutants. Objectivity may be significantly enhanced by combining CAS with the traditional method (Lessman, 2002). Additionally, CAS-CALMS facilitate screening of large numbers of potential mutants throughout the course of development and beyond.

While the zebrafish embryo has been described as "transparent," in fact, it has considerable optical contrast. Live unstained cells generally have some contrast. This is attributed to the optical properties of the inclusions, organelles, membranes, and cytoplasm of the cell. As development proceeds, the zebrafish embryo progressively increases in contrast and becomes increasingly heterogeneous in optical contrast (e.g., the pigmented eyes have higher contrast than the caudal fin). The CAS protocol is based on the principle that, as the embryo changes its position as motility develops, this movement is detectable as a change in optical density within an ROI or fixed window of pixels over each embryo. Different parts of the embryo having varying levels of contrast would lie under the ROI window as the embryo changes position. CALMS uses a similar rationale, but, instead, data are derived from the density-weighted center of the entire well. Therefore, as the larva change position, the mean "center of mass" computed as x-y coordinates will correspondingly change.

Mutagenesis screening techniques should be designed to be simple to operate, reliable, time effective, and low cost. CAS-CALMS allow for many simultaneous analyses to be done on a number of spawns, and thus the combination of these procedures is quite time effective. Computers and flatbed transparency scanners are now relatively inexpensive and therefore cost effective. In fact, CAS-CALMS workstations cost less than research dissecting microscopes. Easy-to-use software makes CAS-CALMS relatively simple and computers are becoming increasingly reliable. Therefore, CAS-CALMS can be designed to be simple to operate, reliable, time effective, and low cost (Lessman, 2002). Design of an effective

CAS-CALMS system involves hardware, software, embryo array, and zebrafish physiological support. We discuss each of these factors in the next sections.

II. Flatbed Transparency Scanners and Computers

A. Scanners

We have successfully used HP ScanJets (Hewlett-Packard Company, Palo Alto, CA) with external transparency adapters (e.g., light source and intensity controller) for most of the development work on CAS-CALMS. Scanners without transparency adapters are limited to the reflected mode. The chorion of the zebrafish embryo is quite reflective. Therefore, reflective-mode scanners produce very low-contrast images containing little useful information for CAS. Other types of scanners to avoid include those which have an internal bay for transparencies. They physically limit the thickness of the object to be scanned (e.g., paper thin) and expose the internal circuitry to the possibility of a water spill! It is also important that the stepper motor driving the CCD scan head be reliable (i.e., it begins the scan at the same position each time). If not, the scans will not be in register when stacked, and interfering frame jumping will occur. It is possible, but tedious, to reregister the frames in the stack using ImageJ or ScionImage (Scion Corporation, Frederick, Maryland).

CAS-CALMS imaging uses a CCD that transduces light into binary computer information. Each CCD pixel scales light intensity into bins, typically 8, 12, or 16 bit (256, 4096, or 65,536 bins, respectively). The linear CCD is scanned under the specimen. Better quality scanners have high-quality CCDs and stepper motors to give true optical resolution of 1200 dpi or higher. In addition, high-quality scanners produce consecutive images in register. Other devices are available that include digital and video cameras; however, these would require a magnification system to image embryo arrays, thereby limiting the size of imaged areas. Therefore, imaging large "arrays of embryo arrays," possible with a scanner, would not be practical with other methods. It should be noted that true optical resolution is a hardware characteristic, while interpolated resolution is really "empty magnification" that involves software generation of subdivided pixels to produce apparent increase in dots per inch. The latter should be avoided. It is recommended that silicone adhesive be used to seal the scanner glass to the body of the scanner, producing a watertight seal. This modification to the scanner will reduce the shock hazard possible in case of accidental spills and protect the scanner circuitry.

Resolution is limited by the size of CCD elements or pixels that will, in turn, allow detection of separate points in the object. Large pixels in the CCD will be unable to separate different points of contrast in an object that is closer than the CCD pixel width. This is a function of the CCD detector. The relationship of optical resolution to pixel size is shown in Fig. 1. Large pixels of 300 dpi allow for considerable density information; however, as pixel size decreases from 1200 to

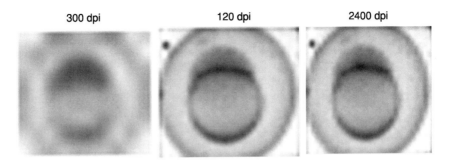

Fig. 1 Zebrafish blastula scanned at 300 dpi (*left*), 1200 dpi (*middle*), and 2400 dpi (*right*) 8-bit grayscale. Each of the original images is approximately 1 × 1 mm and has been enlarged to the size shown on this page. Note that images at 1200 dpi preserve most of the detail found in the higher resolution image. Reprinted (with permission from Elsevier) from Lessman, C. A. (2002). Use of computer-aided screening (CAS) for detection of motility mutants in zebrafish embryos. *Real-Time Imaging* **8**, 189–201.

2400 dpi, localized density information is improved. Nevertheless, for mean density over the entire image, high resolution is less important. Use of a scanner tends to optimize accuracy and precision while reducing noise. This is attributed to the linear CCDs used (i.e., the same pixel elements are used as detectors over a large, linear portion of the image). In addition, unlike a microscope-based system, the fixed relationship of the scan bed to scan head helps produce images of exactly the same size and focus. Files may be compressed in various ways to reduce the need for a large amount of mass storage. We use Joint Photographic Experts Group (JPG) files that produce significant disk-space savings and are compatible with ImageJ. JPG files may lose resolution if opened and re saved as JPG. In addition, we use the least compressed option for JPG since some noticeable loss of detail may otherwise result. Most CAS-CALMS images are saved using JPG at about 10% of the uncompressed file size. It should be noted that (1) not all compression formats are supported by image analysis programs, and (2) files must be uncompressed and loaded into memory for analysis and viewing. Therefore, compressed files may need to be opened and uncompressed by the original compression software before use. Large numbers of uncompressed files may overwhelm the capabilities of most desktop computers because of intrinsic limitations on random access memory (RAM), for example, 4 GB for computers that have the Pentium® processor (Intel®, Santa Clara, CA). As computer processors and associated RAM increase in capacity, this limitation will diminish.

B. Computers

Most Windows-based Pentium® (Microsoft, Redmond, WA) computers and MacIntosh G4/G5s (Apple, Cupertino, CA) are readily capable of handling the multi-megabyte image files produced by flatbed scanners at the upper limit of their optical resolution (i.e., 1200, 2400, or more dots per inch). Actually, rather modest computers are suitable for the data collection phase, especially if JPG or other file compression format is used. However, the computer system used for image analysis would require additional RAM (1 GB or more), a faster central processing unit (CPU) and additional hard drives to process and store large numbers of high-resolution image files. We currently use Dell computers (Dell, Inc., Round Rock, TX). Several older Optiplex GX1 Pentium® computers (Dell, Inc.) are used to collect image data from scanners. For animation and image analysis, higher end Pentium® computers (e.g., Dell Precision 530) with Xeon™ CPUs (Intel®) and 1 GB of RAM are used. Fortunately, the latter types of computers are increasingly commonplace and relatively inexpensive.

III. Software

A. Scanner Programs

Most scanners come bundled with scanner software specific for the particular scanner make and model. We have used HP PrecisionScan Pro with HP ScanJet 5370C scanners and HP Director for HP ScanJet 4570C scanners. The scanner software will scan images to a number of different programs including Microsoft (MS) Imaging (Windows 98), MS Paint, Adobe Photoshop (Adobe, San Jose, CA), and Netscape (Netscape, Mountain View, CA). Usually, the recipient program is selected from a list the scanner program provides at the time of initial scan setup. The files are automatically given temporary file names and numbered consecutively as they are produced. In addition, the files will have a default format [e.g., bit map (BMP), tag image file format (TIF), and JPG], depending on the recipient program. Generally, these files will be placed in a default temporary (Temp) directory under the main Windows directory (if using Windows 98) or under the Documents and Settings directory (if using Microsoft XP).

B. Macro Programs

The keyboard/mouse macro program we use is Macroscheduler, which is available as a share ware program from http://www.mjtnet.com. Macroscheduler uses Windows script language to create macros for automating the scanner operation. The operator sets up the initial scan, selects the area to scan, sets contrast, chooses the scan resolution, and selects the recipient program. Macroscheduler is then used to produce the rest of the scans automatically at preset intervals. Once a macro program is produced, it may be reused for future experiments. Basically, Macroscheduler records keystrokes and mouse operations and produces a script program that is editable. For example, if the scan resolution is increased, the wait time for scan completion may be changed to allow sufficient time for the larger scans. Other keyboard/mouse programs are available and may be adapted for use in CAS.

C. Image Analysis Programs

Currently we use ImageJ for most image stack animations and image analysis procedures carried out on entire stacks. ImageJ is the Java-based version of NIH *Image* and ImageJ is available for Windows, Mac, Linux, and Unix machines at http://rsb.info.nih.gov/ij/download.html. Also useful is the Windows version of NIH *Image* called ScionImage, which is available as free ware from http://www.scioncorp.com/frames/fr_download_now.htm. ScionImage has a useful "average stack" function and has helpful stack cropping and editing tools as plug-ins. Other image analysis programs such as Sigmascan Pro Systat Software, Inc., Point Richmond, California may also be adapted for use in CAS, but these programs may not have image stack capability. It should be stressed here that ImageJ allows image analysis on entire stacks. Therefore image files do not have to be opened individually and measurements made on each image. Most image analysis programs handle 8-, 12-, or 16-bit grayscale images. The ImageJ program uses ROI to define portions of the image to analyze. ROI may be in the form of a template or other complex image overlay to indicate portions of the image for analysis. The most straightforward type of analysis would be to determine the average intensity within an ROI, which gives a measure of light transmittance. For an 8-bit grayscale image $0 =$ no transmittance and $255 = 100\%$ transmittance as in a spectrophotometer. The tabulated data can then be submitted to statistical programs, such as SigmaStat (Jandel Scientific, San Rafael, CA) for determination of significant differences and identification of outlying groups (i.e., mutants) from the normal distribution (i.e., wild type).

IV. Embryos and Larva: Physiological Considerations

A. CAS

To be scanned, the high-density zebrafish embryo arrays used in CAS need to be held such that each embryo is separated from others, yet have continuity of the bathing media to allow fluid exchange. The support must be optically transparent and thin to keep embryos within the focal plane of the scanner. In addition, the support must be physically affixed to the scanner bed so that re scans will represent the same ROI. A product that meets all of these criteria is not commercially available, therefore suitable "chips" need to be specially fabricated. "Chips" are made using commercially available materials, such as tissue culture ware, plastic screening, and silastic adhesive. One feature of CAS is the high-density embryo arrays that are used. These are made possible by the use of plastic canvas available at any needlepoint supply or arts and craft store. The polypropylene sheets have square perforations usually indicated by the number of perforations per inch (2.54 cm), for example, #10 has 10 perforations per 2.54 cm. The sheets are easily cut to fit various culture-ware plates and affixed by small dabs of silicone adhesive to the support vessel. Although the sheets are available in a variety of colors, we have found the natural or clear sheets to work well. Petri plates (100 mm) with shallow quadrant dividers are economical and work well for up to 4 crosses (i.e., each spawn in a separate quadrant). These dishes have good optical properties and will hold more than 400 embryos. For larger numbers of crosses Nunclon rectangular well trays (Sigma-Aldrich, St. Louis, MO) are useful. The 8-well rectangular tray has shallow dividers separating the dish into 26- × 33-mm wells. Each well fitted with #10 plastic canvas would hold about 120 embryos or 960 embryos per dish. We use strips of lab tape to affix the dishes to the scanner glass to reduce accidental movement. Depending on the scanner used, multiple dishes may be positioned on the scanner bed. We have found that it is more practical to place single dishes or trays on each scanner and have several scanner stations to handle increasing numbers of crosses. Scan times increase, as do file size, as the number of dishes per scanner increases. It is important that the embryos be removed from the array before hatching begins, since free-swimming larva are able to move anywhere on the array and mixing of crosses will result. Furthermore, spawns may be asynchronous from batch to batch. Therefore, it is simpler to match development within a group of crosses and place them on one scanner. Other batches, less or more advanced than the first group, are matched to developmental stage and placed together on a second or third scanner.

Since zebrafish embryos will be re scanned a number of times during their development period, they must be maintained in a physiologically relevant state while residing on the scanner bed for up to 48 h. Therefore, environmental factors such as temperature, pH, microbial growth, and salinity must be controlled. Since most scanners generate heat when powered on, the scanner bed will provide some incubator-like conditions. This is supplemented with laminar flow of bathing medium by 2-channel peristaltic pumps, one channel for inflow, the other for outflow. Bathing medium (sterile filtered) is maintained at 28.5 °C in a water bath. Sufficient flow must be maintained to provide adequate medium exchange (Fig. 2). Since water is constantly replenished by the perfusion system, covers for array dishes are optional. We do not use covers for CAS arrays; however, we use covers for CALMS, since water is not perfused and evaporation would be a problem in open dishes. Because the transparency adapter tends to warm the dish cover, condensation is minimized and does not interfere significantly with embryo or larva imaging.

It is necessary to check embryo arrays carefully at the start of incubation to remove those not properly positioned in the chip (e.g., 2 embryos in a well). A simple way to do this is with a boom-mounted stereoscope with ring light attachment positioned over the scanner bed. We have stereoscopes placed adjacent to the scanners for final placement of embryos in arrays. The dish is then carefully moved from the scope to the scanner bed just a short distance away. Loading of arrays is easily done with Pi Pumps fitted with Pasteur pipettes, which takes about 15 m to load 4 crosses into a quadrant plate. To begin the data acquisition phase, the dish containing the array is placed on the scanner bed, input and output lines are fitted from peristaltic pumps, and flow is adjusted to 10 ml/h. The transparency adapter is then positioned over the array and a clear plastic box is placed over the entire assembly to maintain an incubator-like

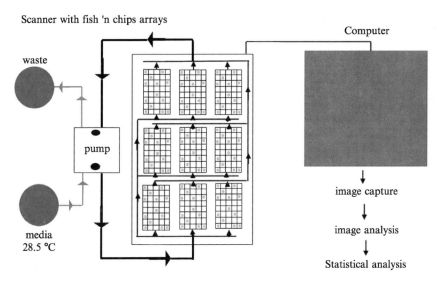

Scanner with fish 'n chips arrays

Fig. 2 Prototype CAS scanner workstation with life support for zebrafish embryos. Scanner bed shown with 9 "chips," each holding a random array of embryos from a single spawn (i.e., an array of arrays). Reprinted (with permission from Elsevier) from Lessman, C. A. (2002). Use of computer-aided screening (CAS) for detection of motility mutants in zebrafish embryos. *Real-Time Imaging* **8**, 189–201.

environment. Thermistor probes are used to monitor the temperature within the array. The water bath containing the vessel of "egg water" is then adjusted to bring the array temperature to 28.5 °C.

B. CALMS

A second tool has been developed to analyze motility in hatched free-swimming larva. The procedure, CALMS, uses the same equipment and software as CAS with the exception that individual larva are placed in single wells of 96-well, round-bottom plates. We use 1-mm capillary tubes fitted with a mouth pipette to load single larva into individual wells containing 100 to 200 μl water. About 15 m are required for 96 wells. The plates are then scanned as in CAS. CAS has been designed to capture and analyze images of embryos within their chorion confines. Because of the nature of the grid and physiological support used in CAS, it is not amenable to analysis of post-hatch larval motility. Preliminary results indicate that single larva, each in a well with 100 to 200 μl water, continue to show vigorous motility for up to 3 days (the longest time tested). Note that, unlike CAS, CALMS uses covered plates without perfusion. Depending on ambient humidity, evaporation may be countered by humidifying the scanner bed. If necessary, partial water changes may be made by 8-channel pipette, using fine gel-loading tips, while the plate resides on the scanner between scans. In CALMS, temperature may be adjusted by venting the clear plastic box enclosing the scanner bed.

V. Data Acquisition

The operator initiates the data acquisition phase by loading the scanner program, selecting transparency mode, setting the scan resolution (e.g., 1200 dpi), scan type (e.g., grayscale) using the program menus, and initiating a prescan. The prescan default is the entire bed surface, after which the operator selects a specific area for the scans. This reduces the file size and eliminates areas of no interest. At this point, the contrast and brightness may be adjusted. We usually use the auto function to set these values automatically. A grayscale "stretch" can be done later on an entire stack in ImageJ or ScionImage. We have found that, as expected, changing the contrast and brightness does affect the magnitude of the changes reflected in the data, but not the general pattern. However, if it should be necessary to compare quantitative changes from more than one experiment (i.e., different stacks), then it becomes important to standardize the brightness and contrast. This can be achieved by placing a standard neutral-density or grayscale filter (e.g., 18% gray) on the scanner bed to set exposure. The operator then initiates the first full scan in HP PrecisionScan Pro, a list of programs is produced, and the operator selects one as the recipient program. We use MS Imaging for Windows 98 machines and Netscape for XP machines, but any program that handles image files may be used as a recipient. After the first scan is complete, Macroscheduler is invoked, the appropriate macro selected, and the "run when" menu is selected. A window containing days, time, and interval to repeat is displayed. The operator selects the appropriate scheduling information (e.g., Monday and Tuesday, start time 9:00 AM, repeat every 10 m). In this example the macro will initiate the scan at 9:00 AM Monday and will repeat the scan every 10 m, stopping at midnight Tuesday. The files are automatically named using sequential numbers and placed in the default Windows Temp directory. Each file also is time stamped when it is saved. After data acquisition is complete, the image files should be moved from the Temp directory into a new directory with an informative name. We usually keep the temporary filenames (e.g., ~LWF0001.bmp), since ImageJ and ScionImage recognize the sequential numbers and load the images sequentially into a stack. Data files may then be archived to CDs or DVDs. One note of caution, if the files are not removed from the Temp directory, running "disk cleanup" programs may erase them unintentionally.

VI. Data Analysis

A. Image Stack Animation

The primary data analysis program we use is ImageJ. Image files are imported as a sequential stack using the "import" selection from the "file" menu. The stack is then manipulated as if it was a single file (e.g., it may be saved or the contrast adjusted using the "adjust" command from the "image" menu). Animation is started by pressing the "=" key and stopped with "esc." Additional animation controls are found in the "image" menu under "stacks." While animating the stack, the zoom tool (magnifying-glass icon) may be used to magnify portions of the array for closer scrutiny (i.e., left mouse button to magnify, right button to zoom out). Animation provides qualitative information about the embryo movements and is a valuable adjunct to performing image analysis. The "average stack" function on the "stacks" menu in ScionImage is useful for producing a single composite image that represents the stack. Embryos or larva appear as "multiple exposures." Therefore, individuals that have no motility or reduced motility appear as relatively distinct images, while wild-type embryos appeared blurred. Another useful tool in ScionImage is the "crop and scale" function from the "stacks" macro. The "stacks" macro must first be loaded using the "special" menu. Once loaded and a selection made from the stack, selecting the "crop and scale" function will produce a stack of just the selected area. A montage of this selected stack may then be made from either ScionImage or ImageJ.

B. Image Analysis

To obtain quantitative data on embryo motility, image analysis is performed on the stack using ImageJ. Select the "Analyze" menu and select "set measurements." A window appears with selectable choices for measurement types. For CAS, the "mean gray value" and "center of mass" are selected, while, for CALMS, only the "center of mass" is used. Next, using the ROI manager found under "tools" in the "Analyze" menu, ROIs are produced. The "oval" drawing tool is used to create a small oval, about 20% of the embryo area over the first embryo in an array (e.g., upper left). The oval is positioned by placing the cursor inside the oval (cursor forms arrow) and by dragging it with the left mouse button down. The oval is placed to one side of the midline and the "add" button is selected in the ROI manager window. The ROI will be identified with its x-y coordinates in pixels. The oval is, in turn, dragged over each embryo and its ROI is added to the ROI manager list. After all embryos in the array have ROIs, they are saved in a directory. We have found it best to save ROIs from a single clutch (e.g., quadrant 1) to a separate directory. It is possible to reuse ROIs if the new stack has exactly the same arrangement of wells. Once the ROIs have been specified, the "select all" choice in the ROI manager will highlight the listed ROIs. Selecting the "measurement" button will produce a pop-up box asking the user whether to process the entire stack. Pressing "yes" will initiate the measurement phase, after which a pop-up box named "results" will appear. The data appear as columns of numbers that are copied to spreadsheet or statistical programs for further manipulation. ImageJ performs measurements on all ROIs for each slice (image) before proceeding to the next slice. Therefore the data are arranged as ROI#1 through ROI#n for slice 1 where n = # embryos in the array. This task is then repeated for subsequent slices. A simple way to rearrange the data in MS Excel is to copy n+1 data cell address to the top of an adjacent empty column.

Then copy this cell to all adjacent empty cells such that sequential data for each embryo are now in a single row. Simple plots of the mean ROI intensity against time for each embryo provide quantitative data; however, this task quickly becomes too complex as additional plots for all embryos are added. Instead, we use the standard deviation of the mean ROI intensity as a quantitative measure of motility. Generally we use 12- or 24-h periods during which to calculate standard deviations for each embryo when hourly scan intervals are used. However, shorter time periods may be used (e.g., 1 or 2 h), especially if short scan intervals were used (e.g., every 10 m). It is also important to note that the "center of mass" function provides x-y coordinates in pixels (or real distance if calibration is used) for each ROI on the array, making it simple to identify and relate the signal to an individual embryo.

C. Example 1: Wild-Type Embryos and CAS

An example of CAS imaging is presented in Fig. 3. Four different wild-type (wt) spawns were placed in an array with each spawn in a separate quadrant. The array was imaged automatically every hour by a macro-controlled program with an HP ScanJet at 1200 dpi, 8-bit grayscale. The array was continuously perfused at a rate of 10 ml/h with fresh dechlorinated tap water that contained methylene blue. Array scans are automatically saved to disk. After data acquisition is complete, files are stacked in ImageJ. At this point, the stack is animated to visually assess the array through time. Next, an ROI is produced covering about 20% of the embryo by using the ROI tool in ImageJ. Examples of ROIs are shown in Fig. 4. The ROI is a fixed subset of pixels that will be analyzed for mean density (optical intensity). As demonstrated in Fig. 4, when embryos begin to move within the confines of the chorion, different portions of the embryo are seen in the fixed ROI. Plots of the mean ROI intensity through time reveal an oscillation of intensity, characteristic of the somersaulting motility of the wt embryo (Fig. 5). The variance observed in Fig. 5 has at least two components: (1) differences in optical properties of the portion of the embryo under the ROI because of embryo movement through time, and (2) changes in overall optical properties of the embryo through time (e.g., increase in pigmentation). The second source of variance is unlikely to produce the oscillatory changes observed and, in any case, may be removed by subtracting the mean intensity for the entire embryo from that of the ROI at each time point. This has been done, and the results are shown in Fig. 6. The oscillations seen in Figs. 5 and 6 then are attributed primarily to embryo movement, and the variance of these oscillations through time may be used as a measure of embryo motility. Therefore, standard deviations of the mean ROIs through time are computed and a frequency distribution plotted (Fig. 7) for all of the wt embryos (n = 163), shown in Fig. 3 (all 4 quadrants taken together). The data were divided into 24-h groupings (e.g., day 1 and day 2). As expected, the majority of movement occurs in day 2 and the greatest amount of variation among individuals also occurs during this period. The day 1 distribution is clearly unimodal, while the day 2

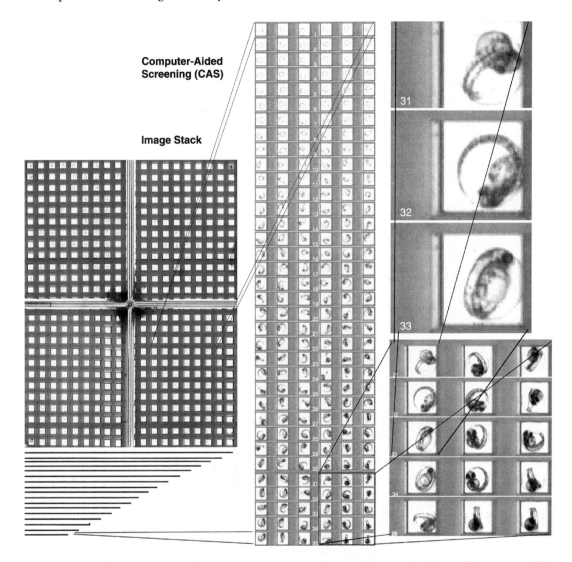

Fig. 3 Panel at left shows an embryo array with 4 quadrants, each holding a different spawn. This array is scanned hourly to produce an image stack. The middle panel shows an enlarged row of 6 embryos, each row of the same embryos from successive hours. Panels on right show progressively enlarged regions of indicated portions of the array.

distribution has multiple peaks indicating a nonhomogeneous population represented by these 4 spawns. Nevertheless, the day 2 distribution has moved well to the right of that for day 1. In CAS analysis, we would expect that motility mutants would have day 2 distributions that overlap significantly with that for day 1.

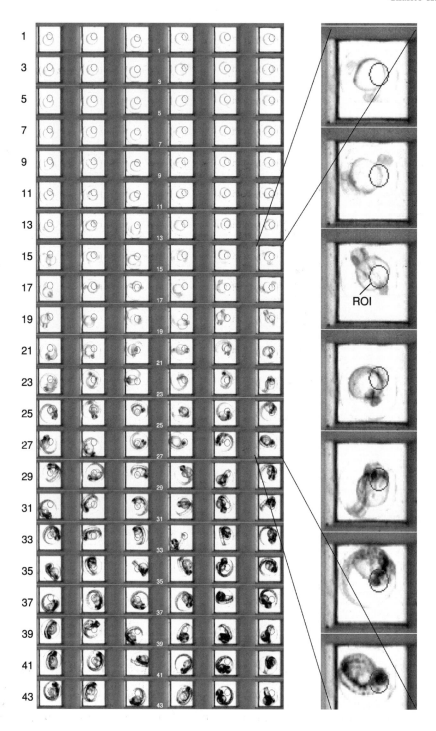

ROI

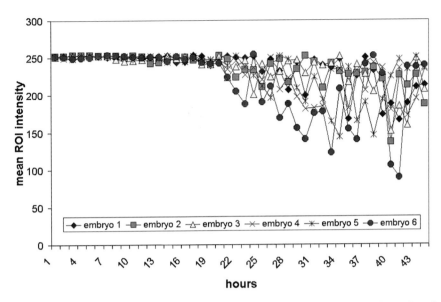

Fig. 5 Plot of mean ROI intensity versus time (hourly) of 6 embryos depicted in Fig. 4. Note that major oscillations in ROI intensity occur after 20 h. An 8-bit grayscale [i.e., 0 (*black*) to 255 (*white*)] is used. The inherent optical properties of the portion of the embryo under the ROI at each time point will determine the range of the oscillations.

Another wt spawn was analyzed by CAS and the data presented as a box plot (Fig. 8). As development occurs, motility increases as denoted by increased median values. The greatest change occurred during day 2. Wild-type motility progressively increases significantly throughout the period covered by this set of scans.

D. Example 2: *Chordino* Mutant and CAS

To provide a "proof of concept" for CAS, two known mutants with suspected motility phenotypes were tested (i.e., pleiotropic mutants). Known *chordino* (allele: din tt250, mutation in the chordin gene) heterozygotes were mated and the resulting offspring subjected to CAS. Homozygous *chordino* embryos have a distinct phenotype including reduced head development and thus are easily recognized. Approximately half of a spawn (n = 51) was placed on a chip and

Fig. 4 Example of region-of-interest (ROI) positioned over each of 6 embryos from an array. The ROIs are shown as circles over each embryo. The panel at the left shows the same row of 6 embryos at odd hours (*left*) from a stack. An enlargement of a single embryo from 15 to 27 h with the fixed ROI is shown in the panel at the right. ImageJ is used to calculate the mean pixel intensity in each ROI throughout the stack. The data are then placed in a spreadsheet and plotted.

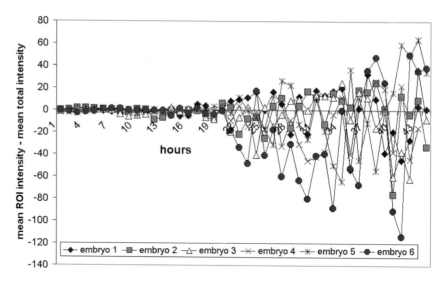

Fig. 6 Plot of mean ROI intensity–mean intensity of the entire embryo at each time point for the 6 embryos depicted in Fig. 4. Note that removal of the variance because of general changes in pigmentation or other optical properties of the whole embryo does not account for the oscillatory changes associated with the ROIs.

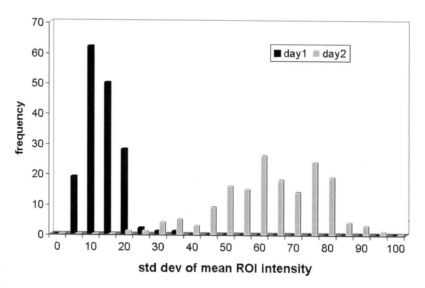

Fig. 7 Frequency distribution of the standard deviation of mean ROI intensity for wild-type zebrafish embryos during day 1 and day 2 of development. Bin = 5, n = 163.

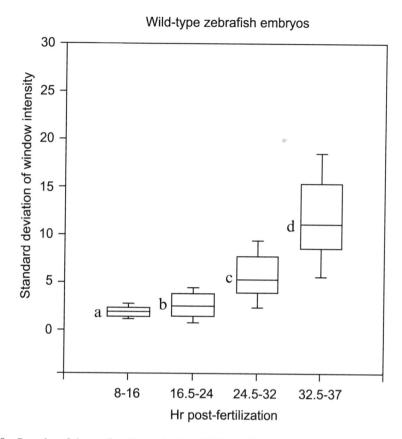

Fig. 8 Box plot of the median (*line in box*) and 95% confidence intervals (error bars) of window (ROI) variation (standard deviation) for wild-type zebrafish during development. The ends of each box represent 25% and 75% values for the distribution. The letters indicate significant differences in the median values in ANOVA, n = 67. Reprinted (with permission from Elsevier) from Lessman, C. A. (2002). Use of computer-aided screening (CAS) for detection of motility mutants in zebrafish embryos. *Real-Time Imaging* **8**, 189–201.

scanned at hourly intervals automatically for two periods: (1) day 1, 11 to 23 hpf and (2) day 2, 34 to 52 hpf. At the end of image collection, the embryo phenotypes were scored by stereoscope. The phenotypes (either *chordino* or wild type) were matched with the x-y coordinates on the array. After this was done, image analysis was performed automatically. Therefore, the results are a blind test. The mean intensity was determined for all ROIs, the standard deviation of the mean ROI intensity was computed, and a frequency distribution constructed for phenotypes scored by microscopy as *chordino* and wild type (Fig. 9).

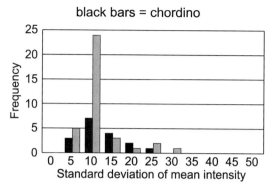

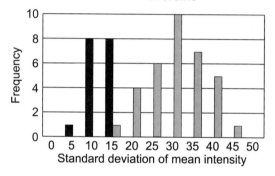

Fig. 9 Frequency distribution plots of the standard deviation of mean intensity for ROIs over 51 embryos of a *chordino* heterozygote mating. Black bars represent *chordino* phenotypes as determined by microscopy, while gray bars are wild type. The upper panel represents day 1 of development, while the lower panel represents day 2. Note that the *chordino* phenotype sorts (significant by ANOVA) from the wild type in the lower panel. Reprinted (with permission from Elsevier) from Lessman, C. A. (2002). Use of computer-aided screening (CAS) for detection of motility mutants in zebrafish embryos. *Real-Time Imaging* **8**, 189–201.

To verify that the distribution in Fig. 9 represents changes in motility, the images were stacked in ScionImage and animated. Observation of the animated stack confirmed the apparent "head-over-tail" tumbling of wild type and the lack of this stereotypic behavior in *chordino*. These results have been replicated several more times and include analysis of matings of F1 offspring from the original parental *chordino* heterozygotes. Thus the CAS method has been useful in identifying new *chordino* heterozygotes from F1 progeny raised here.

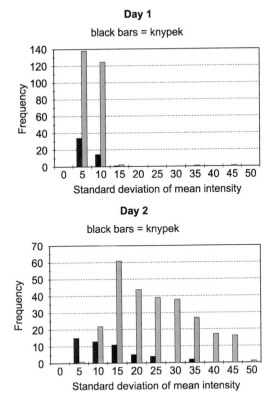

Fig. 10 Frequency distribution plots of the standard deviation of mean intensity for fixed pixel windows over embryos from *knypek* heterozygote matings. Black bars are *knypek* phenotypes as scored by microscope, while gray bars are wild type. Upper panel = day 1, lower panel = day 2. Note that the *knypek* phenotype sorts from the wild type based on CAS motility analysis. Data include 4 crosses: 86 embryos from wild-type cross; 83, 65, and 83 embryos from knypek heterozygote crosses. Thus 50 *knypek* phenotypes were scored from the three *knypek* crosses (i.e., 50:181 or 1:3.6). Reprinted (with permission from Elsevier) from Lessman, C. A. (2002). Use of computer-aided screening (CAS) for detection of motility mutants in zebrafish embryos. *Real-Time Imaging* **8**, 189–201.

E. Example 3: *Knypek* Mutant and CAS

Several pairs of heterozygote *knypek* (allele m119) adults were mated and the resulting embryos subjected to screens using CAS. The results are shown in Fig. 10. The *knypek* phenotype segregates from the wild type in day 2 of development and an analysis of variance (ANOVA) on ranks showed significant difference between frequency distributions of the two CAS-determined phenotypes.

The *knypek* mutant has defects in gastrulation (Topczewski *et al.*, 2001) and in somite formation (Henry *et al.*, 2000). The gene has been identified as a glypican-GPI protein (Topczewski *et al.*, 2001). However, its motility defect was not previously reported until CAS analysis was performed (Lessman, 2002).

F. Example 4: *Dead elvis* Mutant and CAS

A pilot N-ethyl-N-nitrosourea (ENU) mutagenesis project was initiated to generate F2 families to determine whether CAS could detect new motility mutants. Mutagenesis of adult males was carried out as previously described (Solnica-Krezel *et al.*, 1994). The first motility mutant detected by CAS has been named "*dead elvis*" (e.g., does not shake, rattle, or roll). *Dead elvis* has a nonmotile phenotype as determined by CAS, is initially quite normal morphologically until about 36 hpf, then becomes increasingly distinct with pericardial edema, failed or delayed hatching, and swollen yolk cell. *Dead elvis* was found initially in two of six crosses from a single F2 family. These two spawns showed apparent motility defects in a Mendelian ratio on CAS analysis and subsequent stack animation. The crosses showed 1:3 ratio of the same mutant phenotype (Fig. 11) to wt sibs. Adjacent *dead elvis* and wt sibs are shown in Fig. 12 for the interval 24 to 48 h. The data for the incubation period are shown in Fig. 13. The *dead elvis* phenotype remains essentially motionless throughout this and the subsequent period of incubation. The resulting images were subjected to CAS image analysis and the frequency of standard deviations of mean ROI intensity was plotted for 12-h intervals (Fig. 14). It should be noted that re screens with more frequent scans (i.e., every 10 m) differentiate between slow developmental changes that result in contrast differences (e.g., pigment cell migration) and true motility of the embryo itself.

Results of CAS analysis for family 11C are shown in Fig. 14. Note that even with small n-values such as for 11C_I and 11C_J (n = 20 and 24, respectively) CAS analysis is possible. Unimodal distributions are present for 11C_I (assumed wt) while bimodal distributions are evident for 11C_J (*dead elvis*). Nevertheless, more meaningful data are obtained with n = 50 or more. Therefore, very young females that give small spawns need to be re screened as they mature and provide

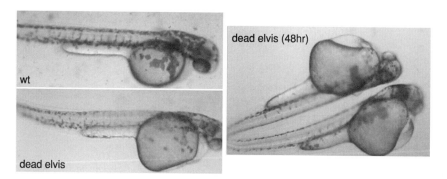

Fig. 11 Comparison of the *dead elvis* phenotype with wild-type sibling (wt) at 48-h incubation. Panel on right is of two *dead elvis* phenotypes. It is necessary to remove *dead elvis* manually from their chorions, while the wt hatches on its own. The *dead elvis* motility phenotype is detectable by CAS well before any morphological differences are apparent.

more progeny per spawn. From the distributions in Fig. 14, 11C_A, 11C_D, 11C_H, and 11C_I appear to be wt and 11C_F and 11C_J are mutants (*dead elvis*).

G. Example 5: *Dead elvis* Mutant and CALMS

To provide a "proof of concept" for the CALMS procedure, embryos from a *dead elvis* heterozygote cross were sorted to motility phenotype by CAS. Embryos were allowed to hatch. (Embryos homozygous for *dead elvis* must be manually dechorionated.) The embryos were then placed in rows of a 96-well plate containing 200 μl system water/well. The covered plate was then scanned at 1200 dpi 8-bit grayscale at 10 m. After scanning 34 images, the frames were stacked in Scion-Image and the "average stack" function was used to produce the image shown in Fig. 15. The wt embryos are seen as "multiple exposures" in their wells while *dead elvis* phenotypes appear distinct because of the averaging of the embryos in their near motionless condition. Surprisingly, small "tail flicks" may be seen in the averaged images of *dead elvis* phenotypes, indicating that these larva are not completely without the ability to move. As in CAS, the stack of images is animated in ImageJ to view the dynamics of movement and, unlike the traditional method of direct microscopic observation at one or two different time points for a minute or two, the entire period of CALMS data acquisition (several hours to days) is compressed into a matter of seconds in the animation. This provides the investigator with much more visual information and also may be used to look for periodicities in movement (e.g., circadian rhythms). Another important difference between CALMS and the traditional screen is the ability to quantify motility in CALMS, while the traditional approach is qualitative only. Since the images obtained by CALMS are digitized, they are immediately amenable to image analysis. In ImageJ, ROIs are produced for each well of a 96-well plate and saved in a directory for subsequent reuse. The "center-of-mass" function is then invoked to compute the x-y coordinates of the pixel in the ROI (whole well) corresponding to the density weighted center. Therefore, as a fish moves away from the geographic center of the well, the x-y coordinates computed are shifted in that direction. The data are plotted as a scattergram (Fig. 16) with standard deviation of the x-coordinate on the x-axis and the standard deviation of the y-coordinate on the y-axis for the *dead elvis* cross shown in Fig. 15.

VII. Potential Problems and Remedies

Absence of motility as determined by the CAS procedure is attributed to (1) true motility mutations, (2) embryo death or paralysis by external agents (e.g., pathogen, toxin), or (3) embryo movements that reposition them in their exact starting position before each scan. Since large numbers of embryos from a spawn (or the entire spawn) will be included in the CAS procedure, statistical analysis will provide definitive evidence expected of Mendelian genetics (i.e., 25% of the spawn

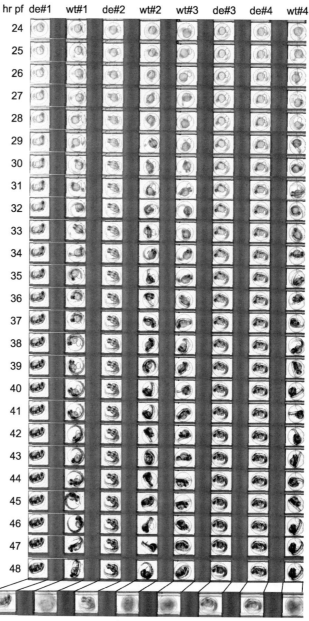

Fig. 12 Hourly images from CAS for adjacent wt sibs and *dead elvis* phenotype for the period 24 to 48 hpf incubation. Note that the wt sibs show characteristic head-over-tail somersaulting behavior, while *dead elvis* embryos are nearly immotile for this entire period. Nevertheless, developmental changes are ongoing. For example pigment cell migration is readily apparent from dorsal to ventral in

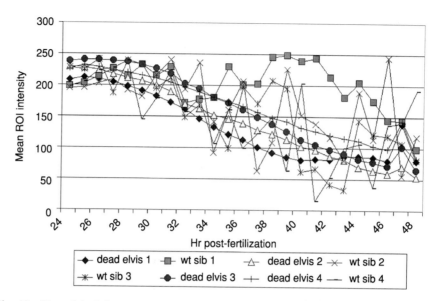

Fig. 13 Plots of *dead elvis* versus wt throughout development for the sibs shown in Fig. 12. Note the typical oscillating pattern of intensity changes as the wt sibs exhibit their normal motility. *Dead elvis* sibs have intensity changes attributed primarily to increased pigmentation.

will show the defect if it is a true Mendelian recessive mutation). Statistically, it is unlikely that exactly 25% of the progeny of a spawn would die or be affected by an external agent or reposition themselves at their individual start point within each well after moving. This is especially true since all potential mutants will be re screened using new progeny from the same putative heterozygote F2 parents. In addition, it has been our experience that embryos that die during CAS undergo rapid cytolysis, which is easily seen in animations of image stacks and on simple observation of the embryo array on the scanner bed with a boom-mounted dissecting microscope. In fact, we have suggested that this phenomenon would allow CAS to be used for toxicity and drug effect screening using zebrafish embryos as a useful vertebrate test system (Lessman, 2002). It is presumed that embryos cytolyze quickly since they are in fresh water and must osmoregulate to survive. After death, adenosine triphosphatase (ATPase) pumps shut down and the cells of the embryo are subjected to the considerable osmotic pressure of the fresh-water environment in which they normally reside. Death or paralysis attributed to an unknown toxin or pathogen in the water supply used for CAS would

the trunk region of *dead elvis*. The "average" function in ScionImage was used to generate the stack average shown at the bottom. The wt embryos appear as blurs because of their normal motility, while *dead elvis* appears rather distinct.

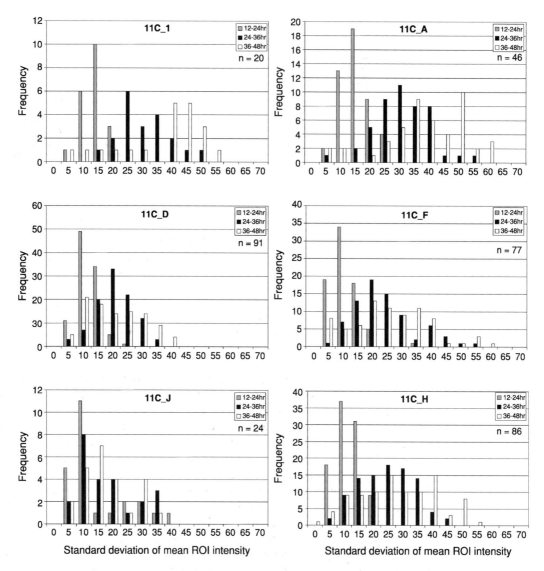

Fig. 14 Frequency distributions of the standard deviation of mean ROI intensity derived from CAS analysis for family 11C F3 progeny (6 different crosses). Number of progeny are indicated by n for each spawn. The bins for standard deviation are 5 and the time intervals are 12 h. 11C_D, 11C_A, 11C_H, and 11C_I show characteristic unimodal distributions for all time intervals. 11C_F and 11C_J show bimodal curves. *Dead elvis* phenotypes are found in these latter two spawns.

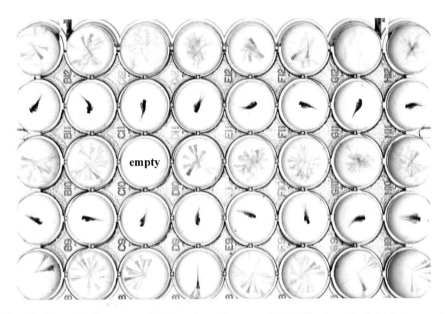

Fig. 15 Example of computer-aided larval motility screen (CALMS) using a *dead elvis* heterozygote cross, in which *dead elvis* phenotypes have been sorted using CAS from wt sibs. The wt sibs were placed in rows 1 (*top*), 3 and 5 (*bottom*). The *dead elvis* phenotypes were placed in rows 2 and 4. Scans were automatically acquired at 10 m (34 frames), stacked in ScionImage and the "average" function used to obtain the single image shown. *Dead elvis* larvae appear distinct while wt sibs show "multiple exposures" because of their motility. Well C10 is empty as a control, other wt wells appear empty (e.g., G12) because of very rapid movements apparent during stack animation, but which produce very low contrast on averaging.

presumably produce some ratio other than 1:3 and would be expected to affect wild-type controls if the system were somehow contaminated.

CAS and CALMS are limited to detecting relatively gross movements as opposed to individual cell migrations, and these are best detected when they are periodic. Therefore, rapid movement analysis is not applicable to these techniques. Nevertheless, CAS and CALMS are especially suited to detecting abnomalities in periodic body movements throughout extended periods, including the entire developmental period of about 48 h. With CALMS, this has been extended to the post-hatch period, at least an additional 2 to 3 days, after which animals require feeding.

Optical properties are critical to both CAS and CALMS. Debris, air bubbles, and wells containing more than one embryo/larva must be carefully avoided. Washing embryos is easily done by placing embryos/larva in fine mesh baskets and rinsing with fresh system water. If parasitic infestation is suspected, an additional wash with 167 ppm formaldehyde may be used. Air bubbles tend to form if water has not equilibrated in temperature with surroundings. The

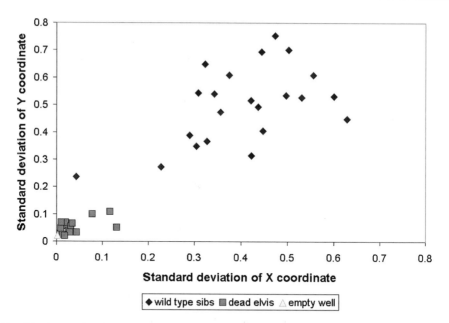

Fig. 16 Scatterplot of the standard deviation of the density weighted center computed for each larva at 10-m intervals from 34 stacked images of a *dead elvis* cross depicted in Fig. 15. Note that the *dead elvis* phenotypes segregate from their wt sibs at the lower left of the plot. The empty well, as expected, sorts near the origin of the plot. Each larva is represented as a single point that corresponds to the standard deviation of the x and y coordinates. CALMS provides quantitative information about motility in these free-swimming larva, using simple, inexpensive, and readily available equipment and software.

formation of air bubbles is easily avoided by prior incubation of water in water baths or incubators set to 28.5 °C. The fish room is thermostatically set to 28.5 °C and powered-on scanners reach an operating temperature on the bed of about 28 °C, thus temperature differentials are avoided.

It is anticipated that mean grayscale density computed for ROIs will change with development of embryonic motility. More importantly, genetic phenotypes will vary from the normal distribution at least at some point in the developmental time sequence. It is expected that motility mutants will show phenotypes such as (1) reduced motility, (2) hypermotility, (3) precocious onset of motility, (4) delayed onset of motility, (5) rhythmicity variants of motility, and (6) aberrations of motility behavior. Generally, it would be expected that abnormal motility development would contribute to lethality as the ability to feed properly as hatchlings would be impaired in most instances. Therefore, many of the motility mutants obtained would likely be lethal mutations, especially those which lack motility development altogether. However, those with other less-severe motility defects may survive as homozygotes and may allow further analyses of the phenotype.

CAS-CALMS identify individual embryos/larva with x-y coordinates in the array, making recovery of particular animals relatively simple.

VIII. Conclusions

A. Advantages of CAS-CALMS

The CAS primary screen is a simple optical density analysis of the "window" or ROI for the overall grayscale channel from white light scans. As part of the primary screen, stacks are also animated and visually inspected for movement aberrations. This latter procedure is quite time effective, since an array spanning many hours is viewed in seconds. Potential mutants picked up by this screen may then be submitted to secondary screens (after verification of the heritability of the mutation by re crosses). Such a secondary screen, to rule out wild type, involves short interval scans (i.e., every 2 to 15 m) to detect any hypermotile mutants that move more frequently than wild type. This latter protocol generates many more image files, and is thus more expensive in terms of both CPU time and work hours for subsequent analysis than the hourly general motility screen. CAS and CALMS may be used sequentially to cover the first 5 days of zebrafish development. Crosses that produce outlying distributions in one or more of these analyses may be submitted to morphological and physiological secondary screens as potential mutants.

The real power of CAS-CALMS is the relentless imaging of embryos and larvae through time that allows determination of onset of any mutated motility gene, including lethal ones. Death involves cytolysis and large increases in optical density, a characteristic easily recorded by the scanner.

At the present time, motility mutants are represented by static images in the ZFIN database and thus clearly fail to provide valuable information about motion. Other methods have yet to detect and characterize fully all of the potential zebrafish motility mutants. This failure is largely attributed to the traditional approach that depends on periodic investigator observation of the potential mutants during screening, and the inability of any investigator to continuously monitor large numbers of spawns throughout the 48-h period of development day after day. CAS-CALMS produce dynamic imaging of embryos and larva, thus providing the needed motility information to allow investigators to narrow the number of candidates for gene identification.

B. Studies on Wild-Type Motility

A few studies have provided background data on development of embryonic motility in wild-type zebrafish. Included is the development of motility behaviors (Bate, 1999; Saint-Amant and Drapeau, 1998), maturation of myoneural junctions (Buss and Drapeau, 2000; Eisen, 1991; Nguyen *et al.*, 1999; Saint-Amant and Drapeau, 2000), and studies of wild-type gene expression (Bolanos-Jimenez *et al.*,

2001; Currie and Ingham, 1996). A time course for development of motility has been reported (Saint-Amant and Drapeau, 1998). The earliest involved spontaneous trunk contractions were recorded at 17 hpf (0.57 Hz), peaking at 19 hpf (0.96 Hz) and then decreasing to less than 0.1 Hz by 27 hpf. The data presented here support the earlier finding that the wild-type embryo develops movements by 17 to 20 hpf.

C. Studies on Motility Mutants

Gene mutations that result in abnormal motility development have been described. Included are descriptions of a mutant without functioning acetylcholine receptors (Westerfield *et al.*, 1990), and mutants with deranged skeletal muscle myofibril development (Felsenfield *et al.*, 1990). A large-scale ethylnitrosourea screen for motility genes in embryos and larva (Granato *et al.*, 1996) identified 166 mutants with specific motility defects among 14 phenotypes and involved up to 48 genes. A number of motility mutants have been described and characterized. These include *vibrato* (Sato and Mishina, 2003), *twitch once* (Ono *et al.*, 2002), and *sofa potato* and *relaxed* (Ono *et al.*, 2001).

CAS-CALMS have an inherent longitudinal temporal design that provides time-based data to better define motility defects. The time component can be very valuable, providing insight into the underlying genetic lesion. Circadian motor activity has been described in larva (Cahill *et al.*, 1998) and circadian clocks have been described in oocytes and embryos of zebrafish (DeLaunay *et al.*, 2000). Melatonin shows rhythmical changes also in zebrafish embryos (Gothilf *et al.*, 1999; Kazimi and Cahill, 1999). In addition, the circler mutants (Ashmore, 1998; Nicolson *et al.*, 1998) have a number of motility defects. Application of CAS-CALMS to these mutants should provide a wealth of time-based data.

The ZFIN database lists about 170 motility mutant phenotypes, more than 40 of which are unnamed. Most of these mutants have only rudimentary descriptions of the motility phenotype present. A handful, such as Nic[tk48d], *twitch once*, *sofa potato*, *relaxed*, and *vibrato* have been more characterized and have defects in the nicotinic acetylcholine receptor (Sepich *et al.*, 1998), rapsyn (Ono *et al.*, 2002), dihydropine receptors (Ono *et al.*, 2001), and Solo (Sato and Mishina, 2003), respectively. The CAS-CALMS approach would greatly augment the number of motility mutants available for study and would provide more meaningful descriptions, as well as quantitative data for specific motility phenotypes.

Acknowledgments

Chordino (allele: din tt250, mutation in the chordin gene) and *Knypek* (allele m119) heterozygote adults were a kind gift of Dr. Lila Solnica-Krezel, Vanderbilt University.

References

Ashmore, J. (1998). Mechanosensation: Swimming round in circles. *Curr. Biol.* **8**, R425–R427.
Bate, M. (1999). Development of motor behavior. *Curr. Opin. Neurobiol.* **9**, 670–675.

Bolanos-Jimenez, F., Bordais, A., Behra, M., Strahle, U., Sahel, J., and Rendon, A. (2001). Dystrophin and Dp71, two products of the DMD gene, show a different pattern of expression during embryonic development in zebrafish. *Mech. Dev.* **102**, 239–241.

Buss, R. R., and Drapeau, P. (2000). Physiological properties of zebrafish embryonic red and white muscle fibers during early development. *J. Neurophysiol.* **84**, 1545–1557.

Cahill, G., Hurd, M. W., and Batchelor, M. M. (1998). Circadian rhythmicity in the locomotor activity of larval zebrafish. *Neuroreport* **9**, 3445–3449.

Currie, P. D., and Ingham, P. W. (1996). Induction of a specific muscle cell type by a hedgehog-like protein in zebrafish. *Nature* **382**, 452–455.

DeLaunay, F., Thisse, C., Marchand, O., Laudet, V., and Thisse, B. (2000). An inherited functional circadian clock in zebrafish embryos. *Science* **289**, 297–300.

Eisen, J. S. (1991). Motoneuronal development in the embryonic zebrafish. *Development* **2**, 141–147.

Felsenfield, A. L., Walker, C., Westerfield, M., Kimmel, C., and Streisinger, G. (1990). Mutations affecting skeletal muscle myofibril structure in the zebrafish. *Development* **108**, 443–459.

Gothilf, Y., Coon, S. L., Toyama, R., Chitnis, A., Namboodiri, M. A., and Klein, D. C. (1999). Zebrafish serotonin N-acetyltransferase-2: Marker for development of pineal photoreceptors and circadian clock function. *Endocrinol.* **140**, 4895–4903.

Granato, M., van Eeden, F. J., Schach, U., Trowe, T., Brand, M., Furutani-Seiki, M., Haffter, P., Hammerschmidt, M., Heisenberg, C. P., Jiang, Y. P., *et al.* (1996). Genes controlling and mediating locomotion behavior of the zebrafish embryo and larva. *Development* **123**, 399–413.

Henry, C. A., Hall, L. A., Hille, M. B., Solnica-Krezel, L., and Cooper, M. S. (2000). Somites in zebrafish doubly mutant for knypek and trilobite form without internal mesenchymal cells or compaction. *Curr. Biol.* **10**, 1063–1066.

Kazimi, N., and Cahill, G. M. (1999). Development of a circadian melatonin rhythm in embryonic zebrafish. *Dev. Brain Res.* **117**, 47–52.

Lessman, C. A. (2002). Use of computer-aided screening (CAS) for detection of motility mutants in zebrafish embryos. *Real-Time Imaging* **8**, 189–201.

Nicolson, T., Rusch, A., Friedrich, R. W., Granato, M., Ruppersberg, J. P., and Nusslein-Volhard, C. (1998). Genetic analysis of vertebrate sensory hair cell mechanosensation: The zebrafish circler mutants. *Neuron* **20**, 271–283.

Nguyen, P. V., Aniksztejn, L., and Catarsi, S. and Drapeau, P. (1999). Maturation of neuromuscular transmission during early development in zebrafish. *J. Neurophysiol.* **81**, 2852–2861.

Ono, F., Higashijima, S., Shcherbatko, A., Fetcho, J., and Brehm, P. (2001). Paralytic zebrafish lacking acetylcholine receptors fail to localize rapsyn clusters to the synapse. *J. Neurosci.* **21**, 5439–5448.

Ono, F., Shcherbatko, A., Higashijima, S., Mandel, G., and Brehm, P. (2002). The zebrafish motility mutant *twitch once* reveals new roles for rapsyn in synaptic function. *J. Neurosci.* **22**, 6491–6498.

Saint-Amant, L., and Drapeau, P. (1998). Time course of the development of motor behaviors in the zebrafish embryo. *J. Neurobiol.* **37**, 622–632.

Saint-Amant, L., and Drapeau, P. (2000). Motoneuron activity patterns related to the earliest behavior of the zebrafish embryo. *J. Neurosci.* **20**, 3964–3972.

Sato, T., and Mishina, M. (2003). Representational difference analysis, high-resolution physical mapping, and transcript identification of the zebrafish genomic region for a motor behavior. *Genomics* **82**, 218–229.

Sepich, D. S., Wegner, J., O'Shea, S., and Westerfield, M. (1998). An altered intron inhibits synthesis of the acetylcholine receptor α-subunit in the paralyzed zebrafish mutant *nic1*. *Genetics* **148**, 361–372.

Solnica-Krezel, L, Schier, A. F., and Driever, W. (1994). Efficient recovery of ENU-induced mutations from the zebrafish germline. *Genetics* **136**, 1401–1420.

Topczewski, J., Sepich, D. S., Myers, D. C., Walker, C., Amores, A., Lele, Z., Hammerschmidt, M., and Solnica-Krezel, L. (2001). The zebrafish glypican Knypek controls cell polarity during gastrulation movements of convergent extension. *Develop. Cell* **1**, 251–264.

Westerfield, M., Liu, D. W., Kimmel, C. B., and Walker, C. (1990). Pathfinding and synapse formation in a zebrafish mutant lacking functional acetylcholine receptors. *Neuron* **4**, 867–874.

CHAPTER 15

Photoreceptor Structure and Development: Analyses Using GFP Transgenes

Brian D. Perkins, *James M. Fadool,[†] and John E. Dowling**

*Department of Molecular and Cellular Biology
Harvard University
Cambridge, Massachusetts 02138

[†]Department of Biological Science
Florida State University
Tallahassee, Florida 32306

Whereas studies in zebrafish have uncovered a number of mechanisms controlling vertebrate retinal development, methods and reagents to study the development of individual cell types *in vivo* and to identify mutations specifically affecting a single cell type are only now beginning to emerge. This is illustrated by the development of several transgenic lines that express fluorescent reporter genes driven by cell-specific promoters. Most prevalent are lines expressing a rod-specific green fluorescent protein (GFP) transgene. With these transgenic lines, questions about rod development, the photoreceptor mosaic, and opsin transport can be addressed. In

Copyright 2004, Elsevier Inc. All rights reserved.
0091-679X/04 $35.00

addition, utilizing these lines in forward genetic screens allows for the direct identification of mutations with specific effects on rod photoreceptors. The ability to monitor and track changes in photoreceptor development over time in a noninvasive manner is an obvious advantage provided by transgenic zebrafish.

I. Introduction

More is known about rod photoreceptor cells than any other cell in the vertebrate retina. From early studies in psychophysics (Hecht *et al.*, 1942) and on visual pigments (Wald, 1955), to the work on mechanisms of dark adaptation (Dowling, 1963), phototransduction (Yau, 1994), and inherited retinal disease (Dryja and Li, 1995), rod photoreceptors have been central to the understanding of retinal function and hereditary blindness disorders. Many of the mutations known to cause hereditary retinal degeneration affect rod-specific genes or otherwise interfere with rod function. Unfortunately, diseases affecting the rods often lead also to the loss of cone photoreceptors and the loss of color and eventually all vision. Future treatments and cures for retinal degeneration will likely depend on therapies designed to replace or regenerate rod photoreceptors, and this will also ensure the survival of cones.

A. Photoreceptor Anatomy and Biochemistry

Both rod and cone photoreceptors are highly specialized cells with a unique morphology consisting of an elongated outer segment, connecting cilium, inner segment, cell body, and synaptic terminal (Fig. 1). The shape and morphology of the outer segments (OS) usually differentiate the rods from the cones and provide the basis for their nomenclature. The OS consists of hundreds of tightly stacked membrane disks that contain the proteins necessary for phototransduction. Protein synthesis occurs in the inner segment and molecules destined for the OS must be transported apically through the connecting cilium. The inner segment also contains numerous mitochondria needed to provide the energy for the demands of protein synthesis, protein trafficking, and phototransduction. The synaptic terminals of rod photoreceptors, known as spherules, are typically smaller than the cone terminals, known as pedicles. Both spherules and pedicles are filled with synaptic vesicles, contain synaptic ribbons, and are presynaptic to the bipolar and horizontal cells.

What most clearly distinguishes rods and cones at the molecular level is the type of light-sensitive visual pigment, or opsin molecule, expressed in a photoreceptor. All vertebrate species possess rods that express a rod opsin, but a variable number of cone types exist among species, each of which express a different cone opsin. Whereas many primates have three cone types, most mammals have just two. On the other hand, cold-blooded vertebrates and birds may have four or more cone types. Zebrafish possess four cone types, which absorb light maximally in the red (570 nm), green (480 nm), blue (415 nm), and ultraviolet (362 nm) regions of the spectrum (Robinson *et al.*, 1993). The four cone types in zebrafish are also

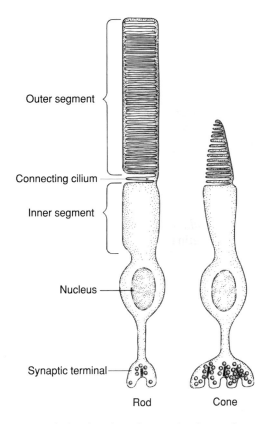

Fig. 1 Schematic drawing of vertebrate rod and cone photoreceptors.

distinguishable morphologically. Short and long single cones contain the ultraviolet (UV) and blue absorbing visual pigments, respectively; whereas the red- and green-sensitive visual pigments are found in the principal and accessory members of the double cones (Robinson *et al.*, 1993).

Opsins are classic examples of G-protein coupled receptors, and the rod visual pigment, rhodopsin, is the prototype for this family of receptors. Opsins, present in the photoreceptor OS, are responsible for light detection. Under dark conditions, a chromophore, 11-cis-retinal, is covalently bound to the opsin. On the absorption of a photon of light, the chromophore isomerizes to all-trans-retinal, initiating a conformational change in opsin. This results in the activation of the heterotrimeric G-protein transducin, which binds to and activates the effector enzyme, a cyclic GMP (cGMP) phosphodiesterase. The phosphodiesterase rapidly hydrolyzes cGMP, thereby lowering the overall concentration of cGMP in the outer segment. The reduction in cGMP concentration results in a closure of cGMP-gated cation channels (Yau and Nakatani, 1985), located in the plasma membrane of the OS, leading to hyperpolarization of the membrane (Tomita, 1970).

B. Transport Mechanisms

Outer segments rapidly turn over throughout the lives of photoreceptor cells. Efficient protein transport mechanisms are thus required to provide photoreceptors with a sufficient supply of protein to replenish that lost during OS turnover. It is estimated that rhodopsin constitutes 90% of the total protein in the rod OS and therefore plays important roles in both the physiology and structural integrity of the photoreceptor (Nathans, 1992). The shedding of the OS disks to the retinal pigment epithelium results in the loss of at least 2000 rhodopsin molecules per minute in some photoreceptors (Young, 1967). Investigations into this process have focused mainly on the transport of rhodopsin. Early work indicated that rhodopsin moves through the inner segment on membrane vesicles that are transported from the Golgi to the connecting cilium (Papermaster *et al.*, 1975, 1985). The proteins involved in the post-Golgi transport of these membranes are not completely known, although they do appear to require the small GTPase, Rab8 (Deretic *et al.*, 1995).

The unconventional myosin VIIA has been thought to play a role in rhodopsin transport through the cilium. For example, myosin VIIA mutations are responsible for the phenotypes in the *shaker-1* mouse, and in Usher Syndrome, a disease that results in deafness and blindness because of photoreceptor degeneration (Gibson *et al.*, 1995; Weil *et al.*, 1995). Interestingly, the *shaker-1* mice do not experience retinal degeneration as humans do. However, rhodopsin transport is affected in the *shaker-1* mutants, as immuno-gold labeling experiments showed rhodopsin molecules being slowed in the connecting cilium (Liu *et al.*, 1999). Studies of opsin transport through the connecting cilium have indicated that kinesin-II mediates opsin movement in association with a process known as intra-flagellar transport, or IFT (Marszalek *et al.*, 2000; Pazour *et al.*, 2002). IFT is a form of microtubule-based motility best understood in *Chlamydomonas* (Scholey, 2003). Do myosin VIIA and the IFT complex both mediate the transport of rhodopsin through the connecting cilium or is one mechanism dominant? Zebrafish mutants exist for several of these genes. The zebrafish *mariner* mutant alleles are mutations in myosin VIIA (Ernest *et al.*, 2000). In addition, a number of mutations in IFT genes have been found by insertional mutagenesis in zebrafish (Svn *et al.*, 2004). Analysis of these mutants may provide insight into the roles of these proteins in opsin transport.

II. Photoreceptor Development

In vertebrates, neurogenesis of the retina begins when proliferating neuroblasts exit the cell cycle and begin the process of differentiation (see Malicki, chapter 16). In mammals, the rods exit the cell cycle in two distinct phases but after most other retinal neurons have differentiated (Morrow *et al.*, 1998a). For both mammalian rods and cones, the exit from mitosis is followed by a delay of several hours before the onset of opsin expression (Morrow *et al.*, 1998a; Szel *et al.*, 1994). In most species, cone differentiation occurs before rod differentiation. This is true of zebrafish as well, though the story is complex. The expression of the zebrafish rod opsin gene begins at approximately 50 hpf (Schmitt and Dowling, 1996) in a region of precocious neurogenesis in the ventral nasal retina referred to as the ventral patch. The red and blue cone opsin genes are also expressed at about 50 hpf while expression of the UV opsin gene is observed about 5 h later (Schmitt *et al.*, 1999). Similarly, small clusters of rods and the red/green double cones are labeled with the monoclonal antibodies ROS-1 and Zpr-1, respectively, in whole-mount at 50 hpf (Raymond *et al.*, 1995). Given that the first cells in the outer nuclear layer of the retina to become postmitotic are seen between 43 and 48 hpf (Hu and Easter, 1999), it is likely that cell cycle exit precedes opsin expression by a few hours. Cone differentiation occurs in a sweeping fashion from the ventronasal side of the choroid fissure to the dorsonasal retina and then to the dorsotemporal and ventrotemporal side of the choroid fissure (Raymond *et al.*, 1995; Schmitt and Dowling, 1996). On the other hand, rod differentiation initiates in the ventral retina on the nasal side of the choroid fissure but soon crosses directly to the temporal side of the fissure. This ventral patch of rods increases symmetrically across the choroid fissure and increases in density while slowly expanding. Between 50 and 72 hpf, rods fill in the dorsal retina in a seemingly random fashion that does not resemble the wave of differentiation that all other retinal cell types exhibit (Raymond *et al.*, 1995; Schmitt and Dowling, 1996). Despite similarities in timing of early rod and cone differentiation, zebrafish vision is predominantly cone-driven for several days. Based on behavioral studies and electroretinography (ERG) recordings, rod function can first be demonstrated only between 14 and 21 dpf (Bilotta *et al.*, 2001; Saszik *et al.*, 1999). Spectral sensitivity appears to be cone dominated through 15 dpf, and the rod contribution is not adult-like until close to 1 month of age (Bilotta *et al.*, 2001; Saszik *et al.*, 1999).

A number of intrinsic and extrinsic factors contribute to the specification, differentiation, and maturation of rods, but the story is far from complete. For example, the transcription factors neural retina leucine zipper (Nrl) and the cone-rod homeobox (Crx) gene have been extensively studied *in vivo* and clearly seem to be involved (Furukawa *et al.*, 1997, 1999; Mears *et al.*, 2001). The signaling molecule *sonic hedgehog* (*shh*) also plays a role in photoreceptor differentiation. Retinas from the zebrafish mutant *sonic-you*, which contains a deletion of the *sonic hedgehog* gene, showed few or no rhodopsin-expressing cells in the ventral patch at 75 hpf, a time point when a significant number of rods have differentiated in wild-type eyes (Stenkamp *et al.*, 2000). Furthermore, injection of antisense morpholinos against the *shh* gene eliminated rod opsin expression at 58 hpf (Stenkamp and Frey, 2003). These results demonstrate a role for *sonic hedgehog* in photoreceptor differentiation. A number of other possible factors have been suggested from studies on rod development in tissue culture systems (see Morrow *et al.*, 1998b) and references therein). In zebrafish, a role for retinoic acid in the development of rods has been demonstrated (Hyatt *et al.*, 1996; Perkins *et al.*, 2002). By adding retinoic acid to the embryo media, an increase of rhodopsin-expressing cells was seen at 4 dpf; whereas the number of cones was decreased. Further experiments are needed to determine whether retinoic acid increases the final number of rods or whether it causes rod progenitors to mature more quickly but does not cause an overall fate shift.

With the ability to modify the external environment during development, as well as modulate gene function by injecting messenger ribonucleic acid (mRNA) or morpholinos, zebrafish are ideally suited to determine whether the effects of various factors seen on mammalian rods in culture can be recapitulated *in vivo*. Furthermore, the ability to easily label specific cell types with fluorescent reporter genes greatly facilitates the examination of a given cell population during development.

III. Transgenic Zebrafish Expressing Photoreceptor-Specific Reporter Genes

Several groups have generated transgenic zebrafish that express GFP in rod photoreceptor cells (Fadool, 2003; Hamaoka *et al.*, 2002; Kennedy *et al.*, 2001; Perkins *et al.*, 2002) and one line that marks ultraviolet cones (Takechi *et al.*, 2003). These lines were generated using standard techniques for producing transgenic zebrafish (Lin, 2000). Briefly, linearized plasmid DNA containing the transgene was injected into the yolks of 1- to 4-cell stage zebrafish embryos. In a small percentage of cases, the DNA randomly integrates into the genome of the germ cells. These "founder fish" produce transgenic progeny.

These lines share many similarities but also have some critical differences (Table I). In all cases, GFP expression is cell specific and restricted to the correct photoreceptor cell type. The lines from Fadool (2003) and Perkins *et al.* (2002) both utilize *Xenopus* promoters (Batni *et al.*, 1996; Knox *et al.*, 1998); whereas the

Table I
Similarities and Differences among Transgenic Zebrafish Lines

Line	Promoter origin (species)	Promoter size (kb)	Localization	Onset of expression (hpf)	Comments	Reference
zops-GFP	Zebrafish	1.2	Cytoplasmic	~72		Kennedy *et al.* (2001)
zops-GFP	Zebrafish	1.1 and 3.7	Cytoplasmic	50	2 separate lines	Hamaoka *et al.* (2002)
xops-GFP-CT44	Xenopus	1.3	Outer segment	~72	Fused to C-terminal 44 amino acids of rhodopsin	Perkins *et al.* (2002)
xops-GFP	Xenopus	5.5	Cytoplasmic	55	Also seen in pineal gland	Fadool (2003)
zsws1-GFP	Zebrafish	5.5	Cytoplasmic	56	UV-cone specific	Takechi *et al.* (2003)

other lines have been generated with constructs containing zebrafish promoters. The onset of expression varies from 50 to 72 hpf, depending on the line. Neither the size nor the origin of the promoter DNA appears to be the cause for this variability as both the 1.1-kb zebrafish promoter (Hamaoka *et al.*, 2002) and the 5.5-kb *Xenopus* promoter (Fadool, 2003) can result in GFP visualization by 55 hpf or earlier, in close agreement to the onset of the endogenous rod opsin gene, which occurs at approximately 50 hpf (Raymond *et al.*, 1995). The other lines show an almost 24-h delay between opsin expression and transgene expression. Why this occurs is not known; although local silencing through nearby suppressor elements or integration into heterochromatic regions of the genome could explain this phenomenon.

By fusing GFP to specific regions of the rhodopsin protein, it is possible to restrict GFP localization within the rod. When GFP is fused to an opsin promoter alone, the protein distributes throughout the cell, including the photoreceptor inner segments and synaptic terminals as well as the outer segments (Fig. 2A). In contrast, the line from Perkins *et al.* (2002) used a transgene that fuses GFP to the C-terminal 44 amino acids of *Xenopus* rhodopsin. In *Xenopus*, this construct was sufficient to direct transport of GFP exclusively to the rod outer segments (Tam *et al.*, 2000). Mutations in the C-terminal region of the rhodopsin gene are known to cause incorrect localization of the protein (Li *et al.*, 1996) and result in retinitis pigmentosa (Berson *et al.*, 1991), thus emphasizing the importance of this

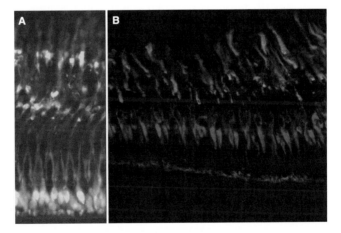

Fig. 2 Analysis of GFP distribution in two photoreceptor-specific transgenic lines. Transverse sections through adult retinas. (A) Fluorescence microscopy image of GFP distribution in the line described by Fadool (2003) demonstrate the cytoplasmic localization and distribution throughout the cell. (B) Confocal analysis of GFP distribution in the line described by Perkins *et al.* (2002). When GFP is fused to the C-terminal 44 amino acids of rhodopsin, GFP fluorescence is restricted to the rod outer segments (*green*). The red/green double cones are stained with Fret-43 (*red*). Fig. 2A was reprinted (with permission from Elsevier) from Fadool, J. M. (2003). Development of a rod photoreceptor mosaic revealed in transgenic zebrafish. *Dev. Biol.* **258**, 277–290. (See Color Insert.)

region of the protein for normal function. In transgenic zebrafish, the GFP is also limited to the rod outer segments (Fig. 2B), indicating that the zebrafish opsin transport machinery recognizes the *Xenopus* peptide. Having a transgenic zebrafish with an actively transported fluorescent marker will enable further investigations into the vectorial sorting of opsin. In contrast, transgenic lines expressing a soluble GFP, which is found ubiquitously in the rod cells, allow for detailed developmental analysis of photoreceptor morphology in living animals, whole-mount embryos, flat mounted retinas, and sections. It is also the case that the signal from GFP in these lines is typically more robust and more uniformly distributed throughout the cell than the labeling patterns from rod-cell-specific antibodies, which are dependent on the subcellular localization of the antigen.

IV. Developmental Studies of Photoreceptors Using Transgenic Zebrafish

A. Photoreceptor Mosaic

It has been known for many years that the cone photoreceptor subtypes in teleosts are arranged in a crystalline-like mosaic and are interdependent on one another (Ali, 1976; Engstrom, 1960). The mosaic of the cone photoreceptors within the zebrafish retina is composed of rows of the blue- and ultraviolet (UV)-sensitive single cones that alternate with parallel rows of red- and green-sensitive double cones (Robinson *et al.*, 1993). It was assumed, however, that the rod photoreceptors were not arranged in a mosaic, in part because rod genesis from rod progenitors located in the inner nuclear layer is sporadic near the proliferative marginal zone and in the central retina (Hagedorn *et al.*, 1998; Marcus *et al.*, 1999; Stenkamp *et al.*, 2000). However, the use of a rod-specific transgenic line demonstrated that a rod mosaic does exist (Fadool, 2003). In this study, flat-mounted adult retinas were imaged through the vitreal surface, using confocal microscopy. Deep scanning through the levels of the cell bodies and rod inner segments showed that the mosaic was maintained (Fig. 3A and B). At the level of the outer plexiform layer, the rod synaptic terminals were seen to be arranged in regularly spaced rows (Fig. 3C). Imaging of the retinal margin in adults showed that newly differentiated rods were arranged around individual UV-sensitive cone nuclei but not around the red and green cone nuclei, which appeared in an alternating fashion with the rod/UV clusters (Fig. 3D through F). These data, combined with spatial pattern analyses of nearest neighbor distances and conformity ratios, support the conclusion of a regularly arranged rod mosaic within the photoreceptor cell layer (Fadool, 2003).

B. Photoreceptor Terminals

Rods form invaginating chemical synapses with horizontal and bipolar cells. During development, neurites of the post-synaptic horizontal and bipolar cells penetrate into the rod spherules, forming the precise geometric arrangement

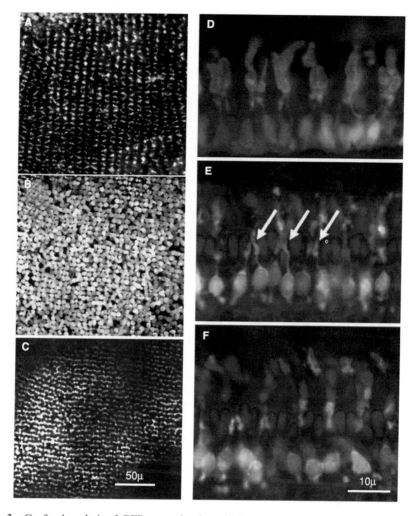

Fig. 3 Confocal analysis of GFP expression in rod photoreceptors reveals a rod mosaic that is established at the retinal margin in adults. (A–C) Whole-mounted retinas in saline were imaged by confocal microscopy from the vitreal side. Images reveal a rod mosaic at the level of the inner segments (A), the cell bodies (B), and the synaptic terminals (C). Note the regularly spaced rows of the rod structures. (E–F) Merged confocal images of the marginal zone in serial, tangential sections through adult retinas that were immuno-labeled (*red*) for rod opsin (D), UV opsin (E), and red opsin (F) with GFP expression (*green*) and 4′,6-diamidino-2-phenyllindole (DAPI) counterstained for nuclei (*blue*). Note the position of clustered rod outer segments at the positions of immature UV cone outer segments (*arrow*). Reprinted (with permission from Elsevier) Fadool, J. M. (2003). Development of a rod photoreceptor mosaic revealed in transgenic zebrafish. *Dev. Biol.* **258**, 277–290. (See Color Insert.)

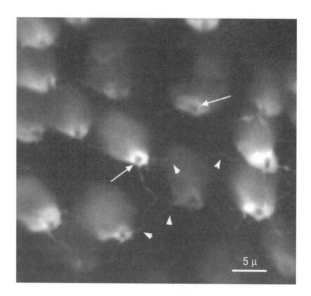

Fig. 4 Confocal image of GFP fluorescence in the dorsal retina of a 5-dpf larva. At the level of the rod terminal, the single invaginating synapse (*arrows*) and numerous telodendria (*arrowheads*) coupling the rod terminals were readily apparent.

of the invaginating synapse (Chen and Witkovsky, 1978). In transgenic GFP zebrafish, the involution of the rod spherule can be observed before the overt differentiation of the rod outer segment (Fig. 4). In addition to invaginating synapses, rod terminals in some species extend numerous telodendria, which form synaptic junctions with neighboring rod photoreceptors (Mariani and Lasansky, 1984; Ohtsuka and Kawamata, 1990; Owen, 1985). It has been proposed that rod coupling may be important for noise reduction under very dim lighting conditions (Witkovsky *et al.*, 2001). In zebrafish larvae, telodendria extend tens of microns to make contact with the neighboring rod terminals (Fig. 4). However, in juvenile and adult retinas, the density of rod terminals is much greater as more rods are incorporated across the retina. The telodendria then extend shorter distances to the nearby spherules, suggesting some level of plasticity in the connections.

1. Genetic Screens for Mutations Affecting Rod Photoreceptor Development

A number of forward genetic screens have identified mutations that affect retinal development and function (Brockerhoff *et al.*, 1995; Fadool *et al.*, 1997; Malicki *et al.*, 1996; Neuhauss *et al.*, 1999). Many of the mutants from these screens were identified using gross morphological criteria, often a small eye phenotype. The nature of the retinal defect was not known until invasive procedures such as histology or immunohistochemistry were performed. As rods represent a relatively small percentage of the total number of photoreceptor cells at 5 dpf, mutations causing an absence or decrease in rods may not result in smaller eyes. Using behavioral tests, such as the optokinetic response (OKR) assay, mutants with normal eye morphology can be found. Of the reported mutants affecting photoreceptors, most appear to affect primarily cone function (Allwardt *et al.*, 2001; Brockerhoff *et al.*, 1997, 2003). This is not surprising since physiological testing indicates that most, if not all, visual function is cone derived during early larval stages (Bilotta *et al.*, 2001). Taken together, it is quite possible that mutations specifically affecting the rods were missed using these screening techniques.

Identifying cell-specific mutations is especially challenging in heterogeneous cell populations such as the retina, where multiple cell types are present in three distinct layers. The dependence on techniques that are invasive or restricted to fixed tissues limits the ability to rapidly identify mutations affecting a specific cell type. However, by using a rod-GFP line as part of the genetic screen, mutants with abnormal rod photoreceptor development can be easily and rapidly identified by simple observation. Embryos to be screened are anesthetized with tricaine and viewed under a fluorescent microscope. Mutants are identified by qualitative differences in the location or total number of GFP-expressing rods. Figure 5 shows an embryo viewed from the lateral side, illustrating that single rods can be easily distinguished in the dorsal retina. As the screen is done on living animals, the embryos can be revived by a simple water change. Mutants can then be viewed during consecutive days to track the changes in the rod populations.

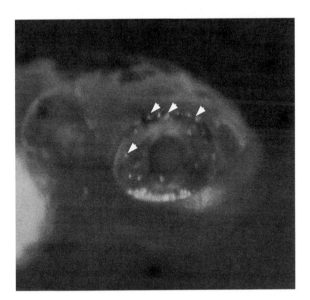

Fig. 5 Individual rod photoreceptors can be identified in living animals. A 4-dpf embryo was imaged with a fluorescent microscope and viewed from the lateral side. Bright autofluorescence is seen in the yolk. Photoreceptors in the dorsal retina are easily seen at this timepoint (*arrowheads*). (See Color Insert.)

Mutations affecting the rods can be separated into three distinct classes. The class I mutants have small eyes and completely eliminate rod development, thus having no GFP expression at any time point. This class would not necessarily be rod specific, as mutants that block differentiation of the entire retina, such as the *young* mutation (Link *et al.*, 2000), would be placed in this category. Class II mutants are categorized as those with small eyes and fewer GFP-expressing cells as in the wild-type siblings. These mutations may slow the rate of rod differentiation, resulting in fewer mature rods at a given age. Alternatively, this phenotype may reflect rod-specific cell death or general retinal degeneration, which would be seen as a loss of fluorescence through several days. The *photoreceptor cell absent* mutant (Fadool *et al.*, 1997) represents a class II mutant. The last category of mutants, the class III mutants, describes mutations most likely to be missed in previous screens. These mutants are expected to have a normal eye size but a significant or complete loss of GFP. This phenotype might reflect a shift away from the rod cell fate by retinal precursors that does not alter the total cell number or density within the retina. While mutants in this category have not yet been identified, this proposed screen would be the most efficient way to find these mutants.

Using this screening approach, the *rods missing* (*rdm*) mutation was identified in a genetic screen for mutants affecting GFP expression (Perkins and Dowling, unpublished results, 2003). The *rods missing* mutant does not fully lack rods, rather fewer GFP cells were observed than in wild-type siblings. When first apparent at 4 dpf, the *rods missing* mutants had a slightly smaller eye and a reduction in GFP expression in the ventral patch. By 6 dpf, the eyes of *rods missing* mutants were considerably smaller than wild type and the GFP expression was still very limited in the dorsal retina. Histological analysis revealed that the retinas of *rods missing* mutants were small, but well patterned. Most striking was the general lack of rod outer segments in the dorsal retina of the mutants. Outer segments from the red/green double cones could be seen and the short UV-cones were present. Interestingly, no evidence for cell death or photoreceptor degeneration was observed throughout the retina. Based on the small-eye phenotype at 4 dpf, the *rods missing* mutant would have been identified using gross morphological criteria, but the identification of this mutant through a screen of transgenic zebrafish rapidly distinguished this mutant as one that directly affected the rods because of the reduction in GFP-expressing cells.

C. Studies of Opsin Transport in *Mariner* Mutants

As noted earlier, the transgenic line by Perkins *et al.* (2002) utilizes a targeted GFP molecule that may be used to study the process of opsin transport. The C-terminal tail of rhodopsin that is fused to the GFP is essential for proper localization of rhodopsin and any changes in subcellular localization of the fusion protein should reflect a specific defect in transport. The zebrafish *mariner* mutant is defective in myosin VIIA and is an obvious candidate in which to study this

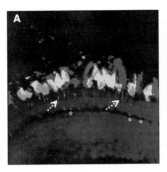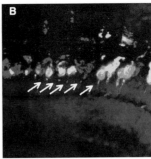

Fig. 6 Confocal analysis of GFP localization in *mariner* mutants that express the GFP-CT44 fusion protein. Tangential sections through 8-dpf animals were immunostained with the rod-specific monoclonal antibody 1D1 (*red*) with GFP expression (*green*) and counterstained with DAPI (*blue*). Colocalization of the 1D1 label and GFP is seen in *yellow*. Note the occasional presence of GFP and 1D1 signal in the inner segments (*dashed arrows*) and the terminals (*asterisks*) of wild-type animals (A). In *mariner* mutants (B), round clusters of colocalized signals were regularly seen at the base of the outer segments, presumably in the connecting cilium (*arrows*). This may represent a mild defect in transport through the connecting cilium. (See Color Insert.)

process. By breeding transgenic fish with *mariner* heterozygotes, fish that carry the mutation and the transgene can be identified in the next generation. Inbreeding of these individuals results in homozygous *mariner* mutants that express the targeted GFP. Transverse sections through wild-type and mutant animals revealed few differences in the location of the GFP signal or in immunolabeling of the rod-specific marker 1D1, which may recognize an epitope on rhodopsin (Fadool *et al.*, 1999). In both wild-type and mutant fish, the 1D1 label and GFP signal was localized predominantly to the OS. Faint GFP signal was observed in the terminals and in the inner segment (Fig. 6A). In *mariner* embryos, reproducible clusters of colocalized GFP and 1D1 labeling were observed at the base of the outer segment, possibly indicating a slight accumulation of protein at or near the connecting cilium (Fig. 6B). These clusters, however, were small and did not result in a significant decrease in GFP or 1D1 localization to the OS. These results are similar to that seen in the *shaker-1* mouse model, where opsin molecules are seen to cluster in the connecting cilium but substantial incorrect localization of opsin was not observed (Liu *et al.*, 1999). These results illustrate the advantages of using a transgenic line with a specifically targeted marker to test a given process.

V. Conclusions

The study of photoreceptor cell structure and development has benefited from the use of transgenic zebrafish expressing cell-specific GFP reporter genes. At the moment, several rod-specific lines have been determined but only one cone-specific GFP line has been determined, namely that for UV-opsin expressing cones. We hope lines will soon exist that express fluorescent reporter genes in other cone cells. The rod-specific lines have revealed the presence of a rod mosaic (Fadool, 2003), provided methods to study protein transport within the rods, and enhanced genetic screens by enabling the search for cell-specific mutations in a noninvasive manner. Yet these types of analyses are certainly not limited to the photoreceptors. A line that expresses GFP in a subset of amacrine cells was used to image the formation of the inner plexiform layer *in vivo* (Kay *et al.*, 2004). With only a subset of amacrine cells expressing GFP, it was possible to resolve the extension of neurites into both the on- and off-sub lamina of the inner plexiform layer.

Transgenic technology will greatly facilitate future studies of neuronal structure and function in zebrafish. Dozens of different transgenic zebrafish lines currently exist (Udvadia and Linney, 2003) and more specialized lines will no doubt appear in the future. The classic studies of Ramón y Cajal used Golgi staining to provide tremendous insights into the structure of retinal neurons in fixed tissue (Ramón y Cajal, 1911). With the increasing power of fluorescent imaging technology, questions about the development and function of neurons *in vivo* and in real time are being addressed. With the appropriate promoters, it should be possible to generate individual transgenic lines that express reporter genes within all subsets of the major classes of retinal cells. These lines could be used for rapid identification of specific cell types before electrophysiological recordings. Additionally, time-lapse imaging of individual cells or whole layers of cells could be used to study the timing and control of synaptogenesis *in vivo*, as seen from the report by Kay *et al.* (2004). Certainly the combination of these techniques with the wide assortment of zebrafish mutants will facilitate our understanding of neural development and function.

References

Ali, M. A. (1976). "Retinas of Fishes: An Atlas." Springer-Verlag, New York.

Allwardt, B. A., Lall, A. B., Brockerhoff, S. E., and Dowling, J. E. (2001). Synapse formation is arrested in retinal photoreceptors of the zebrafish nrc mutant. *J. Neurosci.* **21**, 2330–2342.

Batni, S., Scalzetti, L., Moody, S. A., and Knox, B. E. (1996). Characterization of the Xenopus rhodopsin gene. *J. Biol. Chem.* **271**, 3179–3186.

Berson, E. L., Rosner, B., Sandberg, M. A., Weigel-DiFranco, C., and Dryja, T. P. (1991). Ocular findings in patients with autosomal dominant retinitis pigmentosa and rhodopsin, proline-347-leucine. *Am. J. Ophthalmol.* **111**, 614–623.

Bilotta, J., Saszik, S., and Sutherland, S. E. (2001). Rod contributions to the electroretinogram of the dark-adapted developing zebrafish. *Dev. Dyn.* **222**, 564–570.

Brockerhoff, S. E., Hurley, J. B., Janssen-Bienhold, U., Neuhauss, S. C., Driever, W., and Dowling, J. E. (1995). A behavioral screen for isolating zebrafish mutants with visual system defects. *Proc. Natl. Acad. Sci. USA* **92**, 10545–10549.

Brockerhoff, S. E., Hurley, J. B., Niemi, G. A., and Dowling, J. E. (1997). A new form of inherited red-blindness identified in zebrafish. *J. Neurosci.* **17**, 4236–4242.

Brockerhoff, S. E., Rieke, F., Matthews, H. R., Taylor, M. R., Kennedy, B., Ankoudinova, I., Niemi, G. A., Tucker, C. L., Xiao, M., Cilluffo, M. C., *et al.* (2003). Light stimulates a transducin-independent increase of cytoplasmic Ca2+ and suppression of current in cones from the zebrafish mutant nof. *J. Neurosci.* **23**, 470–480.

Chen, F., and Witkovsky, P. (1978). The formation of photoreceptor synapses in the retina of larval Xenopus. *J. Neurocytol.* **7,** 721–740.

Deretic, D., Huber, L. A., Ransom, N., Mancini, M., Simons, K., and Papermaster, D. S. (1995). rab8 in retinal photoreceptors may participate in rhodopsin transport and in rod outer segment disk morphogenesis. *J. Cell. Sci.* **108,** 215–224.

Dowling, J. E. (1963). Neural and photochemical mechanisms of visual adaptation in the rat. *J. Gen. Physiol.* **46,** 1287–1301.

Dryja, T. P., and Li, T. (1995). Molecular genetics of retinitis pigmentosa. *Hum. Mol. Genet.* **4,** 1739–1743.

Engstrom, K. (1960). Cone types and cone arrangements in retina of some cyprinids. *Acta Zool. (Stockholm)* **41,** 277–295.

Ernest, S., Rauch, G. J., Haffter, P., Geisler, R., Petit, C., and Nicolson, T. (2000). Mariner is defective in myosin VIIA: A zebrafish model for human hereditary deafness. *Hum. Mol. Genet.* **9,** 2189–2196.

Fadool, J. M. (2003). Development of a rod photoreceptor mosaic revealed in transgenic zebrafish. *Dev. Biol.* **258,** 277–290.

Fadool, J. M., Brockerhoff, S. E., Hyatt, G. A., and Dowling, J. E. (1997). Mutations affecting eye morphology in the developing zebrafish (Danio rerio). *Dev. Genet.* **20,** 288–295.

Fadool, J. M., Fadool, D. A., Moore, J. C., and Linser, P. J. (1999). Characterization of monoclonal antibodies against zebrafish retina. *Invest. Opth. Vis. Sci. Suppl.* **40,** 1251.

Furukawa, T., Morrow, E. M., and Cepko, C. L. (1997). Crx, a novel otx-like homeobox gene, shows photoreceptor-specific expression and regulates photoreceptor differentiation. *Cell* **91,** 531–541.

Furukawa, T., Morrow, E. M., Li, T., Davis, F. C., and Cepko, C. L. (1999). Retinopathy and attenuated circadian entrainment in Crx-deficient mice. *Nat. Genet.* **23,** 466–470.

Gibson, F., Walsh, J., Mburu, P., Varela, A., Brown, K. A., Antonio, M., Beisel, K. W., Steel, K. P., and Brown, S. D. (1995). A type VII myosin encoded by the mouse deafness gene shaker-1. *Nature* **374,** 62–64.

Hagedorn, M., Mack, A. F., Evans, B., and Fernald, R. D. (1998). The embryogenesis of rod photoreceptors in the teleost fish retina, Haplochromis burtoni. *Brain Res. Dev. Brain Res.* **108,** 217–227.

Hamaoka, T., Takechi, M., Chinen, A., Nishiwaki, Y., and Kawamura, S. (2002). Visualization of rod photoreceptor development using GFP-transgenic zebrafish. *Genesis* **34,** 215–220.

Hecht, S., Schlaer, S., and Pirenne, M. H. (1942). *J. Optic. Soc. Amer.* **38,** 196–208.

Hu, M., and Easter, S. S. (1999). Retinal neurogenesis: The formation of the initial central patch of postmitotic cells. *Dev. Biol.* **207,** 309–321.

Hyatt, G. A., Schmitt, E. A., Fadool, J. M., and Dowling, J. E. (1996). Retinoic acid alters photoreceptor development in vivo. *Proc. Natl. Acad. Sci. USA* **93,** 13298–13303.

Kay, J. N., Roeser, T., Mumm, J. S., Godinho, L., Mrejeru, A., Wong, R. O., and Baier, H. (2004). Transient requirement for ganglion cells during assembly of retinal synaptic layers. *Development* **131,** 1331–1342.

Kennedy, B. N., Vihtelic, T. S., Checkley, L., Vaughan, K. T., and Hyde, D. R. (2001). Isolation of a zebrafish rod opsin promoter to generate a transgenic zebrafish line expressing enhanced green fluorescent protein in rod photoreceptors. *J. Biol. Chem.* **276,** 14037–14043.

Knox, B. E., Schlueter, C., Sanger, B. M., Green, C. B., and Besharse, J. C. (1998). Transgene expression in Xenopus rods. *FEBS Lett.* **423,** 117–121.

Li, T., Snyder, W. K., Olsson, J. E., and Dryja, T. P. (1996). Transgenic mice carrying the dominant rhodopsin mutation P347S: Evidence for defective vectorial transport of rhodopsin to the outer segments. *Proc. Natl. Acad. Sci. USA* **93,** 14176–14181.

Lin, S. (2000). Transgenic zebrafish. *Methods in Molecular Biology* **136,** 375–383.

Link, B. A., Fadool, J. M., Malicki, J., and Dowling, J. E. (2000). The zebrafish young mutation acts non-cell-autonomously to uncouple differentiation from specification for all retinal cells. *Development* **127,** 2177–2188.

Liu, X., Udovichenko, I. P., Brown, S. D., Steel, K. P., and Williams, D. S. (1999). Myosin VIIa participates in opsin transport through the photoreceptor cilium. *J. Neurosci.* **19,** 6267–6274.

Malicki, J., Neuhauss, S. C., Schier, A. F., Solnica-Krezel, L., Stemple, D. L., Stainier, D. Y., Abdelilah, S., Zwartkruis, F., Rangini, Z., and Driever, W. (1996). Mutations affecting development of the zebrafish retina. *Development* **123**, 263–273.

Marcus, R. C., Delaney, C. L., and Easter, S. S., Jr. (1999). Neurogenesis in the visual system of embryonic and adult zebrafish (Danio rerio). *Vis. Neurosci.* **16**, 417–424.

Mariani, A. P., and Lasansky, A. (1984). Chemical synapses between turtle photoreceptors. *Brain Res.* **310**, 351–354.

Marszalek, J. R., Liu, X., Roberts, E. A., Chui, D., Marth, J. D., Williams, D. S., and Goldstein, L. S. (2000). Genetic evidence for selective transport of opsin and arrestin by kinesin-II in mammalian photoreceptors. *Cell* **102**, 175–187.

Mears, A. J., Kondo, M., Swain, P. K., Takada, Y., Bush, R. A., Saunders, T. L., Sieving, P. A., and Swaroop, A. (2001). Nrl is required for rod photoreceptor development. *Nat. Genet.* **29**, 447–452.

Morrow, E. M., Belliveau, M. J., and Cepko, C. L. (1998a). Two phases of rod photoreceptor differentiation during rat retinal development. *J. Neurosci.* **18**, 3738–3748.

Morrow, E. M., Furukawa, T., and Cepko, C. L. (1999b). Vertebrate photoreceptor cell development and disease. *Trends Cell Biol.* **8**, 353–358.

Nathans, J. (1992). Rhodopsin: Structure, function, and genetics. *Biochemistry* **31**, 4923–4931.

Neuhauss, S. C., Biehlmaier, O., Seeliger, M. W., Das, T., Kohler, K., Harris, W. A., and Baier, H. (1999). Genetic disorders of vision revealed by a behavioral screen of 400 essential loci in zebrafish. *J. Neurosci.* **19**, 8603–8615.

Ohtsuka, T., and Kawamata, K. (1990). Telodendrial contact of HRP-filled photoreceptors in the turtle retina: Pathways of photoreceptor coupling. *J. Comp. Neurol.* **292**, 599–613.

Owen, W. G. (1985). Chemical and electrical synapses between photoreceptors in the retina of the turtle, Chelydra serpentina. *J. Comp. Neurol.* **240**, 423–433.

Papermaster, D. S., Converse, C. A., Siuss, J., Schneider, B. G., and Besharse, J. C. (1975). Membrane biosynthesis in the frog retina: Opsin transport in the photoreceptor cell. *Biochemistry* **14**, 1343–1352.

Papermaster, D. S., Schneider, B. G., and Besharse, J. C. (1985). Vesicular transport of newly synthesized opsin from the Golgi apparatus toward the rod outer segment. Ultrastructural immunocytochemical and autoradiographic evidence in Xenopus retinas. *Invest. Ophthalmol. Vis. Sci.* **26**, 1386–1404.

Pazour, G. J., Baker, S. A., Deane, J. A., Cole, D. G., Dickert, B. L., Rosenbaum, J. L., Witman, G. B., and Besharse, J. C. (2002). The intraflagellar transport protein, IFT88, is essential for vertebrate photoreceptor assembly and maintenance. *J. Cell Biol.* **157**, 103–113.

Perkins, B. D., Kainz, P. M., O'Malley, D. M., and Dowling, J. E. (2002). Transgenic expression of a GFP-rhodopsin COOH-terminal fusion protein in zebrafish rod photoreceptors. *Vis. Neurosci.* **19**, 257–264.

Ramón y Cajal, S. (1911). "Histologie du Systeme Nerveax de l'Homme et des Vertebres." Paris, France.

Raymond, P. A., Barthel, L. K., and Curran, G. A. (1995). Developmental patterning of rod and cone photoreceptors in embryonic zebrafish. *J. Comp. Neuro.* **359**, 537–550.

Robinson, J., Schmitt, E. A., Harosi, F. I., Reece, R. J., and Dowling, J. E. (1993). Zebrafish ultraviolet visual pigment: Absorption spectrum, sequence, and localization. *Proc. Natl. Acad. Sci. USA* **90**, 6009–6012.

Saszik, S., Bilotta, J., and Givin, C. M. (1999). ERG assessment of zebrafish retinal development. *Vis. Neurosci.* **16**, 881–888.

Schmitt, E. A., and Dowling, J. E. (1996). Comparison of topographical patterns of ganglion and photoreceptor cell differentiation in the retina of the zebrafish, Danio rerio. *J. Comp. Neuro.* **371**, 222–234.

Schmitt, E. A., Hyatt, G. A., and Dowling, J. E. (1999). Erratum: Temporal and spatial patterns of opsin gene expression in the zebrafish (Danio rerio): Corrections with additions. *Vis. Neurosci.* **16**, 601–605.

Scholey, J. M. (2003). Intraflagellar transport. *Annu. Rev. Cell Dev. Biol.* **19**, 423–443.

Stenkamp, D. L., and Frey, R. A. (2003). Extraretinal and retinal hedgehog signaling sequentially regulate retinal differentiation in zebrafish. *Dev. Biol.* **258,** 349–363.

Stenkamp, D. L., Frey, R. A., Prabhudesai, S. N., and Raymond, P. A. (2000). Function for Hedgehog genes in zebrafish retinal development. *Dev. Biol.* **220,** 238–252.

Svn, S., Amsterdam, A., Pazour, G. J., Cole, D. G., Miller, M. S., and Hopkins, N. (2004). A genetic screen in zebrafish identifies cilic genes as a principal cause of cystic kidney. *Development* **131,** 4085–4093.

Szel, A., van Veen, T., and Rohlich, P. (1994). Retinal cone differentiation. *Nature* **370,** 336.

Takechi, M., Hamaoka, T., and Kawamura, S. (2003). Fluorescence visualization of ultraviolet-sensitive cone photoreceptor development in living zebrafish. *FEBS Lett.* **553,** 90–94.

Tam, B. M., Moritz, O. L., Hurd, L. B., and Papermaster, D. S. (2000). Identification of an outer segment targeting signal in the COOH terminus of rhodopsin using transgenic Xenopus laevis. *J. Cell Biol.* **151,** 1369–1380.

Tomita, T. (1970). Electrical activity of vertebrate photoreceptors. *Q. Rev. Biophys.* **3,** 179–222.

Udvadia, A. J., and Linney, E. (2003). Windows into development: Historic, current, and future perspectives on transgenic zebrafish. *Dev. Biol.* **256,** 1–17.

Wald, G. (1955). The photoreceptor process in vision. *Am. J. Ophthalmol.* **40,** 18–41.

Weil, D., Blanchard, S., Kaplan, J., Guilford, P., Gibson, F., Walsh, J., Mburu, P., Varela, A., Levilliers, J., Weston, M. D., *et al.* (1995). Defective myosin VIIA gene responsible for Usher syndrome type 1B. *Nature* **374,** 60–61.

Witkovsky, P., Thoreson, W., and Tranchina, D. (2001). Transmission at the photoreceptor synapse. *Prog. Brain. Res.* **131,** 145–159.

Yau, K. W. (1994). Phototransduction mechanism in retinal rods and cones. The Friedenwald Lecture. *Invest. Ophthalmol. Vis. Sci.* **35,** 9–32.

Yau, K. W., and Nakatani, K. (1985). Light-suppressible, cyclic GMP-sensitive conductance in the plasma membrane of a truncated rod outer segment. *Nature* **317,** 252–255.

Young, R. W. (1967). The renewal of photoreceptor cell outer segments. *J. Cell Biol.* **33,** 61–72.

CHAPTER 16

Approaches to Study Neurogenesis in the Zebrafish Retina

Andrei Avanesov and Jarema Malicki

Department of Ophthalmology
Harvard Medical School
Boston, Massachusetts 02114

I. Introduction

The vertebrate central nervous system is enormously complex. The human cerebral cortex alone is estimated to contain in excess of 10^9 neurons (Jacobson, 1991). Each neuron is characterized by the morphology of its soma and processes, its synaptic connections with other cells, the receptors expressed on its surface, the

METHODS IN CELL BIOLOGY, VOL. 76

0091-679X/04 $35.00

neurotransmitters it releases, and numerous other molecular and cellular properties. Together, these characteristics define cell identity. To understand the development of the central nervous system, the multiple steps involved in generation of the numerous cell identities must be determined. One way to approach this enormously complicated task is to choose a region of the central nervous system characterized by a relative simplicity.

The retina is such a region. Several characteristics make the retina more approachable than most other areas of the central nervous system. Cajal (1893) noted that the separation of different cells into distinct layers, the small size of dendritic fields, and the presence of layers consisting almost exclusively of neuronal projections are fortuitous characteristics of the retina. The retina contains a relatively small number of neuronal cell classes characterized by stereotypical positions and distinctive morphologies. Even in very crude histological preparations, the identity of individual cells can be frequently and correctly determined based on their location. In addition, the eye becomes isolated from other parts of the central nervous system early in embryogenesis. Cell migrations into the retina are limited to the optic nerve and the optic chiasm (Burrill and Easter, 1994; Watanabe and Raff, 1988). This relative isolation facilitates the interpretation of developmental events within the retina. Taken together, all these qualities make the retina an excellent model system for the studies of vertebrate neuronal development.

Teleost retinae have been studied for more than a century (Cajal, 1893; Dowling, 1987; Muller, 1857; Rodieck, 1973). The eyes of teleosts in general and zebrafish in particular are large and their neuroanatomy is well characterized. An important advantage of the zebrafish retina for genetic and developmental research is that it forms and becomes functional very early in development. Neurogenesis in the central retina of the zebrafish eye is essentially complete by 60 hpf (Nawrocki, 1985) and, as judged by the startle and optokinetic responses, the zebrafish eye detects light stimuli surprisingly early, starting between 2.5 and 3.5 dpf (Clark, 1981; Easter and Nicola, 1996). Studies of the zebrafish retina benefit from many general qualities of the system: high fecundity, transparency and external development of embryos, the ease of maintenance in large numbers, length of the life cycle, the ability to study haploid development, and most recently from the progress of zebrafish genomics, including the genome sequencing project.

The vertebrate retina has been remarkably conserved in evolution. Early investigators noted that retinae of divergent vertebrate phyla have similar organization (Cajal, 1893; Muller, 1857). Gross morphological and histological features of mammalian and teleost retinae display few differences. Accordingly, human and zebrafish retinae contain the same major cell classes organized in the same layered pattern. Similarities extend beyond histology and morphology. Pax-2/noi and Chx10/Vsx-2 expression patterns, for example, are very similar in mouse and zebrafish eyes (Liu et al., 1994; Macdonald and Wilson, 1997; Nornes et al., 1990; Passini et al., 1997). Likewise, several genetic defects of the zebrafish retina are reminiscent of human disorders. For example, mutations affecting photoreceptor cells in zebrafish (Malicki *et al.*, 1996; Neuhauss *et al.*, 1999) very much resemble retinitis pigmentosa and cone-rod dystrophies (Dryja and Li, 1995; Merin, 1991; Yagasaki and Jacobson, 1989), and mutations in the Pax-2/noi gene produce abnormal optic nerve development in zebrafish, as well as in humans (Macdonald and Wilson, 1997; Sanyanusin *et al.*, 1995, 1996).

The similarity of human and zebrafish retinae is a fortuitous circumstance allowing us to use zebrafish as a model of retinal disorders. Vision is the major sense used by humans in their interactions with the outside world. Throughout the world, diseases of the retina affect millions. In the United States alone, retinitis pigmentosa is estimated to affect 50,000 to 100,000 people (Dryja and Li, 1995), age-related macular degeneration affects more than 15 million people (Seddon, 1994), and in some populations the prevalence of glaucoma can reach more than 4% in people older than 65 years of age (Cedrone *et al.*, 1997). Therefore, in addition to being an excellent model for studies of vertebrate neurogenesis, the zebrafish retina is likely to provide us with insights into the nature of human disorders.

II. Development of the Zebrafish Retina

A. Early Morphogenetic Events

Fate mapping studies indicate that the retina originates from a single field of cells that during early gastrulation is positioned roughly between the telencephalic and the diencephalic precursor fields (Woo and Fraser, 1995). During late gastrulation, the anterior migration of diencephalic precursors is thought to subdivide the retinal field into two separate primordia (Varga *et al.*, 1999). Neurulation in teleosts proceeds somewhat differently than in higher vertebrates. The primordium of the central nervous system does not take the form of a tube (the neural tube), and instead is shaped in the form of a solid rod called the *neural keel* (Fig. 1B and C; Kimmel *et al.*, 1995; Schmitz *et al.*, 1993). Accordingly, in contrast to higher vertebrates, the optic vesicles are not present. The equivalent structures are called optic lobes and these first become evident as bilateral thickenings of the anterior neural keel at about 11.5 hpf and become gradually more and more prominent (Fig. 1A to C; Schmitt and Dowling, 1994). The optic lobes are initially flattened and protrude laterally on both sides of the brain (Fig. 1B and C, *arrows*). At approximately 13 hpf, the posterior portions of optic lobes start to separate from the brain. The anterior portions, on the other hand, remain attached (Fig. 1D). This attachment will persist later in development as the optic stalk. In parallel, the optic lobe turns around its antero-posterior axis so that its lower surface becomes directed towards the brain and the upper surface towards the outside environment (Fig. 1G). Later in development, this outside surface will form the neural retina. Fate mapping studies suggest that starting at around 15 hpf, cells migrate from the medial to lateral epithelial layer of the optic lobe (Fig. 1G; Li *et al.*, 2000b). The medial layer becomes thinner and subsequently

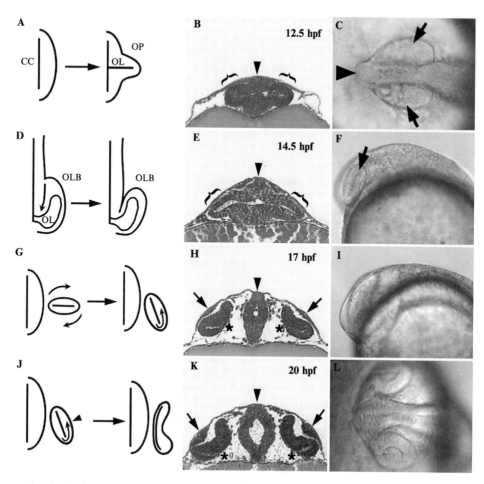

Fig. 1 Early morphogenetic events leading to the formation of the optic cup. (A) A diagram of a transverse section through the anterior neural keel, illustrating the morphogenetic transformation that leads to the formation of optic lobes. The solid horizontal line represents the ventricular lumen of the optic lobe. (B) A transverse plastic section through the anterior portion of the neural keel and optic lobes (*brackets*). (C) Dorsal view of the anterior neural keel and optic lobes (*arrows*) at 12.5 hpf. (D) A schematic representation of the anterior neural keel (dorsal view, anterior down). Wing-shaped optic primordia gradually detach from the neural tube, starting posteriorly (*arrow*). (E) A transverse plastic section through the anterior neural keel and optic lobes (*brackets*) at 14.5 hpf. (F) Lateral view of the anterior neural keel and optic lobe (*arrow*) at the same stage. (G) A diagram of dorso-ventral reorientation of the optic lobe. (H) A transverse plastic section through the neural keel and optic lobes during the reorientation at approximately 17 hpf. At about the same time, lens rudiments start to form (*arrows*) and the medial layer of the optic lobe becomes thinner as it begins to differentiate into the pigmented epithelium (*asterisks*). The lateral surface of the optic lobe starts to invaginate. (I) A lateral view of the anterior neural keel during optic cup formation. (J) A schematic representation of morphogenetic movements that accompany optic cup formation. Cells migrate (*arrow*) from the medial to the lateral cell layer around the ventral edge of the lobe. Simultaneously, the initially flat lobe invaginates (*arrowhead*) to become the concave eye cup. (K) A transverse plastic section through the anterior neural keel during optic cup formation at 20 hpf. Lens rudiments are quite prominent by this

differentiates as the retinal pigmented epithelium (RPE) (Fig. 1H and K, *asterisks*). At about the same time, an invagination forms on the lateral (upper, before turning) surface of the optic lobe (Schmitt and Dowling, 1994). This is accompanied by the appearance of a thickening in the epithelium overlying the optic lobe—the lens rudiment (Fig. 1H, *arrows*). Subsequently, during a period of several hours, both the invagination and the lens placode become increasingly more prominent, transforming the optic lobe into the optic cup (Fig. 1J to L). The choroid fissure forms in the rim of the optic cup next to the optic stalk. The lens placode continues to grow. By 24 hpf, it is detached from the epidermis.

At the beginning of day 2, the optic cup consists of two closely connected sheets of cells: the pseudostratified columnar neuroepithelium (rne) and the cuboidal pigmented epithelium (pe) (Fig. 2A). Starting at about 24 hpf, melanin granules appear in the cells of the pigmented epithelium. In the first half of day 2, concomitant to the expansion of the ventral diencephalon, the eye rotates so that the choroid fissure, which at 24 hpf was pointing above the yolk sack, is now directed towards the heart (Kimmel *et al.*, 1995; Schmitt and Dowling, 1994). Throughout this period, the optic stalk gradually becomes less prominent. In the first half of day 2, as ganglion cells begin to differentiate, the optic stalk provides support for their axons. Later in development, it appears to be entirely replaced by the optic nerve (Macdonald *et al.*, 1997). Rotation is the last major transformation in zebrafish eye development.

B. Neurogenesis

As the second day of development begins, the zebrafish retina still consists of a single sheet of pseudostratified neuroepithelium. Similar to the rest of neural tube epithelia, the retinal neuroepithelium is a highly polarized tissue, characterized by the presence of apico-basal nuclear movements, which correlate with cell cycle phase (Das *et al.*, 2003; Hinds and Hinds, 1974). Nuclei of cells that are about to divide translocate to the apical surface of the neuroepithelium, where both nuclear division and cytokinesis take place. Although it has been assumed for a long time that dividing cells lose their contact with the basal surface of the neuroepithelium (Hinds and Hinds, 1974), more recent two-photon imaging studies in zebrafish indicate that this view may be incorrect, as a tenuous cytoplasmic process extends toward the basal surface during nuclear division of the neuroepithelial cell (Das *et al.*, 2003).

Despite its uniform morphological appearance, the retinal neuroepithelium is the site of transformations, which are obvious in the changes of cell cycle length

stage (*arrows*). Most of the medial cell layer already displays a flattened morphology, except for the ventral-most regions, which still retain columnar appearance (*asterisks*). (L) A dorsal view of the anterior neural keel and optic cups at 20 hpf. Vertical arrowheads in B, C, E, H, and K indicate the midline. CC, central canal; OL, optic lumen; OP, optic primordium; OLB, optic lobe; hpf, hours post fertilization. Except D, C, and L, which show dorsal views, in all panels dorsal is up. Panels A, D, G, J based on Easter and Malicki (2002). The remaining panels reprinted from Pujic and Malicki (2001) with permission from Elsevier.

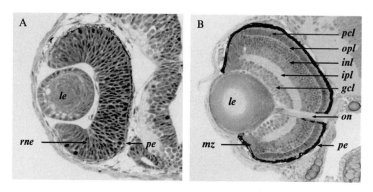

Fig. 2 Neurogenesis in the zebrafish retina. (A) A section through the zebrafish eye during the early stages of neurogenesis at approximately 36 hpf. At this stage, the retina mostly consists of two epithelial layers: the pigmented epithelium and the retinal neuroepithelium. Although some retinal cells are already postmitotic at this stage, they are not numerous enough to form distinct layers. (B) A section through the zebrafish eye at 72 hpf. With the exception of the marginal zone, where cell proliferation will continue throughout the lifetime of the animal, retinal neurogenesis is mostly completed. The major nuclear and plexiform layers, as well as the optic nerve and the pigmented epithelium, are well differentiated. gcl: ganglion cell layer; inl: inner nuclear layer; ipl: inner plexiform layer; le: lens; mz: marginal zone; on: optic nerve; opl: outer plexiform layer; pcl: photoreceptor cell layer; pe: pigmented epithelium; rne: retinal neuroepithelium.

and in the dynamic appearance of gene expression patterns. After a period of a very slow cell cycle progression during early stages of optic cup morphogenesis, cell cycle shortens to approximately 10 h by 24 hpf, and later its duration appears even shorter (Hu and Easter, 1999; Li *et al.*, 2000a; Nawrocki, 1985). The significance of these fluctuations or genetic mechanisms that regulate them are not understood. Similarly, the expression patterns of numerous loci display dramatic changes. While the transcription of some early expressed genes, such as *rx3* or *six3*, is down regulated, other loci become active. The zebrafish *atonal 5* homolog, *lakritz*, is one interesting example of an important genetic regulator characterized by a dynamic expression pattern. The *lakritz* gene becomes active in a small group of cells in the ventral retina by 25 hpf, and from there its expression spreads into the nasal, dorsal, and finally temporal eye (Masai *et al.*, 2000). This gradual advance of expression around the retinal surface is noteworthy because it characterizes many other developmental regulators and neuronal differentiation markers (reviewed in Pujic and Malicki, 2003).

Another noteworthy feature of neuroepithelial cells is the orientation of their mitotic spindles. The position of the mitotic spindle and its role in cell fate determination is an interesting, albeit contentious, issue. It has been proposed that, in some species, the vertical (apico-basal) orientation of the mitotic spindle characterizes asymmetric cell divisions, which produce cells of different identities—a progenitor cell and a postmitotic neuron, for example (Cayouette and Raff, 2003; Cayouette *et al.*, 2001). As such divisions first appear in the neuroepithelium at the onset of neurogenesis, so should vertically oriented mitotic spindles. The analysis of zebrafish neuroepithelial cells found, however, little support for the presence of vertically oriented mitotic spindles: the majority, if not all, of zebrafish neuroepithelial cells divide horizontally (Das *et al.*, 2003).

As the morphogenetic movements that shape and orient the optic cup come to completion, the first retinal cells become postmitotic and begin to differentiate. Gross morphological characteristics of the major retinal cell classes are very well conserved in all vertebrates. Six major classes of neurons arise during neurogenesis: ganglion, amacrine, bipolar, horizontal, interplexiform, and photoreceptor cells. Muller glia are also generated in the same period. Ganglion cell precursors are the first to become postmitotic in a small patch of ventrally located cells between 27 and 28 hpf (Hu and Easter, 1999; Nawrocki, 1985). The early onset of ganglion cell differentiation is again conserved in many vertebrate phyla (Altshuler *et al.*, 1991). Similar to expression patterns that characterize the genetic regulators of retinal neurogenesis, differentiated ganglion cells first appear in the ventral retina, nasal to the optic nerve (Burrill and Easter, 1995; Schmitt and Dowling, 1996). The rudiments of the ganglion cell layer are recognizable in histological sections by 36 hpf. Approximately 10 h after the cell cycle exit of the first ganglion neuron precursors, cells of the inner nuclear layer also become postmitotic, again this first happens in a small ventral group of cells (Hu and Easter, 1999). The first postmitotic photoreceptors appear between 43 and 48 hpf.

By 60 hpf, more than 90% of neurons in the central retina are postmitotic, and the major neuronal layers are distinguishable by morphological criteria (Fig. 2). Cells of different layers become postmitotic in largely nonoverlapping windows of time. This is particularly obvious for ganglion cell precursors, most of which, if not all, are postmitotic before the first inner nuclear layer cells exit the cell cycle (Hu and Easter, 1999). This is different from Xenopus or the mouse, where cell cycle exit times of different cell classes overlap extensively (Holt *et al.*, 1988; Young, 1985). In contrast to mammals, neurogenesis in teleosts and larval amphibians continues at the retinal margin throughout the lifetime of the organism (Marcus *et al.*, 1999). In adult zebrafish, as well as in other teleosts, neurons are also produced in the outer nuclear layer. In contrast to the marginal zone, where many cell types are generated, only rods are added in the outer nuclear layer (Mack and Fernald, 1995; Marcus *et al.*, 1999).

Photoreceptor morphogenesis starts shortly after the exit of photoreceptor precursor cells from the cell cycle (for a review, see Tsujikawa and Malicki, 2003). The photoreceptor cell layer can be distinguished in histological sections by 48 hpf. The photoreceptor outer segments first appear in the ventral patch by 60 hpf, and ribbon synapses of photoreceptor synaptic termini are detectable by 62 hpf (Branchek and Bremiller, 1984; Schmitt and Dowling, 1999). Rods are the first to express opsin, shortly followed by blue and red cones, and somewhat later by short single cones (Raymond *et al.*, 1995; Robinson *et al.*, 1995; Takechi *et al.*, 2003). The photoreceptor cell layer of the zebrafish retina contains five types of photoreceptor cells: rods, short single cones, long single cones, and short and long members of double cone pairs. The differentiation of morphologically distinct photoreceptor types becomes apparent by 4 dpf and, by 12 dpf, all zebrafish photoreceptor classes can be distinguished on the basis of their morphology (Branchek and Bremiller, 1984). The photoreceptor cells of the zebrafish retina are organized in a regular pattern, referred to as the *photoreceptor mosaic*. In the adult retina, rows of cones are separated by rods. Within these rows, double cones are separated from each other by alternating long and short single cones. This pattern is shifted in adjacent rows so that short single cones of one row are positioned between long single cones of the neighboring rows (Fadool, 2003; Larison and Bremiller, 1990). In addition to morphology, individual types of photoreceptors are uniquely characterized by spectral sensitivities and visual pigment expression. Long single cones express blue light-sensitive opsin; short single cones, ultraviolet (UV)-sensitive opsin; double cones, red light-sensitive opsin and green opsins; whereas rods express rod opsin (Hisatomi *et al.*, 1996; Raymond *et al.*, 1993). Recent studies uncovered that the number of opsin genes exceeds the number of different photoreceptor types. Two and four independent loci encode red and green opsins, respectively. All opsin loci appear to be transcribed and their expression levels vary considerably even among closely related genes (Chinen *et al.*, 2003). The spatio-temporal expression patterns of these genes have not yet been described in detail, and it is thus unclear whether any of them are expressed simultaneously in the same cell type.

C. Development of Retinotectal Projection

The neuronal network of the retina is largely self-contained. The only retinal neurons that send their projections outside are ganglion cells. Their axons navigate through the midline of the ventral diencephalon into the dorsal part of the midbrain, the optic tectum. The ganglion cells produce axonal processes shortly after the final mitosis, already when they are migrating toward the vitreal surface (Bodick and Levinthal, 1980). The projections proceed toward the inner surface of the retina and subsequently along the inner limiting membrane toward the optic nerve head. In zebrafish, the first ganglion cell axons exit the eye between 34 and 36 hpf and navigate along the optic stalk and through the ventral region of the brain towards the midline (Burrill and Easter, 1995; Macdonald and Wilson, 1997). At about 2 dpf, the zebrafish optic nerve contains approximately 1800 axons at the exit point from the retina (Bodick and Levinthal, 1980). Cross sections near the nerve head reveal a crescent-shaped optic nerve. Axons of centrally located ganglion cells occupy the outside (dorsal) surface of the crescent; whereas the axons of more peripheral (younger) cells localize to the inside (ventral) surface. With the exception of the axonal trajectories of cells separated by the choroid fissure, axons of neighboring ganglion cells travel together in the optic nerve (Bodick and Levinthal, 1980). In addition to ganglion cell axons, the optic nerve contains retinopetal projections. These appear after 5 dpf and originate in the nucleus olfactoretinalis of the rostral telencephalon (Burrill and

Easter, 1994). After crossing the midline, the axonal projections of ganglion cells split into the dorsal and ventral branches of the optic tract. The ventral branch contains mostly axons of the dorsal retinal ganglion cells, the dorsal branch mostly of the ventral cells (Baier *et al.*, 1996). The growth cones of retinal ganglion cells first enter the optic tectum between 46 and 48 hpf. In addition to the optic tectum, retinal axons innervate nine other much smaller targets in the zebrafish brain (Burrill and Easter, 1994).

Spatial relationships between individual ganglion cells in the retina are precisely reproduced by their projections in the tectum. The exactitude of this pattern has long fascinated biologists and has been a subject of intensive research in many vertebrate species (Drescher *et al.*, 1997; Fraser, 1992; Sanes, 1993). The spatial coordinates of the retina and the tectum are reversed. The ventral-nasal ganglion cells of the zebrafish retina project to the dorsal-posterior optic tectum; whereas the dorsal-temporal cells innervate the ventral-anterior tectum (Stuermer, 1988; Karlstrom *et al.*, 1996; Trowe *et al.*, 1996). By 72 hpf, axons from all quadrants of the retina are in contact with their target territories in the optic tectum.

In summary, development of the zebrafish retina proceeds at a rapid pace. By the end of day 3, all retinal cell classes have been generated and are organized in distinct layers (Fig. 2B), photoreceptor cells have developed outer segments, and ganglion cell axons have innervated the optic tectum. It is also about this time that the zebrafish visual system becomes functional (Clark, 1981; Easter and Nicola, 1996). The brevity of eye morphogenesis and retinal neurogenesis is a major advantage offered by the zebrafish eye as a model system.

III. Analysis of Wild-Type and Mutant Zebrafish Visual Systems

Diverse research approaches have been used to study the zebrafish retina. This chapter provides an overview of the available methods. While some techniques are described in detail, the majority are discussed briefly only because of space constraints. References to more comprehensive discussions are provided. Where applicable, we refer to other chapters of this volume as the source of more complete information. Table I lists some of the most important techniques currently available for the analysis of the zebrafish retina.

Observations of retinal development in the zebrafish embryo after 30 hpf are hampered by pigmentation of the retinal pigmented epithelium. In immunohisto-chemical experiments, for example, the staining pattern is not accessible to visual inspection in whole embryos unless they are sectioned or their pigmentation is inhibited. To inhibit pigmentation, developing zebrafish embryos are raised in media containing 0.003% 1-phenyl-2-thiourea (PTU). In the presence of PTU, however, zebrafish embryonic development does not proceed in the entirely normal way. Starting between 2 and 3 dpf, embryogenesis is somewhat delayed, hatching is inhibited, and pectoral fins are abnormal. Appropriate controls have to be included to account for these deviations from normal embryogenesis. An

Table I
Techniques Available to Study the Zebrafish Retina and their Sources/Examples

Protocol	Goal	Sources/Examples of use*
Histological Analysis		
Electron microscopy	Evaluation of phenotype on a subcellular level	Allwardt et al., 2001; Doerre and Malicki, 2002; Kimmel et al., 1981
Light microscopy	Evaluation of phenotype on a cellular level	Malicki et al., 1996; Schmitt and Dowling, 1994
Molecular Marker Analysis		
Antibody staining (whole mount)	Determination of expression pattern on protein level	Schmitt and Dowling, 1996
Antibody staining (sections)	Determination of expression pattern on protein level	Pujic and Malicki, 2001; Wei and Malicki, 2002
In situ hybridization–double labeling	Parallel determination of two transcript expression patterns	Hauptmann and Gerster, 1994; Jowett, 2001; Jowett and Lettice, 1994; Strahle et al., 1994
In situ hybridization–frozen sections	Determination of expression pattern on transcript level	Barthel and Raymond, 1993; Hisatomi et al., 1996
In situ hybridization–whole mount	Determination of expression pattern on transcript level	Oxtoby and Jowett, 1993; Pujic and Malicki, 2001
Gene Function Analysis		
Implantation	Test of function for a factor (most often diffusible) via the implantation of a bead saturated with this substance	Hyatt et al., 1996
Morpholino knockdown	Test of gene function based on anti-sense inhibition of its activity	Nasevicius and Ekker, 2000; Tsujikawa and Malicki, 2004
Overexpression (DNA injections)	Test of gene function based on enhancement of its activity through DNA injections	reviewed in Malicki et al., 2002; Yoshida and Mishina, 2003

Technique	Description	References
Overexpression (light-mediated RNA/DNA uncaging)	Identification of gene function through enhancement of its activity in selected tissues at specific developmental stages	Ando and Okamoto, 2003; Ando et al., 2001
Overexpression (RNA injections)	Test of gene function based on enhancement of its activity through RNA injections	Macdonald et al., 1995; reviewed in Malicki et al., 2002
Overexpression (UAS-GAL4 system)	Test of gene function in selected tissues using stable transgenic lines	Scheer and Campos-Ortega, 1999; Scheer et al., 2001, 2002
TILLING (Targeting Induced Local Lesions in Genomes)	Identification of chemically induced mutant alleles in a specific genetic locus	Colbert et al., 2001; Wienholds et al., 2002
Embryological Techniques		
Cell labeling (caged fluorophore)	Fate determination for a specific group of cells	Take-uchi et al., 2003
Cell labeling (iontophoretic)	Determination of morphogenetic movements or cell lineage relationships	Li et al., 2000; Varga et al., 1999; Woo and Fraser, 1995
Cell labeling (lipophilic tracers)	Analysis of ganglion cell development (ex. retinotectal projection)	Baier et al., 1996; Malicki and Driever, 1999
Cell labeling (fluorescent protein transgenes)	Determination of cell fate and fine differentiation features in living animals	Fadool, 2003; Neumann and Nuesslein-Volhard, 2000; Yoshida et al., 2002
Mitotic activity detection (BrdU)	Identification of mitotically active cell populations; birthdating	Hu and Easter, 1999; Larison and Bremiller, 1990
Mitotic activity detection (tritiated thymidine)	Identification of mitotically active cell populations; birthdating	Nawrocki, 1985
Tissue ablation	Functional test for a field of cells via their removal by surgical means	Masai et al., 2000
Transplantation (whole eye)	Test whether a defect (in axonal navigation, for example) originates within or outside the eye	Fricke et al., 2001
Transplantation (fragment of tissue)	Functional test for a field of cells via their transplantation to an ectopic position by surgical means	Masai et al., 2000
Transplantation (blastomere)	Test of cell autonomy of a mutant phenotype by generating a genetically mosaic embryo	Ho and Kane, 1990; Jensen et al., 2001; Malicki and Driever, 1999
Behavioral Tests		
Optokinetic response	Test of vision based on eye movements; allows for evaluation of visual acuity	Brockerhoff et al., 1995; Clark, 1981; Neuhauss et al., 1999
Optomotor response	Test of vision based on swimming behavior	Clark, 1981; Neuhauss et al., 1999
Startle response	Simple test of vision based on swimming behavior	Easter and Nicola, 1996

(continues)

Table I (*continued*)

Protocol	Goal	Sources/Examples of use*
Electrophysiological Tests		
ERG	Test of retinal function based on the detection of electrical activity of retinal neurons and glia	Brockerhoff et al., 1995; Neuhauss et al., 1999
Screening Approaches		
Behavioral	Detection of mutant phenotypes using behavioral tests	Brockerhoff et al., 1995; Clark, 1981; Neuhauss et al., 1999
Histological	Detection of mutant phenotypes via analysis of histological analysis of sections	Mohideen et al., 2003
Marker/tracer labeling	Detection of mutant phenotypes via staining with antibodies, RNA probes, or lipophilic tracers	Baier et al., 1996; Guo et al., 1999
Morphological	Detection of mutant phenotypes by morphological criteria	Malicki et al., 1996
Transgene guided	Detection of mutant phenotypes in transgenic lines expressing fluorescent proteins in specific cell populations	reviewed in Tsujikawa and Malicki, 2003

*In this table, we primarily cite experiments performed on the retina. Only where references to work on the eye are not available, we refer to studies of other organs. Most forward genetic approaches such as mutagenesis, mapping, and positional cloning methods do not contain visual system-specific features and thus are not listed in this table. These approaches are discussed in depth in other sections of this volume. Entries are listed alphabetically within each section of the table.

alternative to using PTU is to conduct experiments on pigmentation-deficient animals. Mutations of interest can be crossed into a pigmentation deficient strain. Several loci affecting zebrafish pigmentation, including *albino*, have been described and can be used for this purpose (Kelsh *et al.*, 1996; Streisinger *et al.*, 1986). As crossing a mutation of interest into a pigmentation-deficient background takes two generations, this approach is time-consuming.

A. Histological Analysis

It is safe to assume that a major goal of future eye research in zebrafish will be to characterize new generations of mutant phenotypes. Even in the most comprehensive of genetic screens performed in zebrafish so far, the number of multiple hits per locus was low, indicating that many more genes will be discovered before saturation is achieved (Driever *et al.*, 1996; Golling *et al.*, 2002; Haffter *et al.*, 1996). Therefore, almost certainly, future genetic screens will enrich the already impressive collection of zebrafish eye mutants even further. After morphological description, the first and the simplest step in the analysis of a new mutant phenotype is histological analysis. It allows one to evaluate how a mutation influences the major cell classes in the retina. Because of the exquisitely precise organization of the zebrafish retinal neurons, histological analysis is frequently informative. Plastic sections offer very good tissue preservation for histological analysis. Both epoxy (epon, araldite) and methacrylate (JB4) resins are available for tissue embedding (Polysciences Inc., Warrington, PA). Epoxy resins can be used both for light and electron microscopy. Several fixation methods suitable for plastic sections are routinely used (Li *et al.*, 2000b; Malicki *et al.*, 1996). For light microscopy, plastic sections are frequently prepared at 1-8 μm thickness and stained with an aqueous solution of 1% methylene blue and 1% azure II (Humphrey and Pittman, 1974; Malicki *et al.*, 1996; Schmitt and Dowling, 1999).

Following transmitted light microscopy, histological analysis of mutant phenotypes can be performed at a higher resolution using electron microscopy. This allows one to inspect morphological details of subcellular structures, such as the photoreceptor outer segments, cell junctions, cilia, synaptic ribbons, mitochondria, and many other organelles (Schmitt and Dowling, 1999). These cellular features frequently offer insight into the nature of a mutant phenotype. Electron microscopy can be used in combination with diaminobenzidine (DAB) labeling of specific cell populations. Oxidation of DAB results in the formation of polymers, which are chelated with osmium tetroxide and subsequently observed in the electron microscope (Hanker, 1979). Before electron-microscopic analysis, specific cells can be selectively DAB labeled using several approaches: photo conversion (Burrill and Easter, 1995) antibody staining combined with peroxidase detection (Metcalfe *et al.*, 1990) or retrograde labeling with horseradish peroxidase (HRP) (Metcalfe, 1985).

Details of cell morphology can also be studied using lipophilic carbocyanine dyes, which label cell membranes: DiI and DiO (Honig and Hume, 1986, 1989).

In the retina, these are especially useful in the analysis of ganglion cells. Carbo-cyanine dyes can be used as anterograde, as well as retrograde, tracers. When applied to the retina, DiI and DiO allow one to trace the retinotectal projections (Baier et al., 1996). When applied to the optic tectum or the optic tract, they are used to determine the position of ganglion cell perikarya (Burrill and Easter, 1995; Malicki and Driever, 1999). Since DiI and DiO have different emission spectra, they can be used simultaneously to label two different populations of cells (Baier et al., 1996).

Transgenes that express green fluorescent protein (GFP) and GFP derivatives are also used in the study of cell morphology. Using appropriate promoters, GFP expression can be directed to specific cell populations. Stable transgenic zebrafish lines have been generated that express GFP in ganglion, amacrine, or photorecep-tor cells (Fadool, 2003; Kay et al., 2004; Neumann and Nuesslein-Volhard, 2000; Perkins et al., 2002; Takechi et al., 2003). For photoreceptor cells, transgenic lines that express fluorescent proteins in outer segments or throughout the cell body are available, as well as lines that express GFP in rods or short single cones specifically. These transgenic strains can be used to monitor the morphogenesis of specific photoreceptor cell features, such as outer segments or synaptic termini. Additional dimensions of morphological analysis are opened by membrane-targeted GFP transgenes. These allow one to visualize fine dendritic trees of interneurons (Kay et al., 2004), an informative, as much as esthetically satisfying, endeavor.

B. Molecular Markers

In addition to histology, a variety of molecular markers are used to study the zebrafish retina during and after neurogenesis. Endogenous transcripts and poly-peptides are among the most frequently used markers, although smaller molecules, such as neurotransmitters and neuropeptides, can also be used. During early embryogenesis, the analysis of marker distribution allows one to determine wheth-er the eye field is specified correctly. Several ribonucleic acid (RNA) probes are available to visualize the optic lobe during embryogenesis (Table II). Some of them label all cells of the optic lobe uniformly, while others can be used to monitor the optic stalk area (Table II). After the completion of neurogenesis, cell class-specific markers are used to determine whether particular cell populations are specified and occupy correct positions. Some of these markers are listed in Table II. Many transcript and protein detection methods have been described. Detailed protocols for some of these are available and referenced in Table I.

Antibody staining experiments can be performed in several ways. Staining of whole embryos is the easiest. Many antibodies produce high background in whole-mount experiments, however, and the eye pigmentation needs to be eliminated after 30 hpf as already described. At later stages of development, tissue penetra-tion may become an additional problem. This can be alleviated by increasing detergent concentration above the standard level of 0.5% (2.5% Triton in both blocking and staining solution works well for anti Pax-2 antibody, see Riley et al.,

Table II
Markers Available to Study Neurogenesis in the Zebrafish Retina

Name	Type	Expression pattern	References[a]/Sources
Optic Lobe, Optic Stalk Markers			
efna5b (L4)	RNA probe	Entire OL (12 hpf); nasal neuroepithelium (24 hpf); nasal ganglion cells (≤72 hpf)	Brennan et al., 1997
pax2a (pax 2)	RNA probe & Ab (poly)	Nasal retina, optic stalk (≤24 hpf); ON (2 dpf)	Kikuchi et al., 1997; Macdonald et al., 1997; Covance PRB-276P
rx1 (zrx1)	RNA probe	Anterior neural keel, optic primordia (≤11 hpf)	Chuang et al., 1999; Pujic and Malicki, 2001
rx2 (zrx2)	RNA probe	Anterior neural keel, optic primordia (≤11 hpf)	Chuang et al., 1999; Pujic and Malicki, 2001
rx3 (zrx3)	RNA probe	Anterior neural plate (≤9 hpf); optic primordia (≤12 hpf)	Chuang et al., 1999; Pujic and Malicki, 2001
six3a (six3)	RNA probe	Neural keel, optic primordia (≤11 hpf)	Pujic and Malicki, 2001; Seo et al., 1998
six3b (six6)	RNA probe	Anterior neural keel, optic primordia (≤11 hpf)	Pujic and Malicki, 2001; Seo et al., 1998
vax2	RNA probe	Optic stalk (≤15 hpf); ventral retina (≤18 hpf)	Take-uchi et al., 2003
Ganglion Cell Markers			
alcam[b] (Neurolin)	RNA probe, Ab (mono & poly)	Ganglion cells (28 hpf, RNA; ≤32 hpf protein)	Fashena and Westerfield, 1999; Laessing and Stuermer, 1996; Laessing et al., 1994; Zn-5/Zn-8 DSHB and ZIRC
L3	RNA probe	Ganglion cells (30 hpf)	Brennan et al., 1997
Amacrine Cell Markers			
Calretinin	Ab (poly)	Subset in INL and GCL, IPL (≤5 dpf)	Malicki lab, unpublished data; Chemicon, catalog number AB149
Choline acetyltransferase	Ab (poly)	Subset in INL and GCL, IPL (≤5 dpf)	Malicki lab, unpublished data; Yazulla and Studholme, 2001; Chemicon, catalog number AB144P
GABA	Ab (poly)	Subset in INL and GCL, IPL (2.5 dpf); ON (2 dpf)	Sandell et al., 1994; Chemicon, catalog number AB131; Sigma, catalog number A2052
GAD67	Ab (poly)	Subset in INL and few in GCL, IPL (≤7 dpf)	Connaughton et al., 1999; Kay et al., 2001; Chemicon, catalog number AB108
Hu C/D	Ab (mono)	INL and GCL (≤3 dpf)	Kay et al., 2001; Link et al., 2000; Molecular Probes, catalog number A21271

(continues)

Table II (*continued*)

Name	Type	Expression pattern	References[a]/Sources
Islet-1	Ab (mono)	GCL (36 hpf); INL and GCL (48–72 hpf)	Ericson, 1992; Korzh et al., 1993; Link et al., 2000; *DSHB*
Neuro-peptide Y	Ab (poly)	Subset in INL, IPL (≤4 dpf)	Malicki lab, unpublished data; *ImmunoStar, catalog number 22940*
Parvalbumin	Ab (mono)	Subset in INL and GCL, IPL (≤3 dpf)	Malicki et al., 2003; *Chemicon, catalog number MAB1572*
pax6a (pax6.1)	RNA probe Ab (poly)	Neuroepithelium (12–34 hpf); INL and GCL (2 dpf); INL (5 dpf)	Hitchcock et al., 1996; Macdonald and Wilson, 1997
Serotonin	Ab (poly)	Subset in INL (≤5 dpf)	Malicki lab, unpublished data; *Sigma, catalog number S5545*
Somatostatin	Ab (poly)	Subset in INL (≤5 dpf)	Malicki lab, unpublished data; *ImmunoStar, catalog number 20067*
Substance P	Ab (mono)	Subset in INL (≤5 dpf)	Malicki lab, unpublished data; *AbCam, catalog number AB6338*
Tyrosine hydroxylase	Ab (mono)	Subset in INL (3–3.5 dpf)	Biehlmaier et al., 2003; Pujic and Malicki, 2001; *ImmunoStar, catalog number 22941; Chemicon, catalog number MAB318*
Bipolar Cell Markers			
vsx1	RNA probe	Neuroepithelium (31 hpf); outer INL (50 hpf)	Passini et al., 1997
vsx2	RNA probe	Neuroepithelium (24 hpf); primarily or exclusively in bipolar cells (50 hpf)	Passini et al., 1997
Protein kinase C $\beta1$	Ab (poly)	IPL, OPL (2.5 dpf); bipolar cells somata (≤4 dpf)	Biehlmaier et al., 2003; Kay et al., 2001; *Santa Cruz, catalog number sc-209*
Photoreceptor Markers			
Blue opsin[c]	RNA probe	Blue cones (52 hpf)	Chinen et al., 2003; Raymond et al., 1995; Vihtelic et al., 1999
Blue opsin	Ab (poly)	Blue cones (≤3 dpf)	Doerre and Malicki, 2001; Vihtelic et al., 1999
Green opsin	Ab (poly)	Green cones (≤3 dpf)	Doerre and Malicki, 2001; Vihtelic et al., 1999
Green opsin[d]	RNA probe	Green cones (≤3 dpf)	Chinen et al., 2003; Vihtelic et al., 1999
Red opsin[c]	RNA probe	Red cones (52 hpf)	Chinen et al., 2003; Raymond et al., 1995

Marker	Type	Expression	References
Red opsin	Ab (poly)	Red cones (≤3 dpf)	Doerre and Malicki, 2001; Vihtelic et al., 1999
Rod opsin[c]	RNA probe	Rods (50 hpf)	Chinen et al., 2003; Raymond et al., 1995
Rod opsin	Ab (poly)	Rods (≤3 dpf)	Doerre and Malicki, 2001; Vihtelic et al., 1999
UV opsin	RNA probe	UV cones (56 hpf)	Hisatomi et al., 1996; Takechi et al., 2003
UV opsin	Ab (poly)	UV cones (≤3 dpf)	Doerre and Malicki, 2001; Vihtelic et al., 1999
Zpr1 (Fret 43)	Ab (mono)	Double cones in larvae (48 hpf); double cones & bipolar cell subpopulation in the adult	Larison and Bremiller, 1990; ZIRC
Zpr3 (Fret11)	Ab (mono)	Rods (50 hpf)	Schmitt and Dowling, 1996; ZIRC
Zs-4	Ab (mono)	Rod inner segments (adult), onset unknown	Vihtelic and Hyde, 2000; ZIRC
Muller Glia Markers			
cahz (carbonic anhydrase)	RNA probe Ab (poly)	Mueller glia (≤4 dpf)	Peterson et al., 1997, 2001
GFAP	Ab (poly)	Mueller glia (5 dpf)	Malicki Lab, unpublished data; DAKO, catalog number Z 0034
Glutamine synthetase	Ab (poly)	Mueller glia (60 hpf)	Peterson et al., 2001
Plexiform Layer Markers			
Phalloidin	Fungal alkaloid	IPL, OPL, ON (≤60 hpf)	Malicki et al., 2003; Molecular Probes, catalog number A-12379
Snap-25	Ab (poly)	OPL, IPL (≤2.5 dpf)	Biehlmaier et al., 2003; StressGen, catalog number VAP-SV002
SV2	Ab (mono)	IPL, OPL (≤2.5 dpf)	Biehlmaier et al., 2003; DSHB
Syntaxin-3	Ab (poly)	OPL (2.5 dpf); faint IPL (5 dpf)	Biehlmaier et al., 2003; Alamone labs, catalog number ANR-005

[a] When references to work performed on zebrafish not available, experiments on related fish species are cited.

[b] Zn5 and Zn8 antibodies both recognize neurolin (Kawahara et al., 2002).

[c] Transcript expression onset was estimated by using goldfish probes (Raymond et al., 1995). Two red opsin genes exist in the zebrafish genome (Chinen et al., 2003).

[d] Four green opsin genes are known to exist in the zebrafish genome (Chinen et al., 2003). As their transcript expression patterns have not been yet reported, we provide a predicted timing of transcription onset, based on protein expression data.

Approximate time of expression onset is indicated in parenthesis. Names of markers are listed alphabetically within each section (previous names are indicated in parenthesis). Sources of commercially available reagents are indicated in parenthesis. DSHB = Developmental Studies Hybridoma Bank (http://www.uiowa.edu/~dshbwww/); ZIRC = Zebrafish International Resource Center (http://zfin.org/zirc/home/guide.php); dpf = days post fertilization; hpf = hours post fertilization; mono=monoclonal; poly=polyclonal; GCL = ganglion cell layer; INL = inner nuclear layer; IPL = inner plexiform layer; OPL = outer plexiform layer; OL = optic lobe; ON = optic nerve.

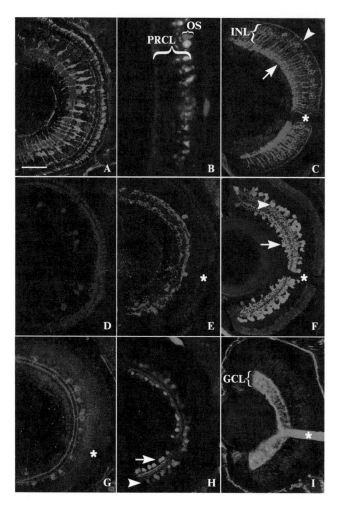

Fig. 3 Transverse sections through the center of the zebrafish eye reveal several major retinal cell classes and their subpopulations. (A) Anti-rod opsin antibody detects rod photoreceptor outer segments (*red*), which are fairly uniformly distributed throughout the outer perimeter of the retina by 5 dpf. On the same section, an antibody to carbonic anhydrase labels cell bodies of Mueller glia (*green*) in the INL, as well as their radially oriented processes. (B) A higher magnification of the photoreceptor cell layer shows the distribution of rod opsin (*red signal*) and UV opsin (*green signal*) in the outer segments (OS) of rods and short single cones, respectively. (C) A subpopulation of bipolar cells is detected using antibody directed to Protein Kinase Cβ (PKC). While cell bodies of PKC-positive bipolar neurons are situated in the central region of the INL, their processes travel radially into the inner (*arrow*) and outer (*arrowhead*) plexiform layers, where they make synaptic connections. (D) Tyrosine hydroxylase-positive interplexiform cells are relatively sparse in the larval retina. (E) Similarly, the distribution of Neuropeptide Y is limited to only a few cells per section. (F) The distribution of GABA, a major inhibitory neurotransmitter. GABA is largely found in amacrine neurons in the INL (*arrowhead*), although some GABA-positive cells are also present in the GCL (*arrow*). (G) Choline acetyltransferase, an enzyme of acetylcholine biosynthetic pathway, is restricted to a relatively small amacrine cell subpopulation. (H) Antibodies directed to a calcium-binding protein, parvalbumin, recognize another fairly large subpopulation of amacrine cells in the INL (*green, arrowhead*). Some parvalbumin-positive

1999), or by enzymatic digestion of embryos (for example collagenase treatment, see Doerre and Malicki, 2002). When background or penetration is a problem, a useful alternative to whole-mounts are frozen or paraffin sections. Confocal microscopy applied to the analysis of sections allows one to reduce background even further.

For cryosectioning, embryos should be fixed as appropriate for a particular antigen and infiltrated in sucrose for cryoprotection. While, for many antigens, simple overnight fixation in 4% paraformaldehyde (PFA) at 4 °C is sufficient, some others require special treatments. Anti-gamma-aminobutyric acid (GABA) staining of amacrine cells, for example, requires fixation in both glutaraldehyde and paraformaldehyde (2% each; see Sandell *et al.*, 1994; Fig. 3F). Proper orientation of fixed specimen can be accomplished in molds prepared from Eppendorf tubes by cutting them transversely into approximately 3- to 4-mm wide rings. These are then placed flat on a glass slide and filled with embedding medium (Tissue-Tek, Sakura Finetek Inc., Torrance, CA). Embryos are placed in the medium, oriented with a probe, and transferred into a cryostat chamber that is cooled to −20 °C. Once the medium solidifies, the rings are removed with a razor blade.

Antibody staining can be effectively performed on 15- to 30-μm sections and analyzed by confocal microscopy. For conventional microscopy, thinner sections may be desired. When modified infiltration and embedding protocols are applied, 3-μm sections of the zebrafish embryos can be prepared and analyzed using a conventional microscope equipped with an UV light source (Barthel and Raymond, 1990). Some antigens require the application of additional steps during staining protocols. Antigen retrieval procedure is necessary, for example, in the case of anti-serotonin or anti-choline Acetyltransferase staining (Fig. 3G and H; Jiao *et al.*, 1999). While staining for the presence of these antigens, sections are immersed in near-boiling solution of 10 mM sodium citrate for 10 min. before the application of blocking solution. Acetone treatment is required for some anti-gamma-tubulin antibodies (Pujic and Malicki, 2001).

In some cases, such as the detection of GABA, antibody staining can be performed on plastic sections. Both epoxy (Epon-812, Electron Microscopy Sciences Inc., Hatfield, PA) and methacrylate (JB-4, Polysciences Inc.) resins can be used as the embedding medium. Plastic sections preserve tissue morphology very well and consequently are more informative than frozen sections. In the GABA protocol, primary antibody can be detected using avidin-HRP conjugate (Vector Laboratories Inc., Burlingame, CA) or a fluorophore-conjugated secondary antibody (Malicki and Driever, 1999; Sandell *et al.*, 1994).

---

cells localize also to the GCL and most likely represent displaced amacrine neurons (*arrow*). By contrast, serotonin-positive neurons (*red*) are exclusively found in the INL. (I) Ganglion cells stain with the Zn-8 antibody directed to neurolin, a cell surface antigen (Fashena and Westerfield, 1999). In addition to neuronal somata, strong Zn8 staining exists in the optic nerve (*asterisk*). In all panels lens is left, dorsal is up. A through H show the retina at 5 dpf. I shows the retina at 3 dpf. Asterisks indicate the optic nerve. Scale bar equals 50 μm in (A), and (C to I) and 10 μm in (B). GCL: ganglion cell layer; INL: inner nuclear layer; OS: outer segments; PRCL: photoreceptor cell layer. Panels D, G, and H reprinted from Pujic and Malicki (2004) with permission from Elsevier. (See Color Insert.)

In situ hybridization with most RNA probes is conveniently done on whole embryos according to standard protocols (Oxtoby and Jowett, 1993). After hybridization, embryos are dehydrated in ethanol and embedded in plastic as described above for antibody staining (Pujic and Malicki, 2001). Expression patterns are subsequently analyzed on 1- to 5-μm sections. Several *in situ* protocols permit one to monitor the expression of two genes simultaneously (Jowett, 1999, 2001; Table II). In the experiment shown in Fig. 4B, expression patterns of two opsins are detected simultaneously using two different chromogenic substrates of alkaline phosphatase (AP) (Hauptmann and Gerster, 1994). *In situ* hybridization can be also combined with antibody staining (Novak and Ribera, 2003; Prince *et al.*, 1998). In embryos older than 5 dpf, *in situ* reagents sometimes do not penetrate to the center of the retina. In such cases, staining can be performed more successfully on sections (Hisatomi *et al.*, 1996).

C. Analysis of Gene Function

A series of mutant alleles of varying severity is arguably the most informative tool of gene function analysis. Although a great variety of mutant lines have been identified, for many loci, chemically induced mutant alleles are not yet available. In these cases, other approaches need to be applied to study gene function. In this section, we briefly discuss the advantages and disadvantages of different loss-of-function and gain-of-function approaches in the context of the zebrafish visual system. In addition, we provide references to more comprehensive discussions of each approach (Table II).

In the absence of chemically induced loss-of-function alleles, antisense-based interference is by far the most common way to obtain information about gene function in the zebrafish embryo (Nasevicius and Ekker, 2000). The reason for this popularity is low cost and low labor expense involved in its use. Although morpholino-modified oligonucleotides have been shown to reproduce chemically induced mutant phenotypes quite well, their use suffers from two main disadvantages. First, they become progressively less effective as development proceeds, presumably because of degradation. Second, some morpholinos produce nonspecific toxicity, which needs to be distinguished from specific features of a morpholino-induced phenotype. Morpholino oligos can be used to interfere either with translation initiation or with splicing. Of note, the efficiency of splice-site morpholinos can be monitored by Reverse-Transcription-Polymerase Chain Reaction (RT-PCR; Draper *et al.*, 2001; Tsujikawa and Malicki, 2004). In general, nearly all splice-site morpholinos reduce the level of wild-type transcript below the level of RT-PCR detection throughout the first 36 h of development, although some have been reported to remain active until 3 or even 5 dpf (Tsujikawa and Malicki, 2004). Most morpholinos are thus sufficient to interfere with genetic pathways involved in retinal neurogenesis but not to study later developmental events or retinal function. Some help designing morpholinos can be obtained from their manufacturer (Gene Tools LLC, Philomath, OR). Detailed protocols for the

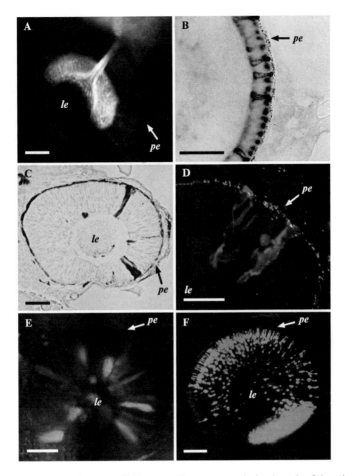

Fig. 4 Examples of techniques available to study neurogenesis in the zebrafish retina. (A) DiI incorporation into the optic tectum retrogradely labels the optic nerve and ganglion cell somata. (B) A transverse plastic section through the zebrafish retina at 3 dpf. *In situ* mRNA hybridization using two probes, each targeted to a different opsin transcript and detected using a different enzymatic reaction, visualizes two types of photoreceptor cells. (C) A plastic section through a genetically mosaic retina at approximately 30 hpf. Biotinylated dextran labeled donor-derived cells incorporate into retinal neuroepithelial sheet of a host embryo and can be detected using HRP staining (*brown precipitate*). (D) A transverse cryosection through a genetically mosaic zebrafish eye at 36 hpf. In this case, donor-derived clones of neuroepithelial cells are detected with fluorophore-conjugated avidin (*red*). The apical surface of the neuroepithelial sheet is visualized with anti-γ-tubulin antibody, which stains centrosomes (*green*). (E) GPF expression in the eye of a zebrafish embryo after injection of a DNA construct containing the GFP gene under the control of a heat-shock promoter. The transgene is expressed in only a small subpopulation of cells. (F) A confocal z-series through the eye of a living transgenic zebrafish, carrying a GFP transgene under the control of the rod opsin promoter (Fadool, 2003). Bright signal is present in rod photoreceptor cells (approximately 3 dpf). Scale bar, 50 μm. pe: pigmented epithelium; le: lens. Panel E reprinted from Malicki *et al.* (2002) with permission from Elsevier. (See Color Insert.)

use of morpholinos, including their sequence requirements, preferable target sites, and injection protocols have been described.

A powerful alternative to the use of morpholinos is Targeting Induced Local Lesions in Genomes (TILLING; Colbert *et al.*, 2001; Wienholds *et al.*, 2002) This approach, essentially a combination of chemical mutagenesis and a PCR-based mutation detection protocol, produces series of variable strength alleles for the target locus. Its main disadvantage is the vast amount of preparatory work that needs to be done to initiate these experiments. A particularly labor-intensive step is the collection of thousands of sperm and DNA samples from mutagenized males. Due to such limitations, TILLING experiments need to be performed by core facilities, which serve a group of laboratories or the entire research community. A detailed discussion of this approach is provided in chapters 5 and 16 of this volume.

To obtain a comprehensive understanding of gene function, one often needs to supplement loss-of-function analysis with overexpression data. In the simplest scenario, this can be accomplished in zebrafish by RNA or DNA injections into the embryo. Several variants of this procedure exist, each with unique advantages and drawbacks (reviewed in Malicki *et al.*, 2002). A frequent limitation of overexpression studies is the pleiotropy of mutant phenotypes: for many loci, early embryonic phenotypes are so severe that they preclude the analysis of late developmental processes, such as retinal neurogenesis. Several experimental tools have been developed to overcome this problem, including the use of heat-shock promoters, the GAL4-Upstream Activator Sequence (UAS) overexpression system, and caged nucleic acids. Similar to invertebrate model systems, the use of heat-shock-induced expression in zebrafish relies on the hsp 70 promoter (Halloran *et al.*, 2000). An interesting variant of this protocol involves the activation of a heat-shock promoter-driven transgene in a small group of living cells by heating them gently with a laser beam, which provides both temporal and spatial control of overexpression pattern (Halloran *et al.*, 2000).

GAL4-UAS system is another method to achieve spatial control of gene expression. Modeled after *Drosophila* (Brand and Perrimon, 1993), the GAL4-UAS overexpression approach takes advantage of two transgenic strains. The activator strain expresses the GAL4 transcriptional activator in a desired subset of tissues, while the effector strain carries the gene of interest under the control of a GAL4 responsive promoter (UAS). The effector transgene is activated by crossing its carrier strain to a line that carries the activator transgene (Scheer *et al.*, 2002). This system has been used in the context of the retina to study *notch* function (Scheer *et al.*, 2001). Finally an interesting method to control gene overexpression patterns is the use of Bhc-caged nucleic acids (Ando *et al.*, 2001). In this approach, embryos are injected with an inactive form of an overexpression construct, which is then later activated in a selected tissue using UV illumination. Both RNA and DNA templates can be used to produce overexpression in this approach.

D. Embryological Techniques

A combination of genetic and embryological approaches is necessary to develop a good understanding of many biological processes. Most embryonic manipulations performed in the zebrafish embryo fall into two categories: cell labeling or transplantation experiments. The first group of techniques is mostly used to monitor cell fate or to determine lineage relationships in wild-type and mutant animals; whereas the second primarily serves to investigate interactions between cells or tissues.

The best established cell labeling approach is iontophoresis. This technique was applied in zebrafish in numerous cell fate studies (Collazo *et al.*, 1994; Devoto *et al.*, 1996; Raible *et al.*, 1992). In the context of visual system development, iontophoretic cell labeling was used to determine cell movements that lead to the formation of the optic primordium (Woo and Fraser, 1995; Varga *et al.*, 1999) and later to study cell rearrangements that accompany optic cup morphogenesis (Li *et al.*, 2000b). Iontophoretic cell labeling has been applied to the study of cell lineage relationships in the developing retina of *Xenopus laevis* (Holt *et al.*, 1988; Wetts and Fraser, 1988). Such studies have not been performed in the zebrafish retina so far, perhaps due to the expectation that they would be unlikely to add much to the results previously obtained in higher vertebrates (Holt *et al.*, 1988; Turner and Cepko, 1987; Turner *et al.*, 1990). Although this may be true, a potentially informative variant of these experiments is to perform them in retinae of mutant animals (Varga *et al.*, 1999).

An alternative cell labeling technique is the activation of caged fluorophores using a laser beam. Caged fluorescein (Molecular Probes, Inc., Eugene, OR) is particularly popular in this type of experiments. This approach has been recently used to study cell fate changes caused by a double knockdown of *vax1* and *vax2* gene function (Take-uchi *et al.*, 2003). Finally, cells of the zebrafish embryo can be labeled using lipophilic dyes or transgenes that express fluorescent proteins (see Histological Analysis in this chapter). These approaches are particularly useful while studying differentiation of specific cell classes. Lipophilic carbocyanine dyes are most frequently used to label axonal trajectories (Baier *et al.*, 1996; Hutson and Chien, 2002). A good example of a transgenic zebrafish line that was used to study cell differentiation is a strain containing an opsin promoter-driven GFP gene (Fadool, 2003).

Transplantation techniques are used to determine cell or tissue interactions. The size of a transplant varies from a small group of cells or even a single cell to the entire organ. In the case of mutations that affect retinotectal projections, it is important to determine whether defects originate in the eye or in brain tissues. This can be accomplished by transplanting the entire optic lobe at 12 hpf and allowing the animals to develop until later stages (Fricke *et al.*, 2001). Smaller size fragments of tissue can be transplanted to document cell-cell signaling events within the optic cup. This approach has been applied to demonstrate inductive properties of the optic stalk tissue (Masai *et al.*, 2000).

Mosaic analysis is a widely used approach that combines genetic and embryological manipulations (Ho and Kane, 1990). The goal of such experiments is to determine the site of the genetic defect responsible for a mutant phenotype. Cell-autonomous phenotypes are caused by gene function defects within the affected cells, while cell-non-autonomous phenotypes are caused by defects in other (frequently neighboring) cells. In contrast to *Drosophila*, zebrafish genetic mosaics are generated by embryological means—blastomere transplantations (Ho and Kane, 1990; Westerfield, 2000). As this technique has been widely used in zebrafish, also in the context of eye development, we provide a more extensive description of how it is applied. In the first step, the donor embryos are labeled at the 2- to 8-cell stage with a tracer. Dextrans conjugated with biotin or a fluorophore are the most commonly used tracers, and frequently a mix of these two tracers is used. Before conducting experiments, it is desirable to purify dextran solution by filtering it through a spin column (Microcon YM-3, Millipore Inc., Billerica, MA). This procedure removes small molecular weight contaminants and thus increases the survival of donor cells. Within a few minutes after injection into the yolk, tracers diffuse throughout the embryo, labeling all blastomeres. Subsequently, starting at about 3 hpf, blastomeres are removed using a glass needle from tracer-labeled donor embryos and transplanted into unlabeled host embryos. The number of transplanted blastomeres usually varies from a few to hundreds, depending on experimental context. One donor embryo is frequently sufficient to supply blastomeres for several hosts. The transplanted blastomeres incorporate into the host embryo and randomly contribute to various tissues including those of experimental interest. To increase the frequency of donor-derived cells in the retina, blastomeres should be transplanted into the animal pole of a host embryo (Moens and Fritz, 1999). Cells in that region will later contribute to eye and brain structures (Woo and Fraser, 1995). Embryos that contain descendants of donor blastomeres in the eye are identified using UV illumination between 24 and 30 hpf, when the retina is only weakly pigmented and contains large radially oriented neuroepithelial cells (see Fig. 4C and D). An elegant control for a cell autonomy test can be generated by transplanting cells from two donor embryos—one wild type, one mutant—into a single host (Ho and Kane, 1990). In such a case, each of the donors has to be labeled with a different tracer. A relatively demanding step of the blastomere transplantation procedure is the preparation of a transplantation needle with an appropriate opening diameter and a sharp, beveled tip, a feature that helps to penetrate the embryo. A transplantation needle preparation method and other technical details of blastomere transplantation protocol have been described (Westerfield, 2000).

Analysis of donor-derived cells in mosaic embryos can proceed in several ways. In the simplest case, donor-derived cells are labeled with a fluorescent tracer only and directly analyzed in whole embryos (Ho and Kane, 1990). Confocal microscopy can be used to achieve this goal. Such analysis is sufficient to provide information about the position and sometimes the morphology of donor-derived cells. When more detailed analysis is necessary, donor-derived cells can be further analyzed on sections. In such cases, donor blastomeres are usually labeled with both fluorophore- and biotin-conjugated dextrans. The fluorophore-conjugated tracer is used to distinguish which embryos contain donor-derived cells in the tissue of interest as described earlier. The biotin-conjugated dextran, on the other hand, is used in detailed analysis at later developmental stages. The HRP-streptavidin version of the ABC kit (Vector Laboratories Inc.) or fluorophore-conjugated avidin (Jackson ImmunoResearch Laboratories Inc., West Grove, PA; Molecular Probes Inc.) can be used to detect biotinylated dextran (see Fig. 4C and D, respectively). HRP detection can be performed in whole mounts and analyzed on plastic sections, as described earlier, for histological analysis. In contrast to that, fluorophore-conjugated tracers are preferably detected after sectioning of frozen tissue, due to fluorophore degradation during plastic embedding procedures. In these experiments, cryosections are prepared as already described for antibody staining. In some experiments, it is desirable to analyze the donor-derived cells for the expression of molecular markers (see Fig. 4D for an example). On frozen sections, avidin detection of donor-derived cells can be combined with antibody staining. Another way to reach this goal is to combine HRP detection of donor-derived cells with *in situ* hybridization or antibody staining (Halpern *et al.*, 1993; Schier *et al.*, 1997). When HRP is used for the detection of donor-derived cells, the resulting reaction product inhibits the detection of the *in situ* probe with alkaline phosphatase (Schier *et al.*, 1997). Because of this, the opposite sequence of enzymatic detection reactions is preferred: *in situ* probe detection first, HRP staining second.

When mosaic analysis is performed in the zebrafish retina at 3 dpf or later, the dilution of a donor-cell tracer can make the interpretation of the results difficult. This is because the descendants of a single transplanted blastomere divide a variable number of times. Therefore, in the donor-derived cells that undergo the highest number of divisions, the tracer dye may be diluted so much that it is no longer detectable. In mosaic animals, such a situation can lead to the appearance of a mutant phenotype or to the rescue of a mutant phenotype in places seemingly not associated with the presence of donor cells, and complicate the interpretation of experimental results. This problem can be sometimes alleviated by increasing the concentration of tracer or in the case of whole-mount experiments by improving the penetration of staining reagents. Similar to antibody staining, collagenase can be used to improve reagent penetration during the detection of donor cells (Doerre and Malicki, 2001). The amount of injected dextran should be increased carefully as concentrations too high are lethal for labeled cells. Mosaic analysis is an important approach that has been used to study many retinal mutants in zebrafish (Doerre and Malicki, 2001, 2002; Goldsmith *et al.*, 2003; Jensen *et al.*, 2001; Link *et al.*, 2000; Malicki and Driever, 1999; Malicki *et al.*, 2003; Pujic and Malicki, 2001; Wei and Malicki, 2002).

Several other embryological techniques are available to study the development of the retina. The level of cell proliferation or the timing of cell-cycle exit (birth date) for the retinal neurons can be determined by H3-thymidine labeling (Nawrocki, 1985) or via 5-bromodeoxyuridine (BrdU) injections into the embryo

(Hu and Easter, 1999). Such studies can be very informative in mutant animals. An excellent example of such an analysis is provided by a study of the *lakritz* mutant, which revealed that the zebrafish *ath5* locus plays a crucial role in the timing of cell cycle exit (Kay *et al.*, 2001).

E. Behavioral Studies

Several vision-dependent behavioral responses have been described in zebrafish larvae and adults: optomotor response (Clark, 1981), optokinetic response (Clark, 1981; Easter and Nicola, 1996), startle response (Easter and Nicola, 1996), photo-taxis (Brockerhoff *et al.*, 1995), escape response (Li and Dowling, 1997), dorsal light reflex (Nicolson *et al.*, 1998). Not surprisingly, larval feeding efficiency also depends on vision (Clark, 1981). While some of these behaviors are already present by 72 hpf, others have been described in adult fish only (for a review, see Neuhauss, 2003). The early onset of vision-dependent behaviors of zebrafish proved to be very useful in genetic screening (as described later in this chapter). The optokinetic response appears to be the most robust and versatile. It is useful both in quick tests of vision and in quantitative estimates of visual acuity. In addition to genetic screens, behavioral tests have recently been used to study the function of the zebrafish optic tectum (Roeser and Baier, 2003).

F. Electrophysiological Analysis

In addition to behavioral tests, measurements of electrical activity in the eye are another, more precise way to evaluate retinal function. Electrical responses of the zebrafish retina can be evaluated by electroretinography (ERG). Similar to other vertebrates, the zebrafish ERG response contains two main waves: a small negative a-wave, which originates from the photoreceptor cells, and a large positive b-wave, which reflects the function of the inner nuclear layer (Dowling, 1987; Makhankov *et al.*, 2004). The goal of an ERG study in zebrafish is no different from that of a similar procedure performed on the human eye. ERG can be used to evaluate the site of retinal defects in mutant animals. Ganglion cell defects do not affect the ERG response (Gnuegge *et al.*, 2001); whereas the alterations of the a-wave or the b-wave suggests a defect in the photoreceptor or in the inner nuclear layers, respectively. The a-wave is relatively small in ERG measurements because of an overlap with the b-wave. To measure the a-wave amplitude, the b-wave can be blocked pharmacologically (Kainz *et al.*, 2003). An additional ERG wave, the d-wave, is produced when longer (approximately 1 sec) flashes of light are used. Referred to as the OFF response, the d-wave is thought to reflect the activity of the OFF-bipolar pathway (Kainz *et al.*, 2003; Makhankov *et al.*, 2004).

Retinal responses are usually elicited using a series of light stimuli that vary in intensity by several orders of magnitude (Allwardt *et al.*, 2001; Kainz *et al.*, 2003). This allows the evaluation of the visual response threshold, a parameter that is sometimes abnormal in mutant animals (Li and Dowling, 1997). Another important variable in ERG measurements is the level of background illumination. ERG measurements can be performed on light-adapted retinae, using background illumination of a constant intensity, or in dark-adapted retinae, which are maintained in total darkness for at least 20 min before measurements (Kainz et al., 2003). Most frequently, recordings are performed on intact anesthetized animals (Makhankov et al., 2004). Alternatively, eyes are gently removed from larvae and bathed in an oxygenated buffer solution. This ensures the oxygen supply to the retina in the absence of blood circulation (Kainz et al., 2003). ERG recordings have become a standard assay to evaluate zebrafish eye mutants (Allwardt et al., 2001; Brockerhoff et al., 1998; Kainz et al., 2003; Makhankov et al., 2004; Neuhauss et al., 1999, 2003).

In addition to ERG, more sophisticated electrophysiological measurements can be used to evaluate zebrafish (mutant) retinae. Ganglion cell function, for example, can be evaluated by recording action potentials from the optic nerve. Such measurements revealed ganglion cell defects in the retinae of *nbb* and *mao* mutants (Gnuegge et al., 2001; Li and Dowling, 2000). Similarly, photoreceptor function has been evaluated by measuring the outer segment currents in isolated cells (Brockerhoff et al., 2003).

IV. Genetic Analysis of the Zebrafish Retina

A. Genetic Screens

The use of zebrafish in genetic studies offers several obvious advantages. The most important of these is the possibility of performing efficient genetic screens. Genetic screening is feasible because adult zebrafish are highly fecund and are easily maintained in large numbers in a fairly small laboratory space. Several screens performed in zebrafish identified many interesting eye phenotypes (Baier et al., 1996; Fadool et al., 1997; Malicki et al., 1996; Neuhauss et al., 1999). As new mutagenesis and screening approaches are being developed, the already impressive collection of zebrafish retinal mutants (Table III) is likely to expand further.

While designing a genetic screen, one has to consider three important variables: the type of mutagen, the design of the breeding scheme, and mutant defect recognition criteria. The majority of screens performed in zebrafish so far involved the use of N-Ethyl-N-nitrosourea (ENU) (Mullins et al., 1994; Solnica-Krezel et al., 1994). This mutagenesis approach is very effective as evidenced by the fact that nearly all mutations that are listed in Table III are ENU induced. A powerful alternative to chemical mutagenesis is insertional retroviral mutagenesis. Although the efficiency of this mutagenesis approach is still lower than that of chemical methods, an obvious advantage of a retroviral mutagen is that it provides means for very rapid identification of mutant genes (Amsterdam et al., 1999; Golling et al., 2002). Retroviral mutagenesis has been recently applied on a large scale to produce hundreds of mutant strains (Golling et al., 2002). The photoreceptor mutant *nrf* is an example of a retinal defect induced using this approach

Table III
Mutations Affecting Zebrafish Eye Development

Locus name	Category and phenotypes *other phenotypes*	Gene/Linkage group	Alleles	References
bozozok (*boz*)	Early patterning MPH: Eye anlage fused *brain, spinal cord*	*dharma*, homeobox transcription factor/LG15	m168	Fekany *et al.*, 1999; Koos and Ho, 1999; Malicki *et al.*, 1996
cyclops (*cyc*)	Early patterning MPH: Eye anlage fused *brain, spinal cord*	*ndr2*, TGF-β related gene/LG12	b16, m101, m122, m294, tf219	Hatta *et al.*, 1991; Rebagliati *et al.*, 1998; Sampath *et al.*, 1998
masterblind (*mbl*)	Early patterning MPH: Optic vesicles absent, eyes not formed MA: Autonomous in the eye *brain, anterior head structures*	*axin1*, Wnt pathway scaffolding factor/LG3	tm13, tm213	Heisenberg *et al.*, 1996, 2001; van de Water *et al.*, 2001
one eyed pinhead (*oep*)	Early patterning MPH: Eye anlage fused *brain, spinal cord*	*oep*, EGF-CFC protein/LG10	m134, tz257, z1, st2	Malicki *et al.*, 1996; Zhang *et al.*, 1998
schmalspur (*sur*)	Early patterning MPH: Eye anlage fused *brain, spinal cord*	*foxh1*, winged-helix transcription factor/LG12	m768, ty68b	Malicki *et al.*, 1996; Pogoda *et al.*, 2000; Sirotkin *et al.*, 2000
silberblick (*slb*)	Early patterning MPH: Eye anlage fused *brain, jaw, spinal cord*	*wnt11*, secreted signaling factor/LG5	tx226, tz216	Heisenberg *et al.*, 1996, 2000
slow muscle omitted (*smu*)	Early patterning MPH: Eye anlage fused HIS: Optic stalk reduced or absent *somites, brain, spinal cord, fins*	*smoh*, seven transmembrane domain receptor/LG4	b577, b641	Barresi *et al.*, 2000; Stenkamp and Frey, 2003; Varga *et al.*, 2001
chokh (*chk*)	early patterning MPH: Optic lobes do not form, lens smaller *none reported*	*rx3*, homeobox transcription factor/LG21	hu499, s399, w29	Loosli *et al.*, 2003
no isthmus (*noi*)	Morphogenesis MPH: Abnormal optic stalk (≤36 hpf) *brain, ear*	*pax2a*, transcription factor/LG13	b593 tb21, tu29a, ty22b, tm243a	Brand *et al.*, 1996a; Lun and Brand, 1998; Macdonald *et al.*, 1997
out of sight (*out*)	Morphogenesis HIS: Reduced size of retina (≤36 hpf) BH: Defective VBA *none described*		m233, m306, m390	Malicki *et al.*, 1996; Malicki Lab, unpublished data

Mutant	Phenotype	Gene/Linkage group	Alleles	References
pandora (pan)	Morphogenesis HIS: Ventral retina absent (3 dpf) *brain, ear, tail, pigmentation*	*supt6h*, transcription elongation factor (*spt6* homology)/LG21	m313, hi1621, hi2505b	Keegan *et al.*, 2002; Malicki *et al.*, 1996
brudas (bru)	Neurogenesis-differentiation-photoreceptors HIS: Central loss of PR (≤3 dpf) EM: Abnormal PR morphology, no OS, IS shortened (≤3 dpf) ERG: Absent BH: OMR, OKR, VBA absent MA: Autonomous in PRCL *touch response, oedema*	LG3	m148, tw212d, s3556	Doerre and Malicki, 2002; Malicki *et al.*, 1996; Neuhauss *et al.*, 1999; Goldsmith *et al.*, 2003
mikre oko (mok)	Neurogenesis-differentiation-photoreceptors HIS: Loss of PR (≤2.5 dpf) EM: Few OS and ST (≤3 dpf) MA: Partially non-autonomous in PRCL *none*		m632	Doerre and Malicki, 2001; Malicki *et al.*, 1996; Malicki Lab, unpublished data
elipsa (eli)	Neurogenesis-differentiation-photoreceptors-OS HIS: Central loss of PR (≤3 dpf) EM: No OS formation ERG: No response (5 dpf) BH: OKR absent MA: Autonomous in PRCL *body axis curled, kidney cysts*	LG9	m649, tp49d	Bahadori *et al.*, 2003; Doerre and Malicki, 2002; Malicki *et al.*, 1996; Neuhauss *et al.*, 1999
fleer (flr)	Neurogenesis-differentiation-photoreceptors-OS HIS: Central loss of PR (≤3 dpf) EM: No OS formation MA: Autonomous in PRCL *body axis curled, kidney cysts*		m477	Doerre and Malicki, 2002; Drummond *et al.*, 1998

(continues)

Table III (continued)

Locus name	Category and phenotypes / *other phenotypes*	Gene/Linkage group	Alleles	References
oval (ovl)	Neurogenesis-differentiation-photoreceptors-OS HIS: Loss of PR (3 dpf) EM: No OS formation ERG: No response BH: OKR absent MA: Autonomous in PRCL body axis *curled, kidney cysts*	*ift88*/LG9	tz288	Bahadori *et al.*, 2003; Doerre and Malicki, 2002; Tsujikawa and Malicki, 2004
no optokinetic response c (nrc)	Neurogenesis-differentiation-photoreceptors-synapse HIS: Thin OPL (6 dpf) EM: ST defect (≤6 dpf) ERG: Delayed and reduced b-wave (6 dpf) BH: OKR absent (6 dpf) *none*	LG10	a14	Allwardt *et al.*, 2001
marginal eye (mre)	Neurogenesis-marginal zone HIS: Cell death at MZ (4 dpf); cell death in INL and PRCL (5 dpf) *none described*		a6	Fadool *et al.*, 1997
piegus (pgu)	Neurogenesis-marginal zone HIS: Cell death predominantly at MZ (≤3 dpf) *pigmentation, ear*		m286, m300	Malicki *et al.*, 1996
round eye (rde)	Neurogenesis-marginal zone HIS: Cell death predominantly at MZ (≤3 dpf); loss of PRCL at MZ (5 dpf) *none described*		a5	Fadool *et al.*, 1997
foggy (fog)	Neurogenesis-patterning HIS: INL and PRCL differentiation delayed or absent (2 dpf) *brain, circulation*	*supt5h*, transcription elongation factor (spt5 homology)/ LG15	m806, sk8	Guo *et al.*, 1999, 2000; Keegan *et al.*, 2002
heart and soul (has)	Neurogenesis-patterning HIS: Neuroepithelium disrupted; all retinal layers are disorganized (3 dpf) *brain, blood, circulation*	*prkci*, atypical PKCι/ LG2	m129, m567, m781	Horne-Badovinac *et al.*, 2001; Malicki *et al.*, 1996; Peterson *et al.*, 2001

mosaic eyes (moe)	Neurogenesis-patterning HIS: Neuroepithelium affected; all retinal cell layers disorganized (3 dpf) MA: Non-autonomous in neuroepithelium *brain, blood circulation*	*moe*, FERM domain protein/LG9	b476, b781, b882	Jensen and Westerfield, 2004; Jensen et al., 2001
nagie oko (nok)	Neurogenesis-patterning HIS: Neuroepithelium affected; all retinal cell layers disorganized (3 dpf) MA: Non-autonomous in neural retina (GCL, PRCL) *brain, blood circulation*	*nok*, MAGUK family scaffolding factor/LG17	m227, m520, ji2	Malicki et al., 1996; Wei and Malicki, 2002
ncad, formerly glass onion, parachute	Neurogenesis-patterning HIS: Retinal neuroepithelium disorganized (≤12 hpf); all retinal cell layers disorganized MA: Non-autonomous in retinal neuroepithelium and PRCL *brain, somites, notochord, tail*	*cdh2*, cadherin adhesion molecule/LG20	m117, tm101b, fr7, rw95, r210	Malicki et al., 2003; Masai et al., 2003; Pujic and Malicki, 2001
oko meduzy (ome)	Neurogenesis-patterning HIS: Neuroepithelium affected, all retinal cell layers disorganized EM: Ectopic cell junctions MA: Non-autonomous in neuroepithelium *brain, blood circulation*		m98, m289, m298, m320	Malicki and Driever, 1999; Malicki et al., 1996
sonic you (syu)	Neurogenesis-patterning HIS: GC (≤40 hpf) and PR (≤58 hpf) differentiation delayed *somites, floor plate, spinal cord, fins*	*shh*, secreted signaling factor/LG7	t4, tq252, tbq70, tbx392	Neumann and Nuesslein-Volhard, 2000; van Eeden et al., 1996
young (yng)	Neurogenesis-patterning HIS: Delayed and disrupted lamination (60 hpf) MA: Non-autonomous in neural retina	*smarca4*, chromatin remodeling factor/LG3	a8, a50	Gregg et al., 2003; Link et al., 2000
you too (yot)	Neurogenesis-patterning (?) pigmentation, heart, body axis curled HIS: PR differentiation delayed (≤75 hpf) *somites, spinal cord*	*gli2*, zinc finger transcription factor/LG9	ty17a, ty119	Karlstrom et al., 1999; Stenkamp et al., 2000; van Eeden et al., 1996
bashful (bal)	Neurogenesis-patterning-ganglion cells HIS: GC disorganized (≤48 hpf) RT projections, notochord, hindbrain		tp86, tp82, tm220a, tr259, tf235	Karlstrom et al., 1996; Odenthal et al., 1996

(continues)

Table III *(continued)*

Locus name	Category and phenotypes *other phenotypes*	Gene/Linkage group	Alleles	References
chameleon (*con*)	Neurogenesis-patterning-ganglion cells HIS: Misrouted axons within the retina (≤48 hpf) *neural tube, curly tail*	*disp1*, multipass transmembrane SSD domain protein/LG20	tf18b, th6, tm15a, tu214, ty60	Brand et al., 1996b; Karlstrom et al., 1996; Nakano et al., 2004
grumpy (*gup*)	Neurogenesis-patterning-ganglion cells HIS: GC disorganized (larvae) ERG: Normal (larvae) BH: OKR defective, VBA normal (larvae) *RT projections, brain, notochord*	*lamb1*, β1 chain of basement membrane protein, laminin/LG25	tx221, m189, tg210, ti228b, tj229a	Karlstrom et al., 1996; Neuhauss et al., 1999; Parsons et al., 2002
lakritz (*lak*)	Neurogenesis-patterning-ganglion-cells HIS: GC not formed, thicker INL BH: OKR, OMR, VBA defective (7 dpf) *none*	*atoh7*, bHLH transcription factor/LG13	th241	Kay et al., 2001; Kelsh et al., 1996; Neuhauss et al., 1999
bergmann (*brg*)	Survival-amacrine cells HIS: Cell loss in inner INL (≤6 dpf) *brain*			Avanesov et al. (in preparation)
chorny (*chy*)	Survival-amacrine cells HIS: Cell loss in inner INL (≤5 dpf) BH: partial VBA (5 dpf) *brain, touch response*			Avanesov et al. (in preparation)
kleks (*kle*)	Survival-amacrine cells HIS: Cell loss in inner INL (5 dpf) BH: VBA absent (5 dpf) *brain*			Avanesov et al. (in preparation)
archie (*arc*)	Survival-ganglion cells HIS: Cell death in inner INL, GCL (≤3 dpf); thin GCL (≤6 dpf) *brain size, body axis curled*	*smarca5*, SWI/SNF related gene/LG1	a3, hi550	Fadool et al., 1997; Golling et al., 2002
night blindness b (*nbb*)	Survival-interplexiform cells HIS: INL abnormal, fewer interplexiform cells (≤11.5 mpf, heterozygotes); non-specific retinal degeneration (≤2.5 dpf, homozygotes) ERG: Normal *brain (homozygotes)*		da15	Li and Dowling, 2000

Mutant	Gene/description	Alleles	Phenotype	References
fade out (fad)	pigmentation	tm63c, tp94c, tc7b, tk224, tg14	Survival-photoreceptors HIS: Loss of PR (≤6 dpf) BH: OKR, OMR defective	Heisenberg et al., 1996; Kelsh et al., 1996; Neuhauss et al., 1999
sleepy (sly)	lamc1, γ1 chain of basement membrane protein, laminin/LG2	ts33a, m466, m86, tp16, te333	Survival-photoreceptors (?) HIS: Shorter OS (larvae) BH: OKR defective, VBA normal (larvae) RT projections, notochord, brain	Karlstrom et al., 1996; Neuhauss et al., 1999; Odenthal et al., 1996; Parsons et al., 2002
bleached (blc)		th204b, ts23, ty89	Survival-photoreceptors HIS: RPE unpigmented and PRCL loss (5 dpf) ERG: No response (5 dpf) BH: Defective OKR, OMR pigmentation, ear, CNS	Haffter et al., 1996; Neuhauss et al., 1999, 2003
discontinuous (dis)		m704	Survival-photoreceptors HIS: Patchy loss of PR (≤3 dpf) brain smaller	Malicki et al., 1996
fading vision (fdv)	defective pigmentation	th236a	Survival-photoreceptors HIS: Loss of PR (≤6 dpf) BH: OKR, OMR	Kelsh et al., 1996; Neuhauss et al., 1999
flathead (fla)		ty76, th5b, ta53c, tf21c, tu255e	Survival-photoreceptors HIS: Loss of PR (6 dpf) ERG: No response (5 dpf) BH: OKR, OMR defective CNS, jaw, branchial arches	Furutani-Seiki et al., 1996; Neuhauss et al., 1999; Schilling et al., 1996
krenty (krt)		m699	Survival-photoreceptors HIS: Patchy loss of PR (≤3 dpf) brain smaller	Malicki et al., 1996
lazy eye (lze)		a101	Survival-photoreceptors HIS: Abnormal Mueller glia morphology, few rods (6 dpf) EM: Fewer rods OS (5–6 dpf) ERG: Reduced a- and b-waves (6 dpf) BH: partial OKR response (5–6 dpf) none	Kainz et al., 2003

(continues)

Table III (*continued*)

Locus name	Category and phenotypes *other phenotypes*	Gene/Linkage group	Alleles	References
niezerka (nie)	Survival-photoreceptors HIS: Peripheral loss of PR (≤3 dpf) EM: Malformed OS and IS (3 dpf) MA: Partially non-autonomous in PRCL *brain smaller*		m743	Doerre and Malicki, 2002; Malicki *et al.*, 1996
not really finished (nrf)	Survival-photoreceptors HIS: Cell death predominantly in PRCL (≤80 dpf) EM: Deformed OS (5 dpf) BH: Partial OKR (5 dpf) *brain smaller*	*nrf1*, transcription factor/LG4	hi399a	Becker *et al.*, 1998
partial optokinetic response b (pob)	Survival-photoreceptors HIS: Few red cones (≤5 dpf) ERG: Enlarged a-wave, delayed and reduced b-wave (6 dpf) BH: No OKR in red light *none*		a1	Brockerhoff *et al.*, 1997
photoreceptors absent (pca)	Survival-photoreceptors HIS: Cental loss of PR (≤5dpf) *none described*		a2	Fadool *et al.*, 1997
punktata (pkt)	Survival-photoreceptors HIS: Cell death predominantly in PRCL, abnormal PR morphology (≤5dpf) *brain, ear, pigmentation*		m288	Malicki *et al.*, 1996
sinusoida (sid)	Survival-photoreceptors HIS: Patchy loss of PR (≤3 dpf) *brain slightly smaller*		m604	Malicki *et al.*, 1996
ivory (ivy)	Survival-photoreceptors, RPE HIS: Loss of PR (5-7 dpf); RPE degeneration (≤5 dpf) BH: Absent OKR (7 dpf) and VBA (5 dpf) MA: Non-autonomous in PRCL *pigmentation reduced*	LG20	tm271a, tp30	Goldsmith *et al.*, 2003; Kelsh *et al.*, 1996; Neuhauss *et al.*, 1999

Mutant	Phenotype	Allele	References
sunbleached (sbl)	Survival-photoreceptors, RPE HIS: Defective RPE and PR (6 dpf) BH: OKR, OMR defective *pigmentation, jaw, branchial arches*	to4	Heisenberg et al., 1996; Kelsh et al., 1996; Neuhauss et al., 1999
night blindness a (nba)	Survival-photoreceptors-late HIS: Patchy rod PR and sometimes cone PR degeneration (5.5 mpf, heterozygotes); INL thinning ($\leq$9.5 mpf, heterozygotes); non-specific retinal cell degeneration ($\leq$2.5 dpf, homozygotes) ERG: Delayed and reduced b-wave (8-13 mpf, heterozygotes) BH: ER reduced (4 mpf) *brain, optic tectum (homozygotes)*	da10 (semi-dominant)	Li and Dowling, 1997
night blindness d (nbd)	Survival-photoreceptors-late HIS: Fewer rod OS ($\leq$2 ypf, heterozygotes); non-specific retinal cell degeneration ($\leq$2.5 dpf, homozygotes) and PR absent ($\leq$5.5 dpf, homozygotes) ERG: Normal BH: ER poor in white or green light ($\leq$2 ypf) *body axis curled (homozygotes)*	da54	Maaswinkel et al., 2003
blass (bls)	Survival-RPE HIS: Defective RPE (6 dpf) BH: Defective OKR, OMR *pigmentation*	tg306	Kelsh et al., 1996; Neuhauss et al., 1999
blurred (blr)	Survival-RPE HIS: RPE degeneration (6 dpf) BH: OKR, OMR defective *pigmentation*	tu29c, tp233a, tm297, tx3, tq262b	Heisenberg et al., 1996; Kelsh et al., 1996; Neuhauss et al., 1999
interface (itf)	Survival-RPE HIS: Vacuolated RPE, partial loss of OS ($\leq$7 dpf) *pigmentation*	nt6	Vihtelic and Hyde, 2002
mirage (mir)	Survival-RPE HIS: Defective RPE (6 dpf) ERG: Defective BH: OKR, OMR defective *pigmentation*	tm79d	Heisenberg et al., 1996; Kelsh et al., 1996; Neuhauss et al., 1999

(continues)

Table III (*continued*)

Locus name	Category and phenotypes *other phenotypes*	Gene/Linkage group	Alleles	References
quasimodo (qam)	Survival-RPE HIS: Defective RPE (6 dpf) ERG: Defective BH: OKR, OMR defective *pigmentation, notochord*	LG14	ta81, tf208, tm138b, tb244c, tw25a	Kelsh *et al.*, 1996; Neuhauss *et al.*, 1999; Odenthal *et al.*, 1996
dropje (drp)	Visual function HIS: Normal ERG: Absence of b-wave BH: Defective OKR, OMR, VBA *none*		tr256	Kelsh *et al.*, 1996; Neuhauss *et al.*, 1999
macho (mao)	Visual function HIS: Normal (larvae) ERG: Normal; lack of GC action potentials (larvae) BH: OKR, OMR, VBA defective (larvae) *RT projections, touch response*	LG2	tt261	Gnuegge *et al.*, 2001; Granato *et al.*, 1996; Neuhauss *et al.*, 1999
no optokinetic nystagmus a (noa)	Visual function HIS: Normal (5 dpf) ERG: Enlarged a-wave, delayed and reduced b-wave (5 dpf) BH: OKR, VBA absent (5 dpf) *pigmentaton*	*dlat*, E2 subunit of pyruvate dehydrogenase complex/LG21	a13, m631	Brockerhoff *et al.*, 1995, 1998; Taylor *et al.*, 2004
no optokinetic response f (nof)	Visual function HIS: Normal (5 and 6 dpf) EM: Normal (5 and 6 dpf) BH: OKR absent *none*	*gnat2*, cone transducin α/ LG8	w21	Brockerhoff *et al.*, 2003
noir (nir)	Visual function HIS: Normal ERG: Absence of b-wave (5 dpf) BH: Defective OKR, OMR, VBA *locomotion*		tc22, tp89	Kelsh *et al.*, 1996; Neuhauss *et al.*, 1999

partial optokinetic nystagmus a (poa)	Visual function HIS: Normal (5 dpf) ERG: Reduced b-wave (6 dpf) BH: Defective OKR, VBA (5 dpf) *none reported*	m724	Brockerhoff *et al.*, 1995
sandy (sdy)	Visual function HIS: RPE pigmentation absent ERG: Normal (larvae) BH: Defective OMR, OKR (larvae) *pigmentation*	te326, ty79, tk20, tm118, to102	Kelsh *et al.*, 1996; Neuhauss *et al.*, 1999
steifftier (ste)	Visual function HIS: Normal BH: Defective OKR, VBA *locomotion*	tf220	Granato *et al.*, 1996; Neuhauss *et al.*, 1999

A large group of zebrafish mutations known to result in a nonspecific degeneration in all retinal cell layers is not listed in this table. Retinotectal projection mutants are not included, unless they produce defects within the retina.

The following criteria were used to form the phenotypic categories included in this table. Mutants in which the primordium is specified correctly (i.e., a defect is not obvious during the first day of embryogenesis) but the eye displays an abnormal shape or size are grouped into morphogenesis category. The morphogenesis category does not, however, include changes of eye size due to cell loss in specific cell populations, as such defects are assigned to the neurogenesis category. Mutations that produce a defect in neuronal specification or differentiation are included in neurogenesis category. Three subdivisions in this category are proposed: neurogenesis–patterning, for mutations that produce a disarrangement of retinal cell pattern, or result in a lack of an element(s) of the retinal pattern, such as a retinal cell class; neurogenesis–differentiation, for mutant lines that display defective differentiation of specific cell features, such as the lack of outer segments in some photoreceptor mutants; and neurogenesis–marginal zone, for genetic defects that predominantly affect cells in the marginal zone. In a large group of mutants, cells appear to first differentiate properly and only later degenerate. These mutant strains are classified as survival. Finally, mutants that display no obvious morphological or histological defects, but do show abnormal function of the visual system are included into visual function group.

This classification almost certainly will have to be modified when more is known about the nature of mutant phenotypes listed. The classification of the mutant phenotypes included in this table is biased by the research emphasis of their discoverers. Mutants of any given category may also contain additional, as yet uncharacterized defects in other aspects of retinal biology. As complementation tests have not been exhaustively performed in this group of mutant loci, it is possible that mutations of single loci are currently represented by two different entries. References are given to papers, which report isolation of mutant alleles as well as mutated genes. More information on specific loci may be obtained from Zebrafish Information Network (www.zfin.org). Nearly all mutations presented in this table were generated via ENU mutagenesis. Among exceptions are a retroviral insertion allele, in the *not really finished* locus and the hi550 allele of *archie*. The absence of the swim bladder phenotype is not listed.

Abbreviations: MPH: morphology; HIS: histology; EM: electron microscopy; ER: escape response; ERG: electroretinogram; OKR: optokinetic response; OMR: optomotor response; GCR: ganglion cell recordings; MA: mosaic analysis; INL: inner nuclear layer; GCL: ganglion cell layer; GC: ganglion cells; MZ: marginal zone; PE: pigmented epithelium; PR: photoreceptors; PRCL: photoreceptor layer; OS: outer segments; IS: inner segments; ST: synaptic terminals; VBA: visual background adaptation; mpf: months postfertilization; ypf: years post fertilization.

(Becker *et al.*, 1998). In addition to chemical mutagens and retroviral vectors, transposons may provide the basis for another effective mutagenesis approach. Transposable elements of the *Tc-1/mariner* and *hAT* families integrate into the zebrafish genome in a transposase-dependent fashion (Fadool *et al.*, 1998; Kawakami *et al.*, 2000; Raz *et al.*, 1998). Although transposons have not yet been used to induce genetic defects in zebrafish, the efficiency of transposition is likely to improve in the future and they too may soon become another mutagenesis tool. The advantages of different mutagenesis options are discussed in detail in other chapters of this volume.

The second important consideration in a genetic screen is the type of breeding scheme that will carry genetic defects from mutagenized animals (G0) to the generation in which the search for mutant phenotypes is performed. The most straightforward option, but also the most space- and time-consuming one, is screening for recessive defects in F3 generation embryos. This procedure was used in the large-scale genetic screens that have been performed to date (Amsterdam *et al.*, 1999; Driever *et al.*, 1996; Haffter *et al.*, 1996). Its main disadvantage is that it requires a very large number of tanks to raise the F2 generation to adulthood. As the majority of laboratories do not have access to several thousands of fish tanks, more space-efficient procedures are frequently required. In this regard, zebrafish offers some possibilities unavailable in other genetically studied vertebrates—haploid and early pressure screens (for a review, see Malicki, 2000). The major asset of these screening strategies is that one generation of animals is omitted and, consequently, time and the amount of laboratory space required are dramatically reduced. Although there are obvious advantages, these two screening strategies also suffer from some limitations. The most significant disadvantage of using haploids is that their development does not proceed in the same way as wild-type embryogenesis. Haploid embryos do not survive beyond 5 dpf, and, even at earlier stages of development, they display obvious defects. Although the eyes of haploid zebrafish appear fairly normal at least until 3 dpf, the architecture of their retinae tends to be disorganized. By 5 dpf, haploid embryos are markedly smaller than the wild type and display numerous abnormalities. In the context of the visual system, haploid screens appear useful to search for early patterning defects before the onset of neurogenesis.

Screening of embryos generated via the application of early pressure (Streisinger *et al.*, 1981) is another strategy that can be used to save both time and space. Similar to haploidization, this technique also allows one to screen for recessive defects in F2 generation embryos. The early pressure technique also involves some shortcomings. Embryos produced via this method display a high frequency of developmental abnormalities which complicate the detection of mutant phenotypes, especially at early developmental stages. Another limitation of early pressure screens is that the fraction of homozygous mutant animals in a clutch of early pressure-generated embryos depends on the distance of a mutant locus from the centromere. For centromeric loci, the fraction of mutant embryos approaches 50%; whereas, for telomeric genes, it decreases below 10% (Streisinger *et al.*,

1986). In other types of screens, mutant phenotypes can be distinguished from non-genetic developmental abnormalities based on their frequencies (25% in the case of screens on F3 embryos). Clearly, this criterion cannot be used in early pressure screens. Despite these limitations, early pressure screens are a very useful approach, especially in small-scale endeavors. Experimental techniques involved in haploid and early pressure screens have been discussed in depth in the previous edition of this volume (Beattie et al., 1999; Walker, 1999).

While the approaches already discussed are used to identify recessive mutant phenotypes, an entirely different breeding scheme is used in searches for dominant defects. These can be already detected in embryos, larvae, or adults of the F1 generation. Although this category of screens requires just a single generation and consequently a very small amount of tank space, few experiments focusing on dominant defects have been performed in zebrafish so far (van Eeden et al., 1999). An example of a search for dominant defects of the visual system is provided by a small behavioral screen of adult animals for the loss of visual perception, which identified a late-onset photoreceptor degeneration phenotype (Li and Dowling, 1997).

The third important consideration while designing a genetic screen is the choice of mutant phenotype detection method. This aspect of screening allows for substantial creativity. Phenotype detection criteria range from very simple to very sophisticated. Ideally, a mutant phenotype recognition strategy should fulfill the following requirements: (1) involve minimal effort; (2) detect gross abnormalities, as well as subtle changes; (3) exclude phenotypes irrelevant to the targeted process. One class of irrelevant phenotypes are nonspecific defects. In large-scale mutagenesis screens performed to date, more than two-thirds of all phenotypes were classified as nonspecific (Driever et al., 1996; Golling et al., 2002; Haffter et al., 1996). The most frequent nonspecific phenotypes in zebrafish are early degeneration that spreads across the entire embryo, and a developmental retardation that affects the brain, eyes, fins, and jaw. Nonspecific phenotypes are not necessarily without value, but are usually considered uninteresting because they are likely to be produced by defects in general cellular mechanisms (such as metabolism), which are not typically targeted by screens performed on vertebrate animals. Another category of irrelevant phenotypes are specific phenotypes of no interest to investigators performing a screen. Such phenotypes are isolated when a screening procedure detects mutations affecting multiple organs, only one of which is of interest. A good example of such a situation is provided by behavioral screens involving the optomotor response. A loss of the optomotor response may be attributed to defects of photoreceptor neurons or skeletal muscles. These two cell types are rarely of interest for a single group of investigators. It is one of the virtues of a well-designed screen that irrelevant phenotypes are efficiently eliminated.

The simplest way to screen for mutant phenotypes is by visual inspection. The most significant disadvantage of this method is that it detects changes only in structures easily recognizable using a microscope (preferably a dissecting scope). Visual inspection screens are thus suitable to search for defects in zebrafish blood vessels, which are easy to see in larvae, but would not detect a loss of a small population of neurons hidden in the depths of the brain. Visual inspection criteria work well when the aim of a screen is the detection of gross morphological changes. Within the eye, such changes may reflect specific defects in a single neuronal lamina. In several mutants, the changes of eye size are caused by a degeneration of photoreceptor cells (Malicki *et al.*, 1996). In this case, the affected cell population is numerous enough to cause a major change of morphology. Most likely, a morphological screen would not detect abnormalities in a less numerous cell class.

Changes confined to small populations of cells usually cannot be identified in a visual inspection screen. To detect them, the target cell population must be somehow made accessible to inspection. Several options exist in this regard: analysis of histological sections, whole-mount antibody staining, *in situ* hybridization, retrograde or anterograde labeling of neurons, and cell class-specific transgenes. One technically simple but rather laborious approach is to embed zebrafish larvae in paraffin and prepare histological sections. This approach was used to screen over 2000 individuals from approximately 50 clutches of F2 early pressure-generated mutagenized larvae and led to the identification of two photoreceptor mutants (Mohideen *et al.*, 2003). In addition to histological analysis, individual cell populations can be visualized in mutagenized animals using antibody staining or *in situ* hybridization. In one screening endeavor, staining of 700 early pressure-generated egg clutches with anti-tyrosine hydroxylase antibody led to the isolation of two retinal mutants (Guo *et al.*, 1999).

An excellent example of a genetic screen that involves labeling of a specific neuronal population has been performed to uncover defects of the retinotectal projection (Baier *et al.*, 1996; Karlstrom *et al.*, 1996; Trowe *et al.*, 1996). In this screen, two subpopulations of retinal ganglion cells were labeled with the carbocyanine tracers, DiI and DiO. Labeling procedures usually make screening much more laborious. To reduce the workload in this screen, DiI and DiO labeling were highly automated. For tracer injection, fish larvae were mounted in a standardized fashion in a temperature-controlled mounting apparatus. After filling the apparatus with liquid agarose and mounting the larvae, the temperature was lowered, allowing the agarose to solidify. Subsequently, the blocks of agarose containing mounted larvae were transferred into the injection setup. Follow injection, the larvae were stored overnight at room temperature to allow for the diffusion of the injected tracer and then transferred to a microscope stage for phenotypic analysis. The authors of this experiment estimate that using this highly automated screening procedure allowed them to inspect more than 2000 larvae per day and to reduce the time spent on the analysis of a single individual to less than 1 min (Baier *et al.*, 1996). Other labeling procedures can also be scaled up to process many clutches of embryos in a single experiment. Antibody or *in situ* protocols, for example, involve multiple changes of staining and washing solutions. To perform these protocols on many embryos in parallel, one can use multi-well staining dishes with stainless steel mesh at the bottom. Such staining dishes can be quickly transferred from one solution to another. Since many labeling procedures are time-consuming, it is essential that during a screen they are performed in parallel on many embryos.

Recent advances provide a way to label specific cell populations in a much less labor-intensive way by using GFP transgenes, such as the one described in the Histological Analysis section. Transgenic GFP lines can be either directly mutagenized or crossed to mutagenized males and the resulting progeny is used to search for defects in fine features of retinal cell populations. In contrast to other cell labeling procedures, the use of GFP transgenes requires very little additional effort, beyond what is required in simple morphological observations of the external phenotype.

Behavioral tests are yet another screening alternative. Several screens based on behavioral criteria have been performed in recent years, leading to the isolation of interesting developmental defects (Brockerhoff et al., 1997, 2003; Li and Dowling, 1997; Neuhauss et al., 1999). Behavioral screens allow one to detect subtle defects of visual function that might evade other search criteria. Similar to many labeling procedures, behavioral screens tend to be laborious. In one instance of a screen involving the optokinetic response, the authors estimate that screening of a single zebrafish larva took, on average, 1 min (Brockerhoff et al., 1995). In addition, since behavioral responses usually involve the cooperation of many cell classes, screens of this type tend to detect a wide range of defects. Optokinetic response screens, for example, may lead to the isolation of defects in the differentiation of lens cells, the specification of retinal neurons or glia, the formation of synaptic connections, the mechanisms of neurotransmitter release, or the development of ocular muscles. Additional tests are usually necessary to ensure that the isolated mutants belong to the desired category. To be useful for screening, a behavioral response should be robust and reproducible and should involve the simplest possible neuronal circuitry. In light of these criteria, the optokinetic response appears to be superior to other behaviors. Both optomotor and startle responses require functional optic tecta while the optokinetic response does not (Clark, 1981; Easter and Nicola, 1996). The optokinetic response also appears to be more robust than the optomotor response and phototaxis (Brockerhoff et al., 1995; Clark, 1981). Behavioral screens can be used to search for both recessive and dominant defects in larvae, as well as in adult fish (Li and Dowling, 1997).

Molecular characterization of defective loci is usually a crucial step that follows the isolation of mutant lines. The development of positional and candidate gene cloning strategies is one of the most significant advances in the field of zebrafish genetics in recent years. These approaches are currently well-established and have played a key role in many important contributions to the understanding of eye development and function (Table III). The positional cloning strategy involves a standard set of steps, such as mapping, chromosomal walking, transcript identification, and the delivery of a proof that the correct gene has been cloned. These steps are largely the same, regardless of the nature of a mutant phenotype and are discussed in depth in other chapters of this volume. An example of a positional cloning strategy, laborious but eventually successful, is the cloning of the *nagie oko* locus (Wei and Malicki, 2002).

B. Mutant Strains Available

Large and small mutagenesis screens identified numerous genetic defects of retinal development in zebrafish. Mutant phenotypes affect a broad range of developmental stages, starting with the specification of the eye primordia, through optic lobe morphogenesis, the specification of neuronal identities, and including the final steps of differentiation, such as outer segment development in photoreceptor cells. A list of mutant lines, excluding the ones that produce nonspecific degeneration of the entire retina, and references to both phenotypic and molecular studies of the underlying loci are provided in Table III.

V. Summary

Similar to other vertebrate species, the zebrafish retina is simpler than other regions of the central nervous system (CNS). Relative simplicity, rapid development, and accessibility to genetic analysis make the zebrafish retina an excellent model system for the studies of neurogenesis in the vertebrate CNS. Numerous genetic screens have led to isolation of an impressive collection of mutations affecting the retina and the retinotectal projection in zebrafish. Mutant phenotypes are being studied using a rich variety of markers: antibodies, RNA probes, retrograde and anterograde tracers, as well as transgenic lines. Particularly impressive progress has been made in the characterization of the zebrafish genome. Consequently, positional and candidate cloning of mutant genes are now fairly easy to accomplish in zebrafish. Many mutant genes have, in fact, already been cloned and their analysis has provided important insights into the gene circuitry that regulates retinal neurogenesis. Genetic screens for visual system defects will continue in the future and progressively more sophisticated screening approaches will make it possible to detect a variety of subtle mutant phenotypes in retinal development. The remarkable evolutionary conservation of the vertebrate eye provides the basis for the use of the zebrafish retina as a model of human disorders. Some of the genetic defects of the zebrafish retina indeed resemble human retinopathies. As new techniques are being introduced and improved at a rapid pace, the zebrafish will continue to be an important organism for the studies of the vertebrate visual system.

Acknowledgments

The authors are grateful to Stefan Heller, Sarah Keller, and Sheila Baker for critical reading of the manuscript and helpful comments. The authors' research on the retina is supported by grants from the National Eye Institute and the Glaucoma Foundation.

Standard safety procedures for the use of some reagents and protocols discussed in this chapter must be followed. For a description of health risks involved in the use of particular chemicals, please contact their manufactures.

References

Allwardt, B. A., Lall, A. B., Brockerhoff, S. E., and Dowling, J. E. (2001). Synapse formation is arrested in retinal photoreceptors of the zebrafish nrc mutant. *J. Neurosci.* **21**, 2330–2342.

Altshuler, D., Turner, D., and Cepko, C. (1991). Specification of cell type in the vertebrate retina. *In* "Development of the Visual System" (D. Lam and C. Shatz, eds.), pp. 37–58. Cambridge MA, MIT Press.

Amsterdam, A., Burgess, S., Golling, G., Chen, W., Sun, Z., Townsend, K., Farrington, S., Haldi, M., and Hopkins, N. (1999). A large-scale insertional mutagenesis screen in zebrafish. *Genes Dev.* **13**, 2713–2724.

Ando, H., Furuta, T., Tsien, R. Y., and Okamoto, H. (2001). Photo-mediated gene activation using caged RNA/DNA in zebrafish embryos. *Nat. Genet.* **28**, 317–325.

Ando, H., and Okamoto, H. (2003). Practical procedures for ectopic induction of gene expression in zebrafish embryos using Bhc-diazo-caged mRNA. *Methods Cell. Sci.* **25**, 25–31.

Bahadori, R., Huber, M., Rinner, O., Seeliger, M. W., Geiger-Rudolph, S., Geisler, R., and Neuhauss, S. C. (2003). Retinal function and morphology in two zebrafish models of oculo-renal syndromes. *Eur. J. Neurosci.* **18**, 1377–1386.

Baier, H., Klostermann, S., Trowe, T., Karlstrom, R. O., Nusslein-Volhard, C., and Bonhoeffer, F. (1996). Genetic dissection of the retinotectal projection. *Development* **123**, 415–425.

Barresi, M. J., Stickney, H. L., and Devoto, S. H. (2000). The zebrafish slow-muscle-omitted gene product is required for Hedgehog signal transduction and the development of slow muscle identity. *Development* **127**, 2189–2199.

Barthel, L. K., and Raymond, P. A. (1990). Improved method for obtaining 3-microns cryosections for immunocytochemistry. *J. Histochem. Cytochem.* **38**, 1383–1388.

Barthel, L. K., and Raymond, P. A. (1993). Subcellular localization of alpha-tubulin and opsin mRNA in the goldfish retina using digoxigenin-labeled cRNA probes detected by alkaline phosphatase and HRP histochemistry. *J. Neurosci. Methods* **50**, 145–152.

Beattie, C. E., Raible, D. W., Henion, P. D., and Eisen, J. S. (1999). Early pressure screens. *Methods Cell. Biol.* **60**, 71–86.

Becker, T. S., Burgess, S. M., Amsterdam, A. H., Allende, M. L., and Hopkins, N. (1998). Not really finished is crucial for development of the zebrafish outer retina and encodes a transcription factor highly homologous to human Nuclear Respiratory Factor-1 and avian Initiation Binding Repressor. *Development* **125**, 4369–4378.

Biehlmaier, O., Neuhauss, S. C., and Kohler, K. (2003). Synaptic plasticity and functionality at the cone terminal of the developing zebrafish retina. *J. Neurobiol.* **56**, 222–236.

Bodick, N., and Levinthal, C. (1980). Growing optic nerve fibers follow neighbors during embryogenesis. *Proc. Natl. Acad. Sci. USA* **77**, 4374–4378.

Branchek, T., and Bremiller, R. (1984). The development of photoreceptors in the zebrafish, Brachydanio rerio. I. Structure. *J. Comp. Neurol.* **224**, 107–115.

Brand, M., Heisenberg, C. P., Jiang, Y. J., Beuchle, D., Lun, K., Furutani-Seiki, M., Granato, M., Haffter, P., Hammerschmidt, M., Kane, D. A., *et al.* (1996a). Mutations in zebrafish genes affecting the formation of the boundary between midbrain and hindbrain. *Development* **123**, 179–190.

Brand, M., Heisenberg, C. P., Warga, R. M., Pelegri, F., Karlstrom, R. O., Beuchle, D., Picker, A., Jiang, Y. J., Furutani-Seiki, M., van Eeden, F. J., *et al.* (1996b). Mutations affecting development of the midline and general body shape during zebrafish embryogenesis. *Development* **123**, 129–142.

Brand, A. H., and Perrimon, N. (1993). Targeted gene expression as a means of altering cell fates and generating dominant phenotypes. *Development* **118**, 401–415.

Brennan, C., Monschau, B., Lindberg, R., Guthrie, B., Drescher, U., Bonhoeffer, F., and Holder, N. (1997). Two Eph receptor tyrosine kinase ligands control axon growth and may be involved in the creation of the retinotectal map in the zebrafish. *Development* **124**, 655–664.

Brockerhoff, S. E., Dowling, J. E., and Hurley, J. B. (1998). Zebrafish retinal mutants. *Vision Res.* **38**, 1335–1339.

Brockerhoff, S. E., Hurley, J. B., Janssen-Bienhold, U., Neuhauss, S. C., Driever, W., and Dowling, J. E. (1995). A behavioral screen for isolating zebrafish mutants with visual system defects. *Proc. Natl. Acad. Sci. USA* **92**, 10545–10549.

Brockerhoff, S. E., Hurley, J. B., Niemi, G. A., and Dowling, J. E. (1997). A new form of inherited red-blindness identified in zebrafish. *J. Neurosci.* **17**, 4236–4242.

Brockerhoff, S. E., Rieke, F., Matthews, H. R., Taylor, M. R., Kennedy, B., Ankoudinova, I., Niemi, G. A., Tucker, C. L., Xiao, M., Cilluffo, M. C., *et al.* (2003). Light stimulates a transducin-independent increase of cytoplasmic Ca2+ and suppression of current in cones from the zebrafish mutant nof. *J. Neurosci.* **23**, 470–480.

Burrill, J., and Easter, S. (1995). The first retinal axons and their microenvironment in zebrafish cryptic pioneers and the pretract. *J. Neurosci.* **15**, 2935–2947.

Burrill, J. D., and Easter, S. S., Jr. (1994). Development of the retinofugal projections in the embryonic and larval zebrafish (Brachydanio rerio). *J. Comp. Neurol.* **346**, 583–600.

Cajal, S. R. (1893). La retine des vertebres. *La. Cellule* **9**, 17–257.

Cayouette, M., and Raff, M. (2003). The orientation of cell division influences cell-fate choice in the developing mammalian retina. *Development* **130**, 2329–2339.

Cayouette, M., Whitmore, A. V., Jeffery, G., and Raff, M. (2001). Asymmetric segregation of Numb in retinal development and the influence of the pigmented epithelium. *J. Neurosci.* **21**, 5643–5651.

Cedrone, C., Culasso, F., Cesareo, M., Zapelloni, A., Cedrone, P., and Cerulli, L. (1997). Prevalence of glaucoma in Ponza, Italy: A comparison with other studies. *Ophthalmic. Epidemiol.* **4**, 59–72.

Chinen, A., Hamaoka, T., Yamada, Y., and Kawamura, S. (2003). Gene duplication and spectral diversification of cone visual pigments of zebrafish. *Genetics* **163**, 663–675.

Chuang, J. C., Mathers, P. H., and Raymond, P. A. (1999). Expression of three Rx homeobox genes in embryonic and adult zebrafish. *Mech. Dev.* **84**, 195–198.

Clark, T. (ed.) (1981). "Visual Responses in Developing Zebrafish (Brachydanio rerio)." University of Oregon, Eugene, OR.

Colbert, T., Till, B. J., Tompa, R., Reynolds, S., Steine, M. N., Yeung, A. T., McCallum, C. M., Comai, L., and Henikoff, S. (2001). High-throughput screening for induced point mutations. *Plant. Physiol.* **126**, 480–484.

Collazo, A., Fraser, S. E., and Mabee, P. M. (1994). A dual embryonic origin for vertebrate mechanoreceptors. *Science* **264**, 426–430.

Connaughton, V. P., Behar, T. N., Liu, W. L., and Massey, S. C. (1999). Immunocytochemical localization of excitatory and inhibitory neurotransmitters in the zebrafish retina [In Process Citation]. *Vis. Neurosci.* **16**, 483–490.

Das, T., Payer, B., Cayouette, M., and Harris, W. A. (2003). *In vivo* time-lapse imaging of cell divisions during neurogenesis in the developing zebrafish retina. *Neuron* **37**, 597–609.

Devoto, S. H., Melancon, E., Eisen, J. S., and Westerfield, M. (1996). Identification of separate slow and fast muscle precursor cells *in vivo*, prior to somite formation. *Development* **122**, 3371–3380.

Doerre, G., and Malicki, J. (2001). A mutation of early photoreceptor development, *mikre oko*, reveals cell-cell interactions involved in the survival and differentiation of zebrafish photoreceptors. *J. Neurosci.* **21**, 6745–6757.

Doerre, G., and Malicki, J. (2002). Genetic analysis of photoreceptor cell development in the zebrafish retina. *Mech. Devel.* **110**, 125–138.

Dowling, J. (1987). "The Retina." Harvard University Press, Cambridge, MA.

Draper, B. W., Morcos, P. A., and Kimmel, C. B. (2001). Inhibition of zebrafish fgf8 pre-mRNA splicing with morpholino oligos: A quantifiable method for gene knockdown. *Genesis* **30**, 154–156.

Drescher, U., Bonhoeffer, F., and Muller, B. K. (1997). The Eph family in retinal axon guidance. *Curr. Opin. Neurobiol.* **7**, 75–80.

Driever, W., Solnica-Krezel, L., Schier, A. F., Neuhauss, S. C., Malicki, J., Stemple, D. L., Stainier, D. Y., Zwartkruis, F., Abdelilah, S., Rangini, Z., *et al.* (1996). A genetic screen for mutations affecting embryogenesis in zebrafish. *Development* **123**, 37–46.

Drummond, I. A., Majumdar, A., Hentschel, H., Elger, M., Solnica-Krezel, L., Schier, A. F., Neuhauss, S. C., Stemple, D. L., Zwartkruis, F., Rangini, Z., *et al.* (1998). Early development of the zebrafish pronephros and analysis of mutations affecting pronephric function. *Development* **125**, 4655–4667.

Dryja, T., and Li, T. (1995). Molecular genetics of retinitis pigmentosa. *Human Molecular Genetics* **4**, 1739–1743.

Easter, S., and Nicola, G. (1996). The development of vision in the zebrafish (Danio rerio). *Dev. Biol.* **180**, 646–663.

Easter, S. S., Jr., and Malicki, J. J. (2002). The zebrafish eye: Developmental and genetic analysis. *Results Probl. Cell Differ.* **40**, 346–370.

Ericson, J. (1992). Early stages of motor neuron differentiation revealed by expression of homeobox gene islet-1. *Science* **256**, 1555–1560.

Fadool, J. M. (2003). Development of a rod photoreceptor mosaic revealed in transgenic zebrafish. *Dev. Biol.* **258**, 277–290.

Fadool, J. M., Brockerhoff, S. E., Hyatt, G. A., and Dowling, J. E. (1997). Mutations affecting eye morphology in the developing zebrafish (Danio rerio). *Dev. Genet.* **20**, 288–295.

Fadool, J. M., Hartl, D. L., and Dowling, J. E. (1998). Transposition of the mariner element from Drosophila mauritiana in zebrafish. *Proc. Natl. Acad. Sci. USA* **95**, 5182–5186.

Fashena, D., and Westerfield, M. (1999). Secondary motoneuron axons localize DM-GRASP on their fasciculated segments. *J. Comp. Neurol.* **406**, 415–424.

Fekany, K., Yamanaka, Y., Leung, T., Sirotkin, H. I., Topczewski, J., Gates, M. A., Hibi, M., Renucci, A., Stemple, D., Radbill, A., *et al.* (1999). The zebrafish bozozok locus encodes Dharma, a homeodomain protein essential for induction of gastrula organizer and dorsoanterior embryonic structures. *Development* **126**, 1427–1438.

Fraser, S. (1992). Patterning of retinotectal connections in the vertebrate visual system. *Curr. Opin. Neurobiol.* **2**, 83–87.

Fricke, C., Lee, J. S., Geiger-Rudolph, S., Bonhoeffer, F., and Chien, C. B. (2001). Astray, a zebrafish roundabout homolog required for retinal axon guidance. *Science* **292**, 507–510.

Furutani-Seiki, M., Jiang, Y. J., Brand, M., Heisenberg, C. P., Houart, C., Beuchle, D., van Eeden, F. J., Granato, M., Haffter, P., Hammerschmidt, M., *et al.* (1996). Neural degeneration mutants in the zebrafish, Danio rerio. *Development* **123**, 229–239.

Gnuegge, L., Schmid, S., and Neuhauss, S. C. (2001). Analysis of the activity-deprived zebrafish mutant macho reveals an essential requirement of neuronal activity for the development of a fine-grained visuotopic map. *J. Neurosci.* **21**, 3542–3548.

Goldsmith, P., Baier, H., and Harris, W. A. (2003). Two zebrafish mutants, ebony and ivory, uncover benefits of neighborhood on photoreceptor survival. *J. Neurobiol.* **57**, 235–245.

Golling, G., Amsterdam, A., Sun, Z., Antonelli, M., Maldonado, E., Chen, W., Burgess, S., Haldi, M., Artzt, K., Farrington, S., *et al.* (2002). Insertional mutagenesis in zebrafish rapidly identifies genes essential for early vertebrate development. *Nat. Genet.* **31**, 135–140.

Granato, M., van Eeden, F. J., Schach, U., Trowe, T., Brand, M., Furutani-Seiki, M., Haffter, P., Hammerschmidt, M., Heisenberg, C. P., Jiang, Y. J., *et al.* (1996). Genes controlling and mediating locomotion behavior of the zebrafish embryo and larva. *Development* **123**, 399–413.

Gregg, R. G., Willer, G. B., Fadool, J. M., Dowling, J. E., and Link, B. A. (2003). Positional cloning of the young mutation identifies an essential role for the Brahma chromatin remodeling complex in mediating retinal cell differentiation. *Proc. Natl. Acad. Sci. USA* **100**, 6535–6540.

Guo, S., Wilson, S. W., Cooke, S., Chitnis, A. B., Driever, W., and Rosenthal, A. (1999b). Mutations in the zebrafish unmask shared regulatory pathways controlling the development of catecholaminergic neurons. *Dev. Biol.* **208**, 473–487.

Guo, S., Yamaguchi, Y., Schilbach, S., Wada, T., Lee, J., Goddard, A., French, D., Handa, H., and Rosenthal, A. (2000). A regulator of transcriptional elongation controls vertebrate neuronal development. *Nature* **408**, 366–369.

Haffter, P., Granato, M., Brand, M., Mullins, M. C., Hammerschmidt, M., Kane, D. A., Odenthal, J., van Eeden, F. J., Jiang, Y. J., Heisenberg, C. P., *et al.* (1996). The identification of genes with unique and essential functions in the development of the zebrafish, Danio rerio. *Development* **123**, 1–36.

Halloran, M. C., Sato-Maeda, M., Warren, J. T., Su, F., Lele, Z., Krone, P. H., and Kuwada, J. Y. and Shoji, W. (2000). Laser-induced gene expression in specific cells of transgenic zebrafish. *Development* **127**, 1953–1960.

Halpern, M., Ho, R., Walker, C., and Kimmel, C. (1993). Induction of muscle pioneers and floor plate is distinguished by the zebrafish no tail mutation. *Cell* **75**, 99–111.

Hanker, J. S. (1979). Osmiophilic reagents in electronmicroscopic histocytochemistry. *Prog. Histochem. Cytochem.* **12**, 1–85.

Hatta, K., Kimmel, C., Ho, R., and Walker, C. (1991). The cyclops mutation blocks specification of the floor plate of the zebrafish central nervous system. *Nature* **350**, 339–341.

Hauptmann, G., and Gerster, T. (1994). Two-color whole-mount *in situ* hybridization to vertebrate and Drosophila embryos. *Trends Genet.* **10**, 266.

Heisenberg, C. P., Brand, M., Jiang, Y. J., Warga, R., Beuchle, D., Eeden, F., Furutani-Seiki, M., Granato, M., Haffter, P., Hammerschmidt, M., *et al.* (1996). Genes involved in forebrain development in the zebrafish, Danio rerio. *Development* **123**, 191–203.

Heisenberg, C. P., Houart, C., Take-Uchi, M., Rauch, G. J., Young, N., Coutinho, P., Masai, I., Caneparo, L., Concha, M. L., Geisler, R., *et al.* (2001). A mutation in the Gsk3-binding domain of zebrafish Masterblind/Axin1 leads to a fate transformation of telencephalon and eyes to diencephalon. *Genes Dev.* **15**, 1427–1434.

Heisenberg, C. P., Tada, M., Rauch, G. J., Saude, L., Concha, M. L., Geisler, R., Stemple, D. L., Smith, J. C., and Wilson, S. W. (2000). Silberblick/Wnt11 mediates convergent extension movements during zebrafish gastrulation. *Nature* **405**, 76–81.

Hinds, J., and Hinds, P. (1974). Early ganglion cell differentiation in the mouse retina: An electron microscopic analysis utilizing serial sections. *Dev. Biol.* **37**, 381–416.

Hisatomi, O., Satoh, T., Barthel, L. K., Stenkamp, D. L., Raymond, P. A., and Tokunaga, F. (1996). Molecular cloning and characterization of the putative ultraviolet-sensitive visual pigment of goldfish. *Vision Res.* **36**, 933–999.

Hitchcock, P. F., Macdonald, R. E., VanDeRyt, J. T., and Wilson, S. W. (1996). Antibodies against Pax6 immunostain amacrine and ganglion cells and neuronal progenitors, but not rod precursors, in the normal and regenerating retina of the goldfish. *J. Neurobiol.* **29**, 399–413.

Ho, R. K., and Kane, D. A. (1990). Cell-autonomous action of zebrafish spt-1 mutation in specific mesodermal precursors. *Nature* **348**, 728–730.

Holt, C., Bertsch, T., Ellis, H., and Harris, W. (1988). Cellular Determination in the Xenopus Retina is Independent of Lineage and Birth Date. *Neuron* **1**, 15–26.

Honig, M. G., and Hume, R. I. (1986). Fluorescent carbocyanine dyes allow living neurons of identified origin to be studied in long-term cultures. *J. Cell. Biol.* **103**, 171–187.

Honig, M. G., and Hume, R. I. (1989). DiI and diO: Versatile fluorescent dyes for neuronal labelling and pathway tracing. *Trends Neurosci.* **12**, 333–335, 340–341.

Horne-Badovinac, S., Lin, D., Waldron, S., Schwarz, M., Mbamalu, G., Pawson, T., Jan, Y., Stainier, D. Y., and Abdelilah-Seyfried, S. (2001). Positional cloning of heart and soul reveals multiple roles for PKC lambda in zebrafish organogenesis. *Curr. Biol.* **11**, 1492–1502.

Hu, M., and Easter, S. S. (1999). Retinal neurogenesis: The formation of the initial central patch of postmitotic cells. *Dev. Biol.* **207**, 309–321.

Humphrey, C., and Pittman, F. (1974). A simple methylene blue-azure II-basic fuchsin stain for epoxy-embedded tissue sections. *Stain. Technol.* **49**, 9–14.

Hutson, L. D., and Chien, C. B. (2002). Pathfinding and error correction by retinal axons: The role of astray/robo2. *Neuron* **33**, 205–217.

Hyatt, G. A., Schmitt, E. A., Marsh-Armstrong, N., McCaffery, P., Drager, U. C., and Dowling, J. E. (1996). Retinoic acid establishes ventral retinal characteristics. *Development* **122**, 195–204.

Jacobson, M. (1991). "Developmental Neurobiology." Plenum Press, New York.

Jensen, A. M., Walker, C., and Westerfield, M. (2001). Mosaic eyes: A zebrafish gene required in pigmented epithelium for apical localization of retinal cell division and lamination. *Development* **128**, 95–105.

Jensen, A. M., and Westerfield, M. (2004). Zebrafish mosaic eyes is a novel FERM protein required for retinal lamination and retinal pigmented epithelial tight junction formation. *Curr. Biol.* **14**, 711–717.

Jiao, Y., Sun, Z., Lee, T., Fusco, F. R., Kimble, T. D., Meade, C. A., Cuthbertson, S., and Reiner, A. (1999). A simple and sensitive antigen retrieval method for free-floating and slide-mounted tissue sections. *J. Neurosci. Methods* **93**, 149–162.

Jowett, T. (1999). Analysis of protein and gene expression. *Methods Cell. Biol.* **59**, 63–85.

Jowett, T. (2001). Double *in situ* hybridization techniques in zebrafish. *Methods* **23**, 345–358.

Jowett, T., and Lettice, L. (1994). Whole-mount *in situ* hybridizations on zebrafish embryos using a mixture of digoxigenin- and fluorescein-labelled probes. *TIG* **10**, 73–74.

Kainz, P. M., Adolph, A. R., Wong, K. Y., and Dowling, J. E. (2003). Lazy eyes zebrafish mutation affects Muller glial cells, compromising photoreceptor function and causing partial blindness. *J. Comp. Neurol.* **463**, 265–280.

Karlstrom, R. O., Talbot, W. S., and Schier, A. F. (1999). Comparative synteny cloning of zebrafish you-too: Mutations in the Hedgehog target gli2 affect ventral forebrain patterning. *Genes Dev.* **13**, 388–393.

Karlstrom, R. O., Trowe, T., Klostermann, S., Baier, H., Brand, M., Crawford, A. D., Grunewald, B., Haffter, P., Hoffmann, H., Meyer, S. U., *et al.* (1996). Zebrafish mutations affecting retinotectal axon pathfinding. *Development* **123**, 427–438.

Kawahara, A., Chien, C. B., and Dawid, I. B. (2002). The homeobox gene mbx is involved in eye and tectum development. *Dev. Biol.* **248**, 107–117.

Kawakami, K., Shima, A., and Kawakami, N. (2000). Identification of a functional transposase of the Tol2 element, an Ac-like element from the Japanese medaka fish, and its transposition in the zebrafish germ lineage. *Proc. Natl. Acad. Sci. USA* **97**, 11403–11408.

Kay, J. N., Finger-Baier, K. C., Roeser, T., Staub, W., and Baier, H. (2001). Retinal ganglion cell genesis requires lakritz, a Zebrafish atonal Homolog. *Neuron* **30**, 725–736.

Kay, J. N., Roeser, T., Mumm, J. S., Godinho, L., Mrejeru, A., Wong, R. O., and Baier, H. (2004). Transient requirement for ganglion cells during assembly of retinal synaptic layers. *Development* **131**, 1331–1342.

Keegan, B. R., Feldman, J. L., Lee, D. H., Koos, D. S., Ho, R. K., Stainier, D. Y., and Yelon, D. (2002). The elongation factors Pandora/Spt6 and Foggy/Spt5 promote transcription in the zebrafish embryo. *Development* **129**, 1623–1632.

Kelsh, R. N., Brand, M., Jiang, Y. J., Heisenberg, C. P., Lin, S., Haffter, P., Odenthal, J., Mullins, M. C., van Eeden, F. J., Furutani-Seiki, M., *et al.* (1996). Zebrafish pigmentation mutations and the processes of neural crest development. *Development* **123**, 369–389.

Kikuchi, Y., Segawa, H., Tokumoto, M., Tsubokawa, T., Hotta, Y., Uyemura, K., and Okamoto, H. (1997). Ocular and cerebellar defects in zebrafish induced by overexpression of the LIM domains of the islet-3 LIM/homeodomain protein. *Neuron* **18**, 369–382.

Kimmel, C. B., Ballard, W. W., Kimmel, S. R., Ullmann, B., and Schilling, T. F. (1995). Stages of embryonic development of the zebrafish. *Dev. Dyn.* **203**, 253–310.

Kimmel, C. B., Sessions, S. K., and Kimmel, R. J. (1981). Morphogenesis and synaptogenesis of the zebrafish Mauthner neuron. *J. Comp. Neurol.* **198**, 101–120.

Koos, D. S., and Ho, R. K. (1999). The nieuwkoid/dharma homeobox gene is essential for bmp2b repression in the zebrafish pregastrula. *Dev. Biol.* **215**, 190–207.

Korzh, V., Edlund, T., and Thor, S. (1993). Zebrafish primary neurons initiate expression of the LIM homeodomain protein Isl-1 at the end of gastrulation. *Development* **118**, 417–425.

Laessing, U., Giordano, S., Stecher, B., Lottspeich, F., and Stuermer, C. A. (1994). Molecular characterization of fish neurolin: A growth-associated cell surface protein and member of the immunoglobulin superfamily in the fish retinotectal system with similarities to chick protein DM-GRASP/SC-1/BEN. *Differentiation* **56**, 21–29.

Laessing, U., and Stuermer, C. A. (1996). Spatiotemporal pattern of retinal ganglion cell differentiation revealed by the expression of neurolin in embryonic zebrafish. *J. Neurobiol.* **29,** 65–74.

Larison, K., and Bremiller, R. (1990). Early onset of phenotype and cell patterning in the embryonic zebrafish retina. *Development* **109,** 567–576.

Li, L., and Dowling, J. E. (1997). A dominant form of inherited retinal degeneration caused by a non-photoreceptor cell-specific mutation. *Proc. Natl. Acad. Sci. USA* **94,** 11645–11650.

Li, L., and Dowling, J. E. (2000). Disruption of the olfactoretinal centrifugal pathway may relate to the visual system defect in night blindness b mutant zebrafish. *J. Neurosci.* **20,** 1883–1892.

Li, Z., Hu, M., Ochocinska, M. J., Joseph, N. M., and Easter, S. S., Jr. (2000a). Modulation of cell proliferation in the embryonic retina of zebrafish (Danio rerio). *Dev. Dyn.* **219,** 391–401.

Li, Z., Joseph, N. M., and Easter, S. S., Jr. (2000b). The morphogenesis of the zebrafish eye, including a fate map of the optic vesicle. *Dev. Dyn.* **218,** 175–188.

Link, B. A., Fadool, J. M., Malicki, J., and Dowling, J. E. (2000). The zebrafish young mutation acts non-cell-autonomously to uncouple differentiation from specification for all retinal cells. *Development* **127,** 2177–2188.

Liu, I. S., Chen, J. D., Ploder, L., Vidgen, D., van der Kooy, D., Kalnins, V. I., and McInnes, R. R. (1994). Developmental expression of a novel murine homeobox gene (Chx10): Evidence for roles in determination of the neuroretina and inner nuclear layer. *Neuron* **13,** 377–393.

Loosli, F., Staub, W., Finger-Baier, K. C., Ober, E. A., Verkade, H., Wittbrodt, J., and Baier, H. (2003). Loss of eyes in zebrafish caused by mutation of chokh/rx3. *EMBO Rep.* **4,** 894–899.

Lun, K., and Brand, M. (1998). A series of no isthmus (noi) alleles of the zebrafish pax2.1 gene reveals multiple signaling events in development of the midbrain-hindbrain boundary. *Development* **125,** 3049–3062.

Maaswinkel, H., Mason, B., and Li, L. (2003). ENU-induced late-onset night blindness associated with rod photoreceptor cell degeneration in zebrafish. *Mech. Ageing Dev.* **124,** 1065–1071.

Macdonald, R., Barth, K. A., Xu, Q., Holder, N., Mikkola, I., and Wilson, S. W. (1995). Midline signalling is required for Pax gene regulation and patterning of the eyes. *Development* **121,** 3267–3278.

Macdonald, R., Scholes, J., Strahle, U., Brennan, C., Holder, N., Brand, M., and Wilson, S. W. (1997). The Pax protein Noi is required for commissural axon pathway formation in the rostral forebrain. *Development* **124,** 2397–2408.

Macdonald, R., and Wilson, S. (1997). Distribution of Pax6 protein during eye development suggests discrete roles in proliferative and differentiated visual cells. *Dev. Genes Evol.* **206,** 363–369.

Mack, A. F., and Fernald, R. D. (1995). New rods move before differentiating in adult teleost retina. *Dev. Biol.* **170,** 136–141.

Makhankov, Y. V., Rinner, O., and Neuhauss, S. C. (2004). An inexpensive device for non-invasive electroretinography in small aquatic vertebrates. *J. Neurosci. Methods* **135,** 205–210.

Malicki, J. (2000). Harnessing the power of forward genetics—analysis of neuronal diversity and patterning in the zebrafish retina. *Trends Neurosc.* **23,** 531–541.

Malicki, J., and Driever, W. (1999). *Oko meduzy* mutations affect neuronal patterning in the zebrafish retina and reveal cell-cell interactions of the retinal neuroepithelial sheet. *Development* **126,** 1235–1246.

Malicki, J., Jo, H., and Pujic, Z. (2003). Zebrafish N-cadherin, encoded by the glass onion locus, plays an essential role in retinal patterning. *Dev. Biol.* **259,** 95–108.

Malicki, J., Jo, H., Wei, X., Hsiung, M., and Pujic, Z. (2002). Analysis of gene function in the zebrafish retina. *Methods* **28,** 427–438.

Malicki, J., Neuhauss, S. C., Schier, A. F., Solnica-Krezel, L., Stemple, D. L., Stainier, D. Y., Abdelilah, S., Zwartkruis, F., Rangini, Z., and Driever, W. (1996). Mutations affecting development of the zebrafish retina. *Development* **123,** 263–273.

Marcus, R. C., Delaney, C. L., and Easter, S. S., Jr. (1999). Neurogenesis in the visual system of embryonic and adult zebrafish (Danio rerio). *Vis. Neurosci.* **16,** 417–424.

Masai, I., Lele, Z., Yamaguchi, M., Komori, A., Nakata, A., Nishiwaki, Y., Wada, H., Tanaka, H., Nojima, Y., Hammerschmidt, M., *et al.* (2003). N-cadherin mediates retinal lamination, maintenance of forebrain compartments and patterning of retinal neurites. *Development* **130,** 2479–2494.

Masai, I., Stemple, D. L., Okamoto, H., and Wilson, S. W. (2000). Midline signals regulate retinal neurogenesis in zebrafish. *Neuron* **27**, 251–263.

Merin, S. (1991). "Inherited Eye Disease." Marcel Dekker, Inc., New York.

Metcalfe, W., Myers, P., Trevarrow, B., Bass, M., and Kimmel, C. (1990). Primary neurons that express the L2/HNK-1 carbohydrate during early development in the zebrafish. *Development* **110**, 491–504.

Metcalfe, W. K. (1985). Sensory neuron growth cones comigrate with posterior lateral line primordial cells in zebrafish. *J. Comp. Neurol.* **238**, 218–224.

Moens, C. B., and Fritz, A. (1999). Techniques in neural development. *Methods Cell Biol.* **59**, 253–272.

Mohideen, M. A., Beckwith, L. G., Tsao-Wu, G. S., Moore, J. L., Wong, A. C., Chinoy, M. R., and Cheng, K. C. (2003). Histology-based screen for zebrafish mutants with abnormal cell differentiation. *Dev. Dyn.* **228**, 414–423.

Muller, H. (1857). Anatomisch-physiologische untersuchungen uber die Retina bei Menschen und Wirbelthieren. *Z. Wiss. Zool.* **8**, 1–122.

Mullins, M. C., Hammerschmidt, M., Haffter, P., and Nusslein-Volhard, C. (1994). Large-scale mutagenesis in the zebrafish: In search of genes controlling development in a vertebrate. *Curr. Biol.* **4**, 189–202.

Nakano, Y., Kim, H. R., Kawakami, A., Roy, S., Schier, A. F., and Ingham, P. W. (2004). Inactivation of dispatched 1 by the chameleon mutation disrupts Hedgehog signalling in the zebrafish embryo. *Dev. Biol.* **269**, 381–392.

Nasevicius, A., and Ekker, S. C. (2000). Effective targeted gene 'knockdown' in zebrafish. *Nat. Genet.* **26**, 216–220.

Nawrocki, W. (ed.) (1985). "Development of the Neural Retina in the Zebrafish, Brachydanio rerio." University of Oregon, Eugene, OR.

Neuhauss, S. C. (2003). Behavioral genetic approaches to visual system development and function in zebrafish. *J. Neurobiol.* **54**, 148–160.

Neuhauss, S. C., Biehlmaier, O., Seeliger, M. W., Das, T., Kohler, K., Harris, W. A., and Baier, H. (1999). Genetic disorders of vision revealed by a behavioral screen of 400 essential loci in zebrafish. *J. Neurosci.* **19**, 8603–8615.

Neuhauss, S. C., Seeliger, M. W., Schepp, C. P., and Biehlmaier, O. (2003). Retinal defects in the zebrafish bleached mutant. *Doc. Ophthalmol.* **107**, 71–78.

Neumann, C. J., and Nuesslein-Volhard, C. (2000). Patterning of the zebrafish retina by a wave of sonic hedgehog activity. *Science* **289**, 2137–2139.

Nicolson, T., Rusch, A., Friedrich, R. W., Granato, M., Ruppersberg, J. P., and Nusslein-Volhard, C. (1998). Genetic analysis of vertebrate sensory hair cell mechanosensation: The zebrafish circler mutants. *Neuron* **20**, 271–283.

Nornes, H. O., Dressler, G. R., Knapik, E. W., Deutsch, U., and Gruss, P. (1990). Spatially and temporally restricted expression of Pax2 during murine neurogenesis. *Development* **109**, 797–809.

Novak, A. E., and Ribera, A. B. (2003). Immunocytochemistry as a tool for zebrafish developmental neurobiology. *Methods Cell Sci.* **25**, 79–83.

Odenthal, J., Haffter, P., Vogelsang, E., Brand, M., van Eeden, F. J., Furutani-Seiki, M., Granato, M., Hammerschmidt, M., Heisenberg, C. P., Jiang, Y. J., *et al.* (1996). Mutations affecting the formation of the notochord in the zebrafish, Danio rerio. *Development* **123**, 103–115.

Oxtoby, E., and Jowett, T. (1993). Cloning of the zebrafish krox-20 gene (krx-20) and its expression during hindbrain development. *NAR* **21**, 1087–1095.

Parsons, M. J., Pollard, S. M., Saude, L., Feldman, B., Coutinho, P., Hirst, E. M., and Stemple, D. L. (2002). Zebrafish mutants identify an essential role for laminins in notochord formation. *Development* **129**, 3137–3146.

Passini, M. A., Levine, E. M., Canger, A. K., Raymond, P. A., and Schechter, N. (1997). Vsx-1 and Vsx-2: Differential expression of two paired-like homeobox genes during zebrafish and goldfish retinogenesis. *J. Comp. Neurol.* **388**, 495–505.

Perkins, B. D., Kainz, P. M., O'Malley, D. M., and Dowling, J. E. (2002). Transgenic expression of a GFP-rhodopsin COOH-terminal fusion protein in zebrafish rod photoreceptors. *Vis. Neurosci.* **19**, 257–264.

Peterson, R. E., Fadool, J. M., McClintock, J., and Linser, P. J. (2001). Muller cell differentiation in the zebrafish neural retina: Evidence of distinct early and late stages in cell maturation. *J. Comp. Neurol.* **429**, 530–540.

Peterson, R. E., Tu, C., and Linser, P. J. (1997). Isolation and characterization of a carbonic anhydrase homologue from the zebrafish (Danio rerio). *J. Mol. Evol.* **44**, 432–439.

Peterson, R. T., Mably, J. D., Chen, J. N., and Fishman, M. C. (2001). Convergence of distinct pathways to heart patterning revealed by the small molecule concentramide and the mutation heart-and-soul. *Curr. Biol.* **11**, 1481–1491.

Pogoda, H. M., Solnica-Krezel, L., Driever, W., and Meyer, D. (2000). The zebrafish forkhead transcription factor FoxH1/Fast1 is a modulator of nodal signaling required for organizer formation. *Curr. Biol.* **10**, 1041–1049.

Prince, V. E., Joly, L., Ekker, M., and Ho, R. K. (1998). Zebrafish hox genes: Genomic organization and modified colinear expression patterns in the trunk [In Process Citation]. *Development* **125**, 407–420.

Pujic, Z., and Malicki, J. (2001). Mutation of the zebrafish glass onion locus causes early cell-nonautonomous loss of neuroepithelial integrity followed by severe neuronal patterning defects in the retina. *Dev. Biol.* **234**, 454–469.

Pujic, Z., and Malicki, J. (2004). Retinal Pattern and the Genetic Basis of its formation in Zebrafish. *Seminars in Cell & Dev. Bio.* **15**, 105–114.

Raible, D. W., Wood, A., Hodsdon, W., Henion, P. D., Weston, J. A., and Eisen, J. S. (1992). Segregation and early dispersal of neural crest cells in the embryonic zebrafish. *Dev. Dyn.* **195**, 29–42.

Raymond, P., Barthel, L., and Curran, G. (1995). Developmental patterning of rod and cone photoreceptors in embryonic zebrafish. *J. Comp. Neurol.* **359**, 537–550.

Raymond, P., Barthel, L., Rounsifer, M., Sullivan, S., and Knight, J. (1993). Expression of rod and cone visual pigments in goldfish and zebrafish: A rhodopsin-like gene is expressed in cones. *Neuron.* **10**, 1161–1174.

Raz, E., van Luenen, H. G., Schaerringer, B., Plasterk, R. H., and Driever, W. (1998). Transposition of the nematode Caenorhabditis elegans Tc3 element in the zebrafish Danio rerio. *Curr. Biol.* **8**, 82–88.

Rebagliati, M. R., Toyama, R., Haffter, P., and Dawid, I. B. (1998). Cyclops encodes a nodal-related factor involved in midline signaling. *Proc. Natl. Acad. Sci. USA* **95**, 9932–9937.

Riley, B. B., Chiang, M., Farmer, L., and Heck, R. (1999). The deltaA gene of zebrafish mediates lateral inhibition of hair cells in the inner ear and is regulated by pax2.1. *Development* **126**, 5669–5678.

Robinson, J., Schmitt, E., and Dowling, J. (1995). Temporal and spatial patterns of opsin gene expression in zebrafish (Danio rerio). *Vis. Neurosci.* **12**, 895–906.

Rodieck, R. W. (1973). "The Vertebrate Retina. Principles of Structure and Function." W. H. Freeman & Co., San Francisco, CA.

Roeser, T., and Baier, H. (2003). Visuomotor behaviors in larval zebrafish after GFP-guided laser ablation of the optic tectum. *J. Neurosci.* **23**, 3726–3734.

Sampath, K., Rubinstein, A. L., Cheng, A. M., Liang, J. O., Fekany, K., Solnica-Krezel, L., Korzh, V., Halpern, M. E., and Wright, C. V. (1998). Induction of the zebrafish ventral brain and floorplate requires cyclops/nodal signalling. *Nature* **395**, 185–189.

Sandell, J., Martin, S., and Heinrich, G. (1994). The development of GABA immunoreactivity in the retina of the zebrafish. *J. Comp. Neurol.* **345**, 596–601.

Sanes, J. R. (1993). Topographic maps and molecular gradients. *Curr. Opin. Neurobiol.* **3**, 67–74.

Sanyanusin, P., Schimmenti, L. A., McNoe, L. A., Ward, T. A., Pierpont, M. E., Sullivan, M. J., Dobyns, W. B., and Eccles, M. R. (1995). Mutation of the PAX2 gene in a family with optic nerve colobomas, renal anomalies and vesicoureteral reflux [published erratum appears in Nat Genet 1996 May;13(1):129]. *Nat. Genet.* **9**, 358–364.

Sanyanusin, P., Schimmenti, L. A., McNoe, T. A., Ward, T. A., Pierpont, M. E., Sullivan, M. J., Dobyns, W. B., and Eccles, M. R. (1996). Mutation of the gene in a family with optic nerve colobomas, renal anomolies and vesicoureteral reflux. *Nat. Genet.* **13**, 129.

Scheer, N., and Campos-Ortega, J. A. (1999). Use of the Gal4-UAS technique for targeted gene expression in the zebrafish. *Mech. Dev.* **80**, 153–158.

Scheer, N., Groth, A., Hans, S., and Campos-Ortega, J. A. (2001). An instructive function for Notch in promoting gliogenesis in the zebrafish retina. *Development* **128**, 1099–1107.

Scheer, N., Riedl, I., Warren, J. T., Kuwada, J. Y., and Campos-Ortega, J. A. (2002). A quantitative analysis of the kinetics of Gal4 activator and effector gene expression in the zebrafish. *Mech. Dev.* **112**, 9–14.

Schier, A. F., Neuhauss, S. C., Helde, K. A., Talbot, W. S., and Driever, W. (1997). The one-eyed pinhead gene functions in mesoderm and endoderm formation in zebrafish and interacts with no tail. *Development* **124**, 327–342.

Schilling, T. F., Piotrowski, T., Grandel, H., Brand, M., Heisenberg, C. P., Jiang, Y. J., Beuchle, D., Hammerschmidt, M., Kane, D. A., Mullins, M. C., *et al.* (1996). Jaw and branchial arch mutants in zebrafish I: Branchial arches. *Development* **123**, 329–344.

Schmitt, E., and Dowling, J. (1994). Early eye morphogenesis in the Zebrafish, Brachydanio rerio. *J. Comp. Neurol.* **344**, 532–542.

Schmitt, E. A., and Dowling, J. E. (1996). Comparison of topographical patterns of ganglion and photoreceptor cell differentiation in the retina of the zebrafish, Danio rerio. *J. Comp. Neurol.* **371**, 222–234.

Schmitt, E. A., and Dowling, J. E. (1999). Early retinal development in the zebrafish, Danio rerio: Light and electron microscopic analyses. *J. Comp. Neurol.* **404**, 515–536.

Schmitz, B., Papan, C., and Campos-Ortega, J. (1993). Neurulation in the anterior trunk of the zebrafish Brachydanio rerio. *Roux's Archives of Developmental Biology* **202**, 250–259.

Seddon, J. (1994). Age-related macular degeneration: Epidemiology. *In* "Principles and Practice of Ophthalmology" (B. Albert and F. Jakobiec, eds.). pp. 1266–1274. W. B. Saunders, Philadelphia.

Seo, H. C., Drivenes, O., Ellingsen, S., and Fjose, A. (1998). Expression of two zebrafish homologues of the murine Six3 gene demarcates the initial eye primordia. *Mech. Dev.* **73**, 45–57.

Sirotkin, H. I., Gates, M. A., Kelly, P. D., Schier, A. F., and Talbot, W. S. (2000). Fast1 is required for the development of dorsal axial structures in zebrafish. *Curr. Biol.* **10**, 1051–1054.

Solnica-Krezel, L., Schier, A., and Driever, W. (1994). Efficient recovery of ENU-induced mutations from the zebrafish germline. *Genetics* **136**, 1–20.

Stenkamp, D. L., and Frey, R. A. (2003). Extraretinal and retinal hedgehog signaling sequentially regulate retinal differentiation in zebrafish. *Dev. Biol.* **258**, 349–363.

Stenkamp, D. L., Frey, R. A., Prabhudesai, S. N., and Raymond, P. A. (2000). Function for Hedgehog genes in zebrafish retinal development. *Dev. Biol.* **220**, 238–252.

Strahle, U., Blader, P., Adam, J., and Ingham, P. W. (1994). A simple and efficient procedure for non-isotopic *in situ* hybridization to sectioned material. *Trends Genet.* **10**, 75–76.

Streisinger, G., Singer, F., Walker, C., Knauber, D., and Dower, N. (1986). Segregation analyses and gene-centromere distances in zebrafish. *Genetics* **112**, 311–319.

Streisinger, G., Walker, C., Dower, N., Knauber, D., and Singer, F. (1981). Production of clones of homozygous diploid zebra fish (Brachydanio rerio). *Nature* **291**, 293–296.

Stuermer, C. A. (1988). Retinotopic organization of the developing retinotectal projection in the zebrafish embryo. *J. Neurosci.* **8**, 4513–4530.

Take-uchi, M., Clarke, J. D., and Wilson, S. W. (2003). Hedgehog signalling maintains the optic stalk-retinal interface through the regulation of Vax gene activity. *Development* **130**, 955–968.

Takechi, M., Hamaoka, T., and Kawamura, S. (2003). Fluorescence visualization of ultraviolet-sensitive cone photoreceptor development in living zebrafish. *FEBS Lett.* **553**, 90–94.

Taylor, M. R., Hurley, J. B., Van Epps, H. A., and Brockerhoff, S. E. (2004). A zebrafish model for pyruvate dehydrogenase deficiency: Rescue of neurological dysfunction and embryonic lethality using a ketogenic diet. *Proc. Natl. Acad. Sci. USA* **101**, 4584–4589.

Trowe, T., Klostermann, S., Baier, H., Granato, M., Crawford, A. D., Grunewald, B., Hoffmann, H., Karlstrom, R. O., Meyer, S. U., Muller, B., *et al.* (1996). Mutations disrupting the ordering and topographic mapping of axons in the retinotectal projection of the zebrafish, Danio rerio. *Development* **123**, 439–450.

Tsujikawa, M., and Malicki, J. (2003). Analysis of photoreceptor development and function in zebrafish retina. *IJDB,* invited review, in press.

Tsujikawa, M., and Malicki, J. (2004). Intraflagellar transport genes are essential for differentiation and survival of vertebrate sensory neurons. *Neuron* **42,** 703–716.

Turner, D., and Cepko, C. (1987). A common progenitor for neurons and glia persists in rat retina late in development. *Nature* **328,** 131–136.

Turner, D., Snyder, E., and Cepko, C. (1990). Lineage-independent determination of cell type in the embryonic mouse retina. *Neuron* **4,** 833–845.

van de Water, S., van de Wetering, M., Joore, J., Esseling, J., Bink, R., Clevers, H., and Zivkovic, D. (2001). Ectopic Wnt signal determines the eyeless phenotype of zebrafish masterblind mutant. *Development* **128,** 3877–3888.

van Eeden, F. J., Granato, M., Odenthal, J., and Haffter, P. (1999). Developmental mutant screens in the zebrafish. *Methods Cell Biol.* **60,** 21–41.

van Eeden, F. J., Granato, M., Schach, U., Brand, M., Furutani-Seiki, M., Haffter, P., Hammerschmidt, M., Heisenberg, C. P., Jiang, Y. J., Kane, D. A., *et al.* (1996). Mutations affecting somite formation and patterning in the zebrafish, Danio rerio. *Development* **123,** 153–164.

Varga, Z. M., Amores, A., Lewis, K. E., Yan, Y. L., Postlethwait, J. H., Eisen, J. S., and Westerfield, M. (2001). Zebrafish smoothened functions in ventral neural tube specification and axon tract formation. *Development* **128,** 3497–3509.

Varga, Z. M., Wegner, J., and Westerfield, M. (1999). Anterior movement of ventral diencephalic precursors separates the primordial eye field in the neural plate and requires cyclops. *Development* **126,** 5533–5546.

Vihtelic, T. S., Doro, C. J., and Hyde, D. R. (1999). Cloning and characterization of six zebrafish photoreceptor opsin cDNAs and immunolocalization of their corresponding proteins. *Vis Neurosci.* **16,** 571–585.

Vihtelic, T. S., and Hyde, D. R. (2002). Zebrafish mutagenesis yields eye morphological mutants with retinal and lens defects. *Vis. Res.* **42,** 535–540.

Vihtelic, T. S., and Hyde, D. R. (2000). Light-induced rod and cone cell death and regeneration in the adult albino zebrafish (Danio rerio) retina. *J. Neurobiol.* **44,** 289–307.

Walker, C. (1999). Haploid screens and gamma-ray mutagenesis. *Methods Cell Biol.* **60,** 43–70.

Watanabe, T., and Raff, M. (1988). Retinal astrocytes are immigrants from the optic nerve. *Nature* **332,** 834–837.

Wei, X., and Malicki, J. (2002). Nagie oko, encoding a MAGUK-family protein, is essential for cellular patterning of the retina. *Nature Genetics* **31,** 150–157.

Westerfield, M. (2000). "The Zebrafish Book." University of Oregon Press, Eugene, OR.

Wetts, R., and Fraser, S. (1988). Multipotent precursors can give rise to all major cell types of the frog retina. *Science* **239,** 1142–1145.

Wienholds, E., Schulte-Merker, S., Walderich, B., and Plasterk, R. H. (2002). Target-selected inactivation of the zebrafish rag1 gene. *Science* **297,** 99–102.

Woo, K., and Fraser, S. E. (1995). Order and coherence in the fate map of the zebrafish nervous system. *Development* **121,** 2595–2609.

Yagasaki, K., and Jacobson, S. G. (1989). Cone-rod dystrophy. Phenotypic diversity by retinal function testing. *Arch. Ophthalmol.* **107,** 701–708.

Yazulla, S., and Studholme, K. M. (2001). Neurochemical anatomy of the zebrafish retina as determined by immunocytochemistry. *J. Neurocytol.* **30,** 551–592.

Yoshida, T., Ito, A., Matsuda, N., and Mishina, M. (2002). Regulation by protein kinase A switching of axonal pathfinding of zebrafish olfactory sensory neurons through the olfactory placode-olfactory bulb boundary. *J. Neurosci.* **22,** 4964–4972.

Yoshida, T., and Mishina, M. (2003). Neuron-specific gene manipulations to transparent zebrafish embryos. *Methods Cell Sci.* **25,** 15–23.

Young, R. W. (1985). Cell differentiation in the retina of the mouse. *Anat. Rec.* **212,** 199–205.

Zhang, J., Talbot, W. S., and Schier, A. F. (1998). Positional cloning identifies zebrafish one-eyed pinhead as a permissive EGF-related ligand required during gastrulation. *Cell* **92,** 241–251.

CHAPTER 17

Instrumentation for Measuring Oculomotor Performance and Plasticity in Larval Organisms

James C. Beck, ★ **Edwin Gilland,** ★ **Robert Baker,** ★ **and David W. Tank**†

★Department of Physiology and Neuroscience
New York University School of Medicine
New York, New York 10016

†Departments of Physics and Molecular Biology
Princeton University
Princeton, New Jersey 08544

METHODS IN CELL BIOLOGY, VOL. 76
Copyright 2004, Elsevier Inc. All rights reserved.
0091-679X/04 $35.00

To study the genetic and developmental basis of sensorimotor processing, behavioral output must be quantified before the neural circuit dynamics can be investigated. The oculomotor system is ideal as it has been extensively utilized for quantitative analysis; however, no complete apparatus exists to both elicit and measure the eye movements in small genetic model organisms in real time. Instrumentation designed for much larger animals must be scaled to accommodate animals only a few millimeters in length, while accurately quantifying the motion of the minuscule eyes. To this end, a video microscope and optokinetic drum were mounted on a miniature, motorized vestibular turntable, servo-controlled with velocity and position feedback and capable of producing sinusoidal motion or position triangles with latencies less than 0.10 s and accelerations greater than $1000\,°/s^2$. The optokinetic drum, also feedback controlled, accommodated a wide range of spatial frequencies that could be nested concentrically to provide visual stimuli for both monocular and binocular testing. Larval and juvenile *Xenopus*, zebrafish, goldfish, and medaka were embedded in agarose with the head free, allowing unrestricted eye movements and normal respiration. Infrared transillumination permitted video imaging of eye movements in either light or dark. Video images were computer processed in real-time (60 Hz), producing accurate ($\pm0.1°$) eye position measurements that, in turn, could be utilized in real-time for visuomotor plasticity paradigms. This instrumentation permits high resolution ontogenetic analysis of oculomotor function in small animals as illustrated for larval zebrafish (5 to 35 dpf).

I. Introduction

The next stage for genomics in neurobiology is to discern how genetic networks operate to regulate neural function and control behavior (Bate, 1998; Gerlai, 2003). A direct investigation of the genetic basis of motor behavior can only be achieved when both the motor dynamics and underlying neural circuit dynamics are measurable; thus, the selection of which motor behavior to study from an animal's unique repertoire is critical. The vertebrate oculomotor system is an ideal candidate because both the stimuli, visual and vestibular motion, and the evoked eye movements are well defined, reproducible, and quantifiable (Robinson, 1968). Measuring oculomotor performance requires not only control of retinal visual slip and head acceleration (both linear and angular); but it also necessitates a means of animal restraint, facilitating uniform presentation of sensory stimuli while accurately measuring eye movements. Available oculomotor instrumentation has been optimized for studies in large, adult animals. Nevertheless, these systems provide a basis for designing a purpose-built apparatus for investigations of oculomotor behavior and plasticity in small genetic model organisms.

The quest for understanding the genetic basis of behavior has prompted several initial studies of oculomotor behavior in larval zebrafish. However, these investigations employed relatively simple recording and stimulating technologies and none explored the oculomotor repertoire with the degree of sophistication that has become routine in studies of adult goldfish (Marsh and Baker, 1997; Pastor et al., 1992), cat (Godaux et al., 1983; Robinson, 1976; Shinoda and Yoshida, 1974), rabbit (Barmack, 1981; Collewijn, 1969; Ito et al., 1979), primate (Fernandez and Goldberg, 1971; Lisberger et al., 1981; Melvill Jones, 1977; Skavenski and Robinson, 1973), or mouse (Boyden and Raymond, 2003; van Alphen et al., 2001).

The only investigation of the angular vestibuloocular reflex (VOR) in larval zebrafish used hand-driven rotation to provide the acceleration stimulus (Easter and Nicola, 1997). With no defined control or measure of table velocity, only a qualitative response was reported. In contrast, previous studies of the zebrafish optokinetic reflex (OKR) have employed motor-driven (Brockerhoff et al., 1995; Carvalho et al., 2002; Easter and Nicola, 1997, 1996; Rick et al., 2000) or video-projected (Roeser and Baier, 2003) striped drums to provide visual stimuli; however, a constant drum velocity was used in all experiments. While useful for genetic screens, this simple stimulus paradigm fails to distinguish visual from oculomotor performance and to quantitatively assess visuomotor maturation (Easter and Nicola, 1997, 1996). To evaluate the developing dynamic and integrative functions of central neurons responsible for the oculomotor behaviors, such as eye position (Robinson, 1989, 1975) and velocity integration (Cohen et al., 1977; Raphan et al., 1979), it is essential to employ dynamic stimuli in the form of changing stimulus frequency or amplitude combined with sinusoids and steps. The motor designs in these previous investigations lacked both the position and velocity feedback necessary to control drum kinetics for the accurate reproduction of complex motion.

Since genetic manipulations of the neural circuits underlying motor behavior may yield subtle differences in oculomotor performance, it is important to begin with an accurate measurement of eye movements. Prior work relied on two methods of measuring oculomotor output in larval animals: either direct observation (Brockerhoff et al., 1995, 1997; Carvalho et al., 2002) or video recordings (Easter and Nicola, 1997; Moorman et al., 1999; Neuhauss et al., 1999; Rick et al., 2000; Riley and Moorman, 2000; Roeser and Baier, 2003). While suitable for large-scale screens of mutant animals, direct observation yielded only a binary answer—movement or no movement. This simple result does not provide sufficient quantitative resolution to link genetic expression and motor behavior. Video recordings of eye movements can permit a more quantitative approach to behavioral analysis. Still, most reports have analyzed the behavior manually, frame-by-frame (e.g., Easter and Nicola, 1997, 1996; Moorman *et al.*, 1999; Neuhauss *et al.*, 1999; Riley and Moorman 2000). This approach was laborious, of low resolution, and practical only for short periods of data. Additional efforts have begun to utilize computer aided video analysis but have been restricted to off-line processing at low frame rates (fewer than 5 Hz; Bahadori *et al.*, 2003; Rick *et al.*, 2000; Roeser and Baier, 2003) and limited accuracy ($\pm 2°$, Roeser and Baier, 2003).

Another major compromise in larval eye measurement recordings has been the necessity to perform experiments in the light. While not a concern for visually driven behaviors, isolation and measurement of vestibular reflexes requires the removal of visual feedback, which can be achieved by measuring VOR with infrared illumination (Wilson and Melvill Jones, 1979). However, investigations into both linear (Moorman, 2001; Moorman *et al.*, 1999; Riley and Moorman, 2000) and angular (Easter and Nicola, 1997) VOR did not use available technology to measure eye movements in the dark. As an alternative, these studies were performed with a featureless drum (*ganzfeld*) to minimize visual feedback during vestibular stimulation. The *ganzfeld* employed by Easter and Nicola (1997) did not eliminate all visual motion cues since their observation of adult-like angular VOR in 5-dpf zebrafish was inconsistent with measurements made using infrared video (Beck *et al.*, 2004).

Overall, what is needed in the assessment of larval oculomotor performance, is a means of accurately measuring eye movements in the dark or light during dynamic visual or vestibular stimulation. Measurements available in real-time would, in addition, allow the use of behavioral-feedback learning paradigms to explore the ontogeny of oculomotor plasticity (Major *et al.*, 2004; Mensh *et al.*, 2004) and to potentially address the molecular mechanisms underlying memory and learning, (i.e., long term potentiation and depression) (Boyden and Raymond, 2003; Hartell, 2002; Ito, 2002).

To satisfy the prerequisites for measuring oculomotor performance in larval animals, a vertical axis vestibular turntable and optokinetic drum were constructed. This apparatus accommodates animals of small sizes (0 to 16 mm) and operates over the wide frequency range (0.01 to 10.00 Hz) necessary for both optokinetic and vestibular stimulation (Wilson and Melvill Jones, 1979). The table and drum were independent, each driven by a feedback servo-controller to provide accurate angular acceleration and visual motion stimulation, required to test visuo-vestibular interactions. In real-time (60 Hz), CCD video images of the subject's head and trunk, illuminated from below with infrared LEDs, were captured to computer and analyzed using vision processing algorithms to determine the relative position of the eye with respect to body axis with high fidelity ($\pm 0.1°$). Eye positions could be used to control drum kinetics in real-time for visuomotor plasticity paradigms. Larval and juvenile zebrafish, goldfish, *Xenopus*, and medaka, were compared under both light and dark conditions using classic oculomotor testing paradigms, demonstrating the powerful utility and flexibility of this instrumentation. Part of this work has appeared elsewhere in abstract form (Beck *et al.*, 2002).

To aid in understanding and visualization of this system in operation, supplemental movies are provided in QuickTime 6.0 format, playable on both Mac and PC computers (http://www.apple.com/quicktime/download/standalone/).

Movies are available to licensed users and can be viewed at the Science Direct Web site (http://www.sciencedirect.com) by clicking on Books and going to chapter 17, volume 76 in *Methods in Cell Biology*.

II. Methods

A. Vestibular Turntable and Drum (Supplemental Movie 1)

The vertical axis turntable comprised an aluminum base, housing the drive motor assembly, and an aluminum top connected to the base via a hollow, stainless steel shaft (Fig. 1A, Movie 1). The surface of the table contained a compact, 3-axis micromanipulator for positioning the specimen, a motorized optokinetic drum, and a CCD camera centered on the table's axis. Table-top rotation was belt driven by a compact and powerful 16-W motor (Faulhaber, Clearwater, FL; model 3042W024C001G) with 40 lb-in, planetary gear head (Faulhaber, model 30/1) inline coupled with the motor shaft. Motor velocity was monitored via an inline tachometer (Faulhaber, model 0436C). Separately connected to the table top was a precision position potentiometer that provided a voltage proportional to the angular position of the table (ETI Systems, Carlsbad, CA; model SP40B). The table motor operated with a servo-controller (Western Servo Design, Hayward, CA; model LDH-S1) that utilized both of these velocity and position feedback signals (see Movie 1).

A small drum centered on the table top provided visual stimuli to induce optokinetic reflexes (OKR; Fig. 1B). Clear, acrylic drums of different sizes, 3.5 to 8 cm, nested concentrically around the specimen holder, provided stationary, patterned, or *ganzfeld* (featureless visual surround) stimuli for both monocular and binocular testing. The moving drum was illuminated from below, through the table shaft, by visible light LEDs and was belt driven by a smaller servomotor than that used in the table base (Faulhaber, model 2251U012S1.5G). Gearing (Faulhaber, model 23/1) and tachometer (Faulhaber, model 0431) arrangement were similar. The drum was servo-controlled (Western Servo Design, model LDH-A1) independently from the table and also employed both position (ETI Systems, model SP22GS) and velocity feedback (see Movie 1). Stimulus waveforms controlling table and drum motion were created with two Agilent (Palo Alto, CA) 33120A phase linked, function generators.

The table and drum were located within a removable, light-tight wooden enclosure with a black felt curtain on the front to facilitate access to the specimen. As needed, the enclosure was maintained between 18 and 30 °C with a YSI

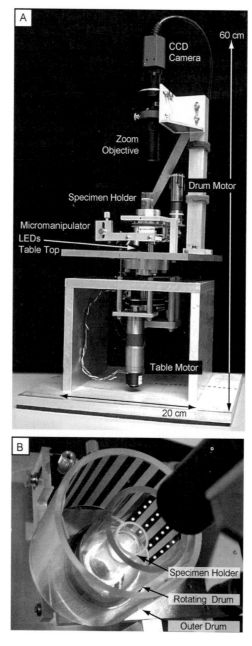

Fig. 1 Vertical axis vestibular turntable and optokinetic drum scaled to accommodate larval and juvenile animals. (A) Side view of vestibular turntable, showing table motor in the base with the optokinetic drum motor, CCD camera with zoom objective, 3-axis micromanipulator, and specimen holder on the table top. The drum has been removed to reveal the specimen holder, lying directly below

temperature controller (Yellow Springs, OH) and a screw base radiant heater (McMaster-Carr, Dayton, NJ).

B. Specimen Holder and Mounting (Supplemental Movie 2)

A transparent glass specimen holder contained the subject in 2.5 mL of water. The specimen holder comprised two pieces: (1) a round, 19-mm diameter, glass chamber that was 12-mm in height, having a floor and roof of 18-mm diameter, round coverslips (Fisher Scientific, Pittsburgh, PA), and (2) a round, 19-mm diameter, glass base 35-mm in height (Figs. 1 and 2B; Movie 2). The specimen-holder base and chamber, as well as the coverslip floor, were held together with clear epoxy (81190; Loctite Corp, Avon, OH). The roof of the specimen chamber was held in place by surface tension of the water and could be left ajar to permit oxygen exchange. The specimen holder was placed into a clear, acrylic mounting stage attached to a 3-axis micromanipulator, allowing the head to be centered on the table and to focus the animal for the camera.

Animals were held in the specimen chamber embedded in a drop of low-melting temperature agarose (2.0%; Sigma, St. Louis, MO; see Movie 2). The head region, rostral to the swim bladder in fish, was freed of agarose with a scalpel, permitting unrestricted movement of the eyes and respiration (Fig. 2A; see Movie 2). Larger and older animals were capable of aspirating the molten agarose, occasionally clogging the pharynx once solidified. This was mitigated by covering the head/pharynx with 6% methylcellulose just before enveloping with agarose. In addition, large animals could also extricate themselves from the agarose once their head was freed, which was reduced by flexing the tail into a "J" configuration before the agarose solidified. Alternatively, an electrolytically sharpened tungsten or platinum 10-μm wire, coated with a 10% benzocaine/ethanol solution, inserted perpendicular to the body through the epithelial ridge in the caudal tail was sufficient to prevent escape (Fig. 2B-3). The solidified agarose block was affixed with several 100-μm minutien pins to a 1-mm thick, clear disk (Sylgard 184; Dow Corning, Midland, MI) that snugly fit within the bottom of the specimen holder (Fig. 2B; see Movie 2). Animals that were removed from the agarose after less than an hour swam normally and could be reared to adult stages.

C. Measuring Eye Movements (Supplemental Movie 3)

A visible light-blocking filter (Wratten 88A; Kodak, Rochester, NY) in the base of the stage provided a dark area below the animal, while permitting the transmission of infrared light. Transillumination of the animal with an infrared (880 nm)

---

are the visible light and infrared LEDs. (B) Top view of specimen holder with surrounding drums, depicting potential visual stimuli arrangements that could be used for *en bloc* or, here, monocular stimulation. The outer drum provides stationary stimuli to one eye while the inner drum rotates (*white dots added for illustration*). Also see Supplemental Movie 1.

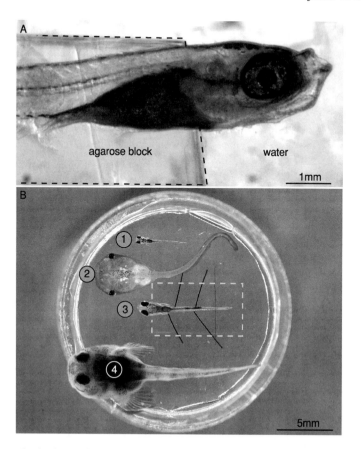

Fig. 2 Larval animals restrained in agarose with head free. (A) Medaka embedded in a block of low-melting temperature agarose (*dashed line*), with the area rostral to the swim bladder freed to allow for unrestricted eye movements and respiration. The upturned mouth of medaka is characteristic of a surface feeding fish. (B) Agarose block (*dashed line*) pinned to a clear Sylgard disk (1-mm thick) and placed in a clear-glass specimen holder (19-mm diameter). Four larval animals that were tested are illustrated, scaled to size: (1) zebrafish, (2) *X. laevis*, (3) goldfish, and (4) midshipman. Animals 1, 2, and 4 were digitally superimposed on the specimen chamber containing the goldfish (3). Also see Supplemental Movie 2 and behavioral movies (5–10).

LED (1) increased the contrast between the eyes and the head and (2) allowed video recording of eye movements in either the light or dark (as required for measuring either VOR or spontaneous fixations in open-loop conditions). The zoom objective (VZM 300i; Edmund Industrial Optics, Barrington, NJ), fitted to a monochrome CCD camera (6412; COHU, San Diego, CA), was adjusted to maximize the head and rostral trunk of the animal within the field of view. Video output was digitized with a National Instruments frame-grabber (PCI-1409; Austin, TX) and analyzed with software developed in National Instruments'

LabVIEW and IMAQ Vision (version 6 for both). The video analysis program was executed on a dual-processor PC running Microsoft Windows XP Professional with two 2.1 GHz Athlon 2400+ MP processors (Movie 3).

Before video analysis, a still image was captured (Fig. 3A-1; see Movie 3) and a region of interest (ROI) was drawn around the body to reduce overall image size used in computation (Fig. 3A-2). In addition, two ROIs were selected, one around the animal's eyes (Fig. 3A-3) and another around a reference on the body axis (Fig. 3A-4). Two video fields, odd and even, each scanned at 60 Hz are interlaced to provide a standard 30 Hz NTSC video frame. To achieve a 60-Hz sampling rate, individual video fields were captured and treated as independent images. This resulted in a reduction in height of the processed video image from 640 pixels in a video frame to 320 pixels in a video field. This change in pixel aspect ratio was taken into account within the software.

Once the real-time video processing began, the two ROIs selected around the eyes and along the body axis were binary thresholded and inverted, producing white eyes (Fig. 3B-5) and body reference (Fig. 3B-6; see Movie 3). The center of mass of the body reference was then calculated (Fig. 3B-7). To select only the eyes within the thresholded ROI (Fig. 3B-5) and to reject pigments (melanophores) in the head around the eyes, the size of each object in the eye ROI was computed. The two largest objects in the eye ROI (Fig. 3A-3) were the eyes and the center of mass of each was then determined (Fig. 3B-8). The center of each eye served as the starting point for a pixel seeding operation ("magic wand") that smoothly filled in the entire shape of each eye by selecting neighboring pixels that were of similar intensity (Fig. 3C-9; see Movie 3). The equivalent ellipse major and minor axes of each filled eye (Fig. 3C-10) and the body axis were then determined (Fig. 3C-11). The body axis connected the midpoint between the eyes to the center of the thresholded body reference (Fig. 3B-6). Eye positions (Fig. 3C-12) were then calculated as the angle of the major axis of each eye (Fig. 3C-10) relative to the body axis (Fig. 3C-11). Left and right directions were established by evaluating the vector cross product of the eye and body axis. Computed eye positions were converted to voltages (16 bit) using D/A converters on a National Instruments computer card (PCI-6052E). Copies of this LabVIEW code will be made available upon request to either J.C.B. or D.W.T.

D. Data Acquisition and Analysis

Data were digitized with an Axon Instruments (Union City, CA) Digidata 1200B and Axoscope 9.0 software executed on a computer separate from the one used to calculate eye positions. Position potentiometer and tachometer outputs for the table and drum, as well as eye positions, and the visible light LED voltage were digitized to hard disk at 200 Hz. Eye position records were subsequently imported into MATLAB and differentiated to produce eye velocity. Both the position and resulting velocity traces were filtered with a sliding-average

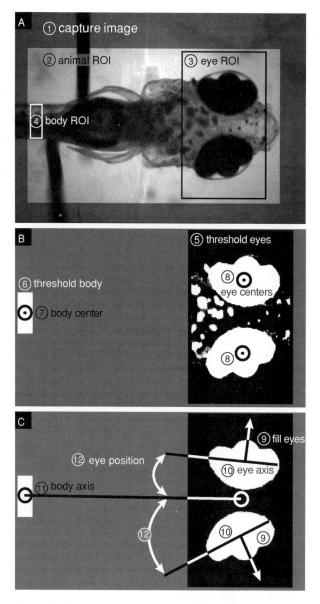

Fig. 3 Eye tracking algorithm. (A) An initial still video image was captured (1) and a region of interest (ROI) drawn around the entire animal (2, *dark border*). ROIs were drawn around the eyes (3, *black rectangle*) and around a fixed point along the body axis (4, *white rectangle*). (B) As video images were captured in real-time, the ROIs around the eyes (5) and body (6) were thresholded and inverted. The centers of the body (7) and the two eyes (8) were determined. (C) From the center of the eye, similar pixel values were filled (9). The body axis (11) was subtracted from the major axis of each eye (10) in order to determine absolute eye position (12). Also see Supplemental Movie 3.

window of 50 ms (10 sample points at 200 Hz). Drum and turntable tachometer and position traces were filtered with a sliding average window of 10 to 30 ms.

E. Experimental Animals

Animals were used in accordance with the *Guide for the Care and Use of Laboratory Animals* (http://www.nap.edu/readingroom/books/labrats/). Specific protocols were approved by the New York University School of Medicine Institutional Animal Care and Use Committee. Larval species tested included established physiological model animals: goldfish (*Carassius auratus*) and midshipmen (*Porichthys notatus*, not illustrated), as well as established developmental and genetic vertebrate models: zebrafish (*Danio rerio*), medaka (*Oryzias latipes*), and larval frogs (*Xenopus laevis* and *X. tropicalis*). Goldfish eggs were obtained from Hunting Creek Fisheries (Thurmont, MD). Zebrafish and medaka adults were obtained from Aquatic Research Organisms, Inc. (Hampton, NH), and larvae were spawned in the lab. *X. laevis* (Movie 5) and *X. tropicalis* larvae were obtained from NASCO (Fort Atkinson, WI). Midshipmen were collected in Bodega Bay, CA, and provided by Dr. Andrew Bass at Cornell University, Ithaca, NY.

III. Results and Discussion

A. Table and Drum (Supplemental Movie 1)

Current technology used to investigate oculomotor behavior in large animals generates angular vestibular stimulation by rotating an adult animal on a vestibular turntable with visual motion stimuli frequently produced by a rotating, striped optokinetic drum (Collewijn, 1991; Wilson and Melvill Jones, 1979). The drum is sufficiently large to physically surround the vestibular platform and thus occupies a large area. In our instrument, the dimensions of the vestibular turntable and optokinetic drum were significantly reduced such that the pair operated as a bench top device (Fig. 1, see Movie 1). The compact size facilitated the ease of recording from millimeter-sized animals.

Powerful motors are required to overcome the inertial mass of the table and drum to faithfully reproduce the command stimulus required for experimental paradigms that cover a wide frequency and amplitude range. Despite their size, the miniature motors chosen to drive the table and drum demonstrated the capability of producing a range of stimuli on par with systems used for oculomotor investigations in much larger animals (Fig. 4). When driven with a waveform generator, the motorized turntable produced sinusoids of up to 10 Hz, as well as position triangles and steps with an average latency of 104 ms (Fig. 4) and accelerations higher than $1000°/s^2$ at 0.5 Hz, $\pm10°/s$. Like the table, the motorized drum produced sinusoids of up to 10 Hz, as well as position steps with latencies of less than 55 ms (Fig. 4). Because of their independent control, the inertial differences between the table and drum could be compensated for by adjusting the drum

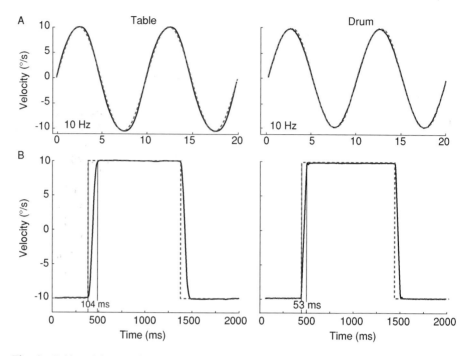

Fig. 4 Table and drum performance during sinusoid and step commands. (A) Despite the small sizes of the drive motors, both the table and drum servo-controllers were able to faithfully reproduce the command signal of 10-Hz sinusoids at ±10°/s. (B) Response to a 0.5-Hz ±10°/s step command demonstrates short latencies for both the table and drum, 104 and 53 ms, respectively, and accurate signal reproduction. Also see Supplemental Movie 1. Table, drum velocity (*solid line*). Command signal (*dashed line*).

phase relative to the table, enabling visuo-vestibular conflict training paradigms. In sum, established behavioral and stimulation paradigms commonly used in the study of the adult oculomotor system could be applied directly to larval and juvenile animals.

B. Animal Immobilization (Supplemental Movie 2)

Another issue that required equal consideration for these investigations was how to properly restrain small animals. A primary concern was that the head remained centered relative to the stimulus. Otherwise, off-axis angular vestibular stimulation would activate both the semicircular canals and the otoliths, making it impossible to distinguish between the linear and angular acceleration systems (Wilson and Melvill Jones, 1979). As reported in behavioral studies of the developing reticulospinal network, larval animals were successfully embedded in low-melting temperature agarose, with the tail subsequently freed to study swimming

(Higashijima *et al.*, 2003; Ritter *et al.*, 2001). Unfortunately, locking the head in agarose also locks the eyes in place, defeating the ability to study eye movements. Alternatively, small animals were also frequently immobilized by immersing them in a highly viscous aqueous mixture of methylcellulose (Brockerhoff *et al.*, 1995; Neuhauss *et al.*, 1999; Rick *et al.*, 2000; Roeser and Baier, 2003) or low concentration agarose (Clark, 1981). However, these methods have several drawbacks as animals immersed in thick liquids cannot respire normally and must exchange oxygen through the skin, which is only effective in the very youngest (approximately 7-dpf) cases (Rombough, 2002). In addition, larger animals can move freely through viscous solutions also limiting its use to early ontogenetic studies (Clark, 1981). More germane, it is likely that the use of a highly viscous mixture to restrain the larval animal would impact oculomotor performance by creating significant drag to eye movement, a possibility tested by placing 5-dpf zebrafish in a 3% methylcellulose solution (see Fig. 8 later in this chapter).

Embedding the animal in a block of low-melting temperature agarose, as done for swimming (Higashijima *et al.*, 2003; Ritter *et al.*, 2001) but with the head and gills exposed, represented an ideal solution to the problems outlined earlier (Fig. 2, see Movie 2). After initial encasement, the agarose covering the head and gills could be quickly removed, and the block trimmed to an appropriate size to fit in the specimen holder. While encased in agarose, the specimen was easily manipulated and could be transferred from Petri dish to holder without injury. Of importance, respiration and eye movements were unencumbered (Fig. 2A). The agarose also acted as a support for minutien (100 μm) pins, allowing the block to be fixed in place (Fig. 2B, see Movie 2). If need be, the animal could be safely removed from the agarose holder and reared to adulthood, providing the opportunity for longitudinal observations and the preservation of isolated mutant animals. Thus, agarose provides a means of gentle, yet effective immobilization with little to no biological perturbation.

By combining a specimen holder constructed entirely of glass with the transparency of the agarose, there was an almost unobstructed visual field encircling the preparation (Fig. 2B). The 19-mm diameter of the specimen holder easily accommodated a wide range of species, including zebrafish, goldfish, medaka, *Xenopus*, and the midshipmen *P. notatus* (Fig. 2B). These animals ranged in size from under 4-mm to greater than 16-mm in length. The largest animals were accommodated by flexing the tail to match the curvature of the holder. By enlarging the diameter of the specimen holder, the oculomotor behavior of even juvenile animals could be studied.

C. Eye Position Measurements (Supplemental Movies 2–4)

The *sine qua non* of any investigation of motor performance is the ability to measure behavioral output. The current standard of eye position measurement is the use of a scleral search coil arrangement (Robinson, 1963). Comprised of a few turns of wire, a coil is attached directly to the eye. Within an alternating magnetic field, coil voltage is proportional to eye position and can be measured with an accuracy of 0.025° at 1000 Hz (Fuchs and Robinson, 1966). Because this approach is clearly intractable for larval animals, video recordings of eye movements have been the preferred alternative (Easter and Nicola, 1997; Neuhauss *et al.*, 1999; Rick *et al.*, 2000; Roeser and Baier, 2003). Video imaging provides a noninvasive yet quantitative approach for measuring eye position and exploits the transparency found in many young animals. Like the scleral search coil system, standard video recordings have a comparatively high temporal resolution that could be employed for real-time measurements. In previous investigations, the limited spatial resolution (±2°; Roeser and Baier, 2003) and low sampling rate used (less than 5 Hz, Bahadori *et al.*, 2003; Rick *et al.*, 2000; Roeser and Baier, 2003) failed to capitalize on this potential.

In contrast, the algorithm presented here (see Fig. 3, Movie 3) operated at 60 Hz by individually capturing the odd and even video fields provided by a standard NTSC CCD camera. The resulting video images were scaled to 320×240 pixels to ensure a square-aspect ratio for eye position measurement, which minimized the complexity in calculating eye position. The efficiency of the algorithm, combined with the computational speed of the dual-processor computer, generated eye position records in real-time with minimal delay (as discussed later in this chapter).

The eye-position measurement algorithm was tested for accuracy and resolution by measuring the rotation angle of computer generated eyes (Fig. 5, see Movie 4). To generate the virtual eye, an ellipse of approximately the same size and location as a larval zebrafish eye was linearly rotated from $-40°$ to $+40°$ in 0.01° increments (Adobe Illustrator, Adobe Systems, San Jose, CA), adequately encompassing the normal ocular range of ±20°. The computer-generated images were processed using the software algorithm (Fig. 5, see Movie 4), and the measured position was compared with the actual position.

Eye movement measurements produced from video systems are typically noisier than those obtained using the well-refined scleral search coil technique. Nevertheless, the algorithm described here demonstrated a remarkable degree of resolution and accuracy throughout the entire range of motion. Linear regression analysis of the measured versus actual position of the virtual eye yielded an excellent linear fit with a slope of 0.9989 ($r = 0.9999$, see Fig. 5B). Because the actual change in virtual eye position was a straight line, the average difference between each step in the measured eye position determined the measurement resolution or mean resolvable step, which was 0.00997°, close to the actual step value of 0.01°. The accuracy was quantified by calculating the standard deviation of the difference between the measured position and the actual position. Measurement accuracy for this virtual eye was determined to be ±0.13°, an order of magnitude improvement over the off-line measurement of ±2° reported (Roeser and Baier, 2003) and closer to that of the search coil (Fuchs and Robinson, 1966).

It is important to note that the measurement accuracy was calculated for only a 320×240-image size, representing a trade-off required to achieve a high frame rate with a standard CCD camera. Indeed, accuracy and image size are closely

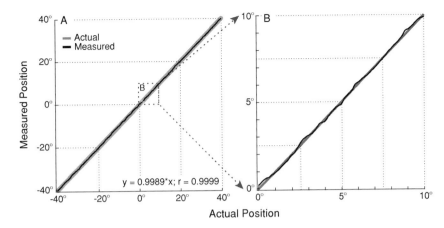

Fig. 5 Accuracy of video eye position measurement system. (A) A computer-generated virtual eye (*ellipse*) was rotated in 0.01° increments over ±40° (actual position, *thick solid grey line*), and measurements of angular rotation determined using the video measurement algorithm, slightly filtered with a 5-point sliding window (measured position, *black line*). Linear regression of the measured positions was well fit (r = 0.9999). (B) Magnification of a smaller range (0 to 10°) highlights the correlation between measured and actual rotation. Also see Supplemental Movie 4.

related. By doubling pixel dimensions (640 × 480), one could improve measurement accuracy to approximately ±0.05 (data not shown). This improvement could be achieved by utilizing a nonstandard, progressive scan camera capturing whole frames (640 × 480) at 60 Hz. Alternatively, accuracy could also be increased by filling the video image with the eyes and a caudal point on the head firmly embedded in agarose.

Accurate timing relationships between the calculated eye position and the table or drum output signals are critical for studying latency during step stimulation or phase during Bode analysis. Digitizing all of the signals through the same apparatus was a simple solution to solve the majority of issues surrounding signal synchrony. Even though the video analysis system operated in real-time, delays were introduced into the eye-position records. To determine the average latency, two ellipses (the eyes) and a square (body) were printed on a piece of clear acetate and directly attached to the optokinetic drum. The position of one ellipse was measured during sinusoidal rotations of the drum (0.065 to 4 Hz). If a processing delay was not produced by the video system, the phase lag between eye velocity and drum velocity would be 0° throughout the frequency range. However, a noticeable phase lag was observed that increased with frequency (Fig. 6). Processing lag was calculated to be 38.1 ms (±1.1), which was then used to shift the eye-position records relative to the table and drum output by 8 sample points (40 ms). Correcting for the phase lag in this manner yielded an average delay of 2.8 ms (±1.2), below the 5-ms sampling resolution for a 200 Hz digitization rate. For

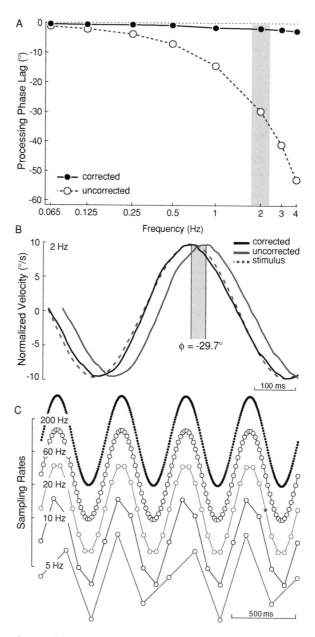

Fig. 6 Impact of processing lag and of sampling rate on signal acquisition. (A) Computing eye positions from video images introduced an average time delay of 38.1 ms (±1.1) into the eye position records that caused a marked phase shift as command frequency increased (*open circles*). (B) At 2 Hz (*dark band in A*), the phase lag was 29.7°. Correcting for the time-delay reduced the lag to only a few degrees even at 4 Hz (*A, filled circles*). (C) Sampling rate has a large impact on the ability to accurately capture high-frequency signals, here 2 Hz.

higher frequencies, more accurate phase correction was achieved by directly subtracting the processing lag from any calculated timing values.

The eye-position measurements presented here were limited by the 60-Hz scan rate of the camera. A comparison of how sample rate affects the capture of a high-frequency (2-Hz) signal can be seen in Fig. 6C. Sampling at 5 or 10 Hz (as in Rick *et al.*, 2000; Roeser and Baier, 2003) produced waveforms that were highly distorted and only vaguely resembling sinusoids. A 20-Hz sampling rate was a large improvement; however, higher sampling rates (60 and 200 Hz) were necessary to capture the signal with fidelity. While a 60-Hz sampling rate was found to be adequate to measure many aspects of oculomotor performance (i.e., slow phase), the temporal resolution of 17 ms limited utility in saccade and step analysis. Advances in imaging technology have made widely available infrared-sensitive cameras that are both high in speed and in resolution yet small in size. For example, the Pulnix PC-640CL is a digital CMOS camera that can achieve greater than 180 Hz at nearly full (640×480) resolution. Therefore, combined with the availability of ever-increasing computational power, it is reasonable to expect newer systems utilizing this measurement algorithm to operate in excess of 180 Hz with a computational accuracy approaching $\pm 0.05°$.

D. Behavioral Recordings (Supplemental Movies 6–10)

1. Optokinetic Measurements

Compensatory eye movements are the result of both vestibular and optokinetic pathways that work together to minimize retinal motion. Each sensory subsystem can be isolated for study by using either vestibular stimuli in the absence of visual input or visual stimuli in the absence of head movement, respectively (Wilson and Melvill Jones, 1979). Significantly, oculomotor performance cannot be fully assessed using only simple constant velocity stimuli (Carvalho *et al.*, 2002; Easter and Nicola, 1997, 1996; Rick *et al.*, 2000; Roeser and Baier, 2003). Instead, more dynamic stimuli are required to determine latencies in pathways, as well as to distinguish positional and velocity components of neuronal signaling. The more common method to dynamically assess the performance of each pathway, independent of the other, is through the use of linear systems frequency analysis (Fernandez and Goldberg, 1971; Melvill Jones and Milsum, 1971). This technique (Bode analysis) often employs sinusoidal stimuli throughout a wide frequency range at a constant velocity amplitude. The Bode plot generated by this analysis provides a measure of eye velocity gain—the ratio of eye velocity to stimulus velocity—and phase relative to the stimulus (see Fig. 9 later in this chapter, Beck *et al.*, 2004).

An example of OKR testing is illustrated with a low-frequency visual stimulus (0.0325 Hz) in a 35-dpf zebrafish (Fig. 7A, see Movie 6). OKR is a combination of slow-phase eye movements (Fig. 7A, top trace, solid lines) and fast, resetting phases (Fig. 7A, top trace, vertical dashed lines). Because stimuli for Bode analysis normally maintain a constant velocity amplitude with increasing frequency, the

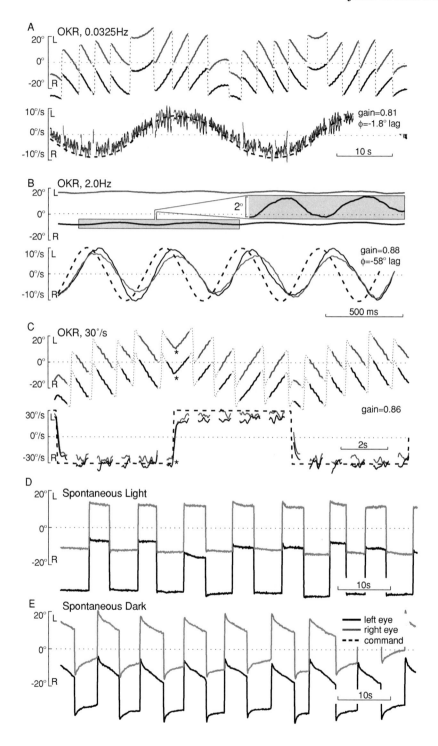

position amplitude of the stimulus actually *decreases* proportionately. Consequently, stimulation with high-frequency sinusoids produces small rotation excursions and often elicits even smaller excursions of eye position. This represents a real test of the accuracy and resolution of the eye tracking system. For example, maintaining a drum velocity amplitude of ±10°/s, as in Fig. 7A, but, at 2 Hz, reduced the drum position amplitude to approximately ±1°. As illustrated in Fig. 7B (Movie 7), a 20-dpf zebrafish was visually stimulated with sinusoids at 2 Hz ±10°/s. Despite the small change in eye position (±1°), the eye tracking system exhibited sufficient resolution (Fig. 7B, top traces, inset) to generate velocity traces for gain and phase analysis with little noise (Fig. 7B, bottom traces). The average gain in this case was 0.88, with a −58° phase lag.

Previous vertebrate studies divided the visuomotor system into direct and indirect components for analysis (Cohen *et al.*, 1977; Marsh and Baker, 1997). The only means of observing these aspects of optokinetic behavior is through the use of abrupt high-acceleration stimuli (e.g., a velocity step). Step performance is determined, in part, by resetting saccades that are of sufficient frequency and amplitude to maintain a compensatory eye velocity. Therefore, velocity steps are also a tool to indirectly assess saccadic performance. In addition, prolonged step stimuli are required to measure the time constant of velocity storage integration. To illustrate, optokinetic drum velocity steps of ±30°/s at 0.1 Hz were employed to induce an OKR in a 15-dpf medaka (Fig. 7C, see Movie 8). The high-speed processing of the eye position tracking algorithm was able to capture both the sharp change in eye position during the drum turn around (Fig. 7C, top traces, asterisks), as well as the fast phases, in sufficient detail (Fig. 7C, top traces, dashed lines). The fact that eye velocity, not eye position, closely correlated with drum velocity (average gain = 0.86) is evidence of the importance to present and analyze data in the velocity domain (Fig. 7A).

Measurements of spontaneous activity often require several minutes of recordings to accumulate enough data for analysis. Such measurements can be easily acquired in either the light or dark. Spontaneous activity in the light provides a measure of how well the oculomotor system utilizes visual feedback to maintain eye position, with position drifts indicating abnormal function. Here, a 10-dpf zebrafish exhibited stable eye positions with no visible drift with time (Fig. 7D).

Fig. 7 Behavioral recordings of optokinetic stimulation. (A) Low-frequency optokinetic stimulation (0.032 Hz, ±10°/s) showing slow and fast phases (upper trace; *solid and dashed lines*, respectively) in a 35-dpf zebrafish. Eye velocity (*bottom*) follows stimulus velocity (Movie 6). (B) At 2 Hz ±10°/s , drum rotation and eye movements were approximately ±1° in a 20-dpf zebrafish (also see Movie 7). (C) OKR stimulation using drum velocity steps of ±30°/s at 0.1 Hz in a 15-dpf medaka (Movie 8). Change in direction of the slow phase (*asterisks*) during reversal of drum velocity (bottom trace, *dashed line*). Recordings of spontaneous activity in a 10-dpf zebrafish both in the light (D) and dark (E). Most animals (A to E) exhibited a 10° nasal bias in eye position. Zero eye position was the eye direction perpendicular to the body axis. Left eye = *black traces*; right eye = *gray traces*; command velocity = *black dashed line*.

Having established a properly functioning visuomotor system, spontaneous activity in the dark can then be assessed. Measurements in the dark remove visual feedback (open loop) and permit an ontogenetic assessment of oculomotor neural integration. In adults, maintaining eye position in the dark with minimal drift is consistent with a properly functioning and tuned oculomotor integrator (Aksay *et al.*, 2000; Major *et al.*, 2004; Mensh *et al.*, 2004; Pastor *et al.*, 1994; Seung, 1996). The same, young zebrafish (Fig. 7E) exhibited a centripetal drift of eye position, which is part of the orderly development of position in holding larval fish (Beck *et al.*, 2003).

2. Negative Effect of Methylcellulose on Optokinetic Performance

Methylcellulose has been extensively used in the oculomotor testing of larval fish (Brockerhoff *et al.*, 1995; Easter and Nicola, 1997, 1996; Moorman, 2001; Moorman *et al.*, 1999; Neuhauss *et al.*, 1999; Rick *et al.*, 2000; Riley and Moorman, 2000; Roeser and Baier, 2003) since, at low concentrations, it is excellent at restricting movement. Therefore, it was logical to assume that methylcellulose might restrict movement of the eyes as well. To test this, 5-dpf larval zebrafish ($n = 5$) were mounted in agarose blocks as described and OKR was measured from 0.065 to 2.0 Hz $\pm 10°/s$ (Fig. 8). Water surrounding the animals was then replaced with a 3% methylcellulose solution, selecting an average of concentrations previously used. When OKR was measured across the same frequency range, oculomotor performance was decreased. With a 0.065-Hz stimulus, control OKR gain was 0.78 (Fig. 8A) and, in the presence of methylcellulose (Fig. 8B), eye-velocity gain dropped to 0.29, returning to control levels upon washout (Fig. 8C). Median drop for all frequencies was 63% (range 37 to 351%). A summary of the results (Fig. 8D) demonstrates that methylcellulose impacted performance at all frequencies, showing the largest drop at the highest.

3. Summary of Optokinetic and Vestibular Performance in Zebrafish

A quantitative, ontogenetic analysis of oculomotor performance in larval zebrafish, medaka, and goldfish has been performed (Beck *et al.*, 2004). The results were in contrast to previous investigations that concluded the oculomotor system of larval zebrafish was adult-like by 4-dpf (Easter and Nicola, 1997 and 1996). Optokinetic performance measured from 5 to 35 dpf (Fig. 9) showed that, while larval zebrafish of all ages have a wide OKR frequency response (Fig. 9A), their ability to follow stimuli of increasing velocity amplitude was initially poor and improved with maturity (Fig. 9B and C). Saccade performance (velocity, amplitude, and frequency) also improved with maturity and significantly impacted eye velocity at higher stimulus velocities (Beck *et al.*, 2004). Although it is tempting to utilize saccade frequency as a measure of OKR performance (Rick *et al.*, 2000), it should be avoided since optokinetic gain is only loosely correlated with saccade frequency (Beck *et al.*, 2004).

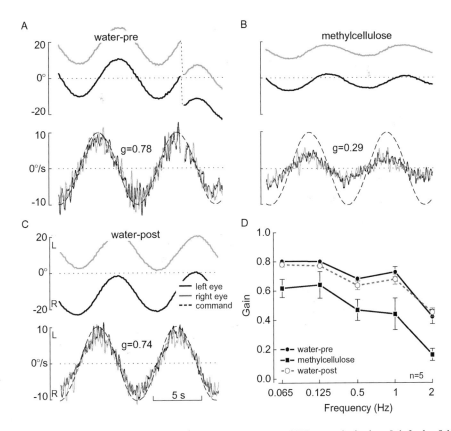

Fig. 8 Effect of methylcellulose on oculomotor performance. OKR eye velocity in a 5-dpf zebrafish at 0.065 Hz $\pm 10°/s$ (gain = 0.78) in water (A) after replacement with a 3% methylcellulose solution (B, gain = 0.29), and after washout of methylcellulose (C, gain = 0.74). *Black trace*: left eye; *gray trace*: right eye; *dashed line*: drum velocity. (D) OKR Bode frequency plot before, during, and after 3% methylcellulose treatment in 5-dpf zebrafish. Bars are standard error, $n = 5$.

Also in contrast to other investigations (Easter and Nicola, 1997), a vestibuloocular reflex was never observed in young larval zebrafish (Movie 9) and only consistently appeared in older animals (Fig. 10, Beck *et al.*, 2004). After 2 weeks of age, a VOR could be observed but was of very low amplitude even with high accelerations (Beck *et al.*, 2004). For example, a typical zebrafish exhibited almost no VOR at low stimulus intensities (Fig. 10A, 0.5 Hz, $\pm 60°/s$,) but demonstrated a slightly stronger response with increased frequency (Fig. 10B) as accelerations became higher (here $\pm 1000°/s^2$). As zebrafish grew in size, their VOR became more sensitive to lower accelerations (Fig. 10C, 0.5 Hz, $\pm 60°/s$, see Movie 10). For comparison, at 0.5 Hz, the average OKR gain in a 5-dpf animal was 0.8 while the VOR gain in a 35-dpf zebrafish barely reaches a gain of 0.2

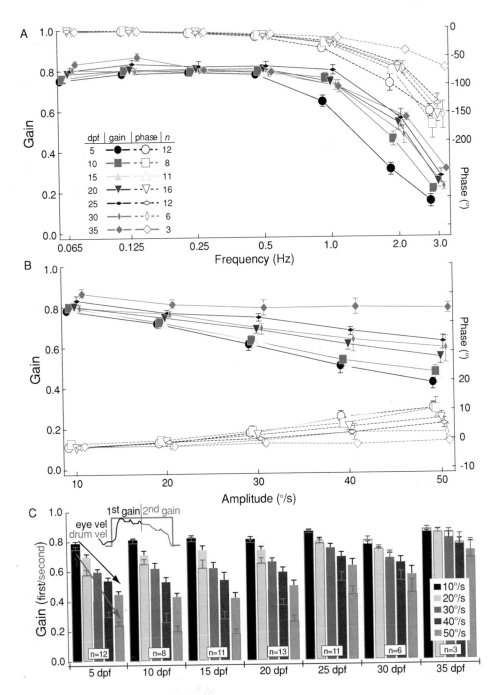

(Fig. 10D, Beck *et al.*, 2004). By adulthood, VOR gain in zebrafish was close to the ideal gain value of 1.0, especially with increasing frequency (Fig. 10E, Gilland *et al.*, 1996). Based on theoretical work relating semicircular canal morphology (i.e., size) and sensitivity (Muller, 1994, 1999; Rabbitt, 1999; Rabbitt *et al.*, 2003), we propose that the lack of an effective angular VOR in larval fish is the result of the small canal lumen diameter, with VOR sensitivity increasing proportional to animal size. Therefore, contrary to the conclusions of Easter and Niccola (1997), who found no difference in the ontogenetic timing of OKR and VOR behaviors, our findings strongly suggest that natural selection has favored the evolution of a system detecting visual over angular motion for early survival in freely swimming larva.

IV. Conclusion

The miniaturized vestibular turntable and optokinetic drum, combined with the novel approach to larval animal immobilization and real-time eye position measurement, represent significant advances in the ability to study eye movements in several species of larval and juvenile animals. Classical physiological methods, combined with imaging approaches, allow structural changes to neuronal architecture as well as genetic mis-expression to be effectively and quantitatively linked with functional changes in oculomotor behavior.

V. Supplemental Movie Descriptions

The supplemental movies provided here are referred to throughout the text and are specifically noted at each section heading. The movies were saved in a QuickTime 6.0 format and are playable on both Mac and PC computers with freely available software (http://www.apple.com/quicktime/download/standalone/). Movies available may be found at the Science Direct Web site (http://www.sciencedirect.com).

Fig. 9 Development of zebrafish OKR. (A) Bode analysis of zebrafish (5 to 35 dpf) from 0.065 to 3.0 Hz at $\pm10°$/s. (B) Phase and gain plot of zebrafish OKR in response to sinusoids (0.125 Hz) of increasing velocity. Traces same as legend in A. (C) Eye velocity gains in response to bidirectional velocity steps (0.1 Hz). Gains were calculated at each drum velocity for (1) the first 2.5 s of the 5-s step (*solid bar*) and (2) the remaining 2.5 s (*open, red bars*), also see inset. Arrows follow the trend in decreasing performance with increasing amplitude between first (*black*) and second (*red*) step halves. All data were averaged through several cycles and offset slightly (A and B) for readability. (See Color Insert.)

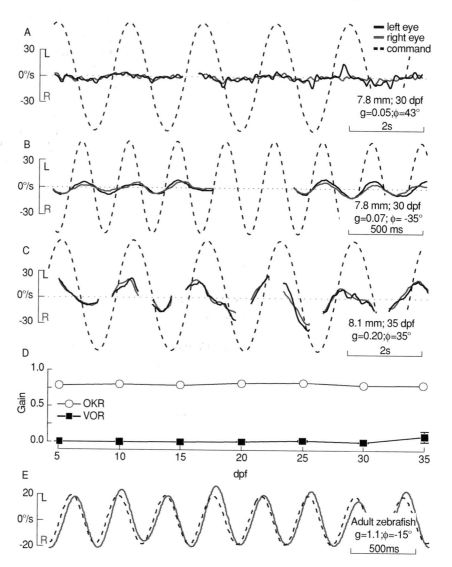

Fig. 10 Zebrafish VOR. VOR was barely detected (A) in a 30-dpf zebrafish at 0.5 Hz, ±60 °/s. At 3.0 Hz, ±45 °/s (B), acceleration forces are higher (±1000 °/s^2) and a slightly larger response was observed. (C) Older and larger animals exhibited increased VOR sensitivity at lower table amplitudes (0.5 Hz, ±60 °/s). (D) Age comparison of OKR and VOR gain at 0.5 Hz (*n* same as legend in Fig. 9A). (E) Adult zebrafish showed an almost perfect compensatory eye velocity with a table stimulus of 3.0 Hz, ±20 °/s. Table velocity (A to E) was inverted to facilitate comparison with eye velocity. Spaces in traces are where fast phases have been removed. See Movies 9 and 10.

A. Movie 1

Operation of motorized vestibular turntable and optokinetic drum. A brief identification of the parts of the apparatus followed by a demonstration of vestibular, optokinetic, and visuo-vestibular stimuli. Vestibular stimuli in order shown are sinusoids at 0.125 Hz, $\pm20°$/s and 0.5 Hz, $\pm60°$/s, as well as velocity steps at 1.0 Hz, $\pm60°$/s and 10.0 Hz, $\pm10°$/s. Optokinetic stimuli are sinusoids at 0.125 Hz, $\pm30°$/s, velocity steps at 0.1 Hz, and high frequency sinusoids (0.5 and 2.0 Hz, $\pm10°$/s). The movie concludes with the turntable and drum producing stimuli for VOR in the light (gain 1), gain up training (gain 2$\times$), and gain down training (gain 0 or cancellation). Length: 2:43; size: 5.7 MB.

B. Movie 2

A step-by-step demonstration of embedding a 5-dpf zebrafish in agarose and preparing the animal for eye movement recordings. Although all steps are shown in their entirety, events not pertaining directly to the process have been edited for time (e.g., changing focus). Length: 4:28; size: 6.0 MB.

C. Movie 3

Step-by-step demonstration of the eye movement algorithm in operation, from initial acquisition to adjusting parameters for optimal eye tracking. Length: 0:58; size: 11.0 MB.

D. Movie 4

An example of the eye tracking algorithm in action, measuring the position of a computer-generated virtual eye (*blue*) that was rotated linearly $\pm40°$ in $0.1°$ increments. The green trace superimposed on the measured position record is the actual position record. Length: 0:34; size: 3.5 MB.

E. Movie 5

Optokinetic response in *Xenopus laevis*. Wild-type *X. laevis* (35 dpf, 13.4-mm total length, stage 48 to 50) was stimulated with a drum velocity of $\pm10°$/s at 0.25 Hz, demonstrating the utility of the apparatus for testing other species. Eye velocity was similar to stimulus velocity (mean gain = 0.82); however, eye movements were jerky and not as smooth as in larval fish. Superimposed on the video image of the animal (*top*) are lines depicting the body axis (*green*) and direction of eye rotation (*blue: left eye; red: right eye*). Below the video image are measurements of eye position, as well as eye velocity and stimulus velocity (*green*). Measurement of time is along the abscissa in seconds. To permit reduced movie file size, sampling rate was reduced to 24 Hz. Eye velocity traces were smoothed with a Gaussian filter off line. Stimulus waveforms were based on original recording. Length: 0:42; size: 9.3 MB.

F. Movie 6

Measurement of OKR in a 25-dpf zebrafish at a low frequency (0.03125 Hz). Traces are same as in Move 5. The abrupt changes in the eye position traces (*middle*) correspond to resetting fast phases in eye velocity. Eye velocity closely matched stimulus velocity and phase (mean: gain = 0.84, phase = 0.4° lag). Movie plays at 2× normal speed. Length: 0:45; size: 9.1 MB.

G. Movie 7

High-frequency OKR in a 40-dpf goldfish. Drum rotation was 2.0 Hz ±10°/s, producing modest changes in drum position and small changes in eye position. Eye velocity nearly matched drum velocity (gain = 0.67). Because of processing delays in the visuomotor system, the eye velocity is out of phase with stimulus velocity and lags by 53°. Length: 0:42; size: 12.3 MB.

H. Movie 8

OKR drum velocity steps of ±30 °/s at 0.1 Hz in a 30-dpf medaka, used to assess both fixed and dynamic neural components of the visuomotor system. A slight buildup in eye velocity during the course of the step was observed. Length: 0:45; size: 10.0 MB.

I. Movie 9

Angular VOR was absent in 4-dpf larvae. A small, young zebrafish larvae (here, 3.8 mm; 4 dpf) was rotated sinusoidally at 0.5 Hz, ±30°/s. A motion tracer within the water column partially reflects the angular acceleration forces the zebrafish larva is experiencing. The inset within the larval video image is the table activity during this stimulus paradigm. Length: 0:28; size: 11 MB.

J. Movie 10

Measurement of the angular VOR in a 25-dpf medaka during sinusoidal rotation in the dark at 1 Hz, ±60 °/s. The angular VOR was robust as eye velocity was nearly the same as the table velocity (gain = 0.9, phase = 9° lead). Table velocity trace was inverted to facilitate comparison with eye velocity. Length: 0:38; size: 7.4 MB.

Acknowledgments

The authors wish to thank Alfred Benedek for constructing the vestibular turntable and drum, Ray Stepnoski (Lucent Technology) for assistance with the construction of an early prototype, and Dr. Sebastian Seung (M.I.T.) for initial advice with eye detection software. This work was supported by grants from the National Institutes of Health (D.W.T and R.B.), as well as a National Research Service Award from the National Eye Institute (J.C.B.).

References

Aksay, E., Baker, R., Seung, H. S., and Tank, D. W. (2000). Anatomy and discharge properties of premotor neurons in the goldfish medulla that have eye-position signals during fixations. *J. Neurophysiol.* **84**, 1035–1049.

Bahadori, R., Huber, M., Rinner, O., Seeliger, M. W., Geiger-Rudolph, S., Geisler, R., and Neuhauss, S. C. (2003). Retinal function and morphology in two zebrafish models of oculo-renal syndromes. *Eur. J. Neurosci.* **18**, 1377–1386.

Barmack, N. H. (1981). A comparison of the horizontal and vertical vestibulo-ocular reflexes of the rabbit. *J. Physiol.* **314**, 547–564.

Bate, M. (1998). Making sense of behavior. *Int. J. Dev. Biol.* **42**, 507–509.

Beck, J. C., Gilland, E., Tank, D. W., and Baker, R. (2004). Quantifying the ontogeny of optokinetic and vestibuloocular behaviors in zebrafish, medaka, and goldfish. *J. Neurophysiol.* **92**, in press.

Beck, J. C., Tank, D. W., and Baker, R. (2003). Ontogeny of persistent neural activity: Maturation of gaze holding in zebrafish. *In* "2003 Abstract Viewer/Itinerary Planner" (Online ed.). Society for Neuroscience, Washington, DC.

Beck, J. C., Tank, D. W., Gilland, E., and Baker, R. (2002). Instrumentation for measuring oculomotor performance and learning in larval and juvenile fish. *In* "2002 Abstract Viewer/Itinerary Planner" (Online ed.). Society for Neuroscience, Washington, DC.

Boyden, E. S., and Raymond, J. L. (2003). Active reversal of motor memories reveals rules governing memory encoding. *Neuron* **39**, 1031–1042.

Brockerhoff, S. E., Hurley, J. B., Janssen-Bienhold, U., Neuhauss, S. C., Driever, W., and Dowling, J. E. (1995). A behavioral screen for isolating zebrafish mutants with visual system defects. *Proc. Natl. Acad. Sci. USA* **92**, 10545–10549.

Brockerhoff, S. E., Hurley, J. B., Niemi, G. A., and Dowling, J. E. (1997). A new form of inherited red-blindness identified in zebrafish. *J. Neurosci.* **17**, 4236–4242.

Carvalho, P. S. M., Noltie, D. B., and Tillitt, D. E. (2002). Ontogenetic improvement of visual function in the medaka *Oryzias latipes* based on an optomotor testing system for larval and adult fish. *Animal Behaviour* **64**, 1–10.

Clark, D. T. (1981). Visual Responses in Developing Zebrafish (*Brachydanio rerio*). Ph. D. Thesis. University of Oregon.

Cohen, B., Matuso, V., and Raphan, T. (1977). Quantitative analysis of the velocity characteristics of the optokinetic nystagmus and optokinetic after-nystagmus. *J. Physiol.* **270**, 321–344.

Collewijn, H. (1991). The optokinetic contribution. *In* "Eye Movements" (R. H. S. Carpenter, ed.), pp. 45–70. CRC Press, Boca Raton, FL.

Collewijn, H. (1969). Optokinetic eye movements in the rabbit: Input-output relations. *Vision Res.* **9**, 117–132.

Easter, S. S., Jr., and Nicola, G. N. (1997). The development of eye movements in the zebrafish (*Danio rerio*). *Dev. Psychobiol.* **31**, 267–276.

Easter, S. S., Jr., and Nicola, G. N. (1996). The development of vision in the zebrafish (*Danio rerio*). *Dev. Biol.* **180**, 646–663.

Fernandez, C., and Goldberg, J. M. (1971). Physiology of peripheral neurons innervating semicircular canals of the squirrel monkey. II. Response to sinusoidal stimulation and dynamics of peripheral vestibular system. *J. Neurophysiol.* **34**, 661–675.

Fuchs, A. F., and Robinson, D. A. (1966). A method for measuring horizontal and vertical eye movement chronically in the monkey. *J. Appl. Physiol.* **21**, 1068–1070.

Gerlai, R. (2003). Zebra fish: An uncharted behavior genetic model. *Behav. Genet.* **33**, 461–468.

Gilland, E., Marsh, E., Suwa, H., and Baker, R. (1996). Oculomotor performance and adaptation in the zebrafish and closely related cyprinids. *Soc. Neuro. Abstr.* **22**, 1833.

Godaux, E., Gobert, C., and Halleux, J. (1983). Vestibuloocular reflex, optokinetic response, and their interactions in the alert cat. *Exp. Neurol.* **80**, 42–54.

Hartell, N. A. (2002). Parallel fiber plasticity. *Cerebellum* **1**, 3–18.

Higashijima, S.-i., Masino, M. A., Mandel, G., and Fetcho, J. R. (2003). Imaging neuronal activity during zebrafish behavior with a genetically encoded calcium indicator. *J. Neurophysiol.* **90,** 3986–3997.

Ito, M. (2002). The molecular organization of cerebellar long-term depression. *Nat. Rev. Neurosci.* **3,** 896–902.

Ito, M., Jastreboff, P. J., and Miyashita, Y. (1979). Adaptive modification of the rabbit's horizontal vestibulo-ocular reflex during sustained vestibular and optokinetic stimulation. *Exp. Brain Res.* **37,** 17–30.

Lisberger, S. G., Miles, F. A., Optican, L. M., and Eighmy, B. B. (1981). Optokinetic response in monkey: Underlying mechanisms and their sensitivity to long-term adaptive changes in vestibulooocular reflex. *J. Neurophysiol.* **45,** 869–890.

Major, G., Baker, R., Aksaya, E., Mensh, B, Seung, H. S., and Tank, D. W. (2004). Plasticity and tuning by visual feedback of the stability of a neural integrator. *Proc. Natl. Acad. Sci. USA* **101,** 7739–7744.

Marsh, E., and Baker, R. (1997). Normal and adapted visuooculomotor reflexes in goldfish. *J. Neurophysiol.* **77,** 1099–1118.

Melvill Jones, G. (1977). Plasticity in the adult vestibulo-ocular reflex arc. *Philos. Trans. R. Soc. Lond. B. Biol. Sci.* **278,** 319–334.

Melvill Jones, G., and Milsum, J. H. (1971). Frequency-response analysis of central vestibular unit activity resulting from rotational stimulation of the semicircular canals. *J. Physiol.* **219,** 191–215.

Mensh, B. D., Aksay, E., Lee, D. D., Seung, H. S., and Tank, D. W. (2004). Spontaneous eye movements in goldfish: Oculomotor integrator performance, plasticity, and dependence on visual feedback. *Vision Research* **44,** 711–726.

Moorman, S. J. (2001). Development of sensory systems in zebrafish (*Danio rerio*). *Ilar. J.* **42,** 292–298.

Moorman, S. J., Burress, C., Cordova, R., and Slater, J. (1999). Stimulus dependence of the development of the zebrafish (*Danio rerio*) vestibular system. *J. Neurobiol.* **38,** 247–258.

Muller, M. (1994). Semicircular duct dimensions and sensitivity of the vertebrate vestibular system. *J. Theor. Biol.* **167,** 239–256.

Muller, M. (1999). Size limitations in semicircular duct systems. *J. Theor. Biol.* **198,** 405–437.

Neuhauss, S. C., Biehlmaier, O., Seeliger, M. W., Das, T., Kohler, K., Harris, W. A., and Baier, H. (1999). Genetic disorders of vision revealed by a behavioral screen of 400 essential loci in zebrafish. *J. Neurosci.* **19,** 8603–8615.

Pastor, A. M., De la Cruz, R. R., and Baker, R. (1992). Characterization and adaptive modification of the goldfish vestibulooocular reflex by sinusoidal and velocity step vestibular stimulation. *J. Neurophysiol.* **68,** 2003–2015.

Pastor, A. M., De la Cruz, R. R., and Baker, R. (1994). Eye position and eye velocity integrators reside in separate brainstem nuclei. *Proc. Natl. Acad. Sci. USA* **91,** 807–811.

Rabbitt, R. D. (1999). Directional coding of three-dimensional movements by the vestibular semicircular canals. *Biol. Cybern.* **80,** 417–431.

Rabbitt, R. D., Damiano, E. R., and Grant, J. W. (2003). Biomechanics of the vestibular semicircular canals and otolith organs. *In* "The Vestibular System" (S. M. Highstein, R. R. Fay, and A. N. Popper, eds.), pp. 153–201. Springer-Verlag, New York.

Raphan, T., Matsuo, V., and Cohen, B. (1979). Velocity storage in the vestibulo-ocular reflex arc (VOR). *Exp. Brain Res.* **35,** 229–248.

Rick, J. M., Horschke, I., and Neuhauss, S. C. (2000). Optokinetic behavior is reversed in a chiasmatic mutant zebrafish larvae. *Curr. Biol.* **10,** 595–598.

Riley, B. B., and Moorman, S. J. (2000). Development of utricular otoliths, but not saccular otoliths, is necessary for vestibular function and survival in zebrafish. *J. Neurobiol.* **43,** 329–337.

Ritter, D. A., Bhatt, D. H., and Fetcho, J. R. (2001). *In vivo* imaging of zebrafish reveals differences in the spinal networks for escape and swimming movements. *J. Neurosci.* **21,** 8956–8965.

Robinson, D. A. (1976). Adaptive gain control of vestibulooocular reflex by the cerebellum. *J. Neurophysiol.* **39,** 954–969.

Robinson, D. A. (1968). Eye movement control in primates. The oculomotor system contains specialized subsystems for acquiring and tracking visual targets. *Science* **161**, 1219–1224.

Robinson, D. A. (1989). Integrating with neurons. *Ann. Rev. of Neurosci.* **12**, 33–45.

Robinson, D. A. (1963). A method of measuring eye movement using a scleral search coil in a magnetic field. *IEEE Trans. Biomed. Electr.* **10**, 137–145.

Robinson, D. A. (1975). Oculomotor signal controls. *In* "Basic Mechanisms of Ocular Motility and Their Clinical Implications" (G. Lennerstrand and P. Bach-y-Rita, eds.), pp. 337–374. Pergamon, New York.

Roeser, T., and Baier, H. (2003). Visuomotor behaviors in larval zebrafish after GFP-guided laser ablation of the optic tectum. *J. Neurosci.* **23**, 3726–3734.

Rombough, P. (2002). Gills are needed for ionoregulation before they are needed for O(2) uptake in developing zebrafish. *Danio rerio. J. Exp. Biol.* **205**, 1787–1794.

Seung, H. S. (1996). How the brain keeps the eyes still. *Proc. Natl. Acad. Sci. USA* **93**, 13339–13344.

Shinoda, Y., and Yoshida, K. (1974). Dynamic characteristics of responses to horizontal head angular acceleration in vestibuloocular pathway in the cat. *J. Neurophysiol.* **37**, 653–673.

Skavenski, A. A., and Robinson, D. A. (1973). Role of abducens neurons in vestibuloocular reflex. *J. Neurophysiol.* **36**, 724–738.

van Alphen, A. M., Stahl, J. S., and De Zeeuw, C. I. (2001). The dynamic characteristics of the mouse horizontal vestibulo-ocular and optokinetic response. *Brain Res.* **890**, 296–305.

Wilson, V. J., and Melvill Jones, G. (1979). "Mammalian Vestibular Physiology." Plenum Press, New York.

CHAPTER 18

Development of Cartilage and Bone

Yashar Javidan and Thomas F. Schilling

Department of Developmental and Cell Biology
University of California, Irvine
Irvine, California 92697

I. Introduction

A. The Zebrafish Model

The skeletons of vertebrates are remarkably similar. Centuries of studies of bone morphology and the fact that bones are preserved as fossils has revealed that identical bones form much of the skull, vertebrae, and appendicular skeleton in fish, amphibians, reptiles, and mammals (De Beer, 1937; Goodrich, 1930). Clear homologies between bones can be drawn based not only on fossils but also on commonalities in their articulations with one another and their patterns of development. Genetic studies in humans, mice, and zebrafish are now revealing that these similarities reflect similar developmental mechanisms underlying skeletal morphogenesis.

Zebrafish develop a simple pattern of early larval cartilages and bones, and it is this early basic skeletal pattern that is most highly conserved among all vertebrates. The last common ancestor of humans and zebrafish existed approximately 420 million years ago, yet both form the same skeletal cell types, including endochondral and dermal bones, as well as cartilages that persist in the adult (Hall and Hanken, 1985). Mutant screens in zebrafish have identified many genes required for early cartilage development, and a number of these mutations are also associated with craniofacial syndromes in humans.

Here we summarize current methods for visualizing cartilage and bone that make zebrafish an excellent genetic model for dissecting the molecular basis of skeletal development. Further, we describe methods of screening for mutants with anomalous craniofacial and other skeletal phenotypes.

B. Zebrafish Skeletal Anatomy

By simply fixing and staining whole zebrafish larvae in Alcian dyes, the entire pattern of cartilage is revealed in whole-mounted specimens (Fig. 1A and B). This pattern emerges between 45 and 72 hpf and includes both the pharyngeal skeleton (Fig. 1A) of the jaw and gills, as well as the neurocranial skeleton, which houses the brain (Fig. 1B; Schilling and Kimmel, 1997). Figure 1C illustrates the simple pattern of neurocranial cartilages, and Fig. 1D shows some of the homologues (e.g., auditory capsule, trabeculae) in the developing human skull (adapted from De Beer, 1937).

In the neurocranium (Fig. 1B and C), the first chondrifications lie parallel to the anterior end of the notochord at 45 to 48 hpf, and these form the parachordal cartilage. The trabecular cartilages chondrify further anteriorly, between the eyes, and eventually fuse in the midline to form the trabeculae communis (at approximately 52 hpf) and flare outwards laterally as the ethmoid plate. At approximately this stage, cartilage of the parachordals fuses anteriorly with the trabeculae to form the basal plate. Early chondrification patterns in mammals are remarkably similar (compare Fig. 1C and D).

Meanwhile, the pharyngeal arch skeleton begins to chondrify (Fig. 1A). Like most teleost fishes, the zebrafish forms a series of seven pharyngeal arches, a mandibular (jaw), hyoid, and five branchials that support the gills. Each pharyngeal arch segment consists of a similar set of serially homologous cartilages. These form a basic pattern of five elements from dorsal to ventral (pharyngobranchial, epibranchial, ceratobranchial, hypobranchial, basibranchial) that is conserved in all vertebrates (reviewed in Schilling, 1997). Only a subset of these is visible in the larval zebrafish (Figs. 1A and 2E). Ceratobranchial elements of the first two arches, Meckel's cartilage and the ceratohyal, are the first to chondrify at approximately 54 hpf and these are followed by the appearance of five more posterior ceratobranchials between 60 and 72 hpf. Here, there are some major differences

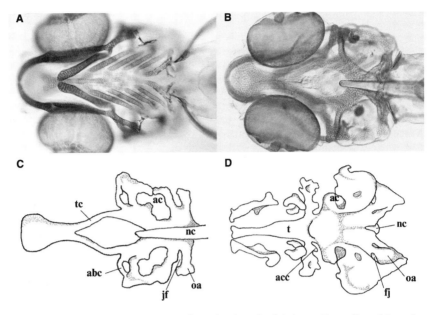

Fig. 1 The simple pattern of early cartilages in the zebrafish larva (A to C) and homologous cartilages in the neurocranium of the human (D). Anterior is to the left in all panels. (A and B) Larvae (approximately 4.0 mm, 5 dpf) are stained with Alcian blue and shown in ventral view (A) to display the pharyngeal arches and dorsal view (B) to show the neurocranium. (C and D) Diagrams illustrating the neurocranium in dorsal view of zebrafish (C; approximately 4.0 mm, 5 dpf) and human (D; approximately 20.0-mm stage; adapted from De Beer, 1937). Abbreviations: abc, anterior basicapsular commissure; ac, auditory capsule; acc, alicochlear commissure; fj, foramen jugulars; jf, jugular foramen; nc, notochord; oa, occipital arch; tc, trabecula cranii (t, trabecular plate).

between fish and mammals in adult morphology. For example, in the first two arches, the dorsal elements in fish (palatoquadrate and hyomandibulae) are involved in jaw articulation, but in mammals form middle ear bones (incus, malleus, and stapes).

As in all bony fishes, zebrafish form several different types of cartilage and bone and their development resembles that of the mammalian skeleton (Hall and Hanken, 1985). The development and chondrification sequence of larval cartilage has been described in the greatest detail (Kimmel *et al.*, 1998; Schilling and Kimmel, 1997). The adult skeletal anatomy and ossification sequence of craniofacial bones in zebrafish has been described by Cubbage and Mabee (1996), and the axial skeleton by Bird and Mabee (2003).

Many bones develop by replacement of cartilage precursors by osteocytes and, in zebrafish, these include bones of the neurocranium (perichondral) and pharyngeal arches (endochondral). While most neurocranial bones ossify relatively late, many endochondral arch bones begin to ossify by 6 to 7 dpf. The first dermal bones (membrane bone) ossify even earlier. These bones form directly from a mesenchymal condensation, without a cartilage intermediate. Examples include the opercle (3 dpf), parasphenoid (4 dpf), and branchiostegal rays (4 to 5 dpf). Dermal plates that cover the skull (parietal, frontal, etc.) develop later, when fish are approximately 8.0 mm in length (3 to 4 weeks) and these have clear homologues in humans. Many teleosts, including ostariophysans zebrafish, are reported to lack acellular bone (Parenti, 1986).

C. Mutations that Disrupt the Skeleton

Mutations provide entry points into a molecular analysis of developmental mechanisms controlling skeletal development, via cloning of the mutated genes. The zebrafish is the only animal in which large-scale screens have been performed to identify mutations that disrupt the skeleton. To date, such screens have focused on finding mutations that disrupt early mouth and pharyngeal morphology during the first 5 to 6 days of development. These have yielded a large collection of mutants that disrupt craniofacial development (Neuhauss *et al.*, 1996; Piotrowski *et al.*, 1996; Schilling *et al.*, 1996b), and for an increasing number, the underlying genetic causes are known (Table I; reviewed in Yelick and Schilling, 2002). Fig. 2B is an example of the pharyngeal defects in *lockjaw*, a mutant in *tfap2a*, which lacks hyoid cartilage and instead develops fusions of mandibular cartilage to the neurocranium (arrowhead).

Several classes of mutations with similar phenotypes have been characterized (Table I). In the neurocranium, mutations that disrupt midline development and either Nodal (*cyclops, one-eyed pinhead*) or Hedgehog signaling (*sonic you, slow muscle omitted*) also disrupt trabecular formation and chondrification of the ethmoid plate (Kimmel *et al.*, 1998; Schilling, 1997). Figure 2D is an example of the neurocranial defects in *sonic you*, a mutation in *sonic hedgehog*, showing characteristic fusion across the midline of structures such as the trabeculae (arrows).

Specific classes of mutations that disrupt pharyngeal arches include *sucker* encoding *endothelin-1 (et-1)*, which disrupts formation of the lower jaw. *sucker* resembles at least four additional "anterior" arch mutants such as *schmerle, sturgeon* and *hoover*, which may be other components of the ET-1 signaling pathway (Piotrowski *et al.*, 1996; Table I). Mutations that disrupt endodermal development, such as *casanova* and *van gogh*, also show pharyngeal cartilage defects, probably secondary consequences of a loss of chondrogenic signals from the endoderm.

The *jellyfish* is one of a large class of mutants called "hammerheads" (Table I). The *jellyfish* mutation disrupts *sox9a*, a transcription factor known to be required for condensation and differentiation of cartilage precursors, suggesting that many hammerhead mutants represent genes involved in this process. The recent identification of the *jekyll* mutation as a *uridine 5′ diphosphate D-glucose dehydrogenase (udgh)* required for proteoglycan production is consistent with this prediction.

Several additional mutations with skeletal defects have been generated by insertional mutagenesis, such as the *hands-off* mutant that lacks pectoral fins, and *chihuahua*, which exhibits skeletal dysplasia, caused by a mutation in type I collagen and was detected through a radiography screen.

II. Cartilage Visualization Techniques

A. Alcian Blue Staining

Alcian blue and green are dyes that stain the proteoglycan components of the extracellular matrix associated with chondrocytes and that are used to visualize cartilage patterns in larvae and in adults (Fig. 1A). Animals are fixed in formalin and stained for several hours to days in the dye, which will label different types of proteoglycans, depending on the pH and aqueous content of the staining solution. Here we describe our standard method for staining in 80% ethanol at a pH of approximately 5.7 and without $MgCl_2$. Under these conditions, Alcian dye labels most anionic tissue components, mucosubstances with carboxyl and sulfate groups, including hyaluronate and chondroitin sulfate. At higher $MgCl_2$ concentrations, only highly sulfated proteoglycans such as heparin sulfate are stained (Wardi and Allen, 1972; Whiteman, 1973).

Alcian Blue first labels the earliest differentiating chondrocytes at 54 hours of development (Schilling and Kimmel, 1997), but the full pattern of early cranial cartilages is best labeled after 72 hpf. Mouth extension has proven to be a useful staging tool for this phase of development (Schilling and Kimmel, 1997).

1. Protocol 1.1

a. Alcian Blue Staining of Cartilage

1. Anesthetize larvae in 5 to 10% methyl ethane sulfonate (i.e., tricaine).

2. Fix in 3.7% neutral buffered formaldehyde (10% formalin in phosphate buffered saline [PBS]) at room temperature for several hours to overnight. Care should be taken not to over fix the preparation.

3. Wash larvae in PBT (PBT: PBS, 0.1% Tween-20) 3 to 5 times for at least 5 minutes per wash.

4. Transfer into a 0.1% solution of Alcian blue (Sigma, St. Louis, MO) dissolved in 80% ethanol and 20% glacial acetic acid for at least 6 to 8 hours to overnight. (Do not overstain). Alcian solution should be syringe filtered (0.22 um) before use.

5. Rinse larvae in ethanol and rehydrate gradually into PBS. Use ethanol, PBT solutions of 70%-30%, 50%-50%, and 30%-70%, respectively.

6. If fish have not been raised in phenyl-thiourea (PTU), remove pigmentation by bleaching in 3% hydrogen peroxide, 1% potassium hydroxide for 1 to 3 hours. This reaction should be monitored carefully and stopped once eye pigmentation

Table I

Cloned and Published Zebrafish Mutations that Disrupt the Skeleton

Mutant Name	Gene	Protein Type (Gene Family)	Skeletal Phenotype	Associated Human Syndrome	Reference
Neurocranial and Midline Defects					
cyclops (cyc^{tf291})	nodal related factor (ndr2);	Secreted ligand (TGFβ)	Absence of anterior and lateral neurocranium		Rebagliati et al. (1998)
one-eye pinhead (oep$^{m134\ \&\ tz57}$)	EGF-CFC	Co-factor activin type II receptor	Severely reduced pharyngeal and neurocranial cartilage		Zhang et al. (1998)
sonic-you (syu^{tq252})	sonic hedgehog (shh)	Secreted ligand (Hedgehog)	Fused anterior neurocranium	HPE and cleft palate	Schauerte et al. (1998)
slow muscle omitted (smu$^{b577\ \&\ b641}$)	smoothened (smu)	Transmembrane signaling protein (Hedgehog family)	Severely reduced pharyngeal and neurocranial cartilage	Basal cell carcinomas	Barresi et al. (2000)
Pharyngeal Cartilage Defects					
lockjaw (low^{ts213})	tfap2a	Transcription factor (Ap2)	Hyoid arch defects	Deafness (Law et al., 98)	Knight et al. (2003)
sucker (suc^{tf216b})	endothelin-1 (et-1)	Secreted ligand (Et)	Loss of lower jaw and other ventral mandibular and hyoid cartilages		Miller et al. (2000)
van-gogh (vgo$^{tm208\ \&\ nu285}$)	tbx1	Transcription factor (T-Box)	Reduced pharyngeal cartilage	DiGeorge deletion (DGS)	Piotrowski et al. (2003)
integrinα5	integrinα5	Cell adhesion molecule	Hyoid arch defects		Crump et al. (2004)

Mutant (gene/allele)	Gene	Protein	Phenotype	Human disease	Reference
monocytic leukemia zincfinger (*moz^b719*)	*moz*	Histone acetyltransferase	Hyoid arch defects (transformation)		Miller *et al.* (2004)
valentino (*val^b337*)	*Mafβ,*	Transcription factor (bZIP)	Ectopic cartilage elements in 3rd branchial arch		Prince *et al.* (1998)
lazarus (*lzr^b557*)	*pbx 4*	Hox co-factor (Pbx)	Fusion of mandibular and hyoid arch cartilages		Popperl *et al.* (2000)
foxi one (*foo*)	*foxi 1*	Transcription factor (Forkhead)	Hyoid arch defects		Nissen *et al.* (2003)
casanova (*cas^ta56 & s4*)	*sox 17 (novel sox gene)*	Transcription factor (Sox)	Loss of pharyngeal cartilage		Kikuchi (2001); Dickmeis (2001)
Cartilage Differentiation Defects					
jellyfish (*jef^thl1134 & tw37*)	*sox 9a*	Transcription factor (HMG)	Lack of neurocranial, pharyngeal, and pectoral girdle cartilages	Camptomelic dysplasia	Yan *et al.* (2002)
jekyll (*jek^m151,m310*)	*ugdh*	Enzyme (Ugdh)	Defect in cartilage differentiation (no Alcian staining)		Walsh and Stainier (2001)
Other					
acerebellar (*ace^ti282a*)	*fgf8*	Secreted ligand (Fgf)	Hyoid arch defects		Reifers *et al.* (1998)
handsoff (*hns^s6 & c99*)	*hand 2*	Transcription factor (bHLH)	Loss of pectoral fins		Yelon *et al.* (2000)
chihuahua (*cht^dc124*)	*collagen 1 (α1)*	Extracellular matrix structural protein	Adult skeletal dysplasia, short thick bones	Osteogenesis imperfecta	Fisher *et al.* (2003)

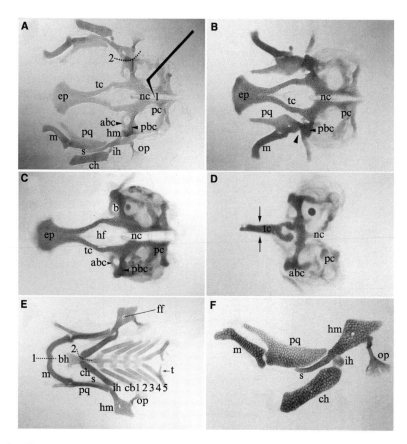

Fig. 2 Flat-mounted preparations of larval cartilages. Early larvae (approximately 3.5 mm, 5 dpf) are stained with Alcian blue, dissected and shown flattened (see Protocol 1.2) for detailed analysis of morphology. Anterior is to the left. (A) Butterfly preparation of the wild-type neurocranium with the mandibular and hyoid arches attached. Numbers indicate important dissection sites. 1—securing needle. 2—joint between anterior basicranial commissure of the neurocranium and hyomandibular. (B) Butterfly preparation of the *lockjaw* (*tfap2α* mutant) skeleton. Large arrowhead indicates the fusion point between the neurocranium and the mandibular arch. (C) Neurocranial preparation of a wild type. (D) Similar preparation of the *Sonic you* (*shh* mutant) neurocranium, showing fusion of skeletal elements across the midline. (E) Seven arch preparation of wild-type mandibular, hyoid, and branchial arches. Numbers indicate important dissection sites for the lateral jaw preparation shown in F. 1—midline fusion point of Meckel's cartilages. 2—midline attachment between ceratohyal and basihyal cartilages. (F) Lateral jaw preparation of wild-type mandibular and hyoid arches. Abbreviations: abc, anterior basicapsular commissure; b, basicapsular fenestra; bh, basihyal cartilage; cb, ceratobranchial; ch, ceratohyal; ep, ethmoid plate; ff, facial nerve foramen (VII); hf, hypophyseal fenestra; hm, hyomandibular; ih, inter hyal; m, Meckel's; nc, notochord; op, opercle; pbc, posterior basicapsular commissure; pc, parachordal cartilage; pq, palatoquadrate; s, symplectic; tc, trabeculae.

has cleared. Bubbles may form within the embryo during clearing but will disappear in subsequent steps. Leave the tube caps open.

7. Clear tissue in 0.05% trypsin dissolved in PBS for 1 to 2 hours. A 0.01% solution may be used for overnight trypsinization; however, we normally perform this at room temperature and monitor tissue clearing closely. Stop the trypsin digestion once the brain tissue has cleared.

8. Wash larvae in PBS 2 to 3 times for 5 minutes per wash. Alizarin red staining for bone visualization should be performed at this stage if desired (see Protocol 1.4).

9. Transfer embryos into glycerol, gradually through a series of 30% and 50% dilutions in PBS, and store in 70% glycerol.

B. Microdissection of Larval Craniofacial Cartilage

Cartilage dissection and flat mounting allows for close observation of the cartilage pattern that is not possible in whole-mounted preparations. Flat mounts allow the specimen to be photographed in one focal plane (see Fig. 2), producing very informative depictions of each cartilage. However, flat mounting disturbs 3D relationships and should be used in combination with whole-mount specimens to develop the most accurate configuration of skeletal structures. Dissection of the larval skeleton is a bit intricate; however, with some patience the following methods allow detailed analyses of cartilage pattern.

1. Protocol 1.2

a. Microdissection and Mounting of Stained Cartilage

1. Stain larval skeleton with Alcian blue (see Protocol 1.1) and store in 70% glycerol.

2. Place stained larvae in depression slides under a dissecting stereomicroscope submerged in a large droplet (approximately 100 μl) of 70% glycerol. Use 40× magnification for sorting and securing embryos and 80 to 100× for dissection.

3. Use two tungsten needles to remove eyes and discard. Remove brain tissue by holding the larva with one needle and scraping dorsal to the skeleton and notochord. Cut the head free from the trunk and tail at the level of the pectoral fins (the notochord can be tough). Use very thin tungsten needles (0.1 mm or 0.2 mm in diameter), which we find work better than glass needles. Secure needles to the ends of glass pipettes by melting glass or with superglue. Sharpen needles either by flaming the tips briefly over a Bunsen burner, or with an electrolytic cell containing NaOH, using the needle as the anode. Use tweezers to bend the tips of the needles (approximately 1 mm) at a 90° angle to provide maximum versatility.

4. Use one needle to secure the larva in place and the second to dissect cartilage with your dominant hand. We prefer to secure the specimen with a needle attached

to a micromanipulator to minimize motion. Place the tip of the securing needle at the base of the notochord and skull, near the posterior end of the head. (Fig. 2A, place needle at position 1).

5. Depending on the desired preparation (discussed later), skeletal structures may be dissected by placing the bent tip at a joint between two cartilage elements and pulling away very gently. The pharyngeal arches (mandibular, hyoid, and the 5 branchial arches) can be separated from the neurocranium and displayed intact (Fig. 2E). To retain neurocranial and arch attachments, we use a "butterfly" preparation (Fig. 2A). We used this type of preparation to display the fusion between the skull and the arches in the *lockjaw* mutant (Knight *et al.*, 2003; Fig. 2B). Lateral preparations of the jaw and hyoid are dissected free from the neurocranium to show detail at higher magnification (Fig. 2F).
Suggested preparations:

   a. Neurocranial preparation (Fig. 2C)
      1. Scrape away brain tissue and separate head from trunk and tail.
      2. After positioning the securing needle, place dissecting needle beneath the hyosymplectic (hs), near the joint between hs and the neurocranium. Now gently pull to detach at the joint. Repeat for the contralateral joint (Fig. 2A, cut along line 2).
      3. Place the dissecting needle at the joint between the tip of the palato-quadrate (pq) and the ethmoid plate (ep) and gently pull to detach.
      4. Gently glide the dissecting needle horizontally, along the plane of the neurocranium and dorsal to the arches to remove any remaining soft tissue. Repeat until the jaw and arches separate from the neurocranium (save arches for the seven arch preparation [b]).
   b. Seven arch preparation (mandibular, hyoid, and five brachial arches, Fig. 2E) Follow procedures for the neurocranial preparation [a] but save the arches.
   c. Butterfly preparation (neurocranium, mandibular, and hyoid arches, Fig. 2A)
      1. Scrape away brain tissue and separate head from trunk and tail.
      2. After positioning the securing needle, place the dissecting needle at the midline of Meckel's cartilage (m) and press down with a cutting motion to separate at the fusion point in the midline (Fig. 2E, cut along line 1).
      3. Place the dissecting needle at the joint between the tip of pq and gently pull to detach.
      4. Place the dissecting needle at the midline between the two ceratohyal (ch) cartilages. Gently pull to separate (Fig. 2E, cut along line 2) and then gently split the more posterior arches and remove them.
      5. For mounting this preparation, position the mandibular and hyoid arches to the sides as displayed in Fig. 2A before securing the coverslip.

     d. Lateral jaw preparation (mandibular and hyoid arches Fig. 2F)

1. Remove the eyes and brain by scraping, and separate the head from trunk and tail.

2. Place the dissecting needle at the midline of m and press down with a cutting motion to detach at the midline (Fig. 2E, cut along line 1).

3. Place the dissecting needle at the joint between the tip of pq and gently pull to detach.

4. Place the dissecting needle at the midline between the two ch cartilages. Gently detach and split the posterior arches at the midline. (Fig. 2E, cut along line 2).

5. Place the needle on hs, near the joint between hs and the neurocranium, and gently pull away (Fig. 2A, cut along line 2).

6. Flat mount the desired structures by transferring them (using a glass pipette) to a clean glass slide containing a small droplet (approximately 50 to 80 μl) of 100% glycerol. Manipulate the specimen with needles under the microscope to achieve the desired orientation before adding the coverslip. Coverslips should be elevated slightly to prevent crushing the specimen. Place a small amount of petroleum jelly or modeling clay at four spots around the specimen, in approximately a 1-cm square array. Use petroleum jelly for flat structures (i.e., neurocranium) and clay for thicker (i.e., arches) structures. Place coverslip on petroleum jelly/clay bridges and gently slide the coverslip to manipulate specimen and obtain the ideal orientation.

C. BrdU Labeling

Bromodeoxyuridine (BrdU) is incorporated into newly synthesized DNA and can be used as a label for dividing chondroblasts and osteoblasts in the skeleton (Fig. 3A). One can label animals with BrdU at any stage, but for incorporation into skeletal condensations one should inject/incubate animals after 54 hpf. It is important to dissect and flat-mount cartilages free of other tissues, which will also be BrdU labeled. We describe an incubation time in BrdU of 24 h, which gives strong labeling, but shorter pulsed treatments are possible. Both injection of BrdU (Larison and Bremiller, 1990) and immersion (Kimmel *et al.*, 1998) work at early larval stages.

1. Protocol 1.3

a. BrdU Labeling of Dividing Cartilage Cells Counter Stained with Alcian Blue

1. Incubate larvae in 1-mM BrdU (Vector Laboratories Inc., Burlingame, CA) in embryo medium for 24 h. Change the medium every 6 h to increase BrdU incorporation.

2. Fix larvae in 4% paraformaldehyde (PFA) overnight at 4 °C. (Store in MeOH and rehydrate through a MeOH/PBS series, if required.)

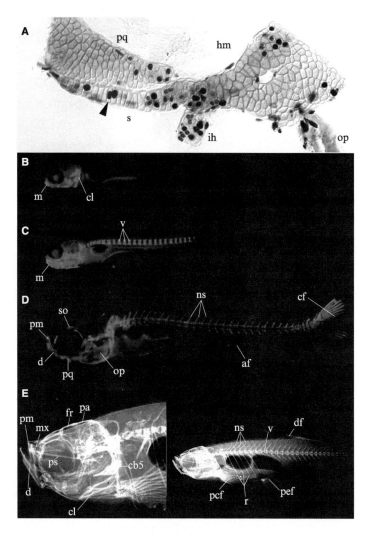

Fig. 3 Techniques for visualizing the zebrafish skeleton. Anterior is to the left in all panels. (A) Partial lateral jaw preparation, showing BrdU-labeled cartilages of the hyoid arch. Arrowhead indicates one of many labeled nuclei that have taken up the BrdU and are labeled with an anti-BrdU antibody. (B through D) Whole larvae (5, 11, and 21 dpf, respectively) in lateral view, showing calcien stained cartilages and bones. (E and F) Radiographs of adult zebrafish. Abbreviations: af, anal fin; cb5, ceratobranchial 5; cf, caudal fin; cl, cleithrum; d, dentary; ds, dorsal fin; fr, frontal; hm, hyomandibular; ih, interhyal; m, Meckel's; mx, maxilla; ns, neural (dorsal) spine; op, opercle; pa, parietal; pcf, pectoral fin; pef, pelvic fin; pm, premaxilla; pq, palatoquadrate; ps, parasphenoid; r, ribs; s, symplectic; so, supraorbital; v, vertebrae.

3. Rinse 3 times for 5 minutes each in PBT (PBT: PBS, 0.1% Tween-20).

4. Rinse 5 minutes in dH_2O.

5. Place larvae in 0.1% trypsin for fewer than 2 minutes (or in acetone at $-20\,°C$ for 7 minutes) to increase antibody penetration.

6. Rinse 1 time in dH_2O and 3 times in PBT for 5 minutes per rinse.

7. Place in blocking solution for up to 1 h (blocking solution: 1×PBS, 1% DMSO, 10% serum, 0.1% Tween-20).

8. Detect BrdU by adding anti-BrdU primary antibody 1:200 (antibody G3G4 developed by S. J. Kaufman and obtained from Developmental Hybridoma Bank, University of Iowa, Iowa City) in blocking solution and incubate overnight at 4 °C with mild agitation on a nutator or shaker.

9. Rinse 5 times for 10 to 15 minutes each in PBT + 1% DMSO.

10. Incubate in biotinylated secondary antibody, diluted 1:500 in blocking solution overnight at 4 °C.

11. Rinse 5 times for 10 to 15 minutes each in PBT (perform Step 12 while waiting).

12. Prepare 'AB' complex: 1 ml of blocking solution, 20 μl Vectastain 'A', 20 μl Vectastain 'B,' and incubate for 30 minutes before applying to specimens.

13. Incubate specimens for 1 h in the 'AB' complex at room temperature.

14. Rinse 5 times for 15 minutes each in PBT.

15. Rinse 3 times for 10 minutes each in 1 × PBS to remove Tween-20. (Tween inhibits subsequent enzymatic reactions.)

16. Preincubate in diaminobenzidine (DAB) solution: 1 ml PBS, 1 ml H_2O, 20 μl DMSO, 25 μl DAB for 15 to 20 minutes (0.5 ml per tube).

17. Add 5 μl of 3% H_2O_2 to each tube (add 5 μl of 3% H_2O_2 for every 500 μl of DAB).

18. Monitor development from 1 up to 30 minutes.

19. Rinse in PBT for at least 3 washes to stop staining.

20. Counterstain preparations with Alcian blue (as in Protocol 1.1) and dissect and flat-mount cartilages (as in Protocol 1.2).

III. Bone Visualization Techniques

A. Alizarin Red Staining

Alizarin red stains calcified matrix associated with bone and can be used to distinguish cartilage and bone by counterstaining with Alcian blue (Cubbage and Mabee, 1996). Alizarin red first labels the earliest differentiating bones at 3 to 4 days of development, and bone development continues for many weeks before achieving the adult pattern. Body length, rather than age, as well as changes in pigmentation and fin morphology are useful staging tools at these ages (Schilling, 2002).

1. Protocol 1.4

a. Alizarin Red Staining of Bone

1. Fix Alcian stain and clear tissue by following Steps 1 through 7 of Protocol 1.1 for Alcian blue staining.
2. Transfer larvae into a 0.05% solution of Alizarin red dissolved in 1% KOH for several hours.
3. Clear larvae in glycerol, gradually through a series of 30% and 50% dilutions in PBS and store in 70% glycerol. For older stages, soft tissues must first be removed by tripsin digestion.

B. Adult/Larval Calcien Staining

The fluorescent chromophore, Calcein ($C_{30}H_{26}N_2O_{13}$), specifically binds to calcium, fluorescently staining the calcified skeletal structures in living zebrafish larvae and juveniles (Fig. 3B to D). This allows analysis of bone in live zebrafish with high sensitivity, though it also labels some cartilages at early stages. Calcein staining can be used to follow the development of skeletal structures from 2 to 21 dpf and is potentially an effective screening tool for identifying skeletal mutants (Du et al., 2001).

1. Protocol 1.5

a. Calcien Staining

1. Immerse living animals (not anesthetized) in 0.2% calcein solution in Petri dishes from 3 to 10 minutes, depending on the size of the larvae or juveniles. Prepare solution by dissolving 2 g of calcein powder (Sigma) in 1 liter of deionized water. To counter calcein's strong acidifying affects, add an appropriate amount of NaOH (0.5 N) to the solution to restore neutral pH (approximately 7.0 to 7.5).
2. Rinse 3 to 4 times in dH_2O and then allow to soak for 10 minutes so that excess, unbound calcein diffuses out of the tissues.
3. Euthanize or anesthetize as desired in 10 to 20% tricaine.
4. Mount on glass depression slides with 3% methylcellulose.
5. View using a microscope with a filter set for green fluorescent protein (GFP) or fluorescein isothiocyanate (FITC). (Excitation wavelength is around 495 nm and the emission wavelength is around 525 nm.)

C. Radiographic Visualization of Adult Zebrafish Skeleton

Radiographic analysis is an excellent technique for visualizing skeletal anatomy and bone morphology (Fig. 3E) and has been previously described by Fisher *et al.* (2003). This technique, in comparison with traditional histological methods such as Alcian blue/Alizarin red staining, is a quick and efficient way to detect subtle skeletal abnormalities in adult zebrafish. This is useful since much of the skeleton matures relatively late in zebrafish development, when the animals are large and difficult to stain and analyze in large numbers. Radiography has been used with zebrafish anywhere between 5 dpf and adulthood and has been shown to be a powerful method for genetic screens to identify subtle phenotypes, such as *chihuahua* (Fisher *et al.*, 2003).

1. Protocol 1.6

a. Radiography of Adult Zebrafish

1. Anesthetize larval or adult fish with 5 to 10% tricaine.
2. Lay animals flat on x-ray platform.
3. Use a small-specimen radiography x-ray machine for higher resolution, such as a Faxitron MX-20 cabinet.
4. Expose 3 to 4 seconds at 17 to 20 kVolts. (Use high-resolution Min-R 2000 film and intensifying screens.)
5. To assign bones, use stained and dissected skeletal preparations as a reference for identification and comparison.

D. Osteoblast and Osteoclast Histology

One can observe bone cell types at a much more detailed level in sections, particularly beyond early larval stages. Here we describe a tartrate-resistant acid phosphatase (TRAP)-staining method modified from Witten and Villwock (1997) and Witten *et al.* (2001). Osteoclasts associated with bone resorption can be identified in light microscopic sections, stained with toluidine blue and labeled by the TRAP method. Alkaline phosphatase (ALP) is visualized (for identification of early osteoblasts). While only mononucleate osteoclasts are present at larval stages, multinucleate osteoclasts appear after approximately 40 dpf. To label osteoclasts, the sections must be decalcified.

1. Protocol 1.7

a. TRAP Staining to Visualize Osteoclasts

1. Fix specimens for 1 h at 4 °C in 10% formaldehyde (methanol-free) in 50 mmol Tris buffer, pH 7.2.

2. Rinse in tap water for 1 h and decalcify in Tris buffer (100 mmol, pH 7.2) containing 10% EDTA for 48 h.

3. Dehydrate for 1.5 h in an acetone series (30, 50, 70, 90, and 100%).

4. Soak specimens in a glycol methacrylate monomer solution for 60 minutes to impregnate bone: 80 ml (2-hydroxyethyl)-methacrylate, 200 ppm p-methoxyphenol, 12 ml ethylene glycol monobutyl ether, 270 mg benzoyl peroxide.

5. Change monomer solution and soak for 24 h to achieve further impregnation.

6. Add 2% catalyst (1 ml N,Ndimethylaniline, 10 ml poly-ethyleneglycol-200) to the monomer solution just before embedding.

7. Embed in polyethylene jars with tight lids.

8. Polymerize at 4 °C for 24 h and another 24 h at room temperature. Tissue blocks can be stored at 4 °C.

9. Section specimens (5-μm thickness), float on demineralized water (25 °C), mount on uncoated slides, and dry at 25 °C.

10. For TRAP-labeling incubate sections again for 30 minutes at 20 °C in 0.1 M acetate buffer+ 50 mM di-sodium tartrate dihydrate, pH 5.5. Incubate in TRAP solution for 30 minutes. This solution is prepared with a naphthol AS-TR phosphate (N-AS-TR-P) as substrate, and hexazotized pararosaniline (PRS): 1 g PRS dissolved in 19.25 ml A. dem., 5.75 ml 32% HCl, and stored in darkness at 4 °C. For hexazotation, add 2 ml 4% $NaNO_2$ (0.58 M) to 1 ml of prepared PRS-solution at 20 °C.

11. Incubate for 30 minutes at 20 °C. Prepare final TRAP incubation solution by adding 1 ml hex azotized PRS, 600 ml 2% $MgCl_2$ solution, 2 ml enzyme substrate solution (2 mg N-ASTR-P dissolved in 2 ml N,N-dimethylformamide), and 100 mM di-sodium tartrate dihydrate to 30 ml of 0.1 M acetate buffer (pH 5.5).

12. Rinse with demineralized water.

13. Counterstain sections with Mayers hematoxylin for 10 minutes, rinse in running tap water for 15 minutes, then once with demineralized water. Dry at 40°C, and mount with DPX.

14. Possible controls for staining specificity include: (1) heat to 90 °C for 10 minutes before incubation, (2) incubate without substrate, (3) incubate without tartrate, or (4) add NaF (10 mmol/l).

2. Protocol 1.8

a. Alkaline Phosphatase Staining to Visualize Osteoblasts

1. Fix in acetone at 4 °C (embed, cut, and mount) as described earlier, but do not perform decalcification.

2. Preincubate in Tris buffer (50 mmol/l, pH 9.5) for 1 h. Incubate in 30 ml Tris buffer (50 mmol/l, pH 9.5) for 2 h at 20 °C. Include hexazotized PRS (as described earlier) and 60 mg Napthol-AS-BI-Phosphate dissolved in 0.2 ml N,N-dimethylformamide in the incubation solution.

3. Visualize fluorescence at 568 nm, following the protocol of Sakakura *et al.* (1998).

4. Possible controls include: (1) heat to 90 °C for 10 minutes before incubation, (2) incubate without substrate, (3) add 10 mmol levamisole, or (4) add EDTA (10 mg/ml).

IV. Molecular Markers of Skeletal Precursors

Gene expression in skeletal precursors and surrounding tissue reveals clues about the tissue interactions and developmental processes that pattern them. Craniofacial mesenchyme in zebrafish is derived from both paraxial mesoderm and migratory neural crest. The latter forms much of the cranial skeleton (Schilling and Kimmel, 1994). We briefly summarize classes of genes expressed in cranial neural crest cells in the zebrafish, which are described in more detail by Yelick and Schilling (2002). These can be used as markers of putative skeletogenic cells at different stages of migration and differentiation, but should be interpreted with caution, because no single marker has been demonstrated to mark only the skeletogenic population or to mark all neural crest at any stage.

A large number of transcription factors are expressed throughout the pre migratory cranial neural crest, including *crestin* (Luo *et al.*, 2001), *foxd3* (Odenthal and Nusslein-Volhard, 1998), *sna2* (Thisse *et al.*, 1995), *sox9a* (Yan *et al.*,1995), *sox10* (Dutton *et al.*, 2001), and *tfap2a* (Knight *et al.*, 2003). Others are more segmentally restricted, such as *hox* group 1-3 genes (Amores *et al.*, 1998; Prince *et al.*, 1998) and *krox20* (Oxtoby and Jowett, 1993). Genes encoding cell surface proteins of the Eph family are also expressed in segment specific domains within the pre migratory neural crest (Xu *et al.*, 1996). For most of these early neural crest markers, expression persists during at least the early stages of neural crest migration into the periphery.

A second set of transcription factors that mark putative skeletogenic neural crest are expressed after migration. These include the homeobox-containing genes *dlx2-7* (Akimenko *et al.*, 1994), *msx* (Ekker *et al.*, 1997), the ETS-domain containing genes *fli1* (Brown *et al.*, 2000), and *pea3* (Brown *et al.*, 1998). Ventral crest cells within each pharyngeal arch express *dhand* (Yelon *et al.*, 2000). Many of these markers continue to be expressed up to the stage of cartilage differentiation in the larvae. By this stage, expression of several markers becomes restricted to skeletogenic populations, including *sox9a*, and new markers become expressed, such as *chondromodulin* and *runx*.

V. Strategy and Potential of Future Screens for Skeletal Mutants

In zebrafish, one can screen large numbers of animals for mutations that disrupt most any aspect of skeletal development. Synchronous development of progeny from single crosses allows efficient recognition of morphological defects by simple comparison between siblings. Previous screens have yielded a large collection of larval cartilage mutants. Simple protocols for staining cartilage and bone and imaging the living skeleton with X-rays, can be used effectively in mutant screens to find more subtle or later phenotypes in the skeleton.

Screens for mutations that disrupt skeletal features after the first week of development, however, require much more careful control of larval health and staging. A larval zebrafish must fill its air bladder by day 5 to 6 to survive, and a majority of the recessive lethal mutations found in previous screens produce defects in this process. In addition, the larva begins to feed, having consumed yolk up to this stage, and underfeeding can lead to delayed development and a failure in skeletal maturation. Mouth extension is an accurate method for staging between 54 and 72 hpf (Schilling and Kimmel, 1997), but for larvae, juveniles and adults, staging requires examining other body features (Schilling, 2002).

Skeletal screens also require long-term maintenance of isolated fish during the analysis of their offspring. At minimum, to examine early larval cartilage formation in a typical screen for recessive lethal mutations, each pair of F2 fish must be kept in a separate container while their progeny are raised for 72 h in addition to time required for staining (see Protocol 1.1) and analysis. Extending the screen to early stages of bone development requires an additional day or two for the larvae to mature, as well as an additional day for Alizarin red, Calcein staining, or x-ray (see Protocols 1.4, 1.5, and 1.6), and this means keeping their parents in separate containers for at least a week. The following is one example of steps involved in a combined screen for cartilage and bone defects at 6 dpf.

1. Protocol 1.9

a. Screening for Recessive, Lethal Mutations That Disrupt Larval Cartilage and Bone

1. Mutagenize male zebrafish (see chapter by van Eeden). Raise founders (3 to 4 months).

2. Intercross F1 founders to create F2 families of mutagenized animals. Raise F2 families (2 to 4 months).

3. Intercross F2 family members and raise embryos in embryo medium to at least 3 dpf for cartilage, and 5 days or more for bone. For growth to these stages, embryos should not be overcrowded (approximately 50 embryos per 30-ml Petri dish), and media changed at least once after hatching (2 to 3 days).

4. Anesthetize at least 25 larvae from each of at least 6 crosses between family members. If possible, first examine living larvae for defects in air bladder formation, head shape, jaw elongation, gill and pectoral fin formation, all of which can be hallmarks of problems with skeletal development.

5. Fix at least 25 larvae in 3.7% neutral buffered formalin and perform Alcian staining (see Protocol 1.1), followed by Alizarin staining (see Protocol 1.4), if desired.

6. Examine stained specimens for cartilage in at least two steps. First in ventral view, examine the large ventral elements (ceratobranchials) of each of the seven pharyngeal arches. These develop in an anterior-to-posterior sequence between 60 and 72 hpf, and screeners should be aware that the posterior most arches are often reduced or lost if animals are developmentally retarded. Then focus dorsally to examine cartilage of the trabeculae and parachordal cartilages of the braincase.

7. As for earlier phenotypes, criteria for keeping mutants should include uniformity of phenotype and Mendelian segregation in more than one cross from the family. Potential mutant carriers should also be screened again immediately.

Acknowledgments

We would like to thank J. Du for comments on specific protocols. We thank members of the Schilling lab for helpful comments on the manuscript. Support was provided by the NIH (NS-41353, DE-13828), March of Dimes (1-FY01-198) and Pew Scholars Foundation (2615SC) to T.S.

References

Akimenko, M. A., Ekker, M., Wegner, J., Lin, W., and Westerfield, M. (1994). Combinatorial expression of three zebrafish genes related to distal-less: Part of a homeobox gene code for the head. *J. Neurosci.* **14,** 3475–3486.

Amores, A., Force, A., Yan, Y. L., Joly, L., Amemiya, C., Fritz, A., Ho, R. K., Langeland, J., Prince, V., Wang, Y. L., *et al.* (1998). Zebrafish hox clusters and vertebrate genome evolution. *Science* **282,** 1711–1714.

Barresi, M. J., Stickney, H. L., and Devoto, S. H. (2000). The zebrafish *slow-muscle-omitted* gene product is required for Hedgehog signal transduction and the development of slow muscle identity. *Development* **127,** 2189–2199.

Bird, N. C., and Mabee, P. M. (2003). Developmental morphology of the axial skeleton of the zebrafish, *Danio rerio* (Ostariophysi: Cyprinidae). *Dev. Dyn.* **228,** 337–357.

Brown, L. A., Amores, A., Schilling, T. F., Jowett, T., Baert, J. L., de Launoit, Y., and Sharrocks, A. D. (1998). Molecular characterization of the zebrafish PEA3 ETS-domain transcription factor. *Oncogene* **17,** 93–104.

Brown, L. A., Rodaway, A. R., Schilling, T. F., Jowett, T., Ingham, P. W., Patient, R. K., and Sharrocks, A. D. (2000). Insights into early vasculogenesis revealed by expression of the ETS-domain transcription factor Fli-1 in wild-type and mutant zebrafish embryos. *Mech. Dev.* **90,** 237–252.

Busch-Nentwich, E., Sollner, C., Roehl, H., and Nicolson, T. (2004). The deafness gene dfna5 is crucial for ugdh expression and HA production in the developing ear in zebrafish. *Development* **131,** 943–951.

Crump, J. G., Swartz, M. E., and Kimmel, C. B. (2004). An integrin-dependent role of pouch endoderm in hyoid cartilage development. *PLoS Biol.* **2,** e244.

Cubbage, C. C., and Mabee, P. M. (1996). Development of the cranium and paired fins in the zebrafish, *Danio rerio* (Ostariophysi, Cyprinidae). *J. Morphol.* **229,** 121–160.

De Beer, G. R. (1937). "The Development of the Vertebrate Skull." Oxford University Press, Oxford. Reprinted 1985, Chicago University Press, Chicago.

Dickmeis, T., Mourrain, P., Saint-Etienne, L., Fischer, N., Aanstad, P., Clark, M., Strahle, U., and Rosa, F. (2001). A crucial component of the endoderm formation pathway, CASANOVA, is encoded by a novel sox-related gene. *Genes Dev.* **15,** 1487–1492.

Du, S. J., Frenkel, V., Kindschi, G., and Zohar, Y. (2001). Visualizing normal and defective bone development in zebrafish embryos using the fluorescent chromophore calcein. *Dev. Biol.* **238**, 239–246.

Dutton, K. A., Pauliny, A., Lopes, S. S., Elworthy, S., Carney, T. J., Rauch, J., Geisler, R., Haffter, P., and Kelsh, R. N. (2001). Zebrafish colourless encodes *sox10* and specifies non-ectomesenchymal neural crest fates. *Development* **128**, 4113–4125.

Ekker, M., Akimenko, M. A., Allende, M. L., Smith, R., Drouin, G., Langille, R. M., Weinberg, E. S., and Westerfield, M. (1997). Relationships among msx gene structure and function in zebrafish and other vertebrates. *Mol. Biol. Evol.* **14**, 1008–1022.

Feldman, B., Gates, M. A., Egan, E. S., Dougan, S. T., Rennebeck, G., Sirotkin, H. I., Schier, A. F., and Talbot, W. S. (1998). Zebrafish organizer development and germ-layer formation require nodal-related signals. *Nature* **395**, 181–185.

Fisher, S., and Halpern, M. E. (1999). Patterning the zebrafish axial skeleton requires early chordin function. *Nat. Genet.* **23**, 442–446.

Fisher, S., Jagadeeswaran, P., and Halpern, M. E. (2003). Radiographic analysis of zebrafish skeletal defects. *Dev. Biol.* **264**, 64–76.

Goodrich, E. S. (1930). "Studies on the Structure and Development of Vertebrates." University of Chicago Press, Chicago.

Hall, B. K., and Hanken, J. (1985). Foreword to reissue of "The Development of the Vertebrate Skull" (G. N. De Beer), pp. vii–xxviii. University of Chicago Press, Chicago.

Kikuchi, Y., Agathon, A., Alexander, J., Thisse, C., Waldron, S., Yelon, D., Thisse, B., and Stainier, D. Y. (2001). Casanova encodes a novel Sox-related protein necessary and sufficient for early endoderm formation in zebrafish. *Genes Dev.* **15**, 1493–1505.

Kimmel, C. B., Miller, C. T., Kruze, G., Ullmann, B., BreMiller, R. A., Larison, K. D., and Snyder, H. C. (1998). The shaping of pharyngeal cartilages during early development of the zebrafish. *Dev. Biol.* **203**, 246–263.

Knight, R. D., Nair, S., Nelson, S., Afshar, A., Javidan, Y., Geisler, R., Rauch, G. J., and Schilling, T. F. (2003). *lockjaw* encodes a zebrafish tfap2a required for early neural crest development. *Development* **130**, 5755–5768.

Larison, K. D., and Bremiller, R. (1990). Early onset of phenotype and cell patterning in the embryonic zebrafish retina. *Development* **109**, 567–576.

Luo, R., An, M., Arduini, B. L., and Henion, P. D. (2001). Specific pan-neural crest expression of zebrafish *crestin* throughout embryonic development. *Dev. Dyn.* **220**, 169–174.

Miller, C. T., Schilling, T. F., Lee, K.-H., Parker, J., and Kimmel, C. B. (2000). *Sucker* encodes a zebrafish Endothelin-1 required for ventral pharyngeal arch development. *Development* **127**, 3815–3828.

Miller, C. T., Maves, L., and Kimmel, C. B. (2004). *moz* regulates Hox expression and pharyngeal segmental identity in zebrafish. *Development* **131**, 2443–2461.

Neuhauss, S. C., Solnica-Krezel, L., Schier, A. F., Zwartkruis, F., Stemple, D. L., Malicki, J., Abdelilah, S., Stainier, D. Y., and Driever, W. (1996). Mutations affecting craniofacial development in zebrafish. *Development* **123**, 357–367.

Nissen, R. M., Yan, J., Amsterdam, A., Hopkins, N., and Burgess, S. M. (2003). Zebrafish *foxi one* modulates cellular responses to Fgf signaling required for the integrity of ear and jaw patterning. *Development* **130**, 2543–2554.

Odenthal, J., and Nusslein-Volhard, C. (1998). Fork head domain genes in zebrafish. *Dev. Genes Evol.* **208**, 245–258.

Oxtoby, E., and Jowett, T. (1993). Cloning of the zebrafish *krox-20* gene (*krx-20*) and its expression during hindbrain development. *Nucleic Acids Res.* **21**, 1087–1095.

Parenti, L. R. (1986). The phylogenetic significance of bone types in EU teleost fishes. *Zool. J. Linn. Soc.* **87**, 37–51.

Piotrowski, T., Ahn, D. G., Schilling, T. F., Nair, S., Ruvinsky, I., Geisler, R., Rauch, G. J., Haffter, P., Zon, L. I., Zhou, Y., *et al.* (2003). The zebrafish *van gogh* mutation disrupts tbx1, which is involved in the DiGeorge deletion syndrome in humans. *Development* **130**, 5043–5052.

Piotrowski, T., Schilling, T. F., Brand, M., Jiang, Y. J., Heisenberg, C. P., Beuchle, D., Grandel, H., Van Eeden, F. J. M., Furutani-Seiki, M., Granato, M., *et al.* (1996). Jaw and branchial arch mutants in zebrafish. II: Anterior arches and cartilage differentiation. *Development* **123**, 345–356.

Popperl, H., Rikhof, H., Chang, H., Haffter, P., Kimmel, C. B., and Moens, C. B. (2000). *lazarus* is a novel *pbx* gene that globally mediates *hox* gene function in zebrafish. *Mol. Cell* **6**, 255–267.

Prince, V. E., Moens, C. B., Kimmel, C. B., and Ho, R. K. (1998). Zebrafish *hox* genes: Expression in the hindbrain region of the wild-type and mutants of the segmentation gene, *valentino*. *Development* **125**, 393–406.

Rebagliati, M. R., Toyama, R., Haffter, P., and Dawid, I. B. (1998). *cyclops* encodes a nodal-related factor involved in midline signaling. *Proc. Natl. Acad. Sci. USA* **95**, 9932–9937.

Reifers, F., Bohli, H., Walsh, E. C., Crossley, P. H., Stainier, D. Y., and Brand, M. (1998). *Fgf8* is mutated in zebrafish *acerebellar* (*ace*) mutants and is required for maintenance of midbrain-hindbrain boundary development and somitogenesis. *Development* **125**, 2381–2395.

Sakakura, Y., Yajima, T., and Tsuruga, E. (1998). Confocal laser scanning microscopic study of tartrate resistant acid phosphatasepositive cells in the dental follicle during early morphogenesis of mouse embryonic molar teeth. *Arch. Oral Biol.* **43**, 353–360.

Schauerte, H. E., van Eeden, F. J., Fricke, C., Odenthal, J., Strahle, U., and Haffter, P. (1998). *sonic hedgehog* is not required for the induction of medial floor plate cells in the zebrafish. *Development* **125**, 2983–2993.

Schilling, T. F., and Kimmel, C. B. (1994). Segment and cell type lineage restrictions during pharyngeal arch development in the zebrafish embryo. *Development* **120**, 2945–2960.

Schilling, T. F., Walker, C., and Kimmel, C. B. (1996a). The *chinless* mutation and neural crest cell interactions during zebrafish jaw development. *Development* **122**, 1417–1426.

Schilling, T. F., Piotrowski, T., Grandel, H., Brand, M., Heisenberg, C.-P., Jiang, Y.-J., Beuchle, D., Hammerschmidt, M., Kane, D. A., Mullins, M. C., *et al.* (1996b). Jaw and branchial arch mutants in zebrafish. I: Branchial arches. *Development* **123**, 329–344.

Schilling, T. F. (1997). Genetic analysis of craniofacial development in the vertebrate embryo. *BioEssays* **19**, 459–468.

Schilling, T. F., and Kimmel, C. B. (1997). Musculoskeletal patterning in the pharyngeal segments of the zebrafish embryo. *Development* **124**, 2945–2960.

Schilling, T. F., Prince, V., and Ingham, P. W. (2001). Plasticity in zebrafish *hox* expression in the hindbrain and cranial neural crest. *Dev. Biol.* **231**, 201–216.

Schilling, T. F. (2002). The morphology of larval and adult zebrafish. *In* "Zebrafish: A Practical Approach" (R. Dahm and C. Nusslein-Volhard, eds.), pp. 59–94. Oxford University Press, Oxford.

Thisse, C., Thisse, B., and Postlethwait, J. H. (1995). Expression of *snail2*, a second member of the zebrafish snail family, in cephalic mesentoderm and presumptive neural crest of wild-type and *spade tail* mutant embryos. *Dev. Biol.* **172**, 86–99.

Walsh, E. C., and Stainier, D. Y. (2001). UDP-glucose dehydrogenase required for cardiac valve formation in zebrafish. *Science* **293**, 1670–1673.

Wardi, A. H., and Allen, W. S. (1972). Alcian blue staining of glycoproteins. *Anal. Biochem.* **48**, 621–623.

Whiteman, P. (1973). The quantitative measurement of Alcian Blue-glycosaminoglycan complexes. *Biochem. J.* **131**, 343–350.

Witten, P. E., and Villwock, W. (1997). Growth requires bone resorption at particular skeletal elements in a teleost fish with acellular bone (*Oreochromis niloticus*, Cichlidae). *J. Appl. Ichtyol.* **13**, 149–158.

Witten, P. E., Hansen, A., and Hall, B. K. (2001). Features of mono- and multinucleated bone resorbing cells of the zebrafish Danio rerio and their contribution to skeletal development, remodeling, and growth. *J. Morphol.* **250**, 197–207.

Xu, Q., Alldus, G., Macdonald, R., Wilkinson, D. G., and Holder, N. (1996). Function of the Eph-related kinase rtk1 in patterning of the zebrafish forebrain. *Nature* **381**, 319–322.

Yan, Y. L., Miller, C. T., Nissen, R. M., Singer, A., Liu, D., Kirn, A., Draper, B., Willoughby, J., Morcos, P. A., Amsterdam, A., *et al.* (2002). A zebrafish *sox9* gene required for cartilage morphogenesis. *Development* **129**, 5065–5079.

Yan, Y. L., Hatta, K., Riggleman, B., and Postlethwait, J. H. (1995). Expression of a type-II collagen gene in the zebrafish embryonic axis. *Dev. Dyn.* **203**, 363–376.

Yelick, P. C., and Schilling, T. F. (2002). Molecular dissection of craniofacial development using zebrafish. *Crit. Rev. Oral. Biol. Med.* **13**, 308–322.

Yelon, D., Ticho, B., Halpern, M. E., Ruvinsky, I., Ho, R. K., Silver, L. M., and Stainier, D. Y. (2000). The bHLH transcription factor hand2 plays parallel roles in zebrafish heart and pectoral fin development. *Development* **127**, 2573–2582.

Zhang, J., Talbot, W. S., and Schier, A. F. (1998). Positional cloning identifies zebrafish *one-eyed pinhead* as a permissive EGF-related ligand required during gastrulation. *Cell* **92**, 241–251.

CHAPTER 19

Morphogenesis of the Jaw: Development Beyond the Embryo

R. Craig Albertson and Pamela C. Yelick

Department of Cytokine Biology
The Forsyth Institute and
Department of Oral and Developmental Biology
Harvard School of Dental Medicine
Boston, Massachusetts 02115

I. Larval Zebrafish Craniofacial Cartilage Development

Since the mid-1990s, the zebrafish model has been used to provide significant insight into the molecular/genetic pathways guiding craniofacial development. Analyses of wild-type zebrafish and the identification and characterization of classes of phenotypically related zebrafish craniofacial mutants have revealed important signaling cascades and tissue interactions that are required for the proper placement, patterning, and differentiation of craniofacial cartilages (Kimmel et al., 1998; Neuhauss et al., 1996; Piotrowski et al., 1996; Schilling et al., 1996; Yelick and Schilling, 2002). As discussed in more detail by Javidan and Schilling in chapter 18 of this volume, the recent molecular characterization of five previously identified zebrafish craniofacial mutants has revealed the importance of signals derived from the endothelium (and of certain transcription factors mediating these signals) for proper craniofacial development. Correlation of the phenotypes of the *suc/et1* zebrafish mutant—caused by a point mutation in the *endothelin-1* gene (Miller *et al.*, 2000); mouse mutants *piebald (s)*, which encode a B-type endothelin receptor (Hosoda *et al.*, 1994); and *lethal spotting (ls)*, which encodes the *endothelin 3* ligand (Baynash *et al.*, 1994)—has demonstrated the importance of the endodermally derived endothelin signaling for proper dorsoventral patterning and differentiation of neural crest cells (NCCs) that give rise to craniofacial cartilage structures. The *suc/et1* phenotype closely resembles human DiGeorge (DGS) and velocardiofacial syndrome phenotypes, thereby providing a useful model for analyses of these human diseases (Kurihara *et al.*, 1995; Paylor *et al.*, 2001). Roles for endothelin signaling pathways in the later developmental events of mineralized craniofacial skeletal elements were also revealed, as discussed later.

Analyses of the craniofacial transcription factor mutants *colourless (cls)/sox10, van Gough (vgo)*, and *lockjaw(low)/tfap2a* have revealed potential genetic interactions of these genes in dorsoventral patterning of the craniofacial complex. The combined pigment and enteric neuronal defects exhibited by *cls/sox10* mutants resemble those of the already mentioned mouse mutants *piebald (s)* and *lethal spotting (ls)* and the *Dominant megacolon (Dom)* mutant, which encodes *Sox10*—the functional homologue of *cls* (Herbarth *et al.*, 1998; Southard-Smith *et al.*, 1998). The *cls/sox10* phenotype also bears a striking resemblance to the human Waardenburg-Shah syndrome and Hirschsprung's disease (Hassinger *et al.*, 1980).

The ventral arch mutant *van Gough (vgo)*, caused by a mutation in the T-Box family member, *tbx1*, (Piotrowski *et al.*, 2003), exhibits severe craniofacial skeletal segmentation defects where cartilages of adjacent arches often fuse. Mutations in the *tbx1* gene are thought to be major contributors to the cardiovascular defects in human DGS patients, a disease affecting several NCC derivatives of the pharyngeal arches (Kurihara *et al*, 1995). Analysis of the *vgo/tbx1* zebrafish mutant demonstrated that *tbx1* acts cell autonomously in the pharyngeal mesendoderm to secondarily influence the development of NCC-derived cartilages. This study also identified regulatory interactions between *vgo/tbx1, edn1*, and *hand2*, genes also implicated in DGS.

An additional mutant, *lockjaw(low)*, caused by a mutation in the *ap2a* gene, exhibits NCC-derived skeletal and pigment defects (Knight *et al.*, 2003). Studies of *low* suggest a model where *tfap2a* functions independently in the specification of subpopulations of NCC-derived pigment cells and in patterning of the pharyngeal skeleton through the regulation of Hox genes.

Together, the successful application of a forward genetic approach to identify and characterize genes essential for early zebrafish craniofacial development has proven extremely beneficial to increasing our understanding of the molecular genetic signals regulating normal human craniofacial development, and in elucidation of molecular defects in human craniofacial syndromes including DGS/VCs, Waardenburg-Shah, and Hirschsprung's disease. In chapter 18 of this volume, many of the techniques that have been perfected to facilitate analyses of early zebrafish craniofacial cartilage growth and patterning up to approximately 5 days of development are described in detail. In this chapter we describe methods used to characterize the molecular/genetic signaling events regulating later stages of craniofacial development, including the growth and differentiation of the craniofacial skeleton, and of replacement tooth development.

II. Analysis of Craniofacial Skeletal and Replacement Tooth Development

Forward genetic mutagenesis screens in zebrafish have provided a detailed knowledge of the molecular signaling cascades regulating *early* embryonic events. The efficiency of this method, however, is simultaneously a limitation. The vast majority of described zebrafish craniofacial mutants exhibit gross qualitative defects and are generally lethal by 5 to 7dpf (Neuhauss *et al.*, 1996; Piotrowski *et al.*, 1996; Schilling *et al.*, 1996; Yelick and Schilling, 2002). At this developmental stage, zebrafish normally have begun to mineralize a bilateral series of 10 to 11 pharyngeal bones and exhibit 1 to 3 teeth in each bilateral ceratobranchial 5 (cb5) arch. Bones and teeth present in the 7-dpf larval zebrafish are illustrated in Fig. 1, Panel C. Two dermal and 2 cartilage replacement bones are derived from the mandibular arch, 3 dermal and 2 endochondral bones arise from the hyoid arch; whereas the only mineralized structure derived from more posterior arches at this stage is cb5. Since these represent only a fraction of the more than 50 bones and 11 bilateral teeth present in adult zebrafish, previously identified homozygous recessive craniofacial mutants are not useful for developmental analyses of the vast majority of skeletal and tooth structures (Cubbage and Mabee, 1996). Therefore, the early lethal phenotypes exhibited by the majority of previously identified craniofacial mutants have resulted in a substantial gap in our knowledge of the genetic control of *later* developmental events including: (1) primary and replacement tooth development, (2) development of the craniofacial skeleton, and (3) growth and remodeling of the craniofacial complex.

In the sections that follow, we discuss experimental approaches that can be used to facilitate analyses of the later developmental events of skeletogenesis and replacement tooth development. To begin, we review the incipient body of literature composing our current understanding of the molecular basis of craniofacial skeletal and pharyngeal tooth development. Next, we propose a new paradigm to use in the characterization and interpretation of the molecular events regulating bone and tooth development in zebrafish.

A. Zebrafish Pharyngeal Tooth Development

Like other cyprinid fishes, zebrafish lack teeth in their oral jaws but rather exhibit a series of bilaterally symmetric teeth on each cb5 arch (Fig. 1). These pharyngeal teeth exhibit all of the characteristics of teeth present in "higher" vertebrates, including a pulp cavity, dentin, and enameloid surface. Recent

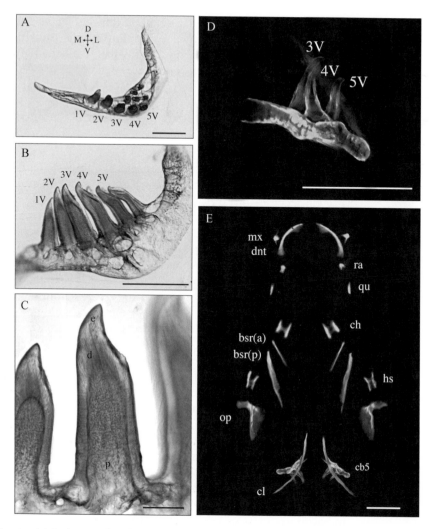

Fig. 1 Adult pharyngeal teeth and larval mineralized skeleton. (A to C) Dissected and Alizarin red–stained cb5 arch and teeth of the adult zebrafish. The large ventral row of pharyngeal teeth is labeled 1V to 5V. Adult zebrafish also possess discrete medial and dorsal rows of teeth (*not labeled*). Dorsal, ventral, medial, and lateral are designated D, V, M, and L, respectively. (C) High magnification of 2V shows the pulp cavity (p), dentin tubules (d), and enameloid (e). (D) Quercetin-stained cb5 in wild-type 7-dpf larvae with three pharyngeal teeth, 3V to 5V. (E) Quercetin-stained mineralized pharyngeal skeleton in 7-dpf larval zebrafish viewed under ultraviolet (UV) illumination. Dermal bones are labeled to the left. Cartilage replacement bones are to the right. Scale bars in A and B represent 0.5 mm. Bars in C to E represent 100 μm. Abbreviations are as follows: mx, maxilla; dnt, dentary; ra, retroarticular; qu, quadrate; ch, ceratohyal; bsr(a), anterior branchiostegal ray; posterior branchiostegal ray; hs, hyosymplectic; op, opercle; cl, cleithrum; cb5, ceratobranchial 5. (See Color Insert.)

morphological and histological studies have described the pattern of tooth development in the larval zebrafish (Perrino and Yelick, 2004; Van der Heyden and Huysseune, 2000; Wautier *et al.*, 2001), and the regulation of adult replacement tooth formation (Huysseune and Sire, 2003; Perrino and Yelick, 2004; Van der Heyden *et al.*, 2001 and 2000). Adult zebrafish teeth closely resemble mammalian teeth, exhibiting distinct pulp cavity, dentin, and enameloid tissues (Fig. 1, Panel B). Zebrafish teeth arise in a synchronized fashion to form three distinct dorsoventral rows (Fig. 1, Panel A). The development of each tooth consists of distinct morphological stages similar to those of mammalian teeth, including initiation, morphogenesis, and cytodifferentiation stages (Thesleff, 2003; Thesleff *et al.*, 1991).

Distinct from mammalian tooth development, zebrafish pharyngeal teeth undergo continuous tooth replacement, characterized by coordinated shedding and attachment stages. Where an older functional tooth is shed, the socket becomes remodeled, and a newly formed replacement tooth is subsequently secured to the cb5 skeletal element. Zebrafish teeth are continuously replaced throughout the life of the zebrafish, providing an opportune model to explore the molecular/genetic signaling cascades regulating both primary *and* replacement tooth development.

Although brief descriptions of tooth defects were included in analyses of zebrafish pharyngeal arch mutants described in the original two large-scale zebrafish mutagenesis screens (Neuhauss *et al.*, 1996; Piotrowski *et al.*, 1996; Schilling *et al.*, 1996), we are unaware of the existence of any tooth-specific mutant. In mammals, the major gene families involved in mammalian tooth specification—Bone morphogenetic protein (Bmp), Wnt, Hedgehog (Shh), Fibroblast growth factor (Fgf), and TNF-alpha families—are reiteratively expressed throughout successive stages of tooth development. Many of these genes are also expressed in the tooth-bearing arches of the zebrafish (for review, see Yelick and Schilling, 2002). Unfortunately, functional characterizations of genes regulating tooth development are less straightforward. In zebrafish, as in all other vertebrates, many of the growth factors that regulate tooth development are also required in very early patterning events of the embryo. Thus, mutations in these genes often result in early lethal phenotypes, resulting in the death of the developing embryo before the initiation or during very early stages, of tooth development. The identification of tooth-specific zebrafish mutants, an ongoing effort in our laboratory, would greatly enhance the understanding of the molecular pathways particularly involved in tooth specification and morphogenesis.

B. Bone Development

Skeletal development in the zebrafish is regulated in much the same way as in other vertebrates, by balanced activities of bone formation by osteoblasts and bone resorption by osteoclasts (Witten *et al.*, 2001). Descriptions of the development of the bony skeleton of zebrafish have been reported (Bird and Mabee, 2003; Cubbage and Mabee, 1996). However, the molecular characterization of this process is virtually nonexistent at this time. Since the mid-1990s, we have seen admirable advances in our understanding of the molecular mechanisms that contribute to the early development of the zebrafish pharyngeal arch cartilages (Kimmel *et al.*, 2001; Miller *et al.*, 2000, 2003; Piotrowski *et al.*, 1996; Schilling *et al.*, 1996). Unfortunately, the relatively late formation of the bony skeleton has precluded its characterization in most analyses of zebrafish homozygous recessive craniofacial mutants. This notable lack of molecular characterization of mineralized tissue formation in the zebrafish has prompted a number of laboratories, including ours, to focus on the elucidation of the molecular/genetic signaling pathways regulating these processes.

Reports of bone development in zebrafish provide insight into certain molecular signals regulating zebrafish skeletal patterning and growth. One study took advantage of the observation that a percentage of *chordin (chd)* homozygous recessive mutants survived to become fertile adults (Fisher and Halpern, 1999). Chordin is a bone morphogenic protein (Bmp) antagonist that plays an important role in the development of the vertebrate gastrula (Miller-Bertoglio *et al.*, 1997; Thomsen, 1997). Mutants deficient in *chordin* exhibit expanded *bmp4* expression in the gastrula, resulting in ventralized phenotypes (Hammerschmidt *et al.*, 1996). While most *chd* mutants exhibited early lethal phenotypes, a small percentage developed into fertile adults (Fisher and Halpern, 1999). Between 4 and 50% of *chd* mutants survive to sexual maturity depending on the allele, providing a unique opportunity to investigate the effects of *chordin* on the later development of bone (Fisher and Halpern, 1999). It was discovered that *chordin*-deficient mutants exhibited defects in axial and caudal skeletal patterning, including the absence, branching, or fusion of the bony processes of the vertebrae and fins (Fisher and Halpern, 1999). These phenotypes were correlated with the ectopic and expanded expression of *bmp4* and its downstream target *msxC*, suggesting a role for *chordin*-mediated BMP activity in skeletal patterning. Interestingly, the craniofacial skeleton was reported as having no noticeable defects.

Another study explored the role of the secreted peptide Endothelin 1 (*edn1*) in regulating pharyngeal bone development in the zebrafish (Kimmel *et al.*, 2003). As already mentioned, the N-ethyl-N-nitrosourea (ENU)-induced mutant *sucker (suc)*, a point mutation in the *endothelin 1 (edn1)* gene, was identified in a genetic screen for pharyngeal cartilage development (Miller *et al.*, 2000; Piotrowski *et al.*, 1996). Zebrafish *edn1* is expressed in a central core of arch paraxial mesoderm, in both surface ectoderm and pharyngeal endodermal epithelia, and not in skeletogenic neural crest. However, zebrafish *edn1* is thought to act directly on post-migratory neural crest cells (Kimmel *et al.*, 2003; Miller *et al.*, 2000). Defects exhibited by *suc* mutants suggest *edn1* is required for dorsal-ventral patterning of the first and second pharyngeal arch cartilages (Miller and Kimmel, 2001; Miller *et al.*, 2000; Piotrowski et al., 1996). *Suc* mutants are characterized by fusion of the dorsal and ventral cartilages in both the first and second pharyngeal arches (Miller *et al.*, 2000). The anterior-posterior polarity of the ventral first arch Meckel's cartilage is also reversed, in that it is directed backwards (Miller *et al.*, 2000). Similar defects are observed in the dermal bones of *suc* larvae (Kimmel *et al.*,

2003). In the first arch, the bones of the upper and lower jaw are fused and reversed in polarity. The effects of *edn1* deficiency on the dermal bones of the second arch are more dynamic, including bone loss, gain, and fusion, in a manner that is dependent on Edn1 levels (Kimmel *et al.*, 2003). When Edn1 was severely reduced (for example using anti-sense morpholino oligomer targeted depletion strategies) second arch bones were generally reduced or missing. Alternatively, when Edn1 was only mildly reduced, second arch bones were more likely to develop, and the dorsal element (opercle) was often increased in size (Kimmel *et al.*, 2003). Mild reduction of Edn1 also led to a fusion of the dorsal and ventral bones of the hyoid arch, revealing defects in dorsoventral patterning. The range of bone defects observed in *suc* mutants suggests that Edn1 patterns the pharyngeal skeleton through morphogenic gradients, and that pharyngeal bones are acutely sensitive to Edn1 protein levels (Kimmel *et al.*, 2003). Unfortunately, because of the severe nature of their defects, *suc* mutant larvae do not survive much beyond one week post fertilization. Thus, the role of Edn1 in regulating bone remodeling and growth remains unknown.

A large-scale screen for skeletal dysplasias in adult ENU-mutagenized zebrafish has been performed (Fisher *et al.*, 2003). Radiography was used to screen living F_1 adults for dominant skeletal mutations. Out of 2000 F_1 fish, only one skeletal mutant was recovered, the *chihuahua*[dc124] *(chi)* mutant. *Chi* was identified as a dominant mutation that transmitted the mutant phenotype to approximately half of its progeny. Adult (*chi*/+) fish appear shorter than their wild-type siblings, but are otherwise morphologically indistinguishable from a gross perspective. In contrast, examination of the skeletal anatomy of adult *(chi/+)* fish using radiographic methods revealed extensive skeletal dysplasia characterized by irregular bone growth, uneven mineralization, and bone weakness as revealed by the presence of multiple bone fractures (Fisher *et al.*, 2003). Many of the defining characteristics of *chi* mutants are similar to individuals exhibiting osteogenesis imperfecta (OI), a dominantly inherited skeletal dysplasia typically caused by mutations in one of two type I collagen genes, COL1A1 or COL1A2 (Benusiene and Kucinskas, 2003). It was therefore not surprising that *chi* mapped to the same interval on LG3 as *col1a1* (Fisher *et al.*, 2003). Although *chi/chi* homozygous mutants failed to develop swim bladders or feed after one week, they were reported to be indistinguishable from age-matched *chi/+* siblings (Fisher *et al.*, 2003). This study by Fisher *et al.* (2003) demonstrates the utility of radiography as a high-throughput screening method for skeletal defects. However, the relatively low frequency that bone-specific mutants were recovered (i.e., 1/2000 or 0.05%), is less encouraging. It will be interesting to further define the effects of the *col1a1* mutation, as present in the *chi* mutant, on bone growth and remodeling.

To date, only one report has focused specifically on characterizing bone growth and remodeling in the zebrafish. Witten *et al.* (2001) endeavored to characterize the manner by which osteoclast activity contributes to skeletal growth and development in zebrafish. Bone remodeling is intimately associated with bone growth (Olsen *et al.*, 2000). Since the zebrafish skeleton continues to grow throughout the life of the fish, osteoclast activity in developing bone was studied from larval to juvenile stages of development to provide an understanding of the mechanisms that regulate this process (Witten et al., 2001). In zebrafish, bone remodeling is initiated between 2 and 3 weeks post fertilization via activation of mononucleated osteoclasts located at the surface of bones poised to undergo the process of remodeling (Witten et al., 2001). As development and growth proceeds, both mono-nucleated and multinucleated osteoclasts contribute to the process of bone resorption. In general, multinucleated cells were found in association with more robust bones and deep absorption pits; whereas mono-nucleated osteoclasts were found in association with thin bones and shallow resorption pits (Witten et al., 2001). While the pattern of bone remodeling in zebrafish appears to be consistent with that in mammals, the regulation of osteoclast activity in zebrafish remains largely unknown and may differ from that in mammals (Witten et al., 2001). Zebrafish, for example, have three parathyroid hormone (PTH) receptors (zPTH1R, zPTH2R, and zPTH3R), and at least two ligands (zPTH1 and zPTH2), suggesting a potentially more complex PTH regulatory system in fishes than in mammals (Gensure et al., 2004). Understanding the fundamental processes of osteoblast- and osteoclast-regulated bone remodeling in teleosts is central to understanding the molecular determinants of bone growth and shape in zebrafish.

C. Paradigm Shift—Analyses of Post-Larval Tooth and Bone Development

Since the mid-1990s, we have seen significant advances in defining molecular mechanisms that pattern the zebrafish embryo. Significantly less is known of zebrafish development beyond early embryonic stages. This is not for lack of interest, but rather is because many of the tools and methods used to analyze early embryonic processes exhibit limited utility for that of later developmental stages. We argue that to better understand the molecular interactions contributing to development beyond the embryo, traditional techniques must be bolstered by the introduction of new methodologies. Specifically, studies in zebrafish aimed at characterizing the development and growth of bones and teeth would be greatly facilitated by the following: (1) genetic screens targeted for later developmental processes (e.g., tooth development, regeneration, and replacement and bone formation, growth, and remodeling); (2) a shift from *qualitative* to *quantitative* characterizations of phenotypes; and (3) transition away from assessment of defects at isolated developmental stages, to the continuous monitoring of mutant phenotypes throughout extended developmental periods.

An elegant example of the first approach was described in the study by Fisher et al. (2003). Similarly, we are currently conducting an ENU-mutagenesis screen for bone and tooth phenotypes in larval and juvenile zebrafish. Our screen begins at 7 dpf, when bones first appear, and continues through 1 month post-fertilization. We are screening F₃ families for homozygous recessive and heterozygous mutations, using both an *in vivo* Quercetin fluorescent stain (Sigma, St. Louis, MO), and Alizarin red/Alcian blue staining methods (modified from Pothoff, 1983).

The Q and Alizarin red stains both bind free Ca^{++}, and therefore label mineralized tissues (e.g., bone and teeth). At 7 dpf, we screen for the appearance of bone and the development of primary teeth. At 1 month, we examine bone growth, as well as the development and patterning of replacement teeth.

Our screening method is simple, robust, and noninvasive. At each developmental stage, F_3 families are placed in a 0.1 mg/ml solution of Quercetin in system water for 1 to 3 hours. The fish are then anesthetized and screened under a Zeiss M2-Bio dissecting microscope fitted with fluorescence. For long-term storage, zebrafish are fixed in buffered 4% paraformaldehyde (PFA), enzymatically cleared with trypsin, stained with Alizarin red in 0.5% KOH, and stored in Glycerol. Using this screening method, we have begun to recover mutants with cartilage, bone, and tooth phenotypes. Some of the newly identified homozygous recessive mutants exhibit severe defects in pharyngeal cartilage development similar to those described in previous mutagenesis screens (Neuhauss *et al.*, 1996; Piotrowski *et al.*, 1996; Schilling *et al.*, 1996; Yelick and Schilling, 2002), making defects in bone and tooth development difficult to interpret. In contrast, analysis of heterozygous mutants has revealed haploinsufficiency bone and tooth phenotypes that appear quite interesting. We have also recovered mutants that exhibit more specific and subtle defects. One such example is shown in Fig. 2. This putative mutant survives well into larval development (i.e., more than 7 dpf), and is distinguished from wild-type siblings by a shortened head and misshapen eyes (Fig. 2, Panels A and D). Examination of dermal bones at 8 dpf revealed defects that appear largely restricted to anterior pharyngeal skeletal elements (Fig. 2, Panels B and E). First-arch bones appear shorter than those of wild-type siblings, while there is no noticeable difference in the length of second-arch bones. Similar defects are observed in pharyngeal arch cartilages (Fig. 2, Panels C and F). In fact, there is a quantifiable difference in the length of Meckel's cartilage (Fig. 2, Panel G), but not of the ceratohyal cartilage (Fig. 2, Panel H). Thus, this mutation appears to specifically affect the cartilages and dermal bones of the mandibular arch. Subtle defects in skeletal anatomy, such as those exhibited by this mutant, can also be quantified using powerful geometric morphometric techniques to reveal regional differences in the size and shape of bony structures.

D. Use of Geometric Morphometric Tools for Quantification of Skeletal Defects

Geometric morphometrics (GM) is an exceedingly powerful tool with which to describe shape variation (Rohlf and Marcus, 1993). A geometric approach to shape analysis is based on the establishment of landmark data, which are homologous anatomical points recorded as a Cartesian coordinate system. Instead of reporting shape as lengths, widths, or angles, a landmark-based approach emphasizes the geometry of a given structure, allowing shape variation to be described relative to other structures. In this way, GM is an ideal method with which to quantify regional defects in experimentally manipulated animals (e.g., genetic mutants). Moreover, geometric descriptors of shape can be reported via pictorial

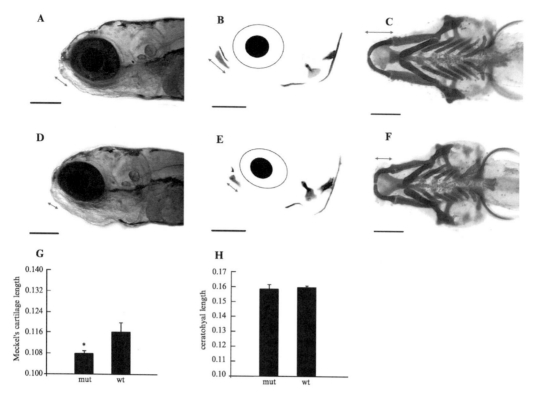

Fig. 2 Characterization of subtle pharyngeal skeletal defects. Eight-dpf wild-type (wt) (A to C) and mutant (D to F) larvae. Mutants are distinguished from wt siblings by a misshaped eye and shortened anterior head structures (A versus D). In the lateral view, first-arch dermal bones appear shorter than those of wt siblings (B versus E). The chondrocranium of mutant zebrafish also exhibits similar defects (C versus F). Meckel's cartilage is significantly shorter in mutant animals (G), while no significant difference in the length of the ceratohyal is detected (H). (See Color Insert.)

representations of the organism/structure, providing an intuitive and precise biological representation of the observed defects.

The interested reader is encouraged to review Bookstein (1991) and Rohlf and Marcus (1993) for a more thorough description of concepts and methods in geometric morphometrics. In general, geometric methods of shape analysis involve three steps: (1) superimposition of landmark data; (2) decomposition of shape variation into a series of geometric variables; and (3) statistical analysis of those variables. The purpose of the first step is to eliminate variance introduced by orientation, position, and size of the specimens as they are measured. Many superimposition algorithms have been developed. A least-squares approach, which is perhaps the most common, superimposes landmark configurations such that the sum of the squared distances between corresponding landmarks is minimized. Thin-plate spline (TPS) analysis is a geometric technique that quantifies

D'Arcy Thompson's concept of Cartesian grid deformations (Thompson, 1917). Deformation grids are constructed from two landmark configurations—the starting form (e.g., mean wild-type shape) and the target form (e.g., mean mutant shape). For morphometrics analysis, the starting form is constrained at some combination of points (i.e., landmarks), but is otherwise free to adopt the target form in a way that minimizes bending energy. This total deformation of the Cartesian grid is then decomposed into a series of geometrically orthogonal variables based on scale (Rohlf and Marcus, 1993). These variables (called partial warps) can be localized to describe precisely what aspects of shape differ between the starting and target forms. When amassing large series of variables, it is often desirable to define major axes of variation. Statistically, partial warps can be treated in the same way as any other variable. Thus, partial warps are perfectly amenable to data reduction analyses such as canonical variate or principal component analysis. The advantage of using geometric variables is that instead of viewing results in terms of graphs and tables, data may be reported as deformation grids. Again, this offers the advantage of direct biological interpretation of the results.

Equipped with these tools and an eye for subtle defects, we can explore cartilage and bone defects in previously identified and in new ENU-induced zebrafish mutants. An example of the application of GM analysis is presented in Fig. 3. Here we have used GM analyses to examine cartilage defects in the larval mutant *acerebellar* (*ace*), which is a mutation in the *fgf8* gene. The *ace/fgf8* mutant was originally classified as a neural mutant based on the lack of midbrain structures (Brand *et al.*, 1996). Since that time, efforts have been made to characterize pharyngeal cartilage defects in *ace* and other *fgf*-deficient mutants (David *et al.*, 2002; Walshe and Mason, 2003). Figure 3, Panel A illustrates the pharyngeal skeleton of wild-type 7-dpf zebrafish larvae in the ventral view. Landmarks that capture variation in first- and second-arch cartilages are shown in red, and are labeled 1 to 9. Age-matched wild-type sibling and *ace/fgf8* mutant larvae are shown (Fig. 3, Panels B and C). Note that the *ace/fgf8* mutant heads apparently shorten (Fig. 3, Panel C). Geometric shape analysis was performed on a family derived from an in-cross of two identified heterozygous *ace/fgf8* adults (Fig. 3, Panels D to G). Landmark variation among specimens is illustrated in Fig. 3, Panel D. Medially placed landmarks (landmark numbers 1, 5, and 6) exhibit distinct anterior-posterior variation; whereas laterally positioned landmark numbers 2, 3, 4, 7, 8, and 9 exhibit distinct variation along the medial-lateral axis. Results of a principal-components analysis on partial warp scores (formally referred to as *relative warp analysis*) are shown (Fig. 3, Panels E to G). Deformation grids representing deviation in shape along the first principal component axis explain 88% of the total phenotypic variation (Fig. 3, Panels E and F). Mutant animals plot on one side of principal component (PC1), while wild-type siblings plot on the opposite side. As expected, mutants exhibit much greater variation along PC1 than do wild-type sibling larvae (Fig. 3, Panel G). Shape variation along PC1 reveals pronounced shortening of both first and second ventral arch cartilages in

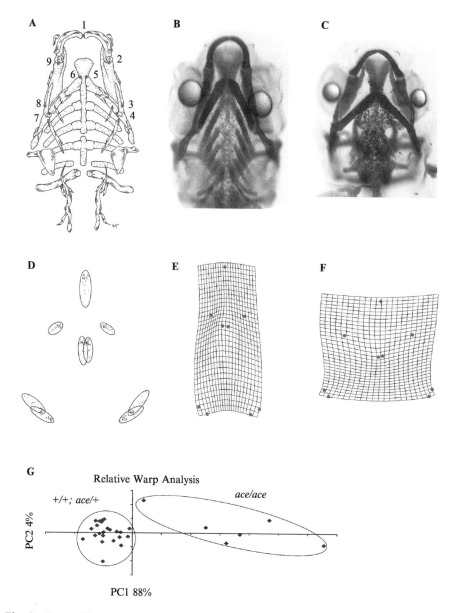

Fig. 3 Shape differences in wild-type (wt) and mutant skeletal elements. Illustration of the 7-dpf wt zebrafish pharyngeal skeleton in the ventral view (A). Seven-dpf wt (B) and *ace* (C) larvae. Note the shortened mutant head. (D to G) Geometric morphometrics analysis of skeletal defects. (D) Variation in landmark configuration after superimposition. (E and F) Deformation of shape along the first principal component axis (PC1). Wt shape is depicted in (E); mutant shape in (F). Mutants exhibit greater variance along PC1 (G). (See Color Insert.)

ace/fgf8 mutants (Fig. 3, Panel F). These elements are also splayed laterally, resulting in a widening of the pharyngeal skeleton. (Note the horizontal displacement of landmarks 2, 3, and 4 and 7, 8, and 9.) As a consequence, first and second arch cartilages have become distanced along the anterior-posterior axis. (Note the vertical displacement of landmarks 5 and 6 relative to landmarks 2 and 9.)

Interestingly, shape variation in the wild-type versus mutant analysis is similar to that exhibited by a comparison of homozygous *ace/fgf8* larvae at 5 and 7 dpf (Fig. 4). Wild-type and *ace/fgf8* mutants at 7 dpf, and 5-dpf *ace/fgf8* mutant larvae are presented (Fig. 4, Panels A, B, and C, respectively). Figure 4, Panels D to F represent deformations of shape along the first principal component axis, which explains 94% of the total shape variation (Fig. 4, Panel G). Deformation grids represent the mean shape of 7-dpf wild-type larvae (Fig. 4, Panel D), 7-dpf *ace/fgf8* mutants (Fig. 4, Panel E) and 5-dpf *ace/fgf8* mutants (Fig. 4, Panel F). The geometry of shape in this analysis is strikingly similar to that presented in the previous example (Fig. 3), in that both exhibit the displacement of the same sets of landmarks relative to one another. Moreover, only one major axis of shape variation (PC1) exists. That is to say, wild-type animals lie along the same trajectory as the mutants from two developmental stages. Altogether, these observations suggest that defects exhibited by *ace/fgf8* mutants are associated with the growth of the pharyngeal skeleton, consistent with the reported roles for fibroblast growth factor (FGF)-signaling in regulating patterning and growth of the anterior region of the vertebrate craniofacial complex (Abu-Issa *et al.*, 2002; Bachler and Neubuser, 2001; Frank *et al.*, 2002; Mina *et al.*, 2002; Tucker *et al.*, 1999).

The examples described demonstrate the utility of GM in quantifying shape defects in mutant zebrafish. Because this approach is dependent on the accurate placement of homologous landmarks among specimens, GM analyses are not useful when defects are so severe that cartilages are missing or grossly misplaced. However, when defects are subtle, this is a consummate technique with which to describe and quantify regional affects of genetic deficiency. As shown in Fig. 4, GM is also a particularly useful tool to describe growth of the craniofacial complex of wild-type and mutant zebrafish. In contrast to the considerable amount of research on larval pharyngeal cartilage development, to date no published studies describe growth of the zebrafish pharyngeal skeleton. A morphometric characterization of wild-type skeletal growth would significantly extend previous studies by providing fundamental knowledge of how adult skeletal form is achieved. Moreover, such a study would provide a reference standard for comparison to that of identified mutants. Discrete changes in the slopes of the growth trajectories of mutants as compared with wild-type siblings could be anatomically localized by GM to identify regional defects of individual gene mutations.

As already mentioned, GM can be used to explore the shape and growth of skeletal elements in previously described heterozygous and nonlethal zebrafish mutants. An appropriate mutant for such analysis is the zebrafish mutant *ikarus* (*ika*), which encodes a *fgf24* mutation (Fischer *et al.*, 2003). Homozygous recessive

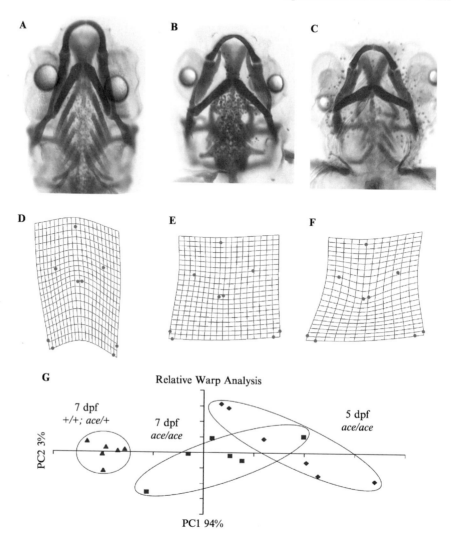

Fig. 4 Geometry and growth of wild-type (wt) and mutant pharyngeal cartilage defects. Representative specimens are shown in (A to C). (A) Wt 7-dpf larvae, (B) mutant 7-dpf larvae, and (C) mutant 5-dpf larvae. Shape deformations are illustrated in (D to E), when pharyngeal skeletal defects in both 7-dpf (E) and 5-dpf (F) *ace* are evaluated against 7-dpf wt larvae (D). The similarity in geometry suggests the *ace* mutation effects growth of the pharyngeal skeleton. All specimens plot on the same principal component axis (PC1), which explains 94% of the total shape variation (G). (See Color Insert.)

ika/fgf24 mutants completely lack pectoral fins, but otherwise appear normal and live to reproductive maturity (Fischer *et al.*, 2003). Another suitable candidate is the *somitabun* (*sbn*) mutant, originally identified in a screen for dorsal-ventral defects in the zebrafish embryo (Mullins *et al.*, 1996). *Sbn/Smad5* mutants exhibits a dominant effect with varying penetrance due to a single amino acid substitution in the Smad5 protein (Hild *et al.*, 1999). Homozygous *sbn/Smad5* embryos are severely dorsalized and die before 3 dpf (Mullins *et al.*, 1996). Embryos that possess one *sbn/Smad5* allele survive to adulthood, but have notable posterior axial skeletal defects including short or no caudal fin (Mullins *et al.*, 1996). The *chi* mutation, discussed earlier, is another example of a dominant mutation with effects on skeletal anatomy. Paradoxically, all of the genes encoded by the mutations delineated earlier are expressed in the developing head, yet no craniofacial defects have been reported. It is possible that compensatory gene expression may alleviate craniofacial defects to a certain extent. We suggest that statistical analyses of the craniofacial complex using sensitive GM tools may reveal previously unidentified defects. Quantification of such phenotypes in heterozygous and nonlethal homozygous mutants could eventually facilitate an understanding of the molecules that regulate pharyngeal skeletal shape and growth.

III. Conclusion

Morphogenesis can be defined as *the processes that are responsible for producing the complex shapes of* adults *from the simple ball of cells that derives from division of the fertilized egg* (On-line Medical Dictionary, © 1997–98 Academic Medical Publishing & CancerWEB). By this definition, a large gap exists in our current knowledge of the morphogenesis of the zebrafish craniofacial complex. In stark contrast to an extensive body of literature that analyzes early craniofacial cartilage development, very few studies in zebrafish have examined stages beyond 5 dpf. Since extensive morphogenetic changes occur after 5 dpf, the existing challenge for developmental biologists is to extend the current genetic paradigm beyond the embryo. Recently developed tools, including *in vivo* mineralized tissue stains combined with sensitive GM analyses described earlier, provide the necessary means to elucidate the molecules and pathways that produce shape in the adult craniofacial skeletal complex. It is likely that the application of these methods for the analyses of later-staged zebrafish development will provide significant insight into the regulation of human craniofacial development.

References

Abu-Issa, R., Smyth, G., Smoak, I., Yamamura, K., and Meyers, E. N. (2002). *Fgf8* is required for pharyngeal arch and cardiovascular development in the mouse. *Development* **129**, 4613–4625.

Bachler, M., and Neubuser, A. (2001). Expression of members of the Fgf family and their receptors during midfacial development. *Mech. Dev.* **100**, 313–316.

Baynash, A. G., Hosoda, K., Giaid, A., Richardson, J. A., Emoto, N., Hammer, R. E., and Yanagisawa, M. (1994). Interaction of *endothelin-3* with *endothelin-B* receptor is essential for development of epidermal melanocytes and enteric neurons. *Cell* **79**, 1277–1285.

Benusiene, E., and Kucinskas, V. (2003). COL1A1 mutation analysis in Lithuanian patients with osteogenesis imperfecta. *J. Appl. Genet.* **44**, 95–102.

Bird, N. C., and Mabee, P. M. (2003). Developmental morphology of the axial skeleton of the zebrafish, *Danio rerio* (Ostariophysi: Cyprinidae). *Dev. Dyn.* **228**, 337–357.

Bookstein, F. L. (1991). "Morphometric Tools for Landmark Data: Geometry and Biology." Cambridge University Press, New York.

Brand, M., Heisenberg, C. P., Jiang, Y. J., Beuchle, D., Lun, K., Furutani-Seiki, M., Granato, M., Haffter, P., Hammerschmidt, M., Kane, D. A., *et al.* (1996). Mutations in zebrafish genes affecting the formation of the boundary between midbrain and hindbrain. *Development* **123**, 179–190.

Cubbage, C. C., and Mabee, P. M. (1996). Development of the cranium and paired fins in the zebrafish *Danio rerio* (Ostariophysi, Cyprinidae). *J. Morph.* **229**, 121–160.

David, N. B., Saint-Etienne, L., Tsang, M., Schilling, T. F., and Rosa, F. M. (2002). Requirement for endoderm and FGF3 in ventral head skeleton formation. *Development* **129**, 4457–4468.

Fischer, S., Draper, B. W., and Neumann, C. J. (2003). The zebrafish *fgf24* mutant identifies an additional level of Fgf signaling involved in vertebrate forelimb initiation. *Development* **130**, 3515–3524.

Fisher, S., and Halpern, M. E. (1999). Patterning the zebrafish axial skeleton requires early chordin function. *Nat. Genet.* **23**, 442–446.

Fisher, S., Jagadeeswaran, P., and Halpern, M. E. (2003). Radiographic analysis of zebrafish skeletal defects. *Dev. Biol.* **264**, 64–76.

Frank, D. U., Fotheringham, L. K., Brewer, J. A., Muglia, L. J., Tristani-Firouzi, M., Capecchi, M. R., and Moon, A. M. (2002). An *Fgf8* mouse mutant phenocopies human 22q11 deletion syndrome. *Development* **129**, 4591–4603.

Gensure, R. C., Ponugoti, B., Gunes, Y., Papasani, M. R., Lanske, B., Bastepe, M., Rubin, D. A., and Juppner, H. (2004). Identification and characterization of two PTH-like molecules in zebrafish. *Endocrinology.* (Epub ahead of print). *Endocrinology* **145**(4), 1634–1639.

Hammerschmidt, M., Pelegri, F., Mullins, M. C., Kane, D. A., van Eeden, F. J., Granato, M., Brand, M., Furutani-Seiki, M., Haffter, P., Heisenberg, C. P., *et al.* (1996). *dino* and *mercedes*, two genes regulating dorsal development in the zebrafish embryo. *Development* **123**, 95–102.

Hassinger, D. D., Mulvihill, J. J., and Chandler, J. B. (1980). Aarskog's syndrome with Hirschsprung's disease, midgut malrotation, and dental anomalies. *J. Med. Genet.* **17**, 235–237.

Herbarth, B., Pingault, V., Bondurand, N., Kuhlbrodt, K., Hermans-Borgmeyer, I., Puliti, A., Lemort, N., Goossens, M., and Wegner, M. (1998). Mutation of the *Sry*-related *Sox10* gene in Dominant megacolon, a mouse model for human Hirschsprung disease. *Proc. Natl. Acad. Sci. USA* **95**, 5161–5165.

Hild, M., Dick, A., Rauch, G. J., Meier, A., Bouwmeester, T., Haffter, P., and Hammerschmidt, M. (1999). The *smad5* mutation *somitabun* blocks Bmp2b signaling during early dorsoventral patterning of the zebrafish embryo. *Development* **10**, 2149–2159.

Hosoda, K., Hammer, R. E., Richardson, J. A., Baynash, A. G., Cheung, J. C., Giaid, A., and Yanagisawa, M. (1994). Targeted and natural (*piebald*-lethal) mutations of *endothelin-B* receptor gene produce megacolon associated with spotted coat color in mice. *Cell* **79**, 1267–1276.

Huysseune, A., and Sire, J. Y. (2003). The role of epithelial remodeling in tooth eruption in larval zebrafish. *Cell Tissue Res.* Oct 29 (epub).

Kimmel, C. B., Miller, C. T., Kruze, G., Ullmann, B., BreMiller, R. A., Larison, K. D., and Snyder, H. C. (1998). The shaping of pharyngeal cartilages during early development of the zebrafish. *Dev. Biol.* **203**, 245–263.

Kimmel, C. B., Miller, C. T., and Moens, C. B. (2001). Specification and morphogenesis of the zebrafish larval head skeleton. *Dev. Biol.* **233**, 239–257.

Kimmel, C. B., Ullmann, B., Walker, M., Miller, C., and Crump, J. G. (2003). Endothelin 1-mediated regulation of pharyngeal bone development in zebrafish. *Development* **130**, 1339–1351.

Knight, R. D., Nair, S., Nelson, S. S., Afshar, A., Javidan, Y., Geisler, R., Rauch, G. J., and Schilling, T. F. (2003). *lockjaw* encodes a zebrafish *tfap2a* required for early neural crest development. *Development* **130**, 5755–5768.

Kurihara, Y., Kurihara, H., Maemura, K., Kuwaki, T., Kumada, M., and Yazaki, Y. (1995). Impaired development of the thyroid and thymus in *endothelin-1* knockout mice. *J. Cardiovasc. Pharmacol.* **3,** S13–S16.

Miller, C. T., and Kimmel, C. B. (2001). Morpholino phenocopies of endothelin 1 (*sucker*) and other anterior arch class mutations. *Genesis.* **30,** 186–187.

Miller, C. T., Schilling, T. F., Lee, K., Parker, J., and Kimmel, C. B. (2000). *sucker* encodes a zebrafish Endothelin-1 required for ventral pharyngeal arch development. *Development* **127,** 3815–3828.

Miller, C. T., Yelon, D., Strainier, D. Y. R., and Kimmel, C. B. (2003). Two *endothelin 1* effectors, *hand2* and *bapx1*, pattern ventral pharyngeal cartilage and the jaw joint. *Development* **130,** 1353–1365.

Miller-Bertoglio, V. E., Fisher, S., Sanchez, A., Mullins, M. C., and Halpern, M. E. (1997). Differential regulation of chordin expression domains in mutant zebrafish. *Dev. Biol.* **129,** 537–550.

Mina, M., Wang, Y. H., Ivanisevic, A. M., Upholt, W. B., and Rodgers, B. (2002). Region- and stage-specific effects of FGFs and BMPs in chick mandibular morphogenesis. *Dev. Dyn.* **223,** 333–352.

Mullins, M. C., Hammerschmidt, M., Kane, D. A., Odenthal, J., Brand, M., van Eeden, F. J., Furutani-Seiki, M., Granato, M., Haffter, P., Heisenberg, C. P., *et al.* (1996). Genes establishing dorsoventral pattern formation in the zebrafish embryo: The ventral specifying genes. *Development* **123,** 81–93.

Neuhauss, S. C., Solnica-Krezel, L., Schier, A. F., Zwartkruis, F., Stemple, D. L., Malicki, J., Abdelilah, S., Stainier, D. Y., and Driever, W. (1996). Mutations affecting craniofacial development in zebrafish. *Development* **123,** 357–367.

Olsen, B. R., Reginato, A. M., and Wang, W. (2000). Bone development. *Annu. Rev. Cell. Dev. Biol.* **16,** 191–220.

Paylor, R., McIlwain, K. L., McAninch, R., Nellis, A., Yuva-Paylor, L. A., Baldini, A., and Lindsay, E. A. (2001). Mice deleted for the DiGeorge/velocardiofacial syndrome region show abnormal sensorimotor gating and learning and memory impairments. *Hum. Mol. Genet.* **10,** 2645–2650.

Perrino, M. A., and Yelick, P. C. (2004). Immunolocalization of *Alk8* during replacement tooth development in zebrafish. *Cells Tissues Organs.* **176,** 17–27.

Piotrowski, T., Schilling, T. F., Brand, M., Jiang, Y. J., Heisenberg, C. P., Beuchle, D., Grandel, H., van Eeden, F. J., Furutani-Seiki, M., Granato, M., *et al.* (1996). Jaw and branchial arch mutants in zebrafish II: Anterior arches and cartilage differentiation. *Development* **123,** 345–356.

Piotrowski, T., Ahn, D. G., Schilling, T. F., Nair, S., Ruvinsky, I., Geisler, R., Rauch, G. J., Haffter, P., Zon, L. I., Zhou, Y., *et al.* (2003). The zebrafish *van gogh* mutation disrupts *tbx1*, which is involved in the DiGeorge deletion syndrome in humans. *Development* **130,** 5043–5052.

Pothoff, T. (1983). Clearing and staining technique. *In* "Ontogeny and Systematics of Fishes. Special Publication No. 1."American Society of Ichthyologists and Herpetologists, Austin, TX.

Rohlf, F. J., and Marcus, L. F. (1993). A revolution in morphometrics. *Trend. Ecol. Evol.* **8,** 129–132.

Schilling, T. F., Piotrowski, T., Grandel, H., Brand, M., Heisenberg, C. P., Jiang, Y. J., Beuchle, D., Hammerschmidt, M., Kane, D. A., Mullins, M. C., *et al.* (1996). Jaw and branchial arch mutants in zebrafish I: branchial arches. *Development* **123,** 329–344.

Southard-Smith, E. M., Kos, L., and Pavan, W. J. (1998). *Sox10* mutation disrupts neural crest development in Dom Hirschsprung mouse model. *Nat. Genet.* **18,** 60–64.

Thesleff, I., Partanen, A. M., and Vainio, S. (1991). Epithelial-mesenchymal interactions in tooth morphogenesis: The roles of extracellular matrix, growth factors, and cell surface receptors. *J. Craniofac. Genet. Dev. Biol.* **11,** 229–237.

Thesleff, I. (2003). Developmental biology and building a tooth. *Quintessence Int.* **34,** 613–620.

Thompson, D'A. W. (1917). "On Growth and Form." Cambridge University Press, Cambridge.

Thomsen, G. H. (1997). Antagonism within and around the organizer: BMP inhibitors in vertebrate body patterning. *Trends Genet.* **13,** 209–211.

Tucker, A. S., Yamada, G., Grigoriou, M., Pachnis, V., and Sharpe, P. T. (1999). Fgf-8 determines rostral-caudal polarity in the first branchial arch. *Development* **126**(1), 51–61.

Van der Heyden, C., Wautier, K., and Huysseune, A. (2001). Tooth succession in the zebrafish (*Danio rerio*). *Arch. Oral. Biol.* **46,** 1051–1058.

Van der Heyden, C., Huysseune, A., and Sire, J. Y. (2000). Development and fine structure of pharyngeal replacement teeth in juvenile zebrafish (*Danio rerio*) (Teleostei, Cyprinidae). *Cell Tissue Res.* **302,** 205–219.

Van der Heyden, C., and Huysseune, A. (2000). Dynamics of tooth formation and replacement in the zebrafish (*Danio rerio*) (Teleostei, Cyprinidae). *Dev. Dyn.* **219,** 486–496.

Walshe, J., and Mason, I. (2003). Fgf signaling is required for formation of cartilage in the head. *Dev. Biol.* **264,** 356–522.

Wautier, K., Van der Heyden, C., and Huysseune, A. (2001). A quantitative analysis of pharyngeal tooth shape in the zebrafish (*Danio rerio*, Teleostei, Cyprinidae). *Arch. Oral. Biol.* **46,** 67–75.

Witten, P. E., Hansen, A., and Hall, B. K. (2001). Features of mono- and multinucleated bone resorbing cells of the zebrafish *Danio rerio* and their contribution to skeletal development, remodeling, and growth. *J. Morph.* **250,** 197–207.

Yelick, P. C., and Schilling, T. F. (2002). Molecular dissection of craniofacial development using zebrafish. *Crit. Rev. Oral. Biol. Med.* **13,** 308–322.

CHAPTER 20

Cardiac Development

Le A. Trinh and Didier Y. R. Stainier

Department of Biochemistry and Biophysics
Programs in Developmental Biology, Genetics, and Human Genetics
University of California, San Francisco
San Francisco, California 94143-0448

I. Introduction

The goal of this chapter is to provide a reference guide for the development of the embryonic heart in zebrafish. Here, we provide a description of the steps of heart development, discussing morphogenetic processes at each stage, as well as regulatory events and gene expression patterns. Although development of the vascular system is integral to cardiovascular function, a description of vascular development is beyond the scope of this chapter (reviewed in Jin *et al.*, 2002; Lawson and Weinstein, 2002).

As in all vertebrates, the zebrafish heart is the first internal organ to form and function. It comprises two layers, an outer muscular layer (the myocardium) and an inner endothelial layer (the endocardium). These two layers are subdivided into two major chambers: the atrium and the ventricle. The zebrafish heart begins to beat at 22 hpf and circulation is initiated by 24 hpf.

Many of the characteristics of the zebrafish have allowed it to emerge as a powerful vertebrate model organism for the study of cardiac development. The external fertilization, rapid development, and optical clarity of the zebrafish provide distinct advantages for the study of organogenesis. The zebrafish heart can be easily observed throughout the stages of its development as it is prominently positioned at the ventral midline of the embryo. In addition, the ability to combine genetics with embryology and cell biology to investigate lineage relationship, cell behavior, and molecular networks has greatly facilitated our understanding of the regulatory processes underlying heart development.

II. Stages of Heart Tube Morphogenesis

Our understanding of heart tube morphogenesis as it occurs in zebrafish comes from a combination of lineage analyses, gene expression studies, and mutations that affect various steps in this process. From these studies, we have divided the stages of heart tube morphogenesis into six discrete phases: heart field formation, migration, heart tube elongation, heart looping, valve formation, and myocardial remodeling. We elaborate on each phase in the following sections by providing a description of the morphological changes that occur at each step. We then discuss the regulatory events.

A. Formation of the Heart Fields

Lineage analyses indicate that cardiac progenitors reside in the first four tiers of marginal blastomeres in the early blastula (Stainier et al., 1993; Warga and Nüsslein-Volhard, 1999). Within the marginal blastomeres, the cardiac progenitors are located bilaterally around 90 to 180° from the dorsal midline (Fig. 1A). This cardiac region encompasses both endocardial and myocardial progenitors with endocardial progenitors positioned more ventrally (Lee et al., 1994).

The endocardial and myocardial cells share a common progenitor before gastrulation as labeling a single cell within the cardiac region in either the early or mid blastula results in progeny contributing to both lineages (Lee et al., 1994). This is not the case with respect to atrial and ventricular lineages (Stainier et al., 1993). Labeling a single cell in the early blastula results in progeny contributing to both the atrium and ventricle, while a single cell labeled in the mid blastula contributes to either the atrium or the ventricle. These results suggest that the cardiac progenitors have acquired positional information by the mid blastula stage embryo. However, it is not known whether this positional information is in the form of a signal that restricts the cells to a chamber-specific fate or attributed to spatial arrangements that limits cell movements during gastrulation.

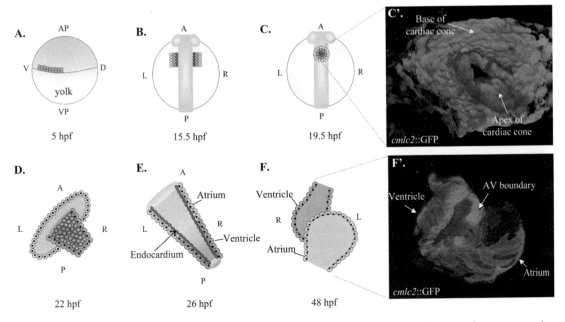

Fig. 1 Zebrafish heart development. (A) In the early blastula, the cardiac progenitors correspond to the ventrolateral margin of the blastoderm. (B) After gastrulation, the cardiac progenitors arrive on either side of the midline. The three rows of circles represent the relative position of the endocardial precursors (*red*), ventricular precursors (*dark green*), and atrial precursors (*light green*) at this stage. By the 13-somite stage (15.5 hpf), the myocardial precursors are patterned mediolaterally, with ventricular precursors positioned medial to the atrial precursors. (C) At the 21-somite stage (19.5 hpf), the myocardial precursors have migrated to the midline and fused to form the cardiac cone. Within the cardiac cone the endocardial precursors (*red*) are located in the central lumen, the ventricular precursors (*dark green*) at the apex, and the atrial precursors (*light green*) at the base. (C') Dorsal view of the cardiac cone as seen by projection of confocal optical sections in a transgenic embryo that expresses GFP in the myocardial precursors (*cmlc2*::GFP). (D) Cardiac cone tilting starts with the apex of the cone bending posteriorly and to the right. Subsequently, the base of the cone, consisting of atrial precursors, will gradually coalesce into a tube. (E) By 26 hpf, elongation of the heart tube results in the leftward positioning of the linear heart tube which is composed of two layers, an inner endocardium (*red*) and an outer myocardium (*green*). (F) By 48 hpf, gradual bending of the heart tube at the atrioventricular boundary forms an S-shaped loop that positions the ventricle (*dark green*) to the right of the atrium (*light green*) as seen in a head-on view. (F') Projection of confocal optical sections of a 56-hpf embryo with GFP expressed in the myocardium. In the projection, a transverse section through the myocardium shows a thick, compact ventricular wall while the atrial wall appears thinner. (See Color Insert.)

During gastrulation, cardiac progenitors are among the first mesodermal cells to involute (Warga and Kimmel, 1990). After involuting, they move toward the animal pole and converge dorsally towards the embryonic axis (Stainier and Fishman, 1992). At the end of gastrulation, the cardiac progenitors arrive on either side of the embryonic midline, where they reside as subpopulations of cells within the anterior lateral plate mesoderm (LPM) (Fig. 1B). At the on-set of somatogenesis, the LPM begins to express a number of myocardial differentiation genes. These differentiations factors include the homeodomain gene, *nkx2.5*, and a number of *gata* genes (4/5/6), which appear to define the cardiac fields within the anterior LPM.

During the early somatogenesis stages (10 to 16 somites), the cardiac fields appear to regulate in response to injury (Serbedzija *et al.*, 1998). Unilateral laser ablation of the cardiac progenitors at the 10-somite stage results in repopulation of the cardiac fields by LPM cells lateral and anterior to the normal cardiac domain. This ability to regulate the heart field persists to the 16-somite stage.

Additionally, the notochord appears to restrict the posterior extent of the cardiac fields in the LPM (Goldstein and Fishman, 1998). Lineage analysis of the LPM indicates that only LPM anterior to the tip of the notochord contributes to the heart. Laser ablation of the notochord results in the posterior expansion of the myocardial expression domain, indicating that the notochord provides an inhibitory signal that limits the extent of the cardiac fields.

B. Migration to the Midline

After the cardiac progenitors arrive on either side of the midline, they undergo a secondary migration toward the midline, where they fuse to form the cardiac cone (Fig. 2A to C). At the 16-somite stage, the myocardial precursors are positioned on either side of the midline, ventral to the anterior endoderm and dorsal to the yolk syncytial layer (YSL) (Stainier *et al.*, 1993). By the 18-somite stage, these bilateral populations have moved toward the midline to make contacts medially (Fig. 2B) (Yelon *et al.*, 1999). The initial contact by the medial myocardial precursors is followed by fusion of the more posterior domain. At the 21-somite stage, the anterior domains fuse to form a cone with a central lumen occupied by endocardial progenitors (Fig. 1C and C'; Stainier *et al.*, 1993; Yelon *et al.*, 1999).

As the myocardial precursors move to the midline, they undergo epithelial maturation (Trinh and Stainier, 2004). At the 16-somite stage, myocardial precursors are cuboidal in shape and show an enrichment of the adherens junction proteins atypical protein kinase (aPKCs) at points of cell-cell contacts (Fig. 2D). As migration proceeds, cell junctional proteins such as aPKCs, zonula occluden-1 (ZO-1), and β-catenin show an increased asymmetric localization in the myocardial precursors (Fig. 2E). The asymmetry in protein localization is accompanied by a change in cell shape, with the medial myocardial precursors becoming columnar. The enrichment of cell junctional proteins in the myocardial precursors throughout the migration stages indicates that these cells are forming tight adhesion to one another and migrate as coherent populations.

The early morphogenesis of endocardial progenitors has not been as extensively examined. Based on expression analyses with the early endothelial progenitor marker, *flk-1,* it is thought that the endocardial progenitors are at the midline by the 18-somite stage (Liao *et al.*, 1997). Similarly, transverse sections of a

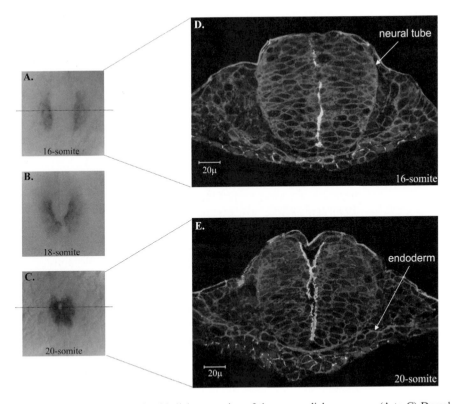

Fig. 2 Midline migration and epithelial maturation of the myocardial precursors. (A to C) Dorsal views of *cmlc2* expression during myocardial migration to the midline. Dash lines indicate the level of the transverse sections in D and E. (A) At the 16-somite stage, the myocardial precursors are positioned on either side of the midline. (B) By the 18-somite stage, the medial myocardial precursors make contacts at the midline. (C) At the 20-somite stage, the myocardial precursors are fusing to form the cardiac cone. (D and E) Transverse confocal images of *cmlc2*::GFP (*false colored blue*) transgenic embryos immunostained with antibodies against β-catenin (*red*) and aPKCs (*green*); dorsal to the top. (D) At the 16-somite stage, the cuboidal myocardial precursors show cortical localization of β-catenin and an enrichment of aPKCs on the apical surface and at points of cell-cell contacts. (E) By the 20-somite stage, the transverse section through the middle of the cardiac cone shows that the myocardial precursors form two U-shaped structures, with the medial cells appearing columnar in shape, while the lateral cells are cuboidal. The aPKCs are restricted to the apicolateral domains, while β-catenin localizes to the basolateral domains of the myocardial precursors. (See Color Insert.)

transgenic line expressing green fluorescent protein (GFP) under the control of the promoter of the *flk-1* gene indicate that the endocardial precursors are at the embryonic midline by the 16-somite stage (Trinh and Stainier, 2004). These results place the endocardial progenitors at the midline before myocardial fusion.

Analyses in *cloche* (*clo*) mutants indicate that endocardial-myocardial interaction is required for proper timing of myocardial migration and heart tube formation. In *clo* mutants, which show a complete absence of endocardial cells (Stainier *et al.*,

1995), myocardial migration is delayed (Trinh and Stainier, 2004). Gene expression analyses demonstrate that the endocardial precursors are a source of Fibronectin that is deposited at the midline on the ventral side of the anterior endoderm (Trinh and Stainier, 2004). In *clo* mutants, Fibronectin deposition is absent at the midline, which may provide the substrate for the temporal regulation of myocardial migration. These data suggest that endocardial progenitors play an important role in the timing of myocardial migration.

Large-scale genetic screens in zebrafish have identified eight mutations—*hands off (han), faust (fau), casanova (cas), bonnie and clyde (bon), one-eye pinhead (oep), natter (nat), miles apart (mil)*, and *two-of-heart (toh)*—that disrupt the medial migration of the myocardial precursors, resulting in the formation of two separate hearts, a phenotype referred to as cardia bifida (Alexander *et al.*, 1998; Chen *et al.*, 1996; Jiang *et al.*, 1996; Stainier *et al.*, 1996). Analyses of these mutants have led to the identification of several requirements for the coordinated movement of myocardial precursors to the midline. First, myocardial differentiation appears to be critical for migration as mutations that disrupt myocardial differentiation (e.g., *han, fau,* and *oep*) all exhibit migration defects (Reiter *et al.*, 1999, 2001; Schier *et al.*, 1997; Yelon *et al.*, 2000). The *han* mutation encodes the basic Helix-Loop-Helix (bHLH) transcription factor Hand2 and appears to regulate both the number of myocardial precursors and myocardial migration (Yelon *et al.*, 2000). *Hand2* is expressed exclusively in the LPM during myocardial migration, suggesting that an aspect of the migration process is autonomous to the myocardial precursors.

Second, the anterior endoderm appears to be essential for myocardial migration as mutants that lack the anterior endoderm (e.g., *cas, bon, fau,* and *oep*) display cardia bifida (Alexander *et al.*, 1999; Kikuchi *et al.*, 2000; Reiter *et al.*, 1999; Schier *et al.*, 1997). Additionally, wild-type endoderm, when transplanted into a subclass of cardia bifida mutants, can rescue myocardial migration (David and Rosa, 2001). Though these studies point to the involvement for the endoderm in myocardial precursor migration, the basis of this requirement remains to be determined.

Third, epithelial organization of the myocardial precursors appears to be critical for the coordinated movement of myocardial precursors (Trinh and Stainier, 2004). The *nat* mutation disrupts adherens junction clustering in the myocardial epithelia and causes cardia bifida. The *nat* mutation encodes Fibronectin, which itself is deposited in the basal substratum around the myocardial precursors throughout the migration stages. Additionally, Fibronectin is deposited at the midline between the endoderm and endocardial precursors. In the complete absence of Fibronectin deposition, myocardial migration is disrupted and adherens junctions between the myocardial precursors do not form properly. These findings suggest that the Fibronectin matrix provides a positional cue for establishing cellular asymmetry in the myocardial precursors and that cell-substratum interaction is required for their epithelial organization and migration.

Finally, signaling mediated by the sphingosine 1-phosphate (S1P) receptor, *miles apart (mil)*, is critical for myocardial migration (Kupperman *et al.*, 2000).

S1P is a bioactive lysophospholipid that regulates a wide range of processes including cell proliferation, differentiation, and survival (Hla, 2003; Panetti *et al.*, 2000). Myocardial differentiation appears unaffected in *mil* mutants. Additionally, cell autonomy studies indicate that *mil* is not required in the migrating myocardial precursors. These results have led to a model in which *mil* functions to provide an environment permissive for myocardial migration to the midline.

C. Heart Tube Elongation

Once the myocardial precursors reach the midline and fuse to form the cardiac cone, the cone will extend to form the linear heart tube. Heart tube elongation begins at the 22-somite stage and proceeds until 26 hpf (Yelon *et al.*, 1999). This process begins with the apex of the cardiac cone tilting posteriorly and toward the right side of the embryo, repositioning the cardiac cone from a dorsal-ventral (D-V) axis to an anterior-posterior (A-P) axis (Fig. 1D). The apex of the cone comprises ventricular cells and establishes the arterial end of the heart tube (Stainier *et al.*, 1993; Yelon *et al.*, 1999). The atrial cells occupy the base of the cone, which coalesces into a tube as the apex tilts (Yelon *et al.*, 1999). By 24 hpf, the ventricular cells are completely repositioned into a leftward slanted tube, while the atrial cells are continuing to telescope into a tube. By 26 hpf, heart tube elongation is completed, resulting in a linear heart tube positioned on the ventral left side of the embryo (Fig. 1E).

The *heart and soul* (*has*) and *heart and mind* (*had*) mutations provide the first molecular insights into our understanding of heart tube elongation. The *has* mutants form cardiac cones that fail to tilt and elongate while *had* mutants exhibit a delay in heart tube extension (Horne-Badovinac *et al.*, 2001; Shu *et al.*, 2003). The *has* mutant encodes aPKCλ, an adherens junction protein that localizes to the apical domain of polarized epithelia (Horne-Badovinac *et al.*, 2001). The *has* mutants exhibit defects in retinal pigmented epithelia and gut tube formation, as well as gut looping, indicating that aPKCλ is critical for the development and morphogenesis of multiple epithelial tissues. Although it is unknown whether aPKCλ is acting autonomously in the myocardial precursors or non-autonomously in neighboring tissues to provide the driving force for heart cone tilting, the role of this gene in controlling the morphogenesis of other epithelia suggests that epithelial polarity of the myocardial precursors may be essential for heart tube elongation. Consistent with this hypothesis is the finding that *had* encodes the α1 isoform of Na,K-ATPase (Shu *et al.*, 2003). In addition to transporting Na+ and K+ across the plasma membrane to establish proper chemical and electrical gradients, Na,K-ATPases are required for septate junction formation in polarized epithelia of *Drosophila* (Paul *et al.*, 2003). Thus, an analysis of epithelial polarity in the myocardial precursors of these two mutants during the heart tube elongation will extend our understanding of this process.

The orderly fusion of the myocardial precursors in forming the cardiac cone has also been implicated as an important regulatory step for heart tube elongation

(Peterson *et al.*, 2001). As previously discussed, the fusion of myocardial precursors at the midline occurs in an orderly fashion with the posterior cells fusing before the anterior cells. In embryos treated with the small molecule concentramide and in *has* mutants, myocardial fusion is defective and heart cone elongation is blocked (Peterson *et al.*, 2001). However, analyses of cardia bifida mutants indicate that myocardial fusion is not required for heart tube elongation (Yelon *et al.*, 1999). In migration-defective mutants, the unfused bifid populations of myocardial precursors undergo tilting and elongation to form tubes as in wild-type embryos, indicating that the fusion of the two myocardial primordia is not a necessary step in heart tube elongation (Yelon *et al.*, 1999). Thus, the fusion defects seen in concentramide-treated and *has* mutants may be coincident with a lack of heart tube elongation rather than causal.

D. Heart Looping

Heart looping can be temporally segregated into two distinct steps: the initial leftward placement of the elongated heart tube, followed by a gradual bending at the atrioventricular boundary to form an S-shape loop that positions the ventricle to the right of the atrium (Fig. 1F; Chen *et al.*, 1997; Chin *et al.*, 2000). The leftward placement of the heart started by heart tube elongation occurs from 22 to 30 hpf and is the first morphological indication of left-right asymmetry in the zebrafish embryo. The subsequent repositioning of the ventricle to the right of the atrium, known as D-looping, occurs between 30 and 48 hpf. Mutant analyses indicate that the initial placement of the heart can be uncoupled from D-looping (Chin *et al.*, 2000). This discordance is seen in mutations that disrupt the formation of the notochord, suggesting that the notochord provides an essential signal either to couple the initial heart position and D-looping or a midline barrier to this coupling signal (Chin *et al.*, 2000; Danos and Yost, 1996).

The cellular processes regulating the morphogenesis of heart looping are poorly understood; however, molecular asymmetries in heart looping have been extensively documented. Before the morphological asymmetry in the heart, a number of genes are expressed asymmetrically in the anterior LPM. At 19 to 20 somites, components of the Nodal signaling pathway, such as *southpaw, cyclop,* and *lefty-1* are expressed in the left LPM (Bisgrove *et al.*, 1999; Long *et al.*, 2003; Rebagliati *et al.*, 1998; Thisse and Thisse, 1999). The homeobox gene *nkx2.5* and the transcription factor *pitx2* are expressed more posteriorly in the left cardiac field than in the right at 20 to 24 somites (Essner *et al.*, 2000; Schilling *et al.*, 1999). The *bmp4* is expressed uniformly in the myocardial precursors at the time of cardiac cone fusion; however, as the cardiac cone begins to tilt and elongate, *bmp4* expression accumulates predominantly on the left side of the heart tube (Chen *et al.*, 1997). Mutant analyses have shown that the patterns of asymmetric gene expression correlate with the direction of heart looping, suggesting that left-right asymmetry signaling pathways regulate this process (Bisgrove *et al.*, 2000). While numerous genes and developmental processes have been implicated in establishing the left-right axis, our understanding of how left-right asymmetry signals are translated into cellular mechanisms regulating heart looping remains limited.

E. Valve Formation

Formation of a functioning valve is an essential step to prevent retrograde blood flow through the heart. Valves form from the endocardial layer of the heart tube at three sites along the developing heart: the outflow tract, atrioventricular (AV) boundary, and sinus venosus. While proper valve formation at these three sites is critical for unidirectional flow, much of our understanding of valve formation in the zebrafish comes from studies focusing on the AV boundary. It is unclear whether the outflow tract and the sinus venosus form valves in a similar fashion as the AV boundary.

The zebrafish heart begins to beat at 22 hpf and circulation is initiated by 24 hpf (Stainier *et al.*, 1993). However, by 36 hpf, only a single layer of endocardial cells line the lumen of the heart (Fig. 3A; Stainier *et al.*, 2002). At these stages, chamber dynamics drive blood flow unidirectionally from the sinus venosus into the atrium, ventricle, and outflow tract. The contraction of the myocardium starts with peristaltic waves that change to coordinated sequential rhythmic beats of the chambers (Warren *et al.*, 2000). By 48 hpf, the endocardial cells at the AV boundary begin to show cellular characteristics that distinguish them from the rest of the endocardium (Stainier *et al.*, 2002; Walsh and Stainier, 2001). The AV endocardial cells, consisting of approximately 5 to 6 cells along the A-P axis of the endocardium, appear cuboidal while non-AV endocardial cells are more

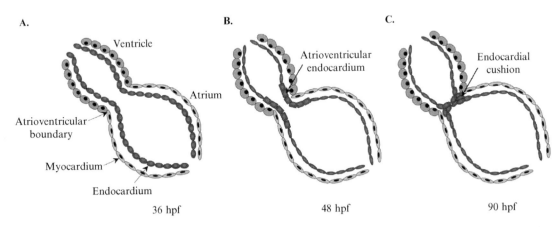

Fig. 3 Atrioventricular (AV) valve formation. (A) At 36 hpf, a single layer of endocardial cells line the lumen of the heart. (B) At 48 hpf, the AV endocardial cells consist of approximately 5 to 6 cuboidal cells along the A-P axis of the endocardium. (C) By 90 hpf, the AV endocardial cells have undergone an epithelial-to-mesenchymal transition, forming endocardial cushions that appear as clusters that are two-cell thick at the AV boundary.

squamous (Fig. 3B). In histological sections, cardiac jelly, a type of extracellular matrix, fills the space between the endocardium and myocardium (Hu *et al.*, 2000). Between 60 and 72 hpf, the AV endocardial cells appear to undergo an epithelial-to-mesenchyme transition (EMT) (Stainier *et al.*, 2002). By 90 hpf, the AV endocardial cells form endocardial cushions that appear as clusters that are two cells thick at the AV boundary (Fig. 3C; Stainier *et al.*, 2002). At 5 dpf, valves at the AV boundary and the outflow tract are visible as two cusps (Hu *et al.*, 2000).

Molecular indications of AV boundary formation occur before morphological changes. In the myocardium, *bmp4* and *versician* are initially expressed throughout the heart and become restricted to the AV myocardial cells at 37 hpf (Walsh and Stainier, 2001). In the endocardium, *notch1b* is initially expressed throughout the entire extent of the heart tube and becomes restricted to the AV endocardial cells at 45 hpf (Walsh and Stainier, 2001). A similar restriction of GFP expression in the endocardium is seen in a GFP-transgenic line that expresses GFP under the control of the *tie2* promoter (*tie2*::GFP; Walsh and Stainier, 2001). The restriction of these molecular markers at the AV boundary suggests that these signaling pathways may be critical regulators of valve formation.

Molecular insights into valve formation in zebrafish first came from identification of the gene disrupted by the *jekyll* (*jek*) mutation, which affects valve formation. The *jek* encodes uridine 5′ diphosphate-glucose dehydrogenase (Ugdh), which is also known as Sugarless in *Drosophila*. The Ugdh is required for the conversion of uridine diphosphate (UDP)-glucose into UDP-glucuronate for the production of heparin sulfate, chondroitin sulfate, and hyaluronic acid (Esko and Selleck, 2002). Proteoglycan biosynthesis has been implicated in a number of signaling pathways including Wnt, BMP, and FGF (Hacker *et al.*, 1997; Jackson *et al.*, 1997; Lin *et al.*, 1999). While, the precise substrate for Jek in cardiac valve formation has yet to be identified, the Wnt signaling pathway has been implicated in valve formation via the identification of a mutation in the *adenomatous polyposis coli* (*apc*) gene of zebrafish (Hurlstone *et al.*, 2003). Apc is a component of the Axin-containing complex involved in phosphorylating β-catenin and targeting it for ubiquitination and degradation by the proteasome. Mutations in *apc* lead to the stabilization of β-catenin and constitutive activation of the Wnt signaling pathway (Fodde *et al.*, 2001). Activated canonical Wnt signaling, as seen by nuclear accumulation of β-catenin, is restricted to the AV endocardial cells in wild-type endocardium. In *apc* mutants, canonical Wnt signaling is activated throughout the endocardium and endocardial cushions form outside of the AV boundary.

In *jek* mutants, the AV endocardial cells fail to form endocardial cushions (Walsh and Stainier, 2001). On the molecular level, expression of AV boundary markers such as *bmp4, versican,* and *notch1b* fail to restrict to the AV boundary, indicating a defect in boundary formation. Interestingly, *clo* mutants, which lack endocardial cells, also fail to restrict AV myocardial genes (Walsh and Stainier, 2001). These observations suggest that myocardial-endocardial interactions are critical for determining the identity of the AV boundary.

While many aspects of zebrafish development are independent of circulation, blood flow appears to be a critical epigenetic factor in valve formation (Hove *et al.*, 2003). Inhibiting blood flow by insertion of a bead in front of the sinus venosus or the back of the ventricle in 37 hpf embryos leads to lack of heart looping and valve formation defects. Additionally, mutations that affect contractility such as *silent heart* and *cardiofunk* exhibit defects in endocardial cushions formation (Bartman *et al.*, 2004). These results underscore the importance of examining the interplay between genetic and epigenetic factors when analyzing cardiovascular defects.

F. Myocardial Remodeling

After assembly into a functioning heart tube, considerable differentiation and morphogenesis continues in the embryo to form the mature heart. In addition to heart looping and valve formation, the myocardium undergoes significant maturation in the form of thickening of the ventricular wall and ventricular trabeculation (Hu *et al.*, 2000). At 36 hpf, morphological differences between the atrium and ventricle are visible by Nomarski optics (Stainier *et al.*, 1996). The ventricle appears thicker than the atrium, which may reflect differences in sarcomere arrangement in the two chambers. Similarly, in transverse sections at 48 hpf, the ventricular myocardial cells appear thick and compact in comparison with the atrium (Fig. 1F' Berdougo *et al.*, 2003; Hu *et al.*, 2000). By 5 dpf, trabeculae, finger-like projections of the myocardium, are visible in the ventricle (Hu *et al.*, 2000). While cardiac function appears to be important for myocardial maturation as mutations in genes encoding the sarcomere scaffold proteins Titin and Tnnt2, and the alpha subunit of the L-type calcium channel result in defective ventricular growth (Rottbauer *et al.*, 2001; Sehnert *et al.*, 2002; Xu *et al.*, 2002), the cellular and molecular mechanisms responsible for these steps in differentiation remain to be elucidated.

III. Gene Expression

Numerous genes have been reported to be expressed in the embryonic heart in zebrafish. This section is not intended to provide a comprehensive list of the genes expressed in the embryonic heart but rather to focus on the expression of a few genes that mark key stages in the patterning and morphogenesis of the embryonic heart. In addition to genes discussed in these sections, we have also delineated genes involved in heart looping and valve formation in the earlier sections on heart morphogenesis.

A. Lateral Plate Mesoderm Gene Expression

Both the myocardial and endocardial precursors arise from the anterior LPM, thus, the patterning and differentiation of the LPM is critical for heart tube formation. Morphologically the anterior LPM starts out as narrow bilateral stripes of tissues on either side of the midline. During mid-somatogenesis (10- to 20-somite stages), these bilateral stripes undergo a spreading process in which they expand along the medial-lateral extent of the embryo. Within the LPM, a number of myocardial differentiation genes are expressed in overlapping patterns.

The bHLH transcription factor *hand2* is expressed throughout the A-P extent of the LPM at the onset of somatogenesis (Yelon *et al.*, 2000). At the 10-somite stage, *hand2* expression segregates between the anterior and posterior domains of the LPM such that a gap forms between the two expression domains. As the anterior LPM spreads, *hand2* expression becomes broader in the anterior expression domain, while the posterior domain remains narrow. The expression pattern of *hand2* is consistent with its role in LPM morphogenesis and myocardial differentiation as mutations in *hand2* result in a lack of LPM spreading and myocardial differentiation defects (Yelon *et al.*, 2000).

Members of the *Gata* family of transcription factor genes (*4/5/6*) are also expressed in the LPM. *Gata* genes are critical regulators of myocardial differentiation and participate in intricate cross-regulatory networks during embryogenesis (Charron *et al.*, 1999; Koutsourakis *et al.*, 1999; Kuo *et al.*, 1997; Molkentin *et al.*, 1997; Reiter *et al.*, 1999). At the 5- to 10-somite stage, *gata4/5/6* expression domains correspond to different A-P extent of the anterior LPM (Fig. 4). The *gata4* expression in the anterior LPM extends to the optic cup and ends posterior to the anterior tip of the notochord (Heicklen-Klein and Evans, 2004; Serbedzija *et al.*, 1998). The *gata5* expression extends posterior to the *gata4* expression domain but does not appear to mark the entire LPM when compared with *hand2* expression (Heicklen-Klein and Evans, 2004; Yelon *et al.*, 2000). Although *gata6* expression extends more posteriorly than *gata4*, its expression domain is shorter than that of the *gata5* expression domain (Heicklen-Klein and Evans, 2004). These overlapping expression patterns in the anterior LPM are consistent with the cross-regulation observed among the Gata family of transcription factors. *Gata6* null mice die at E5.5 and fail to express *gata4* (Koutsourakis *et al.*, 1999). Similarly, *gata5* mutants in zebrafish exhibit a reduction in *gata4* expression in the LPM and overexpression of *gata5* is sufficient to induce ectopic expression of *gata4* and *gata6* (Reiter *et al.*, 1999). Thus, the patterns of overlapping genes may define an overall code for the differentiation of myocardial and endocardial precursors.

The cardiac field, as defined by the expression of the homeodomain gene *nkx2.5*, occupies a sub domain of the anterior LPM (Chen and Fishman, 1996). The *nkx2.5*, the homolog of *Drosophila tinman*, is expressed in the myocardial precursors in all vertebrates (Komuro and Izumo, 1993; Lints *et al.*, 1993; Schultheiss *et al.*, 1995; Tonissen *et al.*, 1994). In *Drosophila*, the *tinman* mutation results in a complete lack of heart formation (Bodmer *et al.*, 1990). In zebrafish, *nkx2.5* expression in the anterior LPM is initially detected at the 5-somite stage (Alexander and Stainier, 1999). At the 10-somite stage, *nkx2.5* expression in the LPM extends posteriorly to the otic vesicle and anteriorly to halfway between the notochord and the eye (Fig. 4; Serbedzija *et al.*, 1998). The anterior tip of the notochord marks the median of the *nkx2.5* expression domain. The A-P boundary

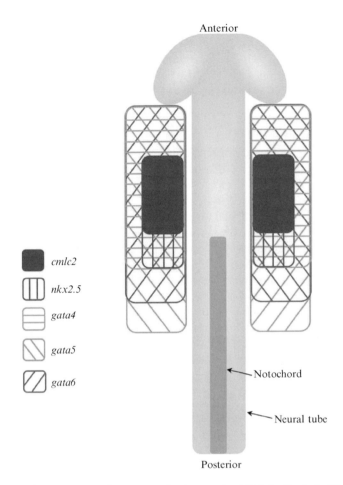

Fig. 4 Overlapping gene expression patterns in the anterior LPM. In the 5- to 10-somite stage embryo, *gata4/5/6* expression domains correspond to different A-P extent of the anterior LPM. The *gata4* expression domain (*green horizontal stripes*) in the anterior LPM extends to the optic cup and ends posterior to the tip of the notochord. The *gata5* expression domain (*orange diagonal stripes*) extends furthest posteriorly, while the *gata6* expression domain (*blue diagonal stripes*) extends more posteriorly than *gata4*, but anterior to that of the *gata5* expression domain. The *nkx2.5* expression domain (*purple vertical stripes*) extends posterior to the tip of the notochord and anteriorly to halfway between the notochord and the eye. *cmlc2* expression (*solid red*) initiates in the 13-somite stage embryo and is limited to the anterior tip of the notochord. (See Color Insert.)

of the *nkx2.5* expression domain is maintained throughout the migration stages. Thereafter, *nkx2.5* is expressed throughout the embryonic myocardium.

Expression of *nkx2.5* anterior to the notochord appears to define the cardiac field. Lineage analyses of the LPM indicate that only cells anterior to the notochord in the *nkx2.5* expression domain contributes progeny to the myocardium

(Goldstein and Fishman, 1998). Additionally, when the cardiac progenitors are ablated, only LPM cells lateral and anterior to the normal cardiac progenitors can compensate for the ablated cells (Serbedzija *et al.*, 1998). Therefore, the ability to contribute to the heart is limited to the region of the LPM anterior to the notochord and not all *nkx2.5*-expressing cells contribute to the myocardium.

B. Myocardial Gene Expression

Myocardial-specific gene expression initiates at the 13-somite stage with the expression of distinct myosin genes in the bilateral populations of myocardial precursors (Yelon *et al.*, 1999). The *cardiac myosin light chain 2* (*cmlc2*) expression overlaps with that of *nkx2.5* but is more restricted. The posterior boundary of *cmlc2* expression aligns with the anterior boundary of the notochord, while *nkx2.5* expression extends beyond the tip of the notochord (Fig. 4; Yelon *et al.*, 1999). As development proceeds, the bilateral stripes of *cmlc2*-expressing cells delineate the migration pattern of the myocardial precursors (see Figure 2 and section on migration to the midline). At 18 somite, *cmlc2* expression resembles a butterfly pattern, with the medial *cmlc2*-positive cells making contact at the midline. By 21 somite, when myocardial fusion is completed, *cmlc2* expression marks the cardiac cone. Subsequently, *cmlc2* expression is maintained in the myocardium throughout development.

Myocardial ventricular-specific gene expression also initiates at the 13-somite stage. At this stage, *ventricular myosin heavy chain* (*vmhc*) expression is restricted to the medially located cells of the bilateral myocardial populations (Yelon *et al.*, 1999). These medial cells compose the leading edge of the migrating myocardial precursors, suggesting that ventricular precursors may play an essential role in the migration process. Additionally, the early expression of *vmhc* indicates that the myocardial precursors exhibit molecular chamber identity before morphological differences.

Known myocardial atrial-specific gene expression begins at the 19-somite stage with the expression of the *atrial myosin heavy chain* (*amhc*) (Berdougo *et al.*, 2003). At this stage, *amhc*-positive cells mark the outer portion of the forming cardiac cone as defined by *cmlc2* expression. The inner portion of the cardiac cone expresses *vmhc*. Thus, the expression patterns of *amhc* and *vmhc* are complementary to each other. As heart tube morphogenesis proceeds, the two complementary expression patterns are maintained. The segregation of *amhc* and *vmhc* expression at these early stages, indicate that the myocardial precursors are patterned before heart tube formation.

C. Endocardial Gene Expression

Many of the genes expressed in the endocardial cells are expressed throughout the vascular endothelium. The endothelial-specific receptor tyrosine kinase gene *flk-1* is expressed in two bilateral stripes of cells within the LPM at the 5-somite stage (Liao *et al.*, 1997). These bilateral stripes of *flk-1*-expressing cells appear more medial than the *nkx2.5* expression domain (Alexander and Stainier, 1999). As development proceeds, the bilateral stripes of *flk-1* expression extend in both the anterior and posterior directions. By the 15-somite stage, at the level of the cardiac field, *flk-1*-expressing cells can be detected medial to the bilateral populations of myocardial precursors. At the 18-somite stage, the medial expression of *flk-1* is seen as a dense cluster and mostly marks the endocardial precursors (Liao *et al.*, 1997). As the heart tube elongates and loops, *flk-1* expression is detected in the inner layer of the tube, as well as in the developing vasculature. Two other receptor tyrosine kinase genes, *tie-2* and *flt-4,* are expressed in a similar pattern as *flk-1* starting at the 18-somite stage (Liao *et al.*, 1997; Thompson *et al.*, 1998). The commonality in gene expression between the endocardial and endothelial cells suggests that two-cell types share a common progenitor.

One gene expressed in the endocardial precursors, but not in the endothelial cells, is the extracellular matrix gene *fibronectin.* The *fibronectin* appears to be one of the earliest markers of the endocardial precursors as it is expressed in the endocardial precursors before the midline migration of the myocardial precursors (Trinh and Stainier, 2004). At the start of somatogenesis, *fibronectin* is initially expressed in the anterior LPM. However, starting between the 11- and 12-somite stages, *fibronectin* expression can be observed in a few cells medial to the LPM. The medial expression spans the midline by the 15-somite stage and is maintained throughout the stages of myocardial migration to the midline. In transverse sections, the *fibronectin*-expressing cells occupy two ventral layers of cells at the midline. This expression pattern is absent in *clo* mutants, which lack endocardial cells, indicating the midline *fibronectin* expression corresponds to endocardial expression. The medial progression of the *fibronectin* expression between the 10- to 15-somite stages may reflect the migration pathway of the endocardial precursors to the midline.

IV. Conclusion and Future Directions

In this chapter, we have focused on the events and processes known to govern heart morphogenesis in zebrafish. While we are gaining a better understanding of some of the processes that regulate zebrafish cardiac development, from the initial specification of the cardiac precursors to the subsequent morphogenesis in forming a functional organ, much remains to be done to gain a full understanding of the various genes and morphogenetic processes that control heart development.

The recent development of transgenic zebrafish will provide powerful tools for the analyses of cardiac morphogenesis on a detailed cellular level. In particular, transgenic lines that use cardiac-specific promoters to drive expression of fluorescent proteins allow for the visualization of cell behavior throughout development in both wild-type and mutant situations. Several cardiac transgenic lines have been created that will further facilitate our understanding of the complex cell behavior and tissue-tissue interactions during different stages in cardiac morphogenesis. In combination with live imaging, these various transgenic lines should provide insights into the cellular processes such as migration, shape changes, and proliferation that are involved in heart tube assembly and remodeling.

Additionally, the identification of other regulators of cardiac development— through the cloning of existing mutations and isolation of new loci—will be instrumental to further our understanding of the genetic networks involved in cardiac development. These studies will provide the framework to clarify the relationship between genetic networks that regulate cardiac formation and the morphogenetic processes that drive organogenesis.

References

Alexander, J., Rothenberg, M., Henry, G. L., and Stainier, D. Y. (1999). *casanova* plays an early and essential role in endoderm formation in zebrafish. *Dev. Biol.* **215**, 343–357.

Alexander, J., Stainier, D. Y., and Yelon, D. (1998). Screening mosaic F1 females for mutations affecting zebrafish heart induction and patterning. *Dev. Genet.* **22**, 288–299.

Bartman, T., Walsh, E. C., Wen, K. K., McKane, M., Ren, J., Alexander, J., Rubenstein, P.A., and Stainier, D.Y.R. (2004). Early myocardial function affects endocardial cushion development in zebrafish. *PLOS* **2**, 673–681.

Berdougo, E., Coleman, H., Lee, D. H., Stainier, D. Y. R., and Yelon, D. (2003). Mutation of *weak atrium/atrial myosin heavy chain* disrupts atrial function and influences ventricular morphogenesis in zebrafish. *Development* **130**, 6121–6129.

Bisgrove, B. W., Essner, J. J., and Yost, H. J. (1999). Regulation of midline development by antagonism of *lefty* and *nodal* signaling. *Development* **126**, 3253–3262.

Bisgrove, B. W., Essner, J. J., and Yost, H. J. (2000). Multiple pathways in the midline regulate concordant brain, heart and gut left-right asymmetry. *Development* **127**, 3567–3579.

Bodmer, R., Jan, L. Y., and Jan, Y. N. (1990). A new homeobox-containing gene, *msh-2* (*tinman*), is transiently expressed during mesoderm formation in Drosophila. *Development* **110**, 661–669.

Charron, F., Paradis, P., Bronchain, O., Nemer, G., and Nemer, M. (1999). Cooperative interaction between GATA4 and GATA6 regulates myocardial gene expression. *Mol. Cell. Biol.* **19**, 4355–4356.

Chen, J. N., and Fishman, M. C. (1996). Zebrafish *tinman* homolog demarcates the heart field and initiates myocardial differentiation. *Development* **122**, 3809–3816.

Chen, J. N., Haffter, P., Odenthal, J., Vogelsang, E., Brand, M., van Eedan, F. J. M., Furutani-Seike, M., Granato, M., Hammerschmidt, M., Heisenberg, C. P., *et al.* (1996). Mutations affecting the cardiovascular system and other internal organs in zebrafish. *Development* **123**, 293–302.

Chen, J. N., van Eeden, F. J. M., Warren, K. S., Chin, A., Nüsslein-Volhard, C., Haffter, P., and Fishman, M. C. (1997). Left-right pattern of cardiac *bmp4* may drive asymmetry of the heart in zebrafish. *Development* **124**, 4373–4382.

Chin, A. J., Tsang, M., and Weinberg, E. S. (2000). Heart and gut chiralities are controlled independently from initial heart position in the developing zebrafish. *Dev. Biol.* **227**, 403–421.

Danos, M. C., and Yost, H. J. (1996). Role of notochord in specification of cardiac left-right orientation in zebrafish and Xenopus. *Dev. Biol.* **177**, 96–103.

David, N. B., and Rosa, F. M. (2001). Cell autonomous commitment to an endodermal fate and behaviour by activation of Nodal signalling. *Development* **128**, 3937–3947.

Esko, J. D., and Selleck, S. B. (2002). Order out of chaos: Assembly of ligand binding sites in heparan sulfate. *Annu. Rev. Biochem.* **71**, 435–471.

Essner, J. J., Branford, W. W., Zhang, J., and Yost, H. J. (2000). Mesentoderm and left-right brain, heart and gut development are differentially regulated by *pitx2* isoforms. *Development* **127,** 1081–1093.

Fodde, R., Smits, R., and Clevers, H. (2001). APC, signal transduction and genetic instability in colorectal cancer. *Nat. Rev. Cancer* **1,** 55–67.

Goldstein, A. M., and Fishman, M. C. (1998). Notochord regulates cardiac lineage in zebrafish embryos. *Dev. Biol.* **201,** 247–252.

Hacker, U., Lin, X., and Perrimon, N. (1997). The *Drosophila sugarless* gene modulates Wingless signaling and encodes an enzyme involved in polysaccharide biosynthesis. *Development* **124,** 3565–3573.

Heicklen-Klein, A., and Evans, T. (2004). T-box binding sites are required for activity of a cardiac GATA4 enhancer. *Dev. Biol.* **267,** 490–504.

Hla, T. (2003). Signaling and biological actions of sphingosine 1-phosphate. *Pharmacol. Res.* **47,** 401–407.

Horne-Badovinac, S., Lin, D., Waldron, S., Schwarz, M., Mbamalu, G., Pawson, T., Jan, J., Stainier, D. Y. R., and Abdelilah-Seyfried, S. (2001). Positional cloning of *heart and soul* reveals multiple roles for PKC lambda in zebrafish organogenesis. *Curr. Biol.* **11,** 1492–1502.

Hove, J. R., Koster, R. W., Forouhar, A. S., Acevedo-Bolton, G., Fraser, S. E., and Gharib, M. (2003). Intracardiac fluid forces are an essential epigenetic factor for embryonic cardiogenesis. *Nature* **421,** 172–177.

Hu, N., Sedmera, D., Yost, H. J., and Clark, E. B. (2000). Structure and function of the developing zebrafish heart. *Anat. Rec.* **260,** 148–157.

Hurlstone, A. F. L., Haramis, A. G., Wienholds, E., Begthel, H., Korving, J., van Eeden, F., Cuppen, E., Zivkovic, D., Plasterk, R. H. A., and Clevers, H. (2003). The Wnt/β-catenin pathway regulates cardiac valve formation. *Nature* **425,** 633–637.

Jackson, S. M., Nakato, H., Sugiura, M., Jannuzi, A., Oakes, R., Kaluza, V., Golden, C., and Selleck, S. B. (1997). Dally, a *Drosophila* glypican, controls cellular responses to the TGFb-related morphogen, Dpp. *Development* **124,** 4113–4120.

Jiang, Y. J., Brand, M., Heisenberg, C. P., Beuchle, D., Furutani-Seike, M., Kelsh, R. N., Warga, R. M., Granato, M., Haffter, P., Hammerschmidt, M., *et al.* (1996). Mutations affecting neurogenesis and brain morphology in the zebrafish, *Danio rerio. Development* **123,** 205–216.

Jin, S. W., Jungblut, B., and Stainier, D. Y. R. (2002). Angiogenesis in zebrafish. *In* "Genetics of Angiogenesis" (J. Honig, ed.), pp. 101–118. BIOS, London.

Kikuchi, Y., Trinh, L. A., Reiter, J. F., Alexander, J., Yelon, D., and Stainier, D. Y. R. (2000). The zebrafish *bonnie and clyde* gene encodes a Mix family homeodomain protein that regulates the generation of endodermal precursors. *Genes Dev.* **14,** 1279–1289.

Komuro, I., and Izumo, S. (1993). Csx: A murine homeobox-containing gene specifically expressed in the developing heart. *Proc. Natl. Acad. Sci. USA* **90,** 8145–8149.

Koutsourakis, M., Langeveld, A., Patient, R., Beddington, R., and Grosveld, F. (1999). The transcription factor GATA6 is essential for early extraembryonic development. *Development* **126,** 723–731.

Kuo, C. T., Morrisey, E. E., Anandappa, R., Sigrist, K., Lu, M. M., Parnacek, M. S., Soudais, C., and Leiden, J. M. (1997). GATA4 transcription factor is required for ventral morphogenesis and heart tube formation. *Genes Dev.* **11,** 1048–1060.

Kupperman, E., An, S., Osborne, N., Waldron, S., and Stainier, D. Y. (2000). A sphingosine-1-phosphate receptor regulates cell migration during vertebrate heart development. *Nature* **406,** 192–195.

Lawson, N. D., and Weinstein, B. M. (2002). Arteries and veins: Making a difference with zebrafish. *Nat. Rev. Genet.* **3,** 674–682.

Lee, R. K. K., Stainier, D. Y. R., Weinstein, B. M., and Fishman, M. C. (1994). Cardiovascular development in the zebrafish. II. Endocardial progenitors are sequestered within the heart field. *Development* **120,** 3361–3366.

Liao, W. B., Bisgrove, W., Sawyer, H., Hug, B., Bell, B., Peters, K., Grunwald, D. J., and Stainier, D. Y. R. (1997). The zebrafish gene *cloche* acts upstream of a *flk-1* homologue to regulate endothelial cell differentiation. *Development* **124**, 381–389.

Lin, X., Buff, E. M., Perrimon, N., and Michelson, A. M. (1999). Heparan sulfate proteoglycans are essential for Fgf receptor signaling during *Drosophila* embryonic development. *Development* **126**, 3715–3723.

Lints, T. J., Parsons, L. M., Hartley, L., Lyons, I., and Harvey, R. P. (1993). *Nkx2.5*: A novel murine homeobox gene expressed in early heart progenitor cells and their myogenic descendants. *Development* **119**, 419–431.

Long, S., Ahmad, N., and Rebagliati, M. (2003). The zebrafish nodal-related gene *southpaw* is required for visceral and diencephalic left-right asymmetry. *Development* **130**, 2303–2316.

Molkentin, J. D., Lin, Q., Duncan, S. A., and Olson, E. N. (1997). Requirement of the transcription factor GATA4 for heart tube formation and ventral morphogenesis. *Genes Dev.* **11**, 1061–1072.

Panetti, T. S., Nowlen, J., and Mosher, D. F. (2000). Sphingosine-1-phosphate and lysophosphatidic acid stimulate endothelial cell migration. *Arterioscler. Thromb. Vasc. Biol.* **20**, 1013–1019.

Paul, S. M., Ternet, M., Salvaterra, P. M., and Beitel, G. J. (2003). The Na+/K+ ATPase is required for septate junction function and epithelial tube-size control in the *Drosophila* tracheal system. *Development* **130**, 4963–4974.

Peterson, R. T., Mably, J. D., Chen, J. N., and Fishman, M. C. (2001). Convergence of distinct pathways to heart patterning revealed by the small molecule concentramide and the mutation heart-and-soul. *Curr. Biol.* **11**, 1481–1491.

Rebagliati, M. R., Toyama, R., Fricke, C., Haffter, P., and Dawid, I. B. (1998). Zebrafish nodal-related genes are implicated in axial patterning and establishing left-right asymmetry. *Dev. Biol.* **199**, 261–272.

Reiter, J. F., Alexander, J., Rodaway, A., Yelon, D., Patient, R., Holder, N., and Stainier, D. Y. R. (1999). *Gata5* is required for the development of the heart and endoderm in zebrafish. *Genes Dev.* **13**, 2983–2995.

Reiter, J. F., Verkade, H., and Stainier, D. Y. R. (2001). *Bmp2b* and *Oep* promote early myocardial differentiation through their regulation of *gata5*. *Dev. Biol.* **234**, 330–338.

Rottbauer, W., Baker, K., Wo, Z. G., Mohideen, M. A., Cantiello, H. F., and Fishman, M. C. (2001). Growth and function of the embryonic heart depend upon the cardiac-specific L-type calcium channel alpha1 subunit. *Dev. Cell* **2**, 265–275.

Schier, A. F., Neuhauss, S. C., Helde, K. A., Talbot, W. S., and Driever, W. (1997). The *one-eyed pinhead* gene functions in mesoderm and endoderm formation in zebrafish and interacts with no tail. *Development* **124**, 327–342.

Schilling, T. F., Concordet, J. P., and Ingham, P. W. (1999). Regulation of left-right asymmetries in the zebrafish by *shh* and *bmp4*. *Dev. Biol.* **210**, 277–287.

Schultheiss, T. M., Xydas, S., and Lassar, A. B. (1995). Induction of avian cardiac myogenesis by anterior endoderm. *Development* **121**, 4203–4214.

Sehnert, A. J., Huq, A., Weinstein, B. M., Walker, C., Fishman, M., and Stainier, D. Y. (2002). Cardiac troponin T is essential in sarcomere assembly and cardiac contractility. *Nat. Genet.* **31**, 106–110.

Serbedzija, G. N., Chen, J. N., and Fishman, M. C. (1998). Regulation of the heart field in zebrafish. *Development* **125**, 1095–1101.

Shu, S., Cheng, K., Patel, N., Chen, F., Joseph, E., Tsai, H. J., and Chen, J. N. (2003). Na,K-ATPase is essential for embryonic heart development in the zebrafish. *Development* **130**, 6165–6173.

Stainier, D. Y. R., Beis, D., Jungblut, B., and Bartman, T. (2002). Endocardial cushion formation in zebrafish. Cold Spring Harb Symp. *Quant. Biol.* **67**, 49–56.

Stainier, D. Y. R., and Fishman, M. C. (1992). Patterning the zebrafish heart tube: Acquisition of anteroposterior polarity. *Dev. Biol.* **153**, 91–101.

Stainier, D. Y., Fouquet, B., Chen, J., Warren, K. S., Weinstein, B. M., Meiler, S. E., Mohideen, M. P. K., Neuhauss, S. C. F., Solnica-Krezel, L., Schier, A. F., *et al.* (1996). Mutations affecting the formation and function of the cardiovascular system in the zebrafish embryo. *Development* **123**, 285–292.

Stainier, D. Y., Lee, R. K., and Fishman, M. (1993). Cardiovascular development in the zebrafish. I. Myocardial fate map and heart tube formation. *Development* **119**, 31–40.

Stainier, D. Y. R., Weinstein, B. M., Detrich, H. W., Zon, L. I., and Fishman, M. C. (1995). *cloche*, an early acting zebrafish gene, is required by both the endothelial and hematopoietic lineages. *Development* **121**, 3141–3150.

Thisse, C., and Thisse, B. (1999). *Antivin*, a novel and divergent member of the TGFbeta superfamily, negatively regulates mesoderm induction. *Development* **126**, 229–240.

Thompson, M. A., Ranson, D. G., Pratt, S. J., MacLennan, H., Kieran, M. W., Detrich, H. W., Vail, B., Huber, T. L., Paw, B., Brownlie, A. J., *et al.* (1998). The *cloche* and *spadetail* genes differentially affect hematopoiesis and vasculogenesis. *Dev. Biol.* **197**, 248–269.

Tonissen, K. F., Drysdale, T. A., Lints, T. J., Harvey, R. P., and Krieg, P. A. (1994). *XNkx2.5*, a *Xenopus* gene related to *Nkx2.5* and *tinman*: Evidence for a conversed role in cardiac development. *Dev. Biol.* **162**, 325–328.

Trinh, L. A., and Stainier, D. Y. R. (2004). Fibronectin regulates epithelial organization during myocardial migration in zebrafish. *Dev. Cell* **3**, 371–382.

Walsh, E. C., and Stainier, D. Y. R. (2001). UDP-Glucose Dehydrogenase required for cardiac valve formation in zebrafish. *Science* **293**, 1670–1673.

Warga, R. M., and Kimmel, C. (1990). Cell movements during epiboly and gastrulation in zebrafish. *Development* **108**, 569–580.

Warga, R. M., and Nüsslein-Volhard, C. (1999). Origin and development of the zebrafish endoderm. *Development* **126**, 827–838.

Warren, K. S., Wu, J. C., Pinet, F., and Fishman, M. C. (2000). The genetic basis of cardiac function: Dissection by zebrafish (*Danio rerio*) screens. *Philos. Trans. R Soc. Lond. B Biol. Sci.* **355**, 939–944.

Xu, X., Meiler, S. E., Zhong, T. P., Mohideen, M., Crossley, D. A., Burggren, W. W., and Fishman, M. C. (2002). Cardiomyopathy in zebrafish due to mutation in an alternatively spliced exon of *titin*. *Nat. Gen.* **30**, 205–209.

Yelon, D., Horne, S. A., and Stainier, D. Y. R. (1999). Restricted expression of cardiac myosin genes reveals regulated aspects of heart tube assembly in zebrafish. *Dev. Biol.* **214**, 23–37.

Yelon, D., Ticho, B., Halpern, M. E., Ruvinsky, I., Ho, R. K., Silver, L. M., and Stainier, D. Y. R. (2000). The bHLH transcription factor *hand2* plays parallel roles in zebrafish heart and pectoral fin development. *Development* **127**, 2573–2582.

CHAPTER 21

Chemical Approaches to Angiogenesis

Joanne Chan* and Fabrizio C. Serluca[†]

*Vascular Biology Program
Children's Hospital of Boston
Boston, Massachusetts 02115

[†]Novartis Institutes for Biomedical Research
Cambridge, Massachusetts 02139

I. Introduction

Targeting blood vessels in anti-angiogenic therapy is a rational and compelling strategy in the treatment of cancer, as well as other angiogenesis-dependent human diseases. The idea that tumor growth is angiogenesis dependent was first proposed by Judah Folkman in the early 1970s (Folkman, 1971). The clinical approval of an anti-angiogenic therapeutic has significantly boosted enthusiasm and support for this approach, as many anti-angiogenic agents are currently in the pipeline. Basic research supporting these ideas came from many studies using cell-based systems, as well as animal models. These studies have focused on the endothelial cell, a specialized cell that lines blood vessels. The identification of key regulators of endothelial cell function, such as the vascular endothelial growth factor (VEGF) and its receptors (VEGFRs) have made them major targets for anti-angiogenic therapy. Identification of additional angiogenic players will enhance our understanding of the molecular mechanisms required to make blood vessels, providing valuable new targets for improved therapies. In this chapter, we discuss methods used in the zebrafish to examine blood vessels. These simple techniques, combined with the ability to perform genetic and chemical screens, make the zebrafish an ideal vertebrate model organism for vascular biology studies.

A. Advantages of the Zebrafish for Angiogenesis

The zebrafish is easily amenable to N-ethyl-N-nitrosourea (ENU) mutagenesis and has thus become a favored vertebrate model system for genetic and developmental studies for a wide range of biological processes. Primordia of the major organ systems are established by the end of the first day of development, circulation begins between 24 hpf and 26 hpf and vascular complexity increases over several days (Isogai et al., 2001). The external development and transparency of the zebrafish embryo allows easy visual analysis of blood flow through a simple dissecting microscope. The zebrafish embryo can survive for the first few days of development without a functional circulation, as its small size facilitates oxygen diffusion (Ransom et al., 1996; Weinstein et al., 1996). This is particularly useful for the examination of genes with essential functions in the development of the vasculature; whereas analogous vascular defects could lead to embryonic lethality in mammalian models. Experimental over-expression or targeted knockdown of gene function can be easily performed in the zebrafish embryo by mRNA, DNA, or antisense morpholino microinjections. External development of the embryos in a simple salt solution also allows for easy addition of chemicals or chemical libraries to embryos (Peterson et al., 2000). Taking advantage of these features of the zebrafish embryo, we and others have developed (or modified) methods for the evaluation of angiogenesis.

B. Vascular Endothelial Growth Factor and Its Receptors

The VEGF and its receptors are key regulators of blood vessel formation in physiological and pathological conditions (reviewed in Yancopoulos et al., 2000). Their specialized regulation of endothelial cell growth, migration, and survival makes VEGF and its receptors the subjects of intense research. Since many diseases and conditions including cancer, diabetic retinopathy, rheumatoid arthritis, and ischemic heart disease are angiogenesis dependent, elucidation of the VEGF receptors' signaling pathways has tremendous therapeutic potential. Three VEGF receptors are found in mammals, with alternative names derived from human or mouse studies: VEGFR1/Flt1, VEGFR2/KDR/Flk-1,

VEGFR3/flt4 (Carmeliet, 2000; Yancopoulos *et al.*, 2000). Cumulative data from cell culture and *in vivo* work in murine models have come to several generalizations:

1. The VEGFR2/KDR/Flk-1 receptor is a major mediator of VEGF signals for most forms of angiogenesis.
2. The VEGFR1/Flt1 acts as a decoy receptor.
3. The VEGFR3/flt4 functions in lymphogenesis and early venous formation.

Although many effectors of the VEGF signaling pathway have been identified, most of our knowledge comes from cell culture studies (reviewed in Matsumoto and Claesson-Welsh, 2001), which cannot fully recapitulate the cell-to-cell interactions that occur in the living animal.

Zebrafish embryonic circulation and developmental progression of blood vessels are described in the detailed work of Isogai *et al.* (2001), an excellent reference for developing blood vessels over the first week of embryonic life. At 24 hpf, the major vessels, the aorta (dorsal artery) and the posterior cardinal vein, are formed. These major vessels form by the process of vasculogenesis (Lawson and Weinstein, 2002b). By 48 hpf, the intersegmental vessels (ISVs) are developed along the trunk of the embryo. One ISV is formed per somite segment along the length of the embryo. These vessels form by sprouting angiogenesis (Childs *et al.*, 2002; Isogai *et al.*, 2003; Lawson and Weinstein, 2002b). The VEGF ligand-receptor signaling pathway plays critical roles in normal physiological and pathological angiogenesis in vertebrate organisms, including the zebrafish and humans. Current efforts are aimed at inhibiting the pathway for the development of anti-angiogenic therapies. Small molecule chemical compounds that can inhibit the VEGF receptor can be tested on the zebrafish embryo (Chan *et al.*, 2002; Serbedzija *et al.*, 1999). Thus, vascular development in the zebrafish can be used to study genes with critical functions in angiogenesis, as well as a read-out for the evaluation of anti-angiogenic effectiveness of chemical compounds.

C. Chemical Genetics and Current New Therapies

The VEGFR, form a small subclass within a large superfamily of receptor tyrosine kinases (RTKs), all sharing structural and functional homology (reviewed in Schlessinger, 2000). As tyrosine kinases, the VEGFRs share a common catalytic kinase domain with many proteins in cancer cell signaling pathways. One such kinase is Abl, a cytoplasmic kinase whose hyperactivity is responsible for the clinical progression of chronic myelogenous leukemia. STI571 is a small molecule kinase inhibitor that selectively blocks the function of the Abl cytoplasmic tyrosine kinase (reviewed in Druker, 2002). In clinical trials, toxicity from therapy was relatively mild (Kantarjian and Talpaz, 2001). The current clinical success of STI571 has energized the field for the development of other inhibitors targeting kinases. The VEGFRs, as tyrosine kinases, have become high-priority targets for this approach. Several inhibitors with high selectivity for VEGFRs are now in clinical trials for the treatment of various human cancers (Manley *et al.*, 2002). We have been able to show that the zebrafish VEGFR, VEGFR2/flk-1, can be inhibited by PTK787/ZK222584 (Chan *et al.*, 2002).

We have taken advantage of the availability of this chemical inhibitor of the VEGFR tyrosine kinases (PTK787/ZK222584) to study endothelial cell function in the zebrafish. PTK787/ZK222584 is a potent inhibitor of the VEGF receptor tyrosine kinases (Wood *et al.*, 2000). Previous studies have shown the effectiveness of this drug in cell culture, murine retina neovascularization, and carcinoma models (Bold *et al.*, 2000; Drevs *et al.*, 2000; Ozaki *et al.*, 2000; Wood *et al.*, 2000). When added to live zebrafish embryos, PTK787/ZK222584 also robustly prevented the formation of all major blood vessels in the embryo (Chan *et al.*, 2002). The compound exhibited high selectivity in the zebrafish for endothelial cells since blood vessel formation was completely inhibited while other tissues appear intact. This potent inhibition is a result of the structural similarity between the zebrafish and human VEGFRs (78% identity in the kinase domain in VEGFR2; Chan *et al.*, 2002). Using this compound, we have been able to dissect the angiogenic signaling pathway downstream of the VEGFR. By upregulating the activity of Akt, a downstream kinase, we have been able to rescue the upstream VEGFR blockade induced by PTK787/ZK222584. This chemical approach provides a rheostatic control of receptor function, permitting the examination of situations with partial or complete blood vessel formation. Since the inhibitor can be applied at any time, it also facilitates analysis at any time point during vascular development in the embryo.

Our protocol involved a simple addition of the PTK787/ZK222584 compound to zebrafish embryo medium (E3 embryo medium, Nusselin-Volhard and Dahm, 2002). Blood vessel formation can be completely inhibited in a dose-dependent manner. We typically choose 2 time points for evaluation of a novel test compound for its anti-angiogenic effects: treatment at 60% epiboly or shield stage, for analysis of aortic blood flow by 24 hpf to 26 hpf; also treatment at 24 hpf, after the establishment of aortic blood flow, for analysis at 48 hpf for the number of functional intersegmental vessels. While it is possible to add test compounds even earlier (e.g., the 1 cell stage) it would be difficult to determine whether unfertilized or unhealthy eggs resulted from the drug treatment. The early gastrulation period provides a convenient time point when no blood vessels are formed yet and a general assessment of embryo health can be obtained, allowing us to quickly sort through embryos before drug treatment. Control embryos are extremely important for analysis of clutch health and accurate staging of blood flow in key vessels so that anti-angiogenic phenotypes in test embryos can be scored with confidence. For the early treatment (addition during gastrulation) and analysis, the number of embryos with aortic blood flow is counted at 26 hpf, at which time the control embryos all have blood flow. Similarly, for the later time point (addition at 24 hpf), the number of functional intersomitic vessels is counted when the control embryos have the complete set of functional vessels at 48 hpf.

For PTK787/ZK222584, embryos treated at 5 μM did not develop any aortic blood flow since the major blood vessels did not form properly (Chan *et al.*, 2002). The PTK787/ZK222584-induced anti-angiogenic phenotype is remarkably similar to the loss of VEGF-A ligand function produced by antisense morpholino targeting (Nasevicius *et al.*, 2000). Embryos treated with PTK787/ZK222584 showed a complete blockage of blood vessel formation as the dorsal artery and the posterior cardinal vein fail to form by 48 hours of development (Chan *et al.*, 2002). This results in severe pericardial edema, with embryonic death by 3 days of development. The lack of deleterious effects on the general morphology of embryonic structures by 24 hpf demonstrates the selectivity of PTK787/ZK222584 for VEGFRs.

For chemical compound treatments and for evaluation of blood vessel function, we typically remove the chorion by careful pronase treatment at 24 hpf. A good indication for gentle pronase treatment is to treat a test batch of embryos so that 10% of the clutch is still in their chorion. These few can be manually dechorionated, if necessary. Several concentrations of the test compound are used to determine the lethal dose, the effective dose, and the range of effectiveness.

II. Screening Tools

We describe here techniques used to visualize the vascular pattern and activity in the zebrafish embryo. We do not include a comprehensive listing of *in situ* probes as this has been recently reviewed in detail (Lawson and Weinstein, 2002a).

A. Alkaline Phosphatase Activity

Endothelial cells possess an endogenous alkaline phosphatase (AP) activity, which can be detected using AP substrates commonly utilized in whole mount *in situ* hybridization protocols (Chen *et al.*, 2001; Schulte-Merker, 2002). This activity was first noted using human arterial endothelial cells and later observed in the amphibian vasculature (Romanul and Bannister, 1962; Stolk, 1963). The enzymatic activity of AP is not restricted to the endothelial lineage. In zebrafish embryos, epithelial cells of the pronephric kidney and the gut, as well as brain tissue are all AP positive. The vascular AP activity can be seen as early as day 2 but is relatively weak compared with AP activity in other tissues. Staining of day 3 embryos or older is preferred as the major vessels, intersomitic, subintestinal, and cranial vessels can be clearly identified (Fig. 1). Since embryos can be fixed in large numbers and screened for vessel abnormalities at later times, the AP assay is particularly well suited for use in genetic screens. It is also possible to section stained embryos to examine deeper vessels without the need for confocal microscopy.

We have tried several published protocols and found the following to give the best results. It is essentially as described by Schulte-Merker (2002). Additional tips have been incorporated to accommodate new zebrafish researchers.

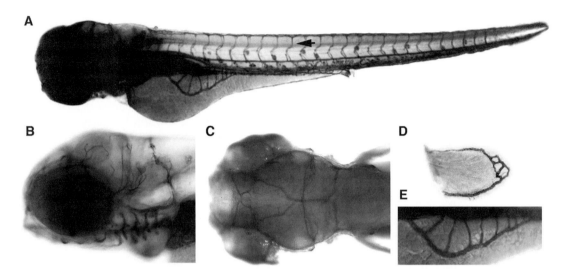

Fig. 1 Alkaline phosphatase staining of blood vessels in a 3-day-old zebrafish larva, using the protocol described. The arrow indicates an intersegmental vessel. (A) Lateral whole view of zebrafish larva. (B) Larval view of head vessels. (C) Dorsal view of head vessels. (D) Pectoral fin vessels. (E) Close-up view of subintestinal vessels.

To prevent natural pigmentation of zebrafish embryos, treat embryos with propylthiouracil (PTU) (0.003% 1-phenyl-2-thiourea) starting at 6 hpf, by simply adding an appropriate amount of 10× PTU stock to the embryo medium. For best results, keep embryo numbers small and remove unhealthy or dead embryos often, refreshing embryo medium/PTU each day. Fix embryos at required stages with 4% paraformaldehyde (PFA) in phosphate buffered saline (PBS) for 30 min at room temperature or overnight at 4 °C. For older stage embryos (e.g., 5 dpf and beyond) some pigmentation is present despite PTU treatment. If embryos are pigmented, they can be cleared after fixation as follows. Remove PFA and rinse twice with PBS + 0.1% Tween-20 (PBT). Add a solution of 1% KOH, 3% H_2O_2, rock 5 to 10 min at room temperature or until the embryos look clear of pigment. Transfer embryos that have had the pigment removed into 50% methanol, 50% PBT for 5 min, then into 100% methanol. Embryos can be stored at −20 °C long term, or up to 1 week at 4 °C. To begin AP staining, treat embryos with ice-cold acetone for 30 min at −20 °C. Rinse twice with PBT (5 min each) and equilibrate embryos with NTMT (100 mM Tris pH 9.5, 100 mM NaCl, 50 mM $MgCl_2$, 0.1% Tween-20) with 3 rinses (5 min each) for a total of 15 min at room temperature. Start the color development reaction by incubating in 500 μl of nitroblue tretrazolium/5-bronio,4-chloro,8-indolylphosphate (NBT/BCIP) staining solution (450 μg NBT and 175 μg BCIP per 1 ml of NTMT, Roche Applied Science, Indianapolis, IN). Staining usually takes about 15 to 30 min for head vessels, but small vessels, such as the intersegmental vessels, may require more time to obtain the right amount of signal. For lightly stained embryos, it is possible to leave embryos overnight in 50% NTMT, 50% stain (NBT/BCIP), or in 100% NTMT at 4 °C until the next day to monitor staining. After staining is completed, wash the embryos 3 times in PBT. (Optional: To allow more time for analysis, we frequently fix embryos a second time in 4% PFA for 20 min at room temperature so that the stained embryos can be kept for months without loss of signal.) After 3 rinses in PBT, transfer embryos to 70% glycerol until all of the embryos sink to the bottom of the tube, about 2 to 3 hours at room temperature. We find that a glycerol series is not necessary for equilibration. For convenience, we frequently transfer embryos into 70% glycerol and keep them at 4 °C overnight.

B. Transgenic Animals

The endogenous AP assay has the advantages of being rapid and simple but it also has a relatively high background since other tissues are also AP positive. Use of transgenic animals, where endothelial cells are genetically marked, provides a method addressing specificity. Two endothelial-specific transgenic lines of zebrafish have been published thus far (Cross *et al.*, 2003; Lawson and Weinstein, 2002b). Lawson and Weinstein (2002b) successfully marked the vasculature by using the friend leukemia virus 1 (fli1) promoter to drive the expression of green fluorescent protein (GFP). This line was used to examine the dynamic process of sprouting angiogenesis in the trunk of the zebrafish embryo during the formation of intersegmental vessels (Isogai *et al.*, 2003) and is available through the Oregon stock center. The VEGFR2/flk promoter was used by Cross *et al.* (2003) to drive expression of a green reef coral fluorescent protein in endothelial cells. Both of these transgenic lines can be used for evaluation of small molecular kinase inhibitors by the presence or absence of endothelial cells lining the blood vessels as shown by Cross *et al.* (2003). The differences in these two transgenic lines are attributed to the promoter chosen and copy number of the transgene. For the TG(fli1:EGFP) line, enhanced green fluorescent protein (EGFP) expression is also seen in some nonvascular neural-crest-derived tissues such as the developing cartilage of the jaw. The copy number of the y1 allele is estimated at more than 25 and thus EGFP is expressed at very high levels (Lawson and Weinstein, 2002b). This also results in a persistent strong expression of the EGFP in the adult animal, which can be seen even in the microvasculature of the tail fin.

C. Microangiography

Blood flow during the first 3 days of embryonic development can easily be assessed using a dissecting microscope. However, for visualization of deeper vessels and to take a pictorial record of functional vessels, it is sometimes necessary to perform microangiography where a fluorescent dye is injected into the embryonic heart. The combined use of microangiography in a transgenic endothelial-GFP line would allow the assessment of blood vessels and blood flow in the same embryo. This proved especially useful for the analysis of blood vessel mutants, where the endothelial cells are in the correct positions but flow is restricted because of a physical constriction or because of subtle patterning defects such as in the *gridlock* mutant (Weinstein *et al.*, 1995; Zhong *et al.*, 2001).

The zebrafish microangiography technique was pioneered by Brant Weinstein (Weinstein et al., 1995). We have made some modifications to increase throughput. We use either a fluorescent lectin or fluorescent microspheres: 5 mg/ml Alexa Fluor® 488 conjugate (Molecular Probes, Eugene, OR) in Danieau's solution [58 mM NaCl, 0.7 mM KCl, 0.4 mM MgSO$_4$, 0.6 mM Ca(NO$_3$)$_2$, 5 mM HEPES, pH 7.6] or we use carboxylate-modified fluorescent microspheres (0.02 μm, FluoSpheres®, Molecular Probes). Embryos from 48 hpf onward are fairly easy to handle with this microangiography protocol. The microinjection needle is the same as for 1-cell stage injections. The fluorescent reagents are first sonicated for 10 to 20 min in a water bath sonicator, then loaded into a microinjection needle. The embryos are anesthetized using 0.02% tricaine, then lined up onto an agarose ramp. The dorsal or ventral placement of the embryos determines the location of the injection. In embryos allowing a ventral view, the site of injection is the cardiac atrium or sinus venosus. Embryos that are dorsal-side up are injected just ventral to the otic vesicle through to the midline of the embryo, gaining access to the aortic arch. After injection, the embryos are immediately placed into embryo medium for recovery. Frequently, the site of needle entry has the highest amount of fluorescence. Optimal views of the heart and head region can also be obtained by injecting the tracer directly into the dorsal artery.

III. Chemical Screening

A. Concentration and Solvent Considerations

The use of the developing zebrafish vasculature as an assay offers several advantages: endogenous vascular development, forward genetics, and the ability to screen through large numbers of compounds in relatively small volumes. Zebrafish embryos can be kept in standard 96-well plates, where a well can easily accommodate 2 to 3 embryos in a 100 to 150-μl volume. Embryos can be distributed either manually before hatching or through the use of automated sorting machinery such as COPAS (Union Biometrica, Somerville, MA) and can develop normally within the well. In contrast to embryos from invertebrate model systems, which possess a tough outer cuticle, the zebrafish chorion does not pose a significant barrier to small molecule entry.

Specific compounds or libraries for use in zebrafish phenotypic screening assays can be stored or diluted into dimethyl sulfoxide (DMSO). Zebrafish embryos can tolerate a range of DMSO concentrations depending on the embryonic stage at the time of addition. In general, higher concentrations are better tolerated when the addition is made at later stages of development and extended incubation times increase the phenotypic noise levels. For example, addition of 2% DMSO at 48 hpf does not cause any gross morphological abnormalities or adverse physiological effects (Milan *et al.*, 2003) during a 24-hour period, however longer incubations may cause pericardial edema in some embryos. Similarly, embryos developing in a 5% DMSO solution added at 8 hpf develop rather normally until the 24 hpf, but are generally necrotic 24 hours later. A final concentration of 1% DMSO does not produce any gross morphological changes even when added as early as the 6-hpf stage.

One caveat to using whole embryos as screening tools is knowledge of the effective concentration of the compounds within the organism. Chemical screening in cell culture systems, where cells have direct contact with the culture medium, has all cells sensing the same concentration. When using zebrafish to screen compounds however, discrepancies between the concentration in the medium and the effective concentration within the embryo certainly exist. This discrepancy is likely dependent on the time of addition where two key variables are compound penetration and metabolism. Direct injection of the compound into the sinus venosus of embryos largely eliminates the penetration problem by having the chemical in circulation but this methodology is not particularly suited for higher throughput analyses. In a pilot screen for selected compounds affecting heart rate, Milan and colleagues noted that molecules known to cause QT-interval prolongation or *torsade de pointes* in humans also cause bradycardia when delivered into the embryo medium (Milan *et al.*, 2003). However, a subset failed to provoke a cardiac phenotype in this fashion and was only able to induce bradycardia on microinjection (with the exception of erythromycin). Such penetration limitations should be taken into consideration, especially when considering screening large chemical collections. Metabolism of exogenous compounds by the cytochrome P450 (CYP) family of enzymes in the liver will undoubtedly affect the compound's stability and may produce a number of biologically active metabolites. A thorough study of zebrafish CYP proteins has not been completed thus far and it would be difficult to predict exactly how exogenous compounds would be metabolized. However public database queries indicate that many of the CYP proteins appear to be conserved in fish and may in fact recapitulate aspects of mammalian hepatic metabolism.

Concentrations within the well for whole embryo assays may need to be up to an order of magnitude greater than the effective concentration observed for cell culture systems, where the compound is directly accessible to all cells. The gamma-secretase inhibitor 7{N-[N-(3,5-difluorophenacetyl)-L-alanyl]-S-phenylglycine} t-butyl ester (DAPT) is active in the nanomolar concentration when added to cell culture systems and has an IC50 of 20 nM in HEK293 cells (Dovey *et al.*, 2001). The gamma-secretase complex is involved in the processing of the amyloid precursor protein (APP) and Notch among other substrates. A 50-μM concentration of DAPT added directly to the embryo media was required to see any effect on zebrafish development. A concentration of 100

μM had a stronger effect and phenocopies Notch-pathway mutations such as *beamter* and *mindbomb* (Geling *et al.*, 2002). A second example is the DNA methyltransferase inhibitor 5-aza-2'-deoxycytidine (5-aza-dC). The inhibitor is active in the nanomolar range in cultured embryonic stem cells (Juttermann *et al.*, 1994) but inhibition of maintenance DNA methylation in zebrafish genomic DNA is only seen in the 50- to 100-μM range when added to whole embryos (Martin *et al.*, 1999).

Both DAPT and 5-aza-deoxyC have profound effects on the early development of embryos. However, similar results are observed with VEGFR inhibitors. PTK787/ZK222584 can inhibit the zebrafish VEGFR expressed in mammalian cells at 100 nM, but low micromolar concentrations are required to block the formation of the major arteries in the embryo at 24 hpf (Chan *et al.*, 2002).

Given the higher concentrations required to see phenotypic effects in whole embryos, as well as unknown variables such as penetration and metabolism, it is not surprising that some chemical screens have utilized several concentrations to effectively screen the chemical space represented by compound libraries. Concentrations of 1, 10, and 100 μg/ml were used in screening compounds for effects on zebrafish embryonic heart rate (Milan *et al.*, 2003). However, this study focused only on 100 compounds and, while this may be feasible when screening hundreds of compounds, screening multiple concentrations for thousands or tens of thousands of compounds may not be a practical approach. In this latter case, setting a high affinity threshold may be in order. Such a strategy was used in the discovery of concentramide, which phenocopies a null mutation of an atypical protein kinase C and is active in the nanomolar range (Peterson *et al.*, 2001).

B. Chemical Genetic Phenotypes

It is likely that the variety of chemically induced phenotypes would somehow parallel genetic phenotypes observed in zebrafish embryos (Peterson *et al.*, 2000). Thus, the large collection of embryonic phenotypes observed in large-scale screens (Haffter *et al.*, 1996) should provide us with a starting point for phenotypic assays. Furthermore as the affected loci in the mutations are being cloned, matching chemical phenotypes to genetic pathways will become easier and develop into a useful tool in the elucidation of the mechanism of action of selected small molecules.

Chemical and genetic phenotypes may not be fully overlapping since the chemical may have several cellular targets of varying affinity; however, the validity of this strategy has been demonstrated for the *heart and soul* mutation (Peterson *et al.*, 2001). Furthermore, PTK787/ZK222584 treatment should inhibit all the VEGFRs, albeit with different affinities, and thus the phenotype closely resembles that of the ligand knockdown rather than a point mutation in one of the high-affinity receptors (Habeck *et al.*, 2002). It is unclear whether more subtle vascular phenotypes such as the ISV patterning defects observed in the *out of bounds* mutant can be phenocopied (Childs *et al.*, 2002). In some cases, knowledge of the affected gene may be necessary.

IV. Sensitized Screens

A complementary approach to direct compound screening is the use of known chemical modulators to sensitize embryonic assays. Genetically sensitized modifier screens have long been used in invertebrates as a means to more efficiently target specific biological processes or signaling pathways (Simon *et al.*, 1991; St Johnston, 2002). In this variant, a chemical can be used to sensitize the embryo to a particular pathway and provide a "first-hit" and thus screen for heterozygous phenotypes or haplo-insufficient loci in the compound's presence. For such screens to be successful however, the cellular target of the chemical modulator should represent a key node in the pathway. Furthermore, the concentration and embryonic stage for chemical addition need to be optimized to screen for heterozygous enhancer or suppressor mutations. The effective range of a compound may be broad and generate a spectrum of phenotypes (Chan *et al.*, 2002) analogous to an allelic series. Weak or strong phenotypes can be used to specifically sensitize the screen for enhancer or suppressors. Alternatively, as more and more defective genes responsible for zebrafish vascular mutants become available, they can be used to test the effects of small molecules on specific pathways.

Acknowledgments

We thank John Mably, Patrick Faloon, and Jeanette Wood for comments on the manuscript and Kimberly Bellavance for the alkaline phosphatase figure.

References

Bold, G., Altmann, K. H., Frei, J., Lang, M., Manley, P. W., Traxler, P., Wietfeld, B., Bruggen, J., Buchdunger, E., Cozens, R., Ferrari, S., Furet, P., Hofmann, F., Martiny-Baron, G., Mestan, J., Rosel, J., Sills, M., Stover, D., Acemoglu, F., Boss, E., Emmenegger, R., Lasser, L., Masso, E., Roth, R., Schlachter, C., and Vetterli, W. (2000). New anilinophthalazines as potent and orally well absorbed inhibitors of the VEGF receptor tyrosine kinases useful as antagonists of tumor-driven angiogenesis. *J. Med. Chem.* **43**, 2310–2323.

Carmeliet, P. (2000). Mechanisms of angiogenesis and arteriogenesis. *Nat. Med.* **6**, 389–395.

Chan, J., Bayliss, P. E., Wood, J. M., and Roberts, T. M. (2002). Dissection of angiogenic signaling in zebrafish using a chemical genetic approach. *Cancer Cell* **1**, 257–267.

Chen, J. N., van Bebber, F., Goldstein, A. M., Serluca, F. C., Jackson, D., Childs, S., Serbedzija, G., Warren, K. S., Mably, J. D., Lindahl, P., Mayer, A., Haffter, P., and Fishman, M. C. (2001). Genetic steps to organ laterality in zebrafish. *Comp. Funct. Genom.* **2**, 60–68.

Childs, S., Chen, J. N., Garrity, D. M., and Fishman, M. C. (2002). Patterning of angiogenesis in the zebrafish embryo. *Development* **129**, 973–982.

Cross, L. M., Cook, M. A., Lin, S., Chen, J. N., and Rubinstein, A. L. (2003). Rapid analysis of angiogenesis drugs in a live fluorescent zebrafish assay. *Arterioscler. Thromb. Vasc. Biol.* **23**, 911–912.

Dovey, H. F., John, V., Anderson, J. P., Chen, L. Z., de Saint Andrieu, P., Fang, L. Y., Freedman, S. B., Folmer, B., Goldbach, E., Holsztynska, E. J., Hu, K. L., Johnson-Wood, K. L., Kennedy, S. L., Kholodenko, D., Knops, J. E., Latimer, L. H., Lee, M., Liao, Z., Lieburg, I. M., Motter,

R. N., Mutter, L. C., Nietz, J., Quinn, K. P., Sacchi, K. L., Seubert, P. A., Shopp, G. M., Thorsett, E. D., Tung, J. S., Wu, J., Yang, S., Yin, C. T., Schenk, D. B., May, P. C., Altstiel, L. D., Bender, M. H., Boggs, L. N., Britton, T. C., Clemens, J. C., Czilli, D. L., Dieckman-McGinty, D. K., Droste, J. J., Fuson, K. S., Gitter, B. D., Hyslop, P. A., Johnstone, E. M., Li, W. Y., Little, S. P., Mabry, T. E., Miller, F. D., and Audia, J. E. (2001). Functional gamma-secretase inhibitors reduce beta-amyloid peptide levels in brain. *J. Neurochem.* **76**, 173–181.

Drevs, J., Hofmann, I., Hugenschmidt, H., Wittig, C., Madjar, H., Muller, M., Wood, J., Martiny-Baron, G., Unger, C., and Marme, D. (2000). Effects of PTK787/ZK 222584, a specific inhibitor of vascular endothelial growth factor receptor tyrosine kinases, on primary tumor, metastasis, vessel density, and blood flow in a murine renal cell carcinoma model. *Cancer Res.* **60**, 4819–4824.

Druker, B. J. (2002). STI571 (Gleevec) as a paradigm for cancer therapy. *Trends Mol. Med.* **8**, S14–S18.

Folkman, J. (1971). Tumor angiogenesis: Therapeutic implications. *N. Engl. J. Med.* **285**, 1182–1186.

Geling, A., Steiner, H., Willem, M., Bally-Cuif, L., and Haass, C. (2002). A gamma-secretase inhibitor blocks Notch signaling *in vivo* and causes a severe neurogenic phenotype in zebrafish. *EMBO Rep.* **3**, 688–694.

Habeck, H., Odenthal, J., Walderich, B., Maischein, H., and Schulte-Merker, S. (2002). Analysis of a zebrafish VEGF receptor mutant reveals specific disruption of angiogenesis. *Curr. Biol.* **12**, 1405–1412.

Haffter, P., Granato, M., Brand, M., Mullins, M. C., Hammerschmidt, M., Kane, D. A., Odenthal, J., van Eeden, F. J., Jiang, Y. J., Heisenberg, C. P., Kelsh, R. N., Furutani-Seiki, M., Vogelsang, E., Beuchle, D., Schach, U., Fabian, C., and Nusslein-Volhard, C. (1996). The identification of genes with unique and essential functions in the development of the zebrafish, Danio rerio. *Development* **123**, 1–36.

Isogai, S., Horiguchi, M., and Weinstein, B. M. (2001). The vascular anatomy of the developing zebrafish: An atlas of embryonic and early larval development. *Dev. Biol.* **230**, 278–301.

Isogai, S., Lawson, N. D., Torrealday, S., Horiguchi, M., and Weinstein, B. M. (2003). Angiogenic network formation in the developing vertebrate trunk. *Development* **130**, 5281–5290.

Juttermann, R., Li, E., and Jaenisch, R. (1994). Toxicity of 5-aza-2'-deoxycytidine to mammalian cells is mediated primarily by covalent trapping of DNA methyltransferase rather than DNA demethylation. *Proc. Natl. Acad. Sci. USA* **91**, 11797–11801.

Kantarjian, H. M., and Talpaz, M. (2001). Imatinib mesylate: Clinical results in Philadelphia chromosome-positive leukemias. *Semin. Oncol.* **28**, 9–18.

Lawson, N. D., and Weinstein, B. M. (2002a). Arteries and veins: Making a difference with zebrafish. *Nat. Rev. Genet.* **3**, 674–682.

Lawson, N. D., and Weinstein, B. M. (2002b). *In Vivo* imaging of embryonic vascular development using transgenic zebrafish. *Dev. Biol.* **248**, 307–318.

Manley, P. W., Martiny-Baron, G., Schlaeppi, J. M., and Wood, J. M. (2002). Therapies directed at vascular endothelial growth factor. *Expert Opin. Investig. Drugs* **11**, 1715–1736.

Martin, C. C., Laforest, L., Akimenko, M. A., and Ekker, M. (1999). A role for DNA methylation in gastrulation and somite patterning. *Dev. Biol.* **206**, 189–205.

Matsumoto, T., and Claesson-Welsh, L. (2001). VEGF receptor signal transduction. *Sci. STKE.* **2001**, RE21.

Milan, D. J., Peterson, T. A., Ruskin, J. N., Peterson, R. T., and MacRae, C. A. (2003). Drugs that induce repolarization abnormalities cause bradycardia in zebrafish. *Circulation* **107**, 1355–1358.

Nasevicius, A., Larson, J., and Ekker, S. C. (2000). Distinct requirements for zebrafish angiogenesis revealed by a VEGF-A morphant. *Yeast* **17**, 294–301.

Nüsslein-Volhard, C., and Dahm, R. (2002). "Zebrafish." Oxford University Press, New York.

Ozaki, H., Seo, M. S., Ozaki, K., Yamada, H., Yamada, E., Okamoto, N., Hofmann, F., Wood, J. M., and Campochiaro, P. A. (2000). Blockade of vascular endothelial cell growth factor receptor signaling is sufficient to completely prevent retinal neovascularization. *Am. J. Pathol.* **156**, 697–707.

Peterson, R. T., Link, B. A., Dowling, J. E., and Schreiber, S. L. (2000). Small molecule developmental screens reveal the logic and timing of vertebrate development. *Proc. Natl. Acad. Sci. USA* **97**, 12965–12969.

Peterson, R. T., Mably, J. D., Chen, J. N., and Fishman, M. C. (2001). Convergence of distinct pathways to heart patterning revealed by the small molecule concentramide and the mutation heart-and-soul. *Curr. Biol.* **11**, 1481–1491.

Ransom, D. G., Haffter, P., Odenthal, J., Brownlie, A., Vogelsang, E., Kelsh, R. N., Brand, M., van Eeden, F. J., Furutani-Seiki, M., Granato, M., Hammerschmidt, M., Heisenberg, C. P., Jiang, Y. J., Kane, D. A., Mullins, M. C., and Nusslein-Volhard, C. (1996). Characterization of zebrafish mutants with defects in embryonic hematopoiesis. *Development* **123**, 311–319.

Romanul, F. C., and Bannister, R. G. (1962). Localized areas of high alkaline phosphatase activity in endothelium of arteries. *Nature* **195**, 611–612.

Schlessinger, J. (2000). Cell signaling by receptor tyrosine kinases. *Cell* **103**, 211–225.

Schulte-Merker, S. (2002). Looking at embryos. *In* "Zebrafish" (C. Nusslein-Volhard and R. Dahm, eds.), pp. 39–58. Oxford University Press, New York.

Serbedzija, G. N., Flynn, E., and Willett, C. E. (1999). Zebrafish angiogenesis: A new model for drug screening. *Angiogenesis* **3**, 353–359.

Simon, M. A., Bowtell, D. D., Dodson, G. S., Laverty, T. R., and Rubin, G. M. (1991). Ras1 and a putative guanine nucleotide exchange factor perform crucial steps in signaling by the sevenless protein tyrosine kinase. *Cell* **67**, 701–716.

St Johnston, D. (2002). The art and design of genetic screens: Drosophila melanogaster. *Nat. Rev. Genet.* **3**, 176–188.

Stolk, A. (1963). Localized areas of high alkaline phosphatase activity in the endothelium of arteries in the axolotl. *Experientia* **19**, 21.

Weinstein, B. M., Schier, A. F., Abdelilah, S., Malicki, J., Solnica-Krezel, L., Stemple, D. L., Stainier, D. Y., Zwartkruis, F., Driever, W., and Fishman, M. C. (1996). Hematopoietic mutations in the zebrafish. *Development* **123**, 303–309.

Weinstein, B. M., Stemple, D. L., Driever, W., and Fishman, M. C. (1995). Gridlock, a localized heritable vascular patterning defect in the zebrafish. *Nat. Med.* **1**, 1143–1147.

Wood, J. M., Bold, G., Buchdunger, E., Cozens, R., Ferrari, S., Frei, J., Hofmann, F., Mestan, J., Mett, H., O'Reilly, T., Persohn, E., Rosel, J., Schnell, C., Stover, D., Theuer, A., Towbin, H., Wenger, F., Woods-Cook, K., Menrad, A., Siemeister, G., Schirner, M., Thierauch, K. H., Schneider, M. R., Drevs, J., Martiny-Baron, G., and Totzke, F. (2000). PTK787/ZK 222584, a novel and potent inhibitor of vascular endothelial growth factor receptor tyrosine kinases, impairs vascular endothelial growth factor-induced responses and tumor growth after oral administration. *Cancer Res.* **60**, 2178–2189.

Yancopoulos, G. D., Davis, S., Gale, N. W., Rudge, J. S., Wiegand, S. J., and Holash, J. (2000). Vascular-specific growth factors and blood vessel formation. *Nature* **407**, 242–248.

Zhong, T. P., Childs, S., Leu, J. P., and Fishman, M. C. (2001). Gridlock signalling pathway fashions the first embryonic artery. *Nature* **414**, 216–220.

CHAPTER 22

Vascular Occlusion and Thrombosis in Zebrafish

Pudur Jagadeeswaran, Matthew Cykowski, and Bijoy Thattaliyath

Department of Cellular and Structural Biology
The University of Texas Health Science Center at San Antonio
San Antonio, Texas 78229

I. Introduction

Hemostasis is a defense mechanism invoked in response to vascular injury to prevent the loss of blood and preserve the integrity of the circulation. It is orchestrated by coagulation proteins, platelets, and components of the vessel wall that function together to form a protective seal composed of platelets and a fibrin network at the site of injury (Mann, 1999). Platelets provide the immediate hemostatic response by adhering to components of the subendothelial matrix (Ruggeri, 2002). This adherence initiates various intracellular signaling pathways in platelets that result in platelet activation and formation of platelet aggregates that seal the damaged vessel in a process termed *primary hemostasis* (Jackson *et al.*, 2003). In addition to this platelet response, the coagulation is initiated by the

METHODS IN CELL BIOLOGY, VOL. 76
0091-679X/04 $35.00

489

exposure of tissue factor on the membrane surfaces of subendothelial cells that bind to VIIa, a coagulation protease. This sets off the activation of a cascade of serine protease zymogens along with cofactors culminating in the ultimate generation of thrombin that generates a fibrin clot from the fibrinogen precursor (Davie, 2003). To control thrombin production and removal of fibrin, a variety of anticoagulant and fibrinolytic pathways are activated (Esmon, 2003; Nesheim, 2003). The vessel wall itself is involved in inhibiting the adverse coagulation and platelet aggregation inside the vessel by providing a layer of endothelial cells that secrete anticoagulant, fibrinolytic, and platelet inhibitory factors such as thrombomodulin, tissue plasminogen activator, and nitric oxide, respectively (Esmon, 1995; Medved and Nieuwenhuizen, 2003; Ritchie *et al.*, 2002).

Human disease conditions in which these processes are not properly maintained result in either excessive bleeding or clotting inside the vessel (Cattaneo, 2003; Levi *et al.*, 2002; Schafer *et al.*, 2003). Formation of clots inside the vessel is referred to as *thrombosis* and is a pathological process caused by disturbance of the vessel wall, blood borne factors, or blood flow. These three components are classically referred to as Virchow's triad (Virchow, 1856), which essentially encompasses all hemostatic components in a global context. Despite extensive investigations into the thrombotic processes, only a limited number of genetic risk factors have been identified to account for venous thrombophilia (Alhenc-Gelas and Aiach, 2003; Heit, 2003). In contrast to these insights into genetic causes of venous thrombosis, no major genetic factor has yet been identified as a definitive risk for arterial thrombosis (Bohm and Al-Khaffaf, 2003; Endler and Mannhalter, 2003). Overall, the genetic factors that influence thrombosis are poorly understood. Thus, alternative approaches to identifying these genetic factors are important for filling the gap in the knowledge of this important life-threatening clinical problem.

A. Zebrafish as a Model for Hemostasis and Thrombosis

Since the mid-1990s, we realized that a classic genetic approach, in which mutations are randomly generated and screened for phenotypic alterations, had greater potential to identify novel factors in a complex problem like hemostasis. For this type of investigation, the choice of the model system was important. Conducting classic mutagenesis and screening for hemostatic mutants in mice would be an arduous task, particularly when recessive mutations have to be identified (Hrabe de Angelis *et al.*, 2000; Justice, 2000; Nolan *et al.*, 2000). Around the millennium, the zebrafish was gaining respect as a genetic model to study vertebrate developmental functions (Driever and Fishman, 1996). However, whether the zebrafish could be used to study mammalian hemostasis was questionable because zebrafish could have primitive hemostatic pathways, and, thus, not be appropriate to study mammalian hemostasis (Doolittle, 1961). In addition, convincing the biochemists, who were accustomed to the bovine model system, of the utility of this miniature model, encouraging them to study hemostasis by developing biochemical assays, was daunting. Even if we were to resolve these issues, the task of developing screening assays to study hemostasis and thrombosis in zebrafish larvae seemed intimidating. To address these problems, first we undertook the characterization of zebrafish hemostasis by developing biochemical methods for analysis of coagulation (Jagadeeswaran and Liu, 1997). By coupling these assay systems with this and molecular genetic methods, we found that the coagulation cascade and anticoagulant pathways in zebrafish were similar to that found in humans (Hanumanthaiah *et al.*, 2002; Jagadeeswaran and Sheehan, 1999; Jagadeeswaran *et al.*, 1999a, 2000a,b; Sheehan *et al.*, 2001). Then, we identified and characterized zebrafish thrombocytes, which are the equivalents of mammalian platelets (Gregory and Jagadeeswaran, 2002; Jagadeeswaran *et al.*, 1999b). By functional analysis assays, as well as immunological methods, we provided evidence that the hemostatic pathways in thrombocytes are similar to those found in mammalian platelets (Gregory and Jagadeeswaran, 2002; Jagadeeswaran *et al.*, 1999b). Thus, it is now reasonably established that the zebrafish system is an excellent model to study mammalian hemostasis and thrombosis.

B. Development of Genetic Screens for Hemostasis

Developing genetic screens for hemostatic pathways has been fraught with problems because of the miniature size of the model, as well as concerns about uniformity and reproducibility of the assays. First, we developed larval bleeding assays as well as microassays for coagulation and thrombocyte function in adult zebrafish (Jagadeeswaran and Liu, 1997; Jagadeeswaran *et al.*, 1999b, 2000a). Unfortunately, these screening tools are labor intensive and did not address all the aspects of Virchow's triad (Virchow, 1856). For example, the larval bleeding assay measures primary hemostasis, as well as coagulation functions. Although this screen is conceptually good, it takes several minutes to perform one assay and is technically demanding. In general, phenotypic screening in larvae are preferred because the parent carrying the recessive mutation could be detected in a few days and further breeding experiments required for analysis of genetic mutation could be planned (Patton and Zon, 2001). Therefore, an assay was required that was simple, had high throughput, and had the ability to screen comprehensively all of the hemostatic pathways in larvae.

II. Vascular Occlusion

A. Chemical Methods

To develop a global, reproducible, and reliable hemostatic assay, uniform wounding in a defined area of a zebrafish larva is the critical feature (Gregory *et al.*, 2002). We will describe our attempts to use chemical and laser injury methods to cause uniform wounding and thrombosis in zebrafish larvae, summarize the advantages of laser injury over the chemical injury methods, and suggest that laser injury is currently the best approach to study thrombosis in zebrafish.

One of the wounding agents, phenylhydrazine, which is known to cause injury in mammals, was used to injure zebrafish larvae (Jain, 1985). Larvae were placed in Tricaine (10 μM) solution and an equal amount of 1% low-melting agarose was added. The agarose-Tricaine mixture, along with a larva (100 μl), was placed in a well formed by placing a plastic ring on a microscopic slide. Phenylhydrazine HCl (100 μl of 40-mM solution) was layered on the surface of the agarose block and vascular occlusion was recorded using a microscope and a digital camera attached to a VCR and a monitor. We found that the caudal artery was always occluded after approximately 10 minutes. This localized occlusion in the caudal artery probably occurs because the caudal area is thinner and may facilitate quick diffusion of phenylhydrazine in contrast to thicker muscular areas in other portions of the larval body. Thus, it was possible to measure time to occlude (TTO) the vessel from the time of addition of this reagent. Though phenylhydrazine treatment gave arterial occlusion in the caudal area, we wanted to test whether the occlusive mass had features of classic thrombosis, including thrombocytes and a fibrin clot. The thrombocyte involvement was assessed by selectively labeling thrombocytes with DiI-C18 (DiI) and testing whether they formed aggregates at the wounding site. We found no aggregates of thrombocytes at the wounding site. We then delivered fluorescein isothiocyanate (FITC)-labeled human fibrinogen into the vascular system of zebrafish larva and tested whether an increase in fluorescence attributed to fibrin formation would occur during phenylhydrazine-induced injury. In these experiments, we did not find any evidence of fibrin formation. Occlusion was found to be irreversible as evidenced by the lack of return of circulation at the site of occlusion. Treatment of larvae with warfarin, a well-known anticoagulant, did not affect the TTO, indicating coagulation was not involved even though erythrocyte membranes were altered, exposing phosphatidylserine (PS), which has the ability to enhance thrombin generation. Thus, the phenylhydrazine-induced arterial occlusion may not be appropriate for thrombosis studies; however, it could be used as a model for understanding the mechanism of occlusion caused by sickle cell disease where sickle cells have PS exposure. Patients with sickle cell anemia are prone to vascular occlusions (Setty *et al.*, 2002; Stuart and Johnson, 1987).

After these difficulties with phenylhydrazine, we attempted to use ferric chloride, another reagent used in mammals to cause vascular injury (injuring the endothelial layer by free radical formation), where a filter paper soaked with ferric chloride is placed on a small area on the surface of the vessel (Kurz *et al.*, 1990). This would be very difficult given the size of the zebrafish larva. Therefore, we added ferric chloride (100 μl of 40-mM solution) to the larva immobilized in agarose as in the case of phenylhydrazine experiment. We found that this treatment also formed an occlusion in the caudal artery similar to that found in the phenylhydrazine experiment. However, the occlusive mass formed in this reaction had fibrin deposition and thrombocytes, as shown by the use of FITC-labeled fibrinogen and DiI labeling, a true sign of the classic thrombus. The TTO for this assay was approximately 6 minutes. Warfarin treatment prolonged the TTO and the occlusion was irreversible. Even though it was a bona fide thrombus, the thrombus formed was not always consistent. Sometimes, there were clumps of cells in circulation generated probably because of systemic activation and a firm arterial occlusion was not noted. Furthermore, it was difficult to control the reaction to cause either arterial or venous thrombus formation, which left us with few remaining choices in the zebrafish model. Therefore, we developed the following laser-induced vascular occlusion assay, which is comprehensive, easy to use, and has a reasonably high throughput.

B. Laser Method

Laser injury has previously been used in animal models to induce thrombus formation (Rosen et al., 2001). In our work we used pulsed nitrogen laser light to cause a lesion in the caudal vein or artery leading to vascular occlusion. We have demonstrated that it is possible to wound the larval vessel uniformly at a defined location both at the arterial side and the venous side. We have shown the formation fibrin in venous occlusive mass by using FITC-fibrinogen and also demonstrated that thrombocytes play a major role in the arterial thrombus formation by using DiI-labeled thrombocytes as already described. The TTO for both arterial and venous occlusions have been established. The validity of this assay has been verified by using the antisense morpholino knockdown technology to knockdown proteins such as prothrombin and factor VII and demonstrating that venous TTO was prolonged. Interestingly, this technology also led to the finding that factor VIIi, a novel factor unique to zebrafish and other fish, is an inhibitor of coagulation and gave shortened TTO. This assay has been applied to verify mutants of hemostasis such as *victoria* (Gregory et al., 2002) and *leopold* (in collaboration with Mary Mullins, University of Pennsylvania, 2003). We will present the detailed procedure of vascular occlusion as performed in our laboratory.

III. Methods

The equipment required for performing laser-induced vascular occlusion is shown in Fig. 1. It is composed of a fluorescent microscope with a fluorescent port designed for detecting green fluorescence through which nitrogen pulsed laser light passed through coumarin 440 dye (445 nm) is delivered (MicroPoint® Laser System, Photonic Instruments Inc., St. Charles, IL). Approximately 10 zebrafish larvae are anesthetized in a microcentrifuge tube containing 500 μl of 10-μM Tricaine solution. Once anesthetized, 500 μl of 1% low-melting agarose (at 40 °C) is added to these larvae. Approximately 50 μl of this Tricaine and agarose mixture, along with a single larva, is taken up with a wide-bore pipette tip and layered in an individual well.

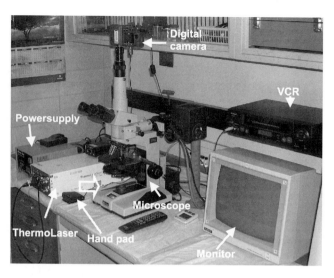

Fig. 1 Powersupply, thermoline laser, digital camera, VCR, monitor, hand pad, and fluorescent microscope are shown by solid arrows. Open arrow shows the point at which a hand is waved towards the monitor in the light path.

The wells are created by punching holes in a rubber sheet of 1.5-mm thickness. The rubber sheet had been placed on a standard glass slide that had been lightly coated with petroleum jelly so the sheet stays in its place (Fig. 2). Usually larvae position themselves sideways such that both the caudal artery and vein are clearly visible. If they are not positioned correctly, they can be carefully moved by a pipette tip to the desired orientation quickly before the agarose solidifies. The larvae embedded in agarose are then placed under the microscope and the laser injury is delivered to an individual larva for 5 sec at 15 pulses/sec, with a laser intensity of setting 16 in the area 5 somites posterior to the anal pore. Because of the thickness of the muscle in older larvae, a slightly more intense laser injury (setting 20) is needed to create the same level of wounding compared with young larva. Thus, the intensity setting can be varied depending on the age of the larva. The hand pedal is pressed with one hand to initiate the laser exposure and after 5 sec the other hand is waved under the microscope (from the position shown in Fig. 1) to block the light so when the images are recorded, using a VCR connected via a digital camera, the time of initiation of laser could be noted as indicated by the interruption of light.

The TTO was taken as the first point after laser treatment when the blood flow through the injured vessel was completely blocked. Because the cells are larger in zebrafish in contrast to human blood cells (such as platelets and red cells), we also estimated the time to adherence (TTA) of the first cell from the time of wounding. Also, in laser injury, since the occlusion was reversible, the time to dissolution (TTD) of thrombus from the TTO could also be measured. Representative images

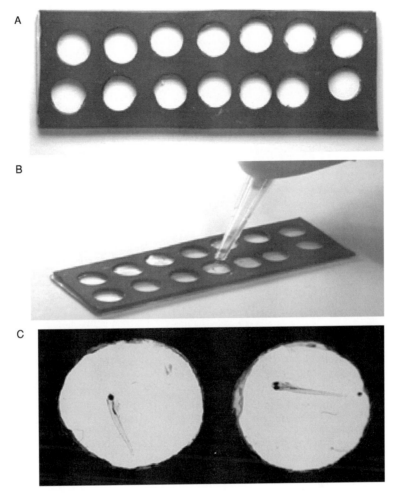

Fig. 2 Method of loading the larvae for laser treatment. (A) Rubber sheet with holes is placed on glass slide after applying a thin coating of petroleum jelly at the bottom of the sheet. (B) Loading of the larvae that have been mixed with 1% low-melting agarose. (C) Magnified view of the larvae in the wells ready for laser ablation.

of thrombus formation in a linear sequence in arterial and venous occlusive events are given in Figs. 3 and 4, respectively. All the three times, TTA, TTO, and TTD for both caudal artery and vein are given in Table I. The values shown in the Table are for 8-day-old larva. At a given laser intensity, the values are prolonged in younger larvae compared with older larvae because the hemostatic mechanisms are not well advanced in young larvae. In fact in 2- to 3-day-old embryos, arterial thrombi will never form because thrombocyte numbers are limited (Gregory and Jagadeeswaran, 2002). Therefore, in our assays we use larvae that are older than

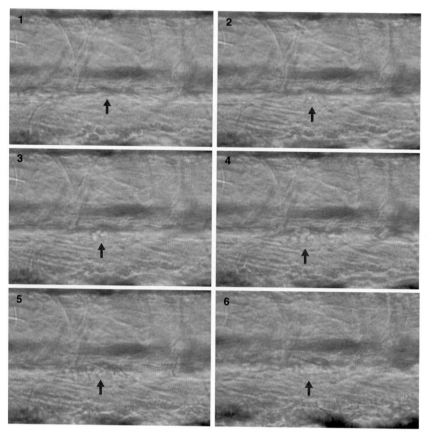

Fig. 3 Sequence of thrombus growth shown by arrows from the time of laser injury 1, and 2 through 6 with increased accumulation of thrombus in the dorsal aorta.

4 days of age. This assay was also used to injure tail arterioles or venules of zebrafish; however, since vessels vary in the diameter and their location in depth, many vessels have to be studied to obtain statistically significant values.

IV. Future Perspectives

The laser-induced vascular injury method is a very reproducible assay to measure thrombus formation using zebrafish larvae because of the advantage that a well-defined target location on the vessel could be easily selected. This is unlike the difficulties in reproducing uniform wounding from one vessel to another in mice that requires analysis of large numbers of vessels to obtain reliable information. In the laser-induced vascular occlusion assay, the entire process of thrombosis and

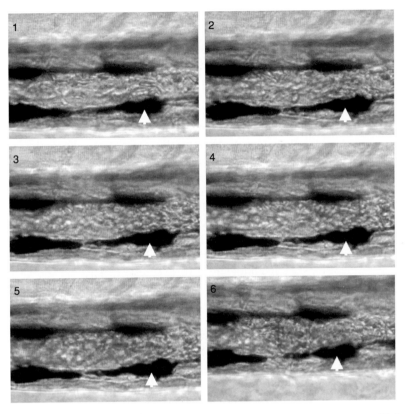

Fig. 4 Sequence of thrombus growth shown by arrows from the time of laser injury 1, and 2 through 6 with increased accumulation of thrombus (*arrow*) in the caudal vein.

Table I
Various Times in Vascular Occlusion

	Arterial	Venous
TTA	11 ± 1 (sec)	4 ± 0.5 (sec)
TTO	75 ± 3 (sec)	21 ± 1 (sec)
TTD	5 ± 0.2 (min)	90 ± 15 (min)

dissolution of thrombus could be followed kinetically. Therefore, coupled with DiI labeling and other fluorescent methods of labeling cells participating in hemostasis, the actual dynamics of participation of cells in thrombus formation could be studied in real time. With the advent of fluorescent labeling of proteins it would be conceivable to study the interactions of the proteins *in vivo* in real time. Furthermore, with the availability of zebrafish with green fluorescent protein (GFP)-labeled endothelium, response of endothelial cells to the thrombus could now be attempted to see the *in vivo* signaling events attributed to thrombus formation.

With the vascular occlusion method developed here, it is possible to conduct larval screens for endothelium and blood components such as thrombocytes and other cells, as well as plasma procoagulant, anticoagulant, and fibrinolytic factors along with factors affecting flow dynamics. Therefore, this assay addresses all the components of Virchow's triad. The assay developed here should facilitate isolation of genes by either saturation mutagenesis (Solnica-Krezel *et al.*, 1994), by morpholino inhibition approaches (Nasevicius and Ekker, 2000), or by Targeting Induced Local Lesions IN Genomes (TILLING) methods (Wienholds *et al.*, 2002, 2003). We have hitherto performed screens for venous occlusion, and in the future other screens for arterial occlusion should be feasible. The dissolution of thrombus assay in both arteries and veins should provide additional screens for the future. Since the methods are now available to measure global hemostasis in larvae, it should be easy to screen for antithrombotic compounds by treating the zebrafish larvae with these compounds and performing TTO assays (Jagadeeswaran *et al.*, 1999b). Thus, the larval vascular occlusion assay presents a powerful tool to discover regulatory elements of blood flow and endothelial cells affecting hemostasis. In addition to the larval screens for thrombus formation, the ability to directly visualize lysis of the thrombus allows the design of assays for mutations affecting fibrinolysis and wound healing. Therefore, in summary, a genetic screen for the components of Virchow's triad is now possible and it is our hope that this may lead to hitherto unidentified components in the complex process of thrombosis. Has genetics ever failed to identify novel pathways?

References

Alhenc-Gelas, M., and Aiach, M. (2003). [Genetics of venous thromboembolism]. *Arch. Mal. Coeur. Vaiss.* **96**, 1111–1115.

Bohm, G., and Al-Khaffaf, H. (2003). Thrombophilia and arterial disease. An up-to-date review of the literature for the vascular surgeon. *Int. Angiol.* **22**, 116–124.

Cattaneo, M. (2003). Inherited platelet-based bleeding disorders. *J. Thromb. Haemost.* **1**, 1628–1636.

Davie, E. W. (2003). A brief historical review of the waterfall/cascade of blood coagulation. *J. Biol. Chem.* **278**, 50819–50832.

Doolittle, R. F. (1961). The Comparative Biochemistry of Blood Coagulation. PhD thesis, Division of Medical Sciences. Harvard University, Cambridge, MA.

Driever, W., and Fishman, M. C. (1996). The zebrafish: Heritable disorders in transparent embryos. *J. Clin. Invest.* **97**, 1788–1794.

Endler, G., and Mannhalter, C. (2003). Polymorphisms in coagulation factor genes and their impact on arterial and venous thrombosis. *Clin. Chim. Acta* **330**, 31–55.

Esmon, C. T. (1995). Thrombomodulin as a model of molecular mechanisms that modulate protease specificity and function at the vessel surface. *FASEB J.* **9**, 946–955.

Esmon, C. T. (2003). The protein C pathway. *Chest* **124**, 26S–32S.

Gregory, M., Hanumanthaiah, R., and Jagadeeswaran, P. (2002). Genetic analysis of hemostasis and thrombosis using vascular occlusion. *Blood Cells Mol. Dis.* **29,** 286–295.

Gregory, M., and Jagadeeswaran, P. (2002). Selective labeling of zebrafish thrombocytes: Quantitation of thrombocyte function and detection during development. *Blood Cells Mol. Dis.* **28,** 418–427.

Hanumanthaiah, R., Day, K., and Jagadeeswaran, P. (2002). Comprehensive analysis of blood coagulation pathways in teleostei: Evolution of coagulation factor genes and identification of zebrafish factor VIIi. *Blood Cells Mol. Dis.* **29,** 57–68.

Heit, J. A. (2003). Risk factors for venous thromboembolism. *Clin. Chest. Med.* **24,** 1–12.

Hrabe de Angelis, M. H., Flaswinkel, H., Fuchs, H., Rathkolb, B., Soewarto, D., Marschall, S., Heffner, S., Pargent, W., Wuensch, K., Jung, M., *et al.* (2000). Genome-wide, large-scale production of mutant mice by ENU mutagenesis. *Nat. Genet.* **25,** 444–447.

Jackson, S. P., Nesbitt, W. S., and Kulkarni, S. (2003). Signaling events underlying thrombus formation. *J. Thromb. Haemost.* **1,** 1602–1612.

Jagadeeswaran, P., Gregory, M., Johnson, S., and Thankavel, B. (2000a). Haemostatic screening and identification of zebrafish mutants with coagulation pathway defects: An approach to identifying novel haemostatic genes in man. *Br. J. Haematol.* **110,** 946–956.

Jagadeeswaran, P., Gregory, M., Zhou, Y., Zon, L., Padmanabhan, K., and Hanumanthaiah, R. (2000b). Characterization of zebrafish full-length prothrombin cDNA and linkage group mapping. *Blood Cells Mol. Dis.* **26,** 479–489.

Jagadeeswaran, P., and Liu, Y. C. (1997). A hemophilia model in zebrafish: Analysis of hemostasis. *Blood Cells Mol. Dis.* **23,** 52–57.

Jagadeeswaran, P., Liu, Y. C., and Sheehan, J. P. (1999a). Analysis of hemostasis in the zebrafish. *Methods Cell Biol.* **59,** 337–357.

Jagadeeswaran, P., and Sheehan, J. P. (1999). Analysis of blood coagulation in the zebrafish. *Blood Cells Mol. Dis.* **25,** 239–249.

Jagadeeswaran, P., Sheehan, J. P., Craig, F. E., and Troyer, D. (1999b). Identification and characterization of zebrafish thrombocytes. *Br. J. Haematol.* **107,** 731–738.

Jain, S. K. (1985). *In Vivo* externalization of phosphatidylserine and phosphatidylethanolamine in the membrane bilayer and hypercoagulability by the lipid peroxidation of erythrocytes in rats. *J. Clin. Invest.* **76,** 281–286.

Justice, M. J. (2000). Capitalizing on large-scale mouse mutagenesis screens. *Nat. Rev. Genet.* **1,** 109–115.

Kurz, K. D., Main, B. W., and Sandusky, G. E. (1990). Rat model of arterial thrombosis induced by ferric chloride. *Thromb. Res.* **60,** 269–280.

Levi, M. M., Vink, R., and de Jonge, E. (2002). Management of bleeding disorders by pro hemostatic therapy. *Int. J. Hematol.* **76**(Suppl. 2), 139–144.

Mann, K. G. (1999). Biochemistry and physiology of blood coagulation. *Thromb. Haemost.* **82,** 165–174.

Medved, L., and Nieuwenhuizen, W. (2003). Molecular mechanisms of initiation of fibrinolysis by fibrin. *Thromb. Haemost.* **89,** 409–419.

Nasevicius, A., and Ekker, S. C. (2000). Effective targeted gene 'knockdown' in zebrafish. *Nat. Genet.* **26,** 216–220.

Nesheim, M. (2003). Thrombin and fibrinolysis. *Chest.* **124,** 33S–39S.

Nolan, P. M., Peters, J., Strivens, M., Rogers, D., Hagan, J., Spurr, N., Gray, I. C., Vizor, L., Brooker, D., Whitehill, E., *et al.* (2000). A systematic, genome-wide, phenotype-driven mutagenesis programme for gene function studies in the mouse. *Nat. Genet.* **25,** 440–443.

Patton, E. E., and Zon, L. I. (2001). The art and design of genetic screens: Zebrafish. *Nat. Rev. Genet.* **2,** 956–966.

Ritchie, J. L., Alexander, H. D., Allen, P., Morgan, D., and McVeigh, G. E. (2002). Effect of nitric oxide modulation on systemic haemodynamics and platelet activation determined by P-selectin expression. *Br. J. Haematol.* **116,** 892–898.

Rosen, E. D., Raymond, S., Zollman, A., Noria, F., Sandoval-Cooper, M., Shulman, A., Merz, J. L., and Castellino, F. J. (2001). Laser-induced noninvasive vascular injury models in mice generate platelet- and coagulation-dependent thrombi. *Am. J. Pathol.* **158,** 1613–1622.

Ruggeri, Z. M. (2002). Platelets in atherothrombosis. *Nat. Med.* **8,** 1227–1234.

Schafer, A. I., Levine, M. N., Konkle, B. A., and Kearon, C. (2003). Thrombotic disorders: Diagnosis and treatment. *Hematology (Am. Soc. Hematol. Educ. Program)* 520–539.

Setty, B. N., Kulkarni, S., and Stuart, M. J. (2002). Role of erythrocyte phosphatidylserine in sickle red cell-endothelial adhesion. *Blood* **99,** 1564–1571.

Sheehan, J., Templer, M., Gregory, M., Hanumanthaiah, R., Troyer, D., Phan, T., Thankavel, B., and Jagadeeswaran, P. (2001). Demonstration of the extrinsic coagulation pathway in teleostei: Identification of zebrafish coagulation factor VII. *Proc. Natl. Acad. Sci. USA* **98,** 8768–8773.

Solnica-Krezel, L., Schier, A. F., and Driever, W. (1994). Efficient recovery of ENU-induced mutations from the zebrafish germline. *Genetics* **136,** 1401–1420.

Stuart, J., and Johnson, C. S. (1987). Rheology of the sickle cell disorders. *Baillieres Clin. Haematol.* **1,** 747–775.

Virchow, R. (1856). "Gesammelte Abhandlungen zur wissenschaftlichen Medicin." Medinger Sohn & Co., Frankfurt.

Wienholds, E., Schulte-Merker, S., Walderich, B., and Plasterk, R. H. (2002). Target-selected inactivation of the zebrafish rag1 gene. *Science.* **297,** 99–102.

Wienholds, E., van Eeden, F., Kosters, M., Mudde, J., Plasterk, R. H., and Cuppen, E. (2003). Efficient target-selected mutagenesis in zebrafish. *Genome Res.* **13,** 2700–2707.

CHAPTER 23

Zebrafish Kidney Development

Iain A. Drummond

Department of Medicine, Harvard Medical School and Renal Unit
Massachusetts General Hospital
Charlestown, Massachusetts 02129

I. Introduction

Kidney development is characterized by the formation of epithelial tubules from mesodermal mesenchymal cells and a subsequent series of interactions of the formed epithelia with vascular tissue to produce an organ that filters blood and regulates body fluid composition. In the course of vertebrate evolution, three distinct kidneys of increasing complexity have been generated: the pronephros, mesonephros, and metanephros (Saxén, 1987). The pronephros is the first kidney to form during embryogenesis. In vertebrates with free-swimming larvae, including amphibians and teleost fish, the pronephros is the functional kidney of early larval life (Howland, 1921; Tytler, 1988; Tytler et al., 1996; Vize et al., 1997) and is required for proper osmoregulation (Howland, 1921). Later, in juvenile stages of zebrafish development, a mesonephros forms around and along the length of the pronephros and later serves as the kidney of adult fish. Despite some differences in organ morphology between the various kidney forms, many common elements exist at the cellular and molecular level that can be exploited to further our understanding of epithelial and vascular differentiation in particular, and organogenesis in general. To date, the zebrafish pronephros has provided a useful model of nephrogenic mesoderm differentiation, kidney cell type differentiation, nephron patterning, kidney: vasculature interactions, glomerular function, and diseases affecting tubule lumen size (i.e., cystic kidney disease). While much remains to be done, the basic features of zebrafish pronephric development and patterning have emerged from studies using simple histology, cell lineage tracing, gene expression patterns, and analysis of zebrafish mutants affecting this process.

II. Pronephric Structure and Function

The kidney has two principal functions: to remove waste from the blood and to balance ion and metabolite concentrations in the blood within physiological ranges that support proper functioning of all other cells (Vize et al., 2002). Kidney function is achieved largely by first filtering the blood and then recovering useful ions and small molecules by directed epithelial transport. This work is performed by *nephrons*, the functional unit of the kidney (Fig. 1). Distinct components or segments of the nephron are specialized for blood filtration and specific solute transport activities. The *glomerulus* is the site of blood filtration and primary urine formation. It is composed of specialized epithelial cells called *podocytes* that form a basket-like extension of cellular processes around a *capillary tuft*. The basement membrane between podocytes and capillary endothelial cells and the specialized junctions between the podocyte cell processes (*slit diaphragms*) function as a blood filtration barrier, allowing passage of small molecules, ions, and blood fluid into the urinary space, while retaining high molecular weight proteins in the vascular system. Blood filtrate is collected first in the epithelial capsule surrounding the capillary tuft (the *nephrocoel* or *Bowman's capsule*) and then processed as it travels down the lumen of the kidney tubules and ducts, where segment specific transport processes recover useful ions and metabolites (Fig. 1; and Vize et al., 2002).

In general, osmoregulation by the fresh water adult teleost kidney is achieved by a high rate of blood filtration, active recovery of essential salts and other blood solutes, and the excretion of copious amounts of dilute urine (Hickman and Trump, 1969). The pronephros is necessary for larval osmoregulation since most mutants with pronephric defects die of edema. Most texts on fish physiology state that nitrogenous waste removal from the blood is not a major function of the fish kidney and is instead carried out primarily by ammonia transport mechanisms in the gills (Hickman and Trump, 1969). Since newly hatched zebrafish larvae do not yet have gills, it is worth considering that the pronephros may also be the primary site of nitrogenous waste removal in the first days of life.

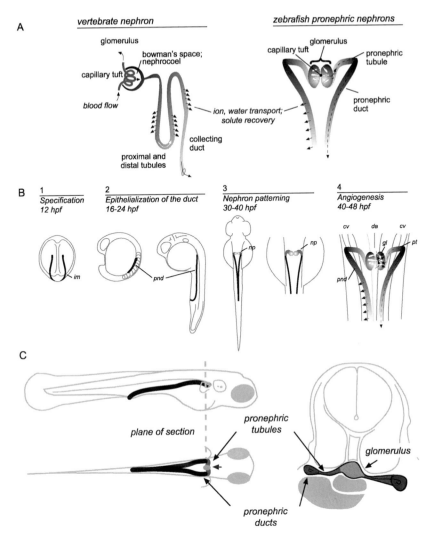

Fig. 1 The zebrafish pronephros. (A) Functional features of the vertebrate nephron and the zebrafish pronephric nephrons. See text for details. (B) Stages in zebrafish pronephric kidney development. (1) Specification of mesoderm to a nephric fate: expression patterns of *pax2.1* and *lim-1* define a posterior region of the intermediate mesoderm (*im*) and suggest that a nephrogenic field is established in early development. (2) Epithelialization of the pronephric duct (*pnd*) follows somitogenesis and is complete by 24 hpf. (3) Patterning of the nephron primordia (*np*) gives rise to the pronephric glomerulus (*gl*) and pronephric tubules (*pt*). (4) Angiogenic sprouts from the dorsal aorta (*da*) invade the glomerulus and form the capillary loop. The cardinal vein (*cv*) is apposed to the tubules and duct and receives recovered solutes. (C) Diagram of the mature zebrafish pronephric kidney in 3 day larva. A midline compound glomerulus connects to the pronephric tubules that run laterally and drain into the pronephric ducts. The ducts are joined at the cloaca, where they communicate with the exterior.

In zebrafish, and several other teleosts, the larval pronephros consists of only two nephrons with glomeruli fused at the embryo midline just ventral to the dorsal aorta (Fig. 1C; Agarwal and John, 1988; Armstrong, 1932; Balfour, 1880; Drummond, 2000; Drummond et al., 1998; Goodrich, 1930; Hentschel and Elger, 1996; Marshall and Smith, 1930; Newstead and Ford, 1960; Tytler, 1988; Tytler et al., 1996). Although simple in form, the pronephric glomerulus is composed of cell types typical of higher vertebrates' kidneys including fenestrated capillary endothelial cells, podocytes, and polarized tubular epithelial cells (Drummond et al., 1998). Two pronephric tubules connect the glomerulus to the pronephric ducts, which run caudally and fuse just before their contact with the outside world at the cloaca. The zebrafish pronephric nephrons form a closed system of blood filtration, tubular resorption, and fluid excretion.

III. Pronephric Development

During the first two days of zebrafish embryonic development, the pronephric kidney forms in a stepwise fashion with sequential addition of the three principal components: the pronephric ducts, tubules, and glomerulus. The pronephric ducts form first and are complete by 24 hpf (Kimmel et al., 1995). The pronephric tubules form slightly later between 30 hpf and 40 hpf and finally the glomerulus becomes a functional blood filter by 48 hpf (Drummond et al., 1998). This stepwise addition of nephron components echoes a general scheme of vertebrate kidney development. Formation of the kidney can be conceptualized in four broadly defined stages: (1) the commitment of undifferentiated mesodermal cells to a nephrogenic fate, (2) the epithelialization and growth of the pronephric duct, (3) the induction and formation of nephrons and the signaling associated with patterning the nephron to form the glomerulus and the different tubular cell types, and (4) the formation of the glomerular capillary tuft by ingrowing endothelial cells and the onset of blood filtration and nephron function (Fig. 1B).

A. Patterning the Intermediate Mesoderm

1. Origins and Patterning of Nephrogenic Mesoderm

Cell labeling and lineage tracing in zebrafish gastrula stage embryos have demonstrated that cells destined to form the pronephros lie just dorsal to the heart progenitors in the shield stage mesodermal germ ring and appear to overlap somewhat with cells fated to form blood (Fig. 2A; Kimmel et al., 1990). This position in the prospective mesoderm is similar to the region defined in amphibian embryos as the source of kidney tissue (Delarue et al., 1997). These cells emerge shortly after gastrulation as a band of tissue, the intermediate mesoderm, at the ventro-lateral edge of the paraxial mesoderm (Fig. 2B and C). Although no data are currently available on when pronephric cells are specified in zebrafish embryos, it is likely that signaling events occurring at this early

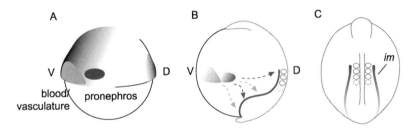

Fig. 2 Origins of the intermediate mesoderm. (A) Approximate positions of cells in a shield stage embryo destined to contribute to the blood/vasculature and pronephric lineages in the ventral (V) germ ring. (D; dorsal shield). (B) migration of cells during gastrulation to populate the intermediate mesoderm (*im*) (C).

stage of development have a major impact on all subsequent differentiation of the pronephros.

The pronephric mesoderm at the blastula and gastrula stages is patterned by morphogen gradients in the early embryo, which are dominated by bone morphogenetic protein (BMP) expression on the ventral side of the embryo and β-catenin/BMP inhibitor expression in the dorsal embryonic shield (Fig. 3A; Hammerschmidt *et al.*, 1996b; Neave *et al.*, 1997; Nikaido *et al.*, 1997). The dorsalized mutants *swirl* (*bmp2b*), *snailhouse* (*bmp7*), *somitabun* (*smad5*), and *lost-a-fin* (*alk8*) all lack signals necessary for ventral mesodermal development and show a reduction or elimination of *pax2.1* positive, presumptive kidney intermediate mesoderm (Hild *et al.*, 1999; Kishimoto *et al.*, 1997; Mullins *et al.*, 1996; Nguyen *et al.*, 1998). Conversely, the ventralized mutant *chordino* lacks the dorsal shield determinant Chordin and shows an expansion of the *pax2.1* expression domain at later stages (Hammerschmidt *et al.*, 1996b). The data from these mutants suggest that BMP signaling, in concert with signals from the shield region, is required to specify kidney mesoderm.

The steps in pronephric mesoderm patterning subsequent to early BMP signaling are less well characterized. However, it is clear that further development of the lateral posterior mesoderm is critically dependent on two functionally related sets of genes: T-box transcription factors and fibroblast growth factors. The T-box transcription factor genes, *notail (ntl), spadetail (spt)*, and *tbx6* function in concert to pattern the trunk and tail mesoderm (Griffin *et al.*, 1998; Hammerschmidt *et al.*, 1996a; Ho and Kane, 1990). In shield stage embryos, all three of these genes are expressed in the ventral and lateral germ ring (Griffin *et al.*, 1998). The *spadetail* and *tbx6* are later expressed broadly in the hypoblast (4- to 6-somite stage) out to the lateral edge of the forming mesoderm and are more specifically involved in trunk development (Griffin *et al.*, 1998). Some work has demonstrated that embryos lacking both *notail* and *spadetail* function do not express the pronephric mesoderm marker *pax2.1* (discussed later) and then show no signs of pronephric tubule development (Amacher *et al.*, 2002). This argues for a critical role for *notail* and *spadetail* in pronephric mesodermal patterning.

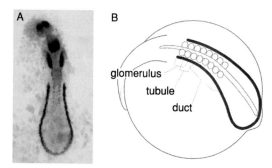

Fig. 3 Derivation of the pronephros from the intermediate mesoderm. (A) The *pax2.1* expression domain in early somitogenesis stage embryos defines a stripe of intermediate mesoderm fated to become the pronephric epithelia. (B) Fate map of the nephrogenic intermediate mesoderm derived from fluorescent dye uncaging lineage experiments.

A dramatic loss of all ventral-posterior embryonic structures is observed in embryos expressing a dominant negative fibroblast growth factor (FGF) receptor (Griffin *et al.*, 1995, 1998). This is most likely attributed to the requirement for FGF signaling in the initiation or maintenance of *notail, spadetail,* and *tbx6* gene expression (Griffin *et al.*, 1995, 1998). In zebrafish, two FGFs, *fgf8* and *fgf24* function redundantly to promote posterior mesoderm development by maintaining *ntl* and *spt* expression (Draper *et al.*, 2003). In these experiments, a combined loss of *fgf8* and either *notail* or *spadetail*, which would be expected to substantially reduce T-box gene activity, did not reduce *pax2.1* expression in the pronephric mesoderm while axial mesoderm was severely affected (Draper *et al.*, 2003). It may be that pronephric mesodermal development requires only low levels of T-box gene function.

Another secreted factor found to influence trunk development and T-box gene expression is Wnt8 (Lekven *et al.*, 2001). The *wnt8* gene is expressed in the germ ring at gastrulation, transiently in the region of the intermediate mesoderm just before somatogenesis, and in a more prolonged fashion in the tail bud (Kelly *et al.*, 1995). Deletion mutants of the bicistronic *wnt8* locus show significant reduction in *tbx6* expression and loss of ventral-posterior mesoderm (Lekven *et al.*, 2001). Inhibition of *wnt8* mRNA translation in *ntl* and *spt* mutant embryos using injected morpholino antisense oligonucleotides has synergistic effects resulting in more profound loss of ventral posterior mesoderm (Lekven *et al.*, 2001). Therefore, trunk development is mediated by at least two secreted signals, *fgf8/24* and Wnt8, which control T-box gene expression and subsequent patterning of trunk mesoderm.

2. Pronephric Nephron Cell Lineage

By the early stages of somite formation, the pronephric intermediate mesoderm is clearly defined by specific pronephric marker gene expression. The Wilms' tumor suppressor, *wt1*, encodes a zinc finger transcriptional regulator that is required for normal mouse kidney development and, when mutated in humans, gives rise to pediatric kidney tumors. Pax-2, a member of the paired-domain-containing homeobox transcriptional regulators, has also been shown to be essential for kidney development in mammals, as well as in zebrafish (as discussed later). Sim-1 is the vertebrate homolog of the Drosophila transcription factor gene *simple-minded*. Based on the overlapping but distinct expression patterns of the zebrafish homologs *wt1, pax2.1*, and *sim1* in the zebrafish intermediate mesoderm (IM), Serluca and Fishman (2001) performed lineage studies by uncaging fluorescent dyes in the expression domains of these three genes before epithelialization. They showed that the areas of the IM fated to become glomerulus, tubule and duct could be defined as sequential anterior to posterior subdomains of the IM that roughly correspond to the expression domains of *wt1, pax2.1* and *sim1* respectively (see Fig. 3; Serluca and Fishman, 2001). In terms of gene function, *pax2.1* has been shown to be required for zebrafish pronephric tubule formation (as discussed later) while currently no data exist on the function of *wt1* and *sim1* in zebrafish. In zebrafish, the mesenchyme to epithelial transition appears to occur *in situ* at all levels of the intermediate mesoderm (Serluca and Fishman, 2001). Ablation of cells within the IM does not result in subsequent absence of kidney tissue, suggesting the existence of a kidney morphogenetic field that can regulate, presumably under the influence of local environmental signals and downstream transcriptional circuits.

3. Gene Expression in the Intermediate Mesoderm

The expression of known signaling molecules and transcription factors in the forming pronephros may provide some entrance points to pathways that regulate early pronephric development. The zebrafish frizzled gene, *frz8a*, presumably encoding a receptor for a locally acting Wnt signal, is expressed throughout the intermediate mesoderm at the 5- to 6-somite stage (Kim *et al.*, 1998). The timing of expression *frz8a* is significantly later than wnt8 expression (as already discussed) suggesting that a different Wnt ligand may signal via *frz8a*. Whereas wnt4 has been reported to be weakly expressed at about this time in bilateral anterior stripes of IM (Ungar *et al.*, 1995), we have not been able to confirm this (Liu *et al.*, 2000; Drummond, unpublished observation, 2000). The *deltaC*, a ligand for notch, is expressed in the IM adjacent to somites 1 to 4 at the 7-somite stage, suggesting that notch signaling may play a role in patterning the pronephric primordium (Smithers *et al.*, 2000). This expression domain corresponds to the precursors of the glomerulus and tubules and extends anterior of the *pax2.1* expression domain (see Fig. 3; Serluca and Fishman, 2001; Smithers *et al.*, 2000). The *foxc1a* gene, encoding a member of the forkhead/winged helix transcription factor family, is expressed strongly in forming somites and also in the IM adjacent to somites 1 to 4 at the 7-somite stage (Topczewska *et al.*, 2001b). Morpholino inhibition of *foxc1a* expression results in loss of *deltaC* positive anterior kidney mesoderm while more posterior *pax2.1* positive mesoderm is unaffected (Topczewska *et al.*, 2001a). The *foxc1a* most likely acts in the anterior pronephric mesoderm and might be expected to play a role in glomerular or tubular development, although this has not been experimentally confirmed. Several other transcription factors are also expressed in the presumptive pronephric mesoderm. As already noted, the paired-domain transcription factor *pax2.1* is expressed in a continuous band of IM from the cloaca up to the posterior boundary of somite 2 (Carroll *et al.*, 1999; Drummond, 2000; Heller and Brandli, 1999; Krauss *et al.*, 1991; Majumdar *et al.*, 2000; Mauch *et al.*, 2000; Pfeffer *et al.*, 1998; Puschel *et al.*, 1992). A second paired-domain transcription factor, pax8, is expressed in a similar pattern. The *pax8* expression is initiated in the IM during gastrulation, slightly earlier than *pax2.1* expression (Pfeffer *et al.*, 1998). The *lim1* gene is expressed during early somitogenesis at all anterior-posterior (A-P) levels of the pronephric mesoderm extending from somite 1 to the future cloaca (Toyama and Dawid, 1997). Posterior *lim1* expression is downregulated by the 12-somite stage while anterior expression (adjacent to somites 1 and 2) just rostral the *pax2.1* expression domain persists until 24 hpf (Fig. 4; Toyama and Dawid, 1997). The lim-1 deficient mouse embryos completely lack the genitourinary tract (Shawlot and Behringer, 1995). The early and extensive expression of *lim1* in the IM suggests that it may play a similarly important role in zebrafish kidney development. The Wilms tumor suppressor gene, *wt1*, is expressed beginning at the 2- to 3-somite stage and later, at the 8-somite stage, in the IM extending from somites 1 to 4 (Serluca and Fishman, 2001). In the mature pronephros, *wt1* is expressed exclusively in podocytes

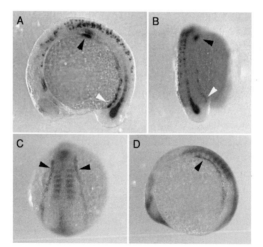

Fig. 4 Gene expression in the intermediate mesoderm. (A and B) *lim1* expression in the anterior pronephric nephron primordium (*black arrows* in A and B) and the posterior pronephric ducts near the forming cloaca (*white arrows* in A and B) at the 15-somite stage. The *lim1* is also expressed in the nervous system and the notochord. (C and D) Est ibd2750 expression marks the entire intermediate mesoderm at the 8-somite stage (*arrows*), as well as the forming somites.

(Drummond *et al.*, 1998; Majumdar and Drummond, 1999, 2000; Majumdar *et al.*, 2000; Serluca and Fishman, 2001).

Along with the expression of these identified genes, data from est *in situ* expression pattern screens is generating a wealth of new intermediate mesoderm markers and potential gene targets for further functional studies (Kudoh *et al.*, 2001; Thisse *et al.*, 2001). Data from two *in situ* screens of this type are available online and can be searched by organ type for candidate markers/genes (ZFIN, 2004). As an example, the novel est ibd2760 (Kudoh *et al.*, 2001) is expressed in the somites and the IM at the 8-somite stage (Fig. 4).

4. Adjacent Tissues and Kidney Cell Specification

In addition to the pronephros, the IM also gives rise to the blood and endothelial cells of the major trunk blood vessels (Detrich *et al.*, 1995; Gering *et al.*, 1998; Horsfield *et al.*, 2002; Weinstein *et al.*, 1996). At the 3- to 5-somite stage, the IM expresses the stem cell leukemia gene (*scl*), a basic helix-loop-helix transcription factor essential for blood cell development in both mouse and fish. The *scl* is expressed most strongly in the medial IM, while kidney markers *pax2.1* and *cdh17* are expressed only in the more lateral IM (Fig. 5; Gering *et al.*, 1998; Horsfield *et al.*, 2002). The formation of distinct kidney and blood lineage cell layers soon after marker gene expression is first detectable suggests that IM cell specification may occur before *pax2.1* and *scl* expression (Davidson *et al.*, 2003). Within the vascular lineage alone, cell fate may be determined in the IM well before major vessel formation, since cells labeled in the IM later give rise to the midline vein or aorta but not both (Zhong *et al.*, 2001). On the other hand, cells of the IM also appear to retain some plasticity in cell fate. Over-expression of *scl* causes excess

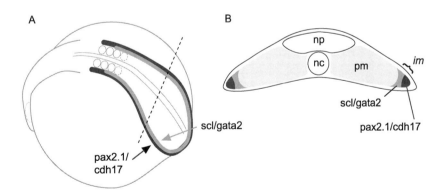

Fig. 5 Blood and kidney progenitor cells in the intermediate mesoderm. (A) During the early stages of somitagenesis, distinct bands of intermediate mesoderm are formed, expressing kidney (*pax2.1/cdh17*) and blood/vascular (*scl/gata2*) markers. Cross-section diagram (B) of lateral nephrogenic mesoderm and more medial blood/vascular mesoderm. *np*, neural plate; *nc*, notochord; *pm*, presomitic mesoderm.

blood cell development at the expense of somite and pronephric duct cell development (Gering *et al.*, 1998; Horsfield *et al.*, 2002). These studies indicate that forced
expression of blood lineage regulators can transfate cells normally destined to
become the kidney. It remains to be determined at what stage genes involved in
specifying the fates of the intermediate mesoderm act and when the commitment
to kidney fate occurs.

Experiments in the chick and frog have suggested that tissues neighboring the
IM may play a role in inducing or maintaining *pax2* expression in the IM and
driving subsequent kidney development (Mauch *et al.*, 2000; Seufert *et al.*, 1999).
In zebrafish, expression of *pax2.1* in the IM and later development of the kidney
occurs independently of the notochord and sonic hedgehog signaling, and does
not require the proper differentiation of the endoderm. These conclusions are
based on studies of *floating head, sonic-you, you-too,* and *one-eyed pinhead* mutant
embryos, where *pax2.1* expression or pronephric duct morphology, if not perfectly
normal, is still observed (Majumdar and Drummond, 2000; Drummond, unpublished observations, 1998). The *spadetail* embryos lacking proper trunk somite
formation still express *pax2.1* in the IM and form the pronephric duct epithelium
(Drummond, unpublished observations, 1998). These observations again suggest
that specification of the kidney IM in zebrafish may occur relatively early, perhaps
as early as gastrulation, and proceed independently of the development of other
axial and paraxial mesoderm.

B. Development of the Pronephric Duct

1. Epithelialization and Tubule Formation

Epithelial tubule formation is a central feature of virtually every organ including the vasculature, the gut, and the liver. Development of the pronephric ducts is
initiated at the boundary of somite 2 and 3 on both sides of the embryo and
temporally follows the anterior to posterior progression of somitogenesis (see
Fig. 1B; Kimmel *et al.*, 1995). By 24 hpf, the duct is complete, existing as an
epithelial tube just dorsal the to yolk extension, from somite 3 to the cloaca
(Drummond *et al.*, 1998; Kimmel *et al.*, 1995). In zebrafish the duct forms
primarily through recruitment of intermediate mesoderm at all axial levels followed by *in situ* epithelial differentiation. Lineage tracing experiments have not
detected any contribution from caudally migrating cells (Serluca and Fishman,
2001). Pronephric duct formation is mediated by a mesenchyme to epithelial
transformation, a process central to kidney formation in all vertebrates (Saxén,
1987). By the end of this transition, the epithelial cells are polarized with an apical
brush border and a basolateral membrane domain containing ion transport
proteins that are essential for the osmoregulatory function of the duct (Fig. 6A;
Drummond *et al.*, 1998). Of interest, duct cells are not morphologically homogeneous. Single, multiflagellated cells are dispersed along the length of the duct,
appearing side by side with the more typical duct cuboidal epithelial cells. The
function of these cells appears to be to drive fluid down the duct by beating their

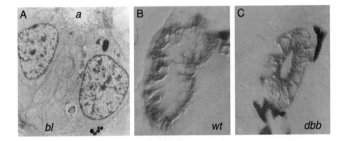

Fig. 6 Epithelial cell polarity in the pronephric duct. (A) Electron micrograph of 2.5-day pronephric duct epithelial cells showing apical (*a*) brush border and basolateral cell surfaces and infoldings. Polarized distribution of the NaK ATPase in 2.5-day pronephric duct epithelial cells visualized by the alpha6F monoclonal antibody. The apical cell surface is devoid of staining while staining is strong on the basolateral cell surface and membrane infoldings (B). Double bubble mutant embryos (C) aberrantly express the NaK ATPase on the apical cytoplasm and cell membrane.

flagella in a corkscrew motion parallel to the axis of the lumen (Kramer-Zucker and Drummond, personal observation, 2004). The pronephric duct forms the basis of the collecting system to which all embryonic and future adult nephrons will attach (Saxén, 1987).

Formation of cell-cell junctions is an essential step in separating apical and basolateral membrane domains and giving an epithelium its vectorial property. Cadherins are the major proteins of the adherens junction that maintain the integrity of epithelial sheets and separate apical and basolateral membrane domains. Cadherin 17 is specifically expressed in the zebrafish IM and later in the pronephric duct epithelium (Horsfield *et al.*, 2002). Loss-of-function studies using antisense morpholino oligonucleotides show that cadherin 17 is essential for duct morphogenesis. Cadherin17-null duct epithelia show a general loss of the normally tight cell-cell adhesion, failure of the bilateral ducts to fuse at the cloaca, and gaps between epithelial cells (Horsfield *et al.*, 2002). Cytoplasmic polarity, in this case the relative placement of nucleus to cytoplasm, is also disrupted. Although it has been thought that cadherins are likely to be essential for mammalian kidney tubule formation, cadherin loss-of-function studies in the mouse kidney have not yielded as clear a phenotype (Dahl *et al.*, 2002). It may be that fewer members of the cadherin gene family are expressed in the more primitive pronephric kidney, potentially avoiding the issue of redundancy in gene function.

Cell polarity and proper targeting of membrane transporters is essential for proper kidney ion transport and function. Several zebrafish mutants have been found to mistarget the NaK ATPase in the tubules and duct from its normal basolateral membrane location to the apical membrane (Fig. 6; Drummond *et al.*, 1998). The activity of the NaK ATPase provides the motive force for many other coupled transport systems (Seldin and Giebisch, 1992). Its mislocalization suggests that severe problems in osmoregulation exist in these mutants. In fact, as discussed later, these mutants later develop cysts in the pronephric tubule and the embryos eventually die of edema (Drummond et al., 1998).

2. Gene Expression in the Pronephric Duct

In addition to the continued expression of pax2.1, pax8, sim1, and to a lesser extent lim1, new expression of the ret1 gene is observed in the forming pronephric duct (Bisgrove et al., 1997; Marcos-Gutierrez et al., 1997). The ret1 is of interest because it encodes a tyrosine kinase cell surface receptor for glial cell-line-derived neurotrophic factor (GDNF), a TGF-β superfamily member that is essential for normal mouse kidney development (Marcos-Gutierrez et al., 1997; Moore et al., 1996; Pichel et al., 1996; Sanchez et al., 1996; Schuchardt et al., 1994, 1995; Shepherd et al., 2001). At 24 hpf, when duct epithelialization is complete, ret1 expression marks the most posterior quarter of the pronephric duct (Marcos-Gutierrez et al., 1997). The function of the ligand GDNF has been studied using antisense oligonucleotides and found to be required for enteric neuron development but dispensable for pronephric development at least at the level of pronephric duct morphology (Shepherd et al., 2001). Other genes implicated in developmental signaling pathways have been shown to be expressed in the pronephric duct. A retinoic-acid response element/green fluorescent protein reporter transgene shows significant activity in the pronephric duct as early as the 18-somite stage (Perz-Edwards et al., 2001). Whether retinoids are involved in cell fate decisions in the pronephros or are only required to support terminal differentiation of pronephric duct epithelial cells is currently unknown. Dietary deficiency of vitamin A and mutations in retinoic acid receptors are both sufficient to severely perturb mouse metanephric development (Batourina et al., 2001; Burrow, 2000; Mendelsohn et al., 1999; Wilson et al., 1953). Retinoic acid (RA)-treated frog embryos show an expansion of the xlim-1 expression domain and in some cases, larger pronephric tubules (Taira et al., 1994). Loss of function of the major RA-synthesizing enzyme RALDH2 in zebrafish neckless mutants (Begemann et al., 2001) does not eliminate pronephric pax2.1 expression (P. Ingham, personal communication, 2000), although other effects on pronephric development remain to be determined.

C. Nephron Differentiation and Segmentation

Nephron segmentation is a general feature of vertebrate kidneys, where osmoregulatory function is dependent on an ordered array of different transporters acting sequentially along the length of the nephron (Seldin and Giebisch, 1992). Nephron segments are also morphologically different, and are composed of distinct epithelial cell types. Although very little is currently known about nephron segmentation in the zebrafish kidney, the linear, anterior-posterior orientation of the zebrafish pronephros makes it an attractive and simplified model system for studies of nephron segmentation and patterning.

1. The Glomerulus-Pronephric Tubule Boundary

Developmental partitioning of the zebrafish pronephros divides it into three structural and functional units: glomerulus, tubules, and ducts (see Fig. 1). Here one of the clearest nephron boundaries exists between the glomerulus and the pronephric tubule. The paired domain transcription factor *pax-2.1* is now known to play a key role in establishing this boundary. The *pax-2.1* is expressed in the early IM and also later, at the time of pronephric tubule formation, in the anterior duct and tubule (Krauss *et al.*, 1991; Majumdar *et al.*, 2000). Fish with mutations in the *pax2.1* gene (*no isthmus, noi*) show a complete absence of the tubule component of the nephron while the glomerulus and duct remain intact (Majumdar *et al.*, 2000). The consequence of loss of *pax2.1* function is even greater when the expression of glomerulus specific genes such as *wt1* and vascular endothelial growth factor (VEGF) is monitored. While in normal embryos the expression of these genes is limited to podocytes (Majumdar *et al.*, 2000; Serluca and Fishman, 2001), in *noi* embryos expression of these podocyte markers is expanded into the anterior pronephric duct epithelial cells (Fig. 7). Although these anterior duct cells maintain an epithelial character (no transdifferentiation to a podocyte morphology is observed), they fail to express normal markers of the duct (NaK ATPase and 3G8,

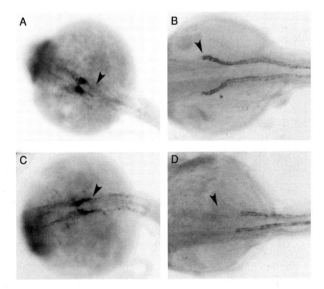

Fig. 7 Formation of the glomerulus-tubule boundary is disrupted in *no isthmus* (*noi; pax2.1*) mutants. Whole-mount *in situ* hybridization with *wt1* marks the presumptive podocytes in wild-type embryos (A). The *wt1* expression is caudally expanded into the anterior pronephric ducts in *noi*[tb21] mutant embryos (C) at 24 hpf (*arrow* in C). Whole-mount antibody staining of wild-type (B) and *noi*[tb21] (D) embryos with mAb •-6F mAb which recognizes a Na[+]/K[+] ATPase ••-subunit. •-6F marks the pronephric duct in wild type (arrows in B) and anterior duct Na[+]/K[+] ATPase expression is missing in *noi*[-/tb21] mutant embryos at 2.5 dpf (D).

a brush border marker), while more posterior duct cells continue to express these two markers despite the loss of *pax2.1* function (Majumdar *et al.*, 2000). The data suggest that *pax2.1* is not only essential for tubule development, but that it also plays an important role in defining the tubule/podocyte boundary by repressing podocyte-specific genes in the anterior duct. It is conceivable that podocyte transcription factors, perhaps including *wt1*, play a similar role in repressing tubule-specific genes in podocytes (Yang *et al.*, 1999). Mutually antagonistic signals acting in concert to define a tissue boundary is a prevalent concept in developmental biology (Briscoe *et al.*, 2000; Li and Joyner, 2001; Schwarz *et al.*, 2000) and is also likely to underlie the patterning of the kidney nephron.

2. Nephron Segments in the Pronephric Tubule and Duct

Although uniform in appearance, the pronephric tubule and duct epithelia can be further subdivided into distinct segments identified on the basis of cell morphology and the expression of membrane transporters. In the mammalian kidney, the proximal tubule is composed of epithelial cells with apical surfaces covered with dense microvillar projections (i.e., a brush border), while in more distal tubule segments, a dense brush border is absent (Vize *et al.*, 2002). An antibody raised against rabbit smooth muscle actin cross reacts with the zebrafish pronephros brush border, presumably recognizing an isoform of actin (Fig. 8C). The monoclonal antibody 3G8, raised against the frog pronephric tubules, also stains the apical brush border in the zebrafish pronephros (Fig. 8B; Drummond, unpublished results, 1998). Both of these antibodies stain only the anterior half of the pronephric duct and also the pronephric tubules. This suggests that (1) the posterior duct lacks a brush border like the mammalian distal nephron segments and (2) the anterior half of the pronephric duct, although differing in developmental origins from the pronephric tubule, may share some functional characteristics with the pronephric tubule.

In addition to morphological differences, membrane transporter expression differs in pronephric epithelial segments (Fig. 9). The anterior pronephric tubule and duct specifically express several transporters associated with acid/base homeostasis: the chloride/bicarbonate anion exchanger AE 2, the sodium/bicarbonate cotransporter NBC1, and the sodium/hydrogen exchanger NHE (Fig. 9A; Zhao and Drummond unpublished, 2004; Davidson *et al.*, unpublished, 2004). Other transporters including the sulphate anion transporter are also expressed in this segment (Davidson *et al.* unpublished, 2004). The expression of these transporters indicates that the anterior duct and tubule may be specialized for tissue fluid pH regulation. It also suggests that the tubule functions equivalent to the mammalian proximal tubule and may extend from the pronephric tubule into the anterior pronephric duct in zebrafish. The mid-segment of the zebrafish pronephric duct expresses *slc12a1* (Na-K-Cl symporter; Fig. 9D), which in mammals is most highly expressed in the distal tubules (loop of Henle, the thick ascending limb, and the distal convoluted tubules). Other markers of the mid-segment include an

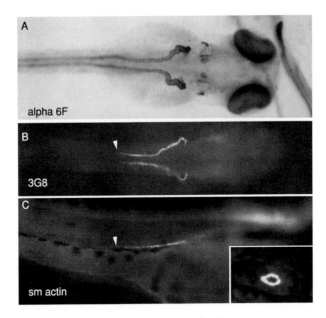

Fig. 8 Segmentation of the pronephric duct. The Na$^+$/K$^+$ ATPase ●●-subunit (A) is expressed uniformly along the pronephric tubules and ducts in a 2.5 dpf embryo as detected by immunohistochemistry with the alpha6F monoclonal antibody. The 3G8 monoclonal antibody stains only the anterior half of the pronephric duct in 2.5-dpf embryos (B). An antibody against rabbit smooth muscle actin (clone 1A4; Sigma, St. Louis, MO) also stains the anterior half of the pronephric duct (C). This monoclonal specifically stains the brush border of pronephric duct epithelial cells (C, inset: cross section of the pronephric duct showing intense apical staining).

aspartoacylase homolog and the zebrafish starmaker gene (Fig. 9; Hukreide *et al.*, personal communication; Sollner *et al.*, 2003; ZFIN, 2004). Finally, the most posterior or distal pronephric nephron segment expresses an est encoding a putative ABC transporter and also the zebrafish c-ret homolog *ret1*. In mammals, c-ret expression is observed in the collecting system, the terminal tubule type associated with water transport and urine concentration. Taken together, this ordered expression of transporters and other genes along the pronephric nephron suggests similarities to the organization of the mammalian nephron.

Signaling systems and genes involved in patterning the nephron segments represent an important area of future investigation. Based on what is known from other organisms, the notch pathway may play a role in this process. Activation of the Notch pathway in Xenopus embryos inhibits pronephric duct development and expression of duct-specific markers while Notch inhibition had the opposite effect (McLaughlin *et al.*, 2000). In zebrafish, *deltaC*, a ligand for Notch, is expressed in the anterior intermediate mesoderm during somitogenesis; however, initial studies of *deltaC* mutants do not show an obvious kidney phenotype

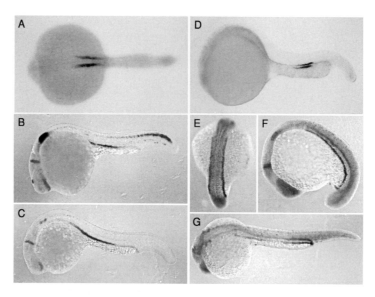

Fig. 9 Transporter mRNA expression defines pronephric nephron segments. (A). The chloride-bicarbonate anion exchanger (AE2) is expressed in the anterior pronephric duct. The mid-segment of the pronephric ducts specifically express the zebrafish starmaker gene (B) and an aspartoacylase homolog (C). Transporters expressed in the duct mid-segment include the Na-K-Cl symporter *slc12a1* (D). Expression of a putative ABC transporter (ibd2207) is observed initially throughout most of the forming pronephric duct at the 15 somite stage (E and F) but becomes restricted primarily to the posterior duct by 24 hpf (G). Embryos in B, C, E, F, and G are counterstained with *pax2.1* probe in red for reference. (A and D courtesy of Alan Davidson *et al.*; B, C, E, F, G courtesy of Neil Hukreide.) (See Color Insert.)

(Smithers *et al.*, 2000). Further exploration of notch and its ligands in the pronephros seems warranted.

3. Pronephric Tubule and Duct Isolation

Historically, the functional aspects of kidney epithelial ion transport have been studied using isolated single epithelial tubules in primary culture. This has not yet been achieved for zebrafish pronephric tubules. However, a useful first step in considering such an approach is larval tissue fractionation and tubule isolation (Fig. 10). Isolation of homogeneous preparations of viable pronephric ducts and tubules is also useful for purifying RNA for use in assaying tissue-specific gene expression. The protocol described next is useful for the isolation of larval pronephroi, as well as for spinal cord, individual myotubes, and anything else that can be identified visually, by transgene expression or by other forms of labeling.

Two- to three-day-old zebrafish larvae show a remarkable resistance to collagenase digestion. Since collagenase is one of the least damaging enzymes used for

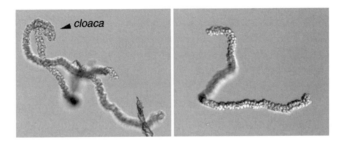

Fig. 10 Pronephric tubules and ducts isolated from 48-hpf zebrafish larvae. Two-day-old larvae were treated with 10 mM DTT followed by incubation in 5 mg/ml collagenase for 3 hours at 28 °C. Individual pronephric ducts can be dissected away from trunk tissue, often with the cloaca intact (*arrowhead*), joining the bilateral ducts at the distal segment.

tissue isolation, we have tested various ways to render larvae collagenase sensitive. We find that a 1-hour pre incubation in a reducing agent (DTT or N-acetyl-cysteine) allows subsequent incubation in collagenase to be effective. These reducing agents were tested because of their known mucolytic activity, which we reasoned would degrade the protective mucous layer of the larval fish skin.

1. Incubate 2- to 3-day-old larvae in 10 mM DTT or N-acetyl-cysteine in egg water for 1 hour at room temperature.
2. Wash the larvae 3 to 4 times with egg water to remove the DTT.
3. Incubate larvae at 28.5 °C in 5 mg/ml collagenase in tissue culture medium or Hanks saline with calcium (Worthington Biochemical Corporation, Lakewood, NJ) for 3 to 6 hours. Collagenase requires calcium for activity.
4. Triturate the larvae gently 5 times with a "blue tip" 1000-μl pipette tip. The larvae should disaggregate into chunks of tissue.
5. Disperse the cell/tissue suspension into a 10-cm Petri dish containing 10 ml of tissue culture medium or Hanks buffer.
6. Collect pronephric ducts by visual identification under a dissecting scope.

Further optimizations of tissue isolation and functional assays of zebrafish pronephric tubule transport would be a significant advance in using fish larvae to study the development of mature organ physiology.

D. Pronephric Vascularization

Proper functioning of the kidney requires a structural integration of pronephric epithelia and blood vessels. The glomerulus is a capillary tuft covered in epithelial cells, the podocytes, and specialized for blood filtration (Drummond *et al.*, 1998; Majumdar and Drummond, 1999; Pavenstadt *et al.*, 2003). Lateral and caudal to the podocytes, the pronephric tubules and ducts lie adjacent to the cardinal veins. Both the glomerular podocytes and the epithelial cells of the tubules and ducts are separated by only a basement membrane from endothelial cells. While the glomerulus is specialized for filtration of blood fluid into the tubules, the close association of tubule and duct epithelia with venous blood vessels facilitates the recovery of ions and metabolites from the glomerular filtrate back to the general circulation.

Vascularization of the glomerulus occurs relatively late in development, after pronephric duct and tubule development is complete (Armstrong, 1932; Drummond *et al.*, 1998; Tytler, 1988). In zebrafish, bilateral glomerular primordia coalesce at 36 to 40 hpf ventral to the notochord, bringing the presumptive podocytes into contact with endothelial cells of the dorsal aorta (Fig. 11; Drummond, 2000; Drummond *et al.*, 1998; Majumdar and Drummond, 1999). Between 40 and 48 hpf, new blood vessels sprout from the aorta and podocytes elaborate foot processes to surround the ingrowing endothelial cells (Fig. 11; Drummond *et al.*, 1998). Further capillary growth and invasion by endothelial cells results in extensive convolution of the glomerular basement membrane, capillary formation, and the onset of glomerular filtration (Drummond *et al.*, 1998).

1. The Role of Podocytes

Evidence that podocytes act to organize vessel ingrowth can be found in (1) the expression patterns of genes known to play an important role in angiogenesis and (2) the recruitment of endothelial cells to ectopic clusters of podocytes in mutant

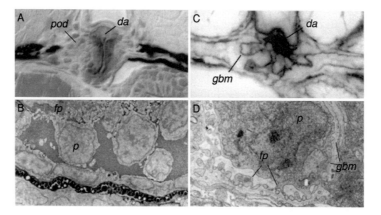

Fig. 11 Interaction of pronephric podocytes with the vasculature. (A) Apposition of nephron primordia at the embryo midline in a 40-hpf zebrafish embryo. Aortic endothelial cells in the cleft separating the nephron primordia are visualized by endogenous alkaline phosphatase activity. Podocytes, *pod*; dorsal aorta, *da*. (B) Ultrastructure of the forming zebrafish glomerulus at 40 hpf. A longitudinal section shows podocytes (*p*) extending foot processes (*fp*) in a dorsal direction and in close contact with overlying capillary endothelial cells. (C) Rhodamine dextran (10,000 M.W.) injected embryos show dye in the dorsal aorta (*da*) and in the glomerular basement membrane (*gbm*) shown here graphically inverted from the original fluorescent image. (D) Podocyte foot process formation does not require signals from endothelial cells as evidenced by the appearance of foot processes (*fp*) in *cloche* mutant embryos which lack all vascular structures. Glomerular basement membrane, (*gbm*); podocyte cell body, (*p*).

embryos that lack the dorsal aorta, the normal blood supply for the pronephric glomerulus. Zebrafish pronephric podocytes express two known mediators of angiogenesis: VEGF and angiopoietin2 (Carmeliet *et al.*, 1996; Ferrara *et al.*, 1996; Majumdar and Drummond, 1999, 2000; Pham *et al.*, 2001; Shalaby *et al.*, 1995). In a complementary manner, capillary-forming endothelial cells express *flk1*, a VEGF receptor, and an early marker of the endothelial differentiation program (Majumdar and Drummond, 1999). In zebrafish embryos at 40 hpf, *flk-1*-positive endothelial cells can be observed invading the glomerular epithelium. In zebrafish *floating head* mutant embryos, the normal source of glomerular blood vessels, the dorsal aorta, is absent (Fouquet *et al.*, 1997). The *floating head* gene encodes a homeobox transcription factor that is required for notochord formation. The absence of a notochord prevents the normal signaling that induces formation of the aorta. Pronephric nephron primordia in *floating head* embryos do not fuse at the midline, instead remaining at ectopic lateral positions in the embryo (Majumdar and Drummond, 2000). These podocytes in the primordia continue to express *wt1* and *vegf* and appear to recruit *flk-1*-positive endothelial cells and go on to form a functional glomerulus (Majumdar and Drummond, 2000). These results support the idea that podocytes, by expressing *vegf*, play a primary role in attracting and assembling the glomerular capillary tuft.

2. The Role of Endothelial Cells

The question of whether complementary signals are emitted from endothelial cells that would stimulate podocyte development has been addressed in studies of the mutant *cloche* (*clo*), which is nearly completely lacking in endothelial cells (Stainier *et al.*, 1995). When assayed by the expression of molecular markers and by an ultrastructural analysis of podocyte morphology, the elaboration of podocyte foot processes in *cloche* embryos occurs normally despite the complete absence of glomerular endothelial cells (Fig. 11; Majumdar and Drummond, 1999). The *cloche* podocytes express *wt1* and *vegf* and form extensive foot processes arranged as pedicels along a glomerular basement membrane (Majumdar and Drummond, 1999). These findings suggest that once initiated, podocyte differentiation proceeds independently of endothelial cells or endothelial cell-derived signals. Other developmental outcomes of signaling between podocytes and endothelial cells remain to be explored. For instance, maturation of the structure and composition of the glomerular basement membrane (Miner and Sanes, 1994) may require a contribution or signal from endothelial cells. Also, proper polarization of podocytes and organization of cell-cell junctions may require the presence of endothelial cells.

3. The Role of Vascular Flow

Zebrafish mutants specifically lacking cardiac function, for instance *silent heart* and *island beat* (*isl*), with mutations in cardiac troponin T and an L-type cardiac calcium channel, respectively (Rottbauer *et al.*, 2001; Sehnert *et al.*, 2002), fail to form a proper glomerular capillary tuft, demonstrating that vascular shear force is required to drive capillary formation. In mutants that lack circulation, endothelial cells fail to invade the encapsulated podocytes and the aorta is instead a broadly dilated vessel surrounding the glomerular primordia (Fig. 12; Serluca *et al.*, 2002). The cells themselves appear normal in several ways. For instance, expression patterns of *wt1* and *vegf* in podocytes and *flk-1* in endothelial cells remain unaltered and individual cell ultrastructure appears normal see (Fig. 6; Serluca *et al.*, 2002). The glomerular morphogenesis defect can be phenocopied by pharmacological and surgical manipulations that disrupt flow through the aorta. This failure

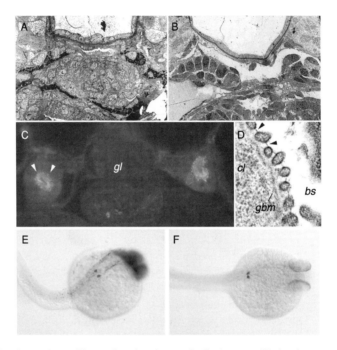

Fig. 12 The glomerular capillary tuft and podocyte slit diaphragms. (A) An electron micrograph of the forming glomerulus at 2.5 dpf with invading endothelial cells from the dorsal aorta shaded in red and podocytes shaded in blue (image false-colored in Adobe Photoshop). (B) A similar stage glomerulus in the mutant island beat which lacks blood flow due to a mutation in an L-type cardiac specific calcium channel. The endothelial cells and podocytes are present but the aorta has a dilated lumen surrounding the podocytes with no sign of glomerular remodeling and morphogenesis. (C) Rhodamine-dextran filtration and uptake by pronephric epithelial cells. 10KD lysine-fixable rhodamine dextran injected into the general circulation can be seen as red fluorescence in glomerular capillaries (*gl*), and filtered dye is seen in apical endosomes of pronephric duct cells (*arrowheads*). Counterstain: FITC wheat germ agglutinin. (D) Electron micrograph of the glomerular basement membrane region in the glomerulus. Individual profiles of podocyte foot processes resting on the glomerular basement membrane (*gbm*) are connected by slit-diaphragms (*arrowheads* at top). *cl*, capillary lumen; *bs*, Bowman's space. Whole-mount *in situ* hybridization shows expression of zebrafish podocin (E) and nephrin (F) specifically in the forming podocytes. (See Color Insert.)

in glomerular morphogenesis is likely be related to the expression of matrix metalloproteinase-2 (MMP-2) in endothelial cells since (1) expression of MMP-2 in the trunk vasculature is modulated by vascular flow and (2) inhibition of MMP activity by TIMP1 (tissue inhibitor of metalloproteinase-1) injections into the vasculature result in a similar dilated aorta and failure to form the glomerulus (Serluca et al., 2002). Studies of blood flow in the zebrafish heart (Hove et al., 2003) confirm that hydrodynamic forces can have a major influence on tissue morphogenesis.

4. A Simple Assay for Glomerular Filtration

Filtration of blood by the pronephric glomerulus can be detected by injections of fluorescent compounds into the general circulation and then monitoring the appearance of fluorescent endosomes in the apical cytoplasm of pronephric duct cells (Fig. 12C; Drummond et al., 1998; Majumdar and Drummond, 2000). From this data, it can be inferred that the fluorescent tracer has passed the glomerular basement membrane and entered the lumen of the pronephric tubules and ducts where it is actively endocytosed. Using this assay, we have established that blood filtration by the zebrafish pronephros begins around 40 hpf (Drummond et al., 1998). Filtered fluorescent dextrans can also be observed directly as they exit the pronephros at the cloaca in live larvae (Kramer-Zucker and Drummond, unpublished, 2004).

IV. The Zebrafish Pronephros as a Model of Human Disease

A. Disease of the Glomerular Filtration Apparatus

A major feature of the glomerular blood filter is the podocyte slit-diaphragm, a specialized adherens junction that forms between the finger-like projections of podocytes, podocyte foot processes (Reiser et al., 2000). Failure of the slit-diaphragm to form results in leakage of high molecular weight proteins into the filtrate, a condition called proteinuria in human patients. Proteinuria is the cardinal feature of several human congenital nephropathies and also a common complication of diabetes (Cooper et al., 2002). Several disease genes' known function in the slit-diaphragm have been cloned. Nephrin is a transmembrane protein present in the slit-diaphragm itself and is thought to contribute to the zipper-like extracellular structure between foot processes (Ruotsalainen et al., 1999). Podocin is a podocyte junction-associated protein (Roselli et al., 2002) that resembles stomatin proteins that play a role in regulating mechanosensitive ion channels (Tavernarakis and Driscoll, 1997). Electron microscopy of the zebrafish pronephric glomerulus reveals that, like mammalian podocytes, zebrafish podocytes form slit-diaphragms between their foot processes (Fig. 12D). Zebrafish homologs of podocin and nephrin are specifically expressed in podocytes as early as 24 hpf (Fig. 12E and F). These functional similarities between mammalian and zebrafish podocytes, coupled with assays for filtration discrimination, point to future applications of fish as a model of human proteinuria pathologies.

B. Disease of the Tubules

One of the most common human genetic diseases is polycystic kidney disease, which affects 1 in 1000 individuals (Calvet and Grantham, 2001). Kidney cysts are the result of grossly expanded kidney tubule lumens and, when present in sufficient size and number, lead to kidney fibrosis and end-stage renal failure. A relatively large set of genetic loci associated with cystic pronephroi have been identified in large-scale N-ethyl-N-nitrosourea (ENU) zebrafish mutant screens (Fig. 13). The 15 genetic loci that we originally identified (Drummond et al., 1998) probably underestimates the number of genes involved since the degree of saturation for this phenotype was low (so far, 12 of the 15 loci are represented by 1 allele). The results of a large-scale retroviral insertional mutagenesis screen have identified 10 zebrafish genes that, when mutated, cause pronephric cysts (Sun et al., 2004). The requirement for a large number of genes is consistent with the idea that maintainance of lumen size and epithelial cell shape is a complex process controlled by many cellular proteins or signaling pathways.

A surprising convergence of data from studies of cystic disease, left-right asymmetry, retinal degeneration, and flagella formation in Chlamydomonas lead to the idea that defects in the formation or function of cilia may underlie the pathology observed in all of these conditions. Cloning the gene responsible for the oakridge polycystic kidney (orpk) mouse was the first link between cilia and

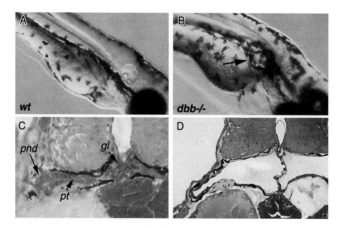

Fig. 13 The zebrafish as a model of polycystic kidney disease. (A and B) Three-day larvae showing wild type (A) and the mutant *double bubble* (*dbb*; B) with a grossly distended pronephric tubule that appears as a bubble (*arrow*) just behind the pectoral fin. (C) Three-day wild-type kidney structure showing the pronephric duct (*pnd*; cross section), pronephric tubule (*pt*), and glomerulus (*gl*). (D) A section of a dbb mutant pronephros shows the distended lumen of the pronephric tubule (*asterisk*) and distended glomerulus at the midline.

kidney cystic disease. The mutant gene, polaris, is a homolog of a Chlamydomonas gene, IFT88, that is required for intraflagellar transport, an essential process in flagellum formation (Pazour *et al.*, 2000; Taulman *et al.*, 2001). Kidney cells are not flagellated but have a single, non-motile apical cilium. The apical cilium had been thought to be a vestigial organelle with uncertain function. The finding that *orpk* mouse kidney epithelial cells have short, malformed apical cilia (Pazour *et al.*, 2000; Taulman *et al.*, 2001; Yoder *et al.*, 2002b) suggested a functional link between cilia and maintenance of epithelial tubule lumen diameter. Subsequent studies revealed that most known cystic mutant genes, including polycystin1, polycystin2, cystin (*cpk* mouse), polaris, inversin, and the C. elegans polycystin homologs lov-1 and pkd2 were, at least in part, localized to cilia (Barr *et al.*, 2001; Morgan *et al.*, 2002; Pazour *et al.*, 2002b; Qin *et al.*, 2001; Yoder *et al.*, 2002a). Cilia function and nodal flow has been implicated in the establishment of left-right asymetry (Tabin and Vogan, 2003). Among the cystic genes, inversin, polaris, and polycystin2 mutations show laterality defects (Morgan *et al.*, 1998; Pennekamp *et al.*, 2002; Taulman *et al.*, 2001). Finally, retinal photoreceptor cells contain a modified cilium that appears to function in membrane transport between the cell body and rod outer membranes. In light of this, the retinal degeneration seen in the orpk/polaris mouse (Pazour *et al.*, 2002a) makes some sense. Of interest, some zebrafish cyst mutants manifest pleiotropic defects, affecting both the kidney and the eye, are referred to as *renal-retinal dysplasia* when observed in human patients (Drummond *et al.*, 1998; Godel *et al.*, 1978). Although the exact function of these genes in vertebrate embryos remains to be worked out, studies of polycystin1 and polycystin2, the genes responsible for autosomal dominant polycystic kidney disease, in epithelial cells suggest that they act together to mediate calcium entry into cells upon flow-induced cilium deflection (Nauli *et al.*, 2003; Praetorius and Spring, 2001). It may be that the cilium acts as a sensor of tubule lumen mechanics and flow, providing a feedback signal that limits lumen diameter. Our own observations indicate that cilia in the zebrafish pronephros are motile and have a "9+2" microtubule doublet organization that is typical of motile cilia and flagella. Motile cilia are often associated with fluid flow and this leads to an alternative hypothesis that cilia act as a fluid pump in the pronephros. When cilia are malformed or immotile, failure to move fluid or particles in the lumen of the pronephric ducts may account for some of the accumulation of fluid in the anterior segments of the pronephros. Motile cilia are a common feature in amphibian pronephric tubules and ducts.

Several genes have been identified so far in zebrafish that are responsible for tubule cyst formation. Nek8 is a member of the NIMA family of serine/threonine kinases and is mutated in the juvenile cystic kidney (*jck*) mouse (Liu *et al.*, 2002). Morpholino antisense oligos that disrupt the function of zebrafish nek8 cause a severe cystic distension of the pronephric tubules. *In vitro* expression of a kinase-dead Nek8 mutant protein or the *jck* Nek8 allele (a missense mutation in a C-terminal putative protein interaction domain) in cultured cells results in multinucleated cells, suggesting a role for nek8 in regulating the cytoskeleton (Liu *et al.*, 2002). Other Nek kinases have links to cytoskeletal functions: Nek2 is localized to centrioles and acts to promote splitting of duplicated centrioles during the cell cycle (Fry, 2002). In Chlamydomonas, FA2 is a Nek kinase necessary for shedding flagella before cell division (Mahjoub *et al.*, 2002). Further studies are needed to test whether Nek8 may have a similar role in cilium or centrosome/basal body function.

Disruption of the zebrafish homolog of the human cystic disease gene polycystin2 also causes pronephric cyst formation (Obara *et al.*, unpublished results, 2004). We have found that co-injected human polycystin2 mRNA can rescue this phenotype, suggesting that the function of polycystin2 has been highly conserved.

Morpholino disruption of the zebrafish inversin gene results in both laterality defects and kidney cysts. The human condition nephronophthisis type 2 (NPHP2) is associated with mutations in the human inversin gene. Both inversin and the mammalian NPHP1 gene nephrocystin are found in basal bodies and cilia, and have been shown to interact biochemically. A splice donor morpholino-induced C-terminal deletion of the putative nephrocystin binding domain in zebrafish inversin results in severe cyst formation, supporting the idea that NPHP proteins act as a multiprotein complex to regulate the function of basal bodies and/or cilia.

The transcription factor HNF1β is required for normal zebrafish pronephric tubule development (Sun and Hopkins, 2001). Cystic tubules are observed in three different retroviral insertional mutants that harbor a disrupted HNF1β gene. The function of HNF1β may be more related to tissue patterning than to cilium function since in the mutants, *pax2.1* expression is lost from the tubules and *wt1* expression in podocyte progenitors appears expanded (Sun and Hopkins, 2001). HNF1β may play a role in specification of the tubule fate and when the process is disturbed, the epithelium becomes cystic. Mutations in HNF1β are associated with glomerulocystic disease and maturity onset diabetes of the young, type V (MODY5) in humans (Froguel and Velho, 1999).

V. Conclusions

The zebrafish pronephric kidney represents one of the many vertebrate kidney forms that have evolved to solve the problem of blood fluid and electrolyte homeostasis in an osmotically challenging environment. Despite differences in organ morphology between the mammalian and teleost kidneys, many parallels exist at the cellular and molecular levels that can be exploited to further our understanding of kidney cell specification, epithelial tubule formation, and the tissue interactions that drive nephrogenesis. The same genes (e.g., *pax2*) and cell types (e.g., podocytes, endothelial cells, and tubular epithelial cells) are employed in the development and function of fish, frog, chicken, and mammalian kidneys. Genes mutated in human disease are also essential for the formation and function of the zebrafish pronephros. The zebrafish therefore presents a useful and relevant model of vertebrate kidney development: its principal strengths lie in the ease with which it can be genetically manipulated and phenotyped so as to rapidly determine the function of genes and cell-cell interactions that underlie the development of all kidney forms.

References

Agarwal, S., and John, P. A. (1988). Studies on the development of the kidney of the guppy, *Lebistes reticulatus*. Part 1. The development of the pronephros. *J. Anim. Morphol. Physiol.* **35**, 17–24.

Amacher, S. L., Draper, B. W., Summers, B. R., and Kimmel, C. B. (2002). The zebrafish T-box genes no tail and spadetail are required for development of trunk and tail mesoderm and medial floor plate. *Development* **129**, 3311–3323.

Armstrong, P. B. (1932). The embryonic origin of function in the pronephros through differentiation and parenchyma-vascular association. *Am. J. Anat.* **51**, 157–188.

Balfour, F. M. (1880). "A Treatise on Comparative Embryology." Macmillan and Co., London.

Barr, M. M., DeModena, J., Braun, D., Nguyen, C. Q., Hall, D. H., and Sternberg, P. W. (2001). The Caenorhabditis elegans autosomal dominant polycystic kidney disease gene homologs lov-1 and pkd-2 act in the same pathway. *Curr. Biol.* **11**, 1341–1346.

Batourina, E., Gim, S., Bello, N., Shy, M., Clagett-Dame, M., Srinivas, S., Costantini, F., and Mendelsohn, C. (2001). Vitamin A controls epithelial/mesenchymal interactions through Ret expression. *Nat. Genet.* **27**, 74–78.

Begemann, G., Schilling, T. F., Rauch, G. J., Geisler, R., and Ingham, P. W. (2001). The zebrafish neckless mutation reveals a requirement for raldh2 in mesodermal signals that pattern the hindbrain. *Development* **128**, 3081–3094.

Bisgrove, B. W., Raible, D. W., Walter, V., Eisen, J. S., and Grunwald, D. J. (1997). Expression of c-ret in the zebrafish embryo: Potential roles in motoneuronal development. *J. Neurobiol.* **33**, 749–768.

Briscoe, J., Pierani, A., Jessell, T. M., and Ericson, J. (2000). A homeodomain protein code specifies progenitor cell identity and neuronal fate in the ventral neural tube. *Cell* **101**, 435–445.

Burrow, C. R. (2000). Retinoids and renal development. *Exp. Nephrol.* **8**, 219–225.

Calvet, J. P., and Grantham, J. J. (2001). The genetics and physiology of polycystic kidney disease. *Semin. Nephrol.* **21**, 107–123.

Carmeliet, P., Ferreira, V., Breier, G., Pollefeyt, S., Kieckens, L., Gertsenstein, M., Fahrig, M., Vandenhoeck, A., Harpal, K., Eberhardt, C., Declercq, C., Pawling, J., Moons, L., and Collen, D. (1996). Abnormal blood vessel development and lethality in embryos lacking a single VEGF allele. *Nature* **380**, 435–439.

Carroll, T. J., Wallingford, J. B., and Vize, P. D. (1999). Dynamic patterns of gene expression in the developing pronephros of Xenopus laevis. *Dev. Genet.* **24**, 199–207.

Cooper, M. E., Mundel, P., and Boner, G. (2002). Role of nephrin in renal disease including diabetic nephropathy. *Semin. Nephrol.* **22**, 393–398.

Dahl, U., Sjodin, A., Larue, L., Radice, G. L., Cajander, S., Takeichi, M., Kemler, R., and Semb, H. (2002). Genetic dissection of cadherin function during nephrogenesis. *Mol. Cell. Biol.* **22**, 1474–1487.

Davidson, A. J., Ernst, P., Wang, Y., Dekens, M. P., Kingsley, P. D., Palis, J., Korsmeyer, S. J., Daley, G. Q., and Zon, L. I. (2003). cdx4 mutants fail to specify blood progenitors and can be rescued by multiple hox genes. *Nature* **425**, 300–306.

Delarue, M., Saez, F. J., Johnson, K. E., and Boucaut, J. C. (1997). Fates of the blastomeres of the 32-cell stage Pleurodeles waltl embryo. *Dev. Dyn.* **210**, 236–248.

Detrich, H. W., 3rd, Kieran, M. W., Chan, F. Y., Barone, L. M., Yee, K., Rundstadler, J. A., Pratt, S., Ransom, D., and Zon, L. I. (1995). Intraembryonic hematopoietic cell migration during vertebrate development. *Proc. Natl. Acad. Sci. USA* **92**, 10713–10717.

Draper, B. W., Stock, D. W., and Kimmel, C. B. (2003). Zebrafish fgf24 functions with fgf8 to promote posterior mesodermal development. *Development* **130**, 4639–4654.

Drummond, I. A. (2000). The zebrafish pronephros: A genetic system for studies of kidney development. *Pediatr. Nephrol.* **14**, 428–435.

Drummond, I. A., Majumdar, A., Hentschel, H., Elger, M., Solnica-Krezel, L., Schier, A. F., Neuhauss, S. C., Stemple, D. L., Zwartkruis, F., Rangini, Z., Driever, W., and Fishman, M. C. (1998). Early development of the zebrafish pronephros and analysis of mutations affecting pronephric function. *Development* **125**, 4655–4667.

Ferrara, N., Carver-Moore, K., Chen, H., Dowd, M., Lu, L., O'Shea, K. S., Powell-Braxton, L., Hillan, K. J., and Moore, M. W. (1996). Heterozygous embryonic lethality induced by targeted inactivation of the VEGF gene. *Nature* **380**, 439–442.

Fouquet, B., Weinstein, B. M., Serluca, F. C., and Fishman, M. C. (1997). Vessel patterning in the embryo of the zebrafish: Guidance by notochord. *Dev. Biol.* **183**, 37–48.

Froguel, P., and Velho, G. (1999). Molecular genetics of maturity-onset diabetes of the young. *Trends Endocrinol. Metab.* **10**, 142–146.

Fry, A. M. (2002). The Nek2 protein kinase: A novel regulator of centrosome structure. *Oncogene* **21**, 6184–6194.

Gering, M., Rodaway, A. R., Gottgens, B., Patient, R. K., and Green, A. R. (1998). The SCL gene specifies haemangioblast development from early mesoderm. *Embo J.* **17**, 4029–4045.

Godel, V., Romano, A., Stein, R., Adam, A., and Goodman, R. M. (1978). Primary retinal dysplasia transmitted as X-chromosome-linked recessive disorder. *Am. J. Ophthalmol.* **86**, 221–227.

Goodrich, E. S. (1930). "Studies on the Structure and Development of Vertebrates." Macmillan, London.

Griffin, K., Patient, R., and Holder, N. (1995). Analysis of FGF function in normal and no tail zebrafish embryos reveals separate mechanisms for formation of the trunk and the tail. *Development* **121**, 2983–2994.

Griffin, K. J., Amacher, S. L., Kimmel, C. B., and Kimelman, D. (1998). Molecular identification of spadetail: Regulation of zebrafish trunk and tail mesoderm formation by T-box genes. *Development* **125**, 3379–3388.

Hammerschmidt, M., Pelegri, F., Mullins, M. C., Kane, D. A., Brand, M., van Eeden, F. J., Furutani-Seiki, M., Granato, M., Haffter, P., Heisenberg, C. P., Jiang, Y. J., Kelsh, R. N., Odenthal, J., Warga, R. M., and Nusslein-Volhard, C. (1996a). Mutations affecting morphogenesis during gastrulation and tail formation in the zebrafish, *Danio rerio. Development* **123**, 143–151.

Hammerschmidt, M., Pelegri, F., Mullins, M. C., Kane, D. A., van Eeden, F. J., Granato, M., Brand, M., Furutani-Seiki, M., Haffter, P., Heisenberg, C. P., Jiang, Y. J., Kelsh, R. N., Odenthal, J., Warga, R. M., and Nusslein-Volhard, C. (1996b). Dino and mercedes, two genes regulating dorsal development in the zebrafish embryo. *Development* **123**, 95–102.

Heller, N., and Brandli, A. W. (1999). Xenopus Pax-2/5/8 orthologues: Novel insights into Pax gene evolution and identification of Pax-8 as the earliest marker for otic and pronephric cell lineages. *Dev. Genet.* **24**, 208–219.

Hentschel, H., and Elger, M. (1996). Functional morphology of the developing pronephric kidney of zebrafish. *J. Amer. Soc. Nephrol.* **7**, 1598.

Hickman, C. P., and Trump, B. F. (1969). The Kidney. *In* "Fish Physiology" (W. S. Hoar and D. J. Randall, eds.), Vol. 1, pp. 91–239. Academic Press, New York.

Hild, M., Dick, A., Rauch, G. J., Meier, A., Bouwmeester, T., Haffter, P., and Hammerschmidt, M. (1999). The smad5 mutation somitabun blocks Bmp2b signaling during early dorsoventral patterning of the zebrafish embryo. *Development* **126**, 2149–2159.

Ho, R. K., and Kane, D. A. (1990). Cell-autonomous action of zebrafish spt-1 mutation in specific mesodermal precursors. *Nature* **348**, 728–730.

Horsfield, J., Ramachandran, A., Reuter, K., LaVallie, E., Collins-Racie, L., Crosier, K., and Crosier, P. (2002). Cadherin-17 is required to maintain pronephric duct integrity during zebrafish development. *Mech. Dev.* **115**, 15–26.

Hove, J. R., Koster, R. W., Forouhar, A. S., Acevedo-Bolton, G., Fraser, S. E., and Gharib, M. (2003). Intracardiac fluid forces are an essential epigenetic factor for embryonic cardiogenesis. *Nature* **421**, 172–177.

Howland, R. B. (1921). Experiments on the effect of the removal of the pronephros of *Ambystoma punctatum. J. Exp. Zool.* **32**, 355–384.

Kelly, G. M., Greenstein, P., Erezyilmaz, D. F., and Moon, R. T. (1995). Zebrafish wnt8 and wnt8b share a common activity but are involved in distinct developmental pathways. *Development* **121**, 1787–1799.

Kim, S. H., Park, H. C., Yeo, S. Y., Hong, S. K., Choi, J. W., Kim, C. H., Weinstein, B. M., and Huh, T. L. (1998). Characterization of two frizzled8 homologues expressed in the embryonic shield and prechordal plate of zebrafish embryos. *Mech. Dev.* **78**, 193–201.

Kimmel, C. B., Ballard, W. W., Kimmel, S. R., Ullmann, B., and Schilling, T. F. (1995). Stages of embryonic development of the zebrafish. *Dev. Dyn.* **203**, 253–310.

Kimmel, C. B., Warga, R. M., and Schilling, T. F. (1990). Origin and organization of the zebrafish fate map. *Development* **108**, 581–594.

Kishimoto, Y., Lee, K. H., Zon, L., Hammerschmidt, M., and Schulte-Merker, S. (1997). The molecular nature of zebrafish swirl: BMP2 function is essential during early dorsoventral patterning. *Development* **124**, 4457–4466.

Krauss, S., Johansen, T., Korzh, V., and Fjose, A. (1991). Expression of the zebrafish paired box gene pax[zf-b] during early neurogenesis. *Development* **113**, 1193–1206.

Kudoh, T., Tsang, M., Hukriede, N. A., Chen, X., Dedekian, M., Clarke, C. J., Kiang, A., Schultz, S., Epstein, J. A., Toyama, R., and Dawid, I. B. (2001). A gene expression screen in zebrafish embryogenesis. *Genome Res.* **11**, 1979–1987.

Lekven, A. C., Thorpe, C. J., Waxman, J. S., and Moon, R. T. (2001). Zebrafish wnt8 encodes two wint8 proteins on a bicistronic transcript and is required for mesoderm and neurectoderm patterning. *Developmental Cell* **1**, 103–114.

Li, J. Y., and Joyner, A. L. (2001). Otx2 and Gbx2 are required for refinement and not induction of mid-hindbrain gene expression. *Development* **128**, 4979–4991.

Liu, A., Majumdar, A., Schauerte, H. E., Haffter, P., and Drummond, I. A. (2000). Zebrafish wnt4b expression in the floor plate is altered in sonic hedgehog and gli-2 mutants. *Mech. Dev.* **91**, 409–413.

Liu, S., Lu, W., Obara, T., Kuida, S., Lehoczky, J., Dewar, K., Drummond, I. A., and Beier, D. R. (2002). A defect in a novel Nek-family kinase causes cystic kidney disease in the mouse and in zebrafish. *Development* **129**, 5839–5846.

Mahjoub, M. R., Montpetit, B., Zhao, L., Finst, R. J., Goh, B., Kim, A. C., and Quarmby, L. M. (2002). The FA2 gene of Chlamydomonas encodes a NIMA family kinase with roles in cell cycle progression and microtubule severing during deflagellation. *J. Cell. Sci.* **115**, 1759–1768.

Majumdar, A., and Drummond, I. A. (1999). Podocyte differentiation in the absence of endothelial cells as revealed in the zebrafish avascular mutant, cloche. *Dev. Genet.* **24**, 220–229.

Majumdar, A., and Drummond, I. A. (2000). The zebrafish floating head mutant demonstrates podocytes play an important role in directing glomerular differentiation. *Dev. Biol.* **222**, 147–157.

Majumdar, A., Lun, K., Brand, M., and Drummond, I. A. (2000). Zebrafish no isthmus reveals a role for pax2.1 in tubule differentiation and patterning events in the pronephric primordia. *Development* **127**, 2089–2098.

Marcos-Gutierrez, C. V., Wilson, S. W., Holder, N., and Pachnis, V. (1997). The zebrafish homologue of the ret receptor and its pattern of expression during embryogenesis. *Oncogene* **14**, 879–889.

Marshall, E. K., and Smith, H. W. (1930). The glomerular development of the vertebrate kidney in relation to habitat. *Biol. Bull.* **59**, 135–153.

Mauch, T. J., Yang, G., Wright, M., Smith, D., and Schoenwolf, G. C. (2000). Signals from trunk paraxial mesoderm induce pronephros formation in chick intermediate mesoderm. *Dev. Biol.* **220**, 62–75.

McLaughlin, K. A., Rones, M. S., and Mercola, M. (2000). Notch regulates cell fate in the developing pronephros. *Dev. Biol.* **227**, 567–580.

Mendelsohn, C., Batourina, E., Fung, S., Gilbert, T., and Dodd, J. (1999). Stromal cells mediate retinoid-dependent functions essential for renal development. *Development* **126**, 1139–1148.

Miner, J. H., and Sanes, J. R. (1994). Collagen IV alpha 3, alpha 4, and alpha 5 chains in rodent basal laminae: Sequence, distribution, association with laminins, and developmental switches. *J. Cell. Biol.* **127**, 879–891.

Moore, M. W., Klein, R. D., Farinas, I., Sauer, H., Armanini, M., Phillips, H., Reichardt, L. F., Ryan, A. M., Carver-Moore, K., and Rosenthal, A. (1996). Renal and neuronal abnormalities in mice lacking GDNF. *Nature* **382**, 76–79.

Morgan, D., Eley, L., Sayer, J., Strachan, T., Yates, L. M., Craighead, A. S., and Goodship, J. A. (2002). Expression analyses and interaction with the anaphase promoting complex protein Apc2 suggest a role for inversin in primary cilia and involvement in the cell cycle. *Hum. Mol. Genet.* **11**, 3345–3350.

Morgan, D., Turnpenny, L., Goodship, J., Dai, W., Majumder, K., Matthews, L., Gardner, A., Schuster, G., Vien, L., Harrison, W., Elder, F. F., Penman-Splitt, M., Overbeek, P., and Strachan, T. (1998). Inversin, a novel gene in the vertebrate left-right axis pathway, is partially deleted in the inv mouse. *Nat. Genet.* **20**, 149–156.

Mullins, M. C., Hammerschmidt, M., Kane, D. A., Odenthal, J., Brand, M., van Eeden, F. J., Furutani-Seiki, M., Granato, M., Haffter, P., Heisenberg, C. P., Jiang, Y. J., Kelsh, R. N., and Nusslein-Volhard, C. (1996). Genes establishing dorsoventral pattern formation in the zebrafish embryo: The ventral specifying genes. *Development* **123**, 81–93.

Nauli, S. M., Alenghat, F. J., Luo, Y., Williams, E., Vassilev, P., Li, X., Elia, A. E., Lu, W., Brown, E. M., Quinn, S. J., Ingber, D. E., and Zhou, J. (2003). Polycystins 1 and 2 mediate mechanosensation in the primary cilium of kidney cells. *Nat. Genet.* **33**, 129–137.

Neave, B., Holder, N., and Patient, R. (1997). A graded response to BMP-4 spatially coordinates patterning of the mesoderm and ectoderm in the zebrafish. *Mech. Dev.* **62**, 183–195.

Newstead, J. D., and Ford, P. (1960). Studies on the development of the kidney of the Pacific Salmon, *Oncorhynchus forbuscha* (Walbaum). 1. The development of the pronephros. *Can. J. Zool.* **36**, 15–21.

Nguyen, V. H., Schmid, B., Trout, J., Connors, S. A., Ekker, M., and Mullins, M. C. (1998). Ventral and lateral regions of the zebrafish gastrula, including the neural crest progenitors, are established by a bmp2b/swirl pathway of genes. *Dev. Biol.* **199**, 93–110.

Nikaido, M., Tada, M., Saji, T., and Ueno, N. (1997). Conservation of BMP signaling in zebrafish mesoderm patterning. *Mech. Dev.* **61**, 75–88.

Pavenstadt, H., Kriz, W., and Kretzler, M. (2003). Cell biology of the glomerular podocyte. *Physiol. Rev.* **83**, 253–307.

Pazour, G. J., Baker, S. A., Deane, J. A., Cole, D. G., Dickert, B. L., Rosenbaum, J. L., Witman, G. B., and Besharse, J. C. (2002a). The intraflagellar transport protein, IFT88, is essential for vertebrate photoreceptor assembly and maintenance. *J. Cell. Biol.* **157**, 103–113.

Pazour, G. J., Dickert, B. L., Vucica, Y., Seeley, E. S., Rosenbaum, J. L., Witman, G. B., and Cole, D. G. (2000). Chlamydomonas IFT88 and its mouse homologue, polycystic kidney disease gene tg737, are required for assembly of cilia and flagella. *J. Cell Biol.* **151**, 709–718.

Pazour, G. J., San Agustin, J. T., Follit, J. A., Rosenbaum, J. L., and Witman, G. B. (2002b). Polycystin-2 localizes to kidney cilia and the ciliary level is elevated in orpk mice with polycystic kidney disease. *Curr. Biol.* **12**, R378–R380.

Pennekamp, P., Karcher, C., Fischer, A., Schweickert, A., Skryabin, B., Horst, J., Blum, M., and Dworniczak, B. (2002). The ion channel polycystin-2 is required for left-right axis determination in mice. *Curr. Biol.* **12**, 938–943.

Perz-Edwards, A., Hardison, N. L., and Linney, E. (2001). Retinoic acid-mediated gene expression in transgenic reporter zebrafish. *Dev. Biol.* **229**, 89–101.

Pfeffer, P. L., Gerster, T., Lun, K., Brand, M., and Busslinger, M. (1998). Characterization of three novel members of the zebrafish Pax2/5/8 family: Dependency of Pax5 and Pax8 expression on the Pax2.1 (noi) function. *Development* **125**, 3063–3074.

Pham, V. N., Roman, B. L., and Weinstein, B. M. (2001). Isolation and expression analysis of three zebrafish angiopoietin genes. *Dev. Dyn.* **221**, 470–474.

Pichel, J. G., Shen, L., Sheng, H. Z., Granholm, A. C., Drago, J., Grinberg, A., Lee, E. J., Huang, S. P., Saarma, M., Hoffer, B. J., Sariola, H., and Westphal, H. (1996). Defects in enteric innervation and kidney development in mice lacking GDNF. *Nature* **382**, 73–76.

Praetorius, H. A., and Spring, K. R. (2001). Bending the MDCK cell primary cilium increases intracellular calcium. *J. Membr. Biol.* **184**, 71–79.

Puschel, A. W., Westerfield, M., and Dressler, G. R. (1992). Comparative analysis of Pax-2 protein distributions during neurulation in mice and zebrafish. *Mech. Dev.* **38**, 197–208.

Qin, H., Rosenbaum, J. L., and Barr, M. M. (2001). An autosomal recessive polycystic kidney disease gene homolog is involved in intraflagellar transport in C. elegans ciliated sensory neurons. *Curr. Biol.* **11**, 457–461.

Reiser, J., Kriz, W., Kretzler, M., and Mundel, P. (2000). The glomerular slit diaphragm is a modified adherens junction. *J. Am. Soc. Nephrol.* **11**, 1–8.

Roselli, S., Gribouval, O., Boute, N., Sich, M., Benessy, F., Attie, T., Gubler, M. C., and Antignac, C. (2002). Podocin localizes in the kidney to the slit diaphragm area. *Am. J. Pathol.* **160**, 131–139.

Rottbauer, W., Baker, K., Wo, Z. G., Mohideen, M. A., Cantiello, H. F., and Fishman, M. C. (2001). Growth and function of the embryonic heart depend upon the cardiac-specific L-type calcium channel alpha1 subunit. *Dev. Cell* **1**, 265–275.

Ruotsalainen, V., Ljungberg, P., Wartiovaara, J., Lenkkeri, U., Kestila, M., Jalanko, H., Holmberg, C., and Tryggvason, K. (1999). Nephrin is specifically located at the slit diaphragm of glomerular podocytes. *Proc. Natl. Acad. Sci. USA* **96**, 7962–7967.

Sanchez, M. P., Silos-Santiago, I., Frisen, J., He, B., Lira, S. A., and Barbacid, M. (1996). Renal agenesis and the absence of enteric neurons in mice lacking GDNF. *Nature* **382**, 70–73.

Saxén, L. (1987). "Organogenesis of the Kidney." Cambridge University Press, Cambridge.

Schuchardt, A., D'Agati, V., Larsson-Blomberg, L., Costantini, F., and Pachnis, V. (1994). Defects in the kidney and enteric nervous system of mice lacking the tyrosine kinase receptor Ret [see comments]. *Nature* **367**, 380–383.

Schuchardt, A., D'Agati, V., Larsson-Blomberg, L., Costantini, F., and Pachnis, V. (1995). RET-deficient mice: An animal model for Hirschsprung's disease and renal agenesis. *J. Intern. Med.* **238**, 327–332.

Schwarz, M., Cecconi, F., Bernier, G., Andrejewski, N., Kammandel, B., Wagner, M., and Gruss, P. (2000). Spatial specification of mammalian eye territories by reciprocal transcriptional repression of Pax2 and Pax6. *Development* **127**, 4325–4334.

Sehnert, A. J., Huq, A., Weinstein, B. M., Walker, C., Fishman, M., and Stainier, D. Y. (2002). Cardiac troponin T is essential in sarcomere assembly and cardiac contractility. *Nat. Genet.* **31**, 106–110.

Seldin, D. W., and Giebisch, G. H. (1992). "The Kidney: Physiology and Pathophysiology." Raven Press, New York.

Serluca, F. C., Drummond, I. A., and Fishman, M. C. (2002). Endothelial signaling in kidney morphogenesis: A role for hemodynamic forces. *Curr. Biol.* **12**, 492–497.

Serluca, F. C., and Fishman, M. C. (2001). Pre-pattern in the pronephric kidney field of zebrafish. *Development* **128**, 2233–2241.

Seufert, D. W., Brennan, H. C., DeGuire, J., Jones, E. A., and Vize, P. D. (1999). Developmental basis of pronephric defects in Xenopus body plan phenotypes. *Dev. Biol.* **215**, 233–242.

Shalaby, F., Rossant, J., Yamaguchi, T. P., Gertsenstein, M., Wu, X. F., Breitman, M. L., and Schuh, A. C. (1995). Failure of blood-island formation and vasculogenesis in Flk-1-deficient mice. *Nature* **376**, 62–66.

Shawlot, W., and Behringer, R. R. (1995). Requirement for Lim1 in head-organizer function. *Nature* **374**, 425–430.

Shepherd, I. T., Beattie, C. E., and Raible, D. W. (2001). Functional analysis of zebrafish GDNF. *Dev. Biol.* **231**, 420–435.

Smithers, L., Haddon, C., Jiang, Y., and Lewis, J. (2000). Sequence and embryonic expression of deltaC in the zebrafish. *Mech. Dev.* **90**, 119–123.

Sollner, C., Burghammer, M., Busch-Nentwich, E., Berger, J., Schwarz, H., Riekel, C., and Nicolson, T. (2003). Control of crystal size and lattice formation by starmaker in otolith biomineralization. *Science* **302**, 282–286.

Stainier, D. Y., Weinstein, B. M., Detrich, H. W., 3rd, Zon, L. I., and Fishman, M. C. (1995). Cloche, an early acting zebrafish gene, is required by both the endothelial and hematopoietic lineages. *Development* **121**, 3141–3150.

Sun, Z., Amsterdam, Pazour, G. J., Cole, D. G., Miller, M. S., and Hapkins, N. (2004). A genetic screen in zebrafish identifies cilia genes as a principal cause of cystic kidney. *Development* **131**(16), 4085–4093.

Sun, Z., and Hopkins, N. (2001). vhnf1, the MODY5 and familial GCKD-associated gene, regulates regional specification of the zebrafish gut, pronephros, and hindbrain. *Genes Dev.* **15**, 3217–3229.

Tabin, C. J., and Vogan, K. J. (2003). A two-cilia model for vertebrate left-right axis specification. *Genes Dev.* **17**, 1–6.

Taira, M., Otani, H., Jamrich, M., and Dawid, I. B. (1994). Expression of the LIM class homeobox gene Xlim-1 in pronephros and CNS cell lineages of Xenopus embryos is affected by retinoic acid and exogastrulation. *Development* **120**, 1525–1536.

Taulman, P. D., Haycraft, C. J., Balkovetz, D. F., and Yoder, B. K. (2001). Polaris, a protein involved in left-right axis patterning, localizes to basal bodies and cilia. *Mol. Biol. Cell* **12**, 589–599.

Tavernarakis, N., and Driscoll, M. (1997). Molecular modeling of mechanotransduction in the nematode Caenorhabditis elegans. *Annu. Rev. Physiol.* **59**, 659–689.

Thisse, B., Pflumio, S., Fürthauer, M., Loppin, B., Heyer, V., Degrave, A., Woehl, R., Lux, A., Steffan, T., Charbonnier, X. Q., and Thisse, C. (2001). Expression of the zebrafish genome during embryogenesis (NIH R01 RR15402). *ZFIN Direct Data Submission* (http://zfin.org).

Topczewska, J. M., Topczewski, J., Shostak, A., Kume, T., Solnica-Krezel, L., and Hogan, B. L. (2001a). The winged helix transcription factor Foxc1a is essential for somitogenesis in zebrafish. *Genes Dev.* **15**, 2483–2493.

Topczewska, J. M., Topczewski, J., Solnica-Krezel, L., and Hogan, B. L. (2001b). Sequence and expression of zebrafish foxc1a and foxc1b, encoding conserved forkhead/winged helix transcription factors. *Mech. Dev.* **100**, 343–347.

Toyama, R., and Dawid, I. B. (1997). lim6, a novel LIM homeobox gene in the zebrafish: Comparison of its expression pattern with lim1. *Dev. Dyn.* **209**, 406–417.

Tytler, P. (1988). Morphology of the pronephros of the juvenile brown trout, Salmo trutta. *J. Morphol.* **195**, 189–204.

Tytler, P., Ireland, J., and Fitches, E. (1996). A study of the structure and function of the pronephros in the lavvae of the turbot (*Scophthalmus maximus*) and the herring (*Clupea harengus*). *Mar. Fresh. Behav. Physiol.* **28**, 3–18.

Ungar, A., Kelly, G. M., and Moon, R. T. (1995). Wnt4 affects morphogenesis when misexpressed in the zebrafish embryo. *Mech. of Dev.* **52**, 1–12.

Vize, P. D., Seufert, D. W., Carroll, T. J., and Wallingford, J. B. (1997). Model systems for the study of kidney development: Use of the pronephros in the analysis of organ induction and patterning. *Dev. Biol.* **188**, 189–204.

Vize, P. D., Woolf, A. S., and Bard, J. B. L. (2002). "The Kidney: From Normal Development to Congenital Diseases." Academic Press, Boston.

Weinstein, B. M., Schier, A. F., Abdelilah, S., Malicki, J., Solnica-Krezel, L., Stemple, D. L., Stainier, D. Y., Zwartkruis, F., Driever, W., and Fishman, M. C. (1996). Hematopoietic mutations in the zebrafish. *Development* **123**, 303–309.

Wilson, J. G., Roth, C. B., and Warkany, J. (1953). An analysis of the syndrome of malformation induced by maternal vitamin. A. deficiency. Effects of restoration of vitamin A at various times during gestation. *Am. J. Anat.* **92**, 189–217.

Yang, Y., Jeanpierre, C., Dressler, G. R., Lacoste, M., Niaudet, P., and Gubler, M. C. (1999). WT1 and PAX-2 podocyte expression in Denys-Drash syndrome and isolated diffuse mesangial sclerosis. *Am. J. Pathol.* **154**, 181–192.

Yoder, B. K., Hou, X., and Guay-Woodford, L. M. (2002a). The polycystic kidney disease proteins, polycystin-1, polycystin-2, polaris, and cystin, are co-localized in renal cilia. *J. Am. Soc. Nephrol.* **13**, 2508–2516.

Yoder, B. K., Tousson, A., Millican, L., Wu, J. H., Bugg, C. E., Jr., Schafer, J. A., and Balkovetz, D. F. (2002b). Polaris, a protein disrupted in orpk mutant mice, is required for assembly of renal cilium. *Am. J. Physiol. Renal. Physiol.* **282**, F541–F552.

ZFIN(2004). ZFIN Gene Expression Data. (http://zfin.org).

Zhong, T. P., Childs, S., Leu, J. P., and Fishman, M. C. (2001). Gridlock signalling pathway fashions the first embryonic artery. *Nature* **414**, 216–220.

CHAPTER 24

Pancreas Development in Zebrafish

Lara Gnügge, Dirk Meyer, and Wolfgang Driever

Developmental Biology
University of Freiburg
D-79104 Freiburg, Germany

I. Pancreas Organogenesis

The vertebrate pancreas has two major tasks: exocrine contributions to food digestion and endocrine control of metabolic homeostasis. These tasks are exerted through two distinct tissue components. The exocrine component is composed of acinar glands that release digestive enzymes into the intestine; whereas the endocrine component is composed of four distinct cell types that secrete hormones into the bloodstream to control glucose homeostasis and digestive functions (reviewed in Slack, 1995). Dysfunction of endocrine beta-cells often leads to diabetes mellitus, a widespread disease affecting more than 150 million people worldwide. Future regenerative or restorative therapies will depend on a better understanding of the development of the endocrine pancreas and specifically its beta-cells. The zebrafish is evolving as an excellent model organism to study vertebrate pancreas organogenesis.

Three major developmental mechanisms contribute to the formation of the pancreas organ primordium and its differentiation into a fully functional organ.

The first is pattern formation of the endoderm, which defines the position of the pancreatic primordium within the endodermal germ layer, and may already contribute to specification of the major pancreatic compartments—the endocrine and exocrine pancreas. The second mechanism involves the control of cell differentiation from pluripotent endodermal precursor cells through hypothetic organ-specific precursors to the specialized pancreatic cell types. The third and least well-understood mechanism is morphogenesis of the pancreas, which involves extensive cellular rearrangements, as well as movements of whole organ parts, and results in the mature asymmetric location of the pancreas in the abdomen and the typical shape and cellular organization of the pancreas.

II. Morphology of the Mature Pancreas

The zebrafish pancreas has assumed its mature shape and position by 6 dpf. The pancreas is located asymmetrically on the right side of the body as an elongated organ extending between the position of somites two to seven. It has a characteristic head-neck-tail morphological organization, where the anterior large portion contains islet tissue surrounded by exocrine tissue, while the small neck and the elongated tail consist of exocrine tissues. In adult fish, upon opening of the abdomen, the pancreas is visible as a very slim (smaller than 1-mm diameter), elongated opaque tissue directly adjacent to the right side of the duodenum.

In higher vertebrates, the Islets of Langerhans constitute the endocrine tissue compartment of the pancreas and are found embedded in the exocrine tissue. The islets are composed of four major cell types: insulin-producing beta-cells form the core of the islet, while somatostatin-producing delta-cells, glucagon-producing alpha-cells, and pancreatic polypeptide secreting PP-cells are located at the periphery. The composition of four different cell types in the endocrine pancreas holds true for most teleostean fishes (i.e., zebrafish); whereas only one (beta-cells), two (beta- and delta-cells), or three (beta-, delta-, and PP-) cell types are usually found in the pancreas of more primitive fishes (Youson and Al-Mahrouki, 1999). This phylogeny is reflected in the ontogeny of the zebrafish pancreas, where beta-cells differentiate before the other hormone-producing cells emerge to form the single principal islet.

In the zebrafish, the spatial arrangement of endocrine cells differs from their mammalian organization in that the core of the islet is composed of both beta- and delta-cells; whereas alpha-cells are found at the periphery (Argenton *et al.*, 1999). However, the lineage relationships between the different endocrine cell types may be conserved. Gene targeting studies of *pax4* and *pax6* (Sosa-Pineda *et al.*, 1997; St-Onge *et al.*, 1997) in mouse suggest a close relationship between beta- and delta-cells, while studies on the origin of alpha- and beta-cells point toward their independent lineage origin (Herrera, 2000). The observation that zebrafish endocrine cells of the 48-hpf (Argenton *et al.*, 1999) and 72-hpf developing islets never co-express insulin and glucagon or somatostatin and glucagon (Biemar *et al.*,

2001) indicates that zebrafish similar to mammalian systems have separate alpha and beta lineages. However, a more comprehensive cell lineage analysis would be required to confirm whether alpha and beta lineages indeed develop separately during early stages of zebrafish embryogenesis.

III. Physiology of the Endocrine Pancreatic Tissues

The function of the islet hormones in fishes is still not well characterized. Current evidence indicates that islet hormones, like in mammals, are required for balancing blood glucose levels between feeding and fasting periods (Wendelaar Bonga, 1993). In teleosts, the release of insulin by elevated blood-glucose levels leads to uptake of glucose by skeletal muscle and adipose tissue, as well as to an increase of glycolysis and glycogenesis in the liver. Insufficient glucose levels or elevated plasma levels of amino acids induce the release of glucagons, the functional antagonist of insulin. Glucagon may increase blood glucose levels by stimulating glycogenolysis in the liver and gluconeogenesis from lactate, amino acids, and glycerol (from lipolysis). In mammals, somatostatin inhibits the release of insulin and glucagon, and therefore the uptake of nutritional substrates from the intestine. Preliminary data obtained for rainbow trout (Eilertson and Sheridan, 1995) suggest that this also holds true for teleosts. Work from our group (Luc St.Onge and Wolfgang Driever, unpublished) indicates that in zebrafish elevated glucose levels stimulate insulin secretion and that insulin lowers blood glucose levels.

IV. Pancreas Development

A remarkable difference between zebrafish and other vertebrates concerns the initial steps of pancreas development, during which the so called "pancreatic buds" are formed in higher vertebrates. In zebrafish embryos, endocrine hormones are expressed significantly before the formation of bud structures (Argenton et al., 1999; Biemar et al., 2001; Ober et al., 2003). The separation of pancreatic tissue from general gut tissue has been observed in the form of a bud-shaped pancreas primordium starting only at about 24 hpf with primordium-expressing endocrine hormone genes, and a second more anterior primordium forming a bud at 34 hpf (Field et al., 2003). In the following, we will summarize the current knowledge of specification of pancreas precursors in the general endoderm of zebrafish (Fig. 1).

V. Formation and Patterning of the Zebrafish Endoderm

Studies in all major vertebrate model organisms have clearly revealed that both the endocrine and exocrine pancreas are derived from endoderm (reviewed in Edlund, 2002; Slack, 1995). In zebrafish, fate map studies have revealed that the

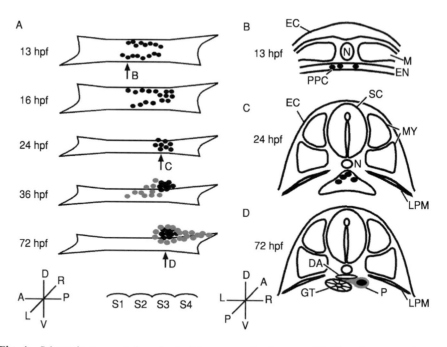

Fig. 1 Schematic representation of zebrafish pancreas development. (A) The *pdx-1*-expressing cells (*black dots*) mark the bilateral domains of the pancreas anlage at the 10-somite stage. These domains fuse during the next 12 hours at the level of the third somite to form a single islet of endocrine precursor cells. The first exocrine precursor cells (*gray dots*) are revealed by *mnr2* expression and appear at 32 hpf anterior to the islet when the epithelial structure of the gut is already established. The exocrine pancreas "buds" off the gut tube, engulfs the endocrine islet, and extends posteriorly to form the "tail" of the pancreas. The relative position of the somites S1 through S4 is indicated at the bottom of (A). (B, C, and D) Schematic representation of cross-sections at the levels indicated by the arrows in (A). The endocrine cells and the islet are represented in black, the exocrine tissue in grey. Abbreviations: DA, dorsal aorta; EC, ectoderm; EN, endoderm; GT, gut tube; LPM, lateral plate mesoderm; MY, myotome; N, notochord; P, pancreas; PPC, pancreatic precursor cells; and SC, spinal cord. The orientation of the drawings is indicated: A-anterior, P-posterior, D-dorsal, V-ventral, L-left, and R-right.

endodermal germ layer forms from the vegetal most cells of the blastoderm adjacent to the margin. Most endoderm derives from the dorsal half of the margin (Warga and Nüsslein-Volhard, 1999; Warga and Stainier, 2002). Endoderm is induced by secreted transforming growth factor beta (TGFbeta) molecules emanating from the yolk syncytial layer underneath, as well as from the vegetal margin of the blastoderm itself (Rodaway *et al.*, 1999; Stainier, 2002). Nodal induces endoderm and mesoderm in a concentration dependent manner (Chen and Schier, 2001; Dougan *et al.*, 2003; Thisse *et al.*, 2000). However, while individual mid-blastula–stage marginal cells may give rise to endoderm and mesoderm, single late-blastula–stage cells give rise to either mesoderm or endoderm. Kikuchi *et al.*

(2004) have found evidence that Notch signaling plays a role in segregation of endoderm from mesoderm. TGFbeta signaling initiates a cascade of several transcription factors and eventually leads to the induction of the HMG transcription factor Sox17. Analysis of several mutants and their effect on expression of other transcription factors in the cascade outlined a molecular pathway of endoderm specification (Alexander and Stainier, 1999; Aoki *et al.*, 2002a; Dickmeis *et al.*, 2001; Kikuchi *et al.*, 2000; Reiter *et al.*, 2001). In the current model, this pathway is initiated by two Nodal-type TGFbeta ligands, which are encoded by the *cyclops (ndr1)* and *squint (ndr2)* genes (Erter *et al.*, 1998; Feldman *et al.*, 1998; Rebagliati *et al.*, 1998; Sampath *et al.*, 1998). Both *cyclops* and *squint* are expressed at the vegetal margin of the blastoderm, the region to which precursors of the endoderm have been mapped. While embryos mutant for either *cyclops* or *squint* alone still develop endoderm, embryos homozygous mutant for both *cyclops* and *squint* completely lack the endodermal germ layer (Dougan *et al.*, 2003). Conversely, ectopic formation of endoderm can be induced by over-expression of Squint or Cyclops, or by blocking Antivin (Lefty) family members, the endogenous antagonists of Nodal factors (Agathon *et al.*, 2001).

High levels of Nodal signaling induce expression of several transcription factors that have been shown to be required for endoderm formation. However, the molecular pathways leading to the activation of these genes are still poorly defined. A TGFbeta receptor that appears to play a central role in Nodal-dependent endoderm formation is Taram-A (Acvr1b). While Nodals are required and sufficient for the formation of endodermal and mesodermal structures, constitutively active or dominant negative forms of Taram-A (Acvr1b) mainly induce or block endodermal differentiation, respectively (Aoki *et al.*, 2002b; Peyrieras *et al.*, 1996; Renucci *et al.*, 1996). Within the cell, Nodal signals are mediated by Smad2 proteins. After phosphorylation by Nodal receptors, Smad2 binds to Smad4 and translocates to the nucleus (Dick *et al.*, 2000; Muller *et al.*, 1999). For the activation of target genes the Smad2-Smad4 complex has to interact with additional cofactors that also define target gene specificity. Currently, only two such Smad2-Smad4 interaction partners, the forkhead domain factor FoxH1 (Fast1) and the mix-related homebox factor Mixer, are known in zebrafish (Kikuchi *et al.*, 2000; Pogoda *et al.*, 2000; Sirotkin *et al.*, 2000). While both factors are present in the early embryo and required for Nodal signaling, genetic studies have shown that additional factors must be involved during the initiation of the transcriptional cascade leading to endoderm formation (Kunwar *et al.*, 2003).

Nodal signaling induces expression of several transcription factors, which appear to act in parallel during endoderm specification. These factors include the mix-like homebox protein Mixer (encoded by *bonnie and clyde, bon*; Alexander and Stainier, 1999), the zinc finger containing factor Gata5 (*faust, fau*; Reiter *et al.*, 2001), and the paired-like homeobox protein Mezzo (Poulain and Lepage, 2002). In a transduction cascade, the activation of these three genes (*mixer, gata5, mezzo*) is required for the expression of the high-mobility-group (HMG) domain protein

Sox32 (*casanova, cas*; Alexander and Stainier, 1999; Aoki *et al.*, 2002a). Sox32 in turn, induces expression of *sox17*, an HMG domain transcription factor acting as the final determinant of endoderm specification.

Mutations in *fau, bon*, and *cas* affect endoderm development. The *bon* mutants develop fewer endodermal cells and eventually lack a functional gut tube including accessory organs of endodermal origin such as the pancreas (Kikuchi *et al.*, 2000). Also, *fau* mutants show defects in endoderm specification, although not as severe as the defects in *bon* mutant embryos (Reiter *et al.*, 2001). The most severe endoderm phenotype is observed in *cas* mutant embryos, in which the loss of active protein leads to a complete loss of *sox17* expression. Bon (Mixer), Faust (Gata5), Mezzo, and Cas (Sox32) all are required to induce normal *sox17* expression. Only Cas (Sox32) can induce *sox17* when nodal signaling is completely abolished in MZ*oep* mutants (Alexander and Stainier, 1999). On the basis of these experiments, the current model for endoderm specification proposes a Nodal-dependent activation of Bon, Mezzo, and Faust, which will then combine to induce *cas*, which in turn induces *sox17*. Also necessary for the induction of *sox17* is Spg (Pou5f1). A mutation in *spg* averts the induction of *sox17* by over-expression of *sox32* (Lunde *et al.*, 2004; Reim *et al.*, 2004) and also causes early downregulation of *sox32* (Lunde *et al.*, 2004). Therefore, *spg* is required twice in the endoderm specification cascade: it maintains—together with *sox32*—the expression of *sox32* in a positive feedback loop and also is required for *sox32* to be able to induce *sox17* (Lunde *et al.*, 2004). Targeted inactivation of *sox17* in the mouse results in a depletion of definitive gut endoderm including the pancreas anlage (Kanai-Azuma *et al.*, 2002). Surprisingly, the differentiation of liver in these mutant mice seems unaffected, arguing for a diversion of endoderm development already at this early stage. The effect of *sox17* on the development of endodermal organs in zebrafish is unknown and still remains to be addressed. Similar to other vertebrates, *foxa* family genes are also expressed in the endoderm during gastrulation and organogenesis in zebrafish (*foxa [fkd4], foxa1[fkd7], foxa2 [fkh1, axial, HNF3b]*, and *foxa3 [fkh2]*; Odenthal and Nüsslein-Volhard, 1998). Further, *hhex* has been shown to be involved in differentiation of some endodermal organs, including liver, thyroid gland, and pancreas (Bort *et al.*, 2004; Rohr and Concha, 2000; Wallace and Pack, 2003; Wallace *et al.*, 2001), and may contribute to regional specification of the duodenum (Sun and Hopkins, 2001).

An important step during the formation of the pancreas anlage is the anterior-posterior regionalization of the endoderm. Retinoic acid (RA), which has been shown to build up a rostrocaudal activity gradient in all three germ layers (Grandel *et al.*, 2002; Perz-Edwards *et al.*, 2001), is important for this anteroposterior patterning (Stafford and Prince, 2002). In embryos mutant for *neckless* (*nls*), which encodes the major retinoic acid synthesis enzyme, RALDH2, neither marker for pancreas precursors nor for hepatic precursors are expressed (Stafford and Prince, 2002). Similar conclusions have been obtained from experiments in which RA signaling was inhibited pharmacologically. Conversely, ectopic application of RA results in an RA-dependent specification of prepancreatic tissue already at the end of gastrulation, and therefore much earlier than the first expression of prepancreatic markers during normal development. RA seems to have the capacity to transfate anterior endoderm to a pancreatic fate (Stafford and Prince, 2002). A similar role in endoderm patterning has been proposed for Bmp2-signaling (Tiso *et al.*, 2002). In mutants with altered BMP2-signaling like *swirl* and *chordino*, the pancreatic primordium is reduced or expanded, respectively. BMP2b signaling, for example, defines the extension of the *her5* expression domain in anterior endoderm (Bally-Cuif *et al.*, 2000).

VI. Formation of the Pancreatic Primordium

One of the earliest specific markers expressed in the endocrine and exocrine primordia in vertebrates is the homeodomain transcription factor Ipf1 (insulin promoter factor 1; in most zebrafish publications named Pdx1—pancreatic and duodenal homeobox1; Milewski *et al.*, 1998; Ohlsson *et al.*, 1993; Stoffers *et al.*, 1997). Early expression of *ipf1 (pdx1)* in zebrafish starts at the 10-somite stage, where *ipf1 (pdx1)* marks cells in the presumptive duodenum, including cells that eventually will give rise to pancreatic tissue (Table I; Biemar *et al.*, 2001). The *ipf1 (pdx1)* expression begins in two bilateral expression domains adjacent to the midline, which coalesce at the midline at the level of the third somite by 16 hpf. Expression of *ipf1 (pdx1)* is decreasing in the epitheliated duodenum at the end of the third day of development, but persists in a subset of cells in the differentiated pancreas (Field *et al.*, 2003). Targeted knockdown of *ipf1 (pdx1)* with morpholinos resulted in a dose-dependent reduction of endocrine and exocrine tissue (Huang *et al.*, 2001a; Yee *et al.*, 2001), but not in a complete lack of endocrine cell types. This suggests that other factors must act in parallel to Ipf1 (Pdx1) during specification of pancreatic precursors (Yee *et al.*, 2001). These data are consistent with results in mouse where initial specification of the pancreatic buds is unaffected in *Ipf1 (Pdx1)* mutant embryos and only the further differentiation and morphogenesis are perturbed (Ahlgren *et al.*, 1996; Jonsson *et al.*, 1994; Offield *et al.*, 1996). A good candidate for a factor acting in parallel to Ipf1 (Pdx1) is the Mnx-homeobox protein Hlxb9 (Hb9). In *Hlxb9* mutant mice, formation of the dorsal pancreatic bud is blocked, while the initial events in ventral bud formation are not affected (Harrison *et al.*, 1999; Li *et al.*, 1999). As expression of *Hlxb9* precedes that of *Ipf1 (Pdx1)* only in the dorsal bud, it had been suggested that *Hlxb9* and *Ipf1 (Pdx1)* might have partially redundant activities during initiation of bud formation (Li *et al.*, 1999). Also in zebrafish, *hlxb9 (hb9)* starts to be expressed in the pancreatic anlage before the onset of *ipf1 (pdx1)* expression (Wendik *et al.*, 2004). Morpholino experiments in zebrafish, however, showed that knockdown of *hlxb9* has no effect on early pancreas morphogenesis. Since the expression of zebrafish *hlxb9* overlaps with that of *ipf1 (pdx1)* during initiation of pancreas development, this situation is similar to the one in the ventral bud of the mouse pancreas. While these data are consistent with the

Table I

Expression of Genes During Pancreas Development

Transcription Factor Genes	Start of Expression	Expression in Pancreas	Reference
hlxb9 (hb9)	Approximately 10 hpf onward	Pre-endocrine cells of pancreatic primordium	Wendik, 2004
hlxb9la (mnr2a)	24 hpf onward	Endodermal cells in anterior part of islet; later also in anterior duodenum, which will migrate and differentiate into exocrine tissue	Wendik, 2004
ipf1 (pdx1)	10-somite stage onwards	Pre-pancreatic cells	Biemar et al., 2001; Argenton et al., 1999
nkx2.2	10-somite stage onwards	Pancreatic primordium	Biemar et al., 2001
pax6.2	Before 12-somite stage	Pancreas primordium	Biemar et al., 2001
isl1	12-somite stage onward	Pancreatic primordium	Biemar et al., 2001
tcf2 (vhnf1)	Approximately 20 hpf onward	Fore- and hindgut	Sun and Hopkins, 2001

Gene	Stage	Expression	Reference
ptf1a (p48)	32 hpf onward	Anterior duodenum, which will migrate and differentiate into exocrine tissue	Zecchin et al., 2004
hhex	21 hpf onwards	Islet primordium (already earlier in liver primordium)	Wallace and Pack, 2003
preproinsulin (ins)	12-somite stage onward	Beta-cells	Sun and Hopkins, 2001
somatostatin2 (sst2)	16-somite stage onward	Delta-cells	Argenton et al., 1999
somatostatin1 (sst1)	24-somite stage onward	Subset of delta-cells	Devos et al., 2002
glucagon (gcg)	24-somite stage onward	alpha-cells	Argenton et al., 1999
pancreatic polypeptide (ppy)	Post 48 hpf	pp-cells (data based on immunohistochemistry, gene not identified)	Argenton et al., 1999
trypsin (try)	Post 48 hpf	Exocrine cells	Biemar et al., 2001
carboxypeptidaseA (cpa)	Post 48 hpf	Exocrine cells	Pack et al., 1996

suggestion of overlapping activities of *ipf1 (pdx1)* and *hlxb9*, direct proof for such an interaction is still missing (Li *et al.*, 1999; Wendik *et al.*, 2004).

In addition to genes expressed in the endoderm, factors emanating from the overlying notochord also seem to play a role in patterning of the endoderm (reviewed in Cleaver and Krieg, 2001; Kim and Hebrok, 2001). In zebrafish *floating head (flh)* mutants, which affect the Xnot homologue, the notochord is missing, and thus signals that may derive from axial mesoderm are not generated. While duodenal expression of *ipf1 (pdx1)* is not affected in *flh* mutant embryos, at 24 hpf only a reduced number of *preproinsulin (ins)*-expressing cells can be detected (Biemar *et al.*, 2001). Further analysis of pancreatic gene expression in *notail (ntl)* mutants, which form chorda mesoderm but do not differentiate a notochord, revealed that the notochord-derived signals required for pancreas development are already present in the chorda mesoderm, as the endocrine pancreas develops in these mutants albeit absence of differentiated notochord.

Sonic hedgehog (shh) appears to be an important signal during formation of the endocrine pancreas in several vertebrate systems, where it has been reported to be a negative regulator of pancreatic development (Hebrok *et al.*, 1998; Kim and Melton, 1998; Litingtung *et al.*, 1998; Thomas *et al.*, 2000). While *Shh* is expressed in endoderm in mouse and *Xenopus, shh*, as well as other zebrafish *hh* genes are not expressed in the zebrafish endoderm during formation of the pancreatic primordium (12-somite stage; Roy *et al.*, 2001). Furthermore, while expression of *shh* is detectable in endoderm at 24 hpf, it is clearly down-regulated in the pancreas anlage. Nevertheless, an influence of *shh* on endocrine pancreas development can clearly be demonstrated in zebrafish too. The lack of Shh-signaling in *smoothened (smu)* mutants (which cannot transduce the Shh signal), as well as inhibition of Shh-signaling by Cyclopamine result in a severe reduction of the endocrine pancreas, while overexpression by means of *shh* mRNA-injection results in an expansion of the endocrine pancreas (Table II; diIorio *et al.*, 2002; Roy *et al.*, 2001). The analysis of the relevant timing of shh signal showed that hh-signaling is required before the 6-somite stage for subsequent differentiation of endoderm into islet tissue (diIorio *et al.*, 2002). This led to a model, in which shh acts during gastrulation via the shield region onto the underlying involuting endodermal cells (diIorio *et al.*, 2002). Interestingly, formation of the exocrine pancreas is not affected by hh signaling, revealing major differences in endocrine and exocrine pancreas formation in zebrafish (Zecchin *et al.*, 2004).

Signals from other sources are also likely to contribute to pancreas development. In *Xenopus* and mice, there is indication that signals from endothelial cells contribute to pancreas and liver development (Lammert *et al.*, 2001; Matsumoto *et al.*, 2001), and may promote and maintain dorsal pancreatic development (Yoshitomi and Zaret, 2004). In zebrafish, however, the analysis of *cloche* mutant embryos, which lack the dorsal aorta, revealed no defect in the formation of insulin-expressing cells (Field *et al.*, 2003), and overexpression or morpholino knockdown of vascular endothelial growth factor-A (VEGF-A) did not significantly affect formation of insulin expressing cells (Ellertsdottir and Driever,

Table II
Zebrafish Mutations with Phenotypes Affecting Pancreas Development

Mutant	Gene and Locus Names	Phenotype	Reference
had	*atp1a1a.1* heart and mind (*Na/K ATPase*)	Primary lateral *pdx*-expressing pancreas primordia do not merge medially, perturbed endoderm morphogenesis	Ellertsdottir and Driever, unpublished; Shu et al., 2003
has	*prkci* heart and soul (*atypical protein kinase C*)	Primary lateral pdx-expressing pancreas primordia do not merge	Field et al., 2003
mib	*mind bomb* (*RING ubiquitin ligase*)	Premature exocrine differentiation	Esni et al., 2004
nls	*aldh1a2 neckless*	No differentiation of pancreas primordium	Stafford and Prince, 2002
smu	*smoh slow muscle omitted* (*smoothened homolog*)	Endocrine cells fail to develop	Roy et al., 2001; diIorio et al., 2002.
syu	*shh sonic hedgehog*	Endocrine cells fail to develop	Roy et al., 2001; diIorio et al., 2002.
tcf2 (*vhnf1*)	*tcf2* (*vhnf1*)	Patterning of gut endoderm disrupted, no differentiation of pancreas primordium	Sun and Hopkins, 2001

unpublished data). Rather, VEGF-C, knockdown of which generates a split, bilateral pancreas, may be involved in the coalescence of anterior endoderm to the midline, and therefore in endoderm morphogenesis (Ober *et al.*, 2004).

Insertional mutations have revealed an important role for the forkhead domain transcription factor Tcf2 (Vhnf1) during endoderm patterning and formation of the pancreas primordium (Sun and Hopkins, 2001). In humans, mutations in *TCF2 (Vhnf1)* can be correlated with a form of human diabetes: maturity onset diabetes of the young, type V (MODY5). The *tcf2 (vhnf1)* mutant zebrafish embryos are characterized, among other phenotypes in organogenesis, by under-development of the pancreas. At a molecular level, *tcf2 (vhnf1)* appears to be required for proper expression of *shh* and *ipf1 (pdx1)* in the endoderm: mutants have reduced *ipf1 (pdx1)* and expanded *shh* expression. These findings made Sun and Hopkins (2001) suggest a model in which *(tcf2) vhnf1* may be required for proper activation of *ipf1 (pdx1)* expression, while Ipf1 (Pdx1) and Shh in turn would act in a reciprocal fashion to down-regulate each others expression during later stages of expression.

VII. Cell Type Specification and Differentiation of the Endocrine Pancreas

A system of lateral inhibition on the basis of the Delta-Notch signaling has been shown to be involved in the decision between exocrine and endocrine fates in mouse (Apelqvist *et al.*, 1999). Mice lacking *ngn3* function (a downstream target of notch signaling) fail to generate any pancreatic endocrine cells and die postnatally from diabetes (Gradwohl *et al.*, 2000). Expression of prepancreatic markers is lost in these embryos and endocrine precursors are lacking in the mutant pancreatic epithelium. Therefore, *ngn3* is required for the specification of a common precursor for the four pancreatic endocrine cell types. Interestingly, in zebrafish, the analysis of the *mind bomb* (*mib*) mutant, in which Notch signaling is severely impaired (Itoh *et al.*, 2003), indicates that Notch signaling may also be required for other pancreatic lineage decision. *mib* mutant embryos develop an increased number of insulin-expressing cells, potentially at the expense of other pancreatic lineages. (Biemar, Epperlein, and Driever, unpublished), However, a late (24 hpf) heat shock driven expression of NotchICD, a dominant active Notch, has been reported to suppress beta-cell differentiation in zebrafish (Esni *et al.*, 2004). While exocrine lineages form in *min* embryos (Zechin *et al.*, 2004), exocrine differentiation appears prematurely, indicating a role of Notch signaling in suppression of differentiation of exocrine precursor cells (Esni *et al.*, 2004). However, a more detailed epistasis analysis is still required to determine at which level of cell fate decisions and during which step of differentiation *mib*, and thus Notch signaling, may act.

Necessary for the differentiation but not for the initial specification of endocrine cells is the homeobox protein Nkx2.2. Mice lacking Nkx2.2 do not maintain properly differentiated alpha-, beta-, and PP-cells and die of diabetes (Sussel *et al.*, 1998). Expression of the zebrafish homolog of *nkx2.2* can be detected as early as the 10-somite stage within the pancreatic primordium and is maintained during is conserved in zebrafish and mice, it is possible that the Nkx2.2 function uncovered in mice may also be conserved in zebrafish.

During early stages in pancreatic endocrine cell differentiation, postmitotic islet cells in mice initiate expression of the LIM-homeodomain protein Isl1 (Ahlgren *et al.*, 1997). Experiments with the mouse *Isl1* knockout suggest that Isl1 is required for the differentiation of all four islet cell lineages in a cell-autonomous way. A second function for Isl1 in mice is in the dorsal pancreatic mesenchyme, which appears to be necessary for the development of the exocrine pancreas (Ahlgren *et al.*, 1997). The *isl1* transcript in the zebrafish pancreatic anlage can be detected as early as the 12-somite stage and continues to be expressed past 24 hpf (Biemar *et al.*, 2001), suggesting that commitment to endocrine cell fate is initiated relatively early in zebrafish, soon after the onset of *pdx-1* expression.

Differentiation of the different subtypes of hormone-expressing cells in the endocrine compartment of the pancreas has been studied extensively in mouse, but so far signals involved in induction or transcription factors involved in specification of defined endocrine fates have not been investigated in zebrafish. The homeobox transcription factor Pax6, for example, is necessary for alpha-cell development, while another family-member, Pax4, is implicated in beta-cell differentiation (reviewed in Dohrmann *et al.*, 2000). The zebrafish pax6.2 gene is expressed within the pancreas primordium in a one-cell thick layer immediately above the yolk, expression is maintained throughout somatogenesis, and still clearly visible by 24 hpf in the region of the developing pancreas (Biemar *et al.*, 2001). The expression of a zebrafish *pax4* has not been reported so far.

A result of the gene expression cascades in the pancreatic primordium is the differentiation of the different endocrine cell types, each expressing a specific pancreatic hormone. Insulin expression starts at the 12-somite stage, roughly 1 hour after the first *ipf1 (pdx1)* transcripts become visible in the region of the developing pancreas (Biemar *et al.*, 2001). A second *preproinsulin* gene, *insb*, has been identified and isolated from the zebrafish genome (Irwin, 2004). While the previously isolated gene (now *insa*) is exclusively expressed in the pancreatic islet, the expression profile for *insb* is still elusive. Potentially the second insulin gene *insb* may be a gene that is expressed in extrapancreatic tissues of fish similar to *preproinsulin* in mammals (Dumonteil and Philippe, 1996).

The *somatostatin* (*sst3*) expression colocalizes with insulin to the core of the pancreatic islet, although expression starts only at the 20-somite stage (Argenton *et al.*, 1999). Two other *somatostatin* genes, isolated from zebrafish are not exclusively linked to pancreas development (Devos *et al.*, 2002). The *sst1* is expressed in the developing pancreas, as well as in the nervous system. The third gene, *sst3*, is a cortistatin-like prohormone and the SS-14 and SS-28 cleavage products of the comparable mammalian gene seem to play a role in neuronal depression and sleep modulation (Spier and de Lecea, 2000).

At 24 hpf, the pancreatic primordium also encompasses the *glucagon (gcg)*-expressing cells—differentiating as the last of the three hormone-expressing cells (Argenton *et al.*, 1999). In the final islet morphology, the glucagon positive cells surround a central group containing insulin- and somatostatin-expressing cells. The fourth cell type in the adult endocrine pancreas would be the ones producing pancreatic polypeptide (Argenton *et al.*, 1999). Cells expressing this protein seem to differentiate much later than the other endocrine cell types, as immunocyto-chemistry with an antibody directed against the mouse homologue reveals absence of pancreatic polypeptide at 48 hpf. This failure to detect the protein cannot be explained by the species variation, as the staining is positive in adult zebrafish pancreas.

VIII. Formation of the Exocrine Pancreas

Until recently, the morphogenic and molecular events underlying the formation of the exocrine pancreas in zebrafish have been poorly understood. Confocal time-lapse analysis of transgenic fish that express green fluorescent protein (GFP) in endodermal cells revealed a detailed picture of the morphogenesis of the developing endodermal organs (for details also see Field *et al.*, 2003). In these studies, the earliest morphological sign of the forming pancreas was found to be a thickening on the dorsal side of the still non-epithelial mono-layered gut endoderm at 24 hpf. Cells in this primordium already express endocrine-specific genes like *insulin* and *somatostatin*. Until 35 hpf, these cells have separated from the gut endoderm to form the endocrine islet on the right side of the now epithelialized gut tube. At the same time, a bud-like protrusion forms on the ventral side of the gut, in a position anterior to the islet. Until the third day of development, this anterior bud extends first in the direction of the islet and then around the islet. Unlike the islet, the anterior protrusion never detaches completely from the intestinal epithelium. Further extension of this protrusion into the posterior direction then leads to the formation of the typical head/tail structure of the exocrine tissue, suggesting that the anterior bud corresponds to the progenitors of the exocrine pancreas (Field *et al.*, 2003). These conclusions are further supported by the recent analysis of the basic Helix-Loop-Helix (bHLH) transcription factor Ptf1a (pancreas transcription factor 1; also known as p48; Kawaguchi *et al.*, 2002; Obata *et al.*, 2001; Zecchin *et al.*, 2004). The first endodermal expression of zebrafish *ptf1a* is detectable at 32 hpf, thus after formation of the endocrine islet and shortly before the formation of the anterior pancreatic bud. This initial expression of *ptf1a* is found anterior to the islet primordium in a pattern overlapping with the anterior expression of *ipf1 (pdx1)*, as well as with that of GFP in the gut:GFP line. In the course of the next 40 hours the *ptf1a* expression domain expands and shifts caudally in a pattern indistinguishable from that of the anterior bud described for the gut:GFP line. Morpholino-induced knockdown of *ptf1a* results in a loss of exocrine pancreas, as revealed by the loss of expression of the exocrine enzyme trypsin (Zecchin *et al.*, 2004). Endocrine gene expression, as well as islet morphology, is not affected in these morphants. The specific expression of *ptf1a* and its requirement for exocrine lineage development reveal major differences in pancreas formation between fish and mammals. In the mouse, genetic loss-of-function studies and lineage tracing experiments revealed requirements for *ptf1a* in converting intestinal to pancreatic lineages, as well as for the development of acinar cells (Kawaguchi *et al.*, 2002; Krapp *et al.*, 1998). Consistent with these functions, mouse *ptf1a* expression initially marks progenitor cells of exocrine and endocrine tissue and is later restricted to the exocrine lineage. In contrast, zebrafish *ptf1a* is expressed late in comparison to endocrine markers, and this expression domain is separated from the endocrine cells. In combination with the different requirement of endocrine and exocrine cells for hh-signaling, this suggests that, different from the situation in amniotes, endocrine and exocrine pancreas tissues develop from separated organ primordia in zebrafish.

However, *ptf1a* appears to have a conserved role in differentiation of exocrine tissue (Kawaguchi *et al.*, 2002). It is interesting to note that epithelial budding is required for the morphogenesis of the endocrine and exocrine pancreas in mouse, but only for the formation of the exocrine pancreas in zebrafish. In this context, *ptf1a* appears to have an additional conserved function upstream of pancreatic budding in converting intestinal fates into pancreatic fates. This could indicate that conserved genetic programs are used in pancreas development of fish and mammals, but that these functions happen in the context of a different temporal organization of pancreas morphogenesis.

A second gene that recently has been shown to function in exocrine pancreas development is the Hlxb9-related factor Hlxb9la (Mnr2a; Wendik *et al.*, 2004). The *hlxb9la* expression marks two populations of endodermal cells, one that emerges from the gut from 30 hpf onward and later marks exocrine pancreas. This expression is very similar to that of *ptf1a*, suggesting that *hlxb9la* expression marks cells of the exocrine lineage too. Expression of *hlxb9la* is further detected from 24 hpf onward in a few cells associated with endocrine tissue. The fate of these has not yet been determined. Notably, these early *hlxb9la*-expressing cells occupy a position close to the islet, which later, at 76 hpf, correlates with the position of individual *insulin*-expressing cells (Field *et al.*, 2003). In other vertebrate systems, duct cells have been indicated to possess stem cell potential in the pancreas (Bonner-Weir *et al.*, 1993; Pictet *et al.*, 1972). In a similar fashion, the region of late beta-cell differentiation marked by *mnr2* expression in zebrafish lies in the proximity of the duct region and therefore appears to harbor bipotential precursors of endocrine and exocrine cells. While *mnr2* genes are not known from mammals, chicken *mnr2* shows a specific expression domain in the bipotential pancreatic precursors of the early pancreatic buds (Grapin-Botton *et al.*, 2001).

Morpholino knockdown of *hlxb9la* results in a reduced size and an abnormal morphology of the exocrine pancreas after 3 dpf. Acinar differentiation is not blocked in the morphants, as revealed by the finding that expression of *trypsin* is reduced, but not missing. Also formation of the endocrine islet and early morphogenesis of the anterior pancreatic bud were not affected after knockdown of *hlxb9la*. This suggests that *hlxb9la* function is required during late morphogenesis of the exocrine pancreas (Wendik *et al.*, 2004). Formation of the mature exocrine pancreas is a complex process that requires extensive proliferation, directed migration, and well-timed terminal differentiation. The molecular mechanisms underlying these processes are poorly characterized. Hlxb9la is involved in these processes, and its further analysis may provide some hints toward location and function of precursor cell populations during the growth phase of the pancreas.

IX. Perspectives

While many of the molecular factors involved are conserved, temporal and morphological differences between pancreas development in zebrafish and mammals promise to help provide insight into basic mechanisms of pancreatogenesis. The early occurrence of endocrine precursors and the spatially and temporally separate appearance of exocrine precursors in zebrafish may reflect the stepwise progress during evolution of the pancreas. In protochordates, insulin-, somatostatin-, and glucagon-producing cells exist solely dispersed in the gut epithelium (reviewed in Youson and Al-Mahrouki, 1999). This distribution reflects the early differentiation of beta- and delta-cells in the zebrafish endodermal sheet during somatogenesis. Larval lampreys develop a one-hormone islet (beta-cells), and adult lampreys, with their three-hormone islet, reveal the trend to the four-hormone islet, which in cartilaginous and bony fish also include glucagon-producing cells. Islets in hagfish are associated with the bile duct, and in adult lamprey with the gut lumen. A pancreatic system composed of duct and acinar cells evolved separately during evolution of the fishes. These steps reflect the mechanisms involved in zebrafish pancreas development as they have become evident from analysis of gene expression and morphogenesis. It may appear that the embryonic phase of pancreas formation reflects four separate processes: (1) specification of endocrine cells; (2) migration of precursors and maturing endocrine cells to form the islet (as the "posterior bud" in zebrafish); (3) specification of exocrine precursors (in an "anterior bud" in zebrafish); and (4) morphogenesis of the exocrine component with ductal and acinar differentiation. In mammals, the more evolved organogenesis of the pancreas may make it difficult to distinguish such evolutionary modules of pancreas development, and thus zebrafish emerge as an opportunity to identify basic molecular mechanisms of cell specification and morphogenesis. Further, after an initial specification of endocrine cells within the endodermal sheet during segmentation, a phase of growth of the islet occurs. Even more extensive growth is observed for the exocrine pancreas. Work in zebrafish (Field *et al.*, 2003; Wendik *et al.*, 2004) may point toward location and function of precursor cell populations acting as "stem-cells," which have been difficult to characterize in mammalian systems. The transparency of the embryos and the availability of GFP-transgenic lines marking defined cell populations during pancreas development (pdx:GFP, ins:GFP—Huang *et al.*, 2001b; gut:GFP—Field *et al.*, 2003) will facilitate the analysis of pancreas morphogenesis, the isolation of additional mutations in genetic screens, and the analysis of gene function.

Acknowledgments

We thank Elin Ellertsdottir for her helpful comments. We are grateful to Francesco Argenton and Bernard Peers for communicating unpublished results. This work was supported in part by a grant from the EU.

References

Agathon, A., Thisse, B., and Thisse, C. (2001). Morpholino knock-down of antivin1 and antivin2 upregulates nodal signaling. *Genesis* **30,** 178–182.

Ahlgren, U., Jonsson, J., and Edlund, H. (1996). The morphogenesis of the pancreatic mesenchyme is uncoupled from that of the pancreatic epithelium in IPF1/PDX1-deficient mice. *Development* **122,** 1409–1416.

Ahlgren, U., Pfaff, S. L., Jessell, T. M., Edlund, T., and Edlund, H. (1997). Independent requirement for ISL1 in formation of pancreatic mesenchyme and islet cells. *Nature* **385,** 257–260.

Alexander, J., and Stainier, D. Y. (1999). A molecular pathway leading to endoderm formation in zebrafish. *Curr. Biol.* **9,** 1147–1157.

Aoki, T. O., David, N. B., Minchiotti, G., Saint-Etienne, L., Dickmeis, T., Persico, G. M., Strahle, U., Mourrain, P., and Rosa, F. M. (2002a). Molecular integration of casanova in the Nodal signalling pathway controlling endoderm formation. *Development* **129,** 275–286.

Aoki, T. O., Mathieu, J., Saint-Etienne, L., Rebagliati, M. R., Peyrieras, N., and Rosa, F. M. (2002b). Regulation of nodal signalling and mesendoderm formation by TARAM-A, a TGFbeta-related type I receptor. *Dev. Biol.* **241,** 273–288.

Apelqvist, A., Li, H., Sommer, L., Beatus, P., Anderson, D. J., Honjo, T., Hrabe de Angelis, M., Lendahl, U., and Edlund, H. (1999). Notch signalling controls pancreatic cell differentiation. *Nature* **400,** 877–881.

Argenton, F., Zecchin, E., and Bortolussi, M. (1999). Early appearance of pancreatic hormone-expressing cells in the zebrafish embryo. *Mech. Dev.* **87,** 217–221.

Bally-Cuif, L., Goutel, C., Wassef, M., Wurst, W., and Rosa, F. (2000). Coregulation of anterior and posterior mesendodermal development by a hairy-related transcriptional repressor. *Genes Dev.* **14,** 1664–1677.

Biemar, F., Argenton, F., Schmidtke, R., Epperlein, S., Peers, B., and Driever, W. (2001). Pancreas development in zebrafish: Early dispersed appearance of endocrine hormone expressing cells and their convergence to form the definitive islet. *Dev. Biol.* **230,** 189–203.

Bonner-Weir, S., Baxter, L. A., Schuppin, G. T., and Smith, F. E. (1993). A second pathway for regeneration of adult exocrine and endocrine pancreas. A possible recapitulation of embryonic development. *Diabetes* **42,** 1715–1720.

Bort, R., Martinez-Barbera, J. P., Beddington, R. S., and Zaret, K. S. (2004). Hex homeobox gene-dependent tissue positioning is required for organogenesis of the ventral pancreas. *Development* **131,** 797–806.

Chen, Y., and Schier, A. F. (2001). The zebrafish Nodal signal Squint functions as a morphogen. *Nature* **411,** 607–610.

Cleaver, O., and Krieg, P. A. (2001). Notochord patterning of the endoderm. *Dev. Biol.* **234,** 1–12.

Devos, N., Deflorian, G., Biemar, F., Bortolussi, M., Martial, J. A., Peers, B., and Argenton, F. (2002). Differential expression of two somatostatin genes during zebrafish embryonic development. *Mech. Dev.* **115**, 133–137.

Dick, A., Mayr, T., Bauer, H., Meier, A., and Hammerschmidt, M. (2000). Cloning and characterization of zebrafish smad2, smad3 and smad4. *Gene* **246**, 69–80.

Dickmeis, T., Rastegar, S., Aanstad, P., Clark, M., Fischer, N., Korzh, V., and Strahle, U. (2001). Expression of the anti-dorsalizing morphogenetic protein gene in the zebrafish embryo. *Dev. Genes Evol.* **211**, 568–572.

diIorio, P. J., Moss, J. B., Sbrogna, J. L., Karlstrom, R. O., and Moss, L. G. (2002). Sonic hedgehog is required early in pancreatic islet development. *Dev. Biol.* **244**, 75–84.

Dohrmann, C., Gruss, P., and Lemaire, L. (2000). Pax genes and the differentiation of hormone-producing endocrine cells in the pancreas. *Mech. Dev.* **92**, 47–54.

Dougan, S. T., Warga, R. M., Kane, D. A., Schier, A. F., and Talbot, W. S. (2003). The role of the zebrafish nodal-related genes squint and cyclops in patterning of mesendoderm. *Development* **130**, 1837–1851.

Dumonteil, E., and Philippe, J. (1996). Insulin gene: Organisation, expression and regulation. *Diabetes Metab.* **22**, 164–173.

Edlund, H. (2002). Pancratic organogenesis—developmental mechanisms and implications for therapy. *Nat. Rev. Genet.* **3**, 524–532.

Eilertson, C. D., and Sheridan, M. A. (1995). Pancreatic somatostatin-14 and somatostatin-25 release in rainbow trout is stimulated by glucose and arginine. *Am. J. Physiol.* **269**, 1017–1023.

Erter, C. E., Solnica-Krezel, L., and Wright, C. V. (1998). Zebrafish nodal-related 2 encodes an early mesendodermal inducer signaling from the extraembryonic yolk syncytial layer. *Dev. Biol.* **204**, 361–372.

Esni, F., Ghosh, B., Biankin, A. V., Lin, J. W., Albert, M. A., Yu, X., MacDonald, R. J., Civin, C. I., Real, F. X., Pack, M. A., Ball, D. W., and Leach, S. D. (2004). Notch inhibits Ptf1 function and acinar cell differentiation in developing mouse and zebrafish pancreas. *Development* **131**, 4213–4224.

Feldman, B., Gates, M. A., Egan, E. S., Dougan, S. T., Rennebeck, G., Sirotkin, H. I., Schier, A. F., and Talbot, W. S. (1998). Zebrafish organizer development and germ-layer formation require nodal-related signals. *Nature* **395**, 181–185.

Field, H. A., Dong, P. D., Beis, D., and Stainier, D. Y. (2003). Formation of the digestive system in zebrafish. II. Pancreas morphogenesis. *Dev. Biol.* **261**, 197–208.

Gradwohl, G., Dierich, A., LeMeur, M., and Guillemot, F. (2000). Neurogenin3 is required for the development of the four endocrine cell lineages of the pancreas. *Proc. Natl. Acad. Sci. USA* **97**, 1607–1611.

Grandel, H., Lun, K., Rauch, G. J., Rhinn, M., Piotrowski, T., Houart, C., Sordino, P., Kuchler, A. M., Schulte-Merker, S., Geisler, R., *et al.* (2002). Retinoic acid signalling in the zebrafish embryo is necessary during pre-segmentation stages to pattern the anterior-posterior axis of the CNS and to induce a pectoral fin bud. *Development* **129**, 2851–2865.

Grapin-Botton, A., Majithia, A. R., and Melton, D. A. (2001). Key events of pancreas formation are triggered in gut endoderm by ectopic expression of pancreatic regulatory genes. *Genes Dev.* **15**, 444–454.

Harrison, K. A., Thaler, J., Pfaff, S. L., Gu, H., and Kehrl, J. H. (1999). Pancreas dorsal lobe agenesis and abnormal islets of Langerhans in Hlxb9-deficient mice. *Nat. Genet.* **23**, 71–75.

Hebrok, M., Kim, S. K., and Melton, D. A. (1998). Notochord repression of endodermal Sonic hedgehog permits pancreas development. *Genes Dev.* **12**, 1705–1713.

Herrera, P. L. (2000). Adult insulin- and glucagon-producing cells differentiate from two independent cell lineages. *Development* **127**, 2317–2322.

Huang, H., Liu, N., and Lin, S. (2001a). Pdx-1 knockdown reduces insulin promoter activity in zebrafish. *Genesis* **30**, 134–136.

Huang, H., Vogel, S. S., Liu, N., Melton, D. A., and Lin, S. (2001b). Analysis of pancreatic development in living transgenic zebrafish embryos. *Mol. Cell. Endocrinol.* **177**, 117–124.

Irwin, D. M. (2004). A second insulin gene in fish genomes. *Gen. Comp. Endocrinol.* **135**, 150–158.

Itoh, M., Kim, C. H., Palardy, G., Oda, T., Jiang, Y. J., Maust, D., Yeo, S. Y., Lorick, K., Wright, G. J., Ariza-McNaughton, L., *et al.* (2003). Mind bomb is a ubiquitin ligase that is essential for efficient activation of Notch signaling by Delta. *Dev. Cell* **4**, 67–82.

Jonsson, J., Carlsson, L., Edlund, T., and Edlund, H. (1994). Insulin-promoter-factor 1 is required for pancreas development in mice. *Nature* **371**, 606–609.

Kanai-Azuma, M., Kanai, Y., Gad, J. M., Tajima, Y., Taya, C., Kurohmaru, M., Sanai, Y., Yonekawa, H., Yazaki, K., Tam, P. P., *et al.* (2002). Depletion of definitive gut endoderm in Sox17-null mutant mice. *Development* **129**, 2367–2379.

Kawaguchi, Y., Cooper, B., Gannon, M., Ray, M., MacDonald, R. J., and Wright, C. V. (2002). The role of the transcriptional regulator Ptf1a in converting intestinal to pancreatic progenitors. *Nat. Genet.* **32**, 128–134.

Kikuchi, Y., Trinh, L. A., Reiter, J. F., Alexander, J., Yelon, D., and Stainier, D. Y. (2000). The zebrafish bonnie and clyde gene encodes a Mix family homeodomain protein that regulates the generation of endodermal precursors. *Genes Dev.* **14**, 1279–1289.

Kikuchi, Y., Verkade, H., Reiter, J. F., Kim, C. H., Chitnis, A. B., Kuroiwa, A., and Stainier, D. Y. (2004). Notch signaling can regulate endoderm formation in zebrafish. *Dev. Dyn.* **229**, 756–762.

Kim, S. K., and Hebrok, M. (2001). Intercellular signals regulating pancreas development and function. *Genes Dev.* **15**, 111–127.

Kim, S. K., and Melton, D. A. (1998). Pancreas development is promoted by cyclopamine, a hedgehog signaling inhibitor. *Proc. Natl. Acad. Sci. USA* **95**, 13036–13041.

Krapp, A., Knofler, M., Ledermann, B., Burki, K., Berney, C., Zoerkler, N., Hagenbuchle, O., and Wellauer, P. K. (1998). The bHLH protein PTF1-p48 is essential for the formation of the exocrine and the correct spatial organization of the endocrine pancreas. *Genes Dev.* **12**, 3752–3763.

Kunwar, P. S., Zimmerman, S., Bennett, J. T., Chen, Y., Whitman, M., and Schier, A. F. (2003). Mixer/Bon and FoxH1/Sur have overlapping and divergent roles in Nodal signaling and mesendoderm induction. *Development* **130**, 5589–5599.

Lammert, E., Cleaver, O., and Melton, D. (2001). Induction of pancreatic differentiation by signals from blood vessels. *Science* **294**, 564–567.

Li, H., Arber, S., Jessell, T. M., and Edlund, H. (1999). Selective agenesis of the dorsal pancreas in mice lacking homeobox gene Hlxb9. *Nat. Genet.* **23**, 67–70.

Litingtung, Y., Lei, L., Westphal, H., and Chiang, C. (1998). Sonic hedgehog is essential to foregut development. *Nat. Genet.* **20**, 58–61.

Lunde, K., Belting, H. G., and Driever, W. (2004). Zebrafish pou5f1/pou2, homolog of mammalian Oct4, functions in the endoderm specification cascade. *Curr. Biol.* **14**, 48–55.

Matsumoto, K., Yoshitomi, H., Rossant, J., and Zaret, K. S. (2001). Liver organogenesis promoted by endothelial cells prior to vascular function. *Science* **294**, 559–563.

Milewski, W. M., Duguay, S. J., Chan, S. J., and Steiner, D. F. (1998). Conservation of PDX-1 structure, function, and expression in zebrafish. *Endocrinology* **139**, 1440–1449.

Muller, F., Blader, P., Rastegar, S., Fischer, N., Knochel, W., and Strahle, U. (1999). Characterization of zebrafish smad1, smad2 and smad5: The amino-terminus of smad1 and smad5 is required for specific function in the embryo. *Mech. Dev.* **88**, 73–88.

Obata, J., Yano, M., Mimura, H., Goto, T., Nakayama, R., Mibu, Y., Oka, C., and Kawaichi, M. (2001). p48 subunit of mouse PTF1 binds to RBP-Jkappa/CBF-1, the intracellular mediator of Notch signalling, and is expressed in the neural tube of early stage embryos. *Genes Cells* **6**, 345–360.

Ober, E. A., Field, H. A., and Stainier, D. Y. (2003). From endoderm formation to liver and pancreas development in zebrafish. *Mech. Dev.* **120**, 5–18.

Ober, E. A., Olofsson, B., Makinen, T., Jin, S. W., Shoji, W., Koh, G. Y., Alitalo, K., and Stainier, D. Y. (2004). Vegfc is required for vascular development and endoderm morphogenesis in zebrafish. *EMBO Rep.* **5**, 78–84.

Odenthal, J., and Nusslein-Volhard, C. (1998). Fork head domain genes in zebrafish. *Dev. Genes Evol.* **208**, 245–258.

Offield, M. F., Jetton, T. L., Labosky, P. A., Ray, M., Stein, R. W., Magnuson, M. A., Hogan, B. L., and Wright, C. V. (1996). PDX-1 is required for pancreatic outgrowth and differentiation of the rostral duodenum. *Development* **122**, 983–995.

Ohlsson, H., Karlsson, K., and Edlund, T. (1993). IPF1, a homeodomain-containing transactivator of the insulin gene. *EMBO J.* **12**, 4251–4259.

Pack, M., Solnica-Krezel, L., Malicki, J., Neuhauss, S. C., Schier, A. F., Stemple, D. L., Driever, W., and Fishman, M. C. (1996). Mutations affecting development of zebrafish digestive organs. *Development* **123**, 321–328.

Perz-Edwards, A., Hardison, N. L., and Linney, E. (2001). Retinoic acid-mediated gene expression in transgenic reporter zebrafish. *Dev. Biol.* **229**, 89–101.

Peyrieras, N., Lu, Y., Renucci, A., Lemarchandel, V., and Rosa, F. (1996). Inhibitory interactions controlling organizer activity in fish. *C R Acad. Sci. III* **319**, 1107–1112.

Pictet, R. L., Clark, W. R., Williams, R. H., and Rutter, W. J. (1972). An ultrastructural analysis of the developing embryonic pancreas. *Dev. Biol.* **29**, 436–467.

Pogoda, H. M., Solnica-Krezel, L., Driever, W., and Meyer, D. (2000). The zebrafish forkhead transcription factor FoxH1/Fast1 is a modulator of nodal signaling required for organizer formation. *Curr. Biol.* **10**, 1041–1049.

Poulain, M., and Lepage, T. (2002). Mezzo, a paired-like homeobox protein is an immediate target of Nodal signalling and regulates endoderm specification in zebrafish. *Development* **129**, 4901–4914.

Rebagliati, M. R., Toyama, R., Haffter, P., and Dawid, I. B. (1998). Cyclops encodes a nodal-related factor involved in midline signaling. *Proc. Natl. Acad. Sci. USA* **95**, 9932–9937.

Reim, G., Mizoguchi, T., Stainier, D. Y., Kikuchi, Y., and Brand, M. (2004). The POU domain protein spg (pou2/Oct4) is essential for endoderm formation in cooperation with the HMG domain protein casanova. *Dev. Cell* **6**, 91–101.

Reiter, J. F., Kikuchi, Y., and Stainier, D. Y. (2001). Multiple roles for Gata5 in zebrafish endoderm formation. *Development* **128**, 125–135.

Renucci, A., Lemarchandel, V., and Rosa, F. (1996). An activated form of type I serine/threonine kinase receptor TARAM-A reveals a specific signalling pathway involved in fish head organiser formation. *Development* **122**, 3735–3743.

Rodaway, A., Takeda, H., Koshida, S., Broadbent, J., Price, B., Smith, J. C., Patient, R., and Holder, N. (1999). Induction of the mesendoderm in the zebrafish germ ring by yolk cell-derived TGF-beta family signals and discrimination of mesoderm and endoderm by FGF. *Development* **126**, 3067–3078.

Rohr, K. B., and Concha, M. L. (2000). Expression of nk2.1a during early development of the thyroid gland in zebrafish. *Mech. Dev.* **95**, 267–270.

Roy, S., Qiao, T., Wolff, C., and Ingham, P. W. (2001). Hedgehog signaling pathway is essential for pancreas specification in the zebrafish embryo. *Curr. Biol.* **11**, 1358–1363.

Sampath, K., Rubinstein, A. L., Cheng, A. M., Liang, J. O., Fekany, K., Solnica-Krezel, L., Korzh, V., Halpern, M. E., and Wright, C. V. (1998). Induction of the zebrafish ventral brain and floorplate requires cyclops/nodal signalling. *Nature* **395**, 185–189.

Shu, X., Cheng, K., Patel, N., Patel, N., Chen, F., Joseph, E, Tsai, H. J., and Chen, J. N. (2003). Na, K-ATPase is essential for embryonic heart development in the zebrafish. *Development* **130**, 6165–6173.

Sirotkin, H. I., Gates, M. A., Kelly, P. D., Schier, A. F., and Talbot, W. S. (2000). Fast1 is required for the development of dorsal axial structures in zebrafish. *Curr. Biol.* **10**, 1051–1054.

Slack, J. M. (1995). Developmental biology of the pancreas. *Development* **121**, 1569–1580.

Sosa-Pineda, B., Chowdhury, K., Torres, M., Oliver, G., and Gruss, P. (1997). The Pax4 gene is essential for differentiation of insulin-producing beta cells in the mammalian pancreas. *Nature* **386**, 399–402.

Spier, A. D., and de Lecea, L. (2000). Cortistatin: A member of the somatostatin neuropeptide family with distinct physiological functions. *Brain Res. Brain Res. Rev.* **33,** 228–241.

St-Onge, L., Sosa-Pineda, B., Chowdhury, K., Mansouri, A., and Gruss, P. (1997). Pax6 is required for differentiation of glucagon-producing alpha-cells in mouse pancreas. *Nature* **387,** 406–409.

Stafford, D., and Prince, V. E. (2002). Retinoic acid signaling is required for a critical early step in zebrafish pancreatic development. *Curr. Biol.* **12,** 1215–1220.

Stainier, D. Y. (2002). A glimpse into the molecular entrails of endoderm formation. *Genes Dev.* **16,** 893–907.

Stoffers, D. A., Zinkin, N. T., Stanojevic, V., Clarke, W. L., and Habener, J. F. (1997). Pancreatic agenesis attributable to a single nucleotide deletion in the human IPF1 gene coding sequence. *Nat. Genet.* **15,** 106–110.

Sun, Z., and Hopkins, N. (2001). vhnf1, the MODY5 and familial GCKD-associated gene, regulates regional specification of the zebrafish gut, pronephros, and hindbrain. *Genes Dev.* **15,** 3217–3229.

Sussel, L., Kalamaras, J., Hartigan-O'Connor, D. J., Meneses, J. J., Pedersen, R. A., Rubenstein, J. L., and German, M. S. (1998). Mice lacking the homeodomain transcription factor Nkx2.2 have diabetes due to arrested differentiation of pancreatic beta cells. *Development* **125,** 2213–2221.

Thisse, B., Wright, C. V., and Thisse, C. (2000). Activin- and Nodal-related factors control antero-posterior patterning of the zebrafish embryo. *Nature* **403,** 425–428.

Thomas, M. K., Rastalsky, N., Lee, J. H., and Habener, J. F. (2000). Hedgehog signaling regulation of insulin production by pancreatic beta-cells. *Diabetes* **49,** 2039–2047.

Tiso, N., Filippi, A., Pauls, S., Bortolussi, M., and Argenton, F. (2002). BMP signalling regulates anteroposterior endoderm patterning in zebrafish. *Mech. Dev.* **118,** 29–37.

Wallace, K. N., and Pack, M. (2003). Unique and conserved aspects of gut development in zebrafish. *Dev. Biol.* **255,** 12–29.

Wallace, K. N., Yusuff, S., Sonntag, J. M., Chin, A. J., and Pack, M. (2001). Zebrafish hhex regulates liver development and digestive organ chirality. *Genesis* **30,** 141–143.

Warga, R. M., and Nüsslein-Volhard, C. (1999). Origin and development of the zebrafish endoderm. *Development* **126,** 827–838.

Warga, R. M., and Stainier, D. Y. (2002). The guts of endoderm formation. *Results Probl. Cell Differ.* **40,** 28–47.

Wendelaar Bonga, S. E. (1993). Endocrinology. *In* "The Physiology of Fishes" (D. E. Evans, ed.), pp. 469–502. CRC Press, Boca Raton, FL.

Wendik, B., Maier, E., and Meyer, D. (2004). Zebrafish mnx genes in endocrin and exocrine pancreas formation. *Devel. Biol.* **268,** 372–383.

Yee, N. S., Yusuff, S., and Pack, M. (2001). Zebrafish pdx1 morphant displays defects in pancreas development and digestive organ chirality, and potentially identifies a multipotent pancreas progenitor cell. *Genesis* **30,** 137–140.

Yoshitomi, H., and Zaret, K. S. (2004). Endothelial cell interactions initiate dorsal pancreas development by selectively inducing the transcription factor Ptf1a. *Development* **131,** 807–817.

Youson, J. H., and Al-Mahrouki, A. A. (1999). Ontogenetic and phylogenetic development of the endocrine pancreas (islet organ) in fish. *Gen. Comp. Endocrinol.* **116,** 303–335.

Zecchin, E., Mavropoulos, A., Devos, N., Filippi, A., Tiso, N., Meyer, D., Peers, B., Bortolussi, M., and Argenton, F. (2004). Evolutionary conserved role of ptf1a in the specification of exocrine pancreatic fates. *Dev. Biol.* **268,** 174–184.

PART III

Disease Models

CHAPTER 25

Mosaic Eyes, Genomic Instability Mutants, and Cancer Susceptibility

Jessica L. Moore,[*,‡] Erin E. Gestl,[*,†] and Keith C. Cheng[*,†]

*Jake Gittlen Cancer Research Institute
Department of Pathology
Pennsylvania State College of Medicine
Hershey, Pennsylvania 17033

†Department of Biochemistry and Molecular Biology
Pennsylvania State College of Medicine
Hershey, Pennsylvania 17033

‡Department of Biology
University of South Florida
Tampa, Florida 33620

The zebrafish mosaic eye assay developed by George Streisinger (1984) takes advantage of the organism's transparency to provide an excellent assay for detecting somatic mutation. We now know that genomic instability contributes to cancer. We therefore took advantage of this assay to identify zebrafish mutants with increased frequencies of somatic mutation and spontaneous cancer. This chapter describes the details of mutagenesis and half-tetrad screening, the basis and practical use of the mosaic eye assay, and the histological methods used to study genomic instability mutants and cancer susceptibility. These techniques should prove useful to other zebrafish researchers, since they are broadly applicable to other biological investigations of zebrafish embryos, larvae and adults.

I. Introduction

Cancer is a multistep process that involves an accumulation of somatic mutations (Lengaur *et al.*, 1998; Loeb *et al.*, 2003). The mosaic eye assay designed by George Streisinger (1984) allowed us to readily score somatic mutation by changes in pigmentation. This was the foundation for a screen in zebrafish for genomic instability mutations to identify novel vertebrate genes that are important in cancer (Moore *et al.*, 2004). In bacteria (Cox, 1976) and yeast (Kolodner *et al.*, 2002), "mutator" phenotypes have been studied in which heritable mutations lead to increased frequencies of somatic mutations. Mutations in genes affecting DNA, including repair, replication, chromosome segregation, and epigenetic control of gene expression have been characterized as mutators (reviewed in Cheng and Loeb, 1997; Table I). In humans, a growing list of heritable cancer phenotypes can be classified as genomic instability syndromes, in which the accumulation of somatic mutations leads to early onset of a variety of spontaneous cancers. These syndromes include Bloom's syndrome, xeroderma pigmentosum, ataxia telangiectasia, Li-Fraumeni syndrome, and hereditary nonpolyposis colon cancer (reviewed in Charames and Bapat, 2003; Hoeijmakers, 2001). Many carriers of genomic instability mutations also display a predisposition to spontaneous cancer.

The ability to perform genetic screens in this vertebrate model system makes it possible to recover mutations in genes that lead to increased frequencies of somatic

Table I
Targets of Genomic Instability[a]

DNA Replication
    Base-selection
    Proofreading
    Mismatch correction
DNA repair
Recombination
Topological management of DNA
Chromosome segregation
Cell cycle regulation
Oncogenes
Tumor-suppressor genes
Metastasis and metastasis-suppressor genes
Senescence genes
Epigenetic regulation of gene expression

[a]Adapted from Cheng and Loeb, 1997.

mutations and cancer susceptibility. We used an embryonic assay for mosaic eye pigmentation to identify genomic instability mutations. The mosaic eye assay was initially described by George Streisinger in a presentation at the National Cancer Institute in which he proposed using zebrafish as a small animal model for carcinogenicity testing (Streisinger, 1984). In this assay, zebrafish embryos, heterozygous for the *golden* pigment locus (*gol/+*), were treated with varying doses of physical and chemical mutagens, including gamma irradiation, ethyl methanesulfonate (EMS), and ethyl nitrosurea (ENU). All experiments described here were done with the *gol^{b1}* allele. Somatic mutations of the wild-type pigmentation allele that resulted from any of these treatments could be seen as clones of "golden" cells in the retinal pigment epithelium (RPE) of the embryonic eye, a monolayer of polygonal cells. These experiments demonstrated the applicability of the mosaic eye assay for determining dose-response of a carcinogen. Heterozygous *gol/+* embryos were treated with increasing doses of gamma irradiation at 5 hpf. The frequency of mosaic eyes in 2-day-old embryos was shown to increase in direct proportion to the amount of radiation received, with a dose of approximately 500 R resulting in a 50% frequency of mosaic eyes. A re-creation of this phenomenon is shown in Fig. 1A. Treatment of blastula stage embryos with 6.0 mM EMS or 0.15 mM ENU also induced 50% mosaic eyes. The frequency of mosaic eyes through a time course of irradiation during the development of *gol/+* embryos was used to determine that 40 cells in the 1000-cell-stage embryo contribute to the final approximately 550 RPE cells per eye (Streisinger *et al.*, 1989). These studies prove that the *golden* gene can serve as a genetic marker for somatic mutations in the live embryo. Our mutant screen was based on the hypothesis that ENU-induced genomic instability mutations would cause somatic mutations detectable with the zebrafish mosaic eye assay

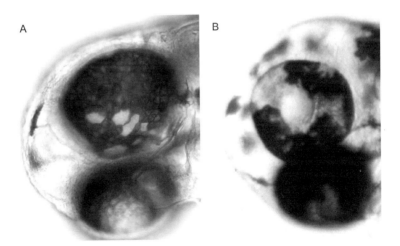

Fig. 1 Examples of mosaic eyes seen in 72 hpf gol/+ embryos. (A) Mosaic eye induced by irradiation at 5 hpf with approximately 510R. (B) Mosaic eye in a *gin-1/gin-1; gol/+* 72-hpf half-tetrad embryo.

(Fig. 1B). In this chapter we describe details of genetic screen and of techniques used to characterize our genomic instability mutants.

II. Mutagenesis and Screening for Mosaic Eyes

Genomic instability (*gin*) and cancer susceptibility mutations can be induced by exposure to the point mutagen ENU, and mutant embryos identified by the mosaic eye phenotype (Fig. 2; Moore *et al.*, 2004). We found it important to use healthy, fertile males between 6 and 9 months of age for ENU mutagenesis, to ensure breeding success several months after ENU treatment. Fertility can be determined by crossing wild-type males with homozygous *gol/gol* females, 1 month before mutagenesis. The resulting clutches of eggs are scored for fertilization and normal morphology at 48 hpf. Embryos are also scored for pigmentation. The presence of *golden* embryos indicates a background mutation at this locus, while an embryo with mosaic eye pigmentation may be attributed to either a background genomic instability mutation or a defect in eye development. We identified highly fertile males by breeding them with fertile females. Only males that successfully fertilized at least 75% of a clutch of 100 embryos, with 90% normal morphology and 100% wild-type pigmentation were used for mutagenesis.

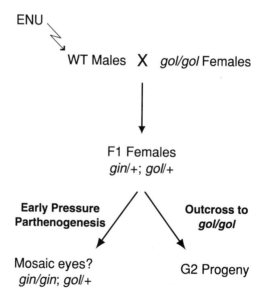

Fig. 2 Wild-type zebrafish males were treated with ethyl nitrosourea (ENU) and then outcrossed to *gol/gol* females. F1 females were screened by early pressure parthenogenesis for *gin* mutations. Half-tetrad embryos were scored using the mosaic eye assay. F1 carrier females were then outcrossed to *gol/gol* males to generate *gin* mutant families.

We found the most productive protocol to be 3 treatments 1 h in length with 2.5 mM ENU (Mullins *et al.*, 1994; Solnica-Krezel *et al.*, 1994). A subset of the pre-tested males can be treated with buffer alone to establish the background frequency of ENU-induced mutations and to generate genetically related wild-type embryos, larvae, or adults for comparison with the mutagenized lines. After mutagenesis, the fish are bred for 2 weeks to void mature sperm that were exposed to the mutagen. ENU-induced base modifications may occur on only 1 of the 2 DNA strands in mature sperm, resulting in genetically mosaic progeny and non-germline inheritance of these mutations (Grunwald and Streisinger, 1992; Mullins *et al.*, 1994). Such strand-specific mutation in the *golden* sequence will produce mosaics that are not the result of genomic instability. ENU-treated males are then bred to homozygous *gol/gol* females bimonthly to generate *gol/+* F1 families. All F1 embryos are screened for mosaic eye pigmentation due to potential dominant genomic instability mutations and for the presence of rare phenotypically *golden* embryos that result from ENU mutations in this gene. These rare *golden* embryos provide an estimate of the average specific-locus hit rate by the mutagen (Russell, 1951). In our experiment, two *golden* mutations were seen among 1818 embryos produced by 10 males mutagenized with 2.5 mM ENU, a single locus mutation frequency of 0.11%. No mosaic eyes or *golden* embryos were seen from the control males from the same experiment, (0/1460 embryos scored) indicating that the frequency of spontaneous mutation at this locus was less than 0.07% (Beckwith *et al.*, 2000; Moore *et al.*, 2004). Previous studies have shown that the specific-locus frequency of mutation at *golden* is about the same as that of the *albino* gene (Mullins *et al.*, 1994; Solnica-Krezel *et al.*, 1994), and therefore may represent an average genetic response to the mutagenic effects of ENU.

We chose to perform a one-generation screen for genomic instability mutations, using early pressure parthenogenesis (EP) to screen F1 females (Gestl *et al.,* 1996; Streisinger *et al.*, 1981; Westerfield, 1995). There are several advantages of a gynogenetic screen for recovery of genomic instability mutants (Cheng and Moore, 1997; Pelegri and Schulte-Merker, 1999):

1. Since the *golden* locus is near the telomere of LG18, 89% of the half-tetrad embryos from a heterozygous female are *gol/+*. Therefore, most of the half-tetrad embryos can be scored for mosaic eyes (Streisinger *et al.*, 1986).

2. Early pressure parthenogenesis allows the generation of a *gin/gin*; *gol/+* genotype for any *gin* mutation. The frequency of *gin/gin* will be between 5 to 50%, depending on the distance from the locus to the centromere (Cheng and Moore, 1997).

3. An F1 gynogenetic screen can begin as early as 4 months after ENU treatment of the males, while an F2 screen entails additional time necessary to raise the second generation. The one-generation screen yields new mutations from fewer fish. Therefore making fewer demands on the fish facility.

Even though two-generation screens have certain advantages over a one-generation screen (Patton and Zon, 2001), it is not practical for the recovery of genomic instability mutations using the mosaic eye phenotype. We note the following caveats for half-tetrad parthenogenesis screens:

1. A one-generation screen requires proficiency in performing early pressure parthenogenesis and other zebrafish *in vitro* fertilization procedures. There is a greater risk of killing valuable carriers during these procedures by breeding naturally.

2. When screening half-tetrad embryos, a relatively large proportion of the clutch may exhibit abnormal morphology because of homozygosity of other mutations that may either mask, or enhance the phenotype of interest (Cheng and Moore, 1997).

3. Gynogenetic screens are biased for detecting mutations that are closer to the centromere than the telomere.

A. Scoring Mosaic Eyes

Day 0: Treat embryos with a 0.5% bleach solution for 90 seconds, then rinse 3× with charcoal-filtered water (Westerfield, 1995). The bleach treatment has the added benefit of hardening the chorion so the embryos do not hatch until after 72 hpf. This allows rolling of the embryos to view the entire RPE, making for efficient and effective scoring of mosaic eyes. Sort developing embryos into bowls or sterile Petri dishes and incubate at 28 °C.

Day 2: At approximately 48 hpf, examine embryos for segregation of *golden* and wild-type pigmentation. Remove any *golden* embryos to another dish. Mosaic eye pigmentation can be scored as follows:

Transfer embryos to a shallow container with 1× Tricaine (MS222, Argent Laboratories, Redmond, WA). Using a high-quality dissecting scope such as a Leica M10 or M16, focus on the eye of a single embryo, using 25 to 40× magnifications. Roll the embryo around with a probe to examine the top, back, and sides of the retina. Mosaic patches on the back of the eye are often more apparent by looking through the head of the embryo. If the embryos have hatched, mosaicism is most easily detected on the lateral sides of the RPE—it will be more difficult to detect mosaic patches on the narrow aspect of the retina. Transfer any mosaic eye embryos to a new container.

Day 3: Reexamine the mosaics that were previously isolated to confirm that they still possess the mosaic phenotype and are not the result of abnormal development. Mosaic patches on the back of the eye can be more difficult to detect at this age because of the increasing opacity of the outer surface of the retina. Rescore the remaining wild-type embryos—some mosaic patches are more obvious at this age because of the darker intensity of the wild-type pigmentation. Last, examine the *golden* embryos that were set aside on day 2. Occasionally an embryo is found that is predominantly *golden* throughout its body with only a few wild-type cells—this is the result of a very early genomic instability event!

Photography: Mosaics can be photographed live on day 2 or 3. Fixation with paraformaldehyde or neutral buffered formalin makes it more difficult to see mosaic patches because opacity is associated with fixation.

III. Mosaic Eyes: Genomic Instability or Retinal Mutations?

It is important to determine whether the mutations causing the mosaic eye phenotype are attributed to somatic mutation events as opposed to mutations that affect retinal morphology such as those reported by Malicki *et al.* (1996). Zebrafish larvae can be embedded for histological analysis of the retina as well as any other tissues of interest. An acrylic mold can be made to create agarose embedding blocks for zebrafish larval sectioning as described in Tsao-Wu *et al.*, (1998) (Fig. 3). We have made molds for arrays of 64 and 96 larvae. Larvae can be fixed at any age of development in 10% neutral buffered formalin (NBF) for 24 hours, and then arranged in an agarose block before paraffin sectioning. We cut 4-μm sections with an average of 8-step sections per block, for 7-day-old larvae. Transverse sections can be made by embedding a single row of larvae in the agarose mold, which is then infiltrated with paraffin. The strip of larvae is then turned upright in the embedding base mold for sectioning. Larval paraffin sections can be stained with regressive Harris hematoxylin and eosin (H & E) for histology (Luna, 1968), or deparaffinized for immunohistochemistry, *in situ* hybridization, RNA or DNA isolation (see Section IV.B).

A. Constructing a Zebrafish Larval Array

1. Treat the acrylic mold with a commercial silicone spray lubricant after you have made several blocks. Make a stock solution of 1% agarose in water. Wrap 1–in. labeling tape around the mold to create a wall approximately 0.5-cm high. For the molds described in Tsao-Wu *et al.* (1998), add 8.5 ml agarose to mold and allow the agarose to completely solidify. Remove tape and gently separate agarose block from the mold. Use a razor blade to cut

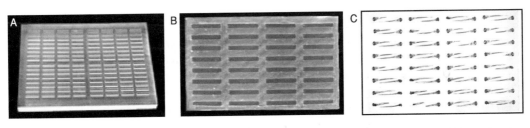

Fig. 3 (A) A zebrafish larval array can be made using an acrylic mold. (B) The mold shown here has 128 teeth in 4 quadrants of 32 teeth each that will form 4 agarose embedding blocks (Tsao-Wu *et al.*, 1998). Sixty-four larvae are arranged in 1 agarose block for processing and sectioning. (C) A single 4-μm section of a larval array stained with hematoxylin and eosin (H & E).

the agarose into 4 equal blocks. Rinse mold and blocks 2× with deionized water to remove any debris. Store unused blocks in water at 4 °C.

2. Pipette fixed larvae in 10% NBF onto the agarose block, removing excess fixative by gently blotting with a tissue. Gently pull 2 embryos into each well, using a dental tool and position so the heads are facing outward and the ventral side is up. The yolk causes the belly to be too rounded to place the embryos consistently on their ventral side. Surface tension will help the tails pull against the sides of the wells.

3. Slowly pipette additional 1% agarose (55 °C) over the larvae so the wells are filled to the same height as the rest of the block. It is important to avoid overlaying the larval arrays with too much agarose, since this will cause the block to warp (Cheng *et al.*, 1999). Trim one corner of the completed agarose block for orientation. The larvae-containing agarose block is then placed in a labeled histology cassette for paraffin embedding using standard tissue processing methods.

IV. Studying Cancer Susceptibility in Adult Zebrafish

Tumor susceptibility results in an increase in spontaneous cancers (Kinzler and Vogelstein, 1998). Fish carrying genomic instability mutations appear to have a predisposition toward developing spontaneous cancers and may be more sensitive to various tumor-promoting agents. A small number of tumors may be visible as outgrowths, asymmetric bulges of the abdomen, or angular bends in the fish's body. However, the majority of zebrafish tumors are most apparent through histology. Whole zebrafish adults can be fixed with 4% paraformaldehyde (PFA) or 10% NBF, followed by decalcification in 0.35 M ethylenediaminetetraacetic acid (EDTA) for paraffin sectioning (Moore *et al.*, 2002). Paraffin sections can be stained with H & E or other stains for histological analysis, with adjacent sections held in reserve for DNA or immunocytochemical experiments (Fig. 4).

A. Fixation and Decalcification of Adult Zebrafish

1. Anesthetize fish in an excess of tricaine (5×, MS222) plus ice until there is no gill movement or startle response. It is critical to keep the fish cold until fixation has occurred, to prevent decomposition of tissues.

2. Make an incision along the ventral midline from the anal pore to just below the gills to ensure access of the fixative to the internal organs. The incision should be made so that none of the internal organs, including the heart, are damaged. Transfer fish to ice cold 4% PFA or 10% NBF and incubate at 4 °C overnight.

3. Pour off fixative and rinse fish twice in 1× PBS. Decalcify by incubating in 0.35 M EDTA, pH 7.8, for 7 days at room temperature. Alternatively, after fixation, fish can be transferred to 70% ethanol and kept for long-term

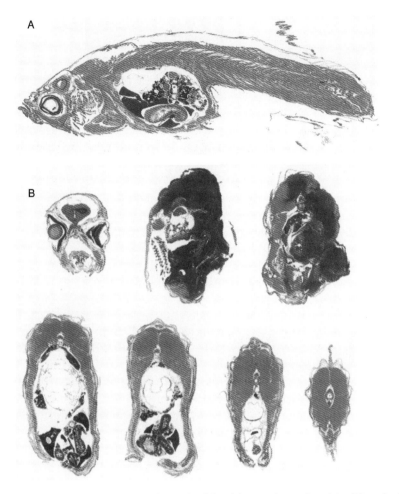

Fig. 4 After fixation and decalcification, zebrafish adults can be sectioned in (A) sagittal, (B) transverse, or (*not shown*) coronal planes. (A) A sagittal section of a zebrafish female stained with hematoxylin and eosin (H & E). (B) Transverse sections of a zebrafish female with a malignant peripheral nerve sheath tumor stained with H & E. The darker stained areas in the top two right sections are the tumor. (See Color Insert.)

storage at 4 °C. When ready to continue with processing for histology, rinse fish twice in 1× PBS and then proceed with decalcification. Fish younger than 1 month of age do not require decalcification.

4. Pour off EDTA and rinse fish twice in 1× PBS. Transfer fish to labeled histology cassettes and proceed with paraffin embedding for sagittal sections (Fig. 4A). For transverse sections, decalcified fish can be cut into 6 or 7 pieces 2 to 3 mm in depth with a single-edge razor blade. India ink can be used to mark the anterior face of these sections that are held in a histology cassette

within a teabag or histology sponges. After paraffin processing, these pieces can be properly oriented in the block for sectioning (Fig. 4B). We typically make 4- to 5-μm step sections of a fish, saving multiple slides per level. One slide is stained with H & E and the others can be saved for additional analysis.

B. DNA Isolation from Paraffin Sections

DNA suitable for PCR can be recovered from paraffin sections of larvae and adults. This can be useful for mapping mutations exhibited by a specific subset of animals from a larval array, or to study gene expression patterns in different adult organs. The following protocol has been used successfully to isolate DNA for a variety of purposes (Moore *et al.*, 2002).

1. Remove wax from slides by transferring them to coplin jars containing two each of the following solutions in this order: xylene, 100% ethanol, 95% ethanol. Allow 2-minute incubations per solution, then air dry slides.

2. Within 1h, hydrate tissues briefly for microdissection by passing the slide quickly through steam generated by a flask of boiling water. Without hydration, scraped tissues will tend to fly off the needle because of static. Microdissection of individual larva or adult tissues can be performed by scraping tissue from the slide using a 25-gauge needle. The dissected sample is transferred to a test tube containing 100 μl of DNA extraction buffer (0.2 mg/mL proteinase K in TE, pH 7.6). Note: Exposure of deparaffinized tissue to air inhibits amplification of the extracted DNA by PCR.

3. Incubate samples at 55 °C for 24 to 48 h, followed by 10-minute incubation at 98 °C to inactivate proteinase K. The scraped slides can be stained with H & E to document the areas removed by microdissection.

V. Transgenic Analysis of Genomic Instability

Transgenic reporter constructs can be designed in many ways to detect mechanisms of genomic instability. An example of this methodology is the development in our laboratory of transgenic constructs to detect frameshift mutations induced during the first 2 weeks of zebrafish development (Fig. 5). A construct was made containing a red-green fluorescent fusion protein (DsRed1-EGFP, Fig. 5A). The red fluorescent domain (DsRed1, Clontech, Palo Alto, CA) is in-frame and serves as an internal control for transgene expression, while reversion of out-of-frame green fluorescent domain (EGFP-1, Clontech) is an indicator of frameshift mutations. Fusions of fluorescent proteins have previously been created that allow the detection of each component separately (Heim and Tsien, 1996; Mitra *et al.*, 1996). To increase the probability of frameshift events occurring between the DsRed1 and EGFP sequences, a run of either 10 or 11 nucleotides was inserted between these domains. As shown in Fig. 5, when the linker region consists of an in-frame run of 10 guanine nucleotides (zG10b), zebrafish larvae show both red

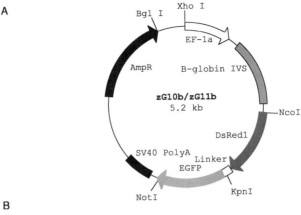

B

zG10b
DsRed1 Ser Gly Leu Arg Ser Thr Gly Arg Gly Gly Gly Thr Asn EGFP
---AAG TCC GGA CTC AGA TCC ACC GGC AGG GGG GGG GGT ACC AAC G---

zG11b
DsRed1 Ser Gly Leu Arg Ser Thr Gly Arg Gly Gly Gly Tyr Gln EGFP[F]
---AAG TCC GGA CTC AGA TCC ACC GGC AGG GGG GGG GGG TAC CAA CG---

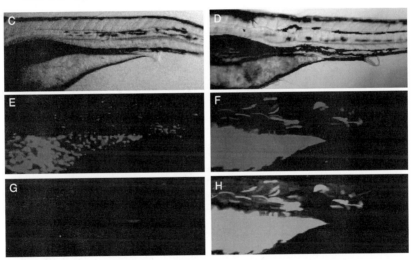

Fig. 5 The zG10b and zG11b transgenic constructs were designed to detect frameshift mutations in zebrafish embryos and larvae. (A) The zG10b/zG11b vector was based on the pXIG vector, a generous gift from Dr. Nancy Hopkins (Amsterdam *et al.*, 1996). A single fusion protein is formed that contains an in-frame red fluorescent protein, a linker region that contains either the G10 or G11 repeat sequence followed by enhanced green fluorescent protein (EGFP). (B) The linker sequence of zG10 maintains the reading frame between the two fluorescent domains, but the linker sequence of zG11 creates a +1 frameshift of the green fluorescent protein (EGFP[F]). (C, E, and G) A 72-hpf embryo injected with zG10b. (D, F, and H) A 72-hpf embryo injected with zG11b. (C and D) Brightfield. (E and F) Red fluorescence seen with a G filter (Leica). The more punctate pattern seen in the zG11-injected embryos may be attributed to the high basic content of the out-of-frame EGFP, resulting in nuclear localization. (G and H) Green fluorescence seen with a GFP3 filter. No green fluorescence is seen in the embryo injected with the out-of-frame zG11b construct. (See Color Insert.)

and green fluorescence. However, when the linker region consists of 11 guanine residues (zG11b), EGFP is out of frame and only red fluorescence can be detected. The zG11b construct was tested for the ability to detect frameshift mutations by treating injected embryos with N-acetoxy-N2-acetylaminofluorene (#005, Midwest Research Institute, Kansas City, MO), which disrupts normal base stacking after forming covalent bonds to the C8 position of guanine residue, causing primarily frameshift mutations (Ames *et al.*, 1972; Shelton and DeMarini, 1995). Green fluorescence was seen in zG11b-injected embryos treated at 24 hpf with 25 μg/ml AAF for 2 h (data not shown). These constructs have been injected into embryos carrying genomic instability mutations to detect point mutations that might occur in the linker regions altering the reading frame of EGFP (Gestl, 2002). So far, none of the genomic instability mutants we have recovered appear to induce frameshift mutations, as tested in a transient assay (Gestl, 2002).

VI. Summary

We now know that genomic instability contributes to cancer. The zebrafish mosaic eye assay developed by George Streisinger takes advantage of the organism's transparency to provide an excellent assay for detecting somatic mutation. This assay allowed us to identify zebrafish mutants with increased frequencies of somatic mutation and spontaneous cancer. Here, we have described details of mutagenesis, the basis and practical use of the mosaic eye assay, and the histological methods used to study genomic instability mutants and cancer susceptibility. These techniques should prove useful to other zebrafish researchers, as they are broadly applicable to many other biological investigations of embryos, larvae, and adult zebrafish.

Acknowledgments

We would like to thank Lynn Budgeon for her excellent histology work, Peggy Hubley for fish maintenance, Rebecca Lamason for help with the gamma experiment and graphics, and the members of the Cheng lab for comments on the manuscript. This work was supported by the NIH grants RO1-CA73935 and RO1-HD40179 to KCC, and NRSA F32-GM119794 to JLM, NSF MCB-93198174 to KCC, NASA grant NGT5-50264 to EGG and the Jake Gittlen Memorial Golf Tournament.

References

Ames, B. N., Gurney, E. G., Miller, J. A., and Bartsch, H. (1972). Carcinogens as frameshift mutagens: Metabolites and derivatives of 2-acetylaminofluorene and other aromatic amine carcinogens. *Proc. Natl. Acad. Sci. USA* **69**, 3128–3132.

Amsterdam, A., Lin, S., Moss, L., and Hopkins, N. (1996). Requirements for green fluorescent protein detection in transgenic zebrafish embryos. *Gene* **173**, 99–103.

Beckwith, L. G., Moore, J. L., Tsao-Wu, G. S., Harshbarger, J. C., and Cheng, K. C. (2000). Ethylnitrosourea induces neoplasia in zebrafish (*Danio rerio*). *Lab. Invest.* **80**, 379–385.

Charames, G. S., and Bapat, B. (2003). Genomic instability and cancer. *Curr. Mol. Med.* **7**, 589–596.

Cheng, K. C., and Loeb, L. A. (1997). Genomic stability and instability: A working paradigm. *Curr. Top. Microbiol. Immunol.* **221**, 5–16.

Cheng, K. C., and Moore, J. L. (1997). Genetic dissection of vertebrate processes in the zebrafish: A comparison of uniparental and two-generation screens. *Biochem. Cell. Biol.* **75,** 525–533.

Cheng, K.C., Beckwith, L., and Wang, X. (1999). Update to: Agarose embedded tissue arrays for histological and genetic analysis. *In* "Expression Genetics: High-Throughput Methods" (M. McClelland and A. Pardee, eds.), p. 37. Eaton Publishing, Westborough, MA.

Cox, E. C. (1976). Bacterial mutator genes and the control of spontaneous mutation. *Ann. Rev. Genet.* **10,** 135–156.

Garcia, A., Lambert, I. B., and Fuchs, R. P. (1993). DNA adduct-induced stabilization of slipped frameshift intermediates within repetitive sequences: Implications for mutagenesis. *Proc. Natl. Acad. Sci. USA* **90,** 5989–5993.

Gestl, E. E. (2002). Development of *in vivo* assays to detect mutation in transgenic zebrafish. Ph.D. Thesis. Department of Biochemistry and Molecular Biology, Pennsylvania State University, Hershey, PA.

Gestl, E. E., Kauffman, E. J., Moore, J. L., and Cheng, K. C. (1996). New conditions for generation of gynogenetic half-tetrad embryos in zebrafish. *J. Hered.* **88,** 76–79.

Grunwald, D. J., and Streisinger, G. (1992). Induction of recessive lethal and specific locus mutations in the zebrafish with ethyl nitrosourea. *Genet. Res. Camb.* **59,** 103–116.

Heim, R., and Tsien, R. (1996). Engineering green fluorescent protein for improved brightness, longer wavelengths and fluorescence resonance energy transfer. *Curr. Biol.* **6,** 178–182.

Hoeijmakers, J. (2001). Genome maintenance mechanisms for preventing cancer. *Nature* **411,** 366–374.

Kinzler, K. W., and Vogelstein, B. (1998). Familial cancer syndromes: The role of caretakers and gatekeepers. *In* "The genetic basis of human cancer" (B. Vogelstein and K. W. Kinzler, eds.), pp. 241–242. McGraw-Hill, New York.

Kolodner, R. D., Putnam, C. D., and Myung, K. (2002). Maintenance of genome stability in *Saccharomyces cerevisiae. Science* **297,** 552–557.

Lengauer, C., Kinzler, K. W., and Vogelstein, B. (1998). Genetic instabilities in human cancers. *Nature* **396,** 643–649.

Loeb, L. A., Loeb, K. R., and Anderson, J. P. (2003). Multiple mutations and cancer. *Proc. Natl. Acad. Sci. USA* **100,** 776–781.

Luna, L. G. (1968). "Manual of Histologic Staining Methods of the Armed Forces Institute of Pathology." McGraw-Hill, New York.

Malicki, J., Neuhauss, S., Schier, A. F., Solnica-Krezel, L., Stemple, D. L., Stainier, D. Y. R., Abdelilah, S., Zwartkruis, F., Rangini, Z., and Driever, W. (1996). Mutations affecting development of the zebrafish retina. *Development* **123,** 263–273.

Mitra, R., Silva, C., and Youvan, D. (1996). Fluorescence resonance energy transfer between blue-emitting and red-shifted derivatives of the green fluorescent protein. *Gene* **173,** 13–17.

Moore, J. L., Aros, M., Steudel, K. G., and Cheng, K. C. (2002). Fixation and decalcification of adult zebrafish for histological, immunocytochemical, and genotypic analysis. *BioTechniques* **32,** 293–298.

Moore, J. L., Breneman, C., Mohideen, M.-A. P. K., and Cheng, K. C. (2004). Zebrafish genomic instability mutants and implications for cancer susceptibility. *Genetics* submitted.

Mullins, M. C., Hammerschmidt, M., Haffter, P., and Nusslein-Volhard, C. (1994). Large-scale mutagenesis in the zebrafish: In search of genes controlling development in a vertebrate. *Curr. Biol.* **4,** 189–202.

Patton, E. E., and Zon, L. I. (2001). The art and design of genetic screens: Zebrafish. *Nat. Rev. Genet.* **12,** 956–966.

Pelegri, F., and Schulte-Merker, S. (1999). A gynogenesis-based screen for maternal-effect genes in the zebrafish, *Danio rerio. Methods Cell Biol.* **60,** 1–20.

Russell, W. L. (1951). X-ray induced mutations in mice. *Cold Spring Harbor Symposium in Quantitative Biology* **16,** 327–336.

Shelton, M. L., and DeMarini, D. M. (1995). Mutagenicity and mutation spectra of 2-acetylaminofluorene at frameshift and base-substitution alleles in four DNA repair backgrounds of Salmonella. *Mut. Res.* **327,** 75–86.

Solnica-Krezel, L., Schier, A. F., and Driever, W. (1994). Efficient recovery of ENU-induced mutations from the zebrafish germline. *Genetics* **136,** 1401–1420.

Streisinger, G., Walker, C., Dower, N., Knauber, D., and Singer, F. (1981). Production of clones of homozygous diploid zebra fish (*Brachydanio rerio*). *Nature* **291,** 293–296.

Streisinger, G. (1984). Attainment of minimal biological variability and measurements of genotoxicity: Production of homozygous diploid zebra fish. *N C I Monograph* **65,** 53–58.

Streisinger, G., Singer, F., Walker, C., Knauber, D., and Dower, N. (1986). Segregation analyses and gene-centromere distances in zebrafish. *Genetics* **112,** 311–319.

Streisinger, G., Coale, F., Taggart, C., Walker, C., and Grunwald, D. J. (1989). Clonal origins of cells in the pigmented retina of the zebrafish eye. *Dev. Biol.* **131,** 60–69.

Tsao-Wu, G. S., Weber, C. H., Budgeon, L. R., and Cheng, K. C. (1998). Agarose-embedded tissue arrays for histologic and genetic analysis. *BioTechniques* **25,** 614–618.

Westerfield, M. (1995). "The Zebrafish Book: A Guide for the Laboratory Use of Zebrafish (*Danio rerio*)." University of Oregon Press, Eugene, Oregon.

CHAPTER 26

Discovery and Use of Small Molecules for Probing Biological Processes in Zebrafish

Randall T. Peterson* and Mark C. Fishman[†]

*Developmental Biology Laboratory
Cardiovascular Research Center
Massachusetts General Hospital
Charlestown, Massachusetts 02129

[†]Novartis Institutes for Biomedical Research
Cambridge, Massachusetts 02139

I. Rationale for Small Molecule Screens in Zebrafish

A. The Strengths and Limitations of Genetic Screens

Of all the virtues of the zebrafish as a model organism, its suitability for large-scale screening is paramount. In no other vertebrate has it been possible to screen for mutations so readily and on such a scale as has been achieved using zebrafish. The earliest genetic screens captured the imaginations of many as wondrous mutant phenotypes were discovered, from the dramatically disrupted to the dramatically subtle (Driever et al., 1996; Haffter et al., 1996). In many cases, these mutants have allowed connections to be drawn between specific genes and their functions, especially for early developmental processes and organogenesis.

Despite the power and elegance of genetic screens, significant limitations prevent their more fruitful use in many circumstances. Most genetic zebrafish mutations discovered thus far are not conditional and do not have the ability, generally, to modulate timing or dose of effect. Consequently, they are best suited for identifying the first developmental function of a gene but are less useful for studying later processes because early-onset effects may obscure, amplify, or complicate later-onset ones. For example, mutation of the sonic hedgehog gene in zebrafish prevents differentiation of medially located muscle pioneer cells (Schauerte et al., 1998; van Eeden et al., 1996). However, many other structures also fail to form in this mutant, and it is unclear from analysis of this mutant alone whether shh is directly involved in muscle pioneer differentiation or whether differentiation is prevented by the failure to form of some other structure, such as the lateral floor plate. In the most extreme cases, gene disruption is embryonic lethal at an early stage of development, making it impossible for traditional genetic screens to identify the role of that gene in any later process. In the absence of an allelic series, no information exists about the effect of different levels of gene product activity.

Additionally, genetic screens may fail to identify mutations in genes for which a functionally redundant isoform exists. In the case of the shh mutant sonic-you, the medial floor plate develops normally (Schauerte et al., 1998), seemingly contradicting the numerous biochemical and embryological studies that have demonstrated a requirement for hedgehog signaling during floor plate induction. The fact that sonic-you mutants form a medial floor plate is likely attributed to the midline expression of two compensatory hedgehog isoforms (tiggy-winkle hedgehog and echidna hedgehog; Dodd et al., 1998). Therefore, a genetic screen for medial floor plate inducers would have failed to identify shh, despite its role in floor plate induction. Functional redundancy of genes may be a more significant problem in zebrafish than in other model organisms, given the apparent genome amplification that occurred in ray-finned fish before the teleost radiation (Postlethwait et al., 1998; Taylor et al., 2003).

Finally, the suppressor and enhancer screens that are valuable tools for identifying upstream and downstream components of genetic pathways in Drosophila *melanogaster*, *Caenorhabditis elegans*, and *Saccharomyces cerevisiae* have not been practical in the zebrafish (St Johnston, 2002).

B. Advantages of Small Molecule Screens

Many of the limitations of traditional genetic screens already outlined can be overcome when genetic screens are complemented with small molecule screens. In fact, *Drosophila* and *C. elegans* are well suited for genetic screening but are not as tractable for small molecule screening because of difficulties in access of small molecules to tissues in these organisms. Zebrafish are amenable to both genetic and small molecule screening, and the ability to combine these approaches in zebrafish is particularly promising. Like genetic mutations, small molecules are a classic means of disrupting biological processes and serve to link genes or gene products with their molecular functions. This approach has different strengths and weaknesses from the genetic one, so chemical and genetic screens are complementary (Fig. 1). For example, small molecules are excellent conditional biological probes, can overcome gene redundancy, and facilitate suppressor and enhancer screens as we will describe later. Furthermore, small molecule screens are generally simpler than are genetic screens.

1. Conditionality

Most zebrafish mutations identified to date are non-conditional and have a fixed allele strength. In an effort to overcome these limitations, some screens for temperature-sensitive alleles have been performed (Johnson and Weston, 1995), but conditional mutants remain the exception. In contrast, small molecules are the ultimate conditional disruptors, allowing both the timing and dosage of pathway disruption to be regulated.

One elegant example of the use of a small molecule to complement genetic mutations is provided by studies of the hedgehog signaling pathway. The small molecule cyclopamine is an antagonist of the hedgehog effector *smoothened* and has been an invaluable tool for dissecting the roles of the hedgehog pathway in many developmental events (Chen *et al.*, 2002). For example, cyclopamine was recently used to determine how three different muscle cell types (the muscle pioneers, the superficial slow fibers, and medial fast fibers) are specified in the zebrafish myotome. Wolff *et al.* (2003) demonstrated that all three muscle cell types are dependent on hedgehog signaling and that the specific identity adopted by a particular cell is governed by the quantity and timing of the hedgehog signal received. By treating embryos with cyclopamine at various times and doses, Wolff *et al.*, were able to alter the cell types present in the developing myotome in a predictable manner and to discover the temporal and quantitative relationships between hedgehog and muscle cell specification. Such an analysis was only possible because of the temporal and quantitative control provided by cyclopamine.

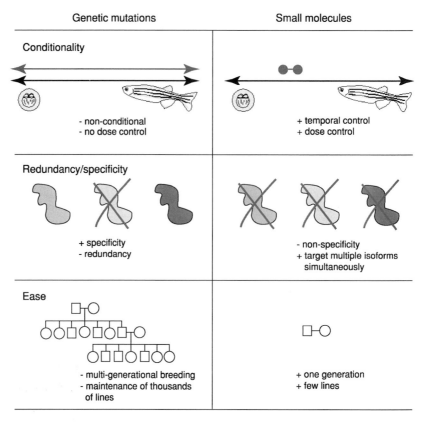

Fig. 1 A comparison between genetic mutations and small molecules. Although mutations and small molecules can both be used to disrupt gene functions, they differ in their conditionality, specificity, and the ease with which they are discovered. See text for details.

The small molecule concentramide was discovered in a zebrafish chemical screen (Peterson *et al.*, 2000, 2001) and has contributed to the biological understanding of the developmental patterning of the heart that is largely derived from genetic studies. One key developmental decision is the axial positioning of the two cardiac chambers, which are normally aligned along the anterior-posterior axis. In the *heart-and-soul* mutant, the ventricle forms within the atrium. When this pattern is established and whether it is because of problems within the cardiac primordia or to cell polarity defects in the adjacent, degenerating brain, was not clear. Concentramide mimics the cardiac effect, without causing cell polarity defects in the central nervous system. By adding concentramide at varying time points, it was possible to identify the precise stage at which cardiac chamber orientation is determined, which is the time (the 14-somite stage) when the bilateral cardiac primordia are preparing to fuse at the midline. In both concentramide-treated embryos and *heart-and-soul* mutants, fusion of the posterior end of the heart field is delayed. This delay prevents the rotation of the heart that transforms the concentrically organized heart cone into a linear structure, with atrium and ventricle adjacent to each other on the anterior-posterior axis (Peterson *et al.*, 2001). As in the case of cyclopamine described earlier, the temporal control with which concentramide could be added was critical for dissecting the process of cardiac patterning and use of the small molecule complemented analysis using existing genetic mutants.

2. Redundancy

Zebrafish chemical screens have identified many phenotypes that are similar to those previously identified using genetic screens, but some of the small molecule-induced phenotypes are unlike any identified by genetic screening (Khersonsky *et al.*, 2003; Moon *et al.*, 2002; Peterson *et al.*, 2000; Spring *et al.*, 2002; Sternson *et al.*, 2001). One potential explanation for this expansion of phenotypes is functional redundancy in the zebrafish genome. When multiple isoforms of a protein play overlapping roles in a biological process, mutation of one isoform may be insufficient to cause an observable phenotype. In contrast, a small molecule may bind to and inhibit multiple isoforms simultaneously, and thereby reveal the importance of those proteins in the biological process. As previously mentioned, the overlapping functions of sonic, tiggy-winkle, and echidna hedgehog genes in establishing the medial floor plate preclude them from being identified via genetic screens as regulators of floor plate induction (Dodd *et al.*, 1998). A small molecule, however, may be capable of inhibiting all three hedgehog proteins and block induction of medial floor plate. In addition, small molecules can modulate both maternal and zygotic gene products and may also have nonprotein targets, including lipids and nucleic acids, and therefore may disrupt processes not able to be disrupted via traditional genetic mutations.

3. Suppressors/Enhancers

No genetic suppressor or enhancer screen has ever been reported in a vertebrate. Small molecule screens make it possible to identify suppressors and enhancers of existing mutations as described in Section III. Chemical Suppressor and Enhancer Screens.

4. Ease

One final advantage of chemical screens is that they are much easier to perform than genetic screens. Whereas to reach any degree of saturation, genetic screens conventionally require large zebrafish facilities for the maintenance of thousands of zebrafish strains and lines, chemical screens typically require at most a few zebrafish lines. And, while the mutagenized fish used for genetic screens are often less fertile, the fish used for chemical screens can be selected in part for fertility.

C. The Potential for Zebrafish-Based Drug Discovery

In addition to their utility for dissection of essential biological processes, zebrafish small molecule screens may be useful for discovering novel therapeutic compounds and drug targets. By modeling human diseases in zebrafish, it may be possible to screen directly for compounds that modify the disease phenotype. Compounds that ameliorate the disease phenotype may serve as lead compounds for drug development, and identification of the compound's protein binding partner may effectively identify novel drug targets for traditional drug discovery efforts.

Many zebrafish models of human diseases have already been developed and are reviewed elsewhere (Amatruda et al., 2002; Rubinstein, 2003; Shin and Fishman, 2002). The majority of these are single-gene mutations that cause zebrafish phenotypes reminiscent of some aspect of human disease. In a number of cases where the genes underlying the human and zebrafish disease are known, orthologous genes are responsible for both conditions (Garrity et al., 2002; Roman et al., 2002; Xu et al., 2002). Recently, it has become possible to identify mutations in virtually any zebrafish gene by target-selected resequencing (Wienholds et al., 2002) or to knockdown the function of a gene using antisense morpholino oligonucleotides (Nasevicius and Ekker, 2000). Therefore, it should be possible to generate zebrafish models for many of the human diseases resulting from a known single-gene mutation. Therapies for many of these human diseases have not been developed because of the difficulty in predicting a priori which proteins should be targeted to reverse the disease phenotype. Significantly, unbiased screening in zebrafish may allow discovery of compounds that reverse the disease, even without knowing what protein is being targeted (MacRae and Peterson, 2003).

In addition to diseases caused by genetic mutation, it may be possible to discover novel drugs for treating infectious diseases. Several zebrafish models of infection have been developed, including models of tuberculosis and Salmonella typhimurium infection (Davis et al., 2002; Van Der Sar et al., 2003). Screening in zebrafish may allow assays to be performed on microbes that cannot be cultured outside of a whole organism. And by screening in the context of a whole organism, it should be possible to identify compounds with antimicrobial activity that have no undue toxicity to the host. Two of the infection models developed thus far use fluorescently labeled microbes for infection (Davis et al., 2002; Van Der Sar et al., 2003), so the efficacy of a small molecule could be measured by quantitating the number of pathogens or by assessing survival of the host.

Will small molecules that reverse a disease phenotype in zebrafish have similar effects in humans? While that question has not been answered, it is clear that many drugs with known effects in humans cause analogous effects in zebrafish. For example, Milan et al. (2003) treated zebrafish with 23 drugs known in humans to lengthen the QT interval on the electrocardiogram, often a harbinger of arrhythmogenesis, an undesirable drug side effect (Milan et al., 2003). Of the 23 drugs, 22 also caused an analogous prolongation of the cardiac cycle in zebrafish. Other drugs that have similar effects in humans and fish include angiogenesis inhibitors, vasodilators, opiates, cholesterol synthesis blockers, and anticoagulants (Langheinrich, 2003). Therefore, tissue access, drug binding sites, and pharmacodynamic effects seem to be generally well conserved between zebrafish and humans.

II. Assay Development

A. General Considerations for Assay Development

In designing an assay for a small molecule screen, two of the most important considerations are assay stringency and reproducibility. A stringent assay will score a small molecule as active only if it causes a phenotype that meets a demanding set of criteria. Obviously, an overly stringent assay may cause valuable small molecules to be overlooked, while an assay that is not stringent enough results in numerous false positives. In our experience, it has been preferable to err on the side of increased stringency, because pursuit of false positives and weakly active compounds can be time consuming. Fortunately, for most screens it is possible to assess the stringency of the assay using a small pilot screen of about 1000 small molecules and increase stringency if an undesirably high hit rate is observed. The optimal hit rate will vary from assay to assay but will likely be in the range of 1 in 100 to 1 in 10,000 molecules screened.

Variability between individual zebrafish can reduce assay reproducibility and lead to unacceptable numbers of false positives. Most reproducibility problems can be eliminated by placing multiple zebrafish embryos in the same well. Three embryos can be raised in a well of a 96-well plate for several days or in a well of a 384-well plate for about 1 day. By requiring that all embryos in a well exhibit the same phenotype, false positives can be eliminated. For example, even if a false positive phenotype is observed in 5% of untreated embryos, the probability of 3 out of 3 embryos in a given microplate well exhibiting the phenotype is about 1/10,000, which is an acceptably low rate of false positives for most assays.

Other factors such as the strain of zebrafish to use, the timing with which small molecules should be added, and the optimal means of imaging or measuring the results must be determined for each assay.

B. Examples of Zebrafish Small Molecule Screens

Most small molecule screens to date have used standard microscopic analysis of morphological defects in developing embryos (Fig. 2). Lesions of the brain, eye, ear, skin, neural crest, blood, heart, vasculature, etc. have been identified (Khersonsky et al., 2003; Moon et al., 2002; Peterson et al., 2000; Spring et al., 2002; Sternson et al., 2001).

Additionally, small molecule perturbants of physiological function have been isolated using assays that take particular advantage of whole animal-based *in vivo*

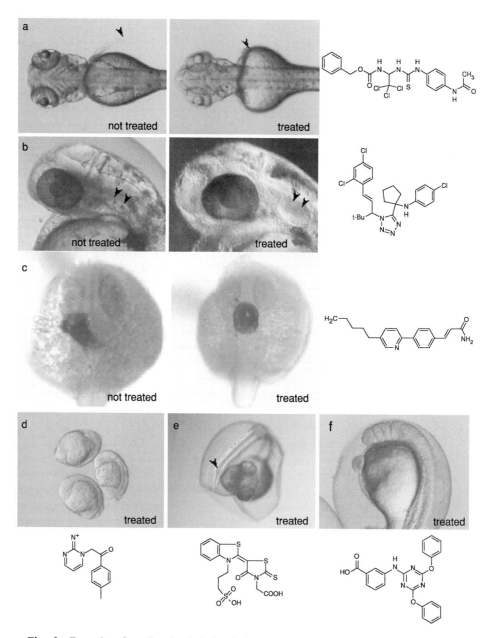

Fig. 2 Examples of small molecule-induced phenotypes. Zebrafish screens have identified small molecules that affect the ontogeny of the fins (a), otoliths (b), heart chambers (c), body axis (d), notochord (e), and eye (f). Arrowheads mark the locations of the pectoral fins (a), otoliths (b), and notochord bulge (e).

screens. For example, an assay has been developed that allows the heart rates of zebrafish larvae in 96- or 384-well plates to be determined automatically using a robotic microscope coupled with a digital video camera and image processing software (Milan *et al.*, 2003). Such an assay can easily be used to identify novel compounds affecting cardiac physiology. Another assay uses a fluorescently quenched phospholipid substrate as a readout of *in vivo* phospholipase activity (Farber *et al.*, 2001), which could be adapted for chemical screening to identify small molecule modifiers of lipid processing or other metabolic processes.

III. Chemical Suppressor and Enhancer Screens

Forward genetic screens have provided novel entrance points into biological pathways, supplemented by suppressor and enhancer screens to identify additional upstream and downstream components of these pathways and to connect them with parallel pathways affecting the process of interest. While the use of modifier screens is widespread in invertebrate genetic models (St Johnston, 2002), genetic modifier screens have thus far been less facile in vertebrates. In part, this is attributed to the lack of balancer chromosomes and the resultant difficulty in phenotypic recognition of a suppressed mutant among its wild-type siblings. For example, in a cross of two fish heterozygous for both the original mutation and its suppressor, only one-sixteenth of the offspring would be expected to be homozygous at both alleles. Therefore, in a recessive suppressor screen, three-sixteenths of a clutch containing a fully penetrant suppressor would still exhibit the mutant phenotype, making it difficult to differentiate from the one-quarter mutant embryos expected in an unsuppressed clutch.

In contrast, chemical modifier screens are feasible because small molecules can be delivered easily and uniformly to all mutant embryos. This makes it possible, in theory, to suppress the mutant phenotype in all treated embryos and eliminates the need to quantify changes in mutant/wild-type ratios. This approach has been used to identify chemical modifiers of the *gridlock* mutation (Peterson *et al.*, 2004). *Gridlock* is a recessive mutation that results in a dysplasia of the dorsal aorta that prevents blood flow to the trunk and tail 2 dpf (Weinstein *et al.*, 1995). By 4 dpf, some homozygous *gridlock* mutants overcome the aortic dysplasia by forming collateral vessels that circumvent the blockage, and these individuals are often able to survive to adulthood. Forty pairs of homozygous *gridlock* adults were mated to generate homozygous *gridlock* embryos, which were arrayed 3 embryos per well in 96-well plates. Mutant embryos were then exposed to small molecules from a diverse chemical library and screened visually for aortic circulation to the tail.

After screening 5000 small molecules, 2 structurally related compounds were identified that completely reverse the *gridlock* phenotype and restore normal circulation to the trunk and tail (Peterson *et al.*, 2004; Fig. 3). No adverse side effects are observed at effective concentrations, and treated embryos survive to

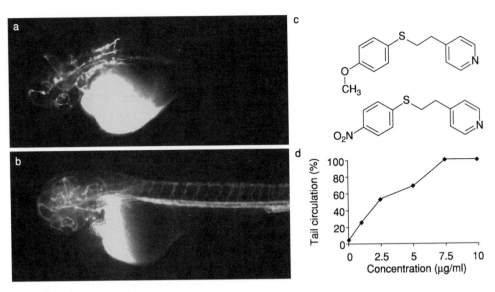

Fig. 3 Chemical suppression of a genetic mutation. Fluorescent microangiograms show the circulation pattern in *gridlock* embryos untreated (a) or treated with a *gridlock* suppressor compound (b). The structures (c) and dose response curve (d) of *gridlock* suppressors identified by zebrafish chemical screening.

adulthood. The compounds appear to function during angioblast specification and migration, and they also promote vessel formation in *in vitro* tubule formation assays using human cells. The precise mechanism of action of these *gridlock* suppressors is not yet known, but they may function in part by inducing expression of vascular endothelial growth factor (VEGF). We are hopeful that, like genetic suppressors, these small molecule suppressors will ultimately lead to the identification of novel components of the *gridlock* signaling pathway and enhance our understanding of how vasculogenesis is regulated in the developing embryo.

The search for chemical suppressors of the *gridlock* mutation was aided by the fact that breeding pairs of homozygous *gridlock* mutants could be generated. Screening for chemical modifiers of recessive lethal mutations is also possible but requires more embryos. Twenty-five percent of embryos from a cross of heterozygotes should exhibit the mutant phenotype. By placing 20 embryos in each well, the probability of encountering a well without a homozygous mutant is 0.003. Therefore, in a screen for chemical suppressors of a recessive mutation, the rate of false positives is acceptably low, and a well containing 20 phenotypically wild-type embryos may be indicative of suppression of the mutant phenotype. A similar strategy has been employed to identify a small molecule suppressor of a recessive embryonic lethal cell cycle mutation (Stern and Zon, 2003).

Although a heterogeneous population of embryos can be used for chemical modifier screens, homogeneous populations may be preferable, and additional methods of generating such homogeneous populations exist. Recently, germ-line replacement was used to create fertile, adult zebrafish whose germ cells were derived completely from a homozygous mutant donor (Ciruna *et al.*, 2002). Because fish generated by this method produce clutches of uniformly mutant embryos, it should be possible to create breeding pairs that will produce homogeneous populations of embryos for almost any mutation. These populations would be ideal for performing chemical modifier screens. In addition, specific pharmacological inhibitors of many proteins exist, and treatment of embryos with these inhibitors often produces a reproducible phenotype. Treatment of large groups of embryos with an inhibitor, followed by screening for compounds that suppress the induced phenotype, may be another means of identifying novel components of the targeted pathway.

Chemical modifier screens have the potential to do for zebrafish what genetic modifier screens have done for invertebrate model organisms, namely moving beyond single gene discovery to the connection of multiple genes into functional pathways. And, given the numerous zebrafish disease models that have been developed (Amatruda *et al.*, 2002; Rubinstein, 2003; Shin and Fishman, 2002), the ability to identify small molecule modifiers of disease phenotypes also presents exciting possibilities for discovering novel therapies and therapeutic targets.

IV. Selection of Small Molecule Libraries

When designing a zebrafish small molecule screen, selection of the compounds to be screened is one of the factors most likely to influence a screen's success. In a genetic screen, one attempts to optimize mutation rate and genomes screened in an effort to efficiently approach "saturation," or the identification of all possible mutations (Mullins *et al.*, 1994). By analogy, a chemical library should be selected that contains a high proportion of biologically active compounds and approaches saturation of chemical space. To have the greatest chance at modulating the activity of every protein in the organism, it is important to ensure that as many library members as possible possess physicochemical properties that are consistent with absorption and bioavailability. In addition, it is important to select a library that possesses as much chemical diversity as possible, rather than a library that is based on a limited number of chemical core structures, or pharmacophores.

The principles that govern absorption, distribution, metabolism, and excretion (ADME) of small molecules have not been studied in detail in zebrafish. However, the effects of small molecule structural features on ADME have been studied extensively for mammalian systems, and generalized principles for predicting the bioavailability of small molecules have been developed. Because these principles have been reviewed extensively (Lin *et al.*, 2003; Lipinski *et al.*, 2001; Poggesi, 2004; Yu and Adedoyin, 2003), they will not be covered here, except to say that such factors as molecular weight, hydrophobicity, and number of hydrogen bond donors and acceptors are considered to be predictive of a small molecule's absorption and potential for biological activity. Many of the collected and combinatorial libraries available have been designed to incorporate only molecules that fall within the accepted ranges for each of these measures (Oprea, 2000; van de Waterbeemd and Gifford, 2003). In addition, a number of chemical moieties have been shown empirically to cause nonspecific toxicity, and molecules containing these moieties are excluded from some libraries (Llorens *et al.*, 2001). Although the conventional predictors of bioavailability were not generated using data from zebrafish, it may be reasonable to presume that similar principles will govern bioavailability in zebrafish, and in the absence of data specific for zebrafish, libraries that have been tailored for use in mammalian cell-based assays are a reasonable starting place for whole organism screens in zebrafish.

One physicochemical factor that has been studied directly in zebrafish, albeit superficially, is the effect of a compound's octanol:water partition coefficient on absorption. The logarithm of the partition ratio between octanol and water (logP) can be measured empirically or calculated for a given chemical structure and correlates with membrane permeability (Table I). In a study of 23 drugs that lengthen a portion of the cardiac cycle in humans known as the QT interval, it was shown that 19 of 19 drugs with a logP higher than 1 were absorbed from the water and 18 of them caused bradycardia in zebrafish, while the 4 drugs with a logP lower than 1 caused bradycardia only after injection into the embryo (Milan *et al.*, 2003). Therefore, logP seems to be predictive of absorption, and small molecules with logP values higher than 1 should generally be selected for zebrafish chemical screens.

Two main types of chemical library are available for use in small molecule screening—collected libraries and combinatorial libraries. Collected libraries are assembled through time and can contain natural products or synthetic compounds synthesized individually or in small groups. Their eclectic nature adds to the diversity of the library. The compounds in a combinatorial library are generally all synthesized in parallel through a series of synthetic reactions that join a small number of building blocks in various combinations to generate large collections of distinct molecules (Fig. 4). The advent of combinatorial chemistry enables a single chemist to generate thousands of novel compounds simultaneously and has been used to generate compounds with interesting biological activities (Batra *et al.*, 2002; Nicolaou and Pfefferkorn, 2001). However, caution should be exercised when using combinatorial libraries for small molecule screens in zebrafish. Despite the large numbers of distinct compounds contained in these libraries, many of the compounds are structurally related and share common core structures. Consequently, it is unclear whether these libraries are able to cover as much chemical space as collected libraries, even when they contain comparable numbers of distinct compounds (Lipinski, 2000). In our experience, combinatorial libraries have generated hit rates in broad screens for developmental defects that are comparable to those of collected libraries but with fewer distinct phenotypes being observed. This observation is consistent with the idea that most combinatorial libraries possess less structural diversity than collected libraries.

Table I
The Relationship Between logP and Absorption in the Zebrafish[a]

Small molecule	logP	Absorbed
Pentamidine	< −1.5	No
Sotalol	0.24	No
Procainamide	0.88	No
NAPA-HCl	0.99*	No
E-4031	2.56*	Yes
Valproic acid	2.75	Yes
Erythromycin	3.06	Yes
Cisapride	3.09*	Yes
Ketanserin	3.29	Yes
Quinidine	3.44	Yes
Droperidol	3.50	Yes
Flecainide	3.78	Yes
Fluoxetine	3.82	Yes
Ibutilide	3.82*	Yes
Haloperidol	4.30	Yes
Amitriptyline	4.92	Yes
Chlorpromazine	5.41	Yes
Thioridazine	5.90	Yes
Pimozide	6.30	Yes
Tamoxifen	6.30*	Yes
Astemizole	6.43*	Yes
Terfenadine	7.62*	Yes
Amiodarone	7.80	Yes

[a] Drugs with logP values greater than 1 were absorbed, while drugs with logP values less than 1 were not absorbed. Where available, experimental logP values were used. Asterisks indicate values calculated using KowWin software (Syracuse Research Corporation, North Syracuse, NY).

Small molecule libraries can be obtained from numerous commercial and noncommercial sources. Commercial sources include ChemDiv, San Diego, CA; Sigma-Aldrich, St. Louis, MO; ChemBridge, San Diego, CA; Peakdale, Chapel en le Firth, Derbyshire, UK; Maybridge, Tintagel, Cornwall, UK; Bionet Camelford, Cornwall, UK; Cerep, Redmond, WA; and Microsource Discovery Systems, Gaylordsville, CT. These companies sell libraries as dry compounds or as pre dissolved stock solutions in micro-format plates. In addition, many academic institutions and research organizations have acquired chemical libraries that are made available at minimal or no cost to affiliated researchers. See http://iccb.med.harvard.edu/index.htm, http://www.hts.wisc.edu/Index.htm, http://www.medicine.mcgill.ca/biochem/htsfacility/, and http://www.hts.ku.edu/index.shtml for examples. Significantly, the National Institutes of Health (NIH), Bethesda, MD, has also made the collection and distribution of small molecule libraries a major priority. Libraries have been created by individual institutes including the National Cancer Institute, which manages a collection of more than 100,000 compounds. Furthermore, one of the major initiatives of the NIH

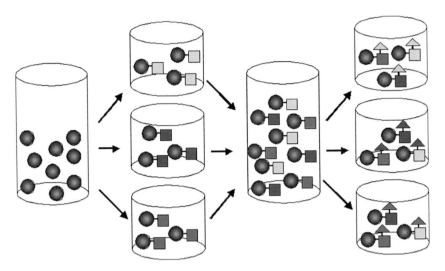

Fig. 4 A split-and-pool approach to combinatorial chemistry. Small molecule libraries for zebrafish chemical screens can be generated by splitting synthesis beads (*circles*) into multiple reaction vessels and attaching different first monomers (*squares*) to the beads in each vessel. The beads are then pooled and split again into separate vessels, where different second monomers (*triangles*) are attached to the first monomers in each vessel. With a relatively small number of monomers and split-and-pool cycles, thousands of distinct small molecules can be synthesized.

Roadmap is to "offer public sector biomedical researchers access to small organic molecules which can be used as chemical probes to study cellular pathways in greater depth," (nihroadmap.nih.gov). Initial libraries generated through this initiative and made available to researchers will contain at least 500,000 compounds. In short, numerous small molecule libraries are available to zebrafish researchers interested in performing chemical screens.

V. Screening Methods

A. Handling Embryos

For most zebrafish chemical screens, embryos should be placed into multi-well plates before addition of small molecules. This task is much easier to perform before hatching, even if the screen design calls for addition of compounds to larval-stage zebrafish. In a basic developmental screen, 96-well plates are prefilled with 200-μL embryo buffer per well using a multichannel pipetter. Phenylthiourea (0.003%) can be added to prevent pigmentation if desired, and antibiotics (160 U/mL penicillin, 160 μg/mL streptomycin) can be added to reduce bacterial contamination if infection becomes problematic. Three embryos are added to each well manually, using a glass Pasteur pipette attached to a manual pipette pump. Hundreds of embryos can be drawn into the pipette at one time and dispensed by gently touching the pipette tip to the surface of the buffer in a well. Surface tension will draw one embryo out of the pipette and into the water without significantly changing the liquid volume in the well. With practice, one person can fill several 96-well plates with embryos in 1 hour. For many screens, the rate-limiting step will be embryo production or phenotyping, not arraying embryos into plates, in which case this manual method should be adequate. However, if manual arraying becomes limiting, a robotic system for arraying zebrafish embryos into 96-well plates has been developed and is available commercially from Union Biometrica, Somerville, MA.

B. Compound Handling

Small molecule stock solutions are prepared in dimethyl sulfoxide (DMSO) at a concentration of about 5 mg/mL and are stored frozen in polypropylene 96- or 384-well plates. After thawing the plates, 100 nL of stock solution should be transferred into the assay plates containing embryos and embryo buffer. This is most reliably performed using a pin transfer robot, but arrays of stainless steel (V&P Scientific, Inc., San Diego, CA) or polypropylene compound transfer pins can also be used to transfer small volumes manually, 96 wells at a time. If a researcher does not have ready access to a pin transfer robot or is only screening a few plates per day, it may be more convenient to perform dispensing for several weeks of screening at once. Compounds can be dispensed into plates containing only 50-μL embryo buffer per well. The plates can then be sealed and frozen at $-80\,^{\circ}$C for several weeks. On each screening day, the plates are thawed, and for each well, the 50-μL embryo buffer containing 100 nL of compound stock solution is transferred to the assay plate, holding the arrayed embryos using a multichannel pipetter. This approach is more convenient for laboratories that do not own compound-handling robots and does not appear to significantly reduce the stability of most compounds.

It is generally impossible to screen all compounds in a library at a full range of concentrations, and so a screening concentration must be selected that is most likely to enable identification of active compounds. Screening at concentrations that are too high can lead to high rates of nonspecific toxicity and death, which obscures interesting results and leads to identification of weakly active compounds. Screening at concentrations that are too low may cause potentially interesting leads to be missed. We have generally used a screening concentration of about 2 μg/mL. Such a dose produces an acceptably high hit rate in our assays but only identifies compounds potent enough to facilitate follow-up studies and efforts to identify protein binding partners.

C. Phenotype Detection

Morphological phenotypes can often be detected by direct observation with a dissecting microscope. Use of screening plates with round-bottom wells facilitates observation by keeping embryos in the center of the well and preventing optical distortion caused by proximity to the side of the well. Observation from below the well is also possible, but most inverted microscopes do not offer stereo views of the embryo. Although the human eye remains the most adept tool for detecting subtle morphological changes, automated screening systems and pattern recognition software can be used to detect changes, especially quantitative changes, in zebrafish morphology or physiology. For example, the Discovery-1 screening system (Universal Imaging Corporation, Downingtown, PA; Fig. 5) has been used to automatically screen through plates of zebrafish embryos and detect subtle changes in heart rate that would be difficult to identify visually (Milan *et al.*, 2003). Many additional screens requiring minimal human intervention can be envisioned, particularly as more fluorescent markers of morphology and physiology are developed for the zebrafish.

VI. Follow-Up Studies for Active Small Molecules

The full benefit of identifying a genetic mutation comes after the mutated gene and mechanism of action have been identified (Talbot and Schier, 1999). Similarly, the full benefit of identifying phenotype-inducing small molecules comes after the

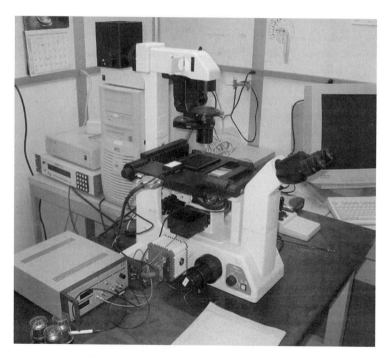

Fig. 5   The Discovery-1™ Screening System. Automated imaging can increase the throughput of zebrafish screens for chemical modifiers of morphology or physiology.

molecular target and mechanism of action are identified. This process allows a phenotype to be converted into molecular understanding of the affected process. Two general approaches have been used successfully to assign mechanisms of action to small molecules. The first approach involves testing whether candidate genes participate in a small molecule's mechanism of action. This "pathway interrogation" approach includes hypothesis-based experiments to test the effect of a small molecule on a single-gene and also genome-wide analyses of the effects of the small molecule on many pathways simultaneously. The second approach is to use affinity between the small molecule and its biological binding partners as a means of purifying and identifying the compound's molecular target. This "affinity purification" approach does not require a priori assumptions about a compound's mechanism of action and includes such techniques as affinity chromatography and expression cloning.

A. Pathway Interrogation

Pathway interrogation experiments aim to identify the molecular pathways affected by a particular small molecule. These experiments may or may not identify a compound's specific molecular target, but they serve to generate mechanistic insight and new, testable hypotheses about how a compound functions. They include:

1. Expression Profiling/Quantitative Polymerase Chain Reaction

Altering a biological process with genetic mutations or small molecules may result in transcriptional changes, which can be detected by quantitative polymerase chain reaction (PCR) for small numbers of genes or by global analyses using microarray transcriptional profiling (Brown and Botstein, 1999). By identifying genes that are up- or down-regulated on treatment with a particular compound, it may be possible to generate testable hypotheses about the compound's mechanism of action. In yeast, transcriptional profiles from drug-treated yeast match profiles from strains in which the drug targets have been deleted (Kuruvilla et al., 2002; Marton et al., 1998). Micro arrays for zebrafish expression profiling are also becoming readily available, including cDNA micro arrays (Clark et al., 2001; Lo et al., 2003; Ton et al., 2002), spotted oligonucleotide micro arrays, and Affymetrix chips. These tools may facilitate the elucidation of small molecule targets, although examples in zebrafish have not yet been reported.

One problem with transcriptional profiling in general is that it is often difficult to separate immediate responses from secondary or adaptive changes. For example, transcriptional changes in embryos with a vascular defect may be complicated by growth inhibition or cell death. Fortunately, with chemical compounds, a treatment time course can be performed to identify the precise time window during which the compound must be applied to have the desired effect, and thereby refine the window for expression profiling. Transcriptional changes observed under these conditions should better reflect rapid alterations in the primary target pathway, rather than downstream responses or adaptations.

2. Overexpression/Knockdown Experiments

Gene overexpression and knockdown experiments in whole zebrafish embryos can be used to test specific hypotheses and to identify epistatic relationships between the small molecule targets and other pathway members. In these experiments, a gene with a hypothesized role as a mediator of a small molecule mechanism of action may be overexpressed or its expression may be "knocked down" by injecting 1- to 4-cell stage embryos with mRNA encoded by the gene or with antisense morpholino oligonucleotides, respectively. Methods for such experiments have been well described (Nasevicius and Ekker, 2000; Nusslein-Volhard and Dahm, 2002). When upregulation of a gene is hypothesized to mediate a small molecule's effect, knockdown of that gene might be expected to block the small molecule from exerting its normal effect. Conversely, overexpression of the gene might be expected to mimic the small molecule's effect. For example, the *gridlock* suppressor GS4012 has been shown to increase expression of the VEGF gene (Peterson *et al.*, 2004). Consistent with a role for VEGF in GS4012's mechanism of action, overexpression of VEGF is sufficient to suppress the *gridlock* phenotype.

Using the logic of epistatic analysis, genes that function upstream of a small molecule target would not normally be expected to modify the effect of a small molecule. Overexpression and knockdown experiments will undoubtedly be important techniques for establishing mechanisms of action for the small molecules under study, just as they are important for assigning functions to genes and genetic mutations.

B. Affinity Purification

Affinity purification experiments seek to identify the specific protein targets to which a small molecule binds. These experiments are unbiased and do not require a priori assumptions about a compound's mechanism of action. They include:

1. Biochemical Purification

Small molecules can be used as affinity probes for biochemical purification of their molecular targets. One classic approach is to radiolabel small molecules and use the radiolabel as a means of following target proteins through the steps of a biochemical purification. Proteins from a cell or embryo lysate treated with the small molecule are fractionated using standard, sequential biochemical purification techniques including ammonium sulphate precipitation, gel filtration, and ion exchange chromatography. Fractions containing target proteins are identified by the presence of the radiolabel. The purified target protein is then identified by tandem mass spectrometry. A small molecule identified in a zebrafish chemical screen can be readily radiolabeled, and several companies will perform custom radiolabeling of compounds on a fee-for-service basis.

2. Affinity Chromatography

The second biochemical approach is to perform affinity chromatography using a matrix coated with the small molecule of interest. Before affinity chromatography, a site for attachment of a chemical linker must be identified. Variants of the compound containing modifications at various potential linker attachment points are purchased or synthesized and then tested for efficacy and potency. Once an attachment point is identified that does not interfere with the compound's activity, a linker (e.g., aminocaproic acid) is appended to the compound. The linker is then covalently attached to a solid support resin such as Affigel (Bio-Rad Laboratories, Hercules, CA), directly or via biotin, following the manufacturer's instructions. A negative control matrix is also synthesized by attaching the linker to the compound of interest at an inactivating site, when possible, or by attaching linker alone to the Affigel. By performing a time course experiment it is usually possible to identify a developmental window during which a compound must be present to cause the expected phenotype. A lysate is prepared from zebrafish embryos of the appropriate stage and exposed to the affinity and control matrices. Non-binding proteins are washed away, and tandem mass spectrometry is used to identify proteins that bind specifically to the affinity matrix. Several excellent references describe this process in further detail (Brown *et al.*, 1994; Chen *et al.*, 1999, 2002; Crews *et al.*, 1994; Harding *et al.*, 1989; Khersonsky *et al.*, 2003; Kwok *et al.*, 2001; Liu *et al.*, 1991; Miller *et al.*, 1990; Shimizu *et al.*, 2000; Taunton *et al.*, 1996).

Many small molecules are known to bind to several targets. As a result, affinity purification may identify proteins that bind a small molecule but are not responsible for its biological effect. This problem can be minimized and the likelihood of identifying the compound's target can be increased by ensuring that the small molecules studied are potent. Compounds with EC50s in the high micromolar range have been used successfully for target identification studies (Nguyen *et al.*, 2003), but generally compounds with such low potency are unsuitable for affinity-based target identification. Small molecules with EC50s in the low micromolar or nanomolar range are more likely to produce successful target identification efforts. For this reason, setting the original screening concentration in the low micromolar range is helpful to ensure that compounds identified can be used for follow-up experiments. Even very potent compounds may bind to proteins that are not responsible for producing the phenotype of interest. Therefore, the function of any protein identified by affinity purification should be confirmed by performing gene knockdowns of the identified targets using antisense morpholino oligonucleotides (Nasevicius and Ekker, 2000). In this way, it should be possible to identify which proteins are relevant to the process being studied, even if multiple binding partners exist.

VII. Conclusions

Since the early 1990s, the zebrafish has become established as a powerful tool for genetic experimentation. In contrast to the other mainstays of metazoan genetics—*Drosophila, C. elegans* and mouse—it has proved equally amenable to chemical screens. Whether the goal is discovery of conditional probes for basic research or discovery of therapeutic targets and lead compounds, the zebrafish enables high-throughput assessment of the biological effects of small molecules. Several challenges remain, particularly in the area of identifying small molecule targets. However, the ability to discover specific, interesting, and useful compounds using zebrafish is clear, and zebrafish-based small molecule screens can provide a novel, accessible approach for launching projects, both in pathway dissection and drug discovery.

References

Amatruda, J. F., Shepard, J. L., Stern, H. M., and Zon, L. I. (2002). Zebrafish as a cancer model system. *Cancer Cell* **1,** 229–231.

Batra, S., Srinivasan, T., Rastogi, S. K., and Kundu, B. (2002). Identification of enzyme inhibitors using combinatorial libraries. *Current Medicinal Chemistry* **9,** 307–319.

Brown, E. J., Albers, M. W., Shin, T. B., Ichikawa, K., Keith, C. T., Lane, W. S., and Schreiber, S. L. (1994). A mammalian protein targeted by G1-arresting rapamycin-receptor complex. *Nature* **369,** 756–758.

Brown, P. O., and Botstein, D. (1999). Exploring the new world of the genome with DNA microarrays. *Nat. Genet.* **21,** 33–37.

Chen, J. K., Lane, W. S., and Schreiber, S. L. (1999). The identification of myriocin-binding proteins. *Chemistry & Biology* **6,** 221–235.

Chen, J. K., Taipale, J., Cooper, M. K., and Beachy, P. A. (2002). Inhibition of Hedgehog signaling by direct binding of cyclopamine to Smoothened. *Genes Dev.* **16,** 2743–2748.

Ciruna, B., Weidinger, G., Knaut, H., Thisse, B., Thisse, C., Raz, E., and Schier, A. F. (2002). Production of maternal-zygotic mutant zebrafish by germ-line replacement. *Proc. Natl. Acad. Sci. USA* **99,** 14919–14924.

Clark, M. D., Hennig, S., Herwig, R., Clifton, S. W., Marra, M. A., Lehrach, H., Johnson, S. L., and Group, T. W. (2001). An oligonucleotide fingerprint normalized and expressed sequence tag characterized zebrafish cDNA library. *Genome Res.* **11,** 1594–1602.

Crews, C. M., Collins, J. L., Lane, W. S., Snapper, M. L., and Schreiber, S. L. (1994). GTP-dependent binding of the antiproliferative agent didemnin to elongation factor 1 alpha. *J. Biol. Chem.* **269,** 15411–15414.

Davis, J. M., Clay, H., Lewis, J. L., Ghori, N., Herbomel, P., and Ramakrishnan, L. (2002). Real-time visualization of mycobacterium-macrophage interactions leading to initiation of granuloma formation in zebrafish embryos. *Immunity* **17,** 693–702.

Dodd, J., Jessell, T. M., and Placzek, M. (1998). The when and where of floor plate induction. *Science* **282,** 1654–1657.

Driever, W., Solnica-Krezel, L., Schier, A. F., Neuhauss, S. C., Malicki, J., Stemple, D. L., Stainier, D. Y., Zwartkruis, F., Abdelilah, S., Rangini, Z., et al. (1996). A genetic screen for mutations affecting embryogenesis in zebrafish. *Development* **123**, 37–46.

Farber, S. A., Pack, M., Ho, S. Y., Johnson, I. D., Wagner, D. S., Dosch, R., Mullins, M. C., Hendrickson, H. S., Hendrickson, E. K., and Halpern, M. E. (2001). Genetic analysis of digestive physiology using fluorescent phospholipid reporters. *Science* **292**, 1385–1388.

Garrity, D. M., Childs, S., and Fishman, M. C. (2002). The heartstrings mutation in zebrafish causes heart/fin Tbx5 deficiency syndrome. *Development* **129**, 4635–4645.

Haffter, P., Granato, M., Brand, M., Mullins, M. C., Hammerschmidt, M., Kane, D. A., Odenthal, J., van Eeden, F. J., Jiang, Y. J., Heisenberg, C. P., et al. (1996). The identification of genes with unique and essential functions in the development of the zebrafish, *Danio rerio*. *Development* **123**, 1–36.

Harding, M. W., Galat, A., Uehling, D. E., and Schreiber, S. L. (1989). A receptor for the immunosuppressant FK506 is a cis-trans peptidyl-prolyl isomerase. *Nature* **341**, 758–760.

Johnson, S. L., and Weston, J. A. (1995). Temperature-sensitive mutations that cause stage-specific defects in Zebrafish fin regeneration. *Genetics* **141**, 1583–1595.

Khersonsky, S. M., Jung, D.-W., Kang, T.-W., Walsh, D. P., Moon, H. S., Jo, H., Jacobson, E. M., Shetty, V., Neubert, T. A., and Chang, Y. T. (2003). Facilitated forward chemical genetics using a tagged triazine library and zebrafish embryo screening. *J. Am. Chem. Soc.* **125**, 11804–11805.

Kuruvilla, F. G., Shamji, A. F., Sternson, S. M., Hergenrother, P. J., and Schreiber, S. L. (2002). Dissecting glucose signalling with diversity-oriented synthesis and small-molecule microarrays. *Nature* **416**, 653–657.

Kwok, B. H., Koh, B., Ndubuisi, M. I., Elofsson, M., and Crews, C. M. (2001). The anti-inflammatory natural product parthenolide from the medicinal herb Feverfew directly binds to and inhibits IkappaB kinase. *Chem. Biol.* **8**, 759–766.

Langheinrich, U. (2003). Zebrafish: A new model on the pharmaceutical catwalk. *Bioessays* **25**, 904–912.

Lin, J., Sahakian, D. C., de Morais, S. M., Xu, J. J., Polzer, R. J., and Winter, S. M. (2003). The role of absorption, distribution, metabolism, excretion and toxicity in drug discovery. *Curr. Top. Med. Chem.* **3**, 1125–1154.

Lipinski, C. A. (2000). Drug-like properties and the causes of poor solubility and poor permeability. *J. Pharmacol. Toxicol. Methods* **44**, 235–249.

Lipinski, C. A., Lombardo, F., Dominy, B. W., and Feeney, P. J. (2001). Experimental and computational approaches to estimate solubility and permeability in drug discovery and development settings. *Adv. Drug Deliv. Rev.* **46**, 3–26.

Liu, J., Farmer, J. D., Jr., Lane, W. S., Friedman, J., Weissman, I., and Schreiber, S. L. (1991). Calcineurin is a common target of cyclophilin-cyclosporin A and FKBP-FK506 complexes. *Cell* **66**, 807–815.

Llorens, O., Perez, J. J., and Villar, H. O. (2001). Toward the design of chemical libraries for mass screening biased against mutagenic compounds. *J. Med. Chem.* **44**, 2793–2804.

Lo, J., Lee, S., Xu, M., Liu, F., Ruan, H., Eun, A., He, Y., Ma, W., Wang, W., Wen, Z., et al. (2003). 15000 unique zebrafish EST clusters and their future use in microarray for profiling gene expression patterns during embryogenesis. *Genome Res.* **13**, 455–466.

MacRae, C. A., and Peterson, R. T. (2003). Zebrafish-based small molecule discovery. *Chem. Biol.* **10**, 901–908.

Marton, M. J., DeRisi, J. L., Bennett, H. A., Iyer, V. R., Meyer, M. R., Roberts, C. J., Stoughton, R., Burchard, J., Slade, D., Dai, H., et al. (1998). Drug target validation and identification of secondary drug target effects using DNA microarrays. *Nat. Med.* **4**, 1293–1301.

Milan, D. J., Peterson, T. A., Ruskin, J. N., Peterson, R. T., and MacRae, C. A. (2003). Drugs that induce repolarization abnormalities cause bradycardia in zebrafish. *Circulation* **107**, 1355–1358.

Miller, D. K., Gillard, J. W., Vickers, P. J., Sadowski, S., Leveille, C., Mancini, J. A., Charleson, P., Dixon, R. A., Ford-Hutchinson, A. W., Fortin, R., et al. (1990). Identification and isolation of a membrane protein necessary for leukotriene production. *Nature* **343**, 278–281.

Moon, H. S., Jacobson, E. M., Khersonsky, S. M., Luzung, M. R., Walsh, D. P., Xiong, W., Lee, J. W., Parikh, P. B., Lam, J. C., Kang, T. W., *et al.* (2002). A novel microtubule destabilizing entity from orthogonal synthesis of triazine library and zebrafish embryo screening. *J. Am. Chem. Soc.* **124,** 11608–11609.

Mullins, M. C., Hammerschmidt, M., Haffter, P., and Nusslein-Volhard, C. (1994). Large-scale mutagenesis in the zebrafish: In search of genes controlling development in a vertebrate. *Curr. Biol.* **4,** 189–202.

Nasevicius, A., and Ekker, S. C. (2000). Effective targeted gene 'knockdown' in zebrafish. [see comments]. *Nature Genetics* **26,** 216–220.

Nguyen, C., Teo, J. L., Matsuda, A., Eguchi, M., Chi, E. Y., Henderson, W. R., Jr., and Kahn, M. (2003). Chemogenomic identification of Ref-1/AP-1 as a therapeutic target for asthma. *Proc. Natl. Acad. Sci. USA* **100,** 1169–1173.

Nicolaou, K. C., and Pfefferkorn, J. A. (2001). Solid phase synthesis of complex natural products and libraries thereof. *Biopolymers* **60,** 171–193.

Nusslein-Volhard, C., and Dahm, R. (2002). Zebrafish. Oxford University Press, Oxford.

Oprea, T. I. (2000). Property distribution of drug-related chemical databases. *J. Comput. Aided Mol. Des.* **14,** 251–264.

Peterson, R. T., Link, B. A., Dowling, J. E., and Schreiber, S. L. (2000). Small molecule developmental screens reveal the logic and timing of vertebrate development. *Proc. Natl. Acad. Sci. USA* **97,** 12965–12969.

Peterson, R. T., Mably, J. D., Chen, J. N., and Fishman, M. C. (2001). Convergence of distinct pathways to heart patterning revealed by the small molecule concentramide and the mutation heart-and-soul. *Current Biology* **11,** 1481–1491.

Peterson, R. T., Shaw, S. Y., Peteron, T. A., Milan, D. J., Zhong, T. P., Schreiber, S. L., MacRae, C. A., and Fishman, M. C. (2004). Chemical suppression of a genetic mutation in a zebrafish model of aortic coarctation. *Nat. Biotechnol.* **22,** 595–599.

Poggesi, I. (2004). Predicting human pharmacokinetics from preclinical data. *Curr. Opin. Drug Discov. Devel.* **7,** 100–111.

Postlethwait, J. H., Yan, Y. L., Gates, M. A., Horne, S., Amores, A., Brownlie, A., Donovan, A., Egan, E. S., Force, A., Gong, Z., *et al.* (1998). Vertebrate genome evolution and the zebrafish gene map. *Nat. Genet.* **18,** 345–349.

Roman, B. L., Pham, V. N., Lawson, N. D., Kulik, M., Childs, S., Lekven, A. C., Garrity, D. M., Moon, R. T., Fishman, M. C., Lechleider, R. J., *et al.* (2002). Disruption of acvrll increases endothelial cell number in zebrafish cranial vessels. *Development* **129,** 3009–3019.

Rubinstein, A. L. (2003). Zebrafish: From disease modeling to drug discovery. *Curr. Opin. Drug Discov. Devel.* **6,** 218–223.

Schauerte, H. E., van Eeden, F. J., Fricke, C., Odenthal, J., Strahle, U., and Haffter, P. (1998). Sonic hedgehog is not required for the induction of medial floor plate cells in the zebrafish. *Development* **125,** 2983–2993.

Shimizu, N., Sugimoto, K., Tang, J., Nishi, T., Sato, I., Hiramoto, M., Aizawa, S., Hatakeyama, M., Ohba, R., Hatori, H., *et al.* (2000). High-performance affinity beads for identifying drug receptors. *Nat. Biotechnol.* **18,** 877–881.

Shin, J. T., and Fishman, M. C. (2002). From Zebrafish to human: Modular medical models. *Annu. Rev. Genomics. Hum. Genet.* **3,** 311–340.

Spring, D. R., Krishnan, S., Blackwell, H. E., and Schreiber, S. L. (2002). Diversity-oriented synthesis of biaryl-containing medium rings using a one bead/one stock solution platform. *J. Am. Chem. Soc.* **124,** 1354–1363.

St Johnston, D. (2002). The art and design of genetic screens: *Drosophila melanogaster. Nat. Rev. Genet.* **3,** 176–188.

Stern, H. M., and Zon, L. I. (2003). Cancer genetics and drug discovery in the zebrafish. *Nat. Rev. Cancer* **3,** 533–539.

Sternson, S. M., Louca, J. B., Wong, J. C., and Schreiber, S. L. (2001). Split–pool synthesis of 1,3-dioxanes leading to arrayed stock solutions of single compounds sufficient for multiple phenotypic and protein-binding assays. *J. Am. Chem. Soc.* **123,** 1740–1747.

Talbot, W. S., and Schier, A. F. (1999). Positional cloning of mutated zebrafish genes. *Methods Cell Biol.* **60,** 259–286.

Taunton, J., Hassig, C. A., and Schreiber, S. L. (1996). A mammalian histone deacetylase related to the yeast transcriptional regulator Rpd3p. [see comments.] *Science* **272,** 408–411.

Taylor, J. S., Braasch, I., Frickey, T., Meyer, A., and Van de Peer, Y. (2003). Genome duplication, a trait shared by 22000 species of ray-finned fish. *Genome Res.* **13,** 382–390.

Ton, C., Stamatiou, D., Dzau, V. J., and Liew, C. C. (2002). Construction of a zebrafish cDNA microarray: Gene expression profiling of the zebrafish during development. *Biochem. Biophys. Res. Commun.* **296,** 1134–1142.

van de Waterbeemd, H., and Gifford, E. (2003). ADMET in silico modelling: Towards prediction paradise? *Nat. Rev. Drug Discov.* **2,** 192–204.

Van Der Sar, A. M., Musters, R. J., Van Eeden, F. J., Appelmelk, B. J., Vandenbroucke-Grauls, C. M., and Bitter, W. (2003). Zebrafish embryos as a model host for the real time analysis of Salmonella typhimurium infections. *Cell Microbiol.* **5,** 601–611.

van Eeden, F. J., Granato, M., Schach, U., Brand, M., Furutani-Seiki, M., Haffter, P., Hammerschmidt, M., Heisenberg, C. P., Jiang, Y. J., Kane, D. A., *et al.* (1996). Mutations affecting somite formation and patterning in the zebrafish, Danio rerio. *Development* **123,** 153–164.

Weinstein, B. M., Stemple, D. L., Driever, W., and Fishman, M. C. (1995). Gridlock, a localized heritable vascular patterning defect in the zebrafish. *Nat. Med.* **1,** 1143–1147.

Wienholds, E., Schulte-Merker, S., Walderich, B., and Plasterk, R. H. (2002). Target-selected inactivation of the zebrafish rag1 gene. *Science* **297,** 99–102.

Wolff, C., Roy, S., and Ingham, P. W. (2003). Multiple muscle cell identities induced by distinct levels and timing of hedgehog activity in the zebrafish embryo. *Curr. Biol.* **13,** 1169–1181.

Xu, X., Meiler, S. E., Zhong, T. P., Mohideen, M., Crossley, D. A., Burggren, W. W., and Fishman, M. C. (2002). Cardiomyopathy in zebrafish due to mutation in an alternatively spliced exon of titin. *Nature Genetics* **30,** 205–209.

Yu, H., and Adedoyin, A. (2003). ADME-Tox in drug discovery: Integration of experimental and computational technologies. *Drug Discov. Today* **8,** 852–861.

CHAPTER 27

Modeling Human Disease by Gene Targeting

Andrew Dodd,★ Stephen P. Chambers,★ Peter E. Nielsen,† and Donald R. Love★

★Molecular Genetics and Development Group
School of Biological Sciences
University of Auckland
Auckland 1001, New Zealand

†Center of Biomolecular Recognition
Department of Medical Biochemistry and Genetics
University of Copenhagen
Copenhagen 2200, Denmark

An accepted practice of studying human disease processes is to establish disease surrogates using model vertebrate and invertebrate species. In many respects, this practice is predicated on the underlying assumption that all species have, to a large extent, the same repertoire of genes, which then leads to the conclusion that model species have common biological pathways. Of importance, mutations in orthologues of human disease-causing genes do not necessarily result in the same pathophysiological outcomes in model organisms. To combat this deficiency, the notion of comparative organism analysis is one that is finding favor among many biologists and it is here that the zebrafish can play a significant role. The need, however, is to effect changes in specific gene expression in the zebrafish to assess the usefulness of this species, among many others, in modeling and studying the underlying biology of specific human disorders. This chapter discusses the means by which this targeting can be achieved in both a transient and heritable manner.

I. Introduction

Since the mid-1990s, the use of zebrafish as a platform for disease analysis has attracted significant attention. This potential has been highlighted by the discovery of mutants with phenotypes similar to known human diseases (Amatruda *et al.*, 2002; Amemiya, 1998; Brownlie *et al.*, 1998; Dodd *et al.*, 2000; Dooley and Zon, 2000; Stainier *et al.*, 1996; Wang *et al.*, 1998; Zon, 1999). These discoveries underscore the role that zebrafish can play in the analysis of disease in conjunction with more commonly used model organisms like mice.

The use of model organisms to study disease is based on the underlying assumption of conservation in biology and that the conservation of pathways would predominate over several pathways arising to perform the same function. While many examples of the conservation of biological pathways exist across diverse and distantly related species, the response of an organism to specific genetic defects can be very difficult to predict. After a decade of disease analysis in mice using knock-out and knock-in techniques we are becoming acutely aware of the difficulties in recreating the complete pathophysiological properties of human disease in other organisms. This awareness has created what has been called the *mousetrap* (Elsea and Lucas, 2002), where one is stuck when mice do not recapitulate the necessary features of the human disease. The unpredictable nature of disease modeling highlights the impact of organism-specific influences on disease processes. With data based on studies from one organism, it can be difficult to distinguish disease-specific processes and responses from organism-specific ones. To try and combat some of these problems, the idea of comparative disease analysis has developed. This idea states that, by combining studies performed in several different organisms, one should be able to distinguish disease-specific responses from organism-specific influences. This type of approach is currently severely hampered by the lack of easily manipulable vertebrate model species. Of importance, the zebrafish has a significant role to play in this approach. Other than the obvious advantages of the zebrafish for studying early development, adopting the zebrafish as a platform for disease analysis would make it an essential partner to mice in the comparative analysis of disease. Clearly, the development of gene-targeting techniques will be important if we are to exploit the possibilities that zebrafish can offer. However, if we are to learn from our experience in studying mice, where it is now appreciated that the limitations imposed by traditional knock-out and knock-in approaches directly limit the ability to study and understand complex processes, then we will need to consider a significant leap in technology.

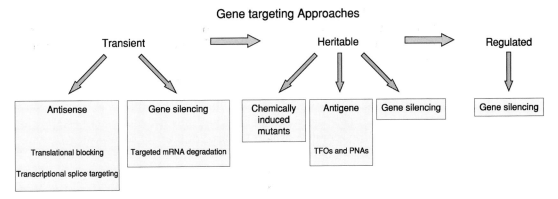

Fig. 1 Gene targeting approaches in the zebrafish. TFO: triplex-forming oligonucleotide; PNA: peptide nucleic acid.

Figure 1 presents a summary of approaches that have been and could be used for gene targeting in the zebrafish that fall into temporal and heritable effects. In terms of the latter, as in the case of mice and *Drosophila*, large collections of random mutants can be screened on the basis of morphological or gene-expression data. This is an excellent resource for identifying genes involved in different processes, and the analysis of these mutants has recently expanded by using methods to screen for mutations in specific genes. These mutants are discussed here, but this approach lacks the directness of gene targeting and the need to effect changes in temporal expression, as well as to analyze the implications of subtle changes in protein structure that could be more specifically engineered.

The following discussion is presented in three complementary parts. The first presents briefly those considerations that should be made concerning an approach that mimics the route to gene targeting in the mouse, namely, the creation of knock-out lines via homologous recombination. The second part concentrates on the means by which temporal changes in gene expression can be achieved in the zebrafish with particular reference to our own studies with gene silencing, and the use of peptide nucleic acids to effect alterations in transcript splicing. The third part examines, by way of a personal perspective, the machinery that is required for site-specific recombination, and the possibility of using peptide nucleic acids (PNAs) to drive a novel approach to achieve gene targeting. This proposition necessitates some appreciation of DNA repair processes.

II. DNA Repair Processes

The generation of knock-out lines of zebrafish, mimicking the protocol developed in the mouse, requires the availability of embryonic stem (ES) cells, the ability to introduce mutations via homologous recombination (HR), and the need to introduce genetically manipulated cells into zebrafish embryos (Altschmied *et al.*, 1998). In support of this process, short-term cultures derived from gastrula-stage zebrafish embryos co-cultured with rainbow trout splenic stromal cells have generated germ-line chimeras after introduction into host embryos, albeit at the low frequency of 4% (Ma *et al.*, 2001). It remains to be determined whether the cultures can be extended long enough to allow for the introduction and selection of colonies with a desired targeting event.

Hagmann *et al.* (1998) have shown that HR can occur in early zebrafish embryos, with a preponderance of non homologous end-joining (NHEJ) over HR. Furthermore, zebrafish exhibit the least number of sequence distortions via the DNA end-joining process compared with the early embryos of *Xenopus* and *Drosophila melanogaster*. The conclusion to this study was that, while the machinery for HR exists in the zebrafish for targeted gene disruption, this process is overwhelmed by NHEJ activity. These authors suggest that targeting proteins involved in NHEJ, such as Ku antigen or poly-ADP-ribose polymerase (PARP), or selectively overexpressing proteins promoting HR, could increase the frequency of HR. With respect to the latter, Cui *et al.* (2003) introduced recombinant plasmids carrying mutated enhanced green fluorescent protein (EGFP) genes together with 236 nucleotide corrective single-stranded DNAs coated with RecA. The 5 to 20% of injected embryos that exhibited EGFP expression could be enhanced with prior UV irradiation of the embryos to stimulate endogenous zebrafish DNA repair systems.

Table I summarizes apparent zebrafish orthologues of those genes that encode proteins involved in DNA repair systems that have been found in *Saccharomyces cerevisiae* and humans. The principal categories of repair are excision repair, single-strand break repair, double-strand break repair that involves HR and NHEJ, and tolerance of DNA damage involving gap repair and lesion bypass.

Of importance, the repair of double-strand breaks (DSBs) can occur via two main pathways: NHEJ and HR. The latter can be further sub-divided into gene conversion and single-strand annealing (Haber *et al.*, 1999). Van Dyck *et al.* (1999) have proposed that the proteins Rad 52 and Ku are competing agents in deciding whether repair proceeds into the competing HR and NHEJ pathways, respectively. It is important to remember that, while NHEJ is generally considered to be mutagenic, conflicting data examining cultured murine fibroblasts shows NHEJ to act strongly in favor of re-ligation to an appropriate partner after a DSB (Ferguson *et al.*, 2000). In addition, while HR is an important pathway in mammalian cells for DNA repair, it also plays an important role in genome rearrangements involving non-allelic homologous sequences (Bishop and Schiestl, 2000). HR can be induced by different types of DNA damage because of the formation of DSBs, which may be replication-independent.

While these scenarios may appear of little relevance to those seeking to follow a homologous recombination route to create targeted mutants, an appreciation of the repair processes plays a pivotal role in understanding how efficient targeting may occur. Indeed, the temporal manipulation of genes expressing proteins

Table I
Proteins Involved in DNA Repair Processes[a]

Human	Yeast	Excision Repair—Nucleotide Excision Repair	Zebrafish Ensemble Gene ID
		Function	
XPC	RAD4	Damage recognition and altering chromatin structure (allows access by damage-processing enzymes) (Rebhan et al., 1997).	ENSDARG00000016503
HR23B	RAD23	Damage recognition and altering chromatin structure (allows access by damage-processing enzymes) (Rebhan et al., 1997).	ENSDARG00000013260
XPA	RAD14	Initiates repair by binding to damaged sites with various affinities, depending on the photoproduct and the transcriptional state of the region (Rebhan et al., 1997).	
RPA70, 32, 14	RPA1, 2, 3	Single-stranded DNA-binding protein. May stabilize open complex with XPA (Rebhan et al., 1997).	ENSDARG00000003938
XPB/ ERCC3	MMS19L/ RAD25	XPB subunit. ATP-dependent 3'-5' DNA helicase. Opens DNA either around the RNA transcription start site or the DNA damage (Rebhan et al., 1997).	ENSDARG00000002402
p62	TFB1	TFIIH basal transcription factor complex (9 proteins). Component of the core-TFIIH basal transcription factor (Rebhan et al., 1997).	ENSDARG00000026701
p52	TFB2	Component of the core-TFIIH basal transcription factor (Rebhan et al., 1997).	ENSDARG00000004772
p44/ GTF2H2	SSL1	Component of the core-TFIIH basal transcription factor. Interacts with XPB, XPD, p62, and P34 (Rebhan et al., 1997).	ENSDARG00000016514
p34/ GTF2H3	TFB4	Component of the core-TFIIH basal transcription factor (Rebhan et al., 1997).	ENSDARG00000005027
XPD/ ERCC2	RAD3	XPD subunit. ATP-dependent 5'-3' DNA helicase. May also have a role in the ageing process (Rebhan et al., 1997).	ENSDARG00000015615
MAT1/ MNAT1	TFB3/ RIG2	CDK assembly factor: stabilizes the cyclin H-CDK7 complex to form a functional CDK-activating kinase (CAK) (Rebhan et al., 1997).	ENSDARG00000002077
CDK7	KIN28	Catalytic subunit of the CAK complex. Complexed to the core-TFIIH basal transcription factor and activates RNA Pol II (Rebhan et al., 1997).	ENSDARG00000007895
CycH	CCL1	Cyclin. Regulates CDK7 (Rebhan et al., 1997).	ENSDARG00000007657
XPG/ ERCC5	RAD2	Single-stranded DNA endonuclease involved in DNA excision repair. Makes the 3' incision in NER (Rebhan et al., 1997).	ENSDARG00000014313
XPF/ ERCC4		Structure-specific DNA repair endonuclease. Responsible for 5' incision during DNA repair. Involved in homologous recombination that assists in removing inter strand cross-link (Rebhan et al., 1997).	ENSDARG00000014161

(continues)

Table I *(continued)*

		Excision Repair—Nucleotide Excision Repair	
Human	Yeast	Function	Zebrafish Ensemble Gene ID
		Strand break repair—single-strand breaks	
LIG1		Seals during DNA replication, DNA recombination and DNA repair nicks in double-stranded DNA (Rebhan et al., 1997).	ENSDARG00000007268
LIG3		Interacts with DNA-repair protein XRCC1. Corrects defective DNA strand-break repair and sister chromatid exchange (Rebhan et al., 1997).	ENSDARG00000021375
LIG4		Efficiently joins single-strand breaks in a double-stranded poly deoxynucleotide. ATP-dependent. Responsible for the ligation step in non-homologous DNA end joining (Rebhan et al., 1997).	ENSDARG00000020823
PARP1		Poly (ADP-Ribose) Polymerase. Modifies various nuclear proteins. Involved in the recovery from DNA damage, possibly acting by protecting the reactive single-strand break ends (Rebhan et al., 1997).	ENSDARG00000015600 ENSDARG00000019529 ENSDARG00000005274
		Strand break repair—double-strand breaks (homologous recombination)	
RAD52	RAD52	Seals during DNA replication and recombination. Repairs nicks in ds-DNA (Rebhan et al., 1997).	
MRE11	MRE11	Possesses single-strand endonuclease activity and double-strand specific 3′-5′ exonuclease activity. Regulates 5′-3′ exonuclease responsible for break-end resectioning (Rebhan et al., 1997).	ENSDARG00000006394
RAD50	RAD50	Involved in DNA recombinational repair and meiosis specific double strand break formation. Complexes with MRE11 and NBS1 (Rebhan et al., 1997).	ENSDARG00000016975
NBS1		Encodes Nibrin (novel DSB repair protein). Mutated in Nijmegen Breakage Syndrome (Rebhan et al., 1997).	ENSDARG00000008320
RAD51	RAD51	Binds to single- and double-stranded DNA. Exhibits DNA-dependent ATPase activity. Unwinds duplex DNA and forms helical nucleoprotein filaments (Rebhan et al., 1997).	ENSDARG00000011751
RAD54 XRCC3	RAD54-like	Recombinational protein associated with RAD51 (Rebhan et al., 1997). Thought to repair chromosomal fragmentation, translocations and deletions (Rebhan et al., 1997).	ENSDARG00000018623 ENSDARG00000017928

Human	Yeast	Description	Zebrafish (ENSDARG)
Strand break repair—double-strand breaks (non-homologous end joining)			
KU70/XRCC6	HDF1	ss-DNA, ATP-dependent helicase. 70kDa subunit. May mediate binding of complex to DNA. Complex binds preferentially to fork-like ends of ds DNA. (Rebhan et al., 1997).	ENSDARG00000020538
KU80/XRCC5	HDF2	80kDa subunit. Works in the 3'-5' direction (Rebhan et al., 1997).	ENSDARG00000015599
DNA-PKcs/XRCC7		Ser/Thr kinase. Forms complex with Ku70/80 autoantigen. Interacts with DNA-PKs Interacting Protein (KIP) (Rebhan et al., 1997).	ENSDARG00000021102
Artemis	PSO2	Endonuclease. Regulated by DNA-PKcs. Prepares for ligation by cutting single-stranded overhangs (Huberman 2003a).	ENSDARG00000005732
LIG4	LIG4	Joins ss-breaks in a ds-poly deoxynucleotide. ATP-dependent. Responsible for the ligation step in non-homologous DNA end joining. Co-operates with XRCC4 (Rebhan et al., 1997).	ENSDARG00000020823
Tolerance of DNA damage-gap formation and recombination			
RAD18	RAD18	Involved in post replication repair of UV-damaged DNA. Gap-fills a daughter strand on replication of damaged DNA. Has ssDNA binding activity (by similarity). Interacts with Rad6 (Rebhan et al., 1997).	ENSDARG00000027938
RAD6/UBE2A	RAD6	Catalyses ubiquitination of other proteins. Required for post replication repair of UV-damaged DNA (Rebhan et al., 1997).	ENSDARG00000001280
UBC13/UBE2N	UBC13	Catalyses ubiquitination of other proteins. Forms a heterodimer with Mms2 (Rebhan et al., 1997).	ENSDARG00000008748
MMS2/UBE2V2	MMS2	Structurally similar to ubiquitin-conjugating enzymes but does not by itself possess ubiquitin-conjugating activity (Rebhan et al., 1997; Huberman 2003b).	ENSDARG00000021406
Tolerance of DNA damage-translesion DNA synthesis: lesion bypass			
Pol Kappa		Incorporates mismatched bases on a non damaged template. Also provides error-prone replication through certain lesions such as abasic sites and DNA adducts (Ohashi et al., 2000)	ENSDARG00000006106
Pol Zeta	REV3	Enables protein-protein interactions with other factors during translesional DNA synthesis. Requires Rev1 protein (Rebhan et al., 1997).	ENSDARG00000020875
REV1	REV1	dCMP transferase; specifically inserts opposite template G, apurinic/apyrimidinic site or uracil residue. Role in mutagenic trans lesion DNA synthesis (Lin et al., 1999).	ENSDARG00000018296
Pol Eta	RAD30	Error-free bypass polymerase. Specifically inserts two adenines opposite cyclobutane thymidine dimers. Mutants show decreased ability to tolerate UV damage. "XP-Variant" (Johnson et al., 1999).	ENSDARG00000026870

[a]Genes encoding for proteins involved in DNA repair processes in *Saccharomyces cerevisiae* (yeast) and their apparent orthologues in *Homo sapiens* (human) and *Danio rerio* (zebrafish) are shown.

involved in repair processes could offer a means of enhancing HR-based gene targeting, but also in the repair of targeted mutagenic insults in zebrafish embryos (as described in Section IV).

III. Transient Means to Effect Altered Gene Expression in the Zebrafish

A. Gene Silencing

The development of gene targeting approaches based on gene silencing or 'RNA interference' has revolutionized gene targeting in many model organisms. In 1998 the groups of Fire and Mello discovered the process of gene expression inhibition by double-stranded RNA (dsRNA), and they termed this process *RNA interference* (Fire *et al.*, 1998). The gene silencing pathway is now recognized to exist in most organisms from plants to humans. The mechanism is thought to be part of an antiviral defense mechanism and also an RNA-based regulation system involved with developmental timing.

This new and powerful technique has become a standard approach in *Caenorhabditis elegans* and *Drosophila*, either through the direct injection of dsRNA or the expression of dsRNA hairpins from recombinant plasmids. However, the introduction of dsRNA to mammalian cells triggers nonspecific responses. These responses highlight an essential difference between how mammals (or vertebrates) and invertebrates respond to dsRNA. In mammals, the introduction of dsRNA induces interferon and a cascade of interferon response genes, resulting in a general repression of transcription and translation leading eventually to apoptosis (Stark *et al.*, 1998). Subsequently, it was found that short dsRNAs (less than 30 bp) could specifically silence genes but did not appear to activate the interferon response (Elbashir *et al.*, 2001). These studies have paved the way for gene silencing becoming an essential tool for manipulating mammalian gene expression.

The exact mechanisms involved in the gene silencing process are not completely understood, but several key steps have been identified (Fjose *et al.*, 2001; Hutvagner and Zamore, 2002; Tijsterman *et al.*, 2002). The central effector molecule in gene silencing is short sections of dsRNA called *short interfering RNAs* (siRNAs). These siRNAs are made from long dsRNAs or RNA hairpins by the enzyme Dicer, a member of the RNase III family of dsRNA-specific endonucleases, in an adenosine triphosphate (ATP)-dependent-step (Bernstein *et al.*, 2001). The siRNA duplexes contain 5′ phosphate and 3′ hydroxyl termini, and two single-stranded nucleotides at their 3′ end (Elbashir *et al.*, 2001). These structural features are important for the entry of siRNAs into the gene silencing pathway. These siRNA duplexes are then incorporated into a protein complex, termed the *RNA-induced silencing complex* (RISC). The ATP-dependent unwinding of the siRNA duplex activates RISC, which can then recognize and cleave a target mRNA complementary to the antisense strand of the siRNA (Fig. 2).

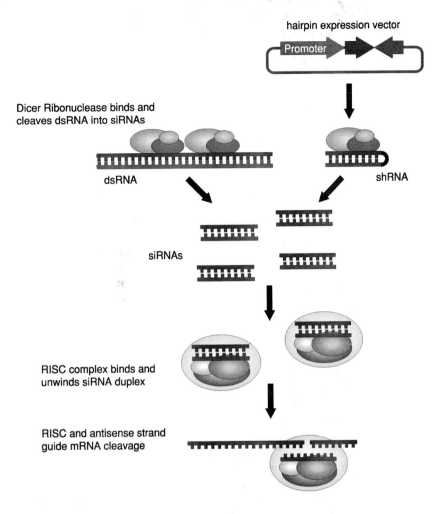

Fig. 2 Gene silencing mechanism. The multi subunit Dicer ribonuclease processes dsRNA and RNA hairpins into siRNAs. The RNA-induced silencing complex (RISC) complex binds siRNAs and activates, guiding the cleavage of complementary mRNA.

Early experiments with dsRNA in zebrafish embryos produced conflicting results. Some groups reported sequence specific gene targeting (Li *et al.*, 2000; Wargelius *et al.*, 1999), while others reported a predominance of non-specific effects (Oates *et al.*, 2000; Zhao *et al.*, 2001). This conflict led to the widely held view that gene silencing was not an appropriate tool for gene targeting in zebrafish embryos. In light of our understanding of the differences between the response of invertebrate and mammalian systems to long dsRNA, the results are not surprising. Of importance, the sensitivity of mammalian systems to long regions of dsRNA is probably a vertebrate issue rather than a mammalian one alone. This conclusion suggests that the siRNA approaches found to be so successful in mammalian systems could be the key to effective gene silencing in other vertebrate systems like the zebrafish. A recent paper by Boonanuntanasarn *et al.* (2003), has demonstrated the utility of siRNAs for gene silencing in trout embryos. These data then add support for this approach in zebrafish.

1. Use of siRNAs in the Zebrafish

To test gene silencing via siRNAs in zebrafish embryos, we decided to target the zebrafish *dmd* gene. The zebrafish has attracted some attention as a model system to study the X-linked recessive disorder Duchenne muscular dystrophy (DMD), a severe neuromuscular disease resulting from mutations in the dystrophin gene. Studies of the zebrafish dystrophin mutant *sapje* have highlighted some potential new roles for the dystrophin protein (Bassett *et al.*, 2003).

We designed several siRNAs to the carboxyl terminal region of the zebrafish dystrophin mRNA (Dodd *et al.*, 2004). These chemically synthesized siRNAs (Xeragon), and a control siRNA (anti-GFP, Dharmacon), were microinjected into single-cell embryos (Fig. 3). In this experiment we were interested in the potential of siRNAs to reduce specifically the levels of target mRNA, while avoiding the induction of non-specific effects. To determine changes in target RNA levels, we measured the levels of dystrophin RNA at various times during development by real-time polymerase chain reaction (PCR) in combination with one reference gene (β-sarcoglycan) that encodes for a protein with which dystrophin interacts, and three unrelated reference genes (Elongation factor 1 alpha, Beta actin, 18s rRNA). This analysis platform enabled changes in gene expression to be assessed more rigorously compared with RNA *in situ* hybridization. Fig. 3a shows a clear reduction in the levels of dystrophin mRNA in siRNA-68 injected embryos (targeting exon 68 of the *dmd* gene). No change in dystrophin mRNA levels was observed in siRNA-GFP injected embryos, and no significant difference in the levels of β-sarcoglycan mRNA was observed (data not shown). No apparent difference was seen in the levels of the three unrelated reference genes used to normalize this experiment. We further tested the effect of siRNA-68 using whole-mount protein immunohistochemistry (Fig. 3b). This analysis showed a reduction in dystrophin staining, indicating that the reduction in mRNA levels translated to a difference in the level of the dystrophin protein.

A comparison between control siRNAs and control long (greater than 300 bp) dsRNAs that are not complementary to any zebrafish genes also highlighted a clear difference in response. Very small amounts of long dsRNA can stimulate a very marked toxic or 'non-specific' response; whereas 10 times the level of a control siRNA exhibited no observable response (data not shown). It remains to be seen whether this difference is attributed to the stimulation of interferon and interferon response genes by long dsRNA. Sledz *et al.* (2003) also indicated

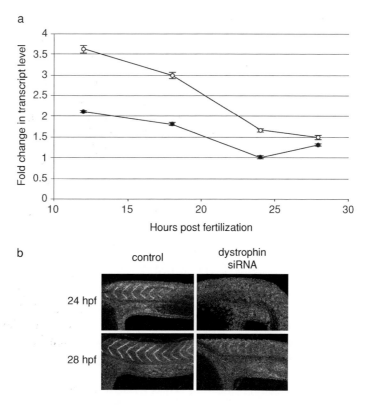

Fig. 3 The siRNA targeting of the zebrafish *dmd* gene. An siRNA (0.3 nl of 50 μM) designed against exon 68 of the zebrafish *dmd* gene was microinjected into zebrafish embryos. Panel a: RNA was extracted using Trizol (Invitrogen) from 10 to 20 embryos at each time point and reverse transcribed using Superscript III (Invitrogen). The cDNAs were subjected to quantitative PCR using an Applied Biosystems model 7900HT platform (Dodd *et al.*, 2004). The fold change in transcript level is based arbitrarily on the lowest transcript level detected in 24 hours siRNA-injected embryos. ($\bigcirc$) control (uninjected) and ($\bullet$) siRNA-68 injected embryos. Vertical bars indicate standard deviations. Panel b: Dystrophin protein was detected using the antibody MANCHO12 (G. E. Morris, NEWI, U.K.) at 1:3 dilution as previously described (Chambers *et al.*, 2003). Rhodamine conjugated secondary antibody was detected by scanning confocal microscopy. Dystrophin protein was absent from the myoseptae (where it is normally expressed) at 24 hpf, but at 28 hpf the recovery of dystrophin expression is detected. Reprinted by permission of Federation of the European Biochemical Societies from "Short interfering RNA-mediated gene targeting in the zebrafish," by Dodd A., *et al.*, (2004). *FEBS Letters* **561**, 89–93.

that siRNAs can stimulate a low level of interferon response. This low-level stimulation must be examined further to determine what impact this could have when using gene silencing in the zebrafish.

B. Oligonucleotide-Based Agents

Synthetic oligonucleotides offer the means of modulating gene expression in an antisense approach (Opalinska and Gewirtz, 2002), or as an antigene strategy (see Section IV). With respect to the former, a variety of backbone chemistries have been developed to increase uptake and biological stability (Dagle and Weeks, 2001; Dove, 2002; Sazani *et al.*, 2001). Of these, morpholino oligonucleotides have been successfully used to block translation, as well as splicing in the zebrafish (Draper *et al.*, 2001; Ekker, 2000; Nasevicius and Ekker, 2000; Xu *et al.*, 2002). However, PNAs have received less attention, with only a single report on their use in blocking translation (Urtishak *et al.*, 2003). PNAs are DNA mimics with the deoxyribose backbone of DNA replaced by an uncharged pseudo peptide. This backbone is resistant to degradation by both nucleases and peptidases (Nielsen, 1999). PNAs bind with extremely high affinity in a sequence-specific manner and they are able to discriminate to one base pair, so careful design is crucial to avoid missing the target. The superior affinity of a PNA to its target has been exploited for other purposes such as a bacteriostat (Nielsen, 1999), and a pre targeting molecule for drug delivery (Ruskowski *et al.*, 1997). In mammalian cells, PNAs are taken up quite poorly, but in the case of the zebrafish, this can be avoided by direct microinjection (Dean, 2000; Urtsihak *et al.*, 2003).

The use of PNAs comes into play at either the level of blocking mRNA translation (Hanvey *et al.*, 1992; Knudsen and Nielsen, 1996; Koppelhus and Nielsen, 2001, 2003), or by targeting hnRNA and binding at splice sites (Karras *et al.*, 2001; Sazani *et al.*, 2001, 2002; Fig. 4). In the case of the former, the assessment of efficacy is protein based with semi-quantitative estimation of protein levels by Western Blotting or immunohistochemical analysis. Binding to splice sites, however, can lead to exon skipping, which can be assessed through the use of real-time polymerase chain reaction (RT-PCR) at the transcript level, as well as protein analysis. RT-PCR is exquisitely sensitive and informative even if not all the transcripts are truncated, as both sets of mRNA would be visible. The use of PNAs in this context is necessarily transient, but the effects can be assessed rapidly.

1. Design

We used 15-mer PNAs designed to bind across the splice donor sites of exons 53, 63, 68, and 71 of the zebrafish *dmd* gene to induce exon skipping. The designs were tested *in silico* using the program Amplify 1.0 to confirm that the PNAs would not form dimers and to confirm binding specificity. The PNAs were supplied as lyophilized pellets and were dissolved in water to an estimated 800-μM stock. The extinction coefficient of each PNA was calculated and used to confirm each stock concentration. These stocks were corrected to pH 6.5 to 7.0 using ammonium bicarbonate buffering, and then resuspended to a working stock of 200 μM. Working dilutions of 10 μM, 25 μM, 50 μM, or 100 μM were made by dilution of the 200-μM working stock in water.

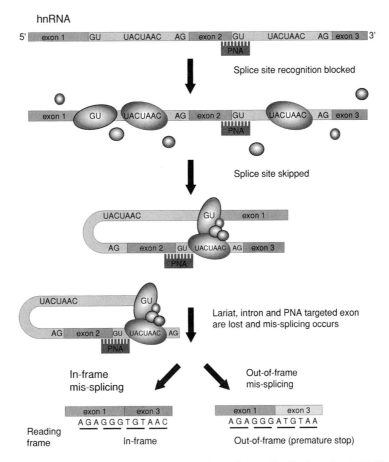

Fig. 4 Means by which PNAs can affect transcript splicing. The binding of a PNA blocks the assembly of splice factors by masking the splice donor site leading to mis-splicing of the primary transcript. This mis-splicing results in the loss of the exon in the processed transcript that is immediately upstream of the splice donor site that has been disrupted by PNA binding.

Routine injections involved the use of approximately 120 embryos per PNA injection, which yielded sufficient embryos for the planned samples, plus a buffer of embryos to account for natural mortality. In the case of an uninjected control population, at least 100 embryos were reserved for matched controls of all samples.

2. Time of Extraction

Dystrophin is detected at 14 hpf (Bolaños-Jiménez *et al.*, 2001), therefore embryos that were microinjected at the 1-cell stage were collected in TRIzol™ (Invitrogen, Carlsbad, CA) at 2, 4, 6, and 13 hpf to assess the onset of transcription (Fig. 5).

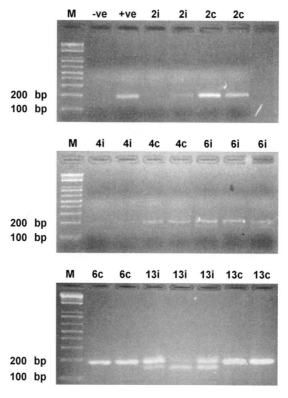

Fig. 5 Time course of RT-PCR amplification of exon 71 of the zebrafish *dmd* gene from PNA-treated embryos. The 900 pl of a 50-μM PNA targeted to the splice donor site of exon 71 of the zebrafish *dmd* gene was introduced into single-cell zebrafish embryos by microinjection, and RNA was extracted during development at the times indicated. The RNAs were subjected to RT-PCR using primers located in exons 70 and 72 of the *dmd* gene transcript. Two RT-PCR amplified fragments were detected corresponding to the correctly spliced (186 bp) and PNA-directed mis-spliced (147 bp) transcripts. It is evident that mis-splicing is induced at 6 hpf in injected embryos, which suggests that embryonic transcription of the *dmd* gene occurs between 4 hpf and 6 hpf. Key: M = 1 kb plus ladder (Invitrogen); +ve = muscle cDNA; −ve = water template; 2i = 2 hpf injection; 2c = 2 hpf control, and so on.

3. Concentration of PNAs for Injections

Nasevicius and Ekker (2000) injected 4.5 ng to 6.0 ng of morpholino oligonucleotides (590 μM to 786 μM, injection volume = 900 pL) into 2-cell embryos. However, initial injections of 900 pL of 200-μM PNAs, resulted in embryos showing delayed development and 100% mortality by 24 hpf. Subsequent injections used approximately 900 pL of 10-μM, 25-μM, 50-μM, or 100-μM PNA solutions.

4. Microinjection of PNA into Zebrafish Embryos

Embryos were microinjected at the one cell stage with glass capillary microinjection needles. Preliminary trials of injections were carried out using PNAs mixed with dextran tetramethyl rhodamine dye (DTMR) at 10 mgmL^{-1}. The fluorescence of rhodamine allowed direct confirmation of injected embryos, and subsequent experiments were performed without dye.

A problem associated with zebrafish microinjection is the mechanical damage that can be brought about by the needle. To account for this, the injection of Holtfreters solution was used as a positive control for needle injury.

5. Analysis of the Effect of PNA Microinjection

Primers were designed to exons flanking the PNA-targeted exon using the design program Primer Premier 5.0 to allow the assessment of exon-skipping by RT-PCR amplification. *In silico* analysis was undertaken using the program Amplify 1.0 to ensure no primer dimerization, hairpins, or false priming would occur.

Primers were tested for amplification efficiency by RT-PCR amplification of the dystrophin transcript from adult zebrafish muscle. After this assay, RNA was extracted from 5 embryos from duplicate samples at each time point. Of each RNA, 1 μg was reverse transcribed using Superscript III (Invitrogen), of which one-fortieth was subjected to PCR amplification with subsequent sizing of the amplicons to detect PNA-directed mis-splicing (Fig. 5).

IV. Heritable Means to Effect Altered Gene Expression in the Zebrafish

While the transient effects of oligonucleotides have been applied in antisense approaches, the less-common approach is their use as antigene agents (i.e., for direct binding to DNA). Two strategies that have been used in this context are triplex-forming oligonucleotides (TFOs) and helix invading PNAs. TFOs bind in the major groove of double-stranded DNA at sites with runs of purines on one strand (Dagle and Weeks, 2001; Knauert and Glazer, 2001). TFOs can act as inhibitors of transcription initiation and elongation, but also as site-specific mutagens, especially when complexed with DNA-damaging agents such as psoralen (Barre *et al.*, 2000; Giovannangeli and Helene, 1997).

Majumdar *et al.* (1998) showed that TFOs complexed with psoralen caused mutation events in CHO-K1 cells that comprised deletions of 4 to 50 bases, some with 1- to 2-base deletion events at the site of the psoralen crosslink, and also base substitution mutations. The authors concluded that the TFO-psoralen-UVA treatment enhanced the frequency of mutagenesis at the target site by a few hundred fold compared with spontaneous mutation events at the target site. It was suggested that deletions might occur as a consequence of gap repair, processing of a

DSB by HR involving single-strand annealing, or NHEJ. TFOs alone can also be mutagenic as a consequence of DNA-repair processes such as nucleotide excision and HR (reviewed by Dagle and Weeks, 2001; Faruqi *et al.*, 2000).

Homo pyridine PNAs invade the helix of double-stranded DNA under low ionic strength or conditions that promote DNA unwinding, such as negative superhelical stress or transcription. This invasion leads to the formation of a D-loop in which the pyrimidine DNA strand is single stranded while a very stable PNA/DNA/PNA triplex, involving Watson-Crick and Hoogsteen base pairing is formed on the purine DNA strand. Often it is advantageous to use bis PNA clamps in which the two PNAs are covalently connected through a flexible linker (Nielsen, 2001). Furthermore, the efficiency of PNA clamps can be significantly improved in particular at physiological ionic strength conditions by conjugation to cationic peptides or to a DNA intercalator (Bentin and Nielsen, 2003; Kaihatsu *et al.*, 2002). These PNA clamps bind to DNA with high specificity and affinity, and have been shown to be mutagenic, at least in mouse cells (Faruqi *et al.*, 1998). Of interest, PNAs have been conjugated to DNA and shown to stimulate repair and recombination in a site-directed manner involving the nucleotide excision repair factor XPA (Rogers *et al.*, 2002). This approach has significant merit in the context of disease modeling where mutations can be targeted to defined protein domains. The outcome of such an approach, when applied to the zebrafish, could be a repertoire of mutants that would enable a better appreciation of the functional consequences of mutation events in the context of a developing organism.

V. Discussion

The accessibility of zebrafish embryos to microinjection make them an ideal platform for gene targeting. In the first instance, transient effectors such as siRNAs offer a novel means of effecting short-term changes in gene expression with readouts that can be assessed at both the transcript and protein levels. Several approaches for generating siRNAs are available. In the case of transient targeting, siRNAs can either be chemically synthesized or produced by *in vitro* digestion of long dsRNAs with recombinant Dicer enzyme (Myers *et al.*, 2003). Stable targeting via the expression of short hairpin RNAs (shRNAs), which are processed to siRNAs (see Fig. 2), is by far the most exciting approach. This method has been gaining popularity as an alternative to homologous recombination in mice (Carmell *et al.*, 2003; Hasuwa *et al.*, 2002; Tiscornia *et al.*, 2003), and offers the ability to target multiple genes in a regulated and tissue-specific manner (Czauderna *et al.*, 2003; van de Wetering *et al.*, 2003).

The possibility of developing targeted mutagenesis strategies using TFOs and PNAs in zebrafish remains promising. The questions surrounding these approaches must concern the availability of the effector molecule to the zebrafish genome at sufficient levels to generate mutants at a frequency in excess of that detected spontaneously, and the efficiency with which any mutagenesis event (direct or indirect) can be repaired. The latter might require some further manipulation such as regulated siRNA-based transcriptional changes in genes encoding for proteins that act in defined repair pathways. Strategies such as this may offer novel routes for manipulating biological pathways in the zebrafish to achieve a disease-modeling outcome. The important corollary to this statement is to know a model when you see it and, for this to be achieved, a greater realization of those parameters that define a disease model, which encompasses more than pheno copying, needs to be adopted.

Acknowledgments

We acknowledge funding support of the University of Auckland Vice Chancellor's Development Fund, University of Auckland Research Committee, Lottery Grants Board of New Zealand, and the Maurice and Phyllis Paykel Trust. We also thank A. Tsai for her assistance in assembling the data presented in Table I and G. Maguire for Fig. 4.

References

Altschmied, J., Hong, Y., and Schartl, M. (1998). Homing in on homologous recombination: Analyzing gene function in fish *in vivo*. *Biol. Chem.* **379**, 631–632.

Amatruda, J. F., Shepard, J. L., Stern, H. M., and Zon, L. I. (2002). Zebrafish as a cancer model system. *Cancer Cell* **1**, 229–231.

Amemiya, C. T. (1998). The zebrafish and haematopoietic justice. *Nat. Genet.* **20**, 222–223.

Barre, F.-X., Ait-Si-Ali, S., Giovannangeli, C., Luis, R., Robin, P., Pritchard, L. L., Helene, C., and Harel-Bellan, A. (2000). Unambiguous demonstration of triple-helix-directed gene modification. *Proc. Natl. Acad. Sci.* **97**, 3084–3088.

Bassett, D. I., Bryson-Richardson, R. J., Daggett, D. F., Gautier, P., Keenan, D. G., and Currie, P. D. (2003). Dystrophin is required for the formation of stable muscle attachments in the zebrafish embryo. *Development* **130**, 5851–5860.

Bentin, T., and Nielsen, P.E. (2003). Superior duplex DNA strand invasion by acridine conjugated peptide nucleic acids. *J. Am. Chem. Soc.* **125**, 6378–6379.

Bernstein, E., Caudy, A. A., Hammond, S. M., and Hannon, G. J. (2001). Role for a bidentate ribonuclease in the initiation step of RNA interference. *Nature* **409**, 363–366.

Bishop, A. J. R., and Schiestl, R. H. (2000). Homologous recombination as a mechanism for genome rearrangements: Environmental and genetic effects. *Hum. Mol. Genet.* **9**, 2427–2434.

Bolaños-Jiménez, F., Bordais, A., Behra, M., Strähle, U., Sahel, J., and Rendón, A. (2001). Dystrophin and Dp71, two products of the DMD gene, show a different pattern of expression during embryonic development in zebrafish. *Mechanisms of Development* **102**, 239–241.

Boonanuntanasarn, S., Yoshizaki, G., and Takeuchi, T. (2003). Specific gene silencing using small interfering RNAs in fish embryos. *Biochem. Biophys. Res. Commun.* **310**, 1089–1095.

Brownlie, A., Donovan, A., Pratt, S. J., Paw, B. H., Oates, A. C., Brugnara, C., Witkowska, H. E., Sassa, S., and Zon, L. I. (1998). Positional cloning of the zebrafish sauternes gene: A model for congenital sideroblastic anaemia. *Nat. Genet.* **20**, 244–250.

Carmell, M. A., Zhang, L., Conklin, D. S., Hannon, G. J., and Rosenquist, T. A. (2003). Germline transmission of RNAi in mice. *Nat. Struct. Biol.* **10**, 91–92.

Chambers, S. P., Anderson, L. V. B., Maguire, G. M., Dodd, A., and Love, D. R. (2003). Sarcoglycans of the zebrafish: Orthology and localization to the sarcolemma and myosepta of muscle. *Biochem. Biophys. Res. Comm.* **303**, 488–495.

Cui, Z., Yang, Y., Kaufman, C. D., Agaliu, D., and Hackett, P. B. (2003). RecA-mediated, targeted mutagenesis in zebrafish. *Mar. Biotechnol. (NY)* **5**, 174–184.

Czauderna, F., Santel, A., Hinz, M., Fechtner, M., Durieux, B., Fisch, G., Leenders, F., Arnold, W., Giese, K., Klippel, A., *et al.* (2003). Inducible shRNA expression for application in a prostate cancer mouse model. *Nucleic Acids Res.* **31**, e127.

Dagle, J. M., and Weeks, D. L. (2001). Oligonucleotide-based strategies to reduce gene expression. *Differentiation* **69**, 75–82.

Dean, D. A. (2000). Peptide nucleic acids: Versatile tools for gene therapy strategies. *Advanced Drug Delivery Reviews* **44**, 81–95.

Dodd, A., Curtis, P. M., Williams, L. C., and Love, D. R. (2000). Zebrafish: Bridging the gap between development and disease. *Hum. Mol. Genet.* **9**, 2443–2449.

Dodd, A, Chambers, S. P., and Love, D. R. (2004). Short interfering RNA-mediated gene targeting in the zebrafish. *FEBS Lett.* **561**, 89–93.

Dooley, K., and Zon, L. (2000). Zebrafish: A model system for the study of human disease. *Current Opinion in Genetics & Development* **10**, 252–256.

Dove, A. (2002). Antisense and sensibility. *Nat. Biotechnol.* **20**, 121–124.

Draper, B. W., Morcos, P. A., and Kimmel, C. B. (2001). Inhibition of zebrafish *fgf8* pre-mRNA splicing with morpholino oligos: A quantifiable method for gene knockdown. *Genesis: The Journal of Genetics & Development* **30**, 154–156.

Ekker, S. C. (2000). Morphants: A new systematic vertebrate functional genomics approach. *Yeast* **17**, 302–306.

Elbashir, S. M., Lendeckel, W., and Tuschl, T. (2001). RNA interference is mediated by 21- and 22-nucleotide RNAs. *Genes Dev.* **15**, 188–200.

Elsea, S. H., and Lucas, R. E. (2002). The mousetrap: What we can learn when the mouse model does not mimic the human disease. *Ilar. J.* **43**, 66–79.

Faruqi, A. F., Egholm, M., and Glazer, P. M. (1998). Peptide nucleic acid-targeted mutagenesis of a chromosomal gene in mouse cells. *Proc. Natl. Acad. Sci.* **95**, 1398–1403.

Faruqi, A. F., Datta, H. J., Carroll, D., Seidman, M. M., and Glazer, P. M. (2000). Triple-helix formation induces recombination in mammalian cells via a nucleotide excision repair-dependent pathway. *Mol. Cell. Biol.* **20**, 990–1000.

Ferguson, D. O., Sekiguchi, J. M., Chang, S., Frank, K. M., Gao, Y., DePinho, R. A., and Alt, F. W. (2000). The non-homologous end-joining pathway of DNA repair is required for genomic stability and the suppression of translocations. *Proc. Natl. Acad. Sci.* **97**, 6630–6633.

Fire, A., Xu, S., Montgomery, M. K., Kostas, S. A., Driver, S. E., and Mello, C. C. (1998). Potent and specific genetic interference by double-stranded RNA in Caenorhabditis elegans. *Nature* **391**, 806–811.

Fjose, A., Ellingsen, S., Wargelius, A., and Seo, H. C. (2001). RNA interference: Mechanisms and applications. *Biotechnol. Annu. Rev.* **7**, 31–57.

Giovannangeli, C., and Helene, C. (1997). Progress in developments of triplex-based strategies. *Antisense Nucleic Acid Drug Dev.* **7**, 413–421.

Haber, J. E. (1999). Gatekeepers of recombination. *Nature* **398**, 665–666.

Hagmann, M., Bruggmann, R., Xue, L., Georgiev, O., Schaffner, W., Rungger, D., Spaniol, P., and Gerster, T. (1998). Homologous recombination and DNA end-joining reactions in zygotes and early embryos of zebrafish (*Danio rerio*) and *Drosophila melanogaster*. *Biol. Chem.* **379**, 673–681.

Hanvey, J. C., Peffer, N. J., Bisi, J. E., Thomson, S. A., Cadilla, R., Josey, J. A., Ricca, D. J., Hassman, C. F., Bonham, M. A., Au, K. G., *et al.* (1992). Antisense and antigene properties of peptide nucleic acids. *Science* **258**, 1481–1485.

Hasuwa, H., Kaseda, K., Einarsdottir, T., and Okabe, M. (2002). Small interfering RNA and gene silencing in transgenic mice and rats. *FEBS Lett.* **532**, 227–230.

Huberman, J. (2003a). DNA Repair lectures—Double Strand Break Repair. Roswell Park Cancer Institute. 3 Nov. http://saturn.roswellpark.org/cmb/huberman/DNA_repair/dsbreak.html.

Huberman, J. (2003b). DNA Repair lectures—DNA Damage Bypass. Roswell Park Cancer Institute. 3 Nov. http://saturn.roswellpark.org/cmb/huberman/DNA_repair/bypass.html.

Hutvagner, G., and Zamore, P. D. (2002). RNAi: Nature abhors a double-strand. *Curr. Opin. Genet. Dev.* **12,** 225–232.

Johnson, R. E., Kondratick, C. M., Prakash, S., and Prakash, L. (1999). hRAD30 mutations in the variant form of xeroderma pigmentosum. *Science* **285,** 263–265.

Kaihatsu, K., Braasch, D. A., Cansizoglu, A., and Corey, D. R. (2002). Enhanced strand invasion by peptide nucleic acid-peptide conjugates. *Biochemistry* **41,** 11118–11125.

Karras, J. G., Maier, M. A., Lu, T., Watt, A., and Manoharan, M. (2001). Peptide nucleic acids are potent modulators of endogenous pre-mRNA splicing of the murine interleukin-5 receptor-alpha chain. *Biochemistry* **40,** 7853–7859.

Knauert, M. P., and Glazer, P. M. (2001). Triplex forming oligonucleotides: Sequence-specific tools for gene targeting. *Hum. Mol. Genet.* **10,** 2243–2251.

Knudsen, H., and Nielsen, P. E. (1996). Antisense properties of duplex- and triplex-forming PNAs. *Nucleic Acids Res.* **24,** 494–500.

Koppelhus, U., and Nielsen, P. E. (2001). Antisense properties of peptide nucleic acid (PNA). *In* "Antisense Drug Technology: Principles, Strategies and Applications" (S. T. Crooke, ed.), pp. 359–374. Marcel Dekker Inc., New York.

Koppelhus, U., and Nielsen, P. E. (2003). Cellular delivery of peptide nucleic acid (PNA). *Advanced Drug Delivery Reviews* **55,** 267–280.

Li, Y. X., Farrell, M. J., Liu, R., Mohanty, N., and Kirby, M. L. (2000). Double-stranded RNA injection produces null phenotypes in zebrafish. *Dev. Biol.* **217,** 394–405.

Lin, W., Xin, H., Zhang, Y., Wu, X., Yuan, F., and Wang, Z. (1999). The human REV1 gene codes for a DNA template-dependent dCMP transferase. *Nucleic Acids Res.* **27,** 4468–4475.

Ma, C., Fan, L., Ganassin, R., Bols, N., and Collodi, P. (2001). Production of zebrafish germ-line chimaeras from embryo cell cultures. *Proc. Natl. Acad. Sci.* **98,** 2461–2466.

Majumdar, A., Khorlin, A., Dyatkina, N., Lin, M., Powell, J., Liu, J., Fei, Z., Khripine, Y., Watanabe, K. A., George, J., *et al.* (1998). Targeted gene knockout mediated by triple helix forming oligonucleotides. *Nat. Genet.* **20,** 212–214.

Myers, J. W., Jones, J. T., Meyer, T., and Ferrell, J. E., Jr. (2003). Recombinant Dicer efficiently converts large dsRNAs into siRNAs suitable for gene silencing. *Nat. Biotechnol.* **21,** 324–328.

Nasevicius, A., and Ekker, S. C. (2000). Effective targeted gene 'knockdown' in zebrafish. *Nature Genetics* **26,** 216–220.

Nielsen, P. E. (1999). Applications of peptide nucleic acids. *Current Opinion in Biotechnology* **10,** 71–75.

Nielsen, P.E. (2001). Targeting double stranded DNA with peptide nucleic acid (PNA). *Current Medicinal Chemistry* **8,** 545–550.

Oates, A. C., Bruce, A. E., and Ho, R. K. (2000). Too much interference: Injection of double-stranded RNA has nonspecific effects in the zebrafish embryo. *Dev. Biol.* **224,** 20–28.

Ohashi, E., Ogi, T., Kusumoto, R., Iwai, S., Masutani, C., Hanaoka, F., and Ohmori, H. (2000). Error-prone bypass of certain DNA lesions by the human DNA polymerase kappa. *Genes Dev.* **14,** 1589–1594.

Opalinska, J. B., and Gewirtz, A. M. (2002). Nucleic-acid therapeutics: Basic principles and recent applications. *Nat. Rev. Drug Discov.* **1,** 503–514.

Rebhan, M., Chalifa-Caspi, V., Prilusky, J., and Lancet, D. (1997). GeneCards: Encyclopedia for genes, proteins and diseases. Weizmann Institute of Science, Bioinformatics Unit and Genome Center (Rehovot, Israel). http://bioinformatics.weizmann.ac.il/cards.

Rogers, F. A., Vasquez, K. M., Egholm, M., and Glazer, P. M. (2002). Site-directed recombination via bifunctional PNA-DNA conjugates. *Proc. Natl. Acad. Sci.* **99,** 16695–16700.

Rusckowski, M., Qu, T., Chang, F., and Hnatowich, D.J. (1997). Pretargeting using peptide nucleic acid. *Cancer* **80,** 2699–2705.

Sazani, P., Kang, S.-H., Maier, M. M., Wei, C., Dillman, J., Summerton, J., Manoharan, M., and Kole, R. (2001). Nuclear antisense effects of neutral, anionic and cationic oligonucleotides analogs. *Nucleic Acids Res.* **29**, 3965–3974.

Sazani, P., Gemignani, F., Kang, S.-H., Maier, M. A., Manoharan, M., Persmark, M., Bortner, D., and Kole, R. (2002). Systematically delivered antisense oligomers upregulate gene expression in mouse tissues. *Nature Biotechnology* **20**, 1228–1233.

Sledz, C. A., Holko, M., de Veer, M. J., Silverman, R. H., and Williams, B. R. (2003). Activation of the interferon system by short-interfering RNAs. *Nat. Cell Biol.* **5**, 834–839.

Stainier, D. Y., Fouquet, B., Chen, J. N., Warren, K. S., Weinstein, B. M., Meiler, S. E., Mohideen, M. A., Neuhauss, S. C., Solnica-Krezel, L., Schier, A. F., *et al.* (1996). Mutations affecting the formation and function of the cardiovascular system in the zebrafish embryo. *Development* **123**, 285–292.

Stark, G. R., Kerr, I. M., Williams, B. R., Silverman, R. H., and Schreiber, R. D. (1998). How cells respond to interferons. *Annu. Rev. Biochem.* **67**, 227–264.

Tijsterman, M., Ketting, R. F., and Plasterk, R. H. (2002). The genetics of RNA silencing. *Annu. Rev. Genet.* **36**, 489–519.

Tiscornia, G., Singer, O., Ikawa, M., and Verma, I. M. (2003). A general method for gene knockdown in mice by using lentiviral vectors expressing small interfering RNA. *Proc. Natl. Acad. Sci.* **100**, 1844–1848.

Urtishak, K. A., Choob, M., Tian, X., Sternehim, N., Talbot, W. S., Wickstrom, E., and Farber, S. A. (2003). Targeted gene knockdown in zebrafish using negatively charged peptide nucleic acid mimics. *Dev. Dyn.* **228**, 405–413.

van de Wetering, M., Oving, I., Muncan, V., Pon Fong, M. T., Brantjes, H., van Leenen, D., Holstege, F. C., Brummelkamp, T. R., Agami, R., and Clevers, H. (2003). Specific inhibition of gene expression using a stably integrated, inducible small-interfering-RNA vector. *EMBO Rep.* **4**, 609–615.

van Dyck, E., Stasiak, A., and West, S. C. (1999). Binding of double-strand breaks in DNA by human Rad52 protein. *Nature* **398**, 728–731.

Wang, H., Long, Q., Marty, S. D., Sassa, S., and Lin, S. (1998). A zebrafish model for hepatoerythropoietic porphyria. *Nat. Genet.* **20**, 239–243.

Wargelius, A., Ellingsen, S., and Fjose, A. (1999). Double-stranded RNA induces specific developmental defects in zebrafish embryos. *Biochem. Biophys. Res. Commun.* **263**, 156–161.

Xu, X., Meiler, S. E., Zhong, T. P., Mohideen, M., Crossley, D. A., Burggren, W. W., and Fishman, M. C. (2002). Cardiomyopathy in zebrafish due to mutation in an alternatively spliced exon of titin. *Nature Genetics* **30**, 205–209.

Zon, L. I. (1999). Zebrafish: A new model for human disease. *Genome Res.* **9**, 99–100.

Zhao, Z., Cao, Y., Li, M., and Meng, A. (2001). Double-stranded RNA injection produces nonspecific defects in zebrafish. *Dev. Biol.* **229**, 215–223.

INDEX

VOLUMES IN SERIES

Founding Series Editor
DAVID M. PRESCOTT

Volume 1 (1964)
Methods in Cell Physiology
Edited by David M. Prescott

Volume 2 (1966)
Methods in Cell Physiology
Edited by David M. Prescott

Volume 3 (1968)
Methods in Cell Physiology
Edited by David M. Prescott

Volume 4 (1970)
Methods in Cell Physiology
Edited by David M. Prescott

Volume 5 (1972)
Methods in Cell Physiology
Edited by David M. Prescott

Volume 6 (1973)
Methods in Cell Physiology
Edited by David M. Prescott

Volume 7 (1973)
Methods in Cell Biology
Edited by David M. Prescott

Volume 8 (1974)
Methods in Cell Biology
Edited by David M. Prescott

Volume 9 (1975)
Methods in Cell Biology
Edited by David M. Prescott

Volume 10 (1975)
Methods in Cell Biology
Edited by David M. Prescott

Volume 11 (1975)
Yeast Cells
Edited by David M. Prescott

Volume 12 (1975)
Yeast Cells
Edited by David M. Prescott

Volume 13 (1976)
Methods in Cell Biology
Edited by David M. Prescott

Volume 14 (1976)
Methods in Cell Biology
Edited by David M. Prescott

Volume 15 (1977)
Methods in Cell Biology
Edited by David M. Prescott

Volume 16 (1977)
Chromatin and Chromosomal Protein Research I
Edited by Gary Stein, Janet Stein, and Lewis J. Kleinsmith

Volume 17 (1978)
Chromatin and Chromosomal Protein Research II
Edited by Gary Stein, Janet Stein, and Lewis J. Kleinsmith

Volume 18 (1978)
Chromatin and Chromosomal Protein Research III
Edited by Gary Stein, Janet Stein, and Lewis J. Kleinsmith

Volume 19 (1978)
Chromatin and Chromosomal Protein Research IV
Edited by Gary Stein, Janet Stein, and Lewis J. Kleinsmith

Volume 20 (1978)
Methods in Cell Biology
Edited by David M. Prescott

Advisory Board Chairman
KEITH R. PORTER

Volume 21A (1980)
**Normal Human Tissue and Cell Culture, Part A: Respiratory, Cardiovascular, and
Integumentary Systems**
Edited by Curtis C. Harris, Benjamin F. Trump, and Gary D. Stoner

Volume 21B (1980)
**Normal Human Tissue and Cell Culture, Part B: Endocrine, Urogenital, and
Gastrointestinal Systems**
Edited by Curtis C. Harris, Benjamin F. Trump, and Gray D. Stoner

Volume 22 (1981)
Three-Dimensional Ultrastructure in Biology
Edited by James N. Turner

Volume 23 (1981)
Basic Mechanisms of Cellular Secretion
Edited by Arthur R. Hand and Constance Oliver

Volume 24 (1982)
The Cytoskeleton, Part A: Cytoskeletal Proteins, Isolation and Characterization
Edited by Leslie Wilson

Volume 25 (1982)
The Cytoskeleton, Part B: Biological Systems and *in Vitro* Models
Edited by Leslie Wilson

Volume 26 (1982)
Prenatal Diagnosis: Cell Biological Approaches
Edited by Samuel A. Latt and Gretchen J. Darlington

Series Editor
LESLIE WILSON

Volume 27 (1986)
Echinoderm Gametes and Embryos
Edited by Thomas E. Schroeder

Volume 28 (1987)
***Dictyostelium discoideum:* Molecular Approaches to Cell Biology**
Edited by James A. Spudich

Volume 29 (1989)
Fluorescence Microscopy of Living Cells in Culture, Part A: Fluorescent Analogs, Labeling Cells, and Basic Microscopy
Edited by Yu-Li Wang and D. Lansing Taylor

Volume 30 (1989)
Fluorescence Microscopy of Living Cells in Culture, Part B: Quantitative Fluorescence Microscopy—Imaging and Spectroscopy
Edited by D. Lansing Taylor and Yu-Li Wang

Volume 31 (1989)
Vesicular Transport, Part A
Edited by Alan M. Tartakoff

Volume 32 (1989)
Vesicular Transport, Part B
Edited by Alan M. Tartakoff

Volume 33 (1990)
Flow Cytometry
Edited by Zbigniew Darzynkiewicz and Harry A. Crissman

Volume 34 (1991)
Vectorial Transport of Proteins into and across Membranes
Edited by Alan M. Tartakoff

Selected from Volumes 31, 32, and 34 (1991)
Laboratory Methods for Vesicular and Vectorial Transport
Edited by Alan M. Tartakoff

Volume 35 (1991)
Functional Organization of the Nucleus: A Laboratory Guide
Edited by Barbara A. Hamkalo and Sarah C. R. Elgin

Volume 36 (1991)
***Xenopus laevis:* Practical Uses in Cell and Molecular Biology**
Edited by Brian K. Kay and H. Benjamin Peng

Series Editors
LESLIE WILSON AND PAUL MATSUDAIRA

Volume 37 (1993)
Antibodies in Cell Biology
Edited by David J. Asai

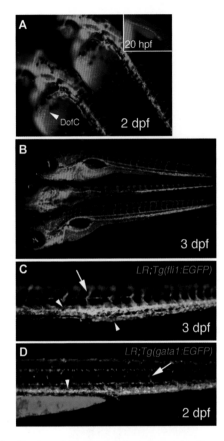

Chapter 1, Fig. 1 Visualization of hematopoietic and vascular tissues using DsRed and EGFP transgenic embryos. (A) In *Tg(lmo2:DsRed)* embryos (abbreviated *LR*), DsRed protein is initially detected at 20 hpf (inset); 2 dpf *LR* embryos labeling hematopoietic and endothelial cells in the ducts of Cuvier (DofC). (B) Labeling of the vascular endothelial network of a 3 dpf *LR* embryo. (C) *LR; Tg(fli1:EGFP)* embryos distinctly label hematopoietic (arrowheads) and endothelial cells (arrow) in 3 dpf embryos. (D) In *LR; Tg(gata1:EGFP)* transgenic embroys, green/red erythrocytes (arrowheads) circulate through vessels (arrow) labeled by DsRed in 2 dpf embryos. (Zhu and Zon, unpublished data.)

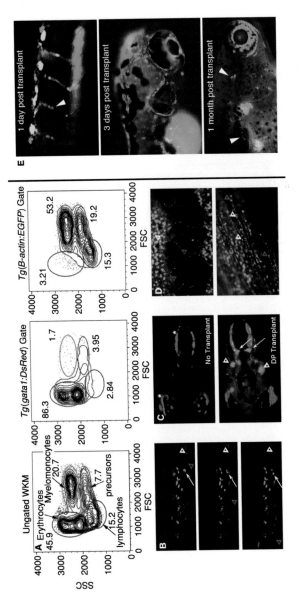

Chapter 1, Fig. 2 Use of multiple fluorescent reporters in transplantation assays. Left panel: Transplantation of whole kidney marrow from double transgenic donors allows independent visualization of leukocytes and erythrocytes in recipient embryos. (A) Scatter profile of ungated WKM in a representative Tg(gata1:DsRed); Tg(β-actin:EGFP) double transgenic adult (left). DsRed+ cells were contained only within the erythrocyte gate (middle), whereas GFP+ cells were non-erythroid (right). (B–C) Transplantation of 48 hpf recipients showed transient reconstitution of donor-derived erythrocytes and leukocytes. (B) Visualization of the tail vessels in a gata1⁻/⁻ transplant recipient showed a slow-moving, round leukocyte (arrowhead), a larger leukocyte displaying an end-over-end tumbling migration (arrow), and a rapidly circulating erythrocyte (red arrowhead) at 1 day post-transplantation. Each frame is separated by 300ms (20× magnification, anterior to the left). (C) Dorsal views comparing untransplanted (upper) and transplanted *bloodless* (*bls*) recipients (lower). *bls* recipients showed rapid and robust engraftment of the pronephros (arrows) and bilateral thymi (arrowheads) by GFP+ leukocytes by day 5 post-transplantation. Asterisks denote autofluorescence of the eyes and swim bladder in the DsRed channel. (D) *bls* recipients display sustained, multilineage hematopoiesis from donor-derived cells. Upper panel shows robust reconstitution of DsRed+ erythrocytes (red arrowheads) and GFP+ leukocytes (white arrowhead) as observed in the dermal capillaries of a *bloodless* recipient at 8 weeks post-transplantation. Lower panel shows similar multilineage reconstitution as observed in the tail capillaries of another *bls* recipient at 8 weeks (20× magnification). From Traver *et al.*, 2003.

Right panel: Transplantation primitive wave hematopoietic progenitors from Tg(lmo2:EGFP) embryos into *vlad tepes* (*vlt*) recipients. (E) The EGFP expressing population from 10–12 somite Tg(lmo2:EGFP); Tg(gata1:DsRed) transgenic embryos were isolated by FACS and transplanted into 48 hpf *vlt* embryos. One day post transplantation, circulating donor-derived cells could be identified by DsRed and EGFP fluorescence (arrowhead points to GFP+ circulating cell). By 3 days post transplantation, most circulating cells were DsRed+, suggesting that the EGFP+ donor progenitors had differentiated into Gata1⁺ erythrocytes. One month after transplantation, each of the surviving recipients carried approximately 10–200 DsRed+ circulating cells (arrowheads). (Zhu and Zon, unpublished data.)

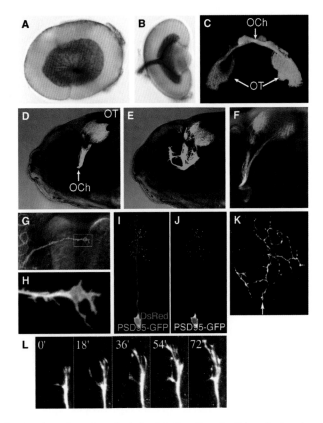

Chapter 2, Fig. 1 A variety of methods for labeling the zebrafish retinotectal system. (A and B) 48 hpf zebrafish eye labeled with the zn-8 antibody, which recognizes the cell surface molecule Alcam/ Neurolin/DM-GRASP, expressed in all RGCs. Courtesy of Arminda Suli. (A) Lateral view, rostral to the right. (B) Dorsal view, rostral up. (C) Dorsal view of 5 dpf retinotectal projection labeled by intraocular injection (Method 1) with DiO (*green*, left eye) and DiI (*red*, right eye). The eyes have been dissected away. Rostral is up. (D and E) Lateral views of optic tracts labeled with DiO in 5 dpf wild type (D) and *astray* (E) larvae. The eye contralateral to the injected eye was removed in order to facilitate imaging. (F) Lateral view of optic tract and tectum of 6 dpf wild-type larva, after topographic labeling of the eye. DiI (*red*) was injected into dorsonasal retina, and DiO (*green*) was injected into ventrotemporal retina. Courtesy of Jeong-Soo Lee. (G and H) Leading growth cone labeled by intraocular injection of DiI (whole eye fill, Method 1). The *boxed area* in G is shown at higher magnification in H. (G) 20× dry objective. (H) A 60× water-immersion objective. (I and J) A single wild-type tectal neuron at 100 hpf, labeled by coinjection of pBSKαtubulin:GAL4 and UAS:PSD95-GFP; UAS:DsRedExpress-1 (Niell *et al.*, 2004; Method 3). The DsRed labels the cytoplasm of the soma and apical dendrite, while PSD95-GFP is preferentially localized to postsynaptic sites. 60×/1.2 water-immersion objective. (I) Merged image. (J) PSD95-GFP image alone. (K) A single retinal axon arbor on the tectum, labeled by injecting plasmid DNA containing the Brn3c promoter driving expression of membrane targeted GFP using GAL4/UAS amplification (Method 3). (L) Time-lapse imaging of a cluster of zebrafish retinal growth cones, labeled with DiI, as they grow through the developing brain over a 72-m period (Method 2). 60× water-immersion objective. (A and B) Obtained on an Olympus BX50WI compound microscope. (C through L) Captured on Olympus FV200 or FV300 confocal microscopes. OCh, optic chiasm; OT, optic tectum.

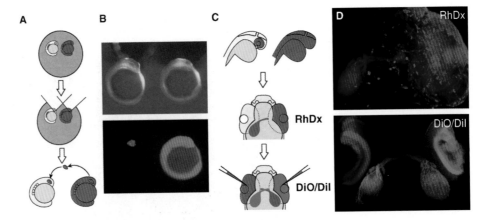

Chapter 2, Fig. 4 Eye transplant procedure, Method 4. (A) At 5 somite stage, host embryo (unlabeled) and donor embryo (labeled with rhodamine dextran) are embedded in a small drop of low-melt agarose. Scalpel cuts are used to remove wedges of agarose, giving access to eye primordial, then the host eye is replaced with the donor eye. (B) Brightfield (*top*) and rhodamine fluorescence (*bottom*) micrographs of embedded embryos after removing agarose wedges. (C) Host and donor embryos are raised until 4 to 5 dpf. Rhodamine fluorescence is used to check that the entire eye has been transplanted without any accompanying brain tissue. Lipophilic dye labeling is then used to assay the retinal projections. (D) Confocal images of transplants. *Top*, rhodamine dextran labeling shows the transplanted eye and retinal axons that have grown out into the host brain. Notice a few cells that have migrated out of the transplanted eye. *Bottom*, lipophilic dye labeling shows the host (*red*) and donor (*green*) projections in a wild type-to-wild type transplant. Note that a few donor axons project ipsilaterally (*yellow* labeling in right tectum).

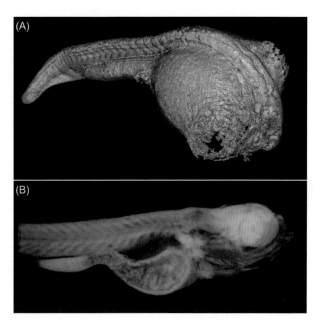

Chapter 3, Fig. 5 Volume and surface renderings. (A) Surface rendering of 28-somite stage embryo labeled with an antiacetylated antibody. The 3D reconstruction can be surface rendered to give a better visualization of the 3D object. The rendered object can be rotated to allow any angle to be viewed. Thresholding can be used to identify the antibody signal (*green*) over that of the embryo. Acetylated tubulin is one of the first components of the neuronal microtubule cytoskeleton to form and is a marker of developing post-mitotic neurons (Piperno and Fuller, 1985; Wilson and Easter, Jr., 1991). (B) Volume rendering of a reconstruction of a 2-week-old juvenile zebrafish. Like surface rendering, volume rendering can be used to give a clear 3D representation of the sample. By reducing the opacity of the rendered object, all of the internal features can be visualized.

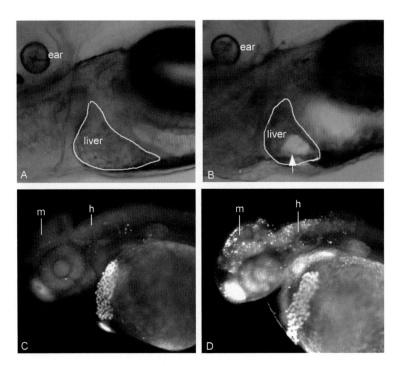

Chapter 5, Fig. 1 Drug-induced apoptosis visualized by acridine orange staining. Acridine orange staining was performed, as described previously (Furutani-Seiki *et al.*, 1996). Untreated zebrafish (A), 4-day-old zebrafish were treated with 1 μg/ml neomycin for 24 h (B). Liver apoptosis can be observed after neomycin treatment. Embryos, (20 hpf) were treated with vehicle alone (C) or with 1 μM retinoic acid (D). Aberrant apoptosis was observed in the midbrain (m) and hindbrain (h) regions.

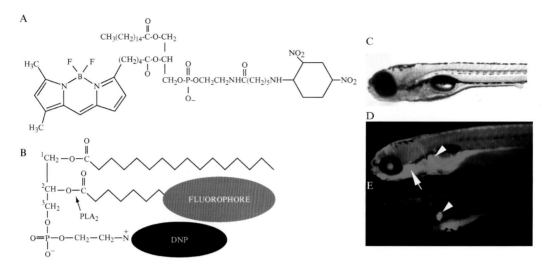

Chapter 6, Fig. 1 PED6 as a biosensor to visualize lipid metabolism in live zebrafish larva. (A) The structure of PED6. The emission of the BODIPY-labeled acyl chain at the *sn*-2 position is quenched by the dinitrophenol group at the *sn*-3 position when this molecule is intact. (B) Upon PLA$_2$ cleavage at the *sn*-2 position, the BODIPY-labeled acyl chain is liberated and can emit a green fluorescence (515 nm). (C) Bright field view of a 5-dpf embryo. (D) Embryo soaked in unquenched phospholipids (2 ug/ml, D3803). Arrowhead marks gall bladder and the arrow marks the pharynx. (E) Embryo (5 dpf) soaked in PED6 (3 ug/ml, 6 h). Arrowhead indicates the gall bladder.

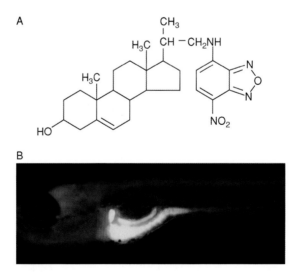

Chapter 6, Fig. 2 NBD-Cholesterol labeled live zebrafish larva. (A) Structure of NBD-Cholesterol. (B) Larval zebrafish (5 dpf) was immersed into embryo media containing NBD-Cholesterol (3 μg/ml, solubilized with fish bile) for 2 h.

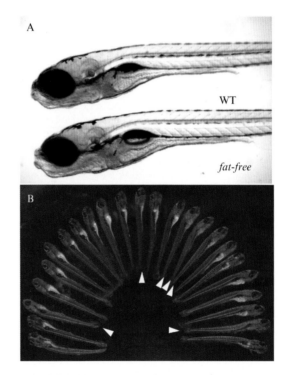

Chapter 6, Fig. 3   PED6 labeled larvae. (A) *Fat-free* is morphologically normal as wild type (WT). (B) *Fat-free* has diminished fluorescence in intestinal lumen and gall bladder. Arrowheads mark mutant embryos.

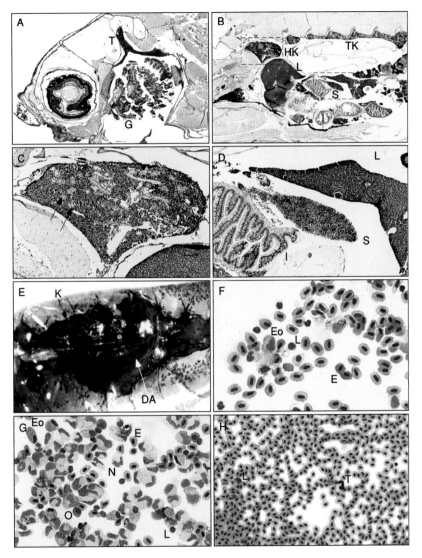

Chapter 8, Fig. 2 Histological analyses of adult hematopoietic sites. (A) Sagittal section showing location of the thymus (T), which is dorsal to the gills (G). (B) Midline sagittal section showing location of the kidney, which is divided into the head kidney (HK) and trunk kidney (TK), and spleen (S). The head kidney shows a higher ratio of blood cells to renal tubules (black arrows), as shown in a close up view of the HK in (C). Close up view of the spleen, which is positioned between the liver (L) and the intestine (I). (E) Light microscopic view of the kidney (K), over which passes the dorsal aorta (DA, white arrow). (F) Cytospin preparation of splenic cells, showing erythrocytes (E), lymphocytes (L), and an eo/basophil (Eo). (G) Cytospin preparation of kidney cells showing cell types as already noted plus neutrophils (N) and erythroid precursors (O, orthochromic erythroblast). (H) Peripheral blood smear showing occasional lymphocytes and thrombocyte (T) clusters amongst mature erythrocytes. (A to D) Hematoxylin and Eosin stains, (F to H) May-Grünwald/Giemsa stains.

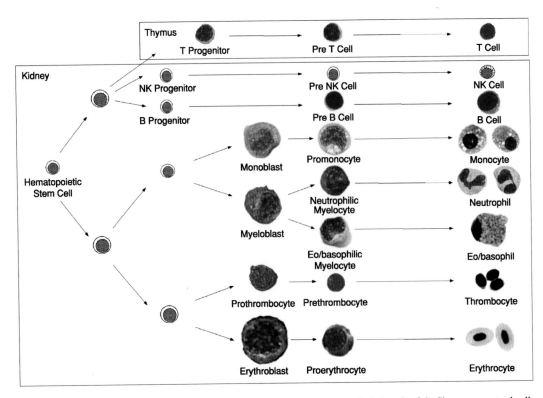

Chapter 8, Fig. 3 Proposed model of definitive hematopoiesis in zebrafish. Shown are actual cells types from adult kidney marrow. All cells were photographed with a 100× oil objective from cytospin preparations. Proposed lineage relationships are based on those demonstrated using clonogenic murine progenitor cells.

Chapter 11, Fig. 3 Analysis of 3D time-lapse recordings. (A) A fast and efficient approach to analyze z-stacks of images that have been recorded through time is to project the individual images of every z-stack/time point into a single plane by either mean value or brightest point projection (shown here). (B) If the 3D nature of the data is to be retained, the recorded signal can be coded with pseudo colors, reflecting the individual z-values within each z-stack. (C) Once every z-stack of each recorded time point is projected into a single plain, these projections can be animated into a movie to allow individual cells to be followed. (D) In the case of color-coded z-levels, movements of cells along the z-axis can be observed by cells changing colors while migrating.

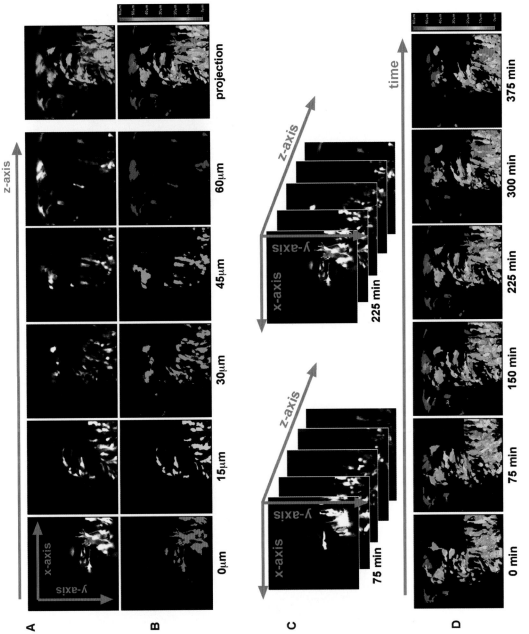

Chapter 11, Fig. 3   (Caption on opposite page)

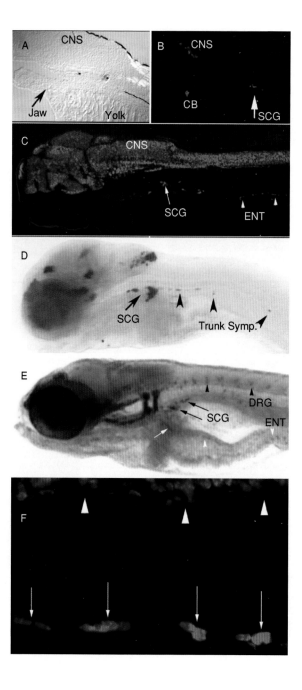

Chapter 12, Fig. 3 Development of the peripheral sympathetic nervous system in zebrafish embryos. (A, B, and C) Parasagittal section of 3.5-dpf embryo. (A) High magnification DIC and (B) fluorescence of the same field, showing TH-IR (*red*) in the SCG (*arrow*), carotid body (CB), and a group of anterior cells in the midbrain (CNS). (C) Low magnification view of 3.5-dpf embryo labeled with anti-Hu to reveal all neurons. A subset of cervical sympathetic neurons are indicated by the arrow and enteric neurons (ENT) by arrowheads. (D) Lateral view of whole-mount *TH* RNA *in situ* preparation at 5 dpf. *TH* RNA is strongly expressed in the SCG (*arrow*) at this stage and is beginning to be expressed in the trunk sympathetic chain (*arrowheads*). A description of *TH* RNA expression in the head is described in Guo *et al.*, 1999. (E) Whole-mount antibody preparation of a 7-dpf larvae labeled with anti-Hu to reveal neurons. Black arrows indicate SCG, black arrowheads indicate dorsal root ganglion (DRG) sensory neurons, and white arrow and white arrowheads indicate enteric neurons (ENT). (F) Parasagittal section in the mid-trunk region of a 17-dpf embryo labeled with anti-Hu. Ventral spinal cord neurons are evident at the top (*arrowheads*). Four segmental sympathetic ganglia (*arrows*) are located ventral to the notochord adjacent to the dorsal aorta.

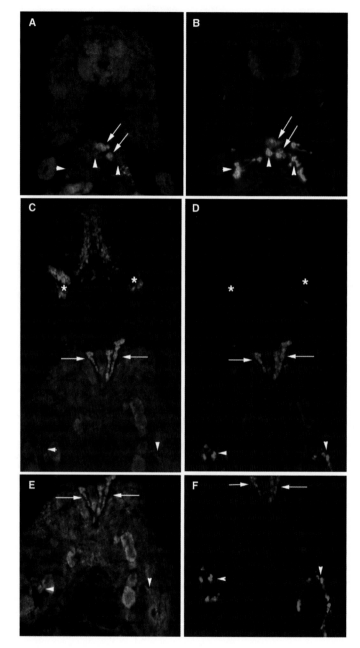

Chapter 12, Fig. 4 Sympathoadrenal derivatives in embryonic and juvenile zebrafish. (A and B) Transverse section of a 3.5-dpf embryo double-labeled with anti-Hu (green) and DβH (red). Arrows indicate Hu$^+$/DβH$^+$ sympathetic neurons of the cervical ganglion. Arrowheads indicate Hu$^-$/DβH$^+$ presumptive chromaffin cells. (C and D) Transverse section through the mid-trunk region at 28 dpf, double-labeled with anti-Hu (green) and anti-TH (red). (E and F) Higher magnification of C and D, including a slightly more ventral region. Arrows indicate sympathetic neurons, arrowheads indicate chromaffin cells, and asterisks denote dorsal root ganglia.

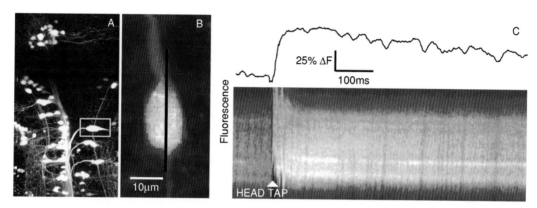

Chapter 13, Fig. 1 Line scan of a Mauthner cell. (A) Mauthner cell (M) in rhombomere 4 is scanned in an awake, restrained larva using a BioRad MRC600 confocal microscope. (B) The image of the cell is rotated 90° and magnified. The laser beam is scanned repeatedly across the cell at a fixed location indicated by the vertical bar. (C) Each successive scan line (acquired at 2 msec intervals) is plotted from left to right. Fluorescence intensity is represented by a color scale with orange to yellow indicating the highest fluorescence and purple to blue indicating the lowest. Relative fluorescence intensity is plotted above the line scan. Line scans reveal the latency of a cell's response to a given stimulus (Gahtan *et al.*, 2002). While the increase in calcium is abrupt, the recovery to resting calcium levels takes several seconds because calcium extrusion mechanisms are slower than initial influx. The calcium indicator itself also slows the response dynamics. Because the Mauthner cell is believed to fire only a single action potential per stimulus (Faber *et al.*, 1989), the size of the calcium response is unexpectedly large. This has been a frequent observation with Mauthner cells and may reflect a calcium potentiation mechanism (O'Malley and Fetcho, 1996). Differences in resting fluorescence values are attributed to regional variation in the intracellular behavior of the indicator, rather than persistent resting calcium gradient (O'Malley, 1994; O'Malley *et al.*, 1999). Relative changes in fluorescence intensity are indicative of the magnitude of the calcium response and, to some extent, the firing rate of action potentials (O'Malley *et al.*, 2003a).

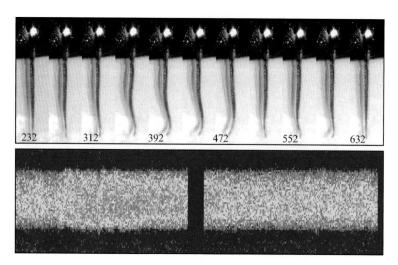

Chapter 13, Fig. 2   Visually evoked swimming (VES). While neuronal activity is being recorded in the brainstem, using an inverted confocal microscope, the behavior of partially restrained larvae can be simultaneously recorded. The larva's head (not visible here) is embedded in agar (*dark region at top of image*). The larva's trunk/tail is free to move only in the lighter region at the bottom. The behavioral sequence was recorded using a Redlake Imaging MotionScope high-speed camera mounted on a dissecting scope positioned above the stage of the inverted microscope. The laser beam, which is scanning neurons within the brainstem, is visible as a white dot at the top of the image. The relative time values within the swim bout are indicated (in milliseconds) for alternating frames in the behavioral sequence. Below the larvae are two consecutive "slow" line scans of an nMLF neuron in the midbrain of the same larvae. The line scans were acquired at 6 msec/scan line. The gap in the scan is attributed to the time-lapse between the end of the first scan and the onset of the second scan (the two line scans fill the frame buffer). By the end of the second scan, about 10 sec after the initial stimulus, calcium levels have returned close to their resting value. The gentle VES-like tail movements in restrained larvae are similar to such movements exhibited in free-swimming larvae and are dramatically different from the vigorous tail movements associated with escape and struggling behaviors as shown in Chapter 13, Fig. 3.

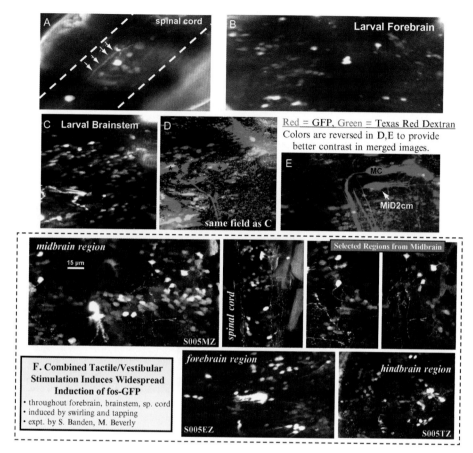

Chapter 13, Fig. 4 Fos-GFP mapping of neural activity. 50 head-taps elicited repetitive escape behaviors and induced fos-GFP expression in the forebrain, brainstem, and spinal cord of some larvae containing fos-GFP DNA. (A) In spinal cord (*between dashed lines*), fos-GFP was induced in several large neurons and is present in neuronal processes as well (*arrows*). Fos-GFP is also induced in forebrain (B) and in brainstem (C). (D) Some larvae were double-labeled by injecting Texas-red dextran into the spinal cord, which labels reticulospinal neurons in brainstem (*green*) and fos-GFP neurons (*red*; same fos-GFP field as shown in C). (E) Another larva shows fewer brainstem neurons with fos-GFP induction (*red*). Two large neurons (Mauthner cell, MiD2cm) were retrogradely labeled but were negative for fos-GFP. Induction is expected to be mosaic in these injected animals, because the fos-GFP DNA is not expected to be present in all neurons. The Mauthner cell normally shows c-fos protein expression in response to this stimulus (Bosch *et al.*, 2001). (F) More widespread induction resulted from combined vestibular/tactile stimulation. Numerous cell bodies, fine processes, and varicosities are labeled. Images acquired pre-stimulation never showed basal expression of fos-GFP. In these larvae, fos-GFP DNA (about 25 ng) was injected into one- or two-cell stage zebrafish embryos and the larvae were imaged at 5 to 7 dpf. Batches of injected larvae were group stimulated by repetitive tapping (and for some animals by periodic swirling) of the 24-well trays in which they were maintained. Typically, 5 to 10% exhibited varying degrees of fos-GFP induction.

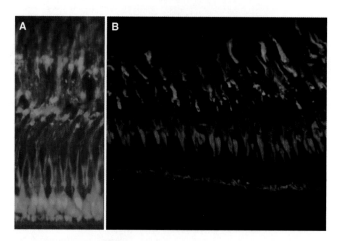

Chapter 15, Fig. 2 Analysis of GFP distribution in two photoreceptor-specific transgenic lines. Transverse sections through adult retinas. (A) Fluorescence microscopy image of GFP distribution in the line described by Fadool (2003) demonstrate the cytoplasmic localization and distribution throughout the cell. (B) Confocal analysis of GFP distribution in the line described by Perkins *et al.* (2002). When GFP is fused to the C-terminal 44 amino acids of rhodopsin, GFP fluorescence is restricted to the rod outer segments (*green*). The red/green double cones are stained with Fret-43 (*red*). Figure 2A was reprinted (with permission from Elsevier) from Fadool, J. M. (2003). Development of a rod photoreceptor mosaic revealed in transgenic zebrafish. *Dev. Biol.* **258,** 277–290.

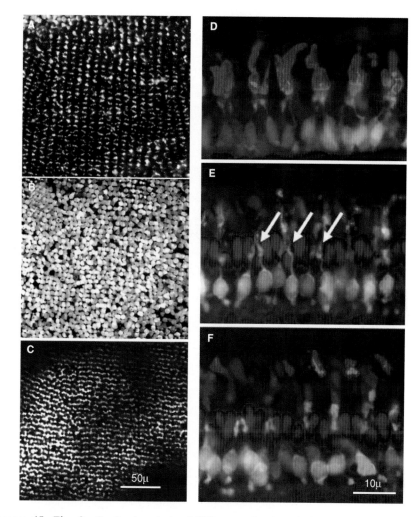

Chapter 15, Fig. 3 Confocal analysis of GFP expression in rod photoreceptors reveals a rod mosaic that is established at the retinal margin in adults. (A–C) Whole-mounted retinas in saline were imaged by confocal microscopy from the vitreal side. Images reveal a rod mosaic at the level of the inner segments (A), the cell bodies (B), and the synaptic terminals (C). Note the regularly spaced rows of the rod structures. (E–F) Merged confocal images of the marginal zone in serial, tangential sections through adult retinas that were immuno labeled (*red*) for rod opsin (D), UV opsin (E), and red opsin (F) with GFP expression (*green*) and 4′,6-diamidino-2-phenyllindole (DAPI) counterstained for nuclei (*blue*). Note the position of clustered rod outer segments at the positions of immature UV cone outer segments (*arrow*). Reprinted (with permission from Elsevier) Fadool, J. M. (2003). Development of a rod photoreceptor mosaic revealed in transgenic zebrafish. *Dev. Biol.* **258**, 277–290.

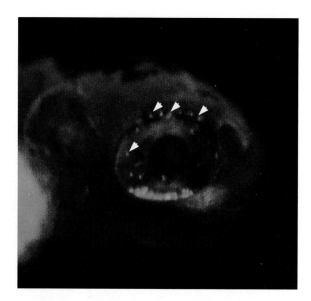

Chapter 15, Fig. 5 Individual rod photoreceptors can be identified in living animals. A 4-dpf embryo was imaged with a fluorescent microscope and viewed from the lateral side. Bright autofluorescence is seen in the yolk. Photoreceptors in the dorsal retina are easily seen at this timepoint (*arrowheads*).

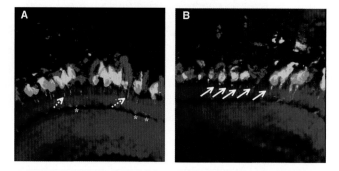

Chapter 15, Fig. 6 Confocal analysis of GFP localization in *mariner* mutants that express the GFP-CT44 fusion protein. Tangential sections through 8-dpf animals were immunostained with the rod-specific monoclonal antibody 1D1 (*red*) with GFP expression (*green*) and counterstained with DAPI (*blue*). Colocalization of the 1D1 label and GFP is seen in *yellow*. Note the occasional presence of GFP and 1D1 signal in the inner segments (*dashed arrows*) and the terminals (*asterisks*) of wild-type animals (A). In *mariner* mutants (B), round clusters of colocalized signals were regularly seen at the base of the outer segments, presumably in the connecting cilium (*arrows*). This may represent a mild defect in transport through the connecting cilium.

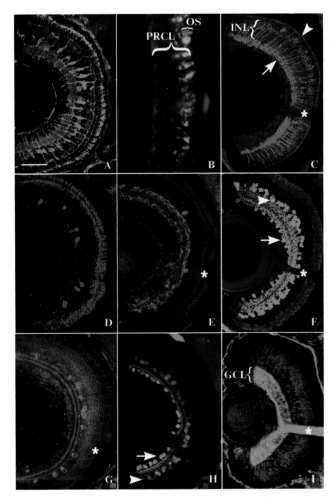

Chapter 16, Fig. 3 Transverse sections through the center of the zebrafish eye reveal several major retinal cell classes and their subpopulations. (A) Anti-rod opsin antibody detects rod photoreceptor outer segments (*red*), which are fairly uniformly distributed throughout the outer perimeter of the retina by 5 dpf. On the same section, an antibody to carbonic anhydrase labels cell bodies of Mueller glia (*green*) in the INL, as well as their radially oriented processes. (B) A higher magnification of the photoreceptor cell layer shows the distribution of rod opsin (*red signal*) and UV opsin (*green signal*) in the outer segments (OS) of rods and short single cones, respectively. (C) A subpopulation of bipolar cells is detected using antibody directed to Protein Kinase Cβ (PKC). While cell bodies of PKC-positive bipolar neurons are situated in the central region of the INL, their processes travel radially into the inner (*arrow*) and outer (*arrowhead*) plexiform layers, where they make synaptic connections. (D) Tyrosine hydroxylase-positive interplexiform cells are relatively sparse in the larval retina. (E) Similarly, the distribution of Neuropeptide Y is limited to only a few cells per section. (F) The distribution of GABA, a major inhibitory neurotransmitter. GABA is largely found in amacrine neurons in the INL (*arrowhead*), although some GABA-positive cells are also present in the GCL (*arrow*). (G) Choline acetyltransferase, an enzyme of acetylcholine biosynthetic pathway, is restricted to a relatively small amacrine cell subpopulation. (H) Antibodies directed to a calcium-binding protein,

(Continued)

Chapter 16, Fig. 3 parvalbumin, recognize another fairly large subpopulation of amacrine cells in the INL (*green, arrowhead*). Some parvalbumin-positive cells localize also to the GCL and most likely represent displaced amacrine neurons (*arrow*). By contrast, serotonin-positive neurons (*red*) are exclusively found in the INL. (I) Ganglion cells stain with the Zn-8 antibody directed to neurolin, a cell surface antigen (Fashena and Westerfield, 1999). In addition to neuronal somata, strong Zn8 staining exists in the optic nerve (*asterisk*). In all panels lens is left, dorsal is up. A through H show the retina at 5 dpf. I shows a retina at 3 dpf. Asterisks indicate the optic nerve. Scale bar equals 50 fm in (A), and (C to I) and 10 fm in (B). GCL: ganglion cell layer; INL: inner nuclear layer; OS: outer segments; PRCL: photoreceptor cell layer.

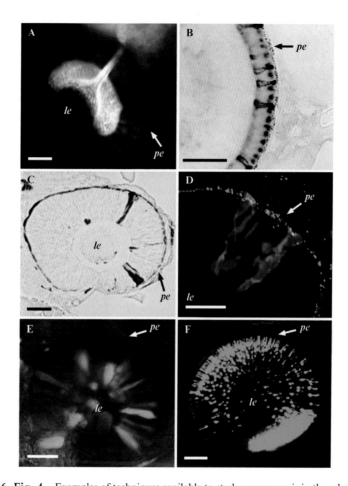

Chapter 16, Fig. 4   Examples of techniques available to study neurogenesis in the zebrafish retina. (A) DiI incorporation into the optic tectum retrogradely labels the optic nerve and ganglion cell somata. (B) A transverse plastic section through the zebrafish retina at 3 dpf. *In situ* mRNA hybridization using two probes, each targeted to a different opsin transcript and detected using a different enzymatic reaction, visualizes two types of photoreceptor cells. (C) A plastic section through a genetically mosaic retina at approximately 30 hpf. Biotinylated dextran labeled donor-derived cells incorporate into retinal neuroepithelial sheet of a host embryo and can be detected using HRP staining (*brown precipitate*). (D) A transverse cryosection through a genetically mosaic zebrafish eye at 36 hpf. In this case, donor-derived clones of neuroepithelial cells are detected with fluorophore-conjugated avidin (*red*). The apical surface of the neuroepithelial sheet is visualized with anti-γ-tubulin antibody, which stains centrosomes (*green*). (E) GPF expression in the eye of a zebrafish embryo after injection of a DNA construct containing the GFP gene under the control of a heat-shock promoter. The transgene is expressed in only a small subpopulation of cells. (F) A confocal z-series through the eye of a living transgenic zebrafish, carrying a GFP transgene under the control of the rod opsin promoter (Fadool, 2003). Bright signal is present in rod photoreceptor cells (approximately 3 dpf). Scale bar, 50 fm. pe: pigmented epithelium; le: lens. Panel E reprinted from Malicki *et al.* (2002) with permission from Elsevier.

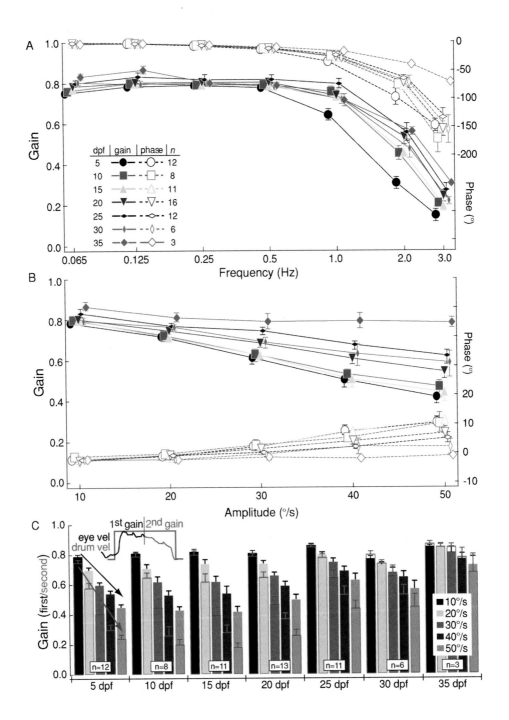

Chapter 17, Fig. 9 Development of zebrafish OKR. (A) Bode analysis of zebrafish (5 to 35 dpf) from 0.065 to 3.0 Hz at ±10°/s. (B) Phase and gain plot of zebrafish OKR in response to sinusoids (0.125 Hz) of increasing velocity. Traces same as legend in A. (C) Eye velocity gains in response to bidirectional velocity steps (0.1 Hz). Gains were calculated at each drum velocity for (1) the first 2.5 s of the 5-s step (*solid bar*) and (2) the remaining 2.5 s (*open, red bars*), also see inset. Arrows follow the trend in decreasing performance with increasing amplitude between first (*black*) and second (*red*) step halves. All data were averaged through several cycles and offset slightly (A and B) for readability.

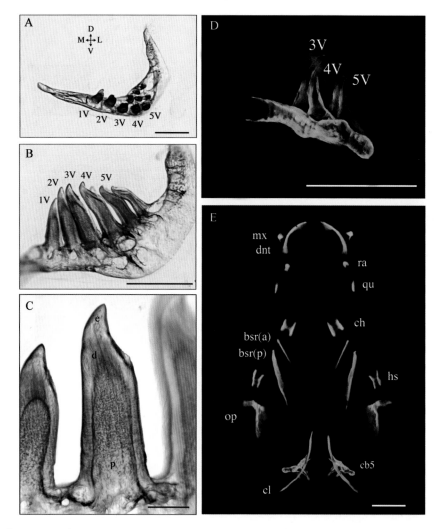

Chapter 19, Fig. 1 Adult pharyngeal teeth and larval mineralized skeleton. (A to C) Dissected and Alizarin red–stained cb5 arch and teeth of the adult zebrafish. The large ventral row of pharyngeal teeth is labeled 1V to 5V. Adult zebrafish also possess discrete medial and dorsal rows of teeth (*not labeled*). Dorsal, ventral, medial, and lateral are designated D, V, M, and L, respectively. (C) High magnification of 2V shows the pulp cavity (p), dentin tubules (d), and enameloid (e). (D) Quercetin-stained cb5 in wild-type 7-dpf larvae with three pharyngeal teeth, 3V to 5V. (E) Quercetin-stained mineralized pharyngeal skeleton in 7-dpf larval zebrafish viewed under ultraviolet (UV) illumination. Dermal bones are labeled to the left. Cartilage replacement bones are to the right. Scale bars in A and B represent 0.5 mm. Bars in C to E represent 100 µm. Abbreviations are as follows: mx, maxilla; dnt, dentary; ra, retroarticular; qu, quadrate; ch, ceratohyal; bsr(a), anterior branchiostegal ray; posterior branchiostegal ray; hs, hyosymplectic; op, opercle; cl, cleithrum; cb5, ceratobranchial 5.

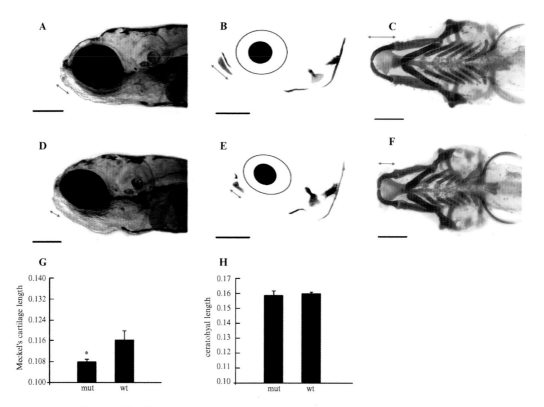

Chapter 19, Fig. 2 Characterization of subtle pharyngeal skeletal defects. Eight-dpf wild-type (wt) (A to C) and mutant (D to F) larvae. Mutants are distinguished from wt siblings by a misshaped eye and shortened anterior head structures (A versus D). In the lateral view, first-arch dermal bones appear shorter than those of wt siblings (B versus E). The chondrocranium of mutant zebrafish also exhibits similar defects (C versus F). Meckel's cartilage is significantly shorter in mutant animals (G), while no significant difference in the length of the ceratohyal is detected (H).

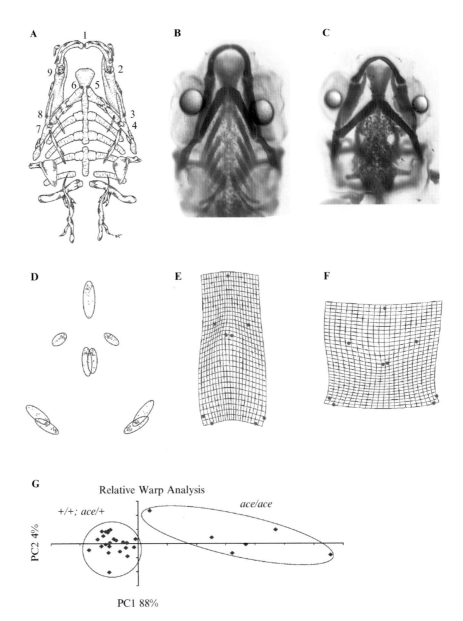

Chapter 19, Fig. 3   Shape differences in wild-type (wt) and mutant skeletal elements. Illustration of the 7-dpf wt zebrafish pharyngeal skeleton in the ventral view (A). Seven-dpf wt (B) and *ace* (C) larvae. Note the shortened mutant head. (D to G) Geometric morphometrics analysis of skeletal defects. (D) Variation in landmark configuration after superimposition. (E and F) Deformation of shape along the first principal component axis (PC1). Wt shape is depicted in (E); mutant shape in (F). Mutants exhibit greater variance along PC1 (G).

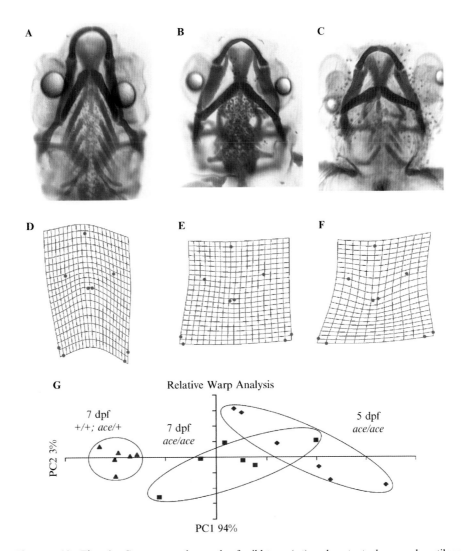

Chapter 19, Fig. 4 Geometry and growth of wild-type (wt) and mutant pharyngeal cartilage defects. Representative specimens are shown in (A to C). (A) Wt 7-dpf larvae, (B) mutant 7-dpf larvae, and (C) mutant 5-dpf larvae. Shape deformations are illustrated in (D to E), when pharyngeal skeletal defects in both 7-dpf (E) and 5-dpf (F) *ace* are evaluated against 7-dpf wt larvae (D). The similarity in geometry suggests the *ace* mutation effects growth of the pharyngeal skeleton. All specimens plot on the same principal component axis (PC1), which explains 94% of the total shape variation (G).

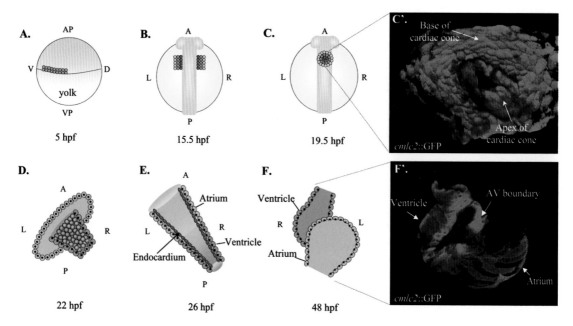

Chapter 20, Fig. 1 Zebrafish heart development. (A) In the early blastula, the cardiac progenitors correspond to the ventrolateral margin of the blastoderm. (B) After gastrulation, the cardiac progenitors arrive on either side of the midline. The three rows of circles represent the relative position of the endocardial precursors (*red*), ventricular precursors (*dark green*), and atrial precursors (*light green*) at this stage. By the 13-somite stage (15.5 hpf), the myocardial precursors are patterned mediolaterally, with ventricular precursors positioned medial to the atrial precursors. (C) At the 21-somite stage (19.5 hpf), the myocardial precursors have migrated to the midline and fused to form the cardiac cone. Within the cardiac cone the endocardial precursors (*red*) are located in the central lumen, the ventricular precursors (*dark green*) at the apex, and the atrial precursors (*light green*) at the base. (C') Dorsal view of the cardiac cone as seen by projection of confocal optical sections in a transgenic embryo that expresses GFP in the myocardial precursors (*cmlc2*::GFP). (D) Cardiac cone tilting starts with the apex of the cone bending posteriorly and to the right. Subsequently, the base of the cone, consisting of atrial precursors, will gradually coalesce into a tube. (E) By 26 hpf, elongation of the heart tube results in the leftward positioning of the linear heart tube which is composed of two layers, an inner endocardium (*red*) and an outer myocardium (*green*). (F) By 48 hpf, gradual bending of the heart tube at the atrioventricular boundary forms an S-shaped loop that positions the ventricle (*dark green*) to the right of the atrium (*light green*) as seen in a head-on view. (F') Projection of confocal optical sections of a 56-hpf embryo with GFP expressed in the myocardium. In the projection, a transverse section through the myocardium shows a thick, compact ventricular wall while the atrial wall appears thinner.

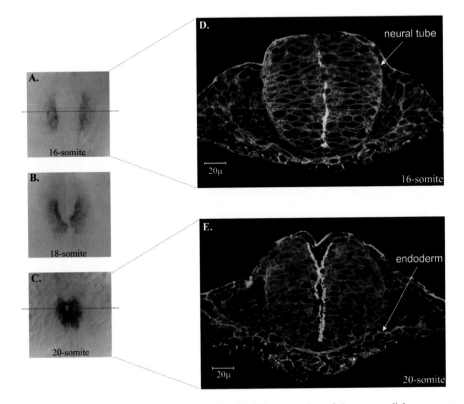

Chapter 20, Fig. 2 Midline migration and epithelial maturation of the myocardial precursors. (A to C) Dorsal views of *cmlc2* expression during myocardial migration to the midline. Dash lines indicate the level of the transverse sections in D and E. (A) At the 16-somite stage, the myocardial precursors are positioned on either side of the midline. (B) By the 18-somite stage, the medial myocardial precursors make contacts at the midline. (C) At the 20-somite stage, the myocardial precursors are fusing to form the cardiac cone. (D and E) Transverse confocal images of *cmlc2*::GFP (*false colored blue*) transgenic embryos immunostained with antibodies against β-catenin (*red*) and aPKCs (*green*); dorsal to the top. (D) At the 16-somite stage, the cuboidal myocardial precursors show cortical localization of β-catenin and an enrichment of aPKCs on the apical surface and at points of cell-cell contacts. (E) By the 20-somite stage, the transverse section through the middle of the cardiac cone shows that the myocardial precursors form two U-shaped structures, with the medial cells appearing columnar in shape, while the lateral cells are cuboidal. The aPKCs are restricted to the apicolateral domains, while β-catenin localizes to the basolateral domains of the myocardial precursors.

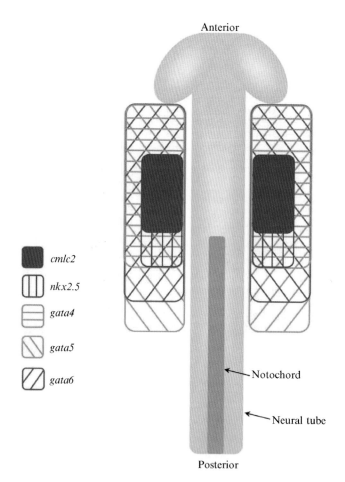

Chapter 20, Fig. 4 Overlapping gene expression patterns in the anterior LPM. In the 5- to 10-somite stage embryo, *gata4/5/6* expression domains correspond to different A-P extent of the anterior LPM. The *gata4* expression domain (*green horizontal stripes*) in the anterior LPM extends to the optic cup and ends posterior to the tip of the notochord. The *gata5* expression domain (*orange diagonal stripes*) extends furthest posteriorly, while the *gata6* expression domain (*blue diagonal stripes*) extends more posteriorly than *gata4*, but anterior to that of the *gata5* expression domain. The *nkx2.5* expression domain (*purple vertical stripes*) extends posterior to the tip of the notochord and anteriorly to halfway between the notochord and the eye. *cmlc2* expression (*solid red*) initiates in the 13-somite stage embryo and is limited to the anterior tip of the notochord.

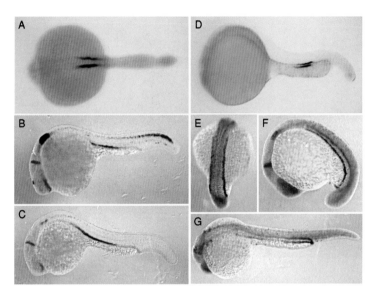

Chapter 23, Fig. 9 Transporter mRNA expression defines pronephric nephron segments. (A) The chloride-bicarbonate anion exchanger (AE2) is expressed in the anterior pronephric duct. The mid-segment of the pronephric ducts specifically express the zebrafish starmaker gene (B) and an aspartoacylase homolog (C). Transporters expressed in the duct mid-segment include the Na-K-Cl symporter *slc12a1* (D). Expression of a putative ABC transporter (ibd2207) is observed initially throughout most of the forming pronephric duct at the 15 somite stage (E and F) but becomes restricted primarily to the posterior duct by 24 hpf (G). Embryos in B, C, E, F, and G are counterstained with *pax2.1* probe in red for reference. (A and D courtesy of Alan Davidson *et al.*; B, C, E, F, G courtesy of Neil Hukreide.)

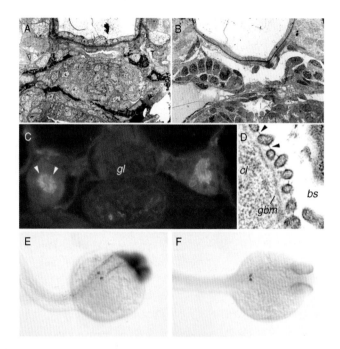

Chapter 23, Fig. 12 The glomerular capillary tuft and podocyte slit diaphragms. (A) An electron micrograph of the forming glomerulus at 2.5 dpf with invading endothelial cells from the dorsal aorta shaded in red and podocytes shaded in blue (image false-colored in Adobe Photoshop). (B) A similar stage glomerulus in the mutant island beat which lacks blood flow due to a mutation in an L-type cardiac specific calcium channel. The endothelial cells and podocytes are present but the aorta has a dilated lumen surrounding the podocytes with no sign of glomerular remodeling and morphogenesis. (C) Rhodamine-dextran filtration and uptake by pronephric epithelial cells. 10KD lysine-fixable rhodamine dextran injected into the general circulation can be seen as red fluorescence in glomerular capillaries (*gl*), and filtered dye is seen in apical endosomes of pronephric duct cells (*arrowheads*). Counterstain: FITC wheat germ agglutinin. (D) Electron micrograph of the glomerular basement membrane region in the glomerulus. Individual profiles of podocyte foot processes resting on the glomerular basement membrane (*gbm*) are connected by slit-diaphragms (*arrowheads* at top). *cl*; capillary lumen, *bs*; Bowman's space. Whole-mount *in situ* hybridization shows expression of zebrafish podocin (E) and nephrin (F) specifically in the forming podocytes.

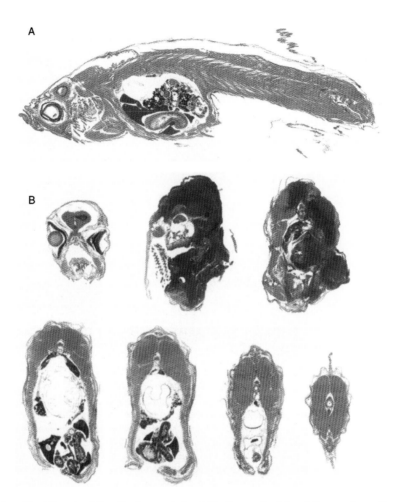

Chapter 25, Fig. 4 After fixation and decalcification, zebrafish adults can be sectioned in (A) sagittal, (B) transverse, or (*not shown*) coronal planes. (A) A sagittal section of a zebrafish female stained with hematoxylin and eosin (H & E). (B) Transverse sections of a zebrafish female with a malignant peripheral nerve sheath tumor stained with H & E. The darker stained areas in the top two right section are the tumor.

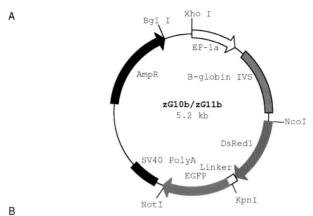

B

zG10b

DsRed1 Ser Gly Leu Arg Ser Thr Gly Arg Gly Gly Gly Thr Asn EGFP
---AAG TCC GGA CTC AGA TCC ACC GGC AGG GGG GGG GGT ACC AAC G---

zG11b

DsRed1 Ser Gly Leu Arg Ser Thr Gly Arg Gly Gly Gly Tyr Gln EGFP^F
---AAG TCC GGA CTC AGA TCC ACC GGC AGG GGG GGG GGG TAC CAA CG---

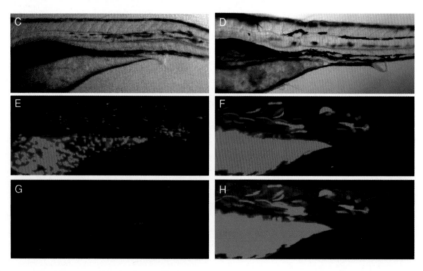

Chapter 25, Fig. 5 The zG10b and zG11b transgenic constructs were designed to detect frameshift mutations in zebrafish embryos and larvae. (A) The zG10b/zG11b vector was based on the pXIG vector, a generous gift from Dr. Nancy Hopkins (Amsterdam *et al.*, 1996). A single fusion protein is formed that contains an in-frame red fluorescent protein, a linker region that contains either the G10 or G11 repeat sequence followed by enhanced green fluorescent protein (EGFP). (B) The linker sequence of zG10 maintains the reading frame between the two fluorescent domains, but the linker sequence of zG11 creates a +1 frameshift of the green fluorescent protein (EGFP^F). (C, E, and G) A 72-hpf embryo injected with zG10b. (D, F, and H) A 72-hpf embryo injected with zG11b. (C and D) Brightfield. (E and F) Red fluorescence seen with a G filter (Leica). The more punctate pattern seen in the zG11-injected embryos may be attributed to the high basic content of the out-of-frame EGFP, resulting in nuclear localization. (G and H) Green fluorescence seen with a GFP3 filter. No green fluorescence is seen in the embryo injected with the out-of-frame zG11b construct.